Lecture Notes in Computer Science 5103

Commenced Publication in 1973
Founding and Former Series Editors:
Gerhard Goos, Juris Hartmanis, and Jan van Leeuwen

Marian Bubak   Geert Dick van Albada
Jack Dongarra   Peter M.A. Sloot (Eds.)

Computational Science – ICCS 2008

8th International Conference
Kraków, Poland, June 23-25, 2008
Proceedings, Part III

 Springer

Volume Editors

Marian Bubak
AGH University of Science and Technology
Institute of Computer Science and
Academic Computer Center CYFRONET
30-950 Kraków, Poland
E-mail: bubak@agh.edu.pl

Geert Dick van Albada
Peter M.A. Sloot
University of Amsterdam
Section Computational Science
1098 SJ Amsterdam, The Netherlands
E-mail: {dick,sloot}@science.uva.nl

Jack Dongarra
University of Tennessee
Computer Science Department
Knoxville, TN 37996, USA
E-mail: dongarra@cs.utk.edu

Library of Congress Control Number: 2008928942

CR Subject Classification (1998): F, D, G, H, I, J, C.2-3

LNCS Sublibrary: SL 1 – Theoretical Computer Science and General Issues

ISSN   0302-9743
ISBN-10   3-540-69388-2 Springer Berlin Heidelberg New York
ISBN-13   978-3-540-69388-8 Springer Berlin Heidelberg New York

Springer is a part of Springer Science+Business Media

springer.com

© Springer-Verlag Berlin Heidelberg 2008
Printed in Germany

Typesetting: Camera-ready by author, data conversion by Scientific Publishing Services, Chennai, India
Printed on acid-free paper   SPIN: 12322213   06/3180   5 4 3 2 1 0

Advancing Science Through Computation

I knock at the stone's front door.
"It's only me, let me come in.
I've come out of pure curiosity.
Only life can quench it.
I mean to stroll through your palace,
then go calling on a leaf, a drop of water.
I don't have much time.
My mortality should touch you."

Wisława Szymborska,
Conversation with a Stone, in Nothing Twice, 1997

The International Conference on Computational Science (ICCS 2008) held in Kraków, Poland, June 23–25, 2008, was the eighth in the series of highly successful conferences: ICCS 2007 in Beijing, China; ICCS 2006 in Reading, UK; ICCS 2005 in Atlanta; ICCS 2004 in Krakow, Poland; ICCS 2003 held simultaneously in Melbourne, Australia and St. Petersburg, Russia; ICCS 2002 in Amsterdam, The Netherlands; and ICCS 2001 in San Francisco, USA.

The theme for ICCS 2008 was "Advancing Science Through Computation," to mark several decades of progress in computational science theory and practice, leading to greatly improved applications in science. This conference was a unique event focusing on recent developments in novel methods and modeling of complex systems for diverse areas of science, scalable scientific algorithms, advanced software tools, computational grids, advanced numerical methods, and novel application areas where the above novel models, algorithms, and tools can be efficiently applied, such as physical systems, computational and systems biology, environment, finance, and others. ICCS 2008 was also meant as a forum for scientists working in mathematics and computer science as the basic computing disciplines and application areas, who are interested in advanced computational methods for physics, chemistry, life sciences, and engineering. The main objective of this conference was to discuss problems and solutions in all areas, to identify new issues, to shape future directions of research, and to help users apply various advanced computational techniques. During previous editions of ICCS, the goal was to build a computational science community; the main challenge in this edition was ensuring very high quality of scientific results presented at the meeting and published in the proceedings.

Keynote lectures were delivered by:

- Maria E. Orłowska: *Intrinsic Limitations in Context Modeling*
- Jesus Villasante: *EU Research in Software and Services: Activities and Priorities in FP7*
- Stefan Blügel: *Computational Materials Science at the Cutting Edge*

- Martin Walker: *New Paradigms for Computational Science*
- Yong Shi: *Multiple Criteria Mathematical Programming and Data Mining*
- Hank Childs: *Why Petascale Visualization and Analysis Will Change the Rules*
- Fabrizio Gagliardi: *HPC Opportunities and Challenges in e-Science*
- Pawel Gepner: *Intel's Technology Vision and Products for HPC*
- Jarek Nieplocha: *Integrated Data and Task Management for Scientific Applications*
- Neil F. Johnson: *What Do Financial Markets, World of Warcraft, and the War in Iraq, all Have in Common? Computational Insights into Human Crowd Dynamics*

We would like to thank all keynote speakers for their interesting and inspiring talks and for submitting the abstracts and papers for these proceedings.

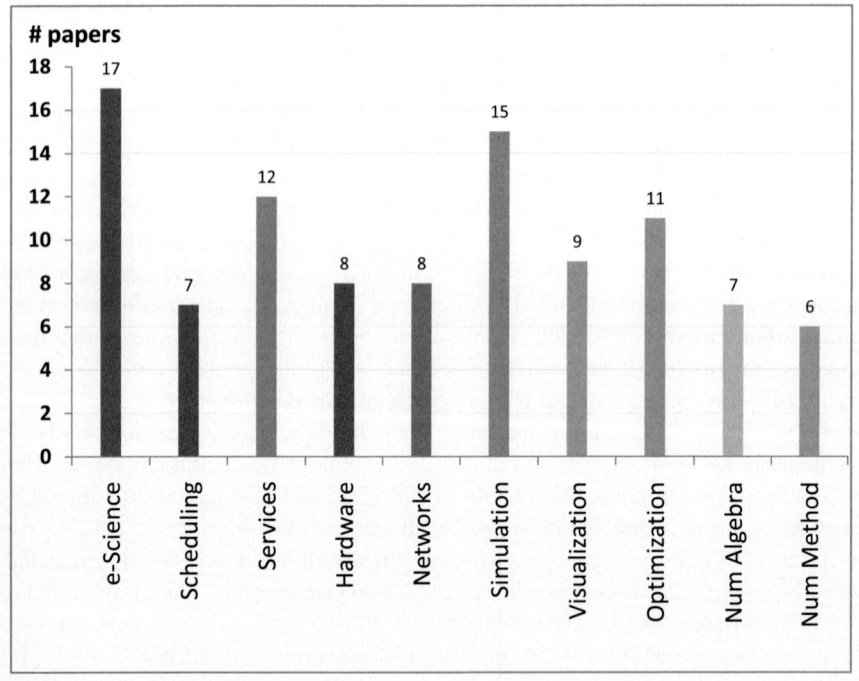

Fig. 1. Number of papers in the general track by topic

The main track of ICSS 2008 was divided into approximately 20 parallel sessions (see Fig. 1) addressing the following topics:

1. e-Science Applications and Systems
2. Scheduling and Load Balancing
3. Software Services and Tools

4. New Hardware and Its Applications
5. Computer Networks
6. Simulation of Complex Systems
7. Image Processing and Visualization
8. Optimization Techniques
9. Numerical Linear Algebra
10. Numerical Algorithms

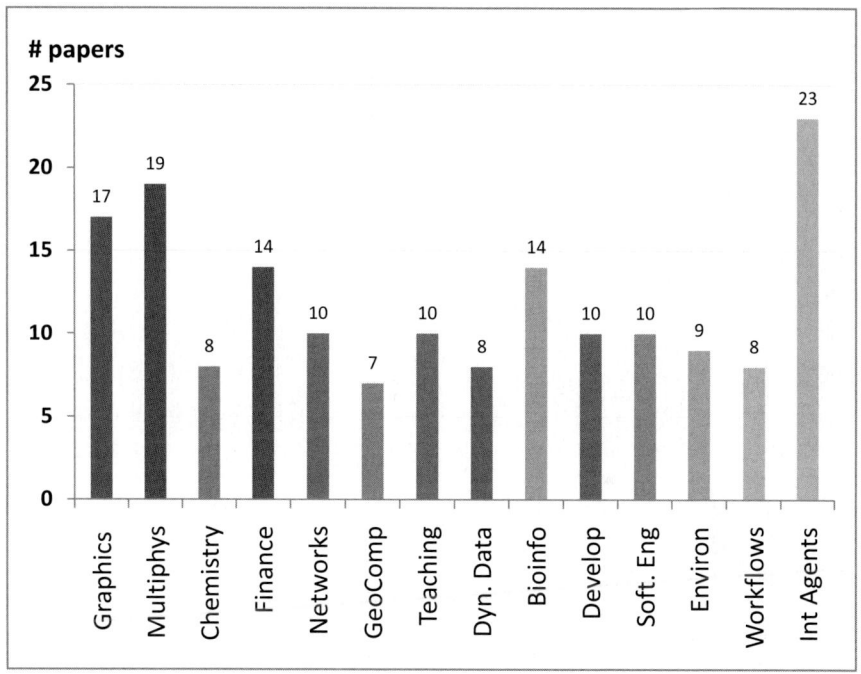

Fig. 2. Number of papers in workshops

The conference included the following workshops (Fig. 2):

1. 7th Workshop on Computer Graphics and Geometric Modeling
2. 5th Workshop on Simulation of Multiphysics Multiscale Systems
3. 3rd Workshop on Computational Chemistry and Its Applications
4. Workshop on Computational Finance and Business Intelligence
5. Workshop on Physical, Biological and Social Networks
6. Workshop on GeoComputation
7. 2nd Workshop on Teaching Computational Science
8. Workshop on Dynamic Data-Driven Application Systems
9. Workshop on Bioinformatics' Challenges to Computer Science
10. Workshop on Tools for Program Development and Analysis in Computational Science

11. Workshop on Software Engineering for Large-Scale Computing
12. Workshop on Collaborative and Cooperative Environments
13. Workshop on Applications of Workflows in Computational Science
14. Workshop on Intelligent Agents and Evolvable Systems

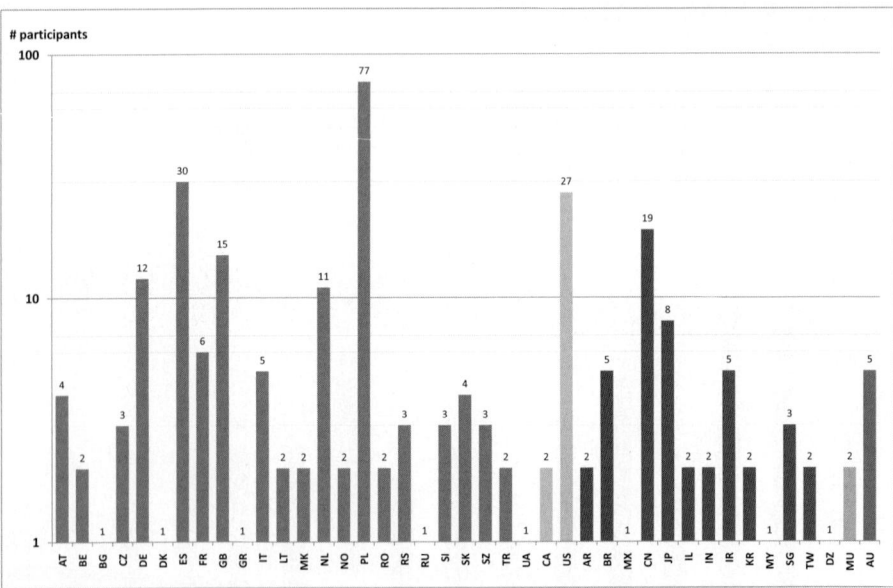

Fig. 3. Number of accepted papers by country

Selection of papers for the conference was possible thanks to the hard work of the Program Committee members and about 510 reviewers; each paper submitted to ICCS 2008 received at least 3 reviews. The distribution of papers accepted for the conference is presented in Fig. 3. ICCS 2008 participants represented all continents; their geographical distribution is presented in Fig. 4.

The ICCS 2008 proceedings consist of three volumes; the first one, LNCS 5101, contains the contributions presented in the general track, while volumes 5102 and 5103 contain papers accepted for workshops. Volume LNCS 5102 is related to various computational research areas and contains papers from Workshops 1–7, while volume LNCS 5103, which contains papers from Workshops 8–14, is mostly related to computer science topics. We hope that the ICCS 2008 proceedings will serve as an important intellectual resource for computational and computer science researchers, pushing forward the boundaries of these two fields and enabling better collaboration and exchange of ideas. We would like to thank Springer for fruitful collaboration during the preparation of the proceedings. At the conference, the best papers from the general track and workshops were nominated and presented on the ICCS 2008 website; awards were funded by Elsevier and Springer. A number of papers will also be published as special issues of selected journals.

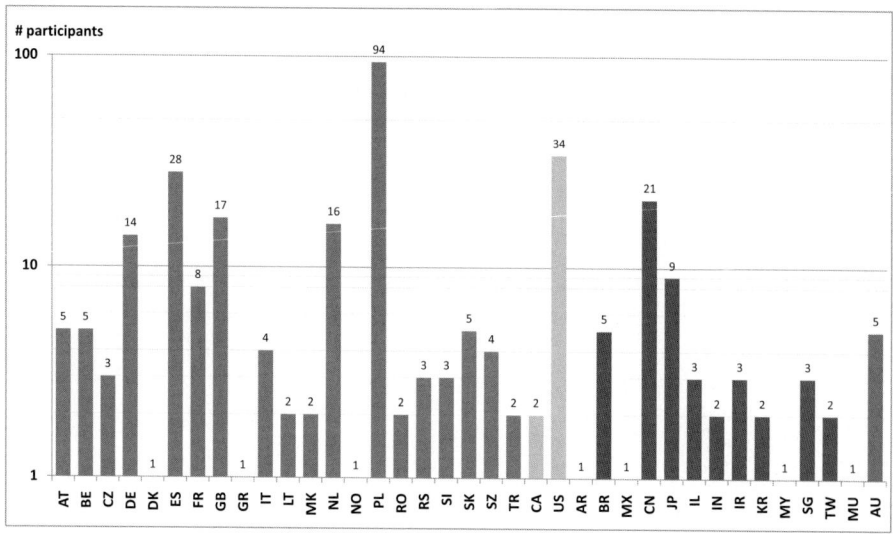

Fig. 4. Number of participants by country

We owe thanks to all workshop organizers and members of the Program Committee for their diligent work, which ensured the very high quality of ICCS 2008. We would like to express our gratitude to the Kazimierz Wiatr, Director of ACC CYFRONET AGH, and to Krzysztof Zieliński, Director of the Institute of Computer Science AGH, for their personal involvement. We are indebted to all the members of the Local Organizing Committee for their enthusiastic work towards the success of ICCS 2008, and to numerous colleagues from ACC CYFRONET AGH and the Institute of Computer Science for their help in editing the proceedings and organizing the event. We very much appreciate the help of the computer science students during the conference. We own thanks to the ICCS 2008 sponsors: Hewlett-Packard, Intel, Qumak-Secom, IBM, Microsoft, ATM, Elsevier (Journal Future Generation Computer Systems), Springer, ACC CYFRONET AGH, and the Institute of Computer Science AGH for their generous support.

We wholeheartedly invite you to once again visit the ICCS 2008 website (http://www.iccs-meeting.org/iccs2008/), to recall the atmosphere of those June days in Kraków.

June 2008

Marian Bubak
G. Dick van Albada
Peter M.A. Sloot
Jack J. Dongarra

Organization

ICCS 2008 was organized by the Academic Computer Centre Cyfronet AGH in cooperation with the Institute of Computer Science AGH (Kraków, Poland), the University of Amsterdam (Amsterdam,The Netherlands) and the University of Tennessee (Knoxville, USA).

All the members of the Local Organizing Committee are staff members of ACC Cyfronet AGH and ICS AGH.

Conference Chairs

Conference Chair — Marian Bubak (AGH University of Science and Technology, Kraków, Poland)

Workshop Chair — Dick van Albada (University of Amsterdam, The Netherlands)

Overall Scientific Co-chair — Jack Dongarra (University of Tennessee, USA)

Overall Scientific Chair — Peter Sloot (University of Amsterdam, The Netherlands)

Local Organizing Committee

Kazimierz Wiatr
Marian Bubak
Zofia Mosurska
Maria Stawiarska
Milena Zając
Mietek Pilipczuk
Karol Frańczak

Sponsoring Institutions

Hewlett-Packard Company
Intel Corporation
Qumak-Sekom S.A. and IBM
Microsoft Corporation
ATM S.A.
Elsevier
Springer

Program Committee

J.H. Abawajy (Deakin University, Australia)
D. Abramson (Monash University, Australia)

V. Alexandrov (University of Reading, UK)
I. Altintas (San Diego Supercomputer Centre, UCSD, USA)
M. Antolovich (Charles Sturt University, Australia)
E. Araujo (Universidade Federal de Campina Grande, Brazil)
M.A. Baker (University of Reading, UK)
B. Baliś (AGH University of Science and Technology, Kraków, Poland)
A. Benoit (LIP, ENS Lyon, France)
I. Bethke (University of Amsterdam, The Netherlands)
J. Bi (Tsinghua University, Beijing, China)
J.A.R. Blais (University of Calgary, Canada)
K. Boryczko (AGH University of Science and Technology, Kraków, Poland)
I. Brandic (Technical University of Vienna, Austria)
M. Bubak (AGH University of Science and Technology, Kraków, Poland)
K. Bubendorfer (Victoria University of Wellington, New Zealand)
B. Cantalupo (Elsag Datamat, Italy)
L. Caroprese (University of Calabria, Italy)
J. Chen (Swinburne University of Technology, Australia)
O. Corcho (Universidad Politcnica de Madrid, Spain)
J. Cui (University of Amsterdam, The Netherlands)
J.C. Cunha (University Nova de Lisboa, Portugal)
S. Date (Osaka University, Japan)
S. Deb (National Institute of Science and Technology, Berhampur, India)
Y.D. Demchenko (University of Amsterdam, The Netherlands)
F. Desprez (INRIA, France)
T. Dhaene (Ghent University, Belgium)
I.T. Dimov (University of Reading, Bulgarian Academy of Sciences, Bulgaria)
J. Dongarra (University of Tennessee, USA)
F. Donno (CERN, Switzerland)
C. Douglas (University of Kentucky, USA)
G. Fox (Indiana University, USA)
W. Funika (AGH University of Science and Technology, Kraków, Poland)
G. Geethakumari (University of Hyderabad, India)
B. Glut (AGH University of Science and Technology, Kraków, Poland)
Y. Gorbachev (St.-Petersburg State Polytechnical University, Russia)
A.M. Gościński (Deakin University, Australia)
M. Govindaraju (Binghamton University, USA)
G.A. Gravvanis (Democritus University of Thrace, Greece)
D.J. Groen (University of Amsterdam, The Netherlands)
T. Gubała (Academic Computer Centre Cyfronet AGH, Kraków, Poland)
M. Hardt (Forschungszentrum Karlsruhe, Germany)
T. Heinis (ETH Zurich, Switzerland)
L. Hluchý (Slovak Academy of Sciences, Slovakia)
W. Hoffmann (University of Amsterdam, The Netherlands)
A. Iglesias (University of Cantabria, Spain)
C.R. Jesshope (University of Amsterdam, The Netherlands)

H. Jin (Huazhong University of Science and Technology, China)
D. Johnson (University of Reading, UK)
B.D. Kandhai (University of Amsterdam, The Netherlands)
S. Kawata (Utsunomiya University, Japan)
W.A. Kelly (Queensland University of Technology, Australia)
J. Kitowski (AGH University of Science and Technology, Kraków, Poland)
M. Koda (University of Tsukuba, Japan)
D. Kranzlmüller (Johannes Kepler University Linz, Austria)
J. Kroc (University of Amsterdam, The Netherlands)
B. Kryza (Academic Computer Centre Cyfronet AGH, Kraków, Poland)
M. Kunze (Forschungszentrum Karlsruhe, Germany)
D. Kurzyniec (Google, Kraków, Poland)
A. Lagana (University of Perugia, Italy)
L. Lefevre (INRIA, France)
A. Lewis (Griffith University, Australia)
H.W. Lim (Royal Holloway, University of London, UK)
E. Lorenz (University of Amsterdam, The Netherlands)
P. Lu (University of Alberta, Canada)
M. Malawski (AGH University of Science and Technology, Kraków, Poland)
A.S. McGough (London e-Science Centre, UK)
P.E.C. Melis (University of Amsterdam, The Netherlands)
E.D. Moreno (UEA-BENq, Manaus, Brazil)
J.T. Mościcki (CERN, Switzerland)
S. Naqvi (CETIC, Belgium)
P.O.A. Navaux (Universidade Federal do Rio Grande do Sul, Brazil)
Z. Nemeth (Hungarian Academy of Science, Hungary)
J. Ni (University of Iowa, USA)
G.E. Norman (Russian Academy of Sciences, Russia)
B.Ó. Nualláin (University of Amsterdam, The Netherlands)
S. Orlando (University of Venice, Italy)
M. Paprzycki (Polish Academy of Sciences, Poland)
M. Parashar (Rutgers University, USA)
C.P. Pautasso (University of Lugano, Switzerland)
M. Postma (University of Amsterdam, The Netherlands)
V. Prasanna (University of Southern California, USA)
T. Priol (IRISA, France)
M.R. Radecki (AGH University of Science and Technology, Kraków, Poland)
M. Ram (C-DAC Bangalore Centre, India)
A. Rendell (Australian National University, Australia)
M. Riedel (Research Centre Jülich, Germany)
D. Rodríguez Garca (University of Alcal, Spain)
K. Rycerz (AGH University of Science and Technology, Kraków, Poland)
R. Santinelli (CERN, Switzerland)
B. Schulze (LNCC, Brazil)
J. Seo (University of Leeds, UK)

A.E. Solomonides (University of the West of England, Bristol, UK)
V. Stankovski (University of Ljubljana, Slovenia)
H. Stockinger (Swiss Institute of Bioinformatics, Switzerland)
A. Streit (Forschungszentrum Jülich, Germany)
H. Sun (Beihang University, China)
R. Tadeusiewicz (AGH University of Science and Technology, Kraków, Poland)
M. Taufer (University of Delaware, USA)
J.C. Tay (Nanyang Technological University, Singapore)
C. Tedeschi (LIP-ENS Lyon, France)
A. Tirado-Ramos (University of Amsterdam, The Netherlands)
P. Tvrdik (Czech Technical University Prague, Czech Republic)
G.D. van Albada (University of Amsterdam, The Netherlands)
R. van den Boomgaard (University of Amsterdam, The Netherlands)
A. Visser (University of Amsterdam, The Netherlands)
D.W. Walker (Cardiff University, UK)
C.L. Wang (University of Hong Kong, China)
A.L. Wendelborn (University of Adelaide, Australia)
Y. Xue (Chinese Academy of Sciences, China)
F.-P. Yang (Chongqing University of Posts and Telecommunications, China)
C.T. Yang (Tunghai University, Taichung, Taiwan)
L.T. Yang (St. Francis Xavier University, Canada)
J. Yu (Renewtek Pty Ltd, Australia)
Y. Zheng (Zhejiang University, China)
E.V. Zudilova-Seinstra (University of Amsterdam, The Netherlands)

Reviewers

J.H. Abawajy
H.H. Abd Allah
D. Abramson
R. Albert
M. Aldinucci
V. Alexandrov
I. Altintas
D. Angulo
C. Anthes
M. Antolovich
E. Araujo
E.F. Archibong
L. Axner
M.A. Baker
B. Bališ
S. Battiato
M. Baumgartner
U. Behn

P. Bekaert
A. Belloum
A. Benoit
G. Bereket
J. Bernsdorf
I. Bethke
B. Bethwaite
J.-L. Beuchat
J. Bi
J. Bin Shyan
B.S. Bindhumadhava
J.A.R. Blais
P. Blowers
B. Boghosian
I. Borges
A.I. Boronin
K. Boryczko
A. Borzi

A. Boutalib
A. Brabazon
J.M. Bradshaw
I. Brandic
V. Breton
R. Brito
W. Bronsvoort
M. Bubak
K. Bubendorfer
J. Buisson
J. Burnett
A. Byrski
M. Caeiro
A. Caiazzo
F.C.A. Campos
M. Cannataro
B. Cantalupo
E. Caron

L. Caroprese
U. Catalyurek
S. Cerbat
K. Cetnarowicz
M. Chakravarty
W. Chaovalitwongse
J. Chen
H. Chojnacki
B. Chopard
C. Choquet
T. Cierzo
T. Clark
S. Collange
P. Combes
O. Corcho
J.M. Cordeiro
A.D. Corso
L. Costa
H. Cota de Freitas
C. Cotta
G. Cottone
C.D. Craig
C. Douglas
A. Craik
J. Cui
J.C. Cunha
R. Custodio
S. Date
A. Datta
D. De Roure
S. Deb
V. Debelov
E. Deelman
Y.D. Demchenko
B. Depardon
F. Desprez
R. Dew
T. Dhaene
G. Di Fatta
A. Diaz-Guilera
R. Dillon
I.T. Dimov
G. Dobrowolski
T. Dokken
J. Dolado

W. Dong
J. Dongarra
F. Donno
C. Douglas
M. Drew
R. Drezewski
A. Duarte
V. Duarte
W. Dubitzky
P. Edmond
A. El Rhalibi
A.A. El-Azhary
V. Ervin
A. Erzan
M. Esseffar
L. Fabrice
Y. Fan
G. Farin
Y. Fei
V. Ferandez
D. Fireman
K. Fisher
A. Folleco
T. Ford
G. Fox
G. Frenking
C. Froidevaux
K. Fülinger
W. Funika
H. Fuss
A. Galvez
R. Garcia
S. Garic
A. Garny
F. Gava
T. Gedeon
G. Geethakumari
A. Gerbessiotis
F. Giacomini
S. Gimelshein
S. Girtelschmid
C. Glasner
T. Glatard
B. Glut
M. Goldman

Y. Gorbachev
A.M. Gościński
M. Govindaraju
E. Grabska
V. Grau
G.A. Gravvanis
C. Grelck
D.J. Groen
J.G. Grujic
Y. Guang Xue
T. Gubała
C. Guerra
V. Guevara
X. Guo
Y. Guo
N.M. Gupte
J.A. Gutierrez de Mesa
P.H. Guzzi
A. Haffegee
S. Hannani
U. Hansmann
M. Hardt
D. Harężlak
M. Harman
R. Harrison
M. Hattori
T. Heinis
P. Heinzlreiter
R. Henschel
F. Hernandez
V. Hernández
P. Herrero
V. Hilaire
L. Hluchý
A. Hoekstra
W. Hoffmann
M. Hofmann-Apitius
J. Holyst
J. Hrusak
J. Hu
X.R. Huang
E. Hunt
K. Ichikawa
A. Iglesias
M. Inda

M. Paszyński
C.P. Pautasso
B. Payne
T. Peachey
S. Pelagatti
J. Peng
Y. Peng
F. Perales
M. Pérez
D. Pfahl
G. Plank
D. Plemenos
A. Pluchino
M. Polak
S.F. Portegies Zwart
M. Postma
B.B. Prahalada
V. Prasanna
R. Preissl
T. Priol
T. Prokosch
M. Py
G. Qiu
J. Quinqueton
M.R. Radecki
B. Raffin
M. Ram
P. Ramasami
P. Ramsamy
O.F. Rana
M. Reformat
A. Rendell
M. Riedel
J.L. Rivail
G.J. Rodgers
C. Rodríguez-Leon
B. Rodríguez
D. Rodríguez
D. Rodríguez Garcia
F. Rogier
G. Rojek
H. Ronghuai
H. Rosmanith
J. Rough
F.-X. Roux

X. Różańska
M. Ruiz
R. Ruiz
K. Rycerz
K. Saetzler
P. Saiz
S. Sanchez
S.K. Khattri
R. Santinelli
A. Santos
M. Sarfraz
M. Satpathy
M. Sbert
H.F. Schaefer
R. Schaefer
M. Schulz
B. Schulze
I. Scriven
E. Segredo
J. Seo
A. Sfarti
Y. Shi
L. Shiyong
Z. Shuai
M.A. Sicilia
L.P. Silva Barra
F. Silvestri
A. Simas
H.M. Singer
V. Sipkova
P.M.A. Sloot
R. Slota
B. Śnieżyński
A.E. Solomonides
R. Soma
A. Sourin
R. Souto
R. Spiteri
V. Srovnal
V. Stankovski
E.B. Stephens
M. Sterzel
H. Stockinger
D. Stokic
A. Streit

B. Strug
H. Sun
Z. Sun
F. Suter
H. Suzuki
D. Szczerba
L. Szirmay-Kalos
R. Tadeusiewicz
B. Tadic
R. Tagliaferri
W.K. Tai
S. Takeda
E.J. Talbi
J. Tan
S. Tan
T. Tang
J. Tao
M. Taufer
J.C. Tay
C. Tedeschi
J.C. Teixeira
D. Teller
G. Terje Lines
C. Te-Yi
A.T. Thakkar
D. Thalmann
S. Thurner
Z. Tianshu
A. Tirado
A. Tirado-Ramos
P. Tjeerd
R.F. Tong
J. Top
H. Torii
V.D. Tran
C. Troyer
P. Trunfio
W. Truszkowski
W. Turek
P. Tvrdik
F. Urmetzer
V. Uskov
G.D. van Albada
R. van den Boomgaard
M. van der Hoef

R. van der Sman	E. Westhof	G. Zhang
B. van Eijk	R. Wismüller	H. Zhang
R. Vannier	C. Wu	J.J. Zhang
P. Veltri	C. Xenophontos	J.Z.H. Zhang
E.J. Vigmond	Y. Xue	L. Zhang
J. Villá i Freixa	N. Yan	J. Zhao
A. Visser	C.T. Yang	Z. Zhao
D.W. Walker	F.-P. Yang	Y. Zheng
C.L. Wang	L.T. Yang	X. Zhiwei
F.L. Wang	X. Yang	A. Zhmakin
J. Wang	J. Yu	N. Zhong
J.Q. Wang	M. Yurkin	M.H. Zhu
J. Weidendorfer	J. Zara	T. Zhu
C. Weihrauch	I. Zelinka	O. Zimmermann
C. Weijun	S. Zeng	J. Zivkovic
A. Weise	C. Zhang	A. Zomaya
A.L. Wendelborn	D.L. Zhang	E.V. Zudilova-Seinstra

Workshops Organizers

7th Workshop on Computer Graphics and Geometric Modeling

A. Iglesias (University of Cantabria, Spain)

5th Workshop on Simulation of Multiphysics Multiscale Systems

V.V. Krzhizhanovskaya and A.G. Hoekstra (University of Amsterdam, The Netherlands)

3rd Workshop on Computational Chemistry and Its Applications

P. Ramasami (University of Mauritius, Mauritius)

Workshop on Computational Finance and Business Intelligence

Y. Shi (Chinese Academy of Sciences, China)

Workshop on Physical, Biological and Social Networks

B. Tadic (Jožef Stefan Institute, Ljubljana, Slovenia)

Workshop on GeoComputation

Y. Xue (London Metropolitan University, UK)

2nd Workshop on Teaching Computational Science

Q. Luo (Wuhan University of Science and Technology Zhongnan Branch, China), A. Tirado-Ramos (University of Amsterdam, The Netherlands), Y.-W. Wu

(Central China Normal University, China) and H.-W. Wang (Wuhan University of Science and Technology Zhongnan Branch, China)

Workshop on Dynamic Data Driven Application Systems

C.C. Douglas (University of Kentucky, USA) and F. Darema (National Science Foundation, USA)

Bioinformatics' Challenges to Computer Science

M. Cannataro (University Magna Gracia of Catanzaro, Italy), M. Romberg (Research Centre Jülich, Germany), J. Sundness (Simula Research Laboratory, Norway), R. Weber dos Santos (Federal University of Juiz de Fora, Brazil)

Workshop on Tools for Program Development and Analysis in Computational Science

A. Knüpfer (University of Technology, Dresden, Germany), J. Tao (Forschungszentrum Karlsruhe, Germany), D. Kranzlmüller (Johannes Kepler University Linz, Austria), A. Bode (University of Technology, München, Germany) and J. Volkert (Johannes Kepler University Linz, Austria)

Workshop on Software Engineering for Large-Scale Computing

D. Rodríguez (University of Alcala, Spain) and R. Ruiz (Pablo de Olavide University, Spain)

Workshop on Collaborative and Cooperative Environments

C. Anthes (Johannes Kepler University Linz, Austria), V. Alexandrov (University of Reading, UK), D. Kranzlmüller, G. Widmer and J. Volkert (Johannes Kepler University Linz, Austria)

Workshop on Applications of Workflows in Computational Science

Z. Zhao and A. Belloum (University of Amsterdam, The Netherlands)

Workshop on Intelligent Agents and Evolvable Systems

K. Cetnarowicz, R. Schaefer (AGH University of Science and Technology, Kraków, Poland) and B. Zheng (South-Central University For Nationalities, Wuhan, China)

Table of Contents – Part III

Workshop on Dynamic Data Driven Application Systems

Bioinformatics' Challenges to Computer Science

Workshop on Tools for Program Development and Analysis in Computational Science

Workshop on Software Engineering for Large-Scale Computing

Workshop on Collaborative and Cooperative Environments

Workshop on Applications of Workflows in Computational Science

Workshop on Intelligent Agents and Evolvable Systems

Workshop on Dynamic Data Driven Application Systems

Dynamic Data Driven Applications Systems – DDDAS 2008

Craig C. Douglas[1,2]

[1] University of Kentucky, Lexington, KY 40506-0046
[2] Yale University, New Haven, CT 06520-8285, USA
douglas-craig@cs.yale.edu

Abstract. This workshop is centered about the recently emerged paradigm of Dynamic Data Driven Applications Systems (DDDAS). The DDDAS concept has already been established as a revolutionary new approach of a symbiotic relation between application and measurement systems, where applications can accept and respond dynamically to new data injected into the executing application, and reversely, the ability of such application systems to dynamically control the measurement processes. The synergistic feedback control-loop between application simulations and measurements can open new domains in the capabilities of simulations with high potential pay-off: create applications with new and enhanced analysis and prediction capabilities and enable a new methodology for more efficient and effective measurement processes. This new paradigm has the potential to transform the way science and engineering are done, with major impact in the way many functions in our society are conducted, such as manufacturing, commerce, transportation, hazard prediction/management, and medicine. The workshop will present such new opportunities, as well as the challenges and approaches in the applications', algorithms' and systems' software technologies needed to enable such capabilities, and will showcase ongoing research in these aspects with examples from several important application areas.

1 Overview

More and more applications are migrating to a data-driven paradigm including contaminant tracking, chemical process plants, petroleum refineries, well bores, and nuclear power plants. In each case sensors produce large quantities of telemetry that are fed into simulations that model key quantities of interest. As data are processed, computational models are adjusted to best agree with known measurements. If properly done, this increases the predictive capability of the simulation system. This allows what-if scenarios to be modeled, disasters to be predicted and avoided with human initiated or automatic responses, and the operation of the plants to be optimized. As this area of computational science grows, a broad spectrum of application areas will reap benefits. Examples include enhanced oil recovery, optimized placement of desalination plants and other water intakes, optimized food production, monitoring the integrity of

M. Bubak et al. (Eds.): ICCS 2008, Part III, LNCS 5103, pp. 3–4, 2008.

engineered structures and thus avoiding failures, and real time traffic advice for drivers. These are but a few of countless examples.

Visualization is used at all stages of DDDAS: setting up data and initial and/or boundary conditions, seeing and analyzing results, and steering computations.

DDDAS is ripe for multidisciplinary research to build applications, algorithms, measurement processes, and software components from which tools can be developed to solve diverse problems of regional and international interest. The advances that will result, including enhanced repositories of re-usable software components and high quality real time applications, support of separation of concerns, and reduced development time, will be of great value to industry and governments, and will set the stage for further valuable research and development and new education paradigms for students. A comprehensive list of ongoing state of the art projects is kept up to date on http://www.dddas.org in the projects area.

Several research thrusts in which advances should significantly enhance the abil ity of data-driven computational science to bring its tremendous benefits to a w ide array of applications. These research thrusts are,

- Effective *assimilation* of streams of data into ongoing simulations.
- *Interpretation, analysis, and adaptation* to assist the analyst and to ensure the most accurate simulation.
- *Cyberinfrastructure* to support data-driven simulations.

These three areas interact with two other research fields symbiotically: (1) forward multiscale modeling and simulation, and (2) deterministic and statistical methods in inverse problems.

Research areas (1) and (2) combined with (3) DDDAS must work within the context of uncertainty and will benefit from the development of statistically sound, unified treatments of uncertainties. For example, in forward multiscale modeling and simulation, input data are uncertain and these uncertainties should be propagated to uncertainties in output quantities of interest. In an inverse problem, proper treatment of measurement uncertainties and errors must be integrated with treatment of uncertainties associated with bforwardb models. be treated systematically. In a data-driven application, all of these uncertainties are present and must

Data management in a DDDASis typically supported by tools for data acquisition, data access, and data dissemination. Data acquisition tools retrieve the real time or near real time data, processing and storing them. Data access tools provide common data manipulation support, e.g., querying, storing, and searching, to upper level models. Data dissemination tools read data from the data store, format them based on requests from data consumers, and deliver formatted data to data consumers.

DDDAS is *the* paradigm of the data rich information age that we live in.

Dynamic Data Driven Applications Systems (DDDAS) – A Transformative Paradigm

Frederica Darema

National Science Foundation, 4201 Wilson Boulevard, Arlington, VA, 22230, USA
fdarema@nsf.gov

Abstract. The Dynamic Data Driven Applications Systems (DDDAS), paradigm entails the ability to dynamically incorporate data into an executing application simulation, and in reverse, the ability of applications to dynamically steer measurement processes. The ability to augment the traditional application modeling with such dynamic data inputs, acquired in real time, online, or existing as archival data, creates the possibility of entirely new capabilities through improved applications modeling, improved instrumentation methods, and systems management methods, enhancing the analysis and prediction capabilities of application simulations, improving the efficiency of simulations and the effectiveness of measurement systems, including providing powerful methodology of management of heterogeneous sensor networks. The dynamic integration of the computational and measurement aspects in DDDAS environments entail a unified of the computational and instrumentation application platform. In this presentation we will discuss the ensuing DDDAS computational model in the context of examples of novel capabilities enabled through its implementation in many application areas.

An overview of the state of the art will be presented. A discussion of how to connect U.S. and European or Asian DDDAS-like projects will be initiated for all present at the DDDAS workshop.

M. Bubak et al. (Eds.): ICCS 2008, Part III, LNCS 5103, p. 5, 2008.
© Springer-Verlag Berlin Heidelberg 2008

Evaluation of Measurement Techniques for the Validation of Agent-Based Simulations Against Streaming Data

Timothy W. Schoenharl and Greg Madey

Department of Computer Science and Engineering
University of Notre Dame
Notre Dame, IN 46556

Abstract. This paper presents a study evaluating the applicability of several different measures to the validation of Agent-Based Modeling simulations against streaming data. We evaluate the various measurements and validation techniques using pedestrian movement simulations used in the WIPER system. The Wireless Integrated Phone-based Emergency Response (WIPER) system is a Dynamic Data-Driven Application System (DDDAS) that uses a stream of cellular network activity to detect, classify and predict crisis events. The WIPER simulation is essential to classification and prediction tasks, as the simulations model human activity, both in movement and cell phone activity, in an attempt to better understand crisis events.[1]

1   Introduction

Crisis events occur without warning on short time scales relative to normal human activities. Simulations designed to model human behavior under crisis scenarios faces several challenges when attempting to model behavior in real time. The DDDAS approach seeks to address that challenge by incorporating real time sensor data into running simulations[1][2]. Dynamic, Data-Driven Application Systems is an approach to developing systems incorporating sensors and simulations where the simulations receive streaming data from the sensors and the sensors receive control information from the simulations.

The WIPER system uses streaming cell phone activity data to detect, track and predict crisis events [3]. The Agent-Based Modeling simulations in the WIPER system are intended to model the movement and cell phone activity of pedestrians. These simulations model crisis events and are intended to be validated in a online fashion against streaming data.

In the WIPER system, ensembles of simulations are created, with each simulation parameterized with a particular crisis scenario and initialized from the streaming data. When the all of the simulations in the ensemble have finished

---

[1] The research on WIPER is supported by an NSF Grant, CISE/CNS-DDDAS, Award #0540348.

M. Bubak et al. (Eds.): ICCS 2008, Part III, LNCS 5103, pp. 6–15, 2008.

running, the results are validated against streaming data from the cell phone network.

Thus the validation technique must provide a method of discriminating between various simulations. In the context of the WIPER project, this means determining which crisis model is the best fit for the phenomenon detected in the streaming data. In this paper we will demonstrate a validation technique and evaluate the technique with several measures.

2 Background

Validation is described as the process of determining whether a given model is an appropriate choice for the phenomenon being modeled. In the model development cycle, validation is normally considered in the context of simulation creation and development, and is done nearly exclusively in an offline fashion. However, the process of selecting an appropriate model for a given phenomenon is precisely what is needed in the dynamic context of the WIPER system.

There exists a large body of work on the topic of simulation validation. A survey of techniques and approaches to offline validation of discrete event simulations can be found in Balci [4]. This work is an essential reference for validation, but many of the techniques are suited to offline validation only, as the interpretation requires human judgement.

This section is divided into three subsections related to the provenance of the techniques we intend to evaluate. The first section deals with canonical offline validation techniques from simulation, the second section presents distance measures and the third section presents work that has been done with online simulations.

2.1 Offline Simulation Validation

Balci presents a thorough evaluation of techniques for validation of models in the context of model and simulation development [4]. The intent for these techniques was to aid in the validation and verification of simulations prior to deployment. Some techniques mentioned by Balci that are useful to our current discussion are predictive validation (also called input-output correlation) and blackbox testing.

Kennedy and Xiang describe the application of several techniques to the validation of Agent-Based Models [5,6]. The authors separate techniques into two categories: subjective, which require human interpretation, and objective, for which success criteria can be determined *a priori*. We focus on objective techniques, as the requirements of a DDDAS system make it impossible to place human decision makers "in the loop".

2.2 Online Simulations

Researchers in the area of discrete event simulations recognize the challenges posed to updating simulations online from streaming data [7]. The need for human interpretation is a serious limitation of traditional validation approaches and limits their usefulness in the context of online validation. Simulation researchers have defined a need for online validation but recognize the challenges to the approach. Davis claims that online validation may be unobtainable due to the difficulty in implementing changes to a model in an online scenario. We present a limited solution to this problem by offering multiple models simultaneously and using validation to select among the best, rather than using online validation to drive a search through model space.

It is important to distinguish between the model, the conceptual understanding of factors driving the phenomenon, and the parameters used to initialize the model. Optimization via simulation is a technique that is similar to canonical optimization and seeks to make optimal choices on selecting input parameters while keeping the underlying model the same and uses a simulation in place of the objective function. These techniques are usually grouped by whether they are appropriate for discrete or continuous input spaces [8]. For simulations with continuous input parameters, the author suggests the use of gradient-based methods. For simulations with discrete input parameters, the author presents approaches using random search on the input space.

3   Measures

We evaluate the following measures in the context of ranking simulations:

- Euclidean distance
- Manhattan distance
- Chebyshev distance
- Canberra distance

The distance measures are used to evaluate the output of the WIPER simulations in the context of agent movement. Agents move on a GIS space and agent locations are generalized to the cell tower that they communicate with. The space is tiled with Voronoi cells [9] that represent the coverage area of each cell tower. Empirical data from the cellular service provider aggregates user locations to the cell tower and the WIPER simulations do the same. Thus we can evaluate the distance measures using a well-defined vector of cell towers where the value for each position in the vector is the number of agents at that tower at each time step.

3.1   Distance Measures

Euclidean distance is the well-known distance metric from Euclidean geometry. The distance measure can be generalized to n dimensions from the common 2 dimensional case. The formula for Euclidean distance in n dimensions is given in Equation 1

$$d(\overline{p}, \overline{q}) = \sqrt{\sum_{i=1}^{n} (p_i - q_i)^2} \tag{1}$$

where

$$\overline{p} = (p_1, p_2, \cdots p_n) \tag{2}$$
$$\overline{q} = (q_1, q_2, \cdots q_n) \tag{3}$$

Manhattan distance, also known as the taxicab metric, is another metric for measuring distance, similar to Euclidean distance. The difference is that Manhattan distance is computed by summing the absolute value of the difference of the individual terms, unlike Euclidean distance which squares the difference, sums over all the differences and takes the square root. From a computational perspective Manhattan distance is significantly less costly to calculate than Euclidean distance, as it does not require taking a square root. The formula for Manhattan distance in n dimensions is given in Equation 4.

$$d(\overline{p}, \overline{q}) = \sum_{i=1}^{n} |p_i - q_i| \tag{4}$$

Chebyshev distance, also called the L_∞ metric, is a distance metric related to Euclidean and Manhattan distances [10]. The formula for the Chebyshev distance is given in Equation 5. The Chebyshev distance returns the maximum distance between elements in the position vectors. For this reason the metric seems appropriate to try on the WIPER simulations, as certain models may produce an output vector with one cell having a large variation from the norm.

$$d(\overline{p}, \overline{q}) = \max_i(|p_i - q_i|) = \lim_{k \to \infty} \left(\sum_{i=1}^{n} |p_i - q_i|^k \right)^{1/k} \tag{5}$$

The Canberra distance metric is used in situations where elements in the vector are always non-negative. In the case of the WIPER simulations, the output vector is composed of the number of agents in each Voronoi cell, which is always non-negative. The formula for Canberra distance is given in Equation 6. As defined, individual elements in the distance calculation could have zero for the numerator or denominator. Thus in cases where $|p_i| = |q_i|$, the element is omitted from the result.

$$d(\overline{p}, \overline{q}) - \sum_{i=1}^{n} \frac{|p_i - q_i|}{|p_i + q_i|} \tag{6}$$

4   Experimental Setup

In order to evaluate the feasibility of our approach, we present three experiments that demonstrate the effectiveness of the measures on validating agent movement models. The first experiment uses output from a particular run of the WIPER simulation as the synthetic data that will be tested against. This output is considered a "target" simulation. For the second movement model experiment we want to examine the effectiveness of measures in ranking models over all model types.

The purpose of these tests are not to demonstrate the effectiveness of the simulation to match the synthetic data but to demonstrate the ability of the measure to differentiate between simulation movement and activity model types. In a DDDAS system models of human behavior will be created and, according to the traditional model development approach, be validated offline. From this set of pre-validated models the system must be able to select, while the system is running, the model that best matches the data.

For the initial movement model experiment, we examine the effectiveness of the various statistical tests and measures in their ability to rank simulations in their closeness to the baseline simulation. The baseline simulation models a crisis scenario where people are fleeing a disaster. All simulations, including the baseline, are started with 900 agents distributed among 20 Voronoi cells. The distribution of agents to Voronoi cells is fixed over all of the simulations. For each of the 5 movement models, 100 replications of the simulation using different random seeds are run. Our evaluation approach is to examine the effectiveness of each measure in ranking output against the baseline. In this experiment, the desired results will show that instances of the Flee model are closer to the target simulation than other model types.

The second movement model experiment considers all of the movement models simultaneously. We use the data from the 500 simulation runs and create a matrix of distance values between every pair of values. Each position m_{ij} in the 500x500 matrix is the value of the distance metric between row i and column j. For consistency we present both the upper and lower halves of the matrix, as well as the diagonal, which is always equal to 0. We create this distance matrix for each of the distance metrics we consider. The outcome of the experiment is determined by examining the matrix and determining if the distance metric used shows low distance for simulation runs of the same model and high distance between simulation runs with differing models.

5   Results

Results of using the Euclidean distance metric to measure the differences in agent locations between movement models is shown in Figure 1(a). At the first time interval the Euclidean metric does an excellent job of ranking the Flee model instances as being the best match to the baseline. All of the Flee model instances have low distance to the baseline and are all lower than any instance of the other models. Interestingly, as the simulations progress, the Euclidean distance of each simulation's output from the Flee model baseline seems to yield good results for classifying the models, as they demonstrate low inter-class distance and high intra-class distance. The exception is the Null and Bounded Flee models. This result is discussed below in the analysis.

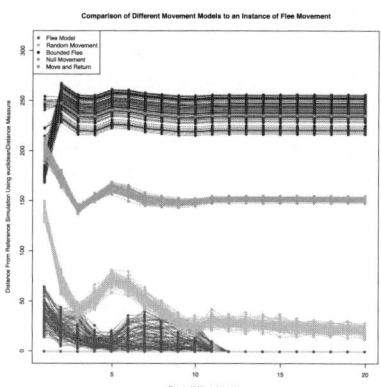

(a) Comparing agent movement in various movement models to flee movement using euclidean distance as the measure.

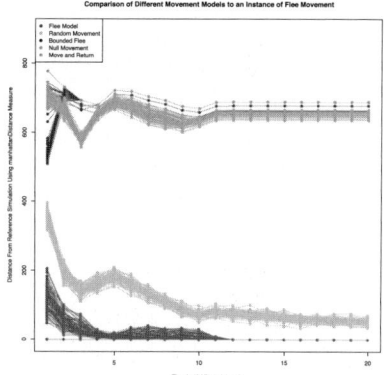

(b) Comparing agent movement in various movement models to flee movement using manhattan distance as the measure.

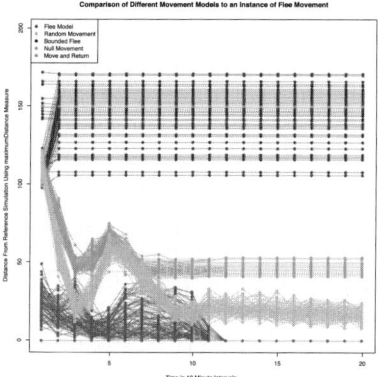

(c) Comparing agent movement in various movement models to flee movement using Chebyshev distance as the measure.

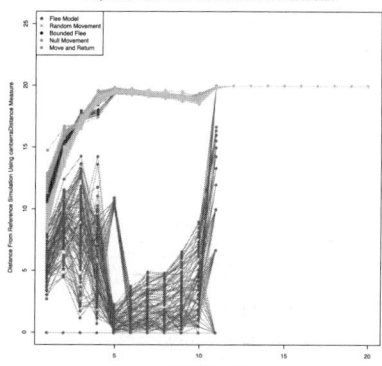

(d) Comparing agent movement in various movement models to flee movement using Canberra distance as the measure.

Fig. 1. Movement models compared periodically over the run of the simulations

Results of using the Manhattan distance metric to measure the differences in agent locations between movement models is shown in Figure 1(b). The Manhattan metric produces similar results to the Euclidean metric, with good results beginning at the first time interval.

Results of using the Chebyshev distance metric to measure the differences in agent locations between movement models is shown in Figure 1(c). As with the other measures in the L family, the Chebyshev distance metric does a good job of differentiating between model types in the early stages of the simulation run and in the late stages.

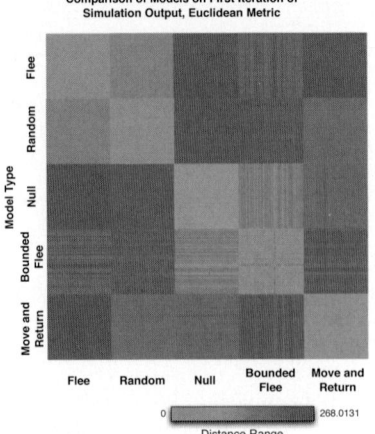

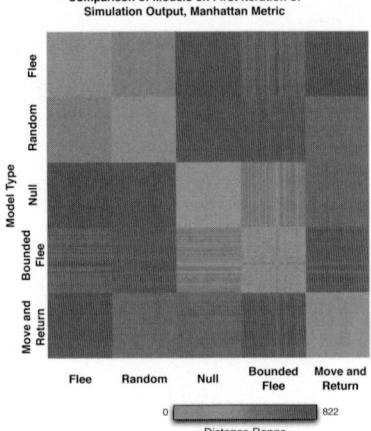

(a) Plot of the euclidean distance between simulation output. Simulations are grouped along x- and y-axis according to movement model in the order Flee movement, Random movement, Null movement, Bounded Flee movement and Move and Return movement.

(b) Plot of the manhattan distance between simulation output. Simulations are grouped along x- and y-axis according to movement model in the order Flee movement, Random movement, Null movement, Bounded Flee movement and Move and Return movement.

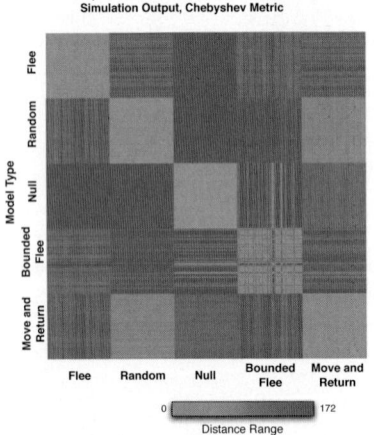

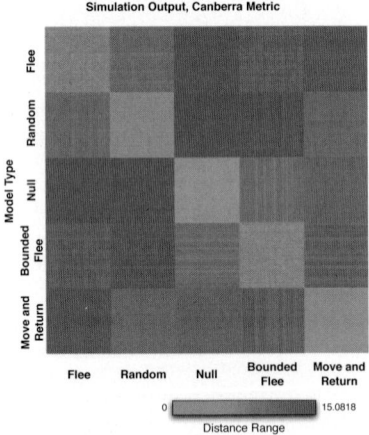

(c) Plot of the Chebyshev distance between simulation output. Simulations are grouped along x- and y-axis according to movement model in the order Flee movement, Random movement, Null movement, Bounded Flee movement and Move and Return movement.

(d) Plot of the canberra distance between simulation output. Simulations are grouped along x- and y-axis according to movement model in the order Flee movement, Random movement, Null movement, Bounded Flee movement and Move and Return movement.

Fig. 2. Distance plots for the simulation outputs

Results of using the Canberra distance metric to measure the differences in agent locations between movement models is shown in Figure 1(d). The Canberra metric appropriately ranks the Flee model simulations as closest, but as the simulations progress the results appear to be unstable and beyond the 11th sampling interval the Canberra metric fails to return valid values for Flee model simulations. Also, unlike the Euclidean and Manhattan metrics, the Canberra metric displays some overlap in the distance results for different model types.

Results of plotting the distance of the output from 500 simulations, 100 runs of each of the 5 movement models, is shown in Figures 2(a), 2(b), 2(c) and 2(d). These results provide a more thorough analysis of the usefulness of the metrics than simply comparing to one run of a simulation. In the ideal scenario the matrix will display low distance between simulations of the same model type (in the figures this would be a bright green square on the diagonal) and high distance when measured against simulations of a different model type (orange or red squares in the remainder of the matrix).

The figures measure the distance in the respective metric of the first time interval of the simulation. The simulations are grouped according to model type in the order, left to right (and top to bottom), Flee Movement, Random Movement, Null Movement, Bounded Flee Movement and Move and Return Movement. Each figure is colored from green to red, with green representing low distance and red high, with the colors scaled to the range of the distance values for the respective metrics.

Figures 2(a), 2(b) and 2(c) present the results for the Euclidean, Manhattan and Chebyshev metrics, respectively. Each of these metrics presents fairly good results in giving simulations of the same model type low distances and simulations with different model types high distance.

The results of the Canberra metric is less clear. The Canberra metric, Figure 2(d) appears to produce high distance values for Flee model simulations against other Flee model simulations and likewise for the Bounded Flee model.

6   Conclusions

Using a distance metric for model selection has several advantages. An experimental study, like that presented in this paper, allows users to calibrate the selection threshold, which makes it possible for the DDDAS to classify a phenomenon based on the distance from the validated model to the event. Alternately, should no model meet the threshold, the system can determine that none of the models are appropriate. In that case the measure may give a suggestion for the "closest fit" model and provides a scenario for new model creation.

In this paper we have presented an evaluation of various tests and measurements for online validation of Agent-Based Models. We have shown that the Euclidean and Manhattan distance metrics work well for validating movement models, however the Canberra distance are significantly less useful.

The Manhattan, Euclidean and Chebyshev metrics produce favorable results when used for measuring the similarity of simulations. Under the conditions we have tested, they produce low inter-model distances with high intra-model distance. Any of these metrics is adequate for an application such as the WIPER project.

The Canberra metric is useful under certain circumstances, but the poor performance measuring Flee model simulations against other Flee model simulations make it less than desirable for use in the WIPER project.

Figures 1(a), 1(b) and 2(b) show the distance metrics failing to differentiate between the Bounded Flee model and the Null Movement model. This result is an artifact of the way movement is measured in the simulations. Since agent locations are aggregated to the level of the Voronoi cell, agent movements below this resolution do not appear in the output. In the Bounded Flee model, agents move 1000 meters from the crisis and then stop moving. Thus, if the crisis is centered in a Voronoi cell that is approximately 2000 meters across, agents in the crisis cell will not appear to have moved at all.

A caveat concerning the use of simple thresholds for model selection in online validation: In the WIPER project, where mislabeling a crisis event as benign could have dire consequences, it is important to factor into the system the cost of false negatives. Crisis event models should be weighted so that when confronted with a crisis event, the chance of labeling it as normal behavior is minimized.

7   Future Work

The work in this paper has focused on measurements for online validation of agent movement models, where validation is selection from among a set of alternatives. Agent behavior in the WIPER simulation is composed of both movement and activity models. It is important for the online validation procedure to treat both movement and activity. In the future we would like to examine measurements for online validation of agent activity models, perhaps in conjunction with work being done to characterize crisis behavior as seen in cell phone activity data [11]. In keeping with our framework, we will need to create not only different input parameters for the activity models, but new models that describe agent behavior under different scenarios (normal activity, crisis, etc). Such work on generating additional agent activity models is currently under way.

Acknowledgements

The graphs in this paper were produced with the R statistics and visualization program and the *plotrix* package for R [12][13].

References

1. Solicitation, N.P.: DDDAS: Dynamic data-driven application systems. NSF Program Solicitation NSF 05-570 (June 2005)
2. Darema, F.: Dynamic Data Driven Application Systems: A new paradigm for application simulations and measurements. In: Bubak, M., van Albada, G.D., Sloot, P.M.A., Dongarra, J. (eds.) ICCS 2004. LNCS, vol. 3038, pp. 662–669. Springer, Heidelberg (2004)

3. Schoenharl, T., Bravo, R., Madey, G.: WIPER: Leveraging the cell phone network for emergency response. International Journal of Intelligent Control and Systems 11(4) (December 2006)
4. Balci, O.: Verification, Validation, and Testing. In: Handbook of Simulation: Principles, Methodology, Advances, Applications, and Practice, John Wiley & Sons, New York (1998)
5. Kennedy, R.C.: Verification and validation of agent-based and equation-based simulations and bioinformatics computing: identifying transposable elements in the aedes aegypti genome. Master's thesis, University of Notre Dame (2006)
6. Xiang, X., Kennedy, R., Madey, G., Cabaniss, S.: Verification and validation of agent-based scientific simulation models. In: Yilmaz, L. (ed.) Proceedings of the 2005 Agent-Directed Simulation Symposium, April 2005. The Society for Modeling and Simulation International, vol. 37, pp. 47–55 (2005)
7. Davis, W.J.: On-line Simulation: Need and Evolving Research Requirements. In: Handbook of Simulation: Principles, Methodology, Advances, Applications, and Practice, John Wiley & Sons, New York (1998)
8. Andradóttir, S.: Simulation Optimization. In: Handbook of Simulation: Principles, Methodology, Advances, Applications, and Practice, John Wiley & Sons, New York (1998)
9. Wikipedia: Voronoi diagram — Wikipedia, the free encyclopedia (2006) (accessed April 25, 2006),
   http://en.wikipedia.org/w/index.php?title=Voronoi_diagram&oldid=47842110
10. Wikipedia: Chebyshev distance — wikipedia, the free encyclopedia (2007) (accessed May 14, 2007)
11. Yan, P., Schoenharl, T., Madey, G.R.: Application of markovmodulated poisson processes to anomaly detection in a cellular telephone network. Technical report, University of Notre Dame (2007)
12. R Development Core Team: R: A Language and Environment for Statistical Computing. R Foundation for Statistical Computing, Vienna, Austria (2006) ISBN 3-900051-07-0
13. Lemon, J., Bolker, B., Oom, S., Klein, E., Rowlingson, B., Wickham, H., Tyagi, A., Eterradossi, O., Grothendieck, G.: Plotrix: Various plotting functions, R package version 2.2 (2007)

Using Intelligent Optimization Methods to Improve the Group Method of Data Handling in Time Series Prediction

Maysam Abbod and Karishma Deshpande

School of Engineering and Design,
Brunel University, West London, UK Uxbridge, UK, UB8 3PH
Maysam.Abbod@Brunel.ac.uk

Abstract. In this paper we show how the performance of the basic algorithm of the Group Method of Data Handling (GMDH) can be improved using Genetic Algorithms (GA) and Particle Swarm Optimization (PSO). The new improved GMDH is then used to predict currency exchange rates: the US Dollar to the Euros. The performance of the hybrid GMDHs are compared with that of the conventional GMDH. Two performance measures, the root mean squared error and the mean absolute percentage errors show that the hybrid GMDH algorithm gives more accurate predictions than the conventional GMDH algorithm.

Keywords: GMDH, GA, PSO, time series, prediction, finance.

1 Introduction

Forecasting future trends of many observable phenomena remains of great interest to a wide circle of people. This requirement has maintained a high rate of activity in various research fields dedicated to temporal prediction methodologies. Two such important application domains are financial markets and environmental systems. Predicting such systems has been attempted for decades but it remains such a challenging task for a wide array of modeling paradigms.

The foreign exchange market is a large business with a large turnover in which trading takes place round the clock and all over the world. Consequently, financial time series prediction has become a very popular and ever growing business. It can be classified as a real world system characterized by the presence of non-linear relations. Modeling real world systems is a demanding task where many factors must be taken into account. The quantity and quality of the data points, the presence of external circumstances such as political issues and inflation rate make the modeling procedure a very difficult task. A survey of the different methods available for modeling non-linear systems is given in Billings [1].

Some researchers tried auto-regressive methods to predict foreign currency exchange rates [2]. Episcopos and Davis [6] used the GARCH and ARIMA statistical methods to identify the model. These methods have not always produced good results which have urged scientists to explore other more effective methods.

M. Bubak et al. (Eds.): ICCS 2008, Part III, LNCS 5103, pp. 16–25, 2008.

During the last two decades adaptive modeling techniques, like neural networks and genetic algorithms, have been extensively used in the field of economics. A list of examples using such methods is given in Deboeck [4] and Chen [3]. An extension to the genetic algorithm paradigm [9] is the versatile genetic programming method introduced by Koza [11]. Scientists were quick to use this method in many aspects of mathematical modeling.

An alternative non-linear modeling method, which was introduced in the late sixties, is the Group Method of Data Handling [11]. Since its introduction researchers from all over the world who used the GMDH in modeling were astonished by its prediction accuracy. Many of these applications were in the field of time series prediction. Parks et al [14] used the GMDH to find a model for the British economy. Ikeda et al [10] used a sequential GMDH algorithm in river flow prediction. Hayashi and Tanaka [8] used a fuzzy GMDH algorithm to predict the production of computers in Japan. Robinson and Mort [15] used an optimized GMDH algorithm for predicting foreign currency exchange rate.

In this paper we describe how the GMDH can be used in conjunction with Genetic Algorithms (GA) and with Particle Swarm Optimization (PSO) to improve the prediction accuracy of the standard GMDH algorithm when it is applied to time series in the form of financial data.

2 The Data

In this paper both the hybrid GMDHs and the conventional GMDH algorithms were used to make one step ahead prediction of the exchange rate from US Dollars to Euros (USD2EURO). Values from 29 September, 2004 to 5 October 2007 were obtained from the website www.oanda.com, a total of 1102 points (Fig. 1). The first 1000 points were used in the training and checking of the GMDH algorithm. The last 102 points were unseen by the algorithm throughout its computation. The performance of the algorithm was evaluated on the last 102 points of the data.

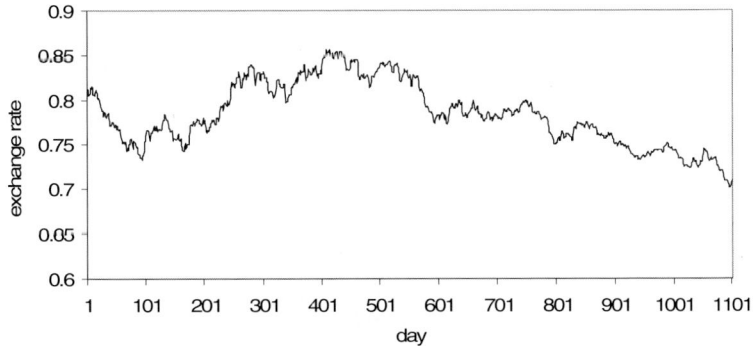

Fig. 1. USD2EURO from 29 Sept, 2004 to 5 Oct, 2007

3 The Group Method of Data Handling (GMDH)

GMDH method was developed by Ivakhnenko [11] as a rival to the method of sto-chastic approximation. The proposed algorithm is based on a multilayer structure using, for each pair of variables, a second order polynomial of the form:

$$Y = A_O + A_1X_I + A_2X_J + A_3X_IX_J + A_4X_I^2 + A_5X_J^2 \tag{1}$$

where x_i and x_j are input variables and y is the corresponding output value. The data points are divided into training and checking sets. The coefficients of the polynomial are found by regression on the training set and its output is then evaluated and tested for suitability using the data points in the checking set. An external criterion, usually the mean squared error (mse), is then used to select the polynomials that are allowed to proceed to the next layer. The output of the selected polynomials becomes the new input values at the next layer. The whole procedure is repeated until the lowest *mse* is no longer smaller than that of the previous layer. The model of the data can be com-puted by tracing back the path of the polynomials that corresponded to the lowest *mse* in each layer.

4 Intelligent Systems Optimization

4.1 Genetic Algorithms

GAs are exploratory search and optimization methods that were devised on the prin-ciples of natural evolution and population genetics [9]. Unlike other optimization techniques, a GA does not require mathematical descriptions of the optimization problem, but instead relies on a cost-function, in order to assess the fitness of a par-ticular solution to the problem in question. Possible solution candidates are repre-sented by a population of individuals (generation) and each individual is encoded as a binary string, referred to as a chromosome containing a well-defined number of alleles (1's and 0's). Initially, a population of individuals is generated and the fittest individuals are chosen by ranking them according to a *priori*-defined fitness-function, which is evaluated for each member of this population. In order to create another better population from the initial one, a mating process is carried out among the fittest individuals in the previous generation, since the relative fitness of each individual is used as a criterion for choice. Hence, the selected individuals are ran-domly combined in pairs to produce two off-springs by *crossing over* parts of their chromosomes at a randomly chosen position of the string. These new off-springs represent a better solution to the problem. In order to provide extra excitation to the process of generation, randomly chosen bits in the strings are inverted (0's to 1's and 1's to 0's). This mechanism is known as *mutation* and helps to speed up convergence and prevents the population from being predominated by the same individuals. All in all, it ensures that the solution set is never naught. A compromise, however, should be reached between too much or too little excitation by choosing a small probability of mutation.

4.2 Practical Swarm Optimization

Particle Swarm Optimization is a global minimization technique for dealing with problems in which a best solution can be represented using position and velocity components. All particles remember the best position they have seen, and communicate this position to the other members of the swarm. The particles will adjust their own positions and velocity based on this information. The communication can be common to the whole swarm, or be divided into local neighborhoods of particles [12].

5   Results and Discussions

The GMDH network, as mentioned earlier, uses a quadratic polynomial as the transfer function in each layer. A question arises: what if the relationship between the input and the output is not best described by a quadratic polynomial? This leads to another question: Could a better global optimization method replace the regression technique to fit the quadratic polynomial and generate a function that described the input-output relationship more accurately, leading to an improvement in the prediction accuracy of the GMDH algorithm?

The GA- and PSO-GMDH algorithm is similar to the conventional GMDH algorithm except that the transfer function polynomials are fitted using better optimization algorithms, namely PSO and GA. The optimization algorithm is applied to the data while the GMDH iterates through in order to find the best function that maps the input to the output. It is important to note that the GA and PSO are applied separately in order to find an exact mapping between the input and the output at different iteration stages.

Two performance measures were used in assessing the accuracy of the conventional GMDH and the hybrid GMDH: the mean absolute percentage error, denoted MAPE, (equation 2) and the widely used root mean squared error, RMSE, (equation 3).

$$\text{MAPE} - \frac{1}{n} \sum_{i=1}^{n} \frac{abs(Y_i - Z_i)}{Y_i} \times 100\% \tag{2}$$

$$\text{RMSE} = \sqrt{\frac{1}{n} \sum_{i=1}^{n} (Y_i - Z_i)^2} \tag{3}$$

where n is the number of variables, Y is the actual output and Z is the predicted output.

The PSO algorithm was set to a population size of 100, while the inertial cognitive and social constants are as follows [5]: $w = 0.7298$; $c1 = 1.49618$; $c2 = 1.49618$

The GA algorithm was set as a binary code of 12 bits for each of the parameters of the polynomial. The polynomial has 6 parameters so that makes the chromosome

12×6=72 bits long. The 12 bit binary number maps to a search space of -2 to +2 units in decimals. The 12 bits settings allows 4096 steps for a search space of 4 units which is a step size of 4/4096= 0.0000976. The GA was set with a mutation rate of 0.1 and a single point crossover at a rate of 0.9.

The results are classified into two types, the training phase and the testing phase. The former phase involves 1000 data points which are used to fit the polynomials at different GMDH iterations. The later stage is to test the GMDH system to data that the system has never experienced before, in this case 102 data points were used. Figs. 2a and 2b shows the predicted against the actual data for a standard GMDH for training and testing respectively. While Figs. 3a and 3b show the same figures but for the PSO-GMDH, and finally Figs. 4a and 4b shows the results of the GA-GMDH version.

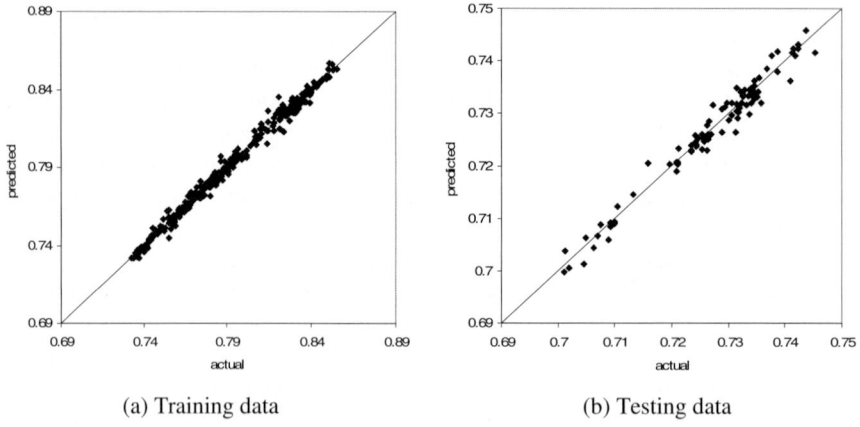

(a) Training data   (b) Testing data

Fig. 2. Graphs of GMDH predicted against actual USD2EURO

The GA-GMDH algorithm usually produces better results due to its global search features. This comes at a price of being slow and computationally intensive. For accurate results, large generation and wide searching space are required. This can be accommodated by creating large chromosomes with bigger constraints for increasing the search accuracy. In contrast, PSO is fast and has no constraints on the search space. But it produces less accurate results globally, but more accurate locally. Combining both algorithms with GMDH provides better accuracy and speed. By starting with a low accuracy GA (fast calculations) for finding the rough global minima, then switching to PSO for finding a more accurate minima in the global GA minima area.

The GA was set to 8 bit settings for each variable, and then preceded by PSO with the same previous settings. Figs. 5a and 5b show the predicted against the actual data for the GA-PSO-GMDH for training and testing respectively

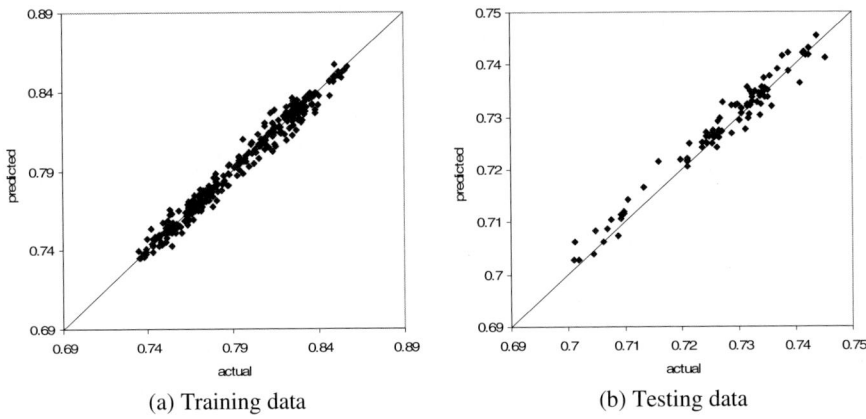

(a) Training data   (b) Testing data

Fig. 3. Graphs of PSO-GMDH predicted against actual USD2EURO

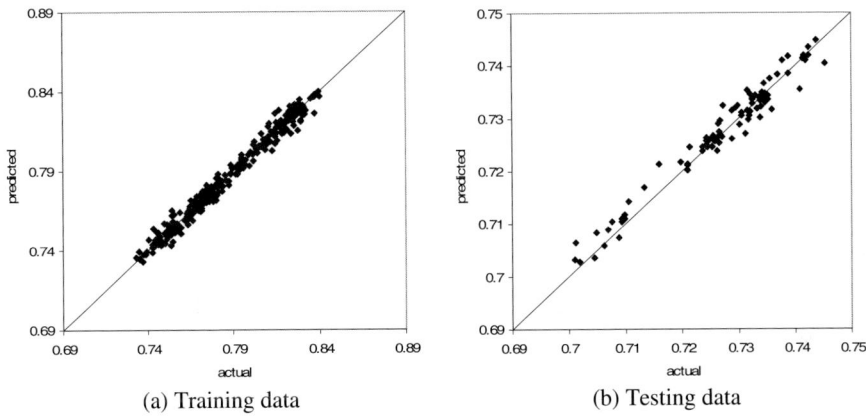

(a) Training data   (b) Testing data

Fig. 4. Graphs of GA-GMDH predicted against actual USD2EURO

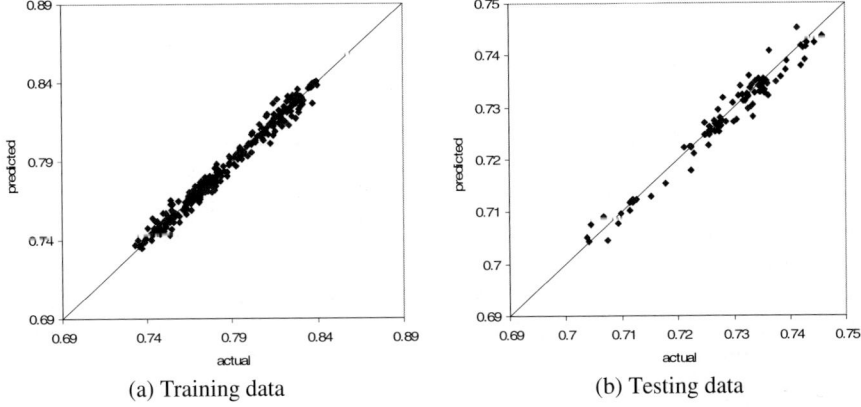

(a) Training data   (b) Testing data

Fig. 5. Graphs of GA-PSO-GMDH predicted against actual USD2EURO

The results for both networks when applied to the entire training and testing data are given in Tables 1 and 2 respectively.

Table 1. Results for USD2EURO training data

	MAPE	RMSE
Linear regression	0.2904	0.003993
GMDH	0.2772	0.003101
PSO-GMDH	0.2604	0.003085
GA-GMDH	0.2602	0.002993
GA-PSO-GMDH	0.2546	0.002989

Table 2. Results for USD2ERO testing data

	MAPE	RMSE
Linear regression	0.1986	0.001841
GMDH	0.1985	0.001831
PSO-GMDH	0.1860	0.001611
GA-GMDH	0.1653	0.001312
GA-PSO-GMDH	0.1416	0.001302

It is evident from the values of both measures that the GA-and PSO-GMDH performs better than the conventional GMDH. Values of the percentage improvement in both performance measures for all the data are given in Table 3.

Table 3. GMDH percentage improvement to the testing data performance

	MAPE Percentage improvement	RMSE Percentage improvement
PSO-GMDH	6.3	12.0
GA-GMDH	16.7	28.3
GA-PSO-GMDH	28.6	28.9

Graphs of the actual against predicted by the conventional GMDH algorithm and predicted by PSO-GMDH algorithm for the USD2EURO exchange rates are shown in Fig. 6. While Fig. 7 shows the predictions of GA-GMDH algorithm and the combinations of GA, PSO and GMDH (GA-PSO-GMDH). The prediction performance of the GMDH network depends on the number of generations over which the algorithm is allowed to evolve. The results given in this paper were produced after only 4 iterations (ambiguous – generations in terms of GA).

Several GMDH runs, using both networks, were carried out each with a different number of generations. It was found that as the individuals were allowed to evolve over a higher number of generations, the value of the minimum of the selection criterion in each generation was decreased. On the other hand, it was found that the performance measures became progressively worse. This was due to the fact that the algorithms became overspecialized for the data they were trained on to the detriment of their ability to predict the unseen data.

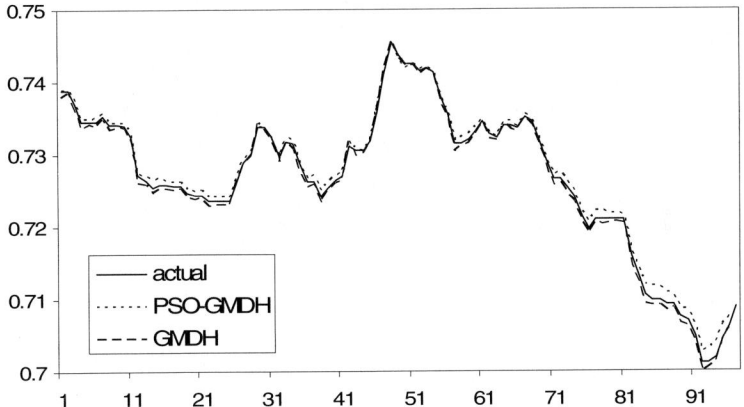

Fig. 6. Graphs of actual testing data, GMDH and PSO-GMDH predictions

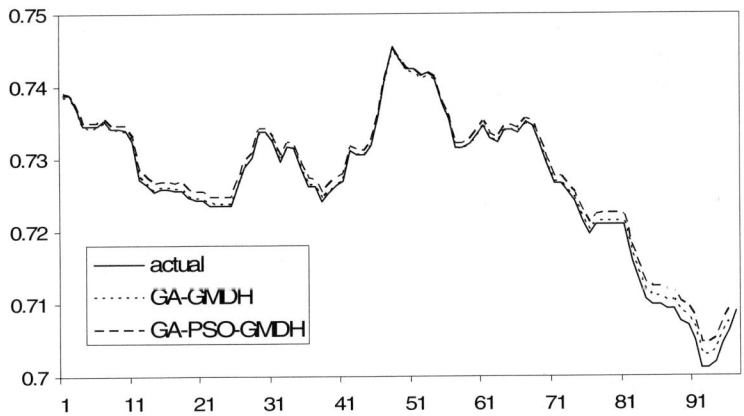

Fig. 7. Graphs of actual testing data, GA-GMDH and GA-PSO-GMDH predictions

6 Conclusions

It was shown in this paper that the performance of the conventional GMDH algorithm can be improved significantly if some information about the type of the relationship between the input and output was available and used in the GMDH run.

The fitness of the best individual in each generation improved while the individuals were allowed to evolve over more generations. This had the reverse effect on the overall performance of the network for the unseen data points. This is an area that is in need of further investigation in order to find a suitable point for terminating the building of the polynomial process.

When the same procedure carried out in this paper was repeated a few times using a different size of training and checking sets, it was noticed that the performance of both GMDH networks, particularly the conventional one, was greatly affected. This leads to the following hypothesis: there might exist a rule for dividing the sample of data that when applied the accuracy of the GMDH algorithm reaches its optimum. Work needs to be carried out to investigate the validity of this hypothesis.

The work carried out in this paper has provided an insight into the way the GMDH deals with time series prediction. While it was shown that it is possible to improve its prediction performance, the GMDH remains a robust and popular method of mathematical modeling.

References

1. Billings, S.A.: Identification of Non-Linear Systems – a Survey. IEE Proceedings on Control Theory and Applications 127(6), 272–285 (1980)
2. Chappell, D., Padmore, J., Mistry, P., Ellis, C.: A Threshold Model for the French Franc/Deutschmark Exchange Rate. Journal of Forecasting 15(3), 155–164 (1995)
3. Chen, C.H.: Neural Networks for Financial Market Prediction. In: IEEE International Conference on Neural Networks, vol. 7, pp. 1199–1202 (1994)
4. Deboeck, G. (ed.): Trading of the Edge: Neural, Genetic and Fuzzy Systems for Chaotic Financial Markets. John Wiley & Sons Inc., Chichester (1994)
5. Eberhart, R.C., Shi, Y.: Comparing Inertia Weights and Constriction Factors in Particle Swarm Optimization. In: Proceedings of the Congress on Evolutionary Computing, San Diego, USA, pp. 84–89 (2000)
6. Episcopos, A., Davis, J.: Predicting Returns on Canadian Exchange Rates with Artificial Neural Networks and EGARCH-M Models. In: Abu-Mostafa, Y., Moody, J., Weigend, A. (eds.) Neural Networks Financial Engineering. Proceedings of the third International Conference on Neural Networks in the Capital Markets, pp. 135–145. World Scientific, Singapore, London (1995)
7. Goldberg, D.E.: Genetic Algorithms in Search, Optimization and Machine Learning. Addison-Wesley, Reading (1989)
8. Hayashi, I., Tanaka, H.: The Fuzzy GMDH Algorithm by Possibility Models and its Application. Fuzzy Sets and Systems 36, 245–258 (1990)
9. Holland, J.H.: Adaption in Natural and Artificial Systems. The University of Michigan Press, Ann Arbor (1975)
10. Ikeda, S., Ochiai, M., Sawaragi, Y.: Sequential GMDH Algorithm and Its Application to River Flow Prediction. IEEE Transaction on Systems, Man and Cybernetics (July 1976)
11. Ivakhnenko, A.G.: The Group Method of Data Handling-A Rival of the Method of Stochastic Approximation. Soviet Automatic Control, vol. 13 c/c of automatika 1(3), 43–55 (1968)

12. Kennedy, J., Eberhart, R.: Particle Swarm Optimization. In: Proc. IEEE Int'l. Conf. on Neural Networks, Perth, Australia, November 1995, pp. 1942–1948 (1995)
13. Koza, J.R.: Genetic Programming: on the programming of computers by means of natural selection. MIT Press, Cambridge (1992)
14. Parks, P., Ivakhnenko, A.G., Boichuk, L.M., Svetalsky, B.K.: A Self-Organizing Model of British Economy for Control with Optimal Prediction using the Balance-of-Variables Criterion. Int. J. of Computer and Information Sciences 4(4) (1975)
15. Robinson, C., Mort, M.: Predicting Foreign Exchange Rates Using Neural and Genetic Models. In: Proceedings of 2nd Asian Control Conference, Seoul, Korea, July 22-25, 1997, vol. 3, pp. 115–118 (1997)

Symbiotic Simulation Control in Semiconductor Manufacturing

Heiko Aydt[1], Stephen John Turner[1], Wentong Cai[1], Malcolm Yoke Hean Low[1], Peter Lendermann[2], and Boon Ping Gan[2]

[1] School of Computer Engineering, Nanyang Technological University,
Singapore 639798
{aydt,assjturner,aswtcai,yhlow}@ntu.edu.sg
[2] D-SIMLAB Technologies Pte Ltd, 30 Biopolis Street #09-02, Matrix,
Singapore 138671
{peter,boonping}@d-simlab.com

Abstract. Semiconductor manufacturing is a highly complex and asset intensive process. Solutions are needed to automate control of equipment and improve efficiency. We describe a symbiotic simulation control system which uses reactive what-if analysis to find a stable configuration of a wet bench tool set. This control system is based on a generic framework for symbiotic simulation. We show that symbiotic simulation can be used to make control decisions in near real-time. Furthermore, we show that our approach yields a notable performance improvement over common practise.

1 Introduction

Semiconductor manufacturing is a complex process which turns pure silicon wafers into integrated circuits with thousands of components. The whole process may require up to several hundred processing steps and up to three months for production [1]. One crucial factor for competitiveness in semiconductor manufacturing is ongoing improvement of the manufacturing process [1]. Automation is a critical factor in this context and the semiconductor industry invests several million US dollars for solutions. According to [2], investments for integrated automation solutions can range between US$ 130 million and US$ 180 million. These integrated solutions also include systems for real-time control of equipment. In addition to high complexity, the semiconductor manufacturing process is also an asset intensive process. A variety of tools are used in semiconductor manufacturing to process wafers and a single tool can cost up to US$ 2 million [3]. Improvement of tool efficiency is therefore important in order to reduce cost for additional tools.

Several performance metrics are considered in semiconductor manufacturing, of which throughput, cycle time, and on-time delivery are considered the most important ones [4]. The performance of most tools depends on their operation mode, further referred to as configuration, and the distribution of different products, further referred to as product mix. If the product mix is changing, it might

M. Bubak et al. (Eds.): ICCS 2008, Part III, LNCS 5103, pp. 26–35, 2008.

be necessary to reconfigure various tools in order to maintain performance targets. Decisions regarding this are currently made by engineers who estimate the required capacity for each product mix. Decisions are then implemented by workers who manually change the configuration of a tool. However, workers do not always strictly follow their instructions. This, and the fact that decision making relies mostly on the experience of engineers, causes the demand for an automated control solution.

Simulations have already been used for offline analysis and optimisation of parameters in the manufacturing process. However, on-line solutions which use simulation for real-time control are still in their infancy. Symbiotic simulation, which is driven by sensor data and capable of making control decisions in real-time, seems to be a promising solution for automation and to improve equipment efficiency. In this paper, we show how a dynamic data driven application can be used for operational decision making regarding the configuration of a wet bench toolset (WBTS), which is used for one of the many critical operations in semiconductor manufacturing. We describe a *symbiotic simulation control system (SSCS)* which is based on a generic framework for symbiotic simulation being developed by us. The application of symbiotic simulation in semiconductor manufacturing also represents the first show-case for the generic framework. The need for such a generic symbiotic framework has been articulated in [5].

The paper is structured as follows: Before presenting our own solution, we give an overview of related work in Sect. 2. This is followed by an introduction to the SSCS and a brief overview of the generic framework in Sect. 3. A more detailed description of the WBTS is given in Sect. 4. We compare common practise with our approach of controlling the WBTS. Therefore, we explain both approaches in Sect. 5 before presenting our experimental results in Sect. 6. Our conclusions are given in Sect. 7.

2   Related Work

The term symbiotic simulation was coined at the Dagstuhl Seminar on Grand Challenges for Modeling and Simulation in 2002 [6]. Symbiotic simulation is closely related to dynamic data-driven application systems (DDDAS). While DDDAS emphasises the symbiotic relationship between an application and a measurement process, symbiotic simulation rather focuses on the symbiotic relationship between a simulation and a physical system. However, symbiotic simulation is also dynamic data-driven and may use sensor steering to improve the accuracy of simulations. Several applications which use symbiotic simulation have been developed since 2002. These include the application of symbiotic simulation for UAV path planning [7], social sciences [8],[9], and business process re-engineering [10]. In the context of semiconductor manufacturing, symbiotic simulation has already been used in [11] to optimise backend operations.

Simulations have been used in several ways to improve the performance of a semiconductor manufacturing process. Efforts in this context mainly aim to improve the cycle time. For example, in [3] a two day reduction of total cycle time could be achieved by investigating a real wafer factory based on simulations. Other work which is concerned with improving cycle times can be found in [12],[13],[14]. Wet bench tools have been the subject of various works. Lee et al. [15] examine the scheduling problem of a wet station for wafer cleaning using Petri net modelling and Gan et al. [16] evaluate the effect of upgrading a bottleneck furnace to the wet bench performance with increasing wafer demand.

3   Symbiotic Simulation Control System (SSCS)

An SSCS is a specific class of symbiotic simulation system. It makes decisions based on the simulation and evaluation of what-if scenarios, each representing an alternative decision, and implements them by means of actuators which are available to the system. A reactive what-if analysis aims to find immediate solutions to sudden, and potentially critical, conditions. For example, the performance of a manufacturing system is observed and a reactive what-if analysis is carried out once the performance drops below a predefined value, i.e., once the performance becomes critical. Various what-if scenarios, each representing alternative decision options (e.g., tool configurations), are simulated and analysed. The decision option of the scenario which performs best in simulation is implemented.

The SSCS is realised using an implementation of our generic framework for symbiotic simulation systems. This framework is based on Jadex[1], a BDI agent system, and represents a library of functional components which are typically required in a symbiotic simulation system. For example, creating and simulating alternative what-if scenarios is an essential concept in symbiotic simulation. Each of these functionalities, among several others, is realised as functional component and provided by the framework.

We use capabilities [17],[18] to implement the various functional components of the framework. The concept of capabilities is used in BDI agent systems to facilitate modularisation of functionality. Agents can be equipped with an arbitrary number of capabilities. This allows flexible design of application-specific symbiotic simulation systems. For example, the design of inherently distributed applications may involve several agents, of which each of them is equipped with only one capability. On the other hand, it is possible to design a symbiotic simulation system which is represented by a single agent only. In this case, the agent is equipped with all required capabilities. The use of capabilities also allows the design of tailor-made symbiotic simulation systems by using only those capabilities which are required in the context of the application.

There are a number of distinct capabilities which are required to realise a symbiotic simulation system. The following description of the SSCS includes only those which are used in the context of the WBTS application. An overview of the WBTS-SSCS capabilities is illustrated in Fig. 1.

The architecture of a symbiotic simulation system consists of three layers for *perception*, *processing*, and *actuation*. The perception layer is responsible for providing measured data to the processing layer where what-if analysis is carried

---

[1] http://vsis-www.informatik.uni-hamburg.de/projects/jadex/

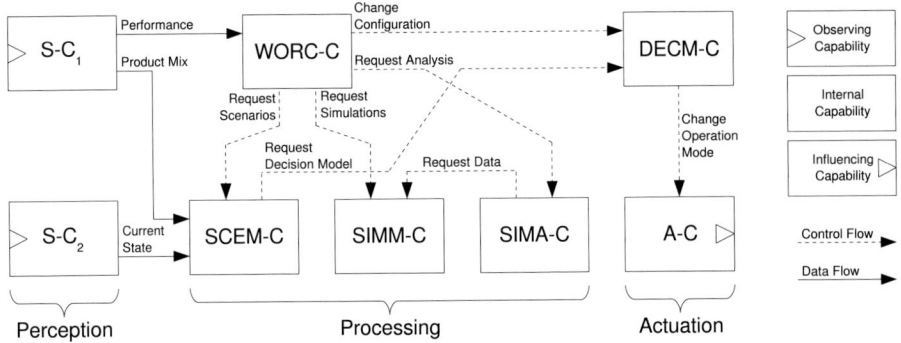

Fig. 1. Overview of various capabilities in the WTBS-SSCS

out. Once a decision is made, the actuation layer is responsible for implementing it. Each layer is associated with a number of specific capabilities.

The workflow control capability WORC-C is the heart of the SSCS. It is responsible for interpreting performance measurements provided by sensor capability S-C_1, triggering the what-if analysis process if the performance becomes critical, and making control decisions regarding the configuration of wet bench tools. If a what-if analysis is triggered, the WORC-C requests a set of scenarios from the scenario management capability SCEM-C, invokes simulations using the simulation management capability SIMM-C, and interprets analysis results provided by the simulation analysis capability SIMA-C. Based on these results, the best decision option is determined by the WORC-C and implemented by using the decision management capability DECM-C.

An SSCS can only implement decision options which are supported by actuators that are available to the SSCS. A decision model is used to reflect possible control options, depending on the available actuators and their abilities to control different aspects of the physical system. In the context of the WBTS application, the decision model reflects the possible operation modes for each wet bench tool. The DECM-C keeps track of available actuators and updates the decision model accordingly.

When requested by the WORC-C, the SCEM-C creates a number of scenarios based on the decision model. Each scenario reflects a different decision option, i.e., a different wet bench configuration. A scenario also consists of an initial state for the simulation run. This initial state is the current state of the physical system, provided by S-C_2. The state of the system represents a snapshot of the WBTS including all wafer lots which are currently being processed. In addition to the decision option and the initial state, a scenario also consists of a given product mix which is used to drive the simulation. The current product mix is provided by S-C_1.

Sensors, actuators, and simulators are inherently application specific. Therefore, the framework cannot provide generic solutions for their corresponding capabilities. Instead, application specific implementations have to be provided by the application developer. The framework provides generic solutions for all other capabilities. However, this does not rule out the possibility of using application specific implementations instead of generic standard solutions.

4   Wet Bench Tool Set (WBTS)

In semiconductor manufacturing, a wet bench is used to clean wafers after certain fabrication steps to yield higher wafer reliability [15],[19]. A wet bench tool consists of a number of baths which contain chemical liquids and process wafer lots according to recipes. Typically, a wafer fab operates a number of wet benches with different bath setups to support a variety of recipes. A recipe specifies constraints regarding the sequence and timing for the cleaning process. Each lot is associated with a recipe and has to be processed strictly according to it. Neither the sequence of chemical liquids in which a lot is processed, nor the time each wafer spends in a particular chemical liquid must be changed. Violation of these requirements would lead to significantly reduced yield.

Particles of contaminants are introduced into the baths during the cleaning process. Therefore, the chemical liquids have to be changed after a certain time. Some recipes introduce particles faster than others. We therefore distinguish between 'dirty' and 'clean' recipes [16]. In order not to compromise the quality of wafers associated with clean recipes, wafers associated with dirty recipes are processed by separate wet benches. Thus, a wet bench operates either in 'clean' mode or 'dirty' mode. By operation mode we mean that a wet bench is dedicated to either one of the two recipe types. Switching the operation mode from 'clean' to 'dirty' does not require any particular activity. However, switching from 'dirty' to 'clean' requires a complete change of liquids. This activity takes up to several hours [16].

5   Control Approaches

5.1   Common Practise Approach

The common practise approach involves engineers who decide upon reconfiguration of wet bench tools. Decision making depends mostly on experience and is therefore difficult to capture. With information from the semiconductor industry we developed the following heuristic: 1) If the number of pending wafer lots exceeds 40, the critical recipe (the most frequent recipe) is determined. 2) In the next step, the set of wet benches is identified which is capable of processing the critical recipe but is not configured to do so yet. 3) One of the wet benches is selected and reconfigured accordingly. This is due to the fact that engineers tend to be careful and change only one wet bench at a time. 4) After changing the operation mode, the system is allowed some time to settle. 5) If, after the settling-in time, the situation did not improve the process is repeated.

We assume a settling-in time of one hour when the operation mode is switched from 'clean' to 'dirty' and four hours otherwise. The latter takes account for a three hour maintenance process where chemicals are changed.

5.2   SSCS Approach

Similar to the common practise approach, the WORC-C of the SSCS observes the number of pending wafer lots. If this number exceeds 40, a reactive what-if analysis is triggered. Each what-if scenario reflects the current state of the physical system with an alternative wet bench configuration. In this particular application, we use eight wet bench tools, of which each can operate either in 'clean' or 'dirty' mode. This results in a total of 256 different possible scenarios, created by the SCEM-C. The performance for the next 24 hours is simulated by the SIMM-C for each scenario. The SIMA-C analyses the simulation results and determines the average performance of each scenario in terms of throughput. The best configuration is determined by the WORC-C and implemented by the DECM-C. To implement a decision, the DECM-C sends corresponding control signals to the A-C, which is capable of changing the operation modes of wet benches in the physical system. Once a decision is implemented, the SSCS is inactive for a fixed settling-in period of four hours.

6   Experiments

For practical reasons, experiments were not performed with a real physical system. Instead, we used a simulation of the WBTS to emulate the physical system. The emulator was paced at a speed 3600 times faster than real-time. This enabled us to perform experiments, reflecting a period of several months, within a few hours. We also used a mechanism to create snapshots of the emulator at any time. Such a snapshot can then be used as initial state for the simulation of a what-if scenario.

The simulation model is deterministic with the only exception being the lot generating process. This process generates arrival events of wafer lots with a distribution according to a given product mix. The SSCS has to consider this when performing a what-if analysis. Therefore, each what-if scenario is simulated using five replications. A whole what-if analysis step, including scenario creation, simulation, analysis, and decision propagation, takes approximately 50-60 seconds on a cluster with 60 processors.

Each experiment was repeated ten times and the mean values over all results are presented here. All results show the performance of the WBTS in terms of cycle time. The cycle time is determined by the ratio of the overall time a wafer lot spent at the toolset to the processing time according to the associated recipe. The overall time includes the time a wafer lot has to wait before being processed.

6.1   Alternating Product Mixes

With the first experiment, we aim to show how well both approaches can maintain performance with an alternating product mix. We use two product mixes with a constant load of 900 lots per day. In this experiment, a period of 20 weeks is considered. Within this period, the product mix is changed every two weeks. We choose the product mixes so that there exists no configuration which is stable for both. This means, a configuration which is stable assuming one of the product mixes, is unstable when assuming the other product mix. An unstable configuration results in a throughput which is below the load. As a consequence, queues of pending lots will build up and an increasing cycle time can be observed. To avoid this, the wet bench configuration has to be changed eventually. Figure 2 illustrates the performance of the SSCS approach and the common practise approach.

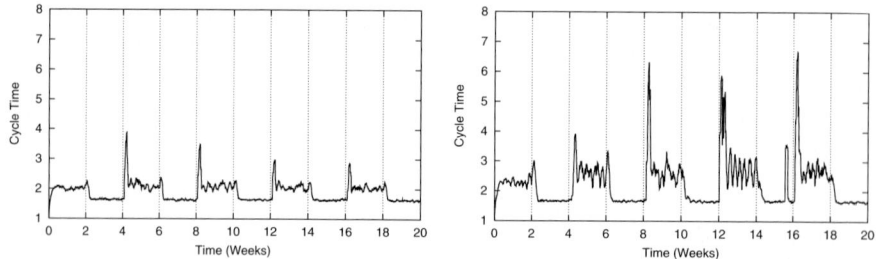

Fig. 2. Cycle time of the WBTS over a period of 20 weeks using the SSCS (left) and the common practise approach (right)

In general, the SSCS approach produces a more homogeneous performance with less variability and lower mean cycle time ($S_{SSCS} = 0.30$, $\overline{X}_{SSCS} = 1.89$) than the common practise approach ($S_{CP} = 0.76$, $\overline{X}_{CP} = 2.26$). One major disadvantage of the common practise approach is the fact that it is a trial and error approach which changes only one tool configuration at a time. Therefore, it takes longer to find a stable configuration, if at all. On the contrary the SSCS is capable of finding a stable solution, if there is one, and making all necessary changes in one step. If no stable solution exists, the solution which performed best in simulations is applied.

6.2   High Workload

At increasing load, it is necessary to oscillate between configurations. This is due to the fact that only two of the eight wet benches used in this application have a bath setup which allows them to process a large variety of recipes. All other wet benches have setups which are limited to a smaller group of recipes. Therefore, these two wet benches represent a bottleneck at high loads. To overcome this, it is necessary to switch the configurations of these wet benches back and forth.

As a consequence, the performance is stabilised. However, above a certain load, the performance cannot be stabilised anymore.

The second experiment aims to determine the critical load of both approaches. The SSCS approach, which is more sophisticated, is capable of stabilising the performance at higher loads than the common practise approach. Figure 3 and 4 illustrate the performance of both approaches over a period of one month with loads of 1000/1200 and 1500/1700 lots per day, respectively.

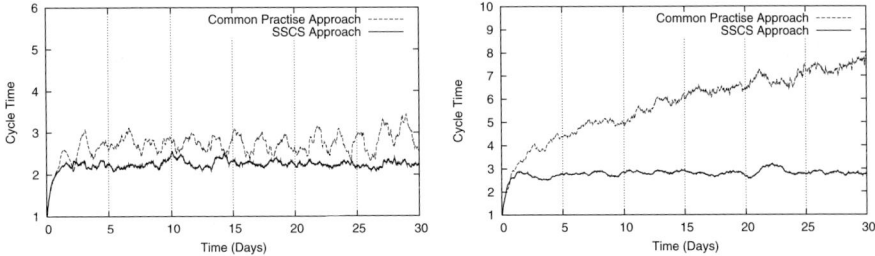

Fig. 3. Cycle time of the WBTS over a period of one month using a load of 1000 (left) and 1200 (right) lots per day

Both approaches are capable of handling a load of 1000 lots per day. At a load of 1200 lots per day, the performance becomes unstable when using the common practise approach, i.e., the cycle time is constantly increasing. In contrast, the SSCS is still capable of maintaining a stable mean cycle time.

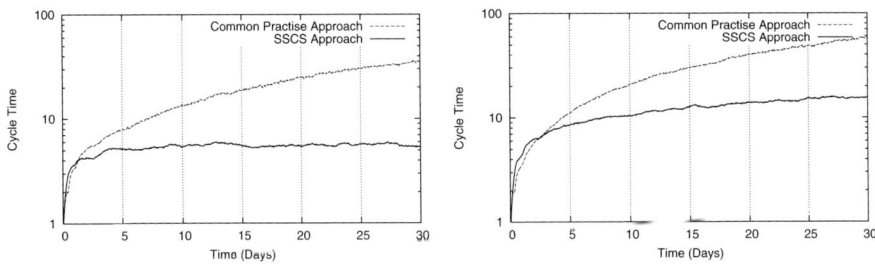

Fig. 4. Cycle time of the WBTS over a period of one month using a load of 1500 (left) and 1700 (right) lots per day. A logarithmic scale was used for cycle time values.

The mean cycle time is elevated at a load of 1500 lots per day but the SSCS approach is still able to handle it. However, at a load of 1700 lots per day the performance finally becomes unstable.

The critical load for the common practise approach and the SSCS approach is at approximately 1000-1200 and 1500-1700 lots per day, respectively. This indicates that the SSCS allows the use of loads which are 25-70% higher as compared to the common practise approach.

7   Conclusions

In this paper, we have described how a DDDAS, in particular symbiotic simulation, can be used for real-time control of semiconductor equipment and to improve performance. Our experimental results show that using the SSCS yields a notable performance improvement over common practise. This is important as higher efficiency reduces demand for additional equipment. Decision making is performed in reasonably short time. All this indicates the applicability of symbiotic simulation in the context of integrated automation solutions. In future work, we will investigate how an SSCS can be applied to an entire semiconductor factory. This will also include investigation of different forms of what-if analysis.

References

1. Potoradi, J., Boon, O., Fowler, J., Pfund, M., Mason, S.: Using simulation-based scheduling to maximize demand fulfillment in a semiconductor assembly facility. In: Proceedings of the Winter Simulation Conference, pp. 1857–1861 (2002)
2. Gan, B.P., Chan, L.P., Turner, S.J.: Interoperating simulations of automatic material handling systems and manufacturing processes. In: Proceedings of the Winter Simulation Conference, pp. 1129–1135 (2006)
3. Scholl, W., Domaschke, J.: Implementation of modeling and simulation in semiconductor wafer fabrication with time constraints between wet etch and furnace operations. In: IEEE Transactions on Semiconductor Manufacturing, August 2000, vol. 13, pp. 273–277 (2000)
4. Pfund, M.E., Mason, S.J., Fowler, J.W.: Semiconductor Manufacturing Scheduling and Dispatching. In: Handbook of Production Scheduling. International Series in Operations Research & Management Science, vol. 89, pp. 213–241. Springer, New York (2006)
5. Huang, S.Y., Cai, W., Turner, S., Hsu, W.J., Zhou, S., Low, M.Y.H., Fujimoto, R., Ayani, R.: A generic symbiotic simulation framework. In: Proceedings of the 20th Workshop on Principles of Advanced and Distributed Simulation, pp. 131–131 (2006)
6. Fujimoto, R., Lunceford, D., Page, E. (eds.): A.M.U.: Grand challenges for modeling and simulation: Dagstuhl report. Technical Report 350, Schloss Dagstuhl. Seminar No 02351 (August 2002)
7. Kamrani, F., Ayani, R.: Using on-line simulation for adaptive path planning of UAVs. In: Proceedings of the 11th IEEE International Symposium on Distributed Simulation and Real-time Applications, Chania, Greece, October 2007, pp. 167–174 (2007)
8. Kennedy, C., Theodoropoulos, G.K., Sorge, V., Ferrari, E., Lee, P., Skelcher, C.: AIMSS: An architecture for data driven simulations in the social sciences. In: Shi, Y., van Albada, G.D., Dongarra, J., Sloot, P.M.A. (eds.) ICCS 2007. LNCS, vol. 4487, pp. 1098–1105. Springer, Heidelberg (2007)
9. Kennedy, C., Theodoropoulos, G.K.: Intelligent management of data driven simulations to support model building in the social sciences. In: Alexandrov, V.N., van Albada, G.D., Sloot, P.M.A., Dongarra, J. (eds.) ICCS 2006. LNCS, vol. 3993, pp. 562–569. Springer, Heidelberg (2006)

10. Low, M.Y.H., Turner, S.J., Ling, D., Peng, H.L., Chai, L.P., Lendermann, P., Buckley, S.: Symbiotic simulation for business process re-engineering in high-tech manufacturing and service networks. In: Proceedings of the Winter Simulation Conference (2007)
11. Low, M.Y.H., Lye, K.W., Lendermann, P., Turner, S.J., Chim, R.T.W., Leo, S.H.: An agent-based approach for managing symbiotic simulation of semiconductor assembly and test operation. In: Proceedings of the 4th International Joint Conference on Autonomous Agents and Multiagent Systems, pp. 85–92. ACM Press, New York (2005)
12. Rulkens, H., van Campen, E., van Herk, J., Rooda, J.: Batch size optimization of a furnace and pre-clean area by using dynamic simulations. In: Advanced Semiconductor Manufacturing Conference and Workshop (IEEE/SEMI), Boston, MA, USA, September 1998, pp. 439–444 (1998)
13. Potoradi, J., Winz, G., Kam, L.W.: Determining optimal lot-size for a semiconductor back-end factory. In: Proceedings of the Winter Simulation Conference, pp. 720–726 (1999)
14. Akçali, E., Uzsoy, R., Hiscock, D.G., Moser, A.L., Teyner, T.J.: Alternative loading and dispatching policies for furnace operations in semiconductor manufacturing: a comparison by simulation. In: Proceedings of the Winter Simulation Conference, pp. 1428–1435 (2000)
15. Lee, T.E., Lee, H.Y., Lee, S.J.: Scheduling a wet station for wafer cleaning with multiple job flows and multiple wafer-handling robots. International Journal of Production Research 45, 487–507 (2007)
16. Gan, B.P., Lendermann, P., Quek, K.P.T., van der Heijden, B., Chin, C.C., Koh, C.Y.: Simulation analysis on the impact of furnace batch size increase in a deposition loop. In: Perrone, L.F., Lawson, B., Liu, J., Wieland, F.P. (eds.) Proceedings of the Winter Simulation Conference, pp. 1821–1828 (2006)
17. Busetta, P., Howden, N., Rönnquist, R., Hodgson, A.: Structuring BDI agents in functional clusters. In: Jennings, N.R. (ed.) ATAL 1999. LNCS, vol. 1757, pp. 277–289. Springer, Heidelberg (2000)
18. Braubach, L., Pokahr, A., Lamersdorf, W.: Extending the capability concept for flexible BDI agent modularization. In: Bordini, R.H., Dastani, M., Dix, J., Seghrouchni, A.E.F. (eds.) PROMAS 2005. LNCS (LNAI), vol. 3862, pp. 139–155. Springer, Heidelberg (2006)
19. Lu, J.K., Ko, F.H., Chu, T.C., Sun, Y.C., Wang, M.Y., Wang, T.K.: Evaluation of cleaning efficiency with a radioactive tracer and development of a microwave digestion method for semiconductor processes. Analytica Chimica Acta 407, 291–300 (2000)

Applying a Dynamic Data Driven Genetic Algorithm to Improve Forest Fire Spread Prediction*

Mónica Denham, Ana Cortés, Tomàs Margalef, and Emilio Luque

Departament d' Arquitectura de Computadors i Sistemes Operatius, E.T.S.E.,
Universitat Autònoma de Barcelona, 08193 - Bellaterra (Barcelona) Spain

Abstract. This work represents the first step toward a DDDAS for Wildland Fire Prediction where our main efforts are oriented to take advantage of the computing power provided by High Performance Computing systems to, on the one hand, propose computational data driven steering strategies to overcome input data uncertainty and, on the other hand, to reduce the execution time of the whole prediction process in order to be reliable during real-time crisis. In particular, this work is focused on the description of a Dynamic Data Driven Genetic Algorithm used as steering strategy to automatic adjust certain input data values of forest fire simulators taking into account the underlying propagation model and the real fire behavior.

1 Introduction

Forest fires are one of the most important threats to forest areas in the whole world. In the lasts years, thousands of hectares were lost by wildfires action. Forest areas losses damage the nature, attempting on ecological balance. Death of different species of animals and plants, profitable areas loss, air pollution, floods, water contamination, are some of the consequences of forest fires. At the same time these problems cause different diseases, famine, animals and vegetables extinction, etc. These facts attempt our standard of living.

Some forest fires occur by nature itself: long dry seasons, elevated temperatures, electric storms, could generate wildland fires. These type of fires can help in the ecological balance: young plants take place where there were old and perhaps unproductive trees. Nature keeps number of fires limited, but in the last years this number was increased by human factors. More than 90% of forest fires are provoked by human hand (accidents, carelessness and negligence) [6]: 16000 hectares were burned in Gran Canaria last summer (2007) during 6 days of strong fires [11]. In Tenerife 15000 hectares were burned at the same time [11]. More than 60 deaths were occurred during forest fire in Greek last July [12].

---

* This work has been supported by the MEC-Spain under contracts TIN 2004-03388 and TIN 2007-64974.

M. Bubak et al. (Eds.): ICCS 2008, Part III, LNCS 5103, pp. 36–45, 2008.

Whereas the number of wildfires had increased in the last years, human work in this area had increased too: studies, strategies and tools to prevent forest fires as well as to reduce the fire during a disaster were developed these last years. Nowadays, several forest fires simulators exist for helping and improving this work. Most of these simulators are based on Rothermel mathematical model [6]. This model describes fire behavior through mathematical equations.

Simulators are used to predict fire behavior, improving the accuracy of actions and reducing fire effects. Several of these simulators use a large number of inputs for describing the environment where fire occurs. It is very difficult to dispose of exact real-time values of these parameters. Several of them change with time: air and fuel humidity change along the day (day-night cycle), weather changes due to elevated temperatures provoked by fire, fires generate very strong gust of winds, etc. Furthermore lots of these parameters have their own behavior pattern and it can change very strongly within a wildland fire. This input data uncertainty causes predictions that are far from the real fire propagation.

During a real hazard such as wildland fire both, the accuracy of the prediction propagation and the prediction response time are crucial key points. For this reason, we propose a fire propagation prediction system, which explores multiple fire propagation scenarios obtained from combining different values of the input parameters by dynamically adapting those scenarios according to observed real fire evolution. These characteristics directly match to the Dynamic Data Driven Application Systems (DDDAS) definition: *DDDAS is a paradigm whereby application/simulations and measurements become a symbiotic feedback control system. DDDAS entails the ability to dynamically incorporate additional data into an executing application, and in reverse, the ability of an application to dynamically steer the measurement process* [7].

An optimal framework for a reliable DDDAS for Wildland Fire Prediction must consider, among others, the following issues: the ability to dynamically couple models from different disciplines; real-time data assimilation strategies for being further injected into the running system; steering strategies for automatic adjusting either models or input data parameters and to have access to enough computational resources to be able to obtain the prediction results under strict real-time deadlines. Some current work on this area could be found in [10,2].

Our currently work consists on the first step toward a DDDAS for Wildland Fire Prediction where our main efforts are oriented to take advantage of the computing power provided by High Performance Computing systems, on the one hand, to propose computational data driven steering strategies to overcome input data uncertainty and, on the other hand, to reduce the execution time of the whole prediction process in order to be reliable during real-time crisis.

In particular, this work is focused on the first working line. A Dynamic Data Driven Genetic Algorithm (GA) is used as steering strategy to automatic adjust certain input data values taking into account the underlying propagation model and the real fire behavior.

This work is organized as follow. Next section describes the proposed dynamic data driven forest fire prediction methodology compared to the classical prediction scheme. Section 3 is focused to describe the Calibration stage of the proposed prediction methodology. The experimental results are shown in section 4 and, finally, the main conclusions are reported in section 5.

2   Forest Fire Spread Prediction

Traditionally, forest fire spread prediction has been performed using a particular forest fire simulator as a prediction tool. Forest fire simulators need to be fed with several input data such as the initial fire line (what we call RFt_0: Real Fire at time t_0), meteorological data, vegetation data, etc. The simulator uses this input data to obtain the simulated fire line for a subsequent instant t_1(we will refer to as SFt_1: Simulated Fire at time t_1). This classical forest fire prediction method is showed in Fig. 1 (a).

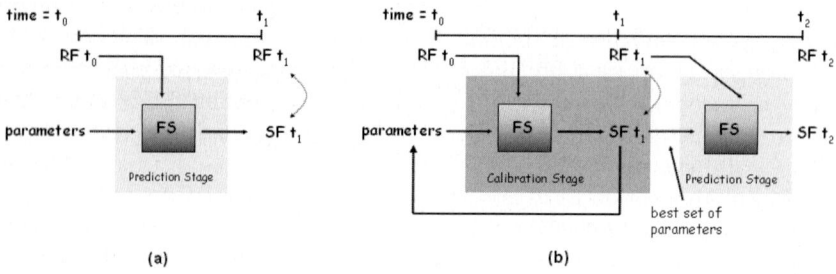

Fig. 1. (a) Classical Forest Fire Propagation Prediction method. (b) Two Stage Propagation Prediction method.

As we can see in this figure, the only information required for the classical prediction method consists of a unique set of values for the input parameters needed by the simulator. Once this input data has been set up, the simulator is executed providing, as part of its output data, the predicted fire line after a certain period of time. Despite of its simplicity, this method has an important drawback: simulation accuracy depends on the unique set of input values, and, as we had mentioned, to dispose their real values is very difficult.

In order to overcome this drawback, we propose a wildfire propagation prediction methodology based on a two stages prediction scheme: Calibration stage and Prediction stage. The DDDAS bases are included in the Calibration stage where taking advantage of the computer power provided by High Performance Computing systems, a GA is used to explore a huge number of input parameters combinations (called scenarios). For this stage being operative, it is necessary to be able to assimilate data about the forest fire front at two consecutive time instants (t_0 and t_1 in Fig. 1(b)).

Each one of the generated scenarios will be fed into the underlying wildland fire simulator, as well as the forest fire front at time t_0. The obtained propagation results will be compared to the real fire propagation at time t_1 and the results obtained from this comparison plus the observed real fire propagation at time t_1 will be used as feedback information at the Calibration stage.

By summarizing, we propose to use a Dynamic Data Driven GA as steering strategy within the calibration stage whose internal operations will be dynamically changed according to the data periodically assimilated from the real fire behavior. As a result of this Dynamic Data Driven Calibration stage, we will obtain a set of input data that minimizes the propagation prediction error from t_0 to t_1. Assuming that the environmental characteristics will keep nearly constant during a certain time (from t_1 to t_2), this set of input data will be used in a classical prediction scheme to predict fire spread at t_2 (see Fig. 1(b)). In the following section Calibration stage will be described.

3   Calibration Stage

GAs are inspired in the evolutionary theory: a population of individuals (scenarios) is evolved by allowing good individual features to survive through out generations. Therefore, the Calibration stage consists of an iterative process where, at each iteration a new generation is obtained taking into account the goodness of the involved scenarios (similarity degree between simulation and real maps).

Consequently, to be able to properly set up the whole Calibration stage we need, on the one hand, to determine the data that must be assimilated from the observed fire propagation for being posteriorly injected in the calibration process. And, on the other hand, the particular implementation of the dynamic data driven GA. The following sections will focus on these topics.

3.1   Fire Spread Model

Several studies have demonstrated that the parameters that most affect forest fire propagation are wind speed, wind direction and slope characteristics. There exist other factors that also influence forest fire spread such as fuel moisture, fuel type and so on, however, it has been shown that the fire rate of spread is less sensitive to these parameters [1]. Therefore, the proposed fire propagation prediction methodology shown in Fig. 1(b) uses this principle to determine what data to assimilate from the observed real fire behavior. According to this fact, we applied reverse engineering to the underlying wildland fire simulator to analytically determine how wind speed, wind direction and slope interact to generate fire propagation. In this work, we use the fire simulator *fireLib* [5].

Fig. 2 shows in a simplified way, how wind and slope are composed to obtain the direction of the maximum propagation spread, being Φ_w the wind effect, Φ_s the slope effect and α the difference between wind and slope directions (Fig. 2 (a)). Both vectors are composed to obtain a third vector, which represents the maximum propagation ratio and its direction (max and β in Fig. 2 (b)). This new vector has a component perpendicular to the slope equal to $\sin(\alpha) * \phi_w$, and a component parallel to the slope equal to $\cos(\alpha) * \phi_w + \phi_s$, and the angle β is the angle between these two vectors.

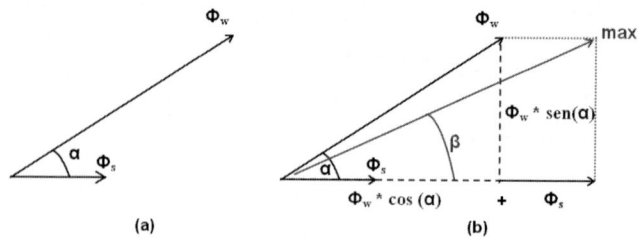

Fig. 2. (a) Wind and slope effect vectors. (b) Composition of spread vector.

Although the scheme shown in Fig. 2 models in a simplify way how *firelib* proceeds to determine fire spread rate, we can conclude that identifying two of the three vectors involved in the process, we have the key to obtain the third one. Therefore, in the next section, we will discuss how from the analysis of the observed fire behavior we assimilate the needed data to identify each one of these three vectors. Finally, in a later section, we will discuss how this information is injected in the steering strategy.

3.2   Data Assimilation

Taking into account slope and wind interactions and their role in fire behavior, we need slope features and real fire progress to obtain the main wind characteristics for performing steering method through Calibration stage. The information related to the slope could easily being obtained from the topography terrain description. Real fire progress is obtained by the analysis of real fire front progress (RFt_1 in Fig. 1 (b)), where point (x, y) in Fig. 3 is calculated and it represents the maximum rate of spread point in the map.

Applying "reverse engineering" to the fire spread model commented in the previous section, and using trigonometry, we can obtain the following equations:

$$x = \Phi_s * \cos(\delta) + \Phi_w \cos(\alpha) \qquad y = \Phi_s * \sin(\delta) + \Phi_w \sin(\alpha) \qquad (1)$$

Then, as x and y values are observed values, it is possible to isolate the wind values from equation 1 as follows:

$$\Phi_w = \frac{x - \Phi_w \cos(\delta)}{\cos(\alpha)} \qquad \alpha = \arctan(\frac{y - \Phi_s \sin(\delta)}{x - \Phi_s \cos(\delta)}) \qquad (2)$$

Once wind factor (Φ_w) is obtained, the "ideal" values for the wind speed and wind direction for the observed x, y point could be deduced. This data will be injected as feedback information inside the dynamic data driven GA to guide the operations applied to a given generation.

It should be recalled that all figures and diagrams are simplified. We are dealing with physics phenomena (fire, wind, slope), where each of them has its own behavior and, in addition, there are physics interactions between them.

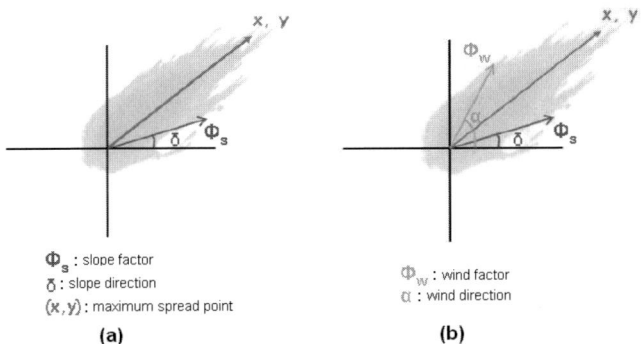

Fig. 3. Available data: slope and point of maximum rate of spread (a) and, the obtained wind vector (b)

3.3 Dynamic Data Driven Genetic Algorithm

The *fireLib* simulator has basically the following input parameters: terrain slope (direction and inclination), wind speed and wind direction, moisture content of the live fuel, moisture content of the dead fuel (at three different times), kind of vegetation (one of the 13 defined through [3]).

Slope and vegetation type change very few in space and time so we assume them as static parameters. Moreover, their values are easy to determine, so we assume them as known and available for our use. Therefore, GA is going to work with the remainder parameters: wind direction and wind speed, moisture content of live fuel and the three moisture content of dead fuel.

Basically, a GA uses three operations to obtain the consecutive generations: *selection*, *crossover* and *mutation*. *Selection* operation selects good quality parents (fitness function) to create children that will inherit parents good characteristics (by *crossover* operation). In order to guarantee nature diversity of individual characteristics, *mutation* phenomenon can occur for each children characteristic (under a very slight probability). *Selection* can include *elitism* where the best j individuals ($j > 0$) are included in the new generation directly. Last two operations intent to guarantee nature diversity of individual characteristics.

In the previous section we had exposed how wind values are obtained in order to steer GA: we modify elitism and mutation operations forcing the assignment of these wind values instead of random values. The main goal of this data changes is to give a correct direction to the GA process, achieving the exploration of promising zones of the whole search space. We try to minimize simulation errors, avoiding the use of individuals with inaccuracy wind values.

In our application, after each simulation (using a particular individual), ideal wind values are calculated and stored together with this individual. After obtaining all the simulations using a certain population, and genetic operations take place (elitism or mutation), the obtained values are assigned to the wind velocity and direction of each individual.

4   Experimental Results

The experimentation main goal is to evaluate the benefit of applying a dynamic data driven GA for fire spread prediction purposes. In particular, the assimilated data will be injected into the mutation and elitism operations independently and compared to applying the prediction methodology without considering any external data. For this purpose, two kind of burns have been used: two synthetic fires (simulation results) and one real prescribe fire. Fig. 4 shows the fire front evolution through time for the three experiments. In all cases, we use the best individual (with minimum error) of calibration stage to obtain the final prediction (at prediction stage).

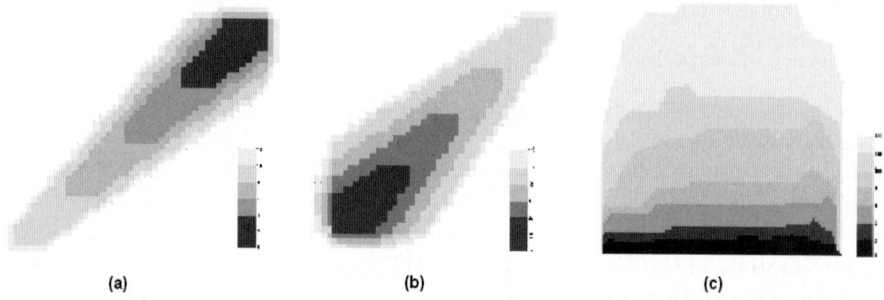

(a)   (b)   (c)

Fig. 4. (a) Experiment 1: synthetic fire case. (b) Experiment 2: synthetic fire case. (c) Experiment 3: real fire case.

All cases had used as initial population a set of 50 individuals with random values. Through GA, 5 iterations are performed (error reductions are insignificant from 3rd or 4th iteration [8]). Experiment results are the average of 5 different initial populations (in order to achieve more descriptive results).

4.1   Experiment 1: Synthetic Case

Experiment 1 concerns with the map depicted in Fig. 4(a) (109 x 89 m^2 and cells of 1 m^2). The terrain had 18° slope and vegetation was fuel model 7 [3].

Figs. 5 (a) and (b) show Calibration and Prediction stages results respectively. In x axis different time intervals are shown and each vertical bar represents one of the three different methods tested: classical GA (non guided), dynamic data driven elitism operation and dynamic data driven mutation.

For both stages (Calibration and Prediction), we can observe that the error has been significantly reduced when any of the two data driven methods have been used. Furthermore, both strategies provide very similar results. Prediction stage results shows errors slightly higher than in Calibration stage, however this is an expected result because we are using the best individual obtained at the Calibration stage to obtain the prediction for a later time interval at the Prediction stage.

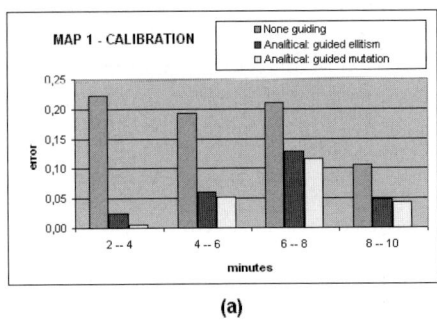 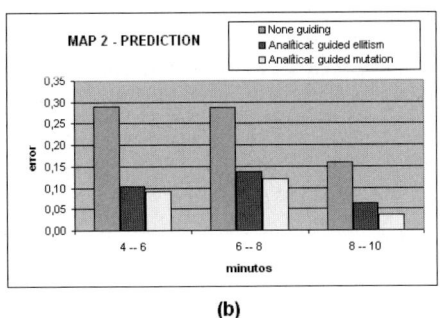

(a)   (b)

Fig. 5. Experiment 1: (a) Calibration stage results, (b) Prediction stage results

4.2 Experiment 2: Synthetic Case

Another synthetic map was used to carry out the second test (concerning with map illustrated in Fig. 4 (b): 33,22 x 27,12m^2 size map and cells of 1 f^2). In this case, we consider 27° slope and vegetation is the same than the first burning case.

The error values obtained for this experiment are shown in Fig. 6. This experiment denotes a similar behavior as the first one. As it happens in the previous experiment, applying any dynamic data driven strategy reduces the calibration error as well as the prediction error.

4.3 Experiment 3: Real Case

The last experiment concerns with Fig. 4(c) (89 x 109m^2 map size and cells size were 1 m^2). This burn has been extracted from a set of prescribe burns performed in the frame of an European project called SPREAD. In particular, these burns were done in Gestosa (Portugal) during 2004 [6].

Fig. 7(a) and 7(b) show the calibration and prediction stages results respectively. The 2 first time terms in Calibration stage we can see that results were similar through all methods, there weren't significant differences between different configurations. But in the 3 last steps, different methods had varied in a significant way. Each method had been adapted to fire features in a different way. For Prediction stage, we can see that in general, all the errors are bigger than observed errors in synthetic fires. Again, the behavior of different methods were different for each slice of time.

We can see that errors from experiment 3 were larger if we compare them with previous experiment errors. This characteristic may be explained considering real fire map features: instead of each step lasts the same amount of time (2 minutes), fire propagation was different in each step. This changing fire behavior may be due to wind or slope or other nature characteristics influences, because of their dynamic behavior. It is important to take into account that the microsystem generated by fires may enlarge the changing characteristics of wind, weather, etc.

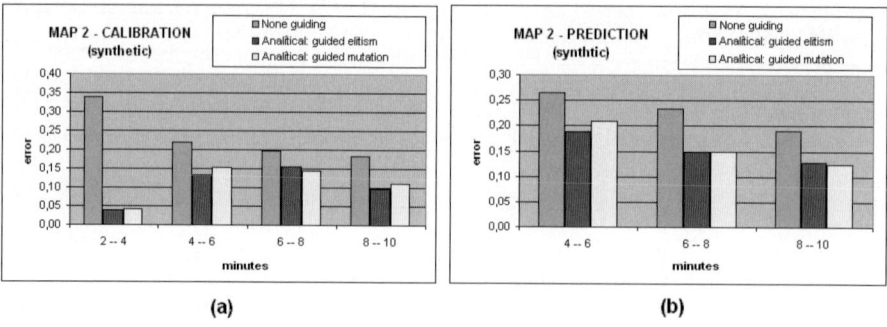

Fig. 6. Experiment 2: (a) Calibration stage results, (b) Prediction stage results

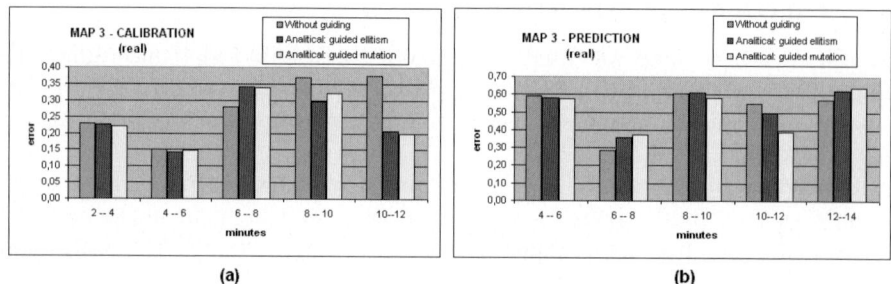

Fig. 7. Experiment 3: (a) Calibration stage results, (b) Prediction stage results

Instead of actual dynamic characteristics of nature, *fireLib* considers slope, wind, fuel type invariant in time [5]. This is an important simulator limit, because it can't adjust its fire advance in favor of actual dynamic behavior of its inputs.

5   Conclusions

This work is focused to improve forest fire spread prediction by using a dynamic data driven GA. For that purpose, certain internal operations of the GA are dynamically adapted according to the observed fire evolution and the underlying fire spread model.

In this work we present three cases of study, two synthetic fires and one real prescribe fire. The obtained results show that the inclusion of the principle of the dynamic data driven systems in the calibration stage improves the quality of the propagation predictions. In particular, we propose two alternative data driven GAs and the results obtained in both cases reduce the prediction error compared to not using any dynamic data driven strategy.

However, we know that this method is strongly coupled to the underlying simulator, therefore, we are currently working in generalize the main idea of the proposed methodology to be completely independent on the underlying fire forest simulator.

References

1. Abdalhaq, B.: A methodology to enhance the Prediction of Forest Fire Propagation. Ph. D Thesis. Universitat Autònoma de Barcelona (Spain) (June 2004)
2. Allen, G.: Building a Dynamic Data Driven Application System for Hurricane Forescasting. In: Shi, Y., van Albada, G.D., Dongarra, J., Sloot, P.M.A. (eds.) ICCS 2007. LNCS, vol. 4487, pp. 1034–1041. Springer, Heidelberg (2007)
3. Anderson, H.E.: Aids to Determining Fuel Models For Estimating Fire Behaviour Intermountain Forest and Range Experiment Station Ogden, UT 84401. General Technical Report INT.122 (April 2002)
4. Andrews, P.L.: BEHAVE: Fire Behavior prediction and modeling systems - Burn subsystem, part 1. General Technical Report INT-194. Odgen, UT, US Department of Agriculture, Forest Service, Intermountain Research Station, p. 130 (1986)
5. Bevins, C.D.: FireLib User Manual & Technical Reference (1996) (accesed January 2006), http://www.fire.org
6. Bianchini, G.: Wildland Fire Prediction based on Statistical Analysis of Multiple Solutions. Ph. D Thesis. Universitat Autònoma de Barcelona (Spain) (July 2006)
7. Dynamic Data Driven Application Systems homepage (November 2007), http://www.dddas.org
8. Denham, M.: Predicción de Incendios Forestales Basada en Algoritmos Evolutivos Guiados por los Datos, MsC thesis. Universitat Autònoma de Barcelona (Spain) (July 2007)
9. FIRE.ORG - Public Domain Software for the Wildland fire Community (acceded May 2007), http://www.fire.org
10. Mandel, J., et al.: A Dynamic Data Driven Wildland Fire Model. In: Shi, Y., van Albada, G.D., Dongarra, J., Sloot, P.M.A. (eds.) ICCS 2007. LNCS, vol. 4487, pp. 1042–1049. Springer, Heidelberg (2007)
11. Canarias 7 (accessed October 2007), http://www.canarias7.es
12. El País (accessed October 2007), http://www.elpais.com

Real-Time Data Driven Wildland Fire Modeling

Jonathan D. Beezley[1,2], Soham Chakraborty[3], Janice L. Coen[2],
Craig C. Douglas[3,5], Jan Mandel[1,2], Anthony Vodacek[4], and Zhen Wang[4]

[1] University of Colorado Denver, Denver, CO 80217-3364, USA
{jon.beezley.math,jan.mandel}@gmail.com
[2] National Center for Atmospheric Research, Boulder, CO 80307-3000, USA
janicec@ucar.edu
[3] University of Kentucky, Lexington, KY 40506-0045, USA
{sohaminator,craig.c.douglas}@gmail.com
[4] Rochester Institute of Technology, Rochester, NY 14623-5603, USA
{vodacek,zxw7546}@cis.rit.edu
[5] Yale University, New Haven, CT 06520-8285, USA

Abstract. We are developing a wildland fire model based on semi-empirical relations that estimate the rate of spread of a surface fire and post-frontal heat release, coupled with WRF, the Weather Research and Forecasting atmospheric model. A level set method identifies the fire front. Data are assimilated using both amplitude and position corrections using a morphing ensemble Kalman filter. We will use thermal images of a fire for observations that will be compared to synthetic image based on the model state.

Keywords: Dynamic data driven application systems, data assimilation, wildland fire modeling, remote sensing, ensemble Kalman filter, image registration, morphing, level set methods, Weather Research and Forecasting model, WRF.

1 Introduction

We describe the current state of a new dynamic data-driven wildland fire simulation based on real-time sensor monitoring. The model will be driven by real-time data from airborne imagery and ground sensors. It will be run on remote supercomputers or clusters.

An earlier summary of this project is in [1], where we have used a regularization approach in an ensemble Kalman filter (EnKF) for wildfires with a fire model by reaction-diffusion-convection partial differential equations [2,3].

Level set methods were applied to empirical fire spread also in [4], without a coupling with the atmosphere or data assimilation. The Morphing EnKF [5] is related to data assimilation by alignment [6].

2 Coupled Fire-Atmosphere Model

The work reported in this section was done in [7], where more details can be found. Our code implements in a simplified numerical framework the mathematical ideas

M. Bubak et al. (Eds.): ICCS 2008, Part III, LNCS 5103, pp. 46–53, 2008.

in [8]. Instead of tracking the fire by a custom ad hoc tracer code, it is now represented using a level set method [9]. Also, the fire model is now coupled with WRF, a community supported numerical weather prediction code. This adds capabilities to the codes used in previous work such as the theoretical experiments of [8] and the reanalysis of a real fire in [10]. We use features of WRF's design to get a natural parallelization of the fire treatment and its support for data assimilation.

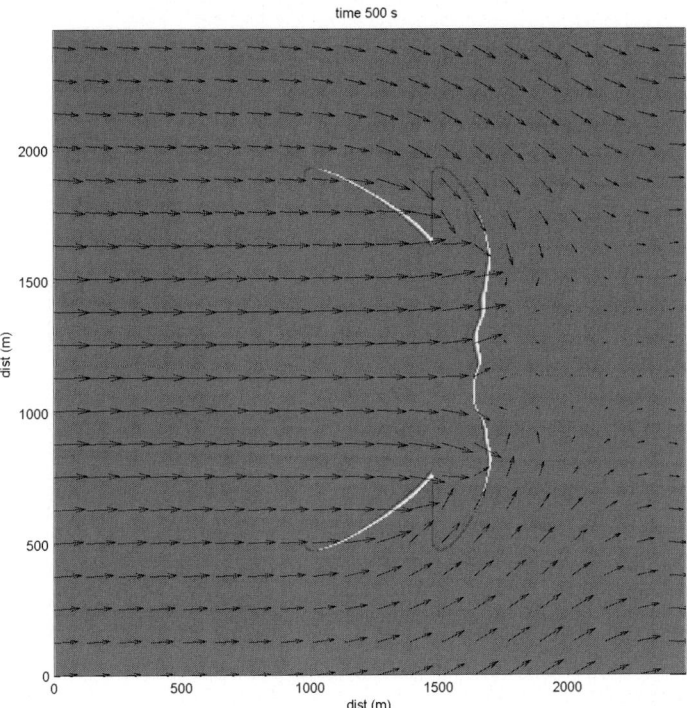

Fig. 1. Coupled fire-atmosphere simulation. Fire propagates from two line ignitions and one circle ignition, which are in the process of merging. The arrows are horizontal wind at ground level. False color is fire heat flux. The fire front on the right has irregular shape and is slowed down because of air being pulled up by the heat created by the fire. This kind of fire behavior cannot be modeled by empirical spread models alone.

We use a semi-empirical fire propagation model to represent a fire spreading on the surface [8,11] . Given the wind speed $\overrightarrow{v}$, terrain height z, and fuel coefficients R_0, a, b, and d, determined from laboratory experiments, the model postulates that the fireline evolves with the spread rate S in the normal direction to the fireline $\overrightarrow{n}$ given by

$$S = R_0 + a\left(\overrightarrow{v} \cdot \overrightarrow{n}\right)^{b} + d\nabla z \cdot \overrightarrow{n}. \tag{1}$$

We further limit the spread rate to satisfy $0 \leq S \leq S_{\max}$, where $S_{\max}$ depends on the fuel. Based on laboratory experiments, the model further assumes that after ignition the fuel amount decreases as an exponential function of time with the characteristic time scale depending on the fuel type, e.g., rapid mass loss in grass and slow mass loss in larger fuel particles (e.g., downed branches). The model delivers the sensible and the latent heat fluxes (temperature and water vapor) from the fire to the atmosphere. The heat fluxes are taken to be proportional to the amount of fuel burned, again with the coefficients determined from laboratory experiments.

We use a level set method for the fire propagation. The burning area at time t is represented by the level set $\{(x,y) : \psi(x,y,t) < 0\}$, where ψ is called the level set function. The fireline at time t is the level set $\{(x,y) : \psi(x,y,t) = 0\}$. The level set function is initialized as the signed distance from the fireline.

The level set function satisfies the differential equation [9]

$$\frac{\partial \psi}{\partial t} + S \left\| \nabla \psi \right\| = 0 \tag{2}$$

which is solved numerically by the Runge-Kutta method or order two,

$$\psi_{n+1} = \psi_n + \Delta t \left[F\left(\psi_n\right) + F\left(\psi_n + \Delta t F\left(\psi_n\right)\right) \right] / 2 \tag{3}$$

with $F(\psi) \approx -S \left\| \nabla \psi \right\|$, using an approximation of $\nabla \psi$ from upwinded approximation of $\nabla \psi$ by Godunov's method [9, p. 58]: each partial derivative is approximated by the left difference if both the left and the central differences are nonnegative, by the right difference if both the right and the central differences are nonpositive, and taken as zero otherwise. It is advantageous that the propagation speed S is defined by (1) with $\vec{n} = \nabla \psi / \left\| \nabla \psi \right\|$ on the whole domain, not just on the fireline, so it can be used directly in the differential equation (2). We use (3) for its better conservation properties, not its accuracy. The explicit Euler method, the obvious first choice, systematically overestimates ψ and thus slows down fire propagation or even stops it altogether while (3) behaves reasonably well.

The fire model is two-way coupled with the atmospheric model, using the horizontal wind velocity components to calculate the fire's heat fluxes, which in turn alter atmospheric conditions such as winds. WRF meshes can be refined only horizontally; only the finest atmospheric mesh interfaces with the fire. In our experiments to date, we have used time step 0.5s with the 60m finest atmospheric mesh step and 6m fire mesh step, which satisfied the CFL stability conditions in the fire and in the atmosphere.

Because WRF does not support flux boundary conditions, the heat flux from the fire model cannot be applied directly as a boundary condition on the derivatives of the corresponding physical field (e.g., air temperature or water vapor contents). Instead, we insert the flux by modifying the temperature and water vapor concentration over a depth of many cells with exponential decay away from the boundary.

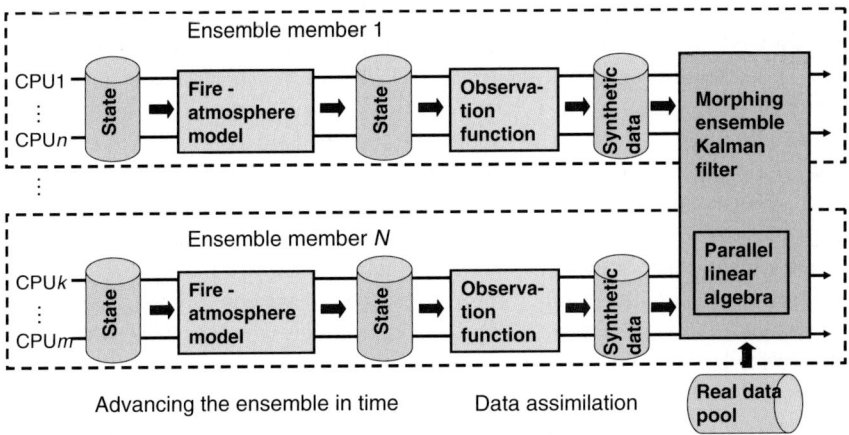

Fig. 2. Parallel implementation of one assimilation cycle. Each ensemble is advanced in time. Using a subset of processors, the observation function is evaluated independently for each ensemble member. The morphing EnKF runs on all processors and adjusts the member states by comparing the synthetic data with real data and balances the uncertainty in the data (given as an error bound) with the uncertainty in the simulation (computed from the spread of the whole ensemble). Currently, the ensemble of model states are kept in disk files. The observation function uses the disk files and delivers synthetic data back to disk files. The EnKF uses the synthetic and real data and modifies the files with the ensemble states. The model, observation function, and EnKF are separate executables.

3   Data Assimilation

Our coding design is based on Fig. 2. Data assimilation is done using an EnKF [12]. We can use data from airborne/space images and land-based sensors, though work remains on making all the data available when needed. In the rest of this section we describe how to treat various kinds of data, how synthetic image data for the fire can be obtained from the system state, and how the model states are adjusted by in response to the data.

Our efforts so far have been to get data from a variety of sources (e.g., weather stations, aerial images, etc.) and to compare it [13] to data from the fire-atmosphere code [8]. The observation function routines receives a state vector that is modified and returned. The state is transferred using disk files, which is slow and must be replaced with a faster mechanism. Individual subvectors corresponding to the most common variables are extracted or replaced in the files. A software layer hides both the fire code and the data transfer method so that the code is not dependent on any particular fire-atmosphere code.

Consider an example of a weather station that reports the location, time, temperature, wind velocity, and humidity. The atmosphere code has multiple nested grids. For a given grid, we have to determine in which cell the weather station is located, which is done using linear interpolation of the location. The data is is determined at relevant grid points using biquadratic interpolation. We compare the computed results with the weather station data and then determine if a fireline is in the cell (or neighboring ones) with the weather station temperature to see if there really is a fire in the cell. In the future, this will be replaced by synthetic data created from the model state, as in Fig. 2.

The model state can be used to produce an image like one from an infrared camera, which can be compared to the actual data in the data assimilation. This process is known as *synthetic scene generation* [14] and depends on the propagation model, parameters such as velocity of the fire movement, as well as its output (the heat flux), and the input (the wind speed). We estimate the 3D flame structure using the parameters from the model, which provides an effective geometry for simulating radiance from the fire scene.

Given the 3D flame structure, we assume we can adequately estimate the infrared scene radiance by including three aspects of radiated energy. These are radiation from the hot ground under the fire front and the cooling of the ground after the fire front passes, which accounts for the heating and cooling of the 2D surface, the direct radiation to the sensor from the 3D flame, which accounts for the intense radiation from the flame itself, and the radiation from the 3D flame that is reflected from the nearby ground. This reflected radiation is most important in the near and mid-wave infrared spectrum. Reflected long-wave radiation is much less important because of the high emissivity (low reflectivity) of burn scar in the long-wave infrared portion of the spectrum [15].

A double exponential is used to estimate the 2D fire front and cooling. We use time constants of 75 seconds and 250 seconds and the peak temperature at the fire front is constrained to be 1075K. The position of the fire front is determined from the propagation model described in the previous section. The 3D flame structure is estimated by using the heat release rate and experimental estimates of flame width and length and the flame is tilted based on wind speed. This 3D structure is represented by a 3D grid of voxels.

We use the ray tracing code Digital Imaging and Remote Sensing Image Generation Model (DIRSIG), which is a first principles based synthetic image

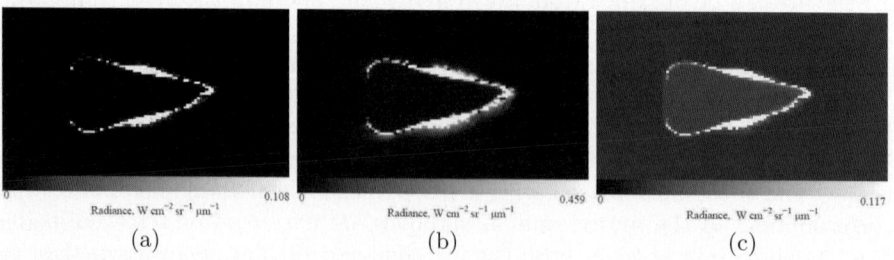

Radiance, W cm^{-2} sr^{-1} μm^{-1} Radiance, W cm^{-2} sr^{-1} μm^{-1} Radiance, W cm^{-2} sr^{-1} μm^{-1}

(a)   (b)   (c)

Fig. 3. Synthetic infrared nighttime radiance scenes of a simulated grassfire. The scenes are rendered as they would be viewed by a multiwavelength camera system on an aircraft flying 3000 m above ground. (a) Shortwave infrared (0.9-1.8 μm). (b) Midwave infrared (3.0-5.0 μm). (c) Longwave infrared (8.0-11.0 μm). Reproduced from [14].

generation model developed by the Digital Imaging and Remote Sensing Laboratory at RIT [16,17]. The 2D ground temperatures and the 3D voxels representing the flame are inputs to DIRSIG, which determines the radiance from those sources as they would be viewed by an airborne remote sensing system. The model can produce multi- or hyper-spectral imagery from the visible through the thermal infrared region of the electromagnetic spectrum. The radiance calculation includes radiance reflected from the ground and the effects of the atmosphere. We validated the resulting synthetic radiance image (Fig. 3) by calculating the fire radiated energy and comparing those results to published values derived from satellite remote sensing data of wildland fires [18]. We are continuing to work on synthetic image generation with the goal of replacing the computationally intensive and accurate ray tracing method with a simpler method of calculating the fire radiance based upon the radiance estimations that are inherent in the fire propagation model.

The rest of this section follows [7] and describes a morphing EnKF. The model state consists of the level set function ψ and the ignition time t_i. Both are given as arrays of values associated with grid nodes. Unlike the tracers in [8], our grid arrays can be modified by data assimilation methods with ease. Data assimilation [19] maintains an approximation of the probability distribution of the state. In each analysis cycle, the probability distribution of the state is advanced in time and then updated from the data likelihood using Bayes theorem. EnKF represents the probability distribution of the state by an ensemble of states and it uses the model only as a black box without any additional coding required. After advancing the ensemble in time, the EnKF replaces the ensemble by its linear combinations with the coefficients obtained by solving a least squares problem to balance the change in the state and the difference from the data.

However, the EnKF applied directly to the model fields does not work well when the data indicates that a fire is in a different location than what is in the state. This is due to the combination of such data having infinitesimally small likelihood and the span of the ensemble does actually contain a state consistent with the data. Therefore, we adjust the simulation state by changing the position of the fire, rather than just an additive correction alone, using the techniques of registration and morphing from image processing. Essentially, we replace the linear combinations of states in the EnKF by intermediate states obtained by morphing, which leads to the morphing EnKF method [5].

Given two functions u_0 and u representing the same physical field (e.g., the temperature or the level set function) from two states of the coupled model, registration can be described as finding a mapping T of the spatial domain so that $u \approx u_0 \circ (I + T)$, where $\circ$ denotes the composition of mappings and I is the identity mapping. The field u and the mapping T are given by their values on a grid. To find the registration mapping T automatically, we solve approximately an optimization problem of the form

$$\|u - u_0 \circ (I + T)\| + \|T\| + \|\nabla T\| \to \min.$$

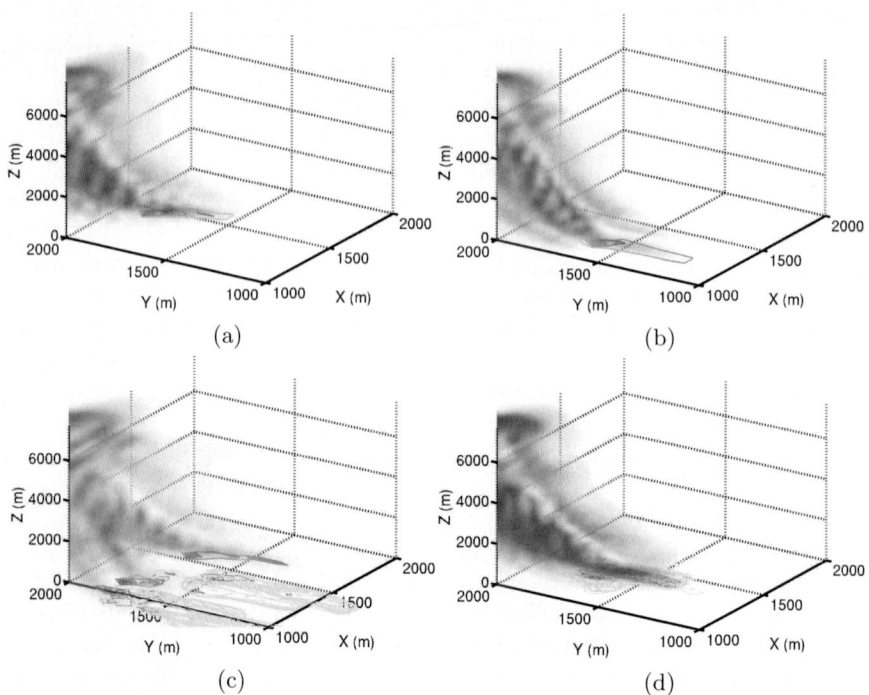

Fig. 4. The morphing EnKF applied to the fireline propagation model coupled with WRF. False color and contour on the horizontal plane is the fire heat flux. The volume shading is the vorticity of the atmosphere. The reference solution (a) is the simulated data. The initial ensemble was created by a random perturbation of the comparison solution (b) with the fire ignited at an intentionally incorrect location. The standard ENKF (c) and the morphing EnKF (d) were applied after 15 minutes. The ensembles have 25 members each with the heat fluxes shown superimposed. The standard EnKF ensembles diverges from the data while the morphing EnKF ensemble keeps closer to the data. Reproduced from [7].

We then construct intermediate functions u_λ between u_0 and u_1 using [5]. For $r = u \circ (I + T)^{-1} - u_0$, we have

$$u_\lambda = (u + \lambda r) \circ (I + \lambda T), \quad 0 \le \lambda \le 1. \tag{4}$$

The morphing EnKF works by transforming the ensemble member into extended states of the form $[r, T]$, which are input into the EnKF. The result is then converted back using (4). Fig. 4 contains an illustrative result.

Acknowledgements. This work was supported by the National Science Foundation under grants CNS-0325314, CNS-0324910, CNS-0324989, CNS-0324876, CNS-0540178, CNS-0719641, CNS-0719626, CNS-0720454, DMS-0623983, EIA-0219627, and OISE-0405349.

References

1. Mandel, J., Beezley, J.D., Bennethum, L.S., Chakraborty, S., Coen, J.L., Douglas, C.C., Hatcher, J., Kim, M., Vodacek, A.: A dynamic data driven wildland fire model. In: Shi, Y., van Albada, G.D., Dongarra, J., Sloot, P.M.A. (eds.) ICCS 2007. LNCS, vol. 4487, pp. 1042–1049. Springer, Heidelberg (2007)
2. Johns, C.J., Mandel, J.: A two-stage ensemble Kalman filter for smooth data assimilation. Environ. Ecological Stat. 15, 101–110 (2008)
3. Mandel, J., Bennethum, L.S., Beezley, J.D., Coen, J.L., Douglas, C.C., Franca, L.P., Kim, M., Vodacek, A.: A wildfire model with data assimilation, arXiv:0709.0086 (2006)
4. Mallet, V., Keyes, D.E., Fendell, F.E.: Modeling wildland fire propagation with level set methodsm arXiv:0710.2694 (2007)
5. Beezley, J.D., Mandel, J.: Morphing ensemble Kalman filters. Tellus 60A, 131–140 (2008)
6. Ravela, S., Emanuel, K.A., McLaughlin, D.: Data assimilation by field alignment. Physica D 230, 127–145 (2007)
7. Mandel, J., Beezley, J.D., Coen, J.L., Kim, M.: Data assimilation for wildland fires: Ensemble Kalman filters in coupled atmosphere-surface models, arXiv:0712.3965 (2007)
8. Clark, T.L., Coen, J., Latham, D.: Description of a coupled atmosphere-fire model. Int. J. Wildland Fire 13, 49–64 (2004)
9. Osher, S., Fedkiw, R.: Level set methods and dynamic implicit surfaces. Springer, New York (2003)
10. Coen, J.L.: Simulation of the Big Elk Fire using coupled atmosphere-fire modeling. Int. J. Wildland Fire 14, 49–59 (2005)
11. Rothermel, R.C.: A mathematical model for predicting fire spread in wildland fires. USDA Forest Service Research Paper INT-115 (1972)
12. Evensen, G.: The ensemble Kalman filter: Theoretical formulation and practical implementation. Ocean Dynamics 53, 343–367 (2003)
13. Chakraborty, S.: Data assimilation and visualization for ensemble wildland fire models. Master's thesis, University of Kentucky, Department of Computer Science, Lexington, KY (2008)
14. Wang, Z.: Modeling Wildland Fire Radiance in Synthetic Remote Sensing Scenes. PhD thesis, Rochester Institute of Technology, Center for Imaging Science (2008)
15. Kremens, R., Faulring, J., Hardy, C.C.: Measurement of the time-temperature and emissivity history of the burn scar for remote sensing applications. In: Paper J1G.5, Proceedings of the 2nd Fire Ecology Congress, Orlando FL, American Meteorological Society (2003)
16. Digital Imaging and Remote Sensing Laboratory: DIRSIG users manual. Rochester Institute of Technology (2006),
    http://www.dirsig.org/docs/manual-2006-11.pdf
17. Schott, J., Brown, S.D., Raqueño, R.V., Gross, H.N., Robinson, G.: An advanced synthetic image generation model and its application to multi/hyperspectral algorithm development. Canadian J. Remote Sensing 25, 99–111 (1999)
18. Wooster, M.J., Zhukov, B., Oertel, D.: Fire radiative energy for quantitative study of biomass burning: derivation from the BIRD experimental satellite and comparison to MODIS fire products. Remote Sensing Environ. 86, 83–107 (2003)
19. Kalnay, E.: Atmospheric Modeling, Data Assimilation and Predictability. Cambridge University Press, Cambridge (2003)

DDDAS Predictions for Water Spills

Craig C. Douglas[1,4], Paul Dostert[2], Yalchin Efendiev[3], Richard E. Ewing[3], Deng Li[1], and Robert A. Lodder[1]

[1] University of Kentucky, Lexington, KY 40506-0046
[2] University of Arizona, Tucson, AZ 85721-0089
[3] Texas A & M University, College Station, TX 77843-3368
[4] Yale University, New Haven, CT 06520-8285, USA
douglas-craig@cs.yale.edu

Abstract. Time based observations are the linchpin of improving predictions in any dynamic data driven application systems. Our predictions are based on solutions to differential equation models with unknown initial conditions and source terms. In this paper we want to simulate a waste spill by a water body, such as near an aquifer or in a river or bay. We employ sensors that can determine the contaminant spill location, where it is at a given time, and where it will go. We estimate initial conditions and source terms using better and new techniques, which improves predictions for a variety of data-driven models.

1 Introduction

In this paper, our goal is to predict contaminant transport, where the contamination is, where the contaminant is going to go, and to monitor the environmental impact of the spill for contaminants in near coastal areas. Sensors measure the contaminant concentration at certain locations. Here, we discuss the quality of the predictions when the initial conditions and the source terms are updated as data is injected.

From a modeling viewpoint, one of the objectives in dynamic data driven application systems (DDDAS) is to improve the predictions as new data is obtained [1]. Data represents the information at different length scales and the precision can vary enormously. Numerous issues are involved in DDDAS, many of which are described in [2,3,4,5,6].

We investigate contaminant transport driven by convection and diffusion. When new sensor based data is obtained, the initial conditions and source terms are updated. The initial conditions are constructed in a finite dimensional space. Measurements often contain errors and uncertainties. We have studied elsewhere approaches where these uncertainties can be taken into account.

The first set of measurements allows recovery of an approximation to the initial conditions. An objective function is used to update initial data in the simulation. The mesh of sensor locations is quite coarse in comparison to the solution's mesh, leading to an ill-posed problem. Prior information about the initial data lets us regularize the problem and then update the initial conditions. We use a penalty

M. Bubak et al. (Eds.): ICCS 2008, Part III, LNCS 5103, pp. 54–63, 2008.

method whose constants depend on time and can be associated with the relative errors in the sensor measurements.

We also update the source terms, which is quite important when there is still a source of the contamination present in the ongoing measurements. We construct an objective functional that lets us perform an update on the source term in a finite dimensional space.

A number of numerical examples are presented that show that the update process is quite important. Using the correct choice of penalty terms demonstratively improves the predictions.

2   Contaminant Concentration Model

Let C be the concentration of a contaminant and assume that the velocity v, obtained from shallow water equations, is known. We assume that the sensors measure the concentration at some known locations. Note that the velocity field and the diffusion coefficients are unknown and need to be estimated in the most general case.

Contaminant transport is modeled using the convection-diffusion equation,

$$\frac{\partial C}{\partial t} + v \cdot \nabla C - \nabla \cdot (D\nabla C) = S(x,t) \text{ in } \Omega.$$

Given measured data, we estimate the initial condition $C(x,0) = C^0(x)$ and the source term $S(x,t)$.

3   Reconstructing Initial Conditions

We begin by obtaining initial data based on sensor readings and forward simulations assuming both a rectangular subdomain $\Omega_c \subset \Omega$ and that our initial condition is in a finite dimensional function space whose support is contained in Ω_c. This finite dimensional space on Ω_c is equipped with a set of linearly independent functions $\{\tilde{C}_i^0(x)\}_{i=1}^{N_c}$. For some $\alpha = (\alpha_1, \alpha_2, \cdots, \alpha_{N_c})$ our initial condition in this space is then represented by

$$\tilde{C}^0(x) = \sum_{i=1}^{N_c} \alpha_i \tilde{C}_i^0(x).$$

We can assume that $\tilde{C}_i(x,t)$ is the solution of our equation using the initial condition $\tilde{C}_i^0(x)$ leading to a solution of the form,

$$\tilde{C}(x,t) = \sum_{i=1}^{N_c} \alpha_i \tilde{C}_i(x,t).$$

Running the forward simulation code for each initial condition $\tilde{C}_i^0(x)$ lets us find $\tilde{C}_i(x,t)$. Let N_c be the number of initial conditions, N_s the number of output concentration locations, and N_t the number of time steps where concentrations are sampled. We save values $\left\{ \tilde{C}_i(x_j, t_k) \right\}_{i=1,j=1,k=1}^{N_c,N_s,N_t}$. The problem is underdetermined for a single set of sensor measurments since $N_s < N_c$ normally. However, $N_s N_t >> N_c$ since data from these sensor locations are collected many times. By using all of the given sensor values at each of the recorded times, we attempt to solve the least squares system to recover the initial conditions and improve the overall predictions made by the model.

If we devise a method so that just a simple least squares problem is solved for each new set of sensor data, then we can solve a much smaller least sqaures problem than in the case when all of the given data is used at once. Further, should not enough data be incorporated at once, solving smaller problems is faster than solving the entire problem over again. We want a model that is improved using smaller quantities of data at any given time.

Once we have collected the data from our forward simulations, the α_i need to be calculated. We want to find the best α such that $\tilde{C}(x,t) \approx C(x,t)$, so we minimize the difference between our simulated concentration and the values at the sensors. Assume there are $N_s < N_c$ sensors in Ω, which leads to minimizing the following objective function:

$$
F(\alpha) = \sum_{j=1}^{N_s} \left(\sum_{i=1}^{N_c} \alpha_i \tilde{C}_i(x_j, t) - \gamma_j(t) \right)^2 + \sum_{i=1}^{N_c} \kappa_i (\alpha_i - \beta_i)^2,
$$

where κ are penalty coefficients for an *a priori* vector β, which will be updated during the simulation to achieve higher accuracy. The minimization of this function gives rise to the linear system $A\alpha = R$, where for $m, n = 1, \cdots, N_c$, and

$$
A_{mn} = \sum_{j=1}^{N_s} \tilde{C}_m(x_j, t) \tilde{C}_n(x_j, t) + \delta_{mn}\kappa_m, \quad \text{and}
$$
$$
R_m = \sum_{j=1}^{N_s} \tilde{C}_m(x_j, t) \gamma_j(t) + \kappa_m \beta_m.
$$

This linear system is clearly symmetric, positive definite and is solved using a direct inversion since it is such a small matrix.

Since we have $\tilde{C}_i(x_j, t)$ at given time steps $\{t_k\}_{k=1}^{N_t}$ we first solve the linear system $A\alpha^1 = R$, where A and R are evaluated at time t_1 and α^1 refers to the values of α_i at t_1. Initially, we begin with a given value of β whose value can be adapted, though generally we initially choose $\beta = 0$. Once α^1 is determined, β is replaqced with α^1 and we solve the linear system again to determine α^2. We continue this process until we get α^{N_t}, which is the most accurate reconstruction of the initial data.

4   Reconstructing the Source Term

Consider the situation when the source term S is given for a particular fixed time interval at a particular location. Assume there is no initial concentration and all of the concentration in the domain comes from the source term so that

$C(x, 0) = 0$ on the whole domain. Now consider a subdomain Ω_C where the source term is defined. Now assume that our region Ω is discretized with N grid points and that the subdomain Ω_C is discretized with N_C grid points. On Ω_C assume that there are basis functions $\{\delta_k\}_{k=1}^{N_C}$ which are nonzero on the kth part of the subdomain and zero otherwise. Assume that $S \approx \tilde{S} = \sum_{k=1}^{N_C} \alpha_k \, \delta_k \, (x, t)$

where $S = \tilde{S} = \delta_k \, (x, t) = 0$ for some $t > \hat{t}$, i.e., the basis functions are nonzero for the same time interval as S and the source is zero after an initial time period. This is the case when there is an immediate spill that takes a short period of time for all of the contaminant to leak into the water. Using this $\tilde{S}$, our equation becomes

$$\frac{\partial \tilde{C}}{\partial t} - L\left(\tilde{C}\right) = \tilde{S} = \sum_{k=1}^{N_C} \alpha_k \, \delta_k \, (x, t)$$

since $\tilde{S}$ is a linear combination of δ_k. We solve

$$\frac{\partial \psi_k}{\partial t} - L\left(\psi_k\right) = \delta_k \, (x, t)$$

for ψ_k and each k. We denote the solution to this equation as $\{\psi_k \, (x, t)\}_{k=1}^{N_s}$ for each k. Under these assumptions, the solution to

$$\frac{\partial C \, (x, t)}{\partial t} - L\left(C \, (x, t)\right) = S \, (x, t), \; C \, (x, 0) = 0 \; x \in \Omega$$

is approximated by

$$C \, (x, t) \approx \tilde{C} \, (x, t) = \sum_{k=1}^{N_C} \alpha_k \, \psi_k \, (x, t).$$

Once again, assume that there are $N_s < N_c$ sensors spread within the domain. Choose a source term $\tilde{S}$ and run the forward problem for this particular source while recording the values of the concentration at each sensor location. These values are given by $\{\gamma_j \, (t)\}_{j=1}^{N_s}$. Once this equation has have solved for each of the source terms, α_k can be reconstructed by solving

$$F \, (\alpha) = \sum_{j=1}^{N_s} \left(\sum_{k=1}^{N_c} \alpha_k \psi_k \, (x, t) - \gamma_k \, (t)\right)^2 + \sum_{k=1}^{N_c} \kappa_k \, (\alpha_k - \beta_k)^2,$$

where κ are penalty coefficients for an a vector β. Minimize this function and solve the corresponding linear system. Note that this is the same exact minimization that was needed for the initial condition recovery problem.

4.1   Solving the Source Term and Initial Condition Problem

We split the solution into two parts in order to predict contaminant transport in the presence of unknown initial conditions and source terms. The first part is due to unknown initial condition and the second one is due to unknown source terms.

We briefly repeat the situation when the source term S is zero, where we assumed there is an initial concentration and solved

$$\frac{\partial C}{\partial t} - L(C) = 0, \ C(x,0) = C^0(x).$$

Consider a discretized subdomain Ω_C where the initial condition is nonzero and assume there is a linearly independent set of functions defined by $\{\varphi_i\}_{i=1}^{N_D}$ on the subdomain given by $C^0(x) \approx \tilde{C}^0(x) = \sum_{i=1}^{N_D} \lambda_i \varphi_i^0(x)$. Now solve

$$\frac{\partial \varphi_i}{\partial t} - L(\varphi_i) = 0, \ \varphi_i(x,0) = \varphi_i^0(x)$$

for each i. The solution to this equation for each i is given by $\varphi_i(x,t)$. We approximate the solution of $\frac{\partial C}{\partial t} - L(C) = 0, \ C(x,0) = C^0(x)$ by $\tilde{C}(x,t) = \sum_{i=1}^{N_D} \lambda_i \varphi_i(x,t)$.

For the second step, we solve the problem for ψ and each k with an unknown source term,

$$\frac{\partial \psi}{\partial t} - L(\psi) = \delta_k(x), \ \psi(x,0) = 0$$

and denote the solution for eazch k as $\{\psi_k(x,t)\}_{k=1}^{N_c}$. Hence, the solution to our original problem with both the source term and initial condition is given by

$$\tilde{C}(x,t) = \sum_{i=1}^{N_D} \lambda_i \varphi_i(x,t) + \sum_{k=1}^{N_c} \alpha_k \psi_k(x,t).$$

We need to verify that this is really the solution. Compute

$$L(\tilde{C}) = \sum_{i=1}^{N_D} \lambda_i L(\varphi_i(x,t)) + \sum_{k=1}^{N_c} \frac{\partial}{\partial t} \alpha_k L(\psi_k(x,t)) \ \text{and}$$

$$\frac{\partial}{\partial t}\tilde{C}(x,t) = \sum_{i=1}^{N_D} \lambda_i \frac{\partial}{\partial t}\varphi_i(x,t) + \sum_{k=1}^{N_c} \frac{\partial}{\partial t}\alpha_k \psi_k(x,t).$$

So

$$\frac{\partial \tilde{C}}{\partial t} - L(\tilde{C}) = \sum_{i=1}^{N_D} \lambda_i \left[\frac{\partial}{\partial t}\varphi_i(x,t) - L(\varphi_i(x,t)) \right] +$$

$$\sum_{k=1}^{N_c} \frac{\partial}{\partial t}\alpha_k [\psi_k(x,t) - L(\psi_k(x,t))]$$

$$= \sum_{k=1}^{N_c} \frac{\partial}{\partial t}\alpha_k [\psi_k(x,t) - L(\psi_k(x,t))] = \sum_{k=1}^{N_c} \alpha_k \delta_k(x).$$

Similarly, $\tilde{C}(x,0) = \sum_{i=1}^{N_D} \lambda_i \varphi_i(x,0) = \tilde{C}^0(x)$. Hence, we have verified that

$$\tilde{C}(x,t) = \sum_{i=1}^{N_D} \lambda_i \varphi_i(x,t) + \sum_{k=1}^{N_c} \alpha_k \psi_k(x,t)$$

really solves our original equation with both an initial condition and source term.

4.2 Reconstruction of Initial Condition and Source Term

After running the forward simulation for each initial basis function and source basis function, we minimize

$$F(\alpha, \lambda) = \sum_{j=1}^{N_s} \left[\left(\sum_{k=1}^{N_c} \alpha_k \psi_k(x_j, t) + \sum_{k=1}^{N_D} \lambda_k \varphi_k(x_j, t) - \gamma_j(t) \right)^2 \right] + \quad (1)$$

$$\sum_{k=1}^{N_c} \tilde{\kappa}_k \left(\alpha_k - \tilde{\beta}_k \right)^2 + \sum_{k=1}^{N_D} \hat{\kappa}_k \left(\lambda_k - \hat{\beta}_k \right)^2.$$

For $N = N_c + N_d$, $\mu = [\alpha_1, \cdots, \alpha_{N_c}, \lambda_1, \cdots, \lambda_{N_D}]$, $\eta(x,t) = [\psi_1, \cdots, \psi_{N_c}, \varphi_1, \cdots, \varphi_{N_D}]$, $\beta = \left[\tilde{\beta}_1, \cdots, \tilde{\beta}_{N_c}, \hat{\beta}_1, \cdots, \hat{\beta}_{N_D} \right]$, $\kappa = [\tilde{\kappa}_1, \cdots, \tilde{\kappa}_{N_c}, \hat{\kappa}_1, \cdots, \hat{\kappa}_{N_D}]$, we minimize

$$F(\mu) = \sum_{j=1}^{N_s} \left[\left(\sum_{k=1}^{N} \mu_k \eta_k(x_j, t) - \gamma_j(t) \right)^2 \right] + \sum_{k=1}^{N} \kappa_k (\mu_k - \beta_k)^2. \quad (2)$$

This is the same minimization that we had previously, which leads to solving a least squares problem of the form $A\mu = R$, where

$$A_{mn} = \sum_{j=1}^{N} \eta_m(x_j, t) \eta_n(x_j, t) + \delta_{mn} \kappa_m \text{ and } R_m = \sum_{j=1}^{N} \eta_m(x_j, t) \gamma_j(t) + \kappa_m \beta_m.$$

Sensor values are recorded only at discrete time steps $t = \{t_j\}_{j=1}^{N_t}$. μ is first estimated using the sensor values at $t = t_1$. Then each successive set of sensor values is used to improve the estimate of μ.

5 Numerical Results

We have performed extensive numerical studies for initial condition and source term estimation [7]. The numerical results convincingly demonstrate that the predictions can be improved by updating initial conditions and source terms.

Each problem has commonality:

- An initial condition is defined on a domain of $[0, 1] \times [0, 1]$.
- Sensor data is recorded at given sensor locations and times and is used to reconstruct the initial condition.
- Biquadratic finite elements are used in both the forward simulation and the reconstruction.
- For our initial condition expansion

$$C^n (x) = \sum_{i=1}^{N} c_i^n \varphi_i (x).$$

We assume φ_i are either piecewise constants or bilinears defined on a given subdomain with its own grid (i.e., there are two different grids): a large (fine) grid where the forward simulation is run and a small (coarse) grid defined only on a given subdomain where we are attempting a reconstruction.
- All velocities are $[2, 2]$ on each cell. Thus our flow is from the lower left corner to the upper right corner of the 2D grid for each problem.
- We sample sensor data every 0.05 seconds for 1.0 seconds at the following five locations: $(.5, .5), (.25, .25), (.25, .75), (.75, .25),$ and $(.75, .75)$.

5.1 Reconstruction Using Piecewise Constants

We attempt to reconstruct a constant initial condition with support on $[0, .2] \times [0, .2]$. We choose an underlying grid defined only on a subregion where we define the basis functions used in the reconstruction.

First, let this grid be exactly where there is support for the function. For example, if we have a 2×2 grid, then we define 4 piecewise constants on $[0, .2] \times [0, .2]$. Hence, support would be on the following four subdomains: $[0, .1] \times [0, .1]$, $[.1, .2] \times [0, .1]$, $[0, .1] \times [.1, .2]$, and $[.1, .2] \times [.1, .2]$. The region is divided similarly for different sized grids.

Second, let the subdomain be larger than the support of the initial condition, e.g., choose $[0, .4] \times [0, .4]$ as the subdomain. Hence, the "effective" area of the basis functions is reduced by a factor of 4 each.

Choose a 2×2 subgrid for the basis functions on $[0, .2] \times [0, .2]$ so that there are only 4 basis functions in the reconstruction. As can be seen in Fig. 1, the reconstructed initial condition are quite good: the initial condition reconstruction only needed two sets of sensor measurements.

Using the same strategy for dividing the domain up into a small number of equal parts we find slightly worse results for larger grids. This seems natural considering that we are using the same amount of information, but we are attempting to reconstruct more functions. Clearly there should be larger errors as the number of functions is increased unless we add more sensor data. Experiments show that this type of folklore is true.

Consider a case with 9 sensor locations instead of 5: $(.25, .25), (.25, .5), (.25, .75), (.5, .25), (.5, .5), (.5, .75), (.75, .25), (.75, .5),$ and $(.75, .75)$. We use

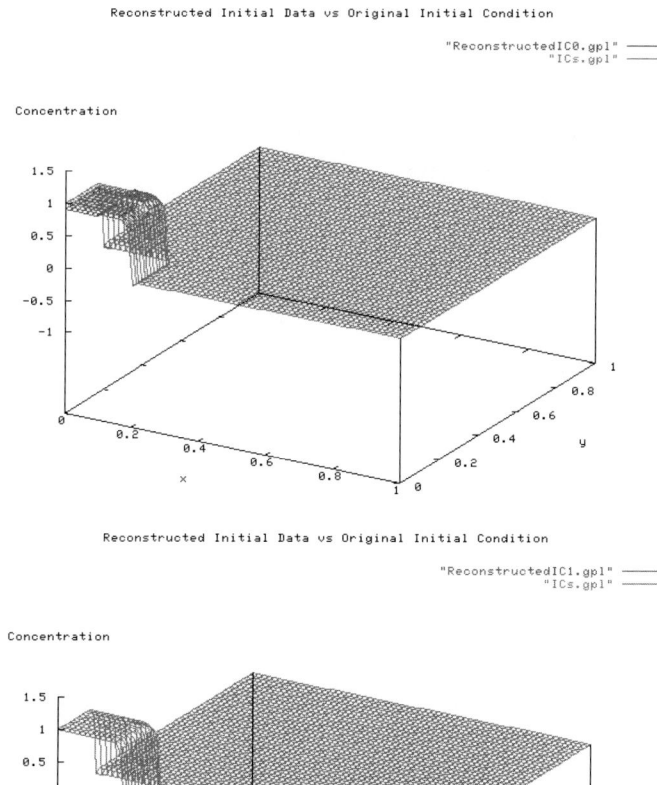

Fig. 1. Five sensors case

bilinears with a 2×2 grid so that there are 16 basis functions instead of 4. The accuracy is far higher than when we used the same parameters with only 5 sensors, as can be seen in Fig. 2.

Consider a case with just 2 sensor locations instead of 5: $(.5, .5)$ and $(.25, .5)$. The accuracy is far lower than when we used the same parameters with only 5 sensors, as can be seen in Fig. 3.

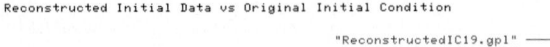

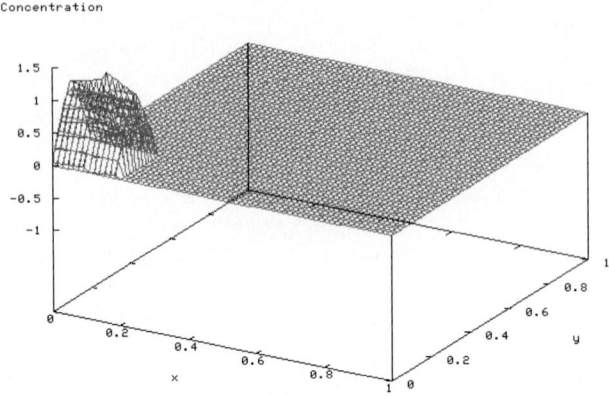

Fig. 2. Nine sensors case

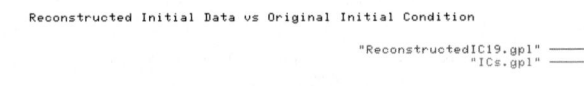

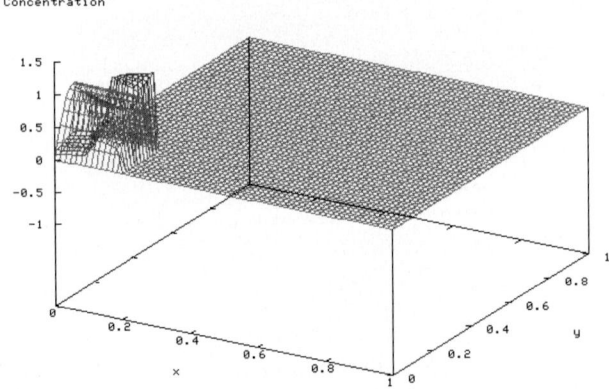

Fig. 3. Two sensors case

For future work, we need to test the proposed methods for numerous other initial conditions and source terms update conditions.

Acknowledgements. This research was supported in part by NSF grants EIA-0219627, EIA-0218229, CNS-0540178, and CNS-0540136.

References

1. Darema, F.: Introduction to the ICCS 2007 Workshop on Dynamic Data Driven Applications Systems. In: Shi, Y., van Albada, G.D., Dongarra, J., Sloot, P.M.A. (eds.) ICCS 2007. LNCS, vol. 4487, pp. 955–962. Springer, Heidelberg (2007)
2. Dostert, P.: Uncertainty Quantification Using Multiscale Methods for Porous Media Flows. PhD thesis, Texas A & M University, College Station, TX (December 2007)
3. Douglas, C., Cole, M., Dostert, P., Efendiev, Y., Ewing, R., Haase, G., Hatcher, J., Iskandarani, M., Johnson, C., Lodder, R.: Dynamic Contaminant Identification in Water. In: Alexandrov, V.N., van Albada, G.D., Sloot, P.M.A., Dongarra, J. (eds.) ICCS 2006. LNCS, vol. 3993, pp. 393–400. Springer, Heidelberg (2006)
4. Douglas, C., Cole, M., Dostert, P., Efendiev, Y., Ewing, R., Haase, G., Hatcher, J., Iskandarani, M., Johnson, C., Lodder, R.: Dynamically identifying and tracking contaminants in water bodies. In: Shi, Y., van Albada, G.D., Dongarra, J., Sloot, P.M.A. (eds.) ICCS 2007. LNCS, vol. 4487, pp. 1002–1009. Springer, Heidelberg (2007)
5. Douglas, C., Efendiev, Y., Ewing, R., Ginting, V., Lazarov, R., Cole, M., Jones, G., Johnson, C.: Multiscale interpolation, backward in time error analysis for data-driven contaminant simulation. In: Sunderam, V.S., van Albada, G.D., Sloot, P.M.A., Dongarra, J. (eds.) ICCS 2005. LNCS, vol. 3515, pp. 640–6470. Springer, Heidelberg (2005)
6. Douglas, C., Efendiev, Y., Ewing, R., Ginting, V., Lazarov, R.: Dynamic data driven simulations in stochastic environments. Computing 77, 321–332 (2006)
7. Dostert, P.: http://math.arizona.edu/~dostert/dddasweb (last visited 2/1/2008)

References

Bioinformatics' Challenges to Computer Science

Bioinformatics' Challenges to Computer Science

Mario Cannataro[1], Mathilde Romberg[2], Joakim Sundnes[3],
and Rodrigo Weber dos Santos[4]

[1] University Magna Graecia, Catanzaro, Italy
cannataro@unicz.it
[2] Research Centre Julich, Germany
m.romberg@fz-juelich.de
[3] Simula Research Laboratory, Norway
sundnes@simula.no
[4] Federal University of Juiz de Fora, Brazil
rodrigo.weber@ufjf.edu.br

Abstract. The workshop Bioinformatics' Challenges to Computer Science covers the topics of data management and integration, modelling and simulation of biological systems and data visualization and image processing. This short paper describes the requirements Bioinformatics has towards computer science, summarizes the papers accepted for the workshop and gives a brief outlook on future developments.

Keywords: Bioinformatics, Data Management and Integration, Modelling and Simulation of Biological Systems, Data Visualization.

1 Bioinformatics - An Overview

Bioinformatics[1], a link between biology and computer science, involves the design and development of advanced algorithms and computational platforms to solve problems in biology and medicine. It also deals with methods for acquiring, storing, retrieving, analyzing, and integrating biological data obtained by experiments, or by querying databases.

Bioinformatics is providing the foundation for fast and reliable analysis of biological and medical data. Genomics, transcriptomics, proteomics, epidemiological, clinical and text mining applications have made essential progress through using bioinformatics tools. Although standard tools are widely offered through the Web, they are no longer sufficient to cope with the increasing demands of the complex analysis and simulation tasks of today's emerging fields of biotechnology research. Moreover, emerging life sciences applications need to use in a coordinated way bioinformatics tools, biological data banks, and patient's clinical data, which require seamless integration, privacy preservation and controlled sharing[2,3,4]. Therefore, new challenges to computer science arise from the sheer problem scale, the huge amounts of data to be integrated and the computing power necessary to analyze large data sets or to simulate complex biological systems.

M. Bubak et al. (Eds.): ICCS 2008, Part III, LNCS 5103, pp. 67–69, 2008.

2   Goals

The aim of the workshop was to bring together scientists from computer and life sciences to discuss future directions of bioinformatics algorithms, applications, and data management. The discussion evolves from the basic building blocks for Bioinformatics applications which are: data sources (e.g. experimental datasets, local and public biological databases); software tools providing specialized services (e.g. searching of protein sequences in protein databases, sequence alignment, biophysical simulations, data classification, etc.); and high level description of the goals and requirements of applications and results produced in past executions. From a computational point of view, bioinformatics applications bring a wide range of challenges and a huge demand for computing power. The challenging issues involve the large number of involved datasets, the size of the datasets, the complexity inherent in the data analysis and simulations, the heterogeneous nature of the data, and the need for a secure infrastructure for processing private data. From another perspective, emerging high performance computer architectures may offer huge computational resources in the trade of specific development of new algorithms for bioinformatics. These are the cases of Grid and Web services, as well as of multicore architectures that demand bioinformatics algorithms to be specifically tailored to these new computational frameworks.

3   Workshop Summary

The papers in the present workshop address a number of the requirements and challenges mentioned above, with special emphasis on data management and integration, modelling and simulation, and data visualization and image processing. A first group of papers discusses data management and integration issues in genomics and proteomics as well as in biomedical applications, including management of provenance data. In particular, Swain *et al.* present a data warehouse and repository for results from protein folding and unfolding simulations together with optimization strategies for the data warehouse that exploit grid tools. The ViroLab virtual laboratory is a common framework discussed by the papers of Balis *et al.* and Assel *et al.*. The first one uses ViroLab to investigate a data provenance model, whereas the second one discusses data integration and security issues for medical data. Mackiewicz *et al.* deal with the analysis of genetic code data and the minimization of prediction errors.

A second group of papers discusses modelling and simulation in systems biology and biosciences. In particular, Cui *et al.* demonstrate a computational model for calcium signalling networks. Cardiac modelling for understanding Chagas' Disease is the focus of Mendoça Costa *et al.* whereas Vega *et al.* present an adaptive algorithm for identifying transcription factor binding regions. The use of DNA microarray expression data for gene selection is exploited by Maciejewski to optimize feature selection for class prediction. Cannataro *et al.* discuss a meta tool for the prediction of possible protein combinations in protein to protein interaction networks.

The third group of papers discusses visualization of genomics data and biomedical images, as well as metadata extraction for e-learning purposes. In particular, Jakubowska *et al.* present a new technique for improving the visual representation of data in genomic browsers. Miranda Teixeira *et al.* discuss methods for automatic segmentation of cardiac Magnetic Resonance Images and Bułat *et al.* deal with algorithms for computer navigation systems assisting bronchoscope positioning. Kononowicz and Wiśniowski exploit the MPEG-7 meta data standard for medical multimedia objects to support e-learning.

The results and current research trends presented in the papers reassure the importance of especially data and meta data management and of computational modelling. In addition, the workshop includes discussion slots on topics not covered by the accepted papers, like bioinformatics middleware for future applications, full semantic integration of biological data banks, and utilization of new visualization techniques.

4   Conclusions and Outlook

Data integration and data management are under research for years but the complexity associated to biological processes and structures demand further investigations and the development of new methods and tools for bioinformatics. The simulation of complex biological phenomena becomes more feasible as the amount of information provided by new experimental techniques and the computing power rapidly increases. On the other hand, the development and use of complex computational models demand new implementations that better exploit the new computer architectures. Research on distributed and Grid computing for bioinformatics applications as well as related workflow modeling will enable emerging life sciences applications to use in a coordinated way bioinformatics tools, biological data banks, and patient's clinical data, that requires seamless integration, privacy preservation and controlled sharing.

Acknowledgements. We would like to thank the members of the program committee for the invaluable assistance on reviewing the papers and the organizers of ICCS for promoting this workshop.

References

1. Cohen, J.: Bioinformatics - an introduction for computer scientists. ACM Computing Surveys 36, 122–158 (2004)
2. Talbi, E.C., Zomaya, A.Y. (eds.): Grid Computing for Bioinformatics and Computational Biology. Wiley, Chichester (2008)
3. Cannataro, M. (ed.): Computational Grid Technologies for Life Sciences, Biomedicine and Healthcare. Information Science Reference. Hershey (in preparation, 2008)
4. Aloisio, G., Breton, V., Mirto, M., Murli, A., Solomonides, T.: Special section: Life science grids for biomedicine and bioinformatics. Future Generation Computer Systems 23, 367–370 (2007)

Grid Computing Solutions for Distributed Repositories of Protein Folding and Unfolding Simulations

Martin Swain[1], Vitaliy Ostropytskyy[1], Cândida G. Silva[2], Frederic Stahl[1], Olivier Riche[1], Rui M.M. Brito[2], and Werner Dubitzky[1]

[1] School of Biomedical Sciences, University of Ulster, Coleraine BT52 1SA,
Northern Ireland, UK
mt.swain@ulster.ac.uk
[2] Chemistry Department, Faculty of Science and Technology, and Center for
Neuroscience and Cell Biology, University of Coimbra, 3004-535 Coimbra, Portugal
brito@ci.uc.pt

Abstract. The P-found protein folding and unfolding simulation repository is designed to allow scientists to perform analyses across large, distributed simulation data sets. There are two storage components in P-found: a primary repository of simulation data and a data warehouse. Here we demonstrate how grid technologies can support multiple, distributed P-found installations. In particular we look at two aspects, first how grid data management technologies can be used to access the distributed data warehouses; and secondly, how the grid can be used to transfer analysis programs to the primary repositories – this is an important and challenging aspect of P-found because the data volumes involved are too large to be centralised. The grid technologies we are developing with the P-found system will allow new large data sets of protein folding simulations to be accessed and analysed in novel ways, with significant potential for enabling new scientific discoveries.

1 Introduction

Protein folding is one of the unsolved paradigms of molecular biology, the understanding of which would provide essential insight into areas as diverse as the therapeutics of neurogenerative diseases or bio-catalysis in organic solvents. Protein folding simulations are time-consuming, data intensive and require access to supercomputing facilities. Once completed, few simulations are made publicly available – this hinders scientists from accessing data reported in publications, performing detailed comparisons, and developing new analytical approaches. A platform to allow access to protein folding and unfolding simulations, and to support data mining functionalities would clearly benefit the field.

The P-found project [1] aims to create a distributed public repository for storing molecular dynamics simulations, particularly those concerned with protein folding and unfolding. It aims to provide the tools needed to support the comparison and analysis of the simulations and thus enable new scientific knowledge to be discovered and shared.

M. Bubak et al. (Eds.): ICCS 2008, Part III, LNCS 5103, pp. 70–79, 2008.

P-found is designed to be a distributed system. This is because protein folding and unfolding simulations require large volumes of storage space: a simulation can easily comprise 1 to 10 Gigabytes of data, with simulation archives consisting of Terabytes. In addition, the groups who created the simulations have often already carefully stored and archived them. Ideally a distributed system would allow scientists to share their data, without undue interference in their existing practices, and without causing them to lose control over their own data.

There are two data storage components to the P-found system: a *primary data store or file repository* for the unaltered protein folding and unfolding simulations; and a *secondary data store or data warehouse* containing, for instance, information about how the simulations were created and derived physical properties of the simulations that are typically used in scientific analysis [1]. The secondary data is approximately 10 to 100 times smaller in volume than the primary data and consists of local and global molecular properties that summarise the simulations.

Grid technology is appropriate for P-found mainly because the user groups involved are based in independent organisations, each with their own IT infrastructures and security policies. Grid technology provides a set of tools whereby these user groups can come together in a virtual organisation, with security components that enable both data and compute servers to be shared across the independent administrative domains [2]. Data grids have been reviewed by Finkelstein *et al.* (2004) [3] and the efficiency of their design has been discussed by Laure *et al.* (2005) [4].

In this paper we consider the use of grid and distributed computing technologies to manage all aspects of data storage and analysis in the P-found system. Many of the results we present here have been adapted from tools and services developed by the DataMiningGrid project [5], [6]. However, P-found is an ongoing project, independent from the DataMiningGrid, and this is the first time that grid solutions for P-found have been presented in detail.

P-found is designed so that the data warehouse facilitates most comparative analyses and data mining operations, with the result that access to the primary data store can be avoided except for particularly novel or complex queries. We assume that every location with a primary data store has an associated data warehouse. However, in practice it is likely that while there will be many primary stores, there may be only a few, centralised data warehouses. This is due to the smaller data volume stored in the data warehouse and the greater administrative effort involved with maintaining such a facility. It is essential that the P-found system supports direct access to the primary data as novel analyses of this data may lead to important scientific discoveries. Therefore, we are investigating the use of grid technologies to solve two aspects of the P-found system:

1. *Distributed data management*: functionality to federate a number of distributed data warehouses so that they may be accessed and managed as if they were a single resource.

2. *Distributed analysis*: functionality for shipping or transferring analysis programs to the distributed primary data stores so that analysis takes place at the data store and thus by-passes the need to transfer large data volumes.

This paper is structured as follows: we briefly describe the DataMiningGrid, and then give more detailed descriptions of how its associated technology can be used to provide solutions to the distributed data management and analysis aspects of P-found. We then discuss the security implications involved with allowing scientists to distribute and execute potentially any kind of program over the P-found infrastructure. Before concluding the paper we outline our on-going and future development plans.

2   Grid Solutions for P-Found

The DataMiningGrid project took P-found's technical requirements into its design, and initial solutions for the problems associated with P-found were developed by the DataMiningGrid. The DataMiningGrid is described in detail elsewhere [5]. In summary, the DataMiningGrid was built using the Globus Toolkit [2], which contains various data management tools, including GridFTP for file transfers and OGSA-DAI version WSRF-2.2 [7], which was used to provide data management functionality for databases. A test-bed was created, based at three sites in the UK, Slovenia and Germany and at each site Globus components were used to interface with local Condor clusters. Condor is an opportunistic, high throughput computing system that uses cycle-stealing mechanisms to create a cluster out of idle computing resources [8].

The P-found system used with the DataMiningGrid is a prototype, and consists of just one installation [1]. This prototype consists of both the file repository and the data warehouse, which is implemented using the Oracle 10g database system. With just one installation it was not possible to fully demonstrate the distributed aspects. However, general grid data management solutions were developed and tested with other distributed database applications in the DataMiningGrid and these are fully applicable to the P-found system. More technical details of components comprising the P-found system are given in Sec. 2.2.

2.1   Data Mining Distributed Data Warehouses

Here we show how OGSA-DAI can be used to federate P-found data warehouses and calculate a summary of the data sets accessed. This summary is then used by different data preprocessing operations, resulting in formatted data sets that are ready for processing with data mining algorithms.

Figure 1 shows how the data summary was developed with the OGSA-DAI APIs. The process begins when the user selects a data service and defines an SQL query to be executed against one of the service's data resources. Then:

1. The client sends the query to the data service and after execution the meta-data associated with the query is retrieved.
2. The meta-data returned by the query is processed by the client, which then sends a request to the data service to create a temporary database table according to the meta-data values.

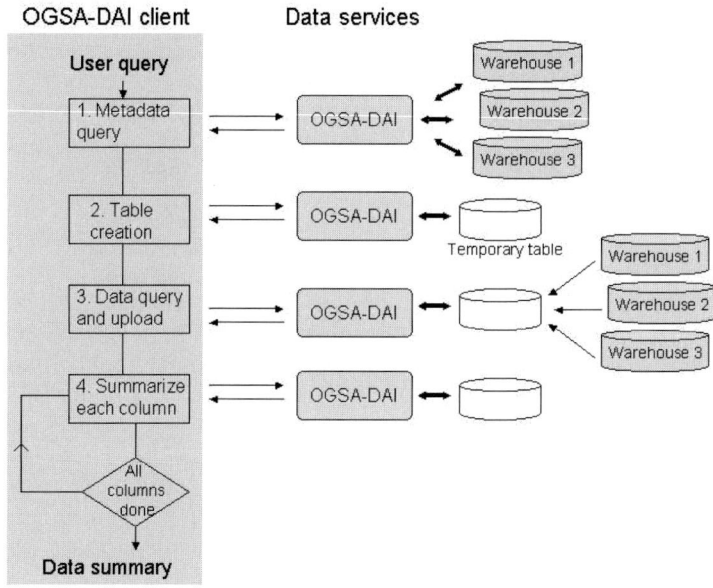

Fig. 1. A P-found application using the OGSA-DAI client API and OGSA-DAI services to calculate a data summary

3. The client retrieves the query data from the data service and loads them into the temporary table.
4. Now it is possible to calculate statistical values of this data set, with the particular values to be calculated depending on the application. Typically, for numeric data the average, standard deviation, maximum, minimum and variance are calculated, while for character data this involves finding all the distinct values in the column. The client performs this by issuing a series of requests, consisting of SQL queries, to summarize the values in each column. The resulting data set summary is stored at the client for use in subsequent requests or routines.

Using software from the DataMiningGrid, OGSA-DAI clients may be used to integrate distributed P-found databases: the clients first access data from a number of repositories, store the data in a temporary table, and then calculate a data summary. In the DataMiningGrid these data were streamed from the temporary table and a number of data filtering operations and data transformations were applied before the data were delivered to a file. The process is shown in Fig. 2:

1. A data summary is calculated, based on data from multiple databases that have already been uploaded to a temporary table. All data are now selected from the table.

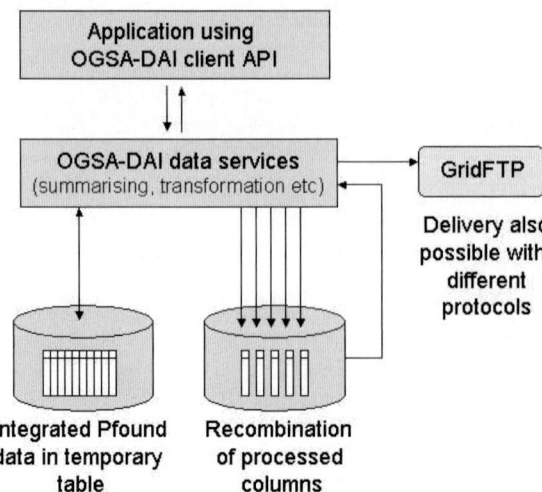

Fig. 2. A P-found application performing data transformation on values generated by a data summary

2. The OGSA-DAI projection operations are used to extract table columns as data are streamed from their source (i.e. the temporary table) to their destination (in this case another temporary database table). As the data is streamed at the data service, columns are extracted and the data in each column is transformed, for example by aggregating values or discretizing continuous numeric values. In this case returning the data to a database table is convenient, as it allows the process to be repeated with a series of different transformations.
3. Now the transformed data is retrieved from the table.
4. Additional functionality may now be applied to convert this transformed data into different formats.
5. In the final step the data are deposited in one or more files on a GridFTP server. These are then available for data mining.

The DataMiningGrid provided some enhancements to OGSA-DAI, for example to perform data mining preprocessing operations such as cross-validation, data transformation, and data formatting i.e. formatting data according to Weka's ARFF structure [9], or the format of the CAREN association algorithm [10]. Such tasks usually required the data summary calculated earlier, and could be used to deliver data transformed and formatted to servers ready for processing with data mining algorithms.

OGSA-DAI supports a variety of data delivery operations, including delivery using GridFTP, HTTP, email or simply returning results from the web service to the client using SOAP messages. There is therefore only a small overhead required to use the clients we have presented here outside of the DataMiningGrid infrastructure.

2.2   Transferring Data Analysis Programs to Distributed File Repositories

Here we present a simplified version of the process for shipping algorithms to multiple, distributed data locations, as implemented by the DataMiningGrid. The main difference is that the DataMiningGrid developed a specialised Resource Broker Service that interfaced with the WS-GRAM utilities from the Globus Toolkit 4 to submit, monitor and cancel jobs on computational resources and local scheduling systems, and to coordinate the input and output staging of data. This Resource Broker Service could ship both data and programs to any location on the grid. It could handle the heterogeneity of grid resources and it provided additional functionality to support data mining applications [5]. Here we assume that the data is never moved from the primary storage repositories, and that there are dedicated compute resources available which are local to the storage repositories.

The system we propose for P-found is shown in Fig. 3 and uses the following components:

1. A client application to analyse, process or data mine the simulations stored in the primary repositories.
2. A system registry: this is a centralised component, listing all the analysis software, simulation repositories, and data warehouses available in the system. It would also contain information to tell either human users or software how to interact with those components.
3. A software repository: there can be one or more of these, and it is here that analysis programs and their associated libraries are stored. This does not require a sophisticated solution: executable files and libraries can be simply stored on a file system and accessed using GridFTP. The software repository could be merged with the client application.
4. Multiple, distributed P-found installations, each installation containing both a primary data store for the simulation files and a secondary data warehouse for computed simulation properties. Compute resources will be available at each installation: these may be a few dedicated processors or a local Condor cluster, connected to the data repositories via a networked file system or high-speed network. There is also a GridFTP server at each installation for performing fast file transfers.

The process for shipping analysis programs to data is shown in Fig. 3 and is described here. Note that there may be a preceding step, not shown in the figure, in which the client application performs a distributed query on the P-found data warehouses, similar to that described in Sec. 2.1. The data warehouses contain all technical information about the simulations i.e. the name of the protein simulated, the techniques used, and so on. In the preceding step this information is queried in order to find a set of interesting simulation files. Then:

1. The client application queries the registry to discover available analysis software. The result will be a list of programs, with a URI giving their location,

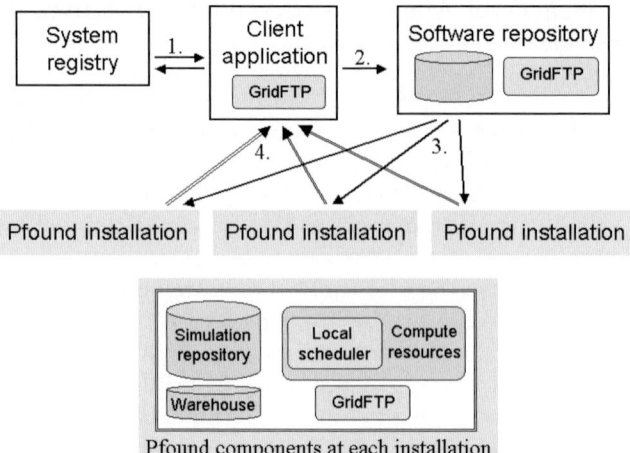

Fig. 3. Shipping programs to the primary P-found data repositories

along with the technical information required to execute these programs, such as how much CPU and memory they need, operating system preferences, etc. This technical information is very important for grid systems which are composed of heterogeneous resources.

2. A program is selected, the correct software repository is contacted, and the program is transferred to the P-found repositories that contain the simulations of interest.

3. Once transferred to the repositories, the programs are executed. This may be quite a complicated procedure and may require a sophisticated component such as the DataMiningGrid's Resource Broker Service, as mentioned earlier in this section. For example, such a component would ensure programs are executed in the correct environment and on machines with the correct specification.

4. Once the analysis is completed, the results are transferred using GridFTP, either back to the client application or to some other location e.g. another P-found repository where they are used in subsequent calculations.

In this scenario there is no explicit support for workflows, unless they are constructed using a scripting language (e.g. Perl or Python). This should be adequate for P-found as many scientists use Perl and Python to coordinate analysis tasks. It does mean, however, that all the executables and associated libraries should be transferred to the primary repositories along with the scripts.

3   Discussion

Grid data management technology is maturing and is able to support relatively sophisticated data analysis scenarios, as has been demonstrated by projects such as Gridminer [11] and the KnowledgeGrid [12], as well as the DataMiningGrid

[5]. Unfortunately, the shipping of programs to data locations has not been investigated in as much detail, even though this is an important approach for distributed data sets that cannot be transferred to centralised processing facilities (due to either their large volume or their confidential nature). There is therefore little "off-the-shelf" software that we can use to provide this functionality.

Allowing users to freely execute arbitrary code on machines outside of their own administrative domain does raise some serious security issues. For example, there is great potential for malicious users to create havoc on the P-found systems by destroying the data storage facilities. To persuade new groups to participate in the P-found system, it is essential for them to be fully confident that their data will remain secure. The Globus Toolkit provides tools for security based on public key cryptography [2], which is sufficient for most purposes. This may also be reinforced by only allowing software that is completely trustworthy to be registered on the system and transferred for execution on different administrative domains. However, this does limit scientists from performing arbitrary analysis on the repositories, one important aim of the P-found system. In the end, the security framework needs to be agreed on by the community, and it must be flexible enough to persuade the most cautious users to participate.

As an alternative to using executables and grid scheduling mechanism, we investigated the Dynasoar framework [13]. Dynasoar provides an architecture for dynamically deploying Web services remotely on a grid or the Internet. A potential use, motivating its design, was to move Web services that access data and perform analysis on it, closer to the data storage facilities, which fits with the requirements of the P-found system. Web services require that an interface is specified, describing the functionality of the Web service and how it should be used – this approach makes a little harder to hide malicious software that is able to damage the system. However, while the Dynasoar has some nice features, it is still a prototype with limited functionality. Moreover, as most scientists from the molecular simulation community typically use scripts (Perl or Python) or executables (C/C++ and Fortran) for analysis and are therefore not familiar with Web service technology, we decided that this approach was currently unsuitable for this project.

The BioSimGrid [14] is a project with similar aims to P-found. It enables data sharing of molecular simulation data between different research groups and universities. A Python scripting environment is used to pre-process, deposit and retrieve data, and a general purpose Python analysis toolkit is available for manipulating deposited data. The authors do not discuss security issues, presumably because the system was designed for use with six UK universities – a relatively small and localised community where all users may easily be held accountable for their actions. This is different to P-found, which we plan to open up to the wider international community.

4   Future Work

P-found is still under development, and project partners are investigating additional approaches to implement the final system. These include using grid services available through the EGEE project (Enabling Grids for E-sciencE), and comparing the approach given here (based on multiple, distributed data warehouses) with a single, centralised data warehouse and multiple storage elements. The P-found partners plan to develop these three prototypes and make them available to the user community. Feedback from the user community will be essential in choosing the final design.

5   Conclusions

To further scientific discoveries there is a need to enable scientists to share and analyse protein folding and unfolding simulations. The P-found protein folding and unfolding simulation repository is designed to fulfill this need, and right now consists of two data storage components: a repository of unprocessed simulation data and a data warehouse containing detailed information regarding how the simulations were generated as well summaries of the simulations in the form of calculated local and global physical properties.

While a centralised version of P-found (`www.p-found.org`) is currently available, P-found is ultimately envisioned as a distributed system due to the massive data volumes involved. Here we have described how grid technologies can be used to realise that vision. We have demonstrated how OGSA-DAI, a grid data management tool can be used to federate distributed P-found data warehouses and prepare data for analysis and data mining tasks, we have also presented a mechanism to allow scientists to ship arbitrary programs to the distributed simulation repositories in order to process this primary data in novel and sophisticated ways. This is an important aspect of P-found, with the potential to generate new scientific discoveries as protein folding and unfolding data sets become more available.

Acknowledgements. This work was supported in part by the European Commission FP6 grants No. 004475 (the DataMiningGrid[1] project), and No. 033437 (the Chemomentum[2] project) to WD, MS and VO; and in part by grant GRID/ GRI/81809/2006 (FCT and FEDER, Portugal) to RMMB, and Doctoral Fellowship SFRH/BD/16888/2004 to CGS.

References

1. Silva, C.G., Ostropytsky, V., Loureiro-Ferreira, N., et al.: P-found: The Protein Folding and Unfolding Simulation Repository. In: Proc. 2006 IEEE Symp. on Computational Intelligence in Bioinformatics and Computational Biology, pp. 101–108 (2006)
2. Foster, I.T.: Globus Toolkit Version 4: Software for Service-Oriented systems. J. Comput. Sci. Technol. 21, 513–520 (2006)

---

[1] http://www.DataMiningGrid.org
[2] http://www.chemomentum.org/

3. Finkelstein, A., Gryce, C., Lewis-Bowen, J.: Relating Requirements and Architectures: A Study of Data-Grids. J. Grid Comput. 2, 207–222 (2004)
4. Laure, E., Stockinger, H., Stockinger, K.: Performance Engineering in Data Grids. Concurrency - Practice and Experience 17, 171–191 (2005)
5. Stankovski, V., Swain, M., Kravtsov, V., et al.: Grid-Enabling Data Mining Applications with DataMiningGrid: An Architectural Perspective. Future Gener. Comput. Syst. 24, 259–279 (2008)
6. Swain, M., Hong, N.P.C.: Data Preprocessing using OGSA-DAI. In: Dubitzky, W. (ed.) Data Mining Techniques in Grid Computing Environments, Wiley, Chichester (in press)
7. Antonioletti, M., Atkinson, M., Baxter, R., et al.: The Design and Implementation of Grid Database Services in OGSA-DAI. Concurr. Comput.: Pract. Exper. 17, 357–376 (2005)
8. Litzkow, M., Livny, M.: Experience with the Condor Distributed Batch System. In: Proc. IEEE Workshop on Experimental Distributed Systems, pp. 97–100 (1990)
9. Witten, I.H., Frank, E.: Practical Machine Learning Tools and Techniques, 2nd edn. Morgan Kaufmann, San Francisco (2005)
10. Azevedo, P.J., Silva, C.G., Rodrigues, J.R., Loureiro-Ferreira, N., Brito, R.M.M.: Detection of Hydrophobic Clusters in Molecular Dynamics Protein Unfolding Simulations Using Association Rules. In: Oliveira, J.L., Maojo, V., Martín-Sánchez, F., Pereira, A.S. (eds.) ISBMDA 2005. LNCS (LNBI), vol. 3745, pp. 329–337. Springer, Heidelberg (2005)
11. Fiser, B., Onan, U., Elsayed, I., Brezany, P., Tjoa, A.: On-Line Analytical Processing on Large Databases Managed by Computational Grids. In: Proc. 15th Int. Workshop on Database and Expert Systems Applications (2004)
12. Congiusta, A., Talia, D., Trunfio, P.: Distributed Data Mining Services Leveraging WSRF. Future Gener. Comput. Syst. 23, 34–41 (2007)
13. Watson, P., Fowler, C.P., Kubicek, C., et al.: Dynamically Deploying Web Services on a Grid using Dynasoar. In: Proc. 9th IEEE Int. Symp. on Object and Component-Oriented Real-Time Distributed Computing (ISORC 2006), Gyeongju, Korea, pp. 151–158. IEEE Computer Society Press, Los Alamitos (2006)
14. Ng, M.H., Johnston, S., Wu, B., et al.: BioSimGrid: Grid-Enabled Biomolecular Simulation Data Storage and Analysis. Future Gener. Comput. Syst. 22, 657–664 (2006)

Provenance Querying for End-Users: A Drug Resistance Case Study

Bartosz Baliś[1,2], Marian Bubak[1,2], Michal Pelczar[2], and Jakub Wach[2]

[1] Institute of Computer Science, AGH, Poland
{balis,bubak}@agh.edu.pl, wach.kuba@gmail.com
[2] Academic Computer Centre – CYFRONET, Poland

Abstract. We present a provenance model based on ontology representation of execution records of in silico experiments. The ontologies describe the execution of experiments, and the semantics of data and computations used during the execution. This model enables query construction in an end-user oriented manner, i.e. by using terms of the scientific domain familiar to researchers, instead of complex query languages. The presented approach is evaluated on a case study Drug Resistance Application exploited in the ViroLab virtual laboratory. We analyze the query capabilities of the presented provenance model. We also describe the process of ontology-based query construction and evaluation.

Keywords: e-Science, Grid, ontology, provenance, ViroLab.

1 Introduction

The importance of provenance tracking in e-Science environments has been pointed out many times. However, providing an adequate end-user support for provenance querying is also an important challenge. It has been recognized that the need for provenance queries goes beyond the lineage of a single data item and searching or mining over many provenance records might be useful [8] [12]. However, there is a technical and conceptual barrier preventing or making it difficult for researchers or specialists who are the end users of e-Science environments, to construct complex queries using the query languages such as XQuery, SQL or SPARQL, or even dedicated APIs built on top of those languages or particular data models. Therefore, there is a need for provenance query methodology which allow end-users to construct powerful queries in an easy way.

The goal of this work is to present a provenance model which enables complex queries over many provenance records. In addition, the model should support end-user oriented querying in order to be usable by non-IT experts. Inspired by the vision of a future e-Science infrastructure, which brings together people, computing infrastructures, data and instruments, and in which semantics and knowledge services are of paramount importance [9] [5], we propose a provenance model based on ontologies which model the execution of scientific workflows. We argue that ontology models describing the execution of experiments (including provenance) are a convenient inter-lingua for: (1) *end-users* who use ontologies as a query language, (2) *query tools* using them to represent and evaluate queries, and *provenance repository* in which the ontologies serve as the data model.

M. Bubak et al. (Eds.): ICCS 2008, Part III, LNCS 5103, pp. 80–89, 2008.

Most existing approaches concentrate on a provenance model sufficient for correct computation of and querying for derivation paths of selected data items [4] [13] [10]. Some approaches introduce simple provenance models that abstract from any computation model, e.g. the Read – Write – State-reset model [4], and the p-assertions model [6]. It has been pointed out that those low-level models limit the provenance query capabilities [13]. A complex RDF and ontology-based provenance model is introduced in myGrid/Taverna [15]. This model has been shown to support complex provenance queries [14]. However, the aspect of query construction in an end-user-friendly manner is not explored in the mentioned works.

The approach to provenance tracking and querying presented in this article is evaluated in the context of the ViroLab Project[1] [11] which provides a virtual laboratory for conducting in silico experiments related to diagnosis of infectious diseases, especially the HIV virus[2] [7].

The remainder of this paper is structured as follows. Section 2 describes the provenance model. Section 3 presents the case study Drug Resistance Workflow. In Section 4, query capabilities of the presented provenance model are investigated. Section 5 introduces the ontology-based provenance query methodology oriented towards end-users. Finally, Section 6 summarizes the current achievements and discusses future work.

2   Provenance Model

Our provenance model is based on the concept of *Experiment Information*, an ontology-based record of experiment execution. The base for this record is an ontology describing in silico experiments in which the workflow model is used as the computation model. The execution stages of the experiment and the data items used in the workflow are linked to domain ontologies of applications and data in order to enhance the semantic description of experiment data and computations. Fig. 1 presents part of the ontology tree in which it is shown how the generic concepts describing the experiment execution are connected do domain ontologies concepts.

Provenance tracking is based on monitoring of the execution of a workflow. The creation of an ontology-based record of the experiment execution is done as a process of translation from low-level monitoring data composed of monitoring events into high-level Experiment Information. This process is depicted in Fig. 2. A workflow enactment engine executes a workflow according to some plan. Monitoring events generated by the instrumentation come from different distributed sources, among others, the workflow enactment engine and the workflow activities. A monitoring system collects and correlates the monitoring events, and passess them to a Semantic Aggregator component which aggregates and translates the monitoring data into an ontology representation, according to an ontology Experiment Information model. The ontology individuals are published into the provenance tracking system (PROToS) [2] and stored in a permanent Experiment Information Store. The process of monitoring, event correlation, aggregation and translation to ontologies is described in detail in [1].

---

[1] ViroLab Project: www.virolab.org
[2] ViroLab virtual laboratory: virolab.cyfronet.pl

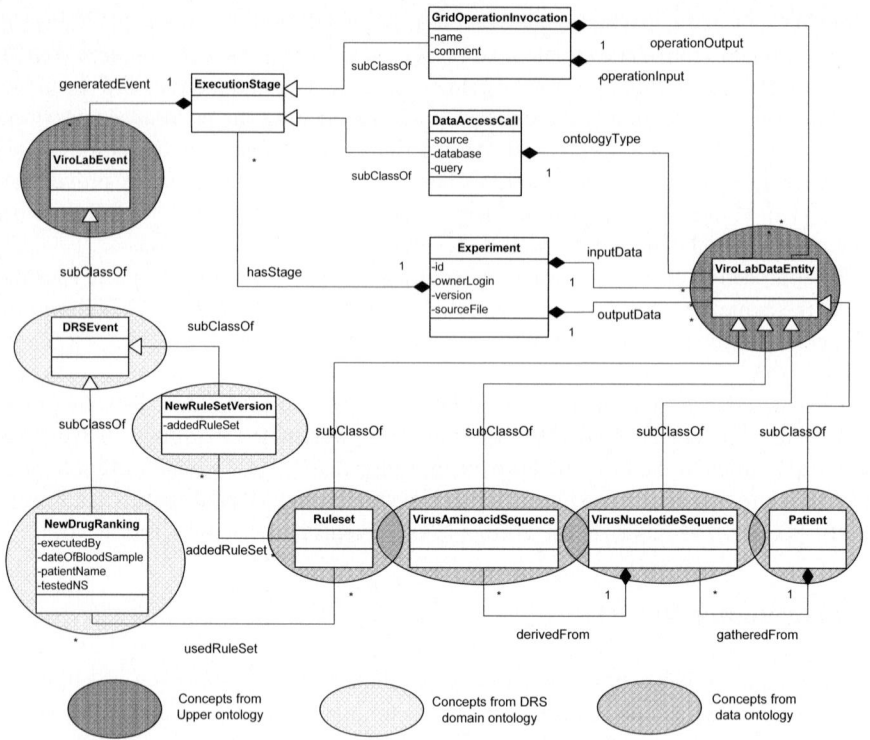

Fig. 1. Generic experiment ontology and related domain ontologies

3 Drug Resistance Workflow

The workflow which is used to illustrate the capabilities of the presented provenance model is the Drug Ranking System (DRS) application exploited in the ViroLab project. The goal of this application is to determine best drug combinations for a particular HIV virus taken from the blood sample of a patient. In the simplest DRS scenario, the inputs for an experiment are *a list of mutations for a given gene of an HIV virus*, and the *name of a ruleset* (chosen out of a few available) to be used to compute drug rankings. A drug ranking tool is executed to compute the drug rankings which are returned as the experiment's results.

We will consider an extended Drug Ranking scenario which contains the following stages:

Experiment Input: HIV virus isolate nucleotide sequence

> **Stage 1.** Align input nucleotide sequence. Result: genes responsible for particular HIV proteins.

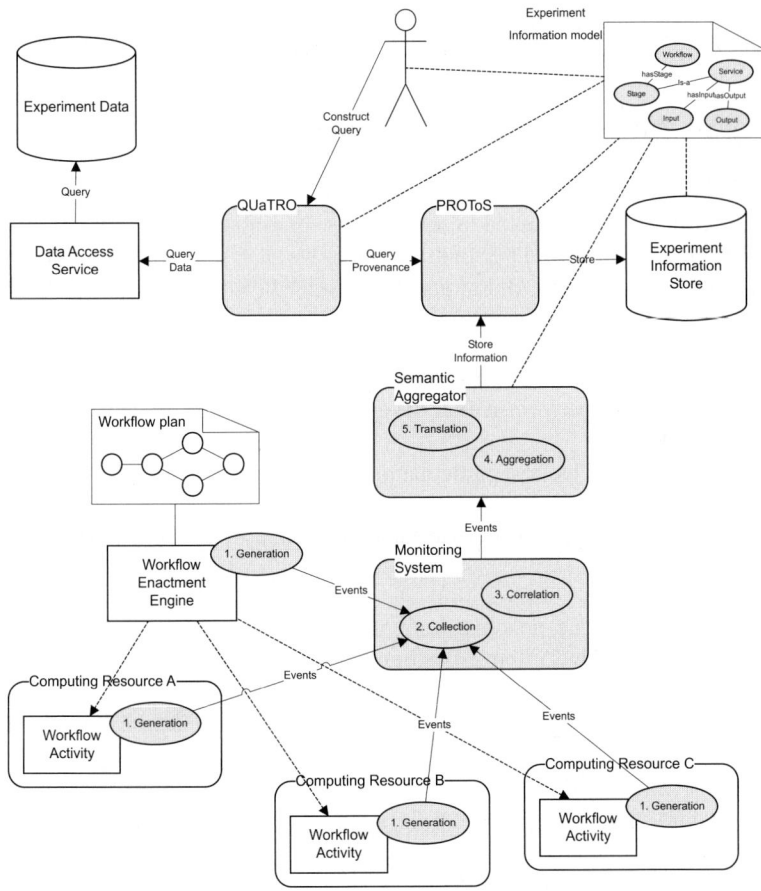

Fig. 2. Provenance tracking and querying in ViroLab

Stage 2. Compare obtained sequences to reference strains in order to determine mutations.

Stage 3. Apply selected rulestes to mutations to determin drug rankings.

In the course of provenance tracking of this simple scenario, multiple events are generated, collected and aggregated into the ontology representation of the experiment's execution. The events include those related to generic experiment enactment (experiment started/ended, operation started/ended, etc.). In addition, *domain-specific* events are generated to enhance the generic information about experiment execution with semantics related to the particular domain. For example, the three steps of the described scenario will generate respective domain events – 'Sequence Alignment', 'Computation of Mutations' and 'New Drug Ranking'. In consequence, appropriate information will be created to denote that the invoked operations were in fact sequence alignments, mutation computations, or drug ranking computations. Also, the input data items will be identified as a nucleotide sequence, a protein sequence, or a mutation list. Those semantic enhancements are achieved by linking generic ontology individuals (e.g. 'Operation Invocation') to domain ontology individuals (e.g. 'New Drug Ranking').

4   Query Capabilities of the Provenance Model

In [8], a number of challenging provenance questions have been formulated. Following were a few articles that attempted to answer those questions on the grounds of several provenance models, e.g. [10] [14]. In this section, we shall define a similar set of questions for our example DRS workflow and demonstrate how they can be answered in our provenance model. The questions are as follows:

Q1 Find the process that led to a particular drug ranking.
Q2 Find all operations performed after the alignment stage that led to a particular drug ranking.
Q3 Find operations that led to a particular drug ranking and were performed as 1st and 2nd ones.
Q4 Find all alignment operations performed on 10.10.2007 which operated on a given nucleotide sequence.

Question Q1 is a basic provenance question which returns a derivation path for a given data result. XQuery implementations of those queries over our provenance model are shown below.
Q1:

```
declare function local:variable-proces($varId as xs:string)
as list {
  for $goi in //GridOperationInvocation
      $dal in //DataAccessLoad
  where $goi//outData contains $varId
  return
    {
      for $input in $goi//inData
      return
        local:variable-proces($input)
    }
    $goi
  where $dal//variableId eq $varId
  return
    local:dataEntity-process($dal//dasId)
    $dal
};

declare function local:dataEntity-proces($dasId as xs:string)
as element {
  for $das in //DataAccessSave
  where $das//dasId eq $dasId
  return
    <process>
```

```
        local:variable-proces($das//variableId)
        $das
     </process>
};


<provenance>
  local:dataEntity-proces({dasId as passed})
</provenance>
```

Q2:

```
<provenance>
for $goi in //GridOperationInvocation,
  $exp in //Experiment
  where
    $exp/outData@[name()='rdf:resource'
    and . eq {drug ranking id}]
    and $exp/stageContext/@rdf:resource = $goi/@rdf:Resource
    and $goi/stageNumber > 1
  return $goi
</provenance>
```

Q3:

```
<provenance>
for $goi in //GridOperationInvocation,
  $exp in //Experiment
  where
    $exp/outData@[name() = 'rdf:resource'
    and . eq {drug ranking id}]
    and $exp/stageContext/@rdf:resource = $goi/@rdf:Resource
    and $goi/stageNumber in {1, 2}
  return $goi
</provenance>
```

Q4:

```
<provenance>
for $goi in //GridOperationInvocation,
  $exp in //Experiment
  where
    $exp/time eq '10.10.2007'
    and $goi/inData@[name() = 'vl-data-protos:dasId'
    and . eq {nucleotide sequence id}]
    and $exp/stageContext/@rdf:resource = $goi/@rdf:Resource
    and $goi/stageNumber = 1
  return $goi
</provenance>
```

5   Ontology-Based Query Construction

On top of the provenance tracking system we have built QUery TRanslation tOols (QUaTRO, Fig. 2), which enable end-user oriented approach to querying both repositories of provenance (through PROToS), and experiment data (through external Data Access Service). Both query construction in QUaTRO and provenance representation in PROToS are based on the ontology model of Experiment Information.

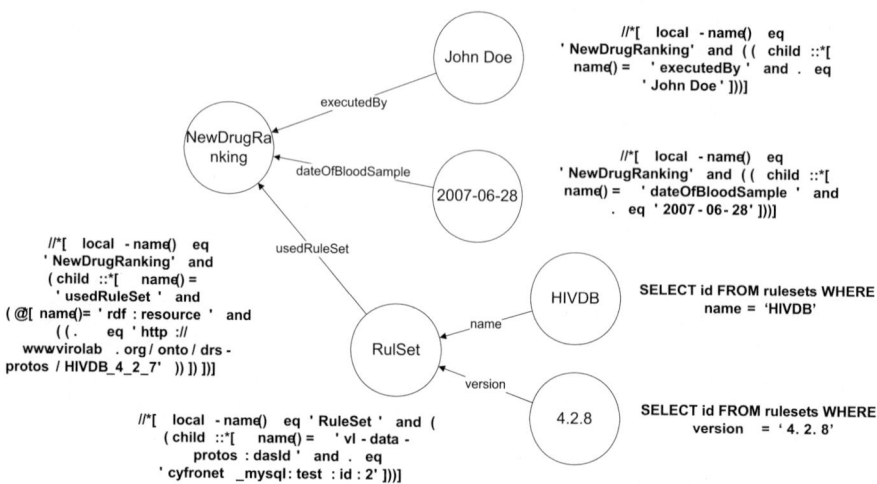

Fig. 3. Sample query tree and its evaluation

QUaTRO provides an easy-to-use graphical user interface. Query construction in QUaTRO amounts to building a query tree which begins with selecting a starting concept which determines the area of interest (for example 'New Drug Ranking', 'Experiment', 'Nucleotide Sequence', 'Ruleset'). The query form automatically expands and allows to follow properties of the starting concept which can lead to (1) other concepts (in which case the tree can be further expanded), (2) literal values, or (3) database mappings, both (2) and (3) becoming query tree leaves. When the query tree construction is completed, the tree is analyzed in a bottom-up fashion, starting from its leaves and following the edges up to the tree root. Subqueries represented in particular subtrees are evaluated and merged in upper nodes, if feasible. In principle, some subtrees might generate queries to relational databases of experiment data, while others – XQuery queries to the provenance repository. This approach enables queries such as *Select all experiments of type 'Drug ranking' whose results contained recommendation for 'Drug X'*. The provenance repository allows to select all experiments of type 'Drug ranking' which resulted in *some* drug recommendations. However, the actual results containing detailed descriptions of what drugs were recommended, are not part of provenance, but of actual experiment data which is stored in a database. Only a combined request to provenance repository and data repository will allow to construct queries of this kind.

Let us consider the following example provenance query: *Select all computations of drug rankings performed by 'John Doe' for blood samples taken on Jun-28-2007, and in which rule set 'HIVDB' in version 4.2.8 was used.* The constructed query tree is shown in Fig. 3, and the corresponding QUaTRO tool web form is presented in

Fig. 4. Query tree nodes denote either concepts ($NewDrugRanking$, $RuleSet$), or literals ($JohnDoe$, $2007 - 06 - 28$, $HIVDB$, 4.2.8). The tree edges are one of the following:

- Object properties which link to another concept ($usedRuleSet$);
- Datatype properties which link to a literal of a given datatype ($executedBy$, $dateOfBloodSample$);
- Database properties which denote that a literal they link to is actually stored in a database (e.g. in table T, column C) ($name$, $version$).

The query is evaluated as shown in Fig. 3. For database values, appropriate SQL queries are constructed. In this case, they return the IDs of rule sets which are compared to IDs stored in the provenance repository (as a property of the $RuleSet$ individuals). This allows us to pick exactly those $RuleSet$ individuals which satisfy the given criteria. Other subtrees are evaluated as XQuery requests to provenance repository and merged in upper-level nodes to minimize the total number of requests.

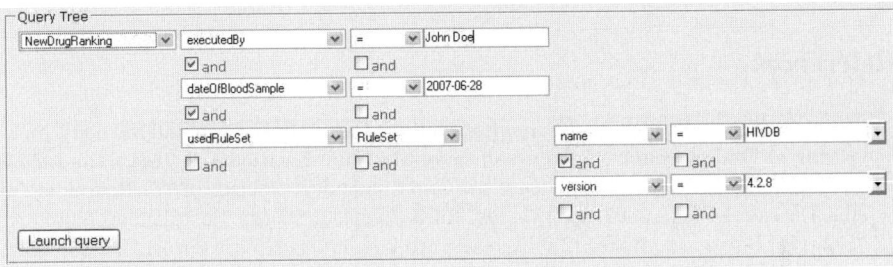

Fig. 4. Query construction in QUaTRO GUI

6 Conclusion and Future Work

We have presented an approach to tracking and querying provenance in which end-users are enabled to construct complex queries in a simple way. Ontologies as a model for provenance proved to be an adequate inter-lingua between: (1) *end users* who use ontologies as a query language while interacting with a graphical user interface; (2) *query tools* using the ontologies to represent and evaluate queries; (3) *provenance repository* in which the ontologies serve as the data model.

A unique feature of our approach is the support for subrequests to databases within a query tree. This enables even more detailed queries in which the structure of data items (not only provenance) is explored [3]. As a matter of fact, QUaTRO tools actually can be used to construct pure SQL-based queries thanks to mappings in the data ontology. This feature is also very useful, since the domain researchers often need to build complex queries to extract interesting input data for in silico experiments.

Currently, prototypes of PROToS and QUaTRO tools are implemented. They allow to record provenance and to construct queries in the described 'wizard-based' manner, starting from an ontology concept and following its properties to build the query tree.

Future work includes the implementation of distributed architecture of PROToS in order to ensure efficient querying. Most importantly, however, we plan several enhancements related to querying capabilities:

- We plan to extend our ontology with reverse relationships. This would allow to issue queries not only in the one-to-many but also in the many-to-one direction. For example, currently one can only query about Experiment which has ExecutionStage, but cannot query directly about ExecutionStages which are part of Experiment.
- Additional operators need to be added to QUaTRO, for example the logical *or*, and the *in* operator denoting that an attribute may have one value from a set thereof.
- Currently, an attribute can be compared only to literal values, but not to the evaluated values of other attributes. We plan to add this enhancement by allowing to explicitly create subqueries within a query, so that attributes could be compared against the result of the evaluated subquery (provided that data types would match).

Acknowledgments. This work was supported by EU project Virolab IST-027446 with the related Polish grant SPUB-M and by the Foundation for Polish Science.

References

1. Balis, B., Bubak, M., Pelczar, M.: From Monitoring Data to Experiment Information – Monitoring of Grid Scientific Workflows. In: Fox, G., Chiu, K., Buyya, R. (eds.) Third IEEE International Conference on e-Science and Grid Computing, e-Science 2007, Bangalore, India, December 10-13, 2007, pp. 187–194. IEEE Computer Society, Los Alamitos (2007)
2. Balis, B., Bubak, M., Wach, J.: Provenance Tracking in the Virolab Virtual Laboratory. In: PPAM 2007. LNCS, Springer, Gdansk, Poland (in Print, 2008)
3. Balis, B., Bubak, M., Wach, J.: User-Oriented Querying over Repositories of Data and Provenance. In: Fox, G., Chiu, K., Buyya, R. (eds.) Third IEEE International Conference on e-Science and Grid Computing, e-Science 2007, Bangalore, India, December 10-13, 2007, pp. 77–84. IEEE Computer Society, Los Alamitos (2007)
4. Bowers, S., McPhillips, T.M., Ludäscher, B., Cohen, S., Davidson, S.B.: A Model for User-Oriented Data Provenance in Pipelined Scientific Workflows. In: Moreau, L., Foster, I. (eds.) IPAW 2006. LNCS, vol. 4145, pp. 133–147. Springer, Heidelberg (2006)
5. Goble, C., Roure, D.D., Shadbolt, N.R., Fernandes, A.A.A.: Enhancing Services and Applications with Knowledge and Semantics. In: Foster, I., Kesselman, C. (eds.) The Grid 2: Blueprint for a New Computing Infrastructure, ch. 23, pp. 432–458. Morgan Kaufmann Publishers, San Francisco (2004)
6. Groth, P., Jiang, S., Miles, S., Munroe, S., Tan, V., Tsasakou, S., Moreau, L.: An Architecture for Provenance Systems, University of Southampton, Tech. Rep. (2006)
7. Gubala, T., Balis, B., Malawski, M., Kasztelnik, M., Nowakowski, P., Assel, M., Harezlak, D., Bartynski, T., Kocot, J., Ciepiela, E., Krol, D., Wach, J., Pelczar, M., Funika, W., Bubak, M.: ViroLab Virtual Laboratory. In: Proc. Cracow Grid Workshop 2007. ACC CYFRONET AGH (2008)
8. Moreau, L., et al.: The first provenance challenge. Concurrency and Computation: Practice and Experience 20(5), 409–418 (2007)
9. De Roure, D., Jennings, N., Shadbolt, N.: The Semantic Grid: A future e-Science infrastructure. In: Berman, F., Fox, G., Hey, A.J.G. (eds.) Grid Computing – Making the Global Infrastructure a Reality, pp. 437–470. John Wiley and Sons, Chichester (2003)

10. Simmhan, Y.L., Plale, B., Gannon, D.: Query capabilities of the Karma provenance frame-work. Concurrency and Computation: Practice and Experience 20(5), 441–451 (2007)
11. Sloot, P.M., Tirado-Ramos, A., Altintas, I., Bubak, M., Boucher, C.: From Molecule to Man: Decision Support in Individualized E-Health. Computer 39(11), 40–46 (2006)
12. Stankovski, V., Swain, M., Kravtsov, V., Niessen, T., Wegener, D., Kindermann, J., Dubitzky, W.: Grid-enabling data mining applications with DataMiningGrid: An architectural perspec-tive. Future Generation Computer Systems 24(4), 259–279 (2008)
13. Zhao, Y., Wilde, M., Foster, I.T.: Applying the virtual data provenance model. In: Moreau, L., Foster, I. (eds.) IPAW 2006. LNCS, vol. 4145, pp. 148–161. Springer, Heidelberg (2006)
14. Zhao, J., Goble, C., Stevens, R., Turi, D.: Mining Taverna's Semantic Web of Provenance. Concurrency and Computation: Practice and Experience 20(5), 463–472 (2007)
15. Zhao, J., Wroe, C., Goble, C.A., Stevens, R., Quan, D., Greenwood, R.M.: Using Semantic Web Technologies for Representing E-science Provenance. In: McIlraith, S.A., Plexousakis, D., van Harmelen, F. (eds.) ISWC 2004. LNCS, vol. 3298, pp. 92–106. Springer, Heidelberg (2004)

Integrating and Accessing Medical Data Resources within the ViroLab Virtual Laboratory

Matthias Assel[1], Piotr Nowakowski[2], and Marian Bubak[2,3]

[1] High Performance Computing Center, University of Stuttgart, D-70550, Germany
[2] Academic Computer Centre CYFRONET AGH, ul. Nawojki 11, 30-950 Kraków, Poland
[3] Institute of Computer Science, AGH, al. Mickiewicza 30, 30-059, Kraków, Poland
assel@hlrs.de, p.nowakowski@cyfronet.pl, bubak@agh.edu.pl

Abstract. This paper presents the data access solutions which have been developed in the ViroLab Virtual Laboratory infrastructure to enable medical researchers and practitioners to conduct experiments in the area of HIV treatment. Such experiments require access to a number of geographically distributed data sets (residing at various hospitals) with heavy focus on integration and security issues. Scientists conducting virtual experiments need to be able to manipulate such distributed data in a consistent and secure manner. We describe the main components of the Virtual Laboratory framework being devoted to data access and explain how data is processed in the presented environment.

Keywords: The Grid, Data access, Virtual laboratories, Data integration, OGSA-DAI, Medical research

1 Introduction and Motivation

The ViroLab Virtual Laboratory is an integrated system of tools for accessing and integrating resources, whose main purpose is to facilitate medical research and treatment in the HIV virology domain as well as other types of research in the general field of medical sciences. The research is carried out in a collaborative working environment using state-of-the-art Grid computing technologies and standards [19] and consisting of distributed computing and data resources deployed at various networked sites. As the complexity of interfacing such resources often presents a steep learning curve for application developers, the main goal of the Virtual Laboratory (VL) is to present a powerful and flexible development environment for virtual experiment developers while preserving ease of use and reusability of the proposed solution, thus allowing transparent and secure access to corresponding underlying infrastructures. A detailed description of the ViroLab Virtual Laboratory design is outside the scope of this paper, but can be found in [21] and [22]. In this paper, we focus on aspects related to data retrieval, integration,, and manipulation in the VL environment.

M. Bubak et al. (Eds.): ICCS 2008, Part III, LNCS 5103, pp. 90–99, 2008.

As one can expect from a virtual laboratory for research and treatment in the area of medical sciences, the system must provide access to a range of medical data including genetic, treatment, and drug information. This data, used to conduct experiments in viral drug resistance interpretation and selecting optimal treatment strategies, comes from various hospitals and medical centers being partners in the ViroLab project and is secured against unauthorized access. From the point of view of experiment developers and users, all data records describing HIV subtypes and mutations are equally valuable and should be treated in an analogous manner. However, a problem emerges with achieving a uniform representation of such data (see next section for details). It is therefore the task of the data access component(s) of the ViroLab Virtual Laboratory to integrate data derived from various sources and to enable experiment developers to manipulate this data in a consistent, efficient and straightforward way.

Due to the sensitivity and confidentiality of data shared within the virtual laboratory, a very critical issue for developing services that allow access to distributed medical databases concerns the overall security including access control to certain resources (who is able to access which information set) as well as data encryption and integrity of relevant data sets processed by the data access infrastructure. ViroLab meets this important issue by introducing a highly dynamic and flexible environment that guarantees security on several levels using established security principles and technologies as described in [2], to protect the confidential information and to keep the patients privacy.

The remainder of this paper is structured as follows: Section 2 contains a description of related work and parallel projects where data access issues are covered. Sections 3 and 4 cover the integration and aggregation of sensitive medical data in the ViroLab project, while section 5 explains how such data can be manipulated by developers of experiments in the presented Virtual Laboratory. Section 6 presents areas of application of the presented technologies and section 7 contains conclusions and closing remarks.

2   Related Work

Data access in the Grid environments has been a subject of study and research for quite some time. Early solutions, such as those employed in batch Grid systems (for instance [16]) relied on replicating data contained in custom-tailored data repositories which were managed by Grid middleware. Prior to performing any calculations, data had to be fetched and staged by a specialized middleware component. Furthermore, when submitting a Grid job, the developer had to specify in advance which data elements (represented by files) were required for the computational task to proceed. Naturally, this was a limiting solution in that it took no notice of structured data storage technologies (such as databases) and did not provide for altering the available input data pool once the job was submitted for processing. Moreover, results had to be collected following the execution of the job and could not typically be stored on the fly as the job progressed. These constraints gave rise to a number of projects aiming at standardization and increased flexibility of data access in Grids, the most important of them being OGSA-DAI [1]. The aim of this project is to develop a middleware system to assist with access and integration of data from separate sources via the Grid. The OGSA-DAI Toolkit supports the smooth exposure of various types of data sources such as relational or XML databases on to grids, and provides easy access to them through common Web Service interfaces. OGSA-DAI is successfully adopted in several research projects such as SIMDAT [20] and BeINGrid [17], and more.

We intend to follow up on the achievements of the aforementioned technologies and further adapt them to the needs of medical researchers in the ViroLab environment. It should be noted that a number of other Grid research initiatives exist, which deal with similar data management issues. Of note is the EUResist project [15], which aims at developing a European integrated system for clinical management of antiretroviral drug resistance. Peer-to-peer data storage solutions are being investigated in specific contexts, such as the the SIGMCC framework [8] or GREDIA [6]. Further discussion on data access and integration solutions applied in medical Grids can be found in both [14] and [7] while similar cluster solutions are discussed in [10]. However, while such technologies are aimed at a narrow group of applications, the solution developed in ViroLab is more generic in nature and allows experiment developers to securely access integrated medical data sets as well as ad-hoc databases used for the purposes of specific experiments instead of being focused on a single type of application or use case.

3 Integration of Heterogeneous (Bio)Medical Data Resources into the Laboratory Infrastructure

Accessing a local database is one of the most common and well-known procedures today but dealing with multiple and distributed systems simultaneously still implies lots of integrational work and results quite often in a real challenge for both administrators and developers. Since descriptions of medical symptoms and their diagnosis vary greatly over different countries, as well as that they may vary in their actual details such as additional circumstances to be considered, e.g. for a pregnant woman vs. for a child etc., elegant and efficient workflows need to be defined in order to integrate those heterogeneous data resources. These inconsistencies together with the sensibility and confidentiality of the information [4] shared make this task not only important but in fact a difficult endeavour. The approach chosen within ViroLab based on the development of a middleware system containing a set of virtualization services that hides the distributed and heterogeneous data resources and their internals from the users and guarantees data access in a transparent, consistent and resource-independent way.

To facilitate information exchange among participants and to ease the storage of (bio)medical data sets, particularly in the field of HIV analysis and treatment, the ViroLab team members decided to use and set up a specifically developed HIV database management system the RegaDB HIV Data and Analysis Management Environment [12] developed by the Rega Institute of the Katholieke

Universiteit Leuven either at each data provider site or at different dedicated locations the so-called collaborative (proxy) databases. RegaDB provides a kind of data storage system including some software tools, which may be installed and managed locally, to store clinical data related to HIV treatment. It aims to support clinicians and researchers in their daily work by delivering a free and open source software solution. For researchers the objective is to offer several tools for HIV sequence analysis and to enable and disburden collaborations between researchers of different hospitals and institutions. Clinicians benefit from this system through the visualization of genetic data, relevant drugs, and algorithms in a well arranged form and the automatic generation of drug resistance interpretation reports. Following the described approaches (having a unified data access point) may alleviate integrational difficulties and ensure beneficial properties for both data providers and medical experts.

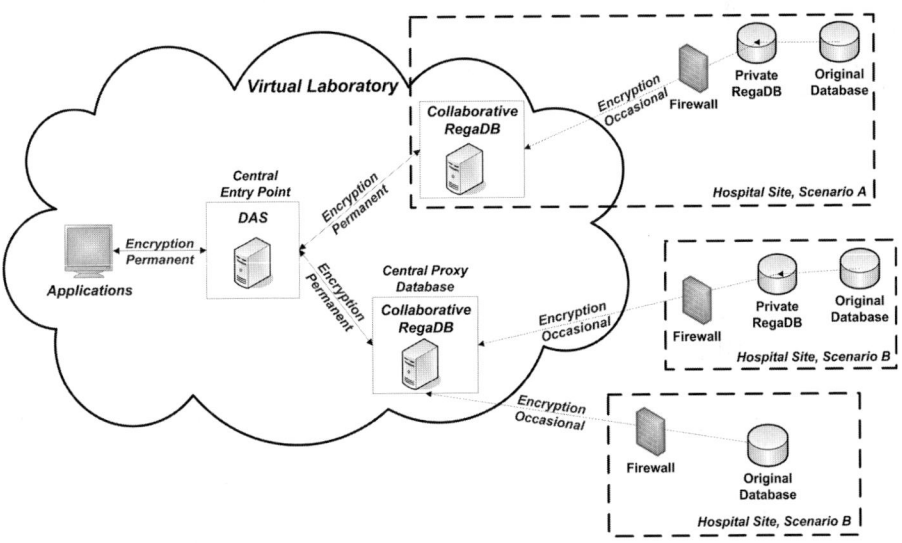

Fig. 1. ViroLab data integration architecture

As depicted in Fig. 1, every data provider can either host a collaborative RegaDB installation within their trusted region, the so-called Demilitarized Zone (DMZ), outside their institutional firewall(s) or upload data onto one of the centrally managed collaborative RegaDB proxies installed at some trusted third parties. Direct external access into the hospitals security regions is not required anymore and the existing database system can still be used independently from sharing any data within the virtual laboratory. Additionally, a hospital can also set up a private RegaDB for its daily purposes and internal data management. To actually contribute data to the virtual environment, the established database schemes need to be converted into the RegaDB schema before data can be stored within any of the RegaDB installations. The migration is done by exporting data from the original database and converting that data extract through a custom script into the latest schema. This procedure can be conducted repeatedly over time at the discretion of the corresponding database administrator(s) and occurs within each hospital. Data anonymization can also occur during that data conversion or alternatively before transferring data from a private RegaDB onto a collaborative one.

4   Exposing the Aggregated Data Sets

Once the integration of heterogeneous data resources has been realized, the queried data sets need to be somehow aggregated and securely published for making them accessible within different applications and/or systems. As mentioned earlier, a particular set of services is required for dealing with multiple resources simultaneously and for ensuring transparent access to corresponding data storages. This set of services, the so-called Data Access Services (DAS), has been designed and implemented according to the specific needs and requirements for biomedical data and confidential information [4] shared among the laboratory users. Basically, for supporting a wide range of end-user applications and allowing a smooth interaction with several virtual laboratory runtime components, the services capabilities implement standard Web Service interfaces that can be independently used of the underlying infrastructure/technologies and that can be easily extended and modified due to the application developers needs. To provide a certain level of flexibility and reliability, DAS has been separated into stand-alone containers, each serving a specific purpose in the overall services functionality. For exposing and querying single or federated databases, the data handling subsystem is responsible for coordinating all data activities including resource identification, database queries, data consultation and transformation, and finally result message preparation and delivery [3]. Observing data confidentiality and ownership but also guaranteeing a secure message transfer, the security handling module takes existing security standards provided amongst others by the Globus Toolkit [13] or Shibboleth [2], and extends those mechanisms with own highly sophisticated features to meet the requirements for sharing single data sets containing confidential patients information. Principally, additional and stronger authorization mechanisms shall be exploited in order to limit, deny or permit access to different resources in a dynamic and flexible way. Attributes of users including their organization, department, role, etc. are used to administer access rights to resources and services. To explain the complex interplay of the major DAS subsystems (data and security handling) we briefly demonstrate a common use case how one can submit a distributed data query to all available data resources concurrently and how individual results are processed before users finally receive the consolidated output. Typically, doctors want to use the virtual environment for requesting patient information including genetic data such as nucleotide sequences or mutations in order to predict any possible drug resistance for their according case. They simply want to retrieve all relevant information without requiring any specific expertise in computer science. Therefore, the way to get to the data must be kept simple and transparent for them but should be as self-explaining as possible. The capability provided by DAS for submitting such queries requires a standard SQL query as input and then automatically performs the following actions: checks which resources are available; requests data resource information such as database type, keywords, etc. of each available resource and compares the resources keywords (pre-defined by resource owner like patients, mutations, sequences, etc.) to the table names stated in the query. If corresponding matches are found, each resource is sequentially queried using the given statement. Finally, the results are merged and the resource ID is added to each new data row as an additional primary key to uniquely identify the origin of each single data set.

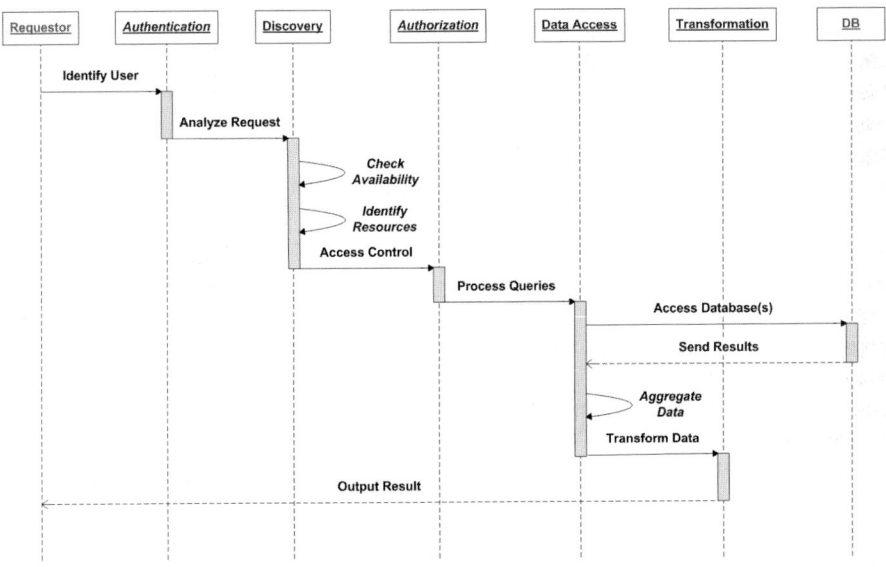

Fig. 2. A typical data access use case within the ViroLab scenario

In Fig. 2, the above-mentioned use case together with the corresponding actions is highlighted again. Each single step - starting with the users request up to the response sent back by the DAS - is depicted within this chain by one specific block. The resource identification, the pure data access, and finally the application-dependent transformation are handled by the data handling module whereas all security operations are carried out by the security handling module in cooperation with the entire security framework deployed within the virtual laboratory. How a doctor can really get to the data and how the DAS is interfaced from other laboratory components in order to interact and in particularly send such queries, is explained in the next section. [11]

5   Interfacing and Manipulating Medical Data in the Virtual Laboratory

The Data Access Services are an integral part of the ViroLab Virtual Laboratory [9], but it cannot be directly interfaced by the ViroLab user layer. In fact, the intricacies associated with the invocation of complex services, manipulating security credentials, submitting queries and processing the received replies would add undue complexity to the VL scripting language, which is designed to be as simple as possible and contain little to no syntax involved with the actual interoperation with the underlying services. Hence, a separate module of the ViroLab runtime system, called the Data Access Client (DAC), is implemented. The DAC has several important functions: carrying out all communications with the Data Access Services, including user authorization, submission of queries and importing results into the context of the ViroLab experiment scripts; presenting data to experiment developers in a way which can be easily manipulated in the experiment host language [18]; interfacing with external "ad hoc" data sources, which are not aggregated under the Data Access Services (for instance, scratchbook databases and P2P storage frameworks), and finally providing a data manipulation layer tool for the submission and storage of ViroLab experiment results.

The Data Access Client is implemented as a JRuby library, which is automatically imported into each experiment script and then, in turn, interfaces with the underlying services such as the Data Access Services. In addition, the interface of the Data Access Client is also exposed directly by the GridSpace Engine (the runtime component of the ViroLab Virtual Library), where it can be utilized by other, external tools which are part of the ViroLab Virtual Laboratory (such as the provenance tracking service).

The basic tenet of the Data Access Client is simplicity of use. Therefore, the API offered by the client to scripting developers is as simple as possible. In order to initiate communication with a data source, all the developer has to do is to instantiate an object of the class DACConnector with the proper arguments. It is only necessary to specify the type of data source and the address (URL) at which the source is located. If there are multiple sources deployed at a given address, it is necessary to select one by providing its schema name.

Fig. 3 presents interaction between the experiment developer and the data access client. Once a data source object is instantiated, the user can use it to import data from the data source, manipulate this data and write data back to the source, if the source supports this functionality. Queries can be formulated in SQL, for standalone data sources and databases aggregated under DAS. Work on integrating XQuery for XML-based data sources is ongoing. The API of the DAC, along with the GSEngine itself, is described in [9]. Thus, the Data Access Client is fully integrated with the JRuby API presented to experiment developers.

As data coming from hospital sources is typically secured against unauthorized access, the Data Access Client must support the authentication methods in use by the Data Access Services (conforming to the overall ViroLab policy on data handling). Since authentication and authorization security in ViroLab

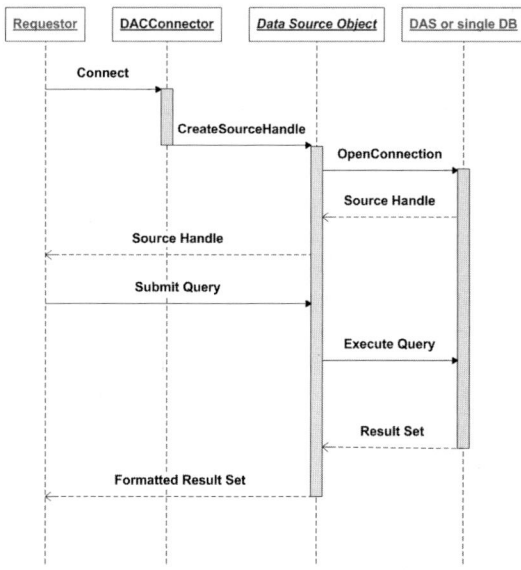

Fig. 3. Interfacing external data resources via DAC

is provided by the Shibboleth attribute-based framework, DAC must authorize itself with DAS prior to retrieval of actual data. In this respect, the DAC relies on the GridSpace Engine (the runtime component of the Virtual Laboratory) to acquire the security handle of the current user, then presents this handle to the Data Access Services so that proper authorization can take place. This process is entirely transparent from the point of view of the experiment user and it does not require the experiment developer to insert additional security-related code in the experiment script.

6   Results

At present, the ViroLab Virtual Laboratory is being applied to a number of applications involving research on the HIV virus. A list of experiments being conducted with the use of the presented solutions can be found at [22]. A representative application in this scope is the "From Genotype to Drug Resistance" framework. This application starts with aggregated data representing viral genotype, then matches this genotype to a given set of rules regarding the susceptibility of the virus to various drugs. In the end, this application is able to determine the most effective course of treatment for a given virus mutation and recommend drugs to be administered to a given patient. In order to ascertain viral susceptibility to various drugs, this application relies on the presented data access subsystem to interface with participating hospitals, securely collect viral genotype data and present it to experiment developers in a uniform manner, using a common schema. This process is further described in [12] and is now being successfully applied in the ViroLab project [21].

7   Summary and Future Work

The presented data access solutions form an integral part of the ViroLab Virtual Laboratory and enable medical researchers to conduct studies in the field of viral diseases treatment as well as other associated areas of medical science.

Current work on the Data Access Client focuses on extending its functionality to provide a backend for the submission, storage and retrieval of VL experiment results. This requires interfacing with WebDAV data storage repositories, as well as with the ProToS Provenance Tracking System [5] which will be used to store metadata describing such results as susceptibility of the HIV virus to various forms of treatment. Once complete, this extension will provide a layer of persistence to all data generated with the use of the ViroLab Virtual Laboratory. We are also conducting research into a potential uniform data schema for all types of data sources used in the ViroLab Virtual Laboratory.

Future developments planned for the Data Access Services will mainly enhance reliability and scalability of the individual services capabilities as well as increase the data submission performance through processing queries in parallel instead of submitting requests one after another. Finally, facilitating the management of access control policies, the corresponding capabilities of the security handling unit will be integrated with a nice and user-friendly graphical user interface allowing the fast and dynamic generation, change, and upload of access control policies for certain data resources in order to provide more flexibility in administering distributed resources within a collaborative working environment.

Acknowledgements: This work is supported by the EU project ViroLab IST-027446 and the related Polish SPUB-M grant, as well as by the EU IST CoreGrid project.

References

1. Antonioletti, M., Atkinson, M.P., Baxter, R., Borley, A., Chue Hong, N.P., Collins, B., Hardman, N., Hume, A., Knox, A., Jackson, M., Krause, A., Laws, S., Magowan, J., Paton, N.W., Pearson, D., Sugden, T., Watson, P., Westhead, M.: The Design and Implementation of Grid Database Services in OGSA-DAI. Concurrency and Computation: Practice and Experience 17(2-4), 357–376 (2005)
2. Assel, M., Kipp, A.: A Secure Infrastructure for Dynamic Collaborative Working Environments. In: Proceedings of the 2007 International Conference on Grid Computing and Applications (GCA 2007), Las Vegas, USA (June 2007)
3. Assel, M., Krammer, B., Loehden, A.: Data Access and Virtualization within ViroLab. In: Proceedings of the 7th Cracow Grid Workshop 2007, Cracow, Poland (October 2007)
4. Assel, M., Krammer, B., Loehden, A.: Management and Access of Biomedical Data in a Grid Environment. In: Proceedings of the 6th Cracow Grid Workshop 2006, Cracow, Poland, October 2006, pp. 263–270 (2006)

5. Balis, B., Bubak, M., Pelczar, M., Wach, J.: Provenance Tracking and Querying in ViroLab. In: Proceedings of Cracow Grid Workshop 2007, Krakow, Poland (December 2007)
6. Bubak, M., Harezlak, D., Nowakowski, P., Gubala, T., Malawski, M.: Appea: A Platform for Development and Execution of Grid Applications e-Challenges - Expanding the Knowledge Economy. IOS Press, Amsterdam (2007) ISBN 978-1-58603-801-4
7. Cannataro, M., Guzzi, P.H., Mazza, T., Tradigo, G., Veltri, P.: Using ontologies for preprocessing and mining spectra data on the Grid. Future Generation Computer Systems 23(1), 55–60 (2007)
8. Cannataro, M., Talia, D., Tradigo, G., Trunfio, P., Veltri, P.: SIGMCC: A system for sharing meta patient records in a Peer-to-Peer environment. Future Generation Computer Systems 24(3), 222–234 (2008)
9. Ciepiela, E., Kocot, J., Gubala, T., Malawski, M., Kasztelnik, M., Bubak, M.: GridSpace Engine of the Virolab Virtual Laboratory. In: Proceedings of Cracow Grid Workshop 2007, Krakow, Poland (December 2007)
10. Frattolillo, F.: Supporting Data Management on Cluster Grids. Future Generation Computer Systems 24(2), 166–176 (2008)
11. Gubala, T., Balis, B., Malawski, M., Kasztelnik, M., Nowakowski, P., Assel, M., Harezlak, D., Bartynski, T., Kocot, J., Ciepiela, E., Krol, D., Wach, J., Pelczar, M., Funika, W., Bubak, M.: ViroLab Virtual Laboratory. In: Proceedings of Cracow Grid Workshop 2007, Krakow, Poland (December 2007)
12. Libin, P., Deforche, K., Van Laethem, K., Camacho, R., Vandamme, A.-M.: RegaDB: An Open Source. In: Community-Driven HIV Data and Analysis Management Environment Fifth European HIV Drug Resistance Workshop, Cascais, Portugal, March 2007, vol. 2007-2, published in Reviews in Antiretroviral Therapy (2007)
13. Foster, I., Kesselman, C., Nick, J., Tuecke, S.: The Physiology of the Grid: An Open Grid Services Architecture for Distributed Systems Integration. Globus Project (2002), http://www.globus.org/research/papers/ogsa.pdf
14. Marovic, B., Jovanovic, Z.: Web-based grid-enabled interaction with 3D medical data. Future Generation Computer Systems 22(4), 385–392 (2006)
15. Zazzi, M., et al.: EuResist: exploration of multiple modeling techniques for prediction of response to treatment. In: Proceedings of the 5th European HIV Drug Resistance Workshop, European AIDS Clinical Society (2007)
16. Enabling Grids for E-sciencE, http://public.eu-egee.org/
17. Gridipedia: The European Grid Marketplace, http://www.gridipedia.eu/
18. JRuby: A reimplementation of the Ruby language in pure Java, http://jruby.codehaus.org/
19. The Open Grid Forum, http://www.gridforum.org/
20. The SIMDAT project, http://www.scai.fraunhofer.de/
21. The ViroLab Project Website, http://www.virolab.org/
22. The ViroLab Virtual Laboratory Web Portal, http://virolab.cyfronet.pl/

Optimisation of Asymmetric Mutational Pressure and Selection Pressure Around the Universal Genetic Code

Paweł Mackiewicz, Przemysław Biecek, Dorota Mackiewicz, Joanna Kiraga, Krystian Baczkowski, Maciej Sobczynski, and Stanisław Cebrat

Department of Genomics, Faculty of Biotechnology, University of Wrocaw, ul. Przybyszewskiego 63/77, 51-148 Wrocaw, Poland
pamac@smorfland.uni.wroc.pl
http://www.smorfland.uni.wroc.pl/

Abstract. One of hypotheses explaining the origin of the genetic code assumes that its evolution has minimised the deleterious effects of mutations in coded proteins. To estimate the level of such optimization, we calculated optimal codes for genes located on differently replicating DNA strands separately assuming the rate of amino acid substitutions in proteins as a measure of code's susceptibility to errors. The optimal code for genes located on one DNA strand was simultaneously worse than the universal code for the genes located on the other strand. Furthermore, we generated 20 million random codes of which only 23 were better than the universal one for genes located on both strands simultaneously while about two orders of magnitude more codes were better for each of the two strands separately. The result indicates that the existing universal code, the mutational pressure, the codon and amino acid compositions are highly optimised for the both differently replicating DNA strands.

Keywords: genetic code, error minimization, adaptation, asymmetric mutational pressure, amino acid usage, leading strand, lagging strand.

1 Introduction

There are three main groups of hypotheses trying to explain the origin and evolution of the genetic code: chemical, historical and adaptive (see for review [1,2,3]). The first one assumes some structural and physicochemical relationships and interactions between stretches of RNA (codons, anticodons, reversed codons, codon-anticodon double helices etc.) and coded amino acids [4,5,6]. So far, a well-confirmed relationship has been found for seven of eight amino acids (see for review: [7]). The second hypothesis states that codons in the simpler, ancestral genetic code coded for only a small subset of amino acids and later, along with the evolution of biochemical organization of primary cells, newly synthesised amino acids took over the codons from the amino acids to which they were related in biosynthetic pathways [8,9,10,11,12]. The third group of hypotheses assumes that the codon assignments could initially vary and it was the selection pressure

M. Bubak et al. (Eds.): ICCS 2008, Part III, LNCS 5103, pp. 100–109, 2008.

which optimized the code to reduce harmful effects of mutations occurring during replication and transcription (lethal-mutation hypothesis) and to minimize errors during translation process (translational-error hypothesis), [6,13,14,15,16,17,18]; see for review: [19].

Primordial organisms whose code reduced the deleterious effects of errors won eventually the competition and survived. During further evolution connected with the increase of genome size, the genetic code was frozen [20] and it was not possible to re-interpret the meaning of any codon because the whole complex translational machinery was already adapted to the code and every such change would have catastrophic consequences for the organisms. Nevertheless, some optimization took place already in the first stages of the genetic code evolution, probably before ,,freezing". One optimization results directly from the simple structural relationships between nucleotides in the double helix - one large and one small nucleotide fit better to form a pair. Thus, transitions which happen with much higher frequency than transversions have much less deleterious mutational effect than transversions. Actually, it was shown that the genetic code is well adapted to the transition/transversion bias [16].

Assuming the adaptive hypothesis of the genetic code evolution we expect that if the genetic code was ,,frozen" at an early stage of evolution when genomes were relatively small, it is the code itself that imposes further restrictions on the mutational pressure, amino acid and codon usage, and the translational machinery in order to minimize the deleterious effects of mutations. Thus, the mutational pressure cannot be completely random, as one could claim, but it is highly biased and it cooperates with the selection pressure on amino acid and codon usage to minimise the harmful effects of nucleotide substitutions. One of the premises is that the most ,,mutable" codons in the genome correspond to the least-represented amino acids [21,22]. Monte Carlo simulations showed that changing of parameters of any of the three counterparts of the coding functions: relative nucleotide substitution rates in the mutational pressure, the way the genetic code is degenerated or the amino acid composition of proteomes increases the deleterious effects of mutations in studied genomes [23].

However, it is not simply to optimise the mutational pressure. The mutational pressures acting on the differently replicating (leading or lagging) DNA strands show different patterns of nucleotide substitutions and leads to the strong bias (asymmetry) in nucleotide composition between the two DNA strands observed in almost all bacterial chromosomes [24,25,26,27,28,29] and long regions of eukaryotic chromosomes [30,31,32]. Therefore, genes are subjected to different mutational pressures depending on their location on the differently replicating DNA strands, which affects their codon usage and amino acid composition of coded proteins [33,34,35,36].

Although several simulation studies about the optimization of the genetic code were carried out [14,15,16,17,18], none of them considered the real and global genomic aspect of this optimization, i.e. the real mutational pressure, gene content, codon and amino acid usage. Zhu *et al.* [37] found that the universal genetic code appears to be less optimised for error minimization when specific codon usage for particular species was considered. However, the authors applied the same and simple mutation pattern for all species in this analysis, which do not fit to the specific codon usage and they concluded that the specific mutation pattern should be taken into account. In this paper we considered the optimization of the genetic code in the context of the two different mutational pressures specific for the leading and lagging DNA strands acting on the asymmetric genome of *Borrelia burgdorferi*. This genome shows the strongest asymmetry between the leading and lagging strands detected so far [33,34,38] thus, it is suitable for such studies.

2   Materials and Methods

All our analyses were performed on the *B. burgdorferi* genome [39] whose sequence and annotations were downloaded from GenBank [40]. Based on these data we calculated the content of codons, amino acids and codon usage for 564 leading strand genes and 286 lagging strand genes. The mutational pressure characteristic for this genome was found by Kowalczuk *et al.* [41]. The pressure was described by the nucleotide substitution matrices (M_n) as follows:

$$M_n = \begin{bmatrix} 1 - pR_A & pR_{AC} & pR_{AG} & pR_{AT} \\ pR_{CA} & 1 - pR_C & pR_{CG} & pR_{CT} \\ pR_{GA} & pR_{GC} & 1 - pR_G & pR_{GT} \\ pR_{TA} & pR_{TC} & pR_{TG} & 1 - pR_T \end{bmatrix}$$

where: p is the overall mutation rate; R_{ij} for $i, j = A, C, G, T$ and $i \neq j$ is the relative rate of substitution of the nucleotide i by the nucleotide j; R_i (in the diagonal) for $i = A, C, G, T$ represents the relative substitution rate of nucleotide i by any of the other three nucleotides.

$$R_i = \sum_{i \neq j} R_{ij}$$

and $R_A + R_C + R_G + R_T = 1$. For $p = 1$ the matrix describing the leading strand mutational pressure is:

$$M_n^{leading} = \begin{bmatrix} 0.808 & 0.023 & 0.067 & 0.103 \\ 0.070 & 0.621 & 0.047 & 0.261 \\ 0.164 & 0.015 & 0.706 & 0.116 \\ 0.065 & 0.035 & 0.035 & 0.865 \end{bmatrix}$$

The matrix represents the most probable pure mutational pressure associated with replication acting on the leading strand. Because DNA strands are complementary, the mutational pressure acting on the lagging strand is a kind of the mirror reflection of the pressure exerted on the leading strand, e.g. R_{GA} for the leading strand corresponds to R_{CT} for the lagging strand etc. In our analyses we have assumed $p = 10^{-8}$ which approximately corresponds to the observed number of substitutions in a bacterial genome per nucleotide per generation [42].

The codon substitution matrix (M_c) containing relative rate of substitutions of one codon by another one was derived from the nucleotide substitution matrix (M_n). The M_c is the Kronecker product of three M_n matrices: $M_c = M_n \otimes M_n \otimes M_n$. For example, the substitution rate of codon GCA to codon CTA equals $R^c_{GCA \rightarrow CTA} = p^2 R_{GC} R_{CT}(1 - pR_A)$. In the M_c each row contains the substitution rates for one of 64 codons to another one:

$$M_c = \begin{bmatrix} R^c_{AAA \rightarrow AAA} & R^c_{AAA \rightarrow AAC} & R^c_{AAA \rightarrow AAG} & \cdots & R^c_{AAA \rightarrow TTT} \\ R^c_{AAC \rightarrow AAA} & R^c_{AAC \rightarrow AAC} & R^c_{AAC \rightarrow AAG} & \cdots & R^c_{AAC \rightarrow TTT} \\ R^c_{AAT \rightarrow AAA} & R^c_{AAT \rightarrow AAC} & R^c_{AAT \rightarrow AAG} & \cdots & R^c_{AAT \rightarrow TTT} \\ \vdots & \vdots & \vdots & \ddots & \vdots \\ R^c_{TTT \rightarrow AAA} & R^c_{TTT \rightarrow AAC} & R^c_{TTT \rightarrow AAG} & \cdots & R^c_{TTT \rightarrow TTT} \end{bmatrix}$$

where: $R^c_{n \rightarrow m}$ for indices of codons $n, m \in \{1..64\}$ represents the relative rate of substitution of codon n by codon m.

Each row of M_c was multiplied by the codon usage of a given codon U_n (i.e. relative frequency of a codon among other synonymous codons coding the same amino acid or stop codon) giving the M_u matrix:

$$M_u = \begin{bmatrix} U_{AAA} R^c_{AAA \rightarrow AAA} & U_{AAA} R^c_{AAA \rightarrow AAC} & \cdots & U_{AAA} R^c_{AAA \rightarrow TTT} \\ U_{AAC} R^c_{AAC \rightarrow AAA} & U_{AAC} R^c_{AAC \rightarrow AAC} & \cdots & U_{AAC} R^c_{AAC \rightarrow TTT} \\ U_{AAT} R^c_{AAT \rightarrow AAA} & U_{AAT} R^c_{AAT \rightarrow AAC} & \cdots & U_{AAT} R^c_{AAT \rightarrow TTT} \\ \vdots & \vdots & \ddots & \vdots \\ U_{TTT} R^c_{TTT \rightarrow AAA} & U_{TTT} R^c_{TTT \rightarrow AAC} & \cdots & U_{TTT} R^c_{TTT \rightarrow TTT} \end{bmatrix}$$

where: U_n stands for codon usage of codon n, where $n \in \{1..64\}$.

To obtain the amino acid substitution matrix (M_a) containing relative rates of substitutions of one amino acid or stop by another, the respective elements of M_u matrix were summed up, which gives the matrix of amino acids (and stops) substitution:

$$M_a = \begin{bmatrix} R^a_{Ala \rightarrow Ala} & R^a_{Ala \rightarrow Arg} & R^a_{Ala \rightarrow Asn} & \cdots & R^a_{Ala \rightarrow Val} \\ R^a_{Arg \rightarrow Ala} & R^a_{Arg \rightarrow Arg} & R^a_{Arg \rightarrow Asn} & \cdots & R^a_{Arg \rightarrow Val} \\ R^a_{Asn \rightarrow Ala} & R^a_{Asn \rightarrow Arg} & R^a_{Asn \rightarrow Asn} & \cdots & R^a_{Asn \rightarrow Val} \\ \vdots & \vdots & \vdots & \ddots & \vdots \\ R^a_{Val \rightarrow Ala} & R^a_{Val \rightarrow Arg} & R^a_{Val \rightarrow Asn} & \cdots & R^a_{Val \rightarrow Val} \end{bmatrix}$$

where: $R^a_{p \rightarrow q}$ for $p, q \in \{1..21\}$ represents the relative rate of substitution of amino acid (or stop) p by amino acid q.

The sum of each row of M_a gives the rate of substitution of amino acid p (or stop) p to another:

$$R^a_p = \sum_{q \neq p} R^a_{p \rightarrow q}.$$

Such calculations were carried out for the leading strand data and for the lagging strand data separately.

3   Results and Discussion

In order to estimate how the genetic code and the mutational pressures are optimized for differently replicating strands, we considered the number of substituted amino acids (and stops). Therefore we multiplied each rate of substitution of a given amino acid R_p^a by the number A_p of this amino acid in the coded proteins and summed the products:

$$S_A = \sum_{p=1}^{21} (A_p R_p^a).$$

In our consideration we applied the number of substituted amino acids instead of fraction because we wanted to analyse the genetic code optimization in the context of the whole genome including the bias between the numbers of the leading and lagging strand genes. For constant R_p^a and A_p, S_A reaches the minimum if $A_p < A_{p+1} < A_{p+N}$ and simultaneously if $R_p^a > R_{p+1}^a > R_{p+N}^a$ i.e. when A_p and S_A are negatively correlated. In other words the total cost of mutations is lower if the rate of mutation is higher for the less frequent residues than for the more frequent ones. Interestingly, A_p and S_A calculated for the real genome data show statistically significant negative correlation (Fig. 1) that suggests a tendency to minimization of amino acid substitutions in the real genome.

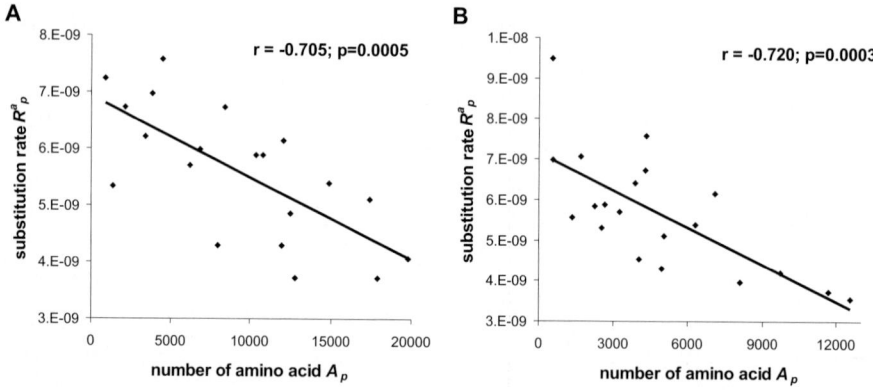

Fig. 1. Correlation between substitution rates R_p^a and the number of amino acids A_p for the leading strand (A) and for the lagging strand (B) data

However, it is possible to find such ascription of amino acids to codons (i.e. to elaborate a new genetic code) which is better optimized than the universal code - according to the minimization of the number of amino acid substitutions. The best way is to rank reversely R_p^a versus A_p. Similarly, one can obtain the worst code giving the highest number of amino acid substitutions by the ranking of A_p and S_A accordingly. The results of such transformations made separately for the leading and for the lagging strand cases are shown in Table 1. Such a transformation did not change the global structure of the genetic code, retains the same degeneracy level and redundancy of the canonical genetic code. It assumes that the groups of codons which code for a single amino acid are fixed and we changed only the assignments between amino acids and the codon blocks. For example, in the case of the leading strand, tryptophane is now coded by four proline codons and proline by one methionine codon. The ascriptions of some amino acids were changed but some of them retained their positions. Because stop codons have special meanings and are represented by only one per gene, we did not change the position of the stops in this transformation.

Table 1. The ascription of codons of a given amino acid that minimizes the number of substitutions separately for the leading and lagging strand cases. Amino acids that have not changed their position are in **bold**.

amino acid	A **R** N D C Q E G H I L **K** M **F** P S T **W** Y V
leading strand	G **R** N S W A K T Q **I** F L H Y M V E P D C
lagging strand	H A L R G D T S M **I** N K C F E Y Q **W** V P

Table 2 shows the expected number of substituted amino acids (including stops) S_A - calculated for the universal genetic code, for the best one and for the worst one for the giving DNA strand. It is possible to find the optimal code for protein coding sequences located on one DNA strand but such a code is not simultaneously the optimal one for genes located on the other DNA strand. In fact it is worse than the universal one. Nevertheless, the S_A values for the universal code and the both classes of genes are much closer to the best code than to the worst one. The value of S_A for the universal code fall between the values for optimal codes for DNA strands.

Table 2. The expected number of missense mutations (including stops) S_A calculated for the universal genetic code, for the best one and for the worst one for the leading and lagging strand proteins

DNA strand	Universal code	Code optimal for:		The worst code for the giving strand
		leading strand	lagging strand	
leading	0.000955	0.000921	0.000974	0.001168
lagging	0.000482	0.000488	0.000465	0.000645

As it was shown above it is easy to find the optimal code for each strand separately but it is difficult to calculate the code that would be optimal for the two strands simultaneously. To solve the problem we have generated 20 million random genetic codes replacing one amino acid by another one as described previously, i.e. retaining the global structure of the genetic code retaining the same degeneracy level and redundancy. Such a transformation corresponds to the method widely used in other studies [14,15,16,17]. For each generated code we calculated the number of substituted amino acids (excluding stops) S_A separately for genes located on the leading and lagging strands. Next we counted for these two sets of genes how many codes produce the S_A value smaller than the value for the universal code and we counted how many random codes are better for both sets. In the last case we considered two conditions:

1. total number of substitutions (i.e. the sum of the S_A for the leading and for the lagging strand) produced by a generated code is smaller than under the universal code;
2. generated code is better simultaneously for each of the two strands.

The first condition treats the leading and the lagging strand genes as one set whereas the second one treats them as separate, independent sets. The results are

Table 3. The number of random (generated) codes (among 20 million) which are better than the universal one according to the number of amino acid substitutions analysed in the aspect of the differently replicating strands. $S^{random}_{A\_leading}$ - the number of substituted amino acids in the leading strand proteins considering the random code; $S^{random}_{A\_lagging}$ - the number of substituted amino acids in the lagging strand proteins considering the random code; $S^{universal}_{A\_leading}$ - the number of substituted amino acids in the leading strand proteins considering the universal code; $S^{universal}_{A\_lagging}$ - the number of substituted amino acids in the lagging strand proteins considering the universal code.

Checked condition	The number of better codes
$S^{random}_{A\_leading} < S^{universal}_{A\_leading}$	6652
$S^{random}_{A\_lagging} < S^{universal}_{A\_lagging}$	733
$S^{random}_{A\_leading} + S^{random}_{A\_lagging} < S^{universal}_{A\_leading} + S^{universal}_{A\_lagging}$	160
$S^{random}_{A\_leading} < S^{universal}_{A\_leading}$ and $S^{random}_{A\_lagging} < S^{universal}_{A\_lagging}$	23

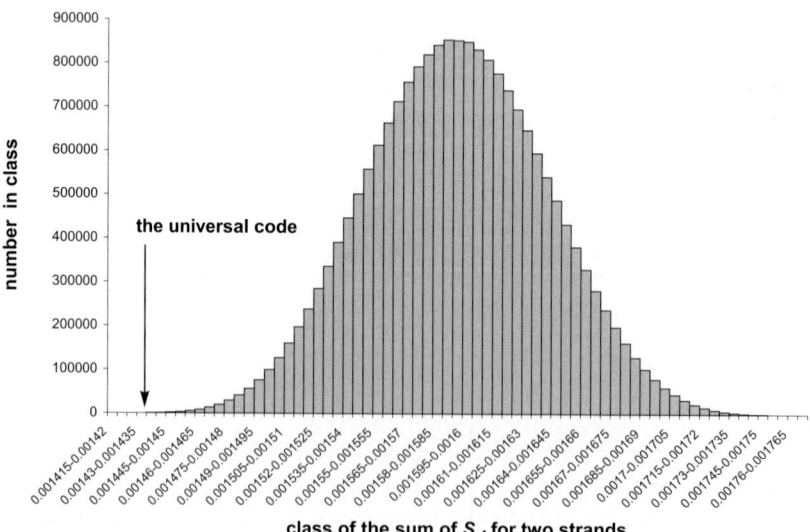

Fig. 2. Distribution of the sum of the number of substituted amino acids S_A for the leading and the lagging strands calculated for 20 million randomly generated genetic codes. The arrow indicates the S_A value for the universal code.

presented in Table 3. We have found that the probability of random generation of a code which would transmit fewer missense mutations in the set of genes located on the leading strand or the lagging strand is relatively high. Nevertheless, we have found much fewer codes which fulfil the first condition (160, i.e. 0.0008%) (Fig. 2) and even fewer, which fulfil the second condition (23, i.e. 0.000115%). The observed optimality of the code is very close to the results obtained by Freeland and Hurst [16], i.e. one per million.

4   Conclusions and Perspectives

The results indicate that the existing universal code, the mutational pressure, the codon and amino acid composition are highly optimised in the context of the two differently replicating DNA strands minimizing the number of substituted amino acids in the coded proteins. In our studies we assumed quite simple measure of a code's susceptibility to errors - number of substituted amino acids - ignoring the differences in their physicochemical properties, e.g. hydrophobicity, polarity or isoelectric point. This simplification enabled to calculate analytically the optimal and the worst assignments of amino acids to codons and to compare them with the result obtained for universal genetic code considering mutational pressure, codon usage and amino acid composition specific for genes lying on differently replicating strands. However, considering of these physicochemical properties would probably decrease the number of random codes better than the universal one and could be further investigated. It would be also interesting to analyze genomic systems of other organisms in this aspect. A better code for one organism could be worse for another organism. If one wanted to look for the optimal code for all organisms, one should check each organism separately - its mutational pressure, amino acid composition and codon usage. It makes no sense to look for a genetic code that would be better for average codon usage or average mutational pressure. There are no average organisms in the biosphere. In the early stages of genetic code evolution the code optimised itself to minimizing harmful effects of various mutational pressures but after it was „frozen" the mutational pressure begun to tune to the universal code independently in different phylogenetic lineages.

Acknowledgements. The work was done in the frame of the ESF program COST Action P10, GIACS and UNESCO chair of interdisciplinary studies.

References

1. Di Giulio, M.: On the origin of the genetic code. J. Theor. Biol. 187, 573–581 (1997)
2. Knight, R.D., Freeland, S.J., Landweber, L.F.: Selection, history and chemistry: the three faces of the genetic code. Trends Biochem. Sci. 24, 241–247 (1999)
3. Knight, R.D.: The origin and evolution of the genetic code: statistical and experimental investigations. Ph.D. Thesis, Department of Ecology and Evolutionary Biology, Princeton University (2001)

4. Dunnill, P.: Triplet nucleotide-amino-acid pairing a stereochemical basis for the division between protein and non-protein amino-acids. Nature 210, 1265–1267 (1966)
5. Pelc, S.R., Welton, M.G.: Stereochemical relationship between coding triplets and amino-acids. Nature 209, 868–872 (1966)
6. Woese, C.R., Dugre, D.H., Saxinger, W.C., Dugre, S.A.: The molecular basis for the genetic code. Proc. Natl. Acad. Sci. USA 55, 966–974 (1966)
7. Yarus, M., Caporaso, J.G., Knight, R.: Origins of the genetic code: the escaped triplet theory. Annu. Rev. Biochem. 74, 179–198 (2005)
8. Wong, J.T.-F.: A Co-Evolution Theory of the Genetic Code. Proc. Natl. Acad. Sci. USA 72, 1909–1912 (1975)
9. Taylor, F.J.R., Coates, D.: The code within the codons. BioSystems 22, 177–187 (1989)
10. Di Giulio, M.: On the relationships between the genetic code coevolution hypothesis and the physicochemical hypothesis. Z. Naturforsch. [C] 46, 305–312 (1991)
11. Amirnovin, R.: An analysis of the metabolic theory of the origin of the genetic code. J. Mol. Evol. 44, 473–476 (1997)
12. Di Giulio, M., Medugno, M.: The Historical Factor: The Biosynthetic Relationships Between Amino Acids and Their Physicochemical Properties in the Origin of the Genetic Code. J. Mol. Evol. 46, 615–621 (1998)
13. Alff-Steinberger, C.: The genetic code and error transmission. Proc. Natl. Acad. Sci. USA 64, 584–591 (1969)
14. Haig, D., Hurst, L.D.: A quantitative measure of error minimization in the genetic code. J. Mol. Evol (Erratum in J. Mol. Evol. 49, 708 (1999)) 33, 412–417 (1991)
15. Ardell, D.H.: On error minimization in a sequential origin of the standard genetic code. J. Mol. Evol. 47, 1–13 (1998)
16. Freeland, S.J., Hurst, L.D.: The genetic code is one in a million. J. Mol. Evol. 47, 238–248 (1998)
17. Freeland, S.J., Knight, R.D., Landweber, L.F., Hurst, L.D.: Early fixation of an optimal genetic code. Mol. Biol. Evol. 17, 511–518 (2000)
18. Sella, G., Ardell, D.H.: The impact of message mutation on the fitness of a genetic code. J. Mol. Evol. 54, 638–651 (2002)
19. Freeland, S.J., Wu, T., Keulmann, N.: The case for an error minimizing standard genetic code. Orig. Life Evol. Biosph. 33, 457–477 (2003)
20. Crick, F.H.: The origin of the genetic code. J. Mol. Evol. 38, 367–379 (1968)
21. Nowicka, A., Mackiewicz, P., Dudkiewicz, M., Mackiewicz, D., Kowalczuk, M., Cebrat, S., Dudek, M.R.: Correlation between mutation pressure, selection pressure, and occurrence of amino acids. In: Sloot, P.M.A., Abramson, D., Bogdanov, A.V., Gorbachev, Y.E., Dongarra, J., Zomaya, A.Y. (eds.) ICCS 2003. LNCS, vol. 2658, pp. 650–657. Springer, Heidelberg (2003)
22. Nowicka, A., Mackiewicz, P., Dudkiewicz, M., Mackiewicz, D., Kowalczuk, M., Banaszak, J., Cebrat, S., Dudek, M.R.: Representation of mutation pressure and selection pressure by PAM matrices. Applied Bioinformatics 3, 31–39 (2004)
23. Dudkiewicz, M., Mackiewicz, P., Nowicka, A., Kowalczuk, M., Mackiewicz, D., Polak, N., Smolarczyk, K., Banaszak, J., Dudek, M.R., Cebrat, S.: Correspondence between mutation and selection pressure and the genetic code degeneracy in the gene evolution. FGCS 21, 1033–1039 (2005)
24. Lobry, J.R.: Asymmetric substitution patterns in the two DNA strands of bacteria. Mol. Biol. Evol. 13, 660–665 (1996)
25. Mrazek, J., Karlin, S.: Strand compositional asymmetry in bacterial and large viral genomes. Proc. Natl. Acad. Sci. USA 95, 3720–3725 (1998)

26. Frank, A.C., Lobry, J.R.: Asymmetric substitution patterns: a review of possible underlying mutational or selective mechanisms. Gene 238, 65–77 (1999)
27. Tillier, E.R., Collins, R.A.: The contributions of replication orientation, gene direction, and signal sequences to base-composition asymmetries in bacterial genomes. J. Mol. Evol. 50, 249–257 (2000)
28. Kowalczuk, M., Mackiewicz, P., Mackiewicz, D., Nowicka, A., Dudkiewicz, M., Dudek, M.R., Cebrat, S.: DNA asymmetry and the replicational mutational pressure. J. Appl. Genet. 42, 553–577 (2001a)
29. Rocha, E.P.: The replication-related organization of bacterial genomes. Microbiology 150, 1609–1627 (2004)
30. Gierlik, A., Kowalczuk, M., Mackiewicz, P., Dudek, M.R., Cebrat, S.: Is there replication-associated mutational pressure in the Saccharomyces cerevisiae genome? J. Theor. Biol. 202, 305–314 (2000)
31. Niu, D.K., Lin, K., Zhang, D.Y.: Strand compositional asymmetries of nuclear DNA in eukaryotes. J. Mol. Evol. 57, 325–334 (2003)
32. Touchon, M., Nicolay, S., Audit, B., Brodie of Brodie, E.B., d'Aubenton-Carafa, Y., Arneodo, A., Thermes, C.: Replication-associated strand asymmetries in mammalian genomes: Toward detection of replication origins. Proc. Natl. Acad. Sci. USA 102, 9836–9841 (2005)
33. McInerney, J.O.: Replicational and transcriptional selection on codon usage in Borrelia burgdorferi. Proc. Natl. Acad. Sci. USA 95, 10698–10703 (1998)
34. Lafay, B., Lloyd, A.T., McLean, M.J., Devine, K.M., Sharp, P.M., Wolfe, K.H.: Proteome composition and codon usage in spirochaetes: species-specific and DNA strand-specific mutational biases. Nucleic Acids Res. 27, 1642–1649 (1999)
35. Mackiewicz, P., Gierlik, A., Kowalczuk, M., Dudek, M.R., Cebrat, S.: How does replication-associated mutational pressure influence amino acid composition of proteins? Genome Res. 9, 409–416 (1999a)
36. Rocha, E.P., Danchin, A., Viari, A.: Universal replication biases in bacteria. Mol. Microbiol. 32, 11–16 (1999)
37. Zhu, C.T., Zeng, X.B., Huang, W.D.: Codon usage decreases the error minimization within the genetic code. J. Mol. Evol. 57, 533–537 (2003)
38. Mackiewicz, P., Gierlik, A., Kowalczuk, M., Szczepanik, D., Dudek, M.R., Cebrat, S.: Mechanisms generating long-range correlation in nucleotide composition of the Borrelia burgdorferi genome. Physica A 273, 103–115 (1999b)
39. Fraser, C.M., Casjens, S., Huang, W.M., Sutton, G.G., Clayton, R., Lathigra, R., White, O., Ketchum, K.A., Dodson, R., Hickey, E.K., et al. (38 co-authors): Genomic sequence of a Lyme disease spirochaete Borrelia burgdorferi. Nature 390, 580–586 (1997)
40. GenBank, ftp://www.ncbi.nlm.nih.gov
41. Kowalczuk, M., Mackiewicz, P., Mackiewicz, D., Nowicka, A., Dudkiewicz, M., Dudek, M.R., Cebrat, S.: High correlation between the turnover of nucleotides un der mutational pressure and the DNA composition. BMC Evol. Biol. 1, 13 (2001b)
42. Drake, J.W., Charlesworth, B., Charlesworth, D., Crow, J.F.: Rates of spontaneous mutation. Genetics 148, 1667–1686 (1998)

Simulating Complex Calcium-Calcineurin Signaling Network

Jiangjun Cui and Jaap A. Kaandorp

Section Computational Science, Faculty of Science, University of Amsterdam,
Kruislaan 403, 1098 SJ Amsterdam, The Netherlands
{jcui,jaapk}@science.uva.nl

Abstract. Understanding of processes in which calcium signaling is involved is of fundamental importance in systems biology and has many applications in medicine. In this paper we have studied the particular case of the complex calcium-calcineurin-MCIP-NFAT signaling network in cardiac myocytes, the understanding of which is critical for treatment of pathologic hypertrophy and heart failure. By including some most recent experimental findings, we constructed a computational model totally based on biochemical principles. The model can correctly predict the mutant (MCIP1$^{-/-}$) behavior under different stress such as PO (pressure overload) and CaN* (activated calcineurin) overexpression.

1 Introduction

In eukaryotic cells, Ca^{2+} functions as a ubiquitous intracellular messenger regulating a diverse range of cellular processes such as cell proliferation, muscle contraction, programmed cell death, etc [5,8,12,20]. Recently it has been recognized that calcium also plays a central role in the control of heart growth through a complex calcium-calcineurin-MCIP-NFAT signaling network (see Fig. 1, please note that the abbreviations used in this paper are listed in the legends of Fig. 1 and in the note of Table 1). The heart responds to physiological and pathological stimuli by hypertrophic growth [20, 21]. Cardiac hypertrophy is a thickening of the heart muscle (myocardium) resulting in a decrease in size of the ventricles. Prolonged pathological hypertrophy may progress to heart failure and significantly increase the risk for sudden death. Thus deciphering the details of the signaling pathways involved in cardiac hypertrophy and understanding their quantitative dynamics through computational modeling will be critical for devising therapeutic drugs for the treatment of heart disease [21].

As shown in the left-up corner of Fig.1, stress incurs the rise of the concentration of cytosolic Ca^{2+} (in normal cardiac myocytes, cytosolic Ca^{2+} concentration rests at a level of less than 200nM and it becomes more than 700 nM under a very strong stress condition), which binds to CaM (4:1). Ca^{2+}-bound CaM binds to CaN to activate it [13]. CaN* (i.e., activated CaN) can bind to MCIP to form Complex1 [21,22]. CaN* can also work as the enzyme to help convert NFAT^P into NFAT [11]. Another enzyme GSK3β works in the reverse conversion of NFAT into

M. Bubak et al. (Eds.): ICCS 2008, Part III, LNCS 5103, pp. 110–119, 2008.

$NFAT^P$, which can bind to 14-3-3 to form Complex3 [2,10]. Such conversion between NFAT and $NFAT^P$ with the help of two enzymes (GSK3β and CaN*) also happens in the nucleus [7]. NFAT in the cytosol will be imported into the nucleus and $NFAT^P$ in the nucleus will be exported into the cytosol. The nuclear NFAT can initiate the transcription of the hypertrophic genes and the gene encoding MCIP (more precisely, MCIP1, a form of MCIP) [7,23]. Both GSK3β and CaN* are shuttled between the nucleus and the cytosol [7, 15]. As shown in the right-up corner of Fig.1, particular stress such as PO can activate BMK1 [19], which catalyzes the conversion of MCIP into $MCIP^P$ [1]. $MCIP^P$ can be converted into $MCIP^{PP}$ by GSK3β . The reverse conversion of $MCIP^{PP}$ into $MCIP^P$ is again catalyzed by CaN* [22]. $MCIP^{PP}$ will bind with 14-3-3 to form Complex2 [1].

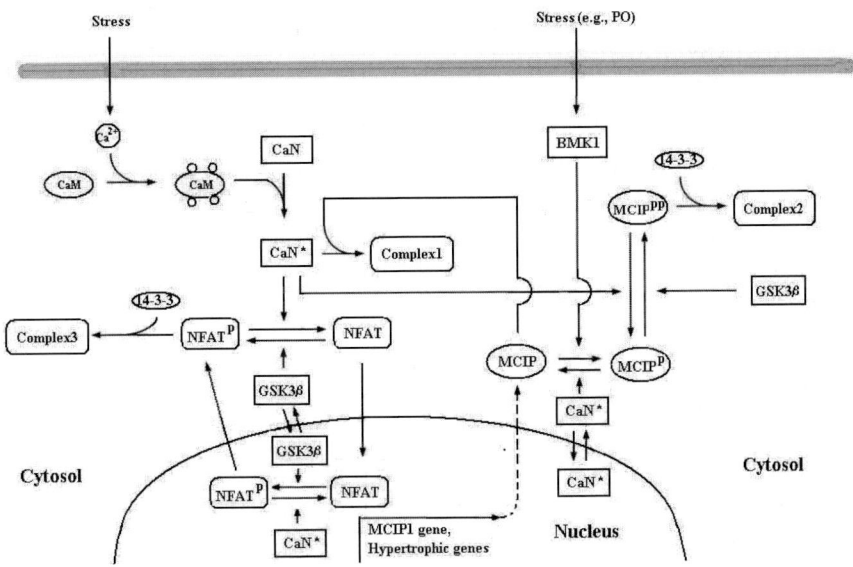

Fig. 1. A schematic graph depicting the Ca^{2+}-calcineurin-MCIP–NFAT signaling networks in cardiac myocytes (for details, please see texts in Introduction). Abbreviations are as follows: calmodulin (CaM); calcineurin (CaN); activated calcineurin (CaN*); nuclear factor of activated T-cells (NFAT); phosphrylated NFAT($NFAT^P$); modulatory calcineurin-interacting protein (MCIP); phosphorylated MCIP on serine 112 ($MCIP^P$); phosphorylated MCIP on both serine 112 and serine 108 ($MCIP^{PP}$); big mitogen-activated protein kinase 1 (BMK1); glycogen synthase 3β (GSK3β); the complex formed by MCIP and calcineurin(Complex1); the complex formed by $MCIP^{PP}$ and protein 14-3-3 (Complex2); the complex formed by $NFAT^P$ and protein 14-3-3 (Complex3); pressure overload (PO); hypertrophic stimuli (stress). The stress of PO is delivered by transverse aortic constriction (TAC).

MCIP1 seems to facilitate or suppress cardiac CaN signaling depending on the nature of the stress (see Fig. 2a). In the case of CaN* transgenic mice, the knock-out of MCIP1 gene (i.e. MCIP1$^{-/-}$ TG mice) exacerbated the hypertrophic response to CaN* overexpression. Paradoxically, however, cardiac hypertrophy in response to PO was blunted in normal MCIP1$^{-/-}$ mice [9,21].

In 2006, Shin *et al.* [18] published a paper in FEBS Letters using switching feedback mechanism to explain this dual role of MCIP in cardiac hypertrophy. The aim of this paper is to propose a much-extended version of Shin's model by including more recent experimental findings (e.g., CaN* is imported into the nucleus to function there [7], MCIP^{PP} will associate with protein 14-3-3 [1] and protein 14-3-3 competes with CaN* to associate with NFAT^P to form Complex3 [10]). The construction of the model is based on biochemical principles and we use an open source software (Cellerator^{TM}) to automatically generate the equations. As we will see later, this model can correctly predict the mutant (MCIP1$^{-/-}$) behavior under different stress such as PO and CaN* overexpression.

2   Computational Model

2.1   Cellerator Software

Cellerator^{TM} is a Mathematica package designed to facilitate biological modeling via automated equation generation [16, 24]. It uses an arrow-based reaction notation to represent biochemical networks and is especially amenable for simulating signal transduction networks.

2.2   Representation of Relevant Reactions

The complex Ca^{2+}-calcineurin signaling network shown in Fig.1 can be represented using 17 reactions grouped into four categories (see Table 1) in addition to a transcription control process of MCIP by NFAT.

2.3   The Equations of the Model

The ODEs (ordinary differential equations) notation of the set of all relevant reactions consists of 28 equations concerning 28 unknowns including $Ca(t)$ which denotes the cytosolic Ca^{2+} concentration. Since calcineurin is unique in its specific responsiveness to sustained, low frequency calcium signals [20], we will assume cytosolic Ca^{2+} concentration as a constant in the following simulations as Shin *et al.* [18] did in their simulations. Then we need to further consider modeling the transcription control process of MCIP by NFAT. We have added the following equation to replace the equation of $Ca(t)$ in our model:

$$MRNA'(t) = k41 \cdot NFATn(t) - k42 \cdot MRNA(t) \tag{1}$$

where $MRNA(t)$ denotes the mRNA concentration of MCIP and $k41$ is the control constant, $k42$ is the degradation constant [28]. Moreover, we need to add an additional production term ($k43*MRNA(t)$) and a degradation term(($ln2/t_{1/2}$)$*$ $MCIP(t)$) in the change rate equation of $MCIP(t)$ where $t_{1/2}$ denotes the half-life time constant of MCIP. Thus eventually we have completed building of our

Table 1. The Representation of Relevant Reactions of The System. Note: Abbreviations and synonyms used in this Table are as follows: $MCIP^P$ ($MCIPp$); $MCIP^{PP}$ ($MCIPpp$); $NFAT^P$ ($NFATp$); cytosolic NFAT ($NFATc$); cytosolic $NFAT^P$ ($NFATpc$); cytosolic inactive CaN ($CaNc$); cytosolic CaN^* ($CaNc^*$); cytosolic GSK3β ($GSK3\beta c$); nuclear NFAT ($NFATn$); nuclear $NFAT^P$ ($NFATpn$); nuclear CaN^* ($CaNn^*$); nuclear GSK3β ($GSK3\beta n$); protein 14-3-3 ($P1433$); Ca^{2+}-bound CaM ($CaMCa$); Complex1 ($Comp1$); Complex2 ($Comp2$); Complex3 ($Comp3$).

Reaction Category	Biochemical Form	Reaction Number	Cellerator Form of Paticular Reactions
Simple Irreversible	$A \xrightarrow{r} B$	(1)	$\{NFATc \to NFATn, k29\}$
		(2)	$\{NFATpn \to NFATpc, k30\}$
Simple Reversible	$A \underset{r2}{\overset{r1}{\rightleftharpoons}} B$	(3)	$\{GSK3\beta c \rightleftharpoons GSK3\beta n, k31, k32\}$
		(4)	$\{CaNc^* \rightleftharpoons CaNn^*, k33, k44\}$
Reversible	$A + B \underset{r2}{\overset{r1}{\rightleftharpoons}} C$	(5)	$\{CaM + Ca^4 \rightleftharpoons CaMCa, k1, k2\}$
		(6)	$\{CaMCa + CaNc \rightleftharpoons CaNc^*, k3, k4\}$
		(7)	$\{CaNc^* + MCIP \rightleftharpoons Comp1, k5, k6\}$
Binding		(8)	$\{P1433 + MCIPpp \rightleftharpoons Comp2, k19, k20\}$
		(9)	$\{NFATpc + P1433 \rightleftharpoons Comp3, k27, k28\}$
Catalytic	$S + E \underset{r2}{\overset{r1}{\rightleftharpoons}} SE$	(10)	$\{MCIP \underset{}{\overset{BMK1}{\rightleftharpoons}} MCIPp, k7, k8, k9\}$
		(11)	$\{MCIPp \underset{}{\overset{CaNc^*}{\rightleftharpoons}} MCIP, k10, k11, k12\}$
		(12)	$\{MCIPp \underset{}{\overset{GSK3\beta c}{\rightleftharpoons}} MCIPpp, k13, k14, k15\}$
		(13)	$\{MCIPpp \underset{}{\overset{CaNc^*}{\rightleftharpoons}} MCIPp, k16, k17, k18\}$
Binding	$SE \xrightarrow{r3} P + E$	(14)	$\{NFATpc \underset{}{\overset{CaNc^*}{\rightleftharpoons}} NFATc, k21, k22, k23\}$
		(15)	$\{NFATc \underset{}{\overset{GSK3\beta c}{\rightleftharpoons}} NFATpc, k24, k25, k26\}$
		(16)	$\{NFATpn \underset{}{\overset{CaNc^*}{\rightleftharpoons}} NFATn, k35, k36, k37\}$
		(17)	$\{NFATn \underset{}{\overset{GSK3\beta n}{\rightleftharpoons}} NFATpn, k38, k39, k40\}$

Table 2. Rate Constants and Other Parameters of the Model. Note: $t_{1/2}$ denotes the half-life time of MCIP1. $[CaN_{tot}]$ denotes the total concentration of calcineurin. $[NFAT_{tot}]$ denotes the total concentration of NFAT.

Parameters	Value	Parameter	Value
$k1$	$0.5\ \mu M^{-4} \cdot min^{-1}$	$k24$	$0.1\ \mu M^{-1} \cdot min^{-1}$
$k2$	$100\ min^{-1}$	$k25$	$0.15\ min^{-1}$
$k3$	$2760\ \mu M^{-1} \cdot min^{-1}$ [13]	$k26$	$0.1\ min^{-1}$
$k4$	$0.072\ min^{-1}$ [13]	$k27$	$0.4\ \mu M^{-1} \cdot min^{-1}$
$k5$	$50\ \mu M^{-1} \cdot min^{-1}$	$k28$	$0.1\ min^{-1}$
$k6$	$0.0567\ min^{-1}$ [18]	$k29$	$0.4\ min^{-1}$ [15]
$k7$	$5\ \mu M^{-1} \cdot min^{-1}$	$k30$	$0.1\ min^{-1}$ [15]
$k8$	$0.1\ min^{-1}$	$k31$	$0.1\ min^{-1}$
$k9$	$0.5\ min^{-1}$	$k32$	$0.05\ min^{-1}$
$k10$	$0.1\ \mu M^{-1} \cdot min^{-1}$	$k33$	$0.114\ min^{-1}$ [17]
$k11$	$0.1\ min^{-1}$	$k34$	$0.0552\ min^{-1}$ [17]
$k12$	$0.1\ min^{-1}$	$k35$	$0.15\ \mu M^{-1} \cdot min^{-1}$
$k13$	$0.5\ \mu M^{-1} \cdot min^{-1}$	$k36$	$0.1\ min^{-1}$
$k14$	$0.5\ min^{-1}$	$k37$	$0.2\ min^{-1}$
$k15$	$0.1\ min^{-1}$	$k38$	$0.1\ \mu M^{-1} \cdot min^{-1}$
$k16$	$0.1\ \mu M^{-1} \cdot min^{-1}$	$k39$	$0.1\ min^{-1}$
$k17$	$0.1\ min^{-1}$	$k40$	$0.1\ min^{-1}$
$k18$	$0.1\ min^{-1}$	$k41$	$0.02\ min^{-1}$
$k19$	$0.5\ \mu M^{-1} \cdot min^{-1}$	$k42$	$0.03\ min^{-1}$
$k20$	$0.1\ min^{-1}$	$k43$	$0.03\ min^{-1}$
$k21$	$0.15\ \mu M^{-1} \cdot min^{-1}$	$t_{1/2}$	$15\ min$ [14]
$k22$	$0.15\ min^{-1}$	$[CaN_{tot}]$	$1\ \mu M$ [4,18]
$k23$	$0.1\ min^{-1}$	$[NFAT_{tot}]$	$0.017\ \mu M$ [3,18]

computational model which consists of 28 ODEs for 28 unknowns. The relevant parameters except parameter Ca, which denotes cytosolic calcium concentration, are listed in Table 2.

3   Results and Discussion

Now we have the model, then a natural ensuing question is how to simulate the stress. Similar as Shin *et al.* did in their paper, we simulate the mild stress (i.e., PO) by setting parameter Ca to a smaller constant ($0.2 \mu M$) (please note that we use a selected steady state as the initial condition for simulating the normally growing heart cells for parameter $Ca = 0.05 \mu M$. This means that we simulate the increase of the cytosolic Ca^{2+} concentration incurred by PO by increasing parameter Ca from $0.05 \mu M$ to $0.2 \mu M$). Moreover, at the same time we increase the initial value of $BMK1(t)$ from $0.012 \mu M$ to $1.2\ \mu M$ because PO activates BMK1 [19]. The strong stress (i.e., CaN* overexpression) is simulated by setting parameter Ca to a bigger constant ($0.4 \mu M$) and simultaneously increasing the initial value of $CaNc^*(t)$ from $0.0275 \mu M$ to $0.825 \mu M$.

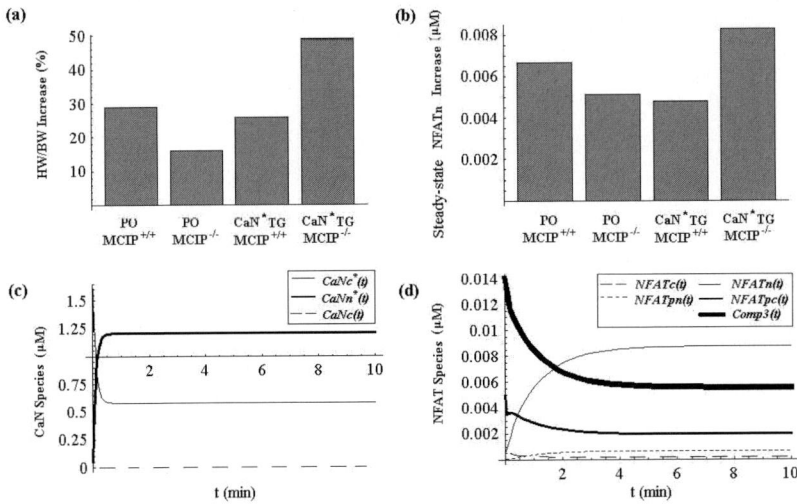

Fig. 2. Hypertrophic response and simulated transient curves for CaN* TG animals. (a) The stimuli of pressure overload (PO) caused more severe hypertrophy in normal animals (the first bar) than in MCIP$^{-/-}$ animals (the second bar) whereas the stimuli of CaN* overexpression (expressed from a muscle-specific transgene) incurred much more HW/BW (heart weight normalized to body weight) increase in MCIP$^{-/-}$ transgenic (TG) mice (the fourth bar) than in normal transgenic mice (the third bar). (b) The simulated value increase of steady-state nuclear NFAT under the different stimuli of PO and CaN* overexpression. (c) Simulated main CaN species concentration as a function of t in the case of CaN* overexpression for MCIP$^{-/-}$ animals. Thin solid line: $CaNc^*(t)$; thick solid line: $CaNn^*(t)$; Dashed line: $CaNc(t)$. (d) Simulated main NFAT species concentration as a function of t in the case of CaN* overexpression for MCIP$^{-/-}$ animals. Thin solid line: $NFATn(t)$; thick solid line: $NFATpc(t)$; extremely thick solid line: $Comp3(t)$;sparsely dashed line: $NFATc(t)$; densely dashed line: $NFATpn(t)$.

3.1 Steady-State Properties

By numerically solving the equations using the parameters listed in Table 2, simulations show that the system does evolve to a steady state. In Fig. 2b, the steady-state value increase of nuclear NFAT (i.e.,$NFATn^*(t)$) under the different stimuli for simulated MCIP$^{+/+}$ and MCIP$^{-/-}$ heart cells are shown. By comparison of the first two bars in this Figure, we can see that PO causes greater increase of the steady-state value of nuclear NFAT in simulated normal cells than in simulated MCIP mutant cells. However, the comparison of the third bar with the fourth bar tells us that CaN* overexpression incurs much less increase of the steady state value of nuclear NFAT in simulated normal cells than in simulated MCIP$^{-/-}$ cells.

3.2 Transients and Mutant Behavior

In Fig. 3, critical transient curves in the case of CaN* overexpression for simulated MCIP$^{+/+}$ cells are shown. From Fig. 3d, we can see that the concentration

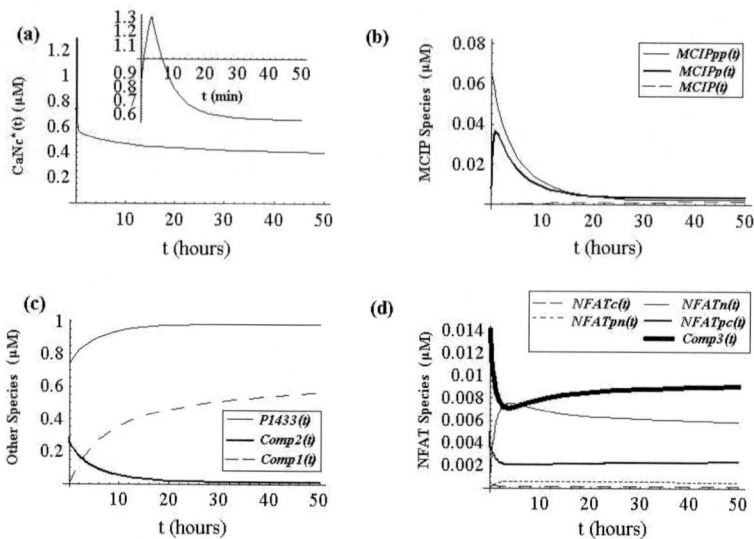

Fig. 3. Simulated transient curves for normal animals under the stimulus of CaN* overexpression. (a) Simulated $CaNc^*(t)$ (i.e., cytosolic CaN*) as a function of t. The small figure in the right-up corner shows the detailed change of CaNc*(t) during the first 50 minutes. (b) Simulated concentration of main MCIP species as a function of t. Thin solid line: $MCIPpp(t)$; thick solid line: $MCIPp(t)$; dashed line: $MCIP(t)$; (c) Simulated concentration of some other species as a function of t. Thin solid line: $P1433(t)$; thick solid line: $Comp2(t)$; dashed line: $Comp1(t)$. (d) Simulated main NFAT species concentration as a function of t. Thin solid line: $NFATn(t)$; thick solid line: $NFATpc(t)$; extremely thick solid line: $Comp3(t)$; sparsely dashed line: $NFATc(t)$; densely dashed line: $NFATpn(t)$.

of nuclear NFAT quickly rises to a peak value of 7.6 nM and then gradually declines (it will eventually rests at 5.3 nM). Similarly, we can perform numerical simulations for the MCIP$^{-/-}$ animals. Fig. 2c and Fig. 2d show the transient curves of the main CaN species and NFAT species under the stimulus of CaN* overexpression. From Fig. 2d, we can see that nuclear NFAT steadily increases from 0.5nM to a resting level of 8.8 nM.

3.3   Discussion

The decrease of Complex2 shown in Fig. 3C means the accelerated dissociation of Complex2 which should produce more MCIP^{PP} (see Reaction 8 in Table 1). However, from Fig. 3b, we can see that MCIP^{PP} is actually decreasing. Since only two reactions (Reaction 8 and 13) can possibly cause the decrease of MCIP^{PP}, the only feasible explanation is that the initial sudden rise of cytosolic CaN* catalyzes the conversion of MCIP^{PP} to MCIP^{P} and then to MCIP and the resultant depletion of MCIP^{PP} promotes the dissociation of Complex2 which also causes the concentration rise of protein 14-3-3 as seen in Fig. 3c. The increasing formation of Complex1 shown in Fig. 3c indicates that the MCIP converted from $MCIP^P$ and the newly expressed MCIP associate with cytosolic CaN*. Since in first 50 hours, the total increase of Complex1 concentration (about 0.5 μM) is much greater than the total decrease (less than 0.09 μM) of the concentration of three MCIP species shown in Fig.3b, we can conclude that the greatest source of MCIP consumed in the process of increasing formation of Complex1 is from newly expressed MCIP. The increasing formation of Complex1 also consumes cytosolic CaN* and realizes the inhibition of MCIP on calcineurin (CaN) activity. In simulated $MCIP^{-/-}$ cells, due to the lack of MCIP, nuclear NFAT increases to a much higher level (8.8nM) than in $MCIP^{+/+}$ cells (5.3 nM) as shown in Fig.2d. Also it takes much less time for the system to evolve to new steady state.

Similarly, the simulations in the case of PO stress (results not shown here due to the limit of space) show that in normal cells, activated BMK1 promotes the conversion of MCIP to $MCIP^P$ then to $MCIP^{PP}$ which causes the increase of Comlex2 formation (from 0.28 μM to 0.96 μM) and the decrease of free 14-3-3 concentration (from 0.71 μM to 0.048 μM). In simulated $MCIP^{-/-}$ cells, due to the existence of large quantity of free 14-3-3, nuclear NFAT increases from 0.5 nM to a less higher level (5.6 nM) than in $MCIP^{+/+}$ cells (7.2 nM) as shown in Fig.2b.

By comparison of the experimental and simulated hypertrophic response to different stress shown in Fig. 2a and Fig.2b, we can see both two second bars are lower than the corresponding first bars which means that our model does reproduce the seemingly facilitating function of MCIP on hypertrophic response under the stimuli of PO. Moreover, The two fourth bars are higher than the corresponding third bars which means that our model does reproduce the inhibitory function of MCIP on hypertrophic response under the stimuli of CaN* overexpression.

4   Conclusion

We have built a computational model for the complex calcium-calcineurin-MCIP-NFAT signaling network in cardiac myocytes. Our model can correctly predict the mutant ($MCIP1^{-/-}$) behavior under different stress such as PO and CaN* overexpression. Our simulation results suggest that in the case of PO, the seeming facilitating role of MCIP is due to that activated BMK1 promotes the conversion of MCIP to $MCIP^P$ then to $MCIP^{PP}$ which associates with 14-3-3 to relieve the inhibitory effect of 14-3-3 on hypertrophic response. In the case of CaN* TG mice, the overexpressed CaN* causes the dissociation of Complex2 by promoting the conversion of $MCIP^{PP}$ to $MCIP^P$ then to MCIP, which associates with CaN* to inhibit its activity. Moreover, the feedback loop of MCIP expression controlled by NFAT contributes greatly to this inhibition.

Acknowledgements. J. Cui sincerely thanks his group leader Prof. P.M.A. Sloot for sustaining support for his research. We would like to thank Prof. Kyle

W. Cunningham (Johns Hopkins University, USA) for suggesting the topic. J. Cui was firstly funded by the Dutch Science Foundation on his project 'Mesoscale simulation paradigms in the silicon cell' and later funded by EU on MORPHEX project.

References

1. Abbasi, S., Lee, J.D., Su, B., Chen, X., Alcon, J.L., Yang, J., Kellems, R.E., Xia, Y.: Protein kinase-mediated regulation of calcineurin through the phosphorylation of modulatory calcineurin-interacting protein 1. J. Biol. Chem. 281, 7717–7726 (2006)
2. Antos, C.L., McKinsey, T.A., Frey, N., Kutschke, W., McAnally, J., Shelton, J.M., Richardson, J.A., Hill, J.A., Olson, E.N.: Activated glycogen synthase-3β suppresses cardiac hypertrophy in vivo. Proc. Natl. Acad. Sci. USA 99, 907–912 (2002)
3. Arron, J.R., Winslow, M.M., Polleri, A., Chang, C.P., Wu, H., Gao, X., Neilson, J.R., Chen, L., Heit, J.J., Kim, S.K., Yamasaki, N., Miyakawa, T., Francke, U., Graef, I.A., Crabtree, G.R.: NFAT dysregulation by increased dosage of DSCR1 and DYRK1A on chromosome 21. Nature 441, 595–600 (2006)
4. Bhalla, U.S., Iyengar, R.: Emergent properties of networks of biological signaling pathways. Science 203, 381–387 (1999)
5. Cui, J., Kaandorp, J.A.: Mathematical modeling of calcium homeostasis in yeast cells. Cell Calcium 39, 337–348 (2006)
6. Dixon, M.: Enzymes, Longman Group Limited London, 3rd edn (1979)
7. Hallhuber, M., Burkard, N., Wu, R., Buch, M.H., Engelhardt, S., Hein, L., Neyses, L., Schuh, K., Ritter, O.: Inhibition of nuclear import of calcineurin prevents myocardial hypertrophy. Circ. Res. 99, 626–635 (2006)
8. Hilioti, Z., Gallagher, D.A., Low-Nam, S.T., Ramaswamy, P., Gajer, P., Kingsbury, T.J., Birchwood, C.J., Levchenko, A., Cunningham, K.W.: GSK-3 kinases enhance calcineurin signaling by phosphorylation of RCNs. Genes & Dev 18, 35–47 (2004)
9. Hill, J.A., Rothermel, B.A., Yoo, K.D., et al.: Targeted inhibition of calcineurin in pressure-overload cardiac hypertropy: preservation of systolic function. J. Biol. Chem. 277, 10251–10255 (2002)
10. Liao, W., Wang, S., Han, C., Zhang, Y.: 14-3-3 Proteins regulate glycogen synthase 3β phosphorylation and inhibit cardiomyocyte hypertrophy. FEBS Journal 272, 1845–1854 (2005)
11. Molkentin, J.D., Lu, J.R., Antos, C.L., Markham, B., Richardson, J., Robbins, J., Grant, S.R., Olson, E.N.: A calcineurin-dependent transcriptional pathway for cardiac hypertrophy. Cell 93, 215–228 (1998)
12. Putney, J.W.: Calcium signaling, 2nd edn. CRC Press, Boca Raton (2005)
13. Quintana, A.R., Wang, D., Forbes, J.E., Waxham, M.N.: Kinetics of calmodulin binding to calcineurin. Biochem. Biophys. Res. Commun. 334, 674–680 (2005)
14. Rothermel, B.A., Vega, R.B., Williams, R.S.: The role of modulatory calcineurin-interacting proteins in calcineurin signaling. Trends Cardiovasc. Med. 13, 15–21 (2003)
15. Salazar, C., Hfer, T.: Allosteric regulation of the transcription factor NFAT1 by multiple phosphorylation sites: a mathematical analysis. J. Mol. Biol. 327, 31–45 (2003)
16. Shapiro, B.E., Levchenko, A., Meyerowitz, E.M., Wold, B.J., Mjolsness, E.D.: Cellerator: extending a computer algebra system to include biochemical arrows for signal transduction simulations. Bioinformatics 19, 677–678 (2003)

17. Shibasaki, F., Price, E.R., Milan, D., McKeon, F.: Role of kinases and the phosphatase calcineurin in the nuclear shuttling of transcription factor NF-AT4. Nature 382, 370–373 (1996)
18. Shin, S.-Y., Choo, S.-M., Kim, D., Baek, S.J., Wolkenhauer, O., Cho, K.-H.: Switching feedback mechanisms realize the dual role of MCIP in the regulation of calcineurin activity. FEBS Letters 580, 5965–5973 (2006)
19. Takeishi, Y., Huang, Q., Abe, J.-i., Glassman, M., Che, W., Lee, J.-D., Kawakatsu, H., Lawrence, E.G., Hoit, B.D., Berk, B.C., Walsh, R.A.: Src and multiple MAP kinase activation in cardiac hypertrophy and congestive heart failure under chronic pressure-overload: Comparison with acute mechanical stretch. Journal of Molecular and Cellular Cardiology 33, 1637–1648 (2001)
20. Vega, R.B., Bassel-Duby, R., Olson, E.N.: Control of Cardiac growth and function by calcineurin signaling. J. Biol. Chem. 278, 36981–36984 (2003)
21. Vega, R.B., Rothermel, B.A., Weinheimer, C.J., Kovacs, A., Naseem, R.H., Bassel-Duby, R., Williams, R.S., Olson, E.N.: Dual roles of modulatory calcineurin-interacting protein 1 in cardiac hypertrophy. Proc. Natl. Acad. Sci. USA 100, 669–674 (2003)
22. Vega, R.B., Yang, J., Rothermel, B.A., Bassel-Duby, R., Williams, R.S.: Multiple domains of MCIP1 contribute to inhibition of calcineurin activity. J. Biol. Chem. 277, 30401–30407 (2002)
23. Yang, J., Rothermel, B., Vega, R.B., Frey, N., McKinsey, T.A., Olson, E.N., Bassel-Duby, R., Williams, R.S.: Independent signals control expression of the calcineurin inhibitory proteins MCIP1 and MCIP2 in striated muscles. Circ. Res. 87, e61–e68 (2000)
24. http://www.cellerator.info/

Web Applications Supporting the Development of Models of Chagas' Disease for Left Ventricular Myocytes of Adult Rats

Caroline Mendonça Costa, Ricardo Silva Campos, Fernando Otaviano Campos, and Rodrigo Weber dos Santos

FISIOCOMP, the Laboratory of Computational Physiology,
Dept. of Computer Science and Master Program in Computational Modeling
Federal University of Juiz de Fora
Campus Martelos,
Juiz de Fora, MG 36036-330, Brazil
mendonca.carol@yahoo.com.br, ricardosilvacampos@terra.com.br,
fcampos1981@gmail.com, rodrigo.weber@ufjf.edu.br
http://www.fisiocomp.ufjf.br

Abstract. Chagas' Disease is an endemic infection in many areas of South and Central America that may cause a fatal type of myocarditis. In the acute phase of the disease, ventricular extrasystoles and tachycardia have been reported. Experiments with cardiac myocytes in the acute stage of Chagas' Disease have suggested a depression of I_{to}, the Ca^{2+}-independent K^+ Transient Outward current. In this work we use computational models of left ventricular myocytes of adult rats to qualify the effects of I_{to} reduction that appear during the acute phase of Chagas' Disease. The simulations are carried out by Web applications based on the CellML language, a recently emerged international standard for the description of biological models. The computational framework supports the development of mathematical models, simulations and visualization of the results. Our preliminary simulation results suggest that the reduction of I_{to} density elicits modifications of electrophysiological mechanisms that are strongly related to the phenomena of arrhythmia.

Keywords: Cardiac Modeling, Web-based Simulations, Problem Solving Environment, Myocarditis, Arrhythmia, Chagas' Disease.

1   Introduction

Chagas' Disease is an endemic infection in many areas of South and Central America causing a distinctive, often fatal myocarditis. Approximately 10 to 20 million individuals are infected, and 50,000 deaths annually are associated with this disease [1].

In the acute period ventricular extrasystoles and tachycardia have been reported in the cases of acute myocarditis [2,3]. Experiments with cardiac myocytes in the acute stage of Chagas' Disease [4] have suggested a depression of I_{to},

M. Bubak et al. (Eds.): ICCS 2008, Part III, LNCS 5103, pp. 120–129, 2008.

the Ca^{2+}-independent K^+ Transient Outward current in canine hearts. Animal models of myocarditis have also reported the depression of I_{to} in rat heart. The availability of many different genetically altered mouse models of known defects in the human cardiovascular system, the recent completion of the rat genome [5] and improvements in methods for genetically modifying rat progeny [6] create a strong demand for a more quantitative understanding of murine cardiovascular physiology.

In this work we use computational models of the left ventricular myocytes of adult rats to qualify the effects of the I_{to} reduction that appear during the acute phase of Chagas' Disease. Due to the multi-scale and multi-physics nature of these biological models, their development becomes particularly challenging not only from a biological or biophysical viewpoint, but also from a mathematical and computational perspective. A Web-based computational framework that provides support for cardiac electrophysiology modeling is under development [7,8]. Such framework integrates different tools and allows one to bypass many complex steps during the development and use of cardiac models that are based on systems of ordinary differential equations. The implementation of cardiac cell models is supported by a code generator named AGOS [7] that generates C++ code by translating models described in the CellML language [9], a recently emerged international open standard for the description of biological models. The generated code allows one to manipulate and solve the mathematical models numerically. The set up and use of the computational models is supported by a user-friendly graphical interface that offers the tasks of simulation configuration, execution in a computer cluster, storage of results and visualization. All these tools are integrated in a Web portal [10]. The Web portal allows one to develop and simulate cardiac models efficiently via this user-friendly integrated environment. As a result, the complex techniques behind cardiac modeling are taken care of by the Web distributed applications.

A series of in-silico experiments were performed in order to better understand how the depression of I_{to} relates to the reported extrasystoles and tachycardia in the acute phase of the disease. The simulations focused on obtaining insights to better describe three important landmarks of arrhythmia: APD (Action Potential Duration) restitution curves; dispersion of repolarization; and the dynamics of intracellular Ca^{2+} concentration. The APD restitution curve seeks an indication of pro-arrhythmic behavior. During pacing and reentrant cardiac arrhythmias, APD restitution slope has been shown to be an important determinant of wave stability. In addition of promoting APD alternans, a steep (modulus greater than 1) APD restitution slope can promote breakup of electrical waves into a fibrillation-like state [11]. It is well known that amplification of dispersion of repolarization underlies the development of life-threatening ventricular arrhythmias [12]. The dispersion of repolarization is influenced by the dispersion of APDs of the ventricular myocytes and the propagation velocity of the depolarization wave. In this work we have taken the difference between endocardial and epicardial APDs as an indication of transmural APD dispersion. Finally, alterations in intracellular calcium handling play a prominent role in the generation of arrhythmias in the failing heart. Abnormal intracellular calcium dynamics are usually correlated to the occurrence of Early (EAD) and Delayed (DAD) Afterdepolarization, i.e. the strong candidates for triggers of arrhythmia [13].

In this work, we use a modified version of the Pandit *et al* [14] model to simulate action potentials (APs) in adult rat left ventricular myocytes. The mathematical description of the Pandit model was obtained from the CellML repository [15]. Our Web Portal supported the modification of the model, the simulation of many experiments and the visualization of results. Our simulation results suggest that the reduction of I_{to} density observed during the acute phase of Chagas' Disease elicits the modification of many important electrophysiological mechanisms and features at the cellular level. The reduction of I_{to} was followed by $[Ca^{2+}]i$ overload and promoted the occurrence of EADs, acting thus in a pro-arrhythmic fashion. However, reduction of I_{to} also decreased the transmural dispersion of APDs, indicating an anti-arrhythmic response. Therefore, the observed modifications do not converge to the same direction in what concerns the prevention or induction of arrhythmia.

2   Methods

We have used a modified version of the Pandit *et al.* [14] model to simulate action potentials (APs) in adult rat left ventricular myocytes. The Pandit model was used to simulate both epicardial and endocardial myocytes as described in [14]. The original model exhibits Early Afterdepolarizations (EAD) at pacing rates greater than 2Hz due to insufficient inactivation and excessive reactivation of the L-type Ca^{2+} current $I_{Ca,L}$ [16]. To solve these limitations the modifications proposed in [16] were implemented to avoid the occurrence of EAD at physiological rates: the time constant of fast inactivation related to the Ca^{2+}-independent K^+ Transient Outward Current I_{to} was reduced by 5%; and the inactivation time constant f11 of $I_{Ca,L}$ was modified according to [16] in order to accelerate its kinetics.

These modifications were performed on the Web via a CellML editor which is one of the applications of a Web-based computational framework that provides support for cardiac electrophysiology modeling [7,8]. In addition, all simulations and the visualization of results were performed on the web with the support of our computational framework. After applying the above modifications to the original CellML model, the AGOS tool [7] translated the new description to an Application Program Interface (API) which is an object oriented C++ code. Functions are created for initialization of parameters like the number of iterations, discretization interval and initial values of variables. There are functions for the numerical solution of the system of ordinary differential equations and for results storage. In addition, the API offers functions which allow the automatic creation of model-specific interfaces. This feature is used by another web application that enables one to set any model initial condition or parameter of the model, displaying their actual names, as documented in the CellML input file, as shown in Figure 1. The Web server is currently connected to a small cluster via the SGE engine [17] which enables the simulations to be executed.

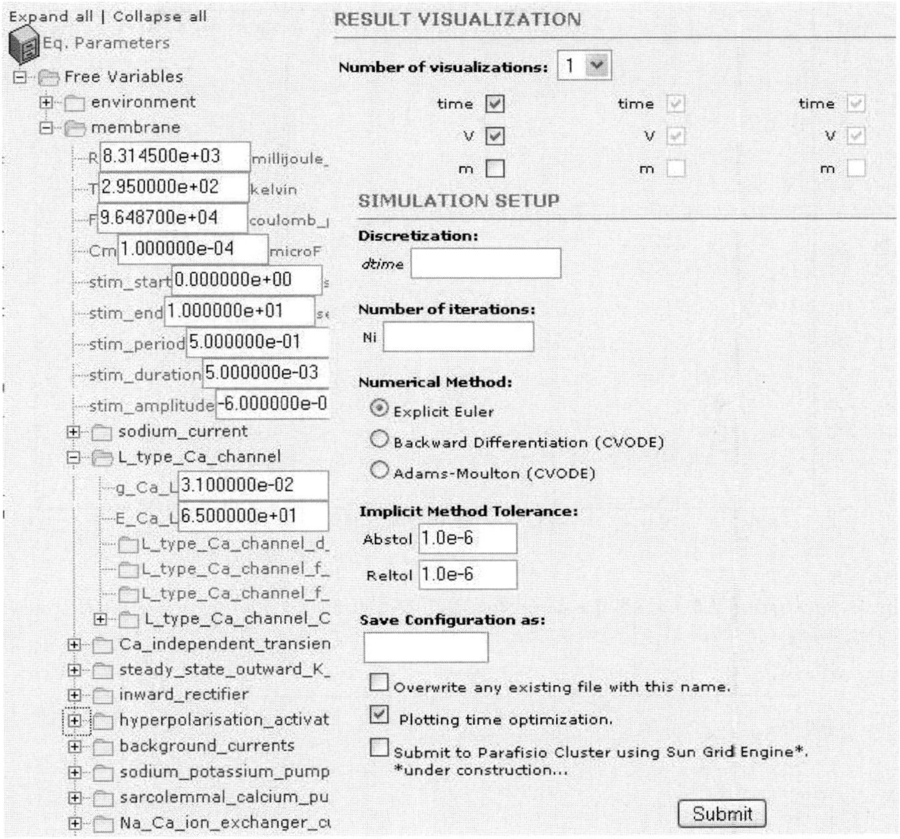

Fig. 1. Web form automatically generated by the AGOS tool for the Pandit model

The in-silico myocyte based on the Pandit model was stimulated by injecting a transmembrane current of 0.6nA during 5 ms. A train of stimuli was simulated in order to reproduce the dynamic restitution protocol described in [18]. Cells were initially paced at a constant BCL (Basic Cycle Length) of 150 ms. After 50 stimuli pacing was stopped and action potential duration (APD90) and the maximal upstroke velocity ($max\ dv/dt$) were measured. APD90, APD at 90% of repolarization, was calculated using the difference between the activation time and repolarization time. Activation times were obtained as the time of maximum rate of rise of the simulated transmembrane potentials. Repolarization times were calculated as the time of crossing a level corresponding to 90% of repolarization to the transmembrane resting potential. Pacing was then restarted at a longer BCL. At every 50 stimuli BCL was increased by 25 ms for the simulation with BCLs ranging from 150 ms to 300 ms; by 50 ms for BCLs ranging from 300 ms to 500 ms; by 250 ms from 500 ms to 1000 ms; and by 500 ms from 1000 ms to 3000 ms.

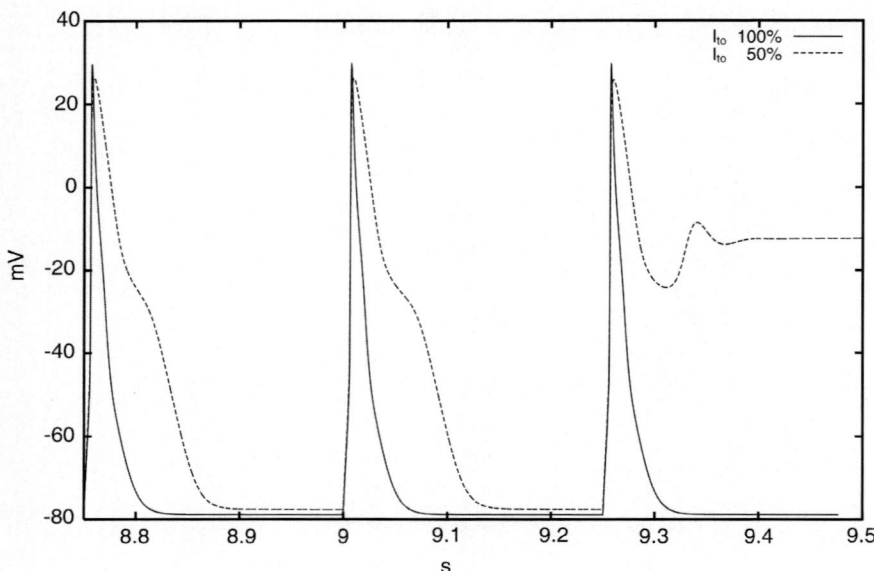

Fig. 2. Epicardial AP paced at a constant BCL of 250 ms for 50% reduction and control I$_{to}$ densities

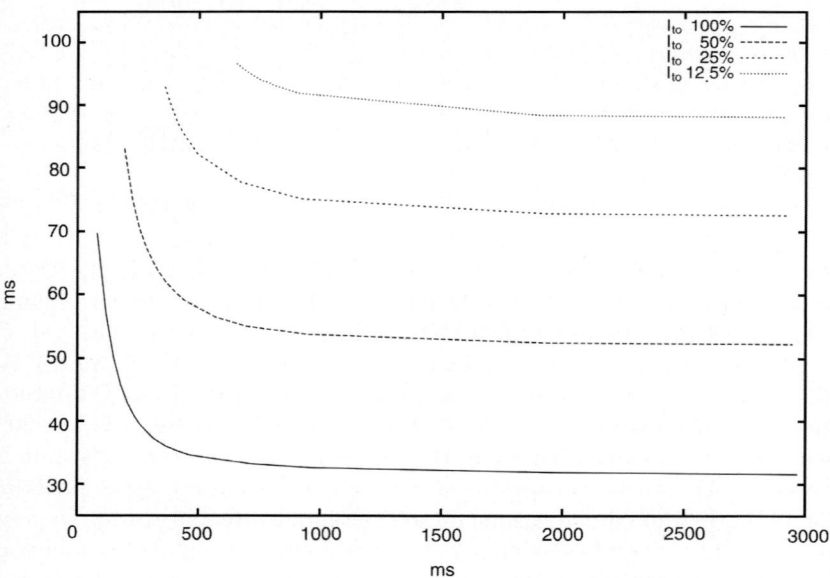

Fig. 3. Epicardial APD for different DIs and I$_{to}$ densities

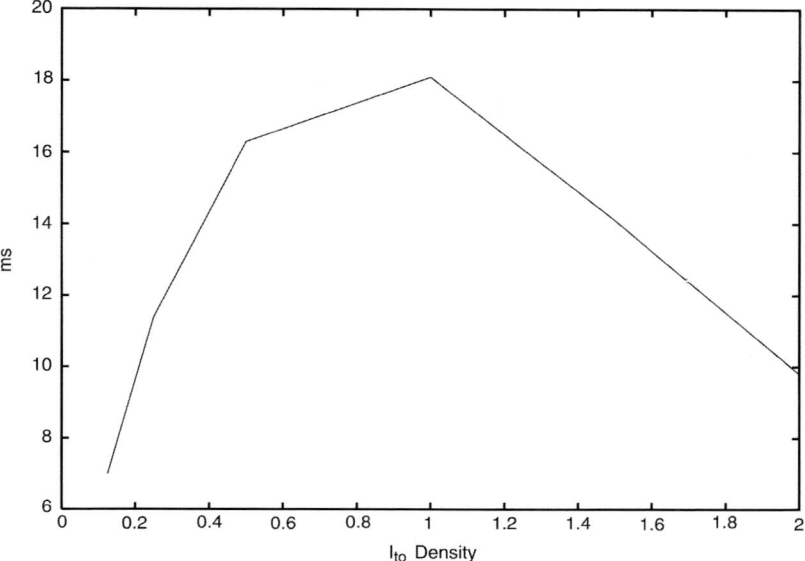

Fig. 4. Difference between endocardial and epicardial APDs for different I_{to} densities

We have also analyzed the dynamics of intracellular Ca^{2+} concentration by using the same steady-state restitution protocol described above. Calcium intracellular concentration was also recorded during 100 pulses at a constant BCL of 750 ms. The density of the I_{to} current was reduced to 50%, 25% and 12.5% in order to investigate the occurrence of EAD and its relation to APD. Transmembranic currents and ion concentrations were computed, stored, visualized and used in subsequent analysis.

3   Results

3.1   Restitution Curves and EADs

Experiments with cardiac myocytes in the acute stage of Chagas' Disease [4] indicate that phase 1 repolarization is attenuated and suggest a depression of I_{to}. In this work we investigated the influence of I_{to} in the APD restitution. For this task we changed the maximum conductance of the I_{to} current varying its density from 12.5% to 100% (control condition).

Figure 2 shows APs using the Modified Pandit Model paced at a BCL of 250 ms for both control (100% of I_{to}) and I_{to} reduced to 50% conditions. At this BCL, Figure 2 shows that a pattern similar to EAD appears after 9 s of simulation when I_{to} is reduced. At BCLs longer than 275 ms we have not observed EADs, i.e. with I_{to} reduced to 50%. However, after further reducing I_{to} density to 25% (12.5%) EAD-like patterns show up again for BCLs shorter than 450 ms

(750 ms). Therefore, these simulations suggest that the reduction of I_{to} density increases the probability of EAD occurrence at physiological rates.

APD restitution curves were calculated and are presented in Figure 3, where APD is plotted versus the diastolic interval (DI). DI was taken as the difference of the BCL and the measured APD90. One may note from Figure 3 that APD decreases for longer BCLs reaching a steady-state condition after 1000 ms. In addition, reducing the maximum conductivity of I_{to} monotononically increases APD. The APD restitution curve may be used as an indication of pro-arrhythmic behavior. During pacing and reentrant cardiac arrhythmias, APD restitution slope has been shown to be an important determinant of wave stability. In addition to promoting APD alternans, a steep (modulus greater than 1) APD restitution slope can promote breakup of electrical waves into a fibrillation-like state [11]. The slopes of the restitution curves presented by Figure 3 were significantly smaller than 1. This agrees with the fact that alternans were not observed in any of the simulation results.

3.2   Dispersion of Repolarization

It is well known that amplification of dispersion of repolarization underlies the development of life-threatening ventricular arrhythmias [12]. The dispersion of repolarization is influenced by the dispersion of APDs of the ventricular myocytes and the propagation velocity of the depolarization wave. In this work we have taken the difference between endocardial and epicardial APDs as an indication of transmural APD dispersion. In addition, since the propagation velocity is proportional to the maximum rise of the action potential, we measured $max\ dv/dt$ and considered this as an indication of changes on the propagation velocity.

Figure 4 presents the difference of APDs between simulated endocardial and epicardial cells for different densities of I_{to} with a pacing rate of 1Hz. Note the biphasic behavior. APD dispersion is greatly decreased by reducing I_{to} density (a near 4-fold decrease from control to 12.5% of I_{to} density). However, increasing I_{to} beyond the control level also reduces APD dispersion.

Since $max\ dv/dt$ is related with the velocity of the wave propagation in the cardiac tissue, we simulated the influence of the I_{to} in the upstroke of the AP. Figure 5 shows that the longer the cycle length the greater the upstroke. Figure 5 also shows that I_{to} marginally affects $max\ dv/dt$. The reduction of I_{to} was followed by a 10% decrease of $max\ dv/dt$.

3.3   Changes in the Intracellular Ca^{2+} Concentration

Alterations in intracellular calcium handling play a prominent role in the generation of arrhythmias in the failing heart [13]. Figure 6 presents the $[Ca^{2+}]i$ versus time for a BCL of 750 ms. By reducing I_{to} density we observe that $[Ca^{2+}]i$ reaches a larger steady-state value. The results thus suggest that reduction of I_{to} density is followed by $[Ca^{2+}]i$ overload.

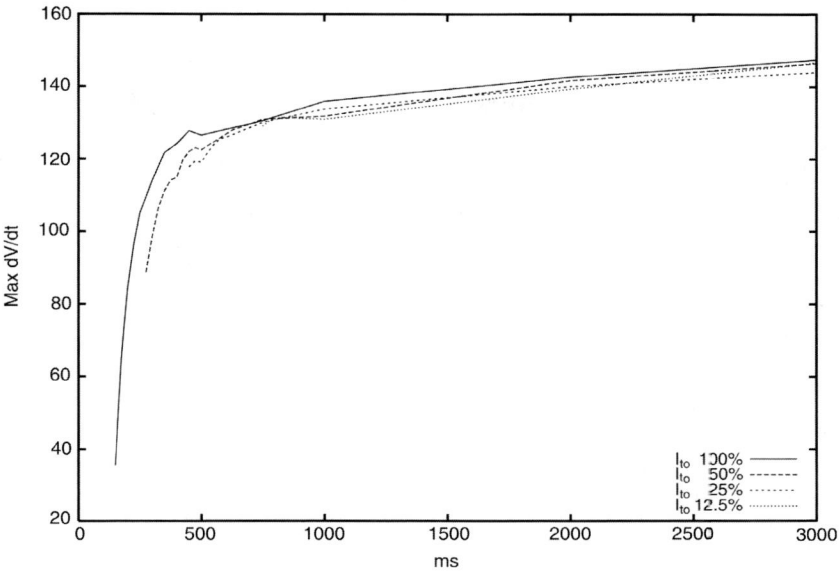

Fig. 5. The upstroke velocity *max dv/dt* for different DIs

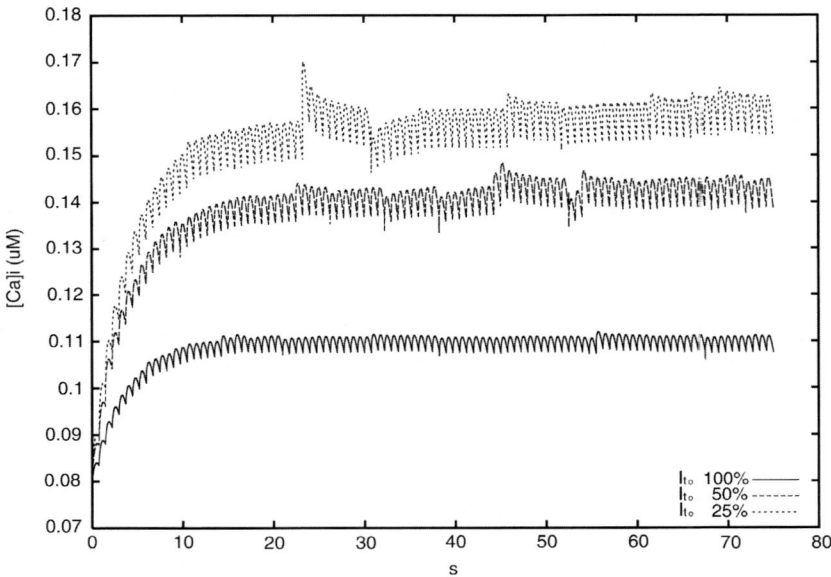

Fig. 6. $[Ca^{2+}]$i for different densities of I_{to} versus time

4   Discussion

In this work we have used a Web-based cardiac computational framework based on the CellML language to simulate mathematical models of the left ventricular myocytes of adult rats. The simulations seek a better understanding of the effects of I_{to} reduction during the acute phase of Chagas' Disease. APD restitution curves were generated, the transmural dispersion of repolarization was investigated and the dynamics of $[Ca^{2+}]i$ analyzed. We have not observed the phenomenon of alternans during the steady-state pacing protocol adopted. The APD restitution curves were calculated as presented in Figure 3. It is known that a steep APD restitution slope may suggest the breakup of electrical waves into a fibrillation-like state. The absolute values of the slopes of the restitution curves presented by Figure 3 were significantly smaller than 1. This agrees with the fact that alternans were not observed in any of the simulation results.

However, during fast pacing a pattern similar to Early Afterdepolarization (EAD) was present (see Figure 2). In addition, EADs occurred in considerably slower pacing rates after I_{to} density was reduced in the models. Therefore, the results suggest that the reduction of I_{to} density increases the probability of EAD occurrence at physiological rates. The premature occurrence of EADs seems to be in accordance with the accumulation of $[Ca^{2+}]i$ as a consequence of reduction of I_{to} density (see Figure 6). Calcium entry into myocytes drives myocyte contraction. To prepare for the next contraction, myocytes must extrude calcium from intracellular space via the Na^+/Ca^{2+} exchanger (NCX1) or sequester it into the sarcoplasmic reticulum. Defective calcium extrusion correlates with increased intracellular calcium levels and may be relevant to heart failure and to the generation of arrhythmias [19]. The results presented in Figure 6 suggest that reduction of I_{to} density may be followed by $[Ca^{2+}]i$ overload.

We have addressed the effects of I_{to} density reduction on the transmural dispersion of repolarization by calculating the changes on endocardial to epicardial APD difference and on the *max dv/dt*, which is related to the velocity of the wave propagation in the cardiac tissue. In our simulations, *max dv/dt* was marginally affected (see Figure 4). However, the transmural dispersion of APD, as presented in Figure 5, is decreased after I_{to} is reduced. The results indicate an anti-arrhythmic effect associated to the reduction of I_{to} density.

In summary, the computational framework simplified the required modifications and usage of the mathematical model studied in this work. The combination of the XML-based automatic code generator, the CellML editor and the Web portal provided an user-friendly environment for our simulations. Our results suggest that the reduction of I_{to} density observed during the acute phase of Chagas' Disease elicits the modification of many important electrophysiological mechanisms and features. The reduction of I_{to} was followed by $[Ca^{2+}]i$ overload and promoted the occurrence of EADs, acting thus in a pro-arrhythmic fashion. However, reduction of I_{to} also decreased the dispersion of APDs, indicating an anti-arrhythmic effect. Therefore, the observed modifications do not converge to the same direction in what concerns the prevention or induction of arrhythmia. Further studies are necessary in order to better characterize and understand the electrophysiological changes that underly the acute phase of Chagas' Disease.

Acknowledgments. The authors would like to thank the support provided by UFJF, CNPQ and FAPEMIG, project TEC-1358/05. Caroline Costa, Ricardo Campos and Fernando Campos are FAPEMIG scholarship holders.

References

1. Control of Chagas' Disease. Report of WHO Expert Committee. World Health Organization, Geneva, p. 1 (1991)
2. Laranja, F.S., Dias, E., Nobrega, G., Miranda, A.: Chagas' Disease. A clinical, epidemiologic and pathologic study. Circulation 14, 1035–1060 (1956)
3. Rosenbaum, M.B.: Chagasic myocardiopathy. Prog. Cardiovasc. Dis. 7, 199–225 (1964)
4. Pacioretty, L.M., Barr, S.C., Han, W.P., Gilmour Jr., R.F.: Reduction of the transient outward potassium current in a canine model of Chagas' Disease. Am. J. Physiol. 268(3 Pt 2), H1258–H1264 (1995)
5. Rat Genome Sequencing Project Consortium: Genome sequence of the Brown Norway rat yields insights in to mammalian evolution. Nature 428, 493–521 (2004)
6. Abbot, A.: The renaissance rat. Nature 428, 464–466 (2004)
7. Barbosa, C.B., Santos, R.W., Amorim, R.M., Ciuffo, L.N., Manfroi, F.M., Oliveira, R.S., Campos, F.O.: A Transformation Tool for ODE based models. In: Alexandrov, V.N., van Albada, G.D., Sloot, P.M.A., Dongarra, J. (eds.) ICCS 2006. LNCS, vol. 3991, pp. 69–75. Springer, Heidelberg (2006)
8. Martins, D., Campos, F.O., Ciuffo, L.N., Oliveira, R.S., Amorim, R.M., Vieira, V.F., Ebecken, N.F.F., Barbosa, C.B., Santos, R.W.: A Computational Framework for Cardiac Modeling Based on Distributed Computing and Web Applications. In: Daydé, M., Palma, J.M.L.M., Coutinho, Á.L.G.A., Pacitti, E., Lopes, J.C. (eds.) VECPAR 2006. LNCS, vol. 4395, pp. 544–555. Springer, Heidelberg (2007)
9. CellMl 1.1 (2006), http://www.cellml.org/specifications/cellml_1.1/
10. Laboratory of Computational Physiology, http://www.fisiocomp ufjf.br
11. Zemlin, C.W., Panfilov, A.V.: Spiral waves in excitable media with negative restitution. Phys. Rev. E Stat. Nonlin. Soft Matter. Phys. 63(4 Pt 1), 041912 (2001)
12. Antzelevitch, C.: Role of spatial dispersion of repolarization in inherited and acquired sudden cardiac death syndromes. Am. J. Physiol. Heart Circ. Physiol. 293(4), H2024–H2038 (2007)
13. Burashnikov, A., Antzelevitch, C.: Late-phase 3 EAD. A unique mechanism contributing to initiation of atrial fibrillation. Pacing Clin. Electrophysiol. 29(3), 290–295 (2006)
14. Pandit, S.V., Clark, R.B., Giles, W.R., Demir, S.S.: A mathematical model of action potential heterogeneity in adult rat left ventricular myocytes. Biophys. J. 81(6), 3029–3051 (2001)
15. CellMl Repository, http://www.cellml.org/models/
16. Kondratyev, A.A., Ponard, J.G., Munteanu, A., Rohr, S., Kucera J.P.: Dynamic changes of cardiac conduction during rapid pacing. Am. J. Physiol. Heart Circ. Physiol. 292(4), H1796–H1811 (2007)
17. SGE, The Sun Grid Engine, http://gridengine.sunsource.net/
18. Koller, M.L., Riccio, M.L., Gilmour Jr., R.F.: Dynamic restitution of action potential duration during electrical alternans and ventricular fibrillation. Am. J. Physiol. 275(5 Pt 2), H1635–H1642 (1998)
19. Bers, D.M., Despa, S., Bossuyt, J.: Regulation of Ca2+ and Na+ in normal and failing cardiac myocytes. Ann. N Y Acad. Sci. 1080, 165–177 (2006)

A Streamlined and Generalized Analysis of Chromatin ImmunoPrecipitation Paired-End diTag Data

Vinsensius B. Vega[1], Yijun Ruan[2], and Wing-Kin Sung[1]

[1] Computational and Mathematical Biology Group
[2] Clone and Sequencing Group
Genome Institute of Singapore
{vegav,ruanyj,sungk}@gis.a-star.edu.sg

Abstract. Comprehensive, accurate and detailed maps of transcription factor binding sites (TFBS) help to unravel the transcriptional regulatory relationship between genes and transcription factors. The recently developed sequencing-based genome-wide approach ChIP-PET (Chromatin ImmunoPrecipitation coupled with Paired-End diTag analysis) permits accurate and unbiased mapping of TF-DNA interactions. In this paper we outline a methodical framework to analyze ChIP-PET sequence data to identify most likely binding regions. Mathematical formulations were derived to streamline and strengthen the analysis. We established a more faithful noise distribution estimation that leads to the adaptive threshold scheme. The algorithms were evaluated using three real-world datasets. Using motif enrichment as indirect evidence and additional ChIP-qPCR validations, the overall performance was consistently satisfactory.

1 Introduction

Transcription factors (TF) play a pivotal role in controlling gene expression, and thus directing the cellular behavior and mechanisms. Part of the effort to fully decipher the intricate regulatory networks and pathway, a key intermediate goal would be to identify the relevant Transcription Factor Binding Sites (TFBS). Such is also one of the goals set out for the human genome[1].

Chromatin Immuoprecipitation (ChIP) is a powerful approach to study in vivo protein-DNA interactions. It consists of five major steps: (i) cross-link the DNA binding proteins to the DNA in vivo, (ii) shear the chromatin fibers using sonication or otherwise, (iii) immunoprecipitate the chromatin fragments using specific antibody against given protein targets, (iv) reverse the cross-linking of protein-bound DNA, and (v) analyze the ChIP enriched DNA fragments. These DNA fragments can then be characterized using high throughput approaches, such as hybridization-based ChIP-chip analysis [2],[3],[4],[5] or direct DNA sequencing. ChIP sequencing can be performed by sequencing individually cloned fragments [6],[7], concatenations of single tags derived from fragments (STAGE) [8],[9],[10],[11] or concatenations of pair-end-ditags to infer the linear structure

M. Bubak et al. (Eds.): ICCS 2008, Part III, LNCS 5103, pp. 130–139, 2008.

of ChIP DNA fragments (ChIP-PET)[12],[13]. The sequencing approaches have their advantages over the hybridization-based approaches by elucidating the exact nucleotide content of target DNA sequences.

In a ChIP-PET experiment, 5' (18bp) and 3' (18bp) signatures for each of the ChIP enriched DNA fragments were extracted and joined to form the paired end tag structure (PET) that were then concatenated for efficient sequencing analysis. The PET sequences were then mapped to the reference genome to infer the full content of each of the ChIP DNA fragments. The protein-DNA interaction regions enriched by ChIP procedure will have more DNA fragments representing the target regions than the non-specific regions. Therefore, with sufficient sequence sampling in the DNA pool of a ChIP experiment, multiple DNA fragments originated from the target regions will be observed.

In previous analyses of ChIP-PET data [12],[13], Monte Carlo simulations were employed to distinguish true signals from noise. This paper proposes mathematical formulations for performing the similar assessment as the Monte Carlo simulation in an efficient manner, and further generalizes the approach to resolve potential irregular noise arising from anomalous chromosomal copies and configuration.

2   Results and Discussion

2.1   PET Clusters as the Readout for TFBS

Presence of PET clusters is clearly an initial indication of genomic loci enriched for TF-bound fragments. The more PETs that a cluster has, the more probable the TF binds to the region. We can set a minimum cut-off criterion, say

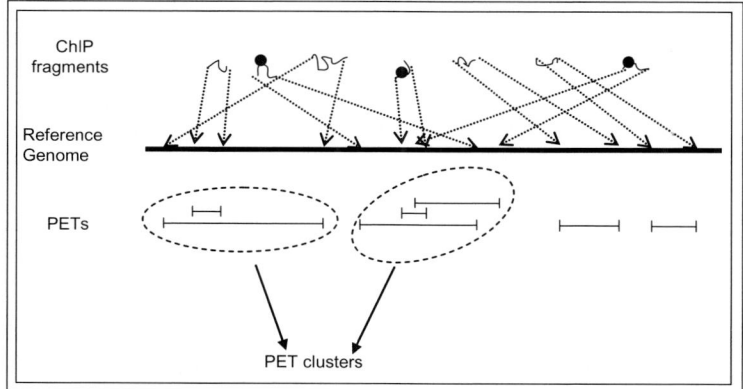

Fig. 1. ChIP fragments, PETs, and ChIP-PET clusters. ChIP fragments might be TF-bound (shaded circles) or simply noise. Mapped ChIP fragments are called PETs. Overlapping PETs are grouped into ChIP-PET clusters.

z, and classify clusters with at least z PETs (i.e. PETz+ clusters) to be the highly probable clusters with TF binding. A Monte Carlo approach has been successfully employed to determine this threshold [12].

More analytically, if we assume that the noisy PETs are randomly and uniformly distributed along the genome, then the distance, D, between any two consecutive random PETs is expected to follow the exponential distribution with rate $\lambda = \frac{F}{G}$, where F is the total number of PETs and G is the genome length. The probability of two PETs overlapping (i.e. the distance between them is less than or equal the (expected) PET length) by chance alone is $P_{\exp}(X \leq k; \lambda)$ where $P_{\exp}$ is the cumulative exponential distribution function whose rate is λ and k is the expected length of a PET. Two overlapping PETs can be found in a PET2 cluster and beyond. Thus, the probability $P_{\exp}(X \leq k; \lambda)$ is the probability of a PET2+ cluster to happen simply by chance. More generally, the probability of the occurrence of a PETn+ cluster by random is:

$$\Pr_{\text{PET}}(Y \geq n; \lambda) \approx (P_{\exp}(X \leq k; \lambda))^{(n-1)} = \left(1 - e^{-\lambda k}\right)^{(n-1)} \qquad (1)$$

In place of the Monte Carlo simulations, one can readily compute the p-value of random PETn+ clusters using the above equation to determine the appropriate threshold for a given ChIP PET library. Additional empirical evidence for Eq. (1) will be provided later in the text.

2.2   Counting on Maximum Support for Defining Binding Regions

While number of PETs in a cluster is useful for assessing whether the cluster is likely to be a true signal, clusters with seemingly good number of PETs can still be generated by random noise. For a true binding region with many PETs, the exact position of the actual protein-DNA site will be more refined and appear as a sharp peak in the cluster. However, it is not uncommon to find big clusters whose overlapping regions are not well concentrated. This is an indication that they might have been formed simply by chance. Figure 2 shows a snapshot of two clusters from real libraries, contrasting a typical good cluster (left) with well defined core, to a configuration with scattered overlap region (right) which was most likely formed by random PETs.

We call a cluster as a moPETn (*maximum overlap* PET n) cluster if all of its sub-region is supported by at most n PETs. Similar to the previous definition, moPETn+ clusters represent the set of moPETm clusters where $m \geq n$. The left cluster in Fig.2 is PET5/moPET5, while the right cluster is PET5/moPET2. To some extent, PETn/moPETm (where $m < n$) clusters are doubtful. There are 96 of such clusters in the p53 library, while the Oct4 and ER libraries contain 4,929 and 910 such clusters.

The probability of a moPETn to be initiated by an arbitrary PET $< s, l >$ can be estimated by the probability of observing additional $(n-1)$ PET starting sites

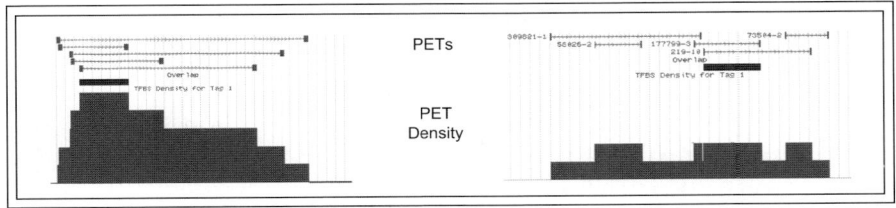

Fig. 2. A comparison of high fidelity cluster and noisy cluster. Good clusters are generally well-defined (left cluster), containing a strong overlapping region. Dispersed ChIP PET segments (right cluster) hint at the possibility of cluster formation purely at random and by chance alone.

at most l-bp away from s. This probability follows that of Poisson distribution for observing $(n-1)$ events whose rate is λ within the interval k (=expected PET length). More formally, the probability of an arbitrary PET to initiate a moPETn cluster:

$$\mathrm{Pr}_{\mathrm{moPET}}(Y = n; \lambda) \approx P_{\mathrm{Poisson}}(X = (n-1); \lambda k) = \frac{e^{-\lambda k}(\lambda k)^{(n-1)}}{(n-1)!} \quad (2)$$

Using $\mathrm{Pr}_{\mathrm{moPET}}(n)$ and given the acceptable p-value level, we can determine the appropriate cut-off of moPETn for identifying true TF-binding regions. Comparison with simulation results is presented below.

2.3   Comparing Empirical Results and Analytical Formulations

To evaluate the correctness of our statistical formulations in Eqs. (1) and (2), we compared the analytical estimations of PETn+ and moPETn clusters distributions to the empirical ones generated through a 100,000 runs of Monte Carlo simulations (see Methods) with different sets of parameter as listed in Table 1. The collected statistics were used to construct empirical distributions which were then compared with the proposed analytical framework. Figure 3(a) and 3(b) contrasts the empirical probability of PETn+ and moPETn occurrence (points) against the analytical estimations (dashed lines). The analytical curves tracks the empirical values very well, reconfirming the validity of the analytical distributions.

2.4   Adaptive Approach for Biased Genomes

The estimation of rate λ, i.e. the expected number of PETs per nucleotide, plays a critical role in Eqs. (1) and (2). This rate signifies the expected noise level of the dataset. A single global rate λ reflects the assumption that the

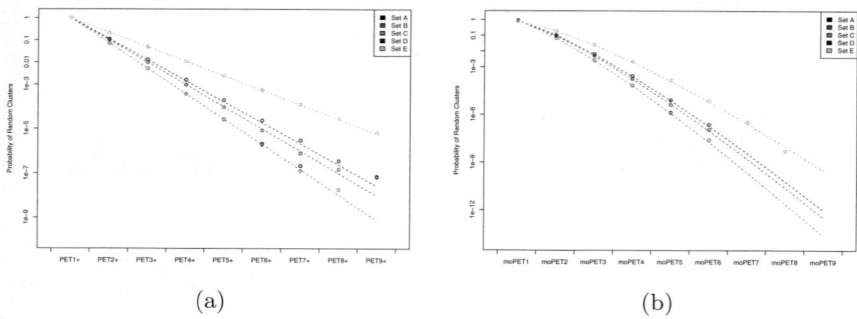

(a)   (b)

Fig. 3. Comparing simulation and analytical results. Probability of a random
(**A**) PETn+ or (**B**) moPETn cluster produced by chance, as estimated by Monte Carlo
simulations (points) and computed analytically (dashed lines).

noisy PETs are randomly uniformly distributed across the genome. Although
may be acceptable in general, libraries generated from aberrant genome require
more refined estimation. Amplified genomic regions will have higher PET counts
than the overall genome, making their purely random clusters bear stronger
signal than those of normal regions. On the other hand, regions with significant deletions will contain less than expected PETs and their true binding
loci will be much weaker. Using single global λ would result in higher false
positive rates in amplified regions and higher false negative rates in deleted
regions.

We devised a two-phase adaptive approach that takes account of local biases
(see Fig.4) in predicting the most probable source (true binding vs. noise) of
each PET cluster. Given a cluster c, the first phase considers the local window
of some predefined size L centered on the cluster c, and, estimates the total
number of noise PETs. The second phase computes the local λ and calculates a
local moPET (or PET) cut-off T_{moPET}. Clusters c is considered to be binding
region if its moPET (or PET) count is greater than T_{moPET}.

The noise estimation step (phase 1) counts the number of potentially noisy
PETs within the window. Overestimation of noise would increase false negatives,
while underestimation would add false positives. We adhere to two heuristics: (i)
the current cluster should not be assumed as real and (ii) other clusters within
the windows that seem to be real clusters should, as much as possible, not be
counted as noise. The first rule stems from the fact that most of the clusters
(especially PET1 clusters) are noise. Observations that binding sites are sometimes located proximal to each other motivated the second rule. For the window
size L, we set it to be at least twice the expected distance between two PETs
(i.e. λ^{-1}). Noise estimation starts by identifying the probable noisy clusters. Assuming that the current cluster c is noisy, clusters within the window L with
higher moPET counts than the current cluster c are contextually considered
non-noise (line 3 in Fig.4). Next, the rate of noisy PETs per cluster is estimated by taking the geometric mean [14],[15] of PET counts of (locally) noisy

GOODCLUSTER(c, p, L)

1   Let D be the set of clusters that are located at most $\frac{L}{2}$ basepairs away (upstream or downstream) from c

2   $G \leftarrow \{\}$ ▷ **Start of local noise estimation**

3   **for** each cluster $d \in D$: **if** MO(d) $\leq$ MO(c) **then** $G = G \cup \{$PET(d)$\}$

4   $g \leftarrow$ GEOMEAN(G)

5   $S \leftarrow \sum_{d \in D} \min(\text{PET}(d), g)$ ▷ S is the estimated local noise

6   $\lambda \leftarrow \frac{S}{L}$ ▷ **Start of threshold determination**

7   $T_{\text{moPET}} \leftarrow \min(\{n | \text{Pr}_{\text{moPET}}(Y \geq n; \lambda) \leq p\})$

8   **if** MO(c) $> T_{\text{moPET}}$

      then return TRUE

      else return FALSE

Fig. 4. Pseudocode of the adaptive thresholding algorithm

clusters (line 4 in Fig.4). The final sum of noisy PETs, S, is calculated by adding the noisy PET counts of all the clusters within the current window (line 5 in Fig.4).

The second phase is quite straightforward through performing sufficient iterations of Monte Carlo simulations or the application of the Eqs. (1) or (2) using the local rate λ ($=S/L$) and considering the window length L.

2.5 Performance on Real Datasets

Three real datasets were used in our evaluation: p53 ChIP-PET [12], Oct4 ChIP-PET [13], and Estrogen Receptor (ER) ChIP-PET [16]. For each dataset, we applied our proposed algorithms to predict TF-bound regions. The predicted regions were then evaluated indirectly by enrichment of putative relevant binding motifs and (whenever available) directly through further ChIP qPCR validation assays (see Methods).

The p53 library was the first and the smallest dataset (65,714 PETs, average length: 625bp) and was constructed using the human HCT116 cancer cell lines. The ER ChIP PET library comprised 136,152 PETs (average length: 72bp), was assayed on human MCF-7 breast cancer cell lines. The largest library among the three, the Oct4 ChIP PET, was based on mouse E14 cell lines (366,639 PETs, average length: 627bp). The non-gapped genome lengths for human and mouse are estimated at 2.8Gbp (UCSC hg17) and 2.5Gbp (UCSC mm5).

Setting the cut-off of at $p = 10^{-3}$, the selected clusters for p53 is PET3+ or moPET3+, for ER is PET4+ or moPET3+, and for Oct4 is PET4+ or moPET4+. Table 2 gives the validations of each PET cluster group in each library and PET or moPET cluster group. Sharp motif enrichment can be seen at the selected cut-offs in all libraries, compared to the PET2 or moPET2 group, which is expected to be noisy. Table 2 also shows high success rate of ChIP-qPCR validations. The p53 library had 100% of the ChIP-qPCR tested sites showing enrichment of p53 binding. The high ChIP-qPCR success rate ($> 95\%$) for the selected Oct4 moPET4+ clusters also increased our confidence of the validity of the cluster selection approach.

Prior to running the ChIP-qPCR validation for the ER library, we noticed unusual concentrations of PETs in regions, which correlated well with the regions previously reported to be amplified in the underlying MCF-7 cell lines [17]. This prompted us to employ the adaptive moPET thresholding algorithm to 'normalize' the amplified regions. We also applied the adaptive approach on the other two datasets, to see its effect on other libraries from relatively normal cell lines (i.e. the p53 and Oct4 libraries). The result is summarized in Table 3.

Adaptive thresholding might exclude clusters selected under the global thresholding and re-include clusters which would otherwise be excluded because they were below the global threshold. Global and adaptive moPET thresholding produced the same results for p53 (see Table 3). Interestingly, application of adaptive thresholding on the Oct4 library re-included some of the moPET3 clusters, with a higher motif enrichment. Only a tiny fraction of the moPET4 was rejected, with no significant impact on motif enrichment. The ChIP qPCR success rates for the adaptive-selected clusters were higher than before. A sizeable portion of the moPET3+ in ER ChIP PET library was rejected and the selected clusters had better motif enrichment, indicative of true binding. ChIP-qPCR assays on random samples of the selected clusters confirmed that further.

3   Methods

3.1   ChIP-PET Clustering

The primary ChIP-PET data is the locations and lengths of the ChIP-PET fragments. The tuple $< s, l >$ represents an l-bp long PET fragment mapped into location s. Two PET fragments $< s_1, l_1 >$ and $< s_2, l_2 >$, where $s_1 \leq s_2$, are said to be overlapping if $s_1 + l_1 \geq s_2$. A ChIP-PET cluster is defined as the largest set of cascading overlapping PET fragments.

3.2   Simulation Procedures

To generate a random PET library, we preformed a Monte Carlo simulation while taking into account the overall genome length (G), the total number of PETs (M), and the desired PETs' lengths (minimum and maximum lengths). In each Monte Carlo simulation, M points were randomly picked along the G-bp genome, mimicking the generation of a PET library containing completely random fragments. For each picked point, a random length was sampled from a uniform distribution within the given minimum and maximum bounds. Simulated PETs were clustered accordingly. Statistics of PETn+ and moPETn clusters were collected and averaged over a sufficient number of Monte Carlo iterations.

3.3   Evaluations of Selected Clusters from Real Datasets

The goodness of these clusters (i.e. whether these clusters were truly bound by the TF) was then doubly-assessed: (i) indirectly from the enrichment of putative relevant binding motifs among the selected clusters and (ii) directly through further ChIP qPCR validation assays. The putative motifs for p53 and Oct4 were identified based on the binding site models described in their respective papers [12],[13]. Putative ER binding motif were based on the consensus sequence, GGTCAnnnTGACC [18], and allowing for up to 2nt mismatches. Additional ChIP qPCR validations have also been carried out on some of the selected clusters [12],[13],[16].

4   Conclusions

We have described a more principled framework for analyzing ChIP-PET data to identify transcription factor binding regions. To dichotomize the PET clusters into potentially binding regions and likely non-binding regions, we utilized a random PET generation model and estimated the improbable concentration of PETs generated at random. The adaptive thresholding framework was introduced to handle aberrant genomes, e.g. due to amplifications and deletions in cancer cell lines, or other experimental conditions. These analyses might be further improved by taking into account other known inherent properties of the genome (e.g. prevalence of repeats). We also noticed a potential utility of the adaptive technique for identifying intrinsic features of the genome (e.g. for delineating amplified or deleted segments).

Acknowledgments. We would like to thank Jane Thomsen for providing the ChIP-qPCR validation data of the ER library. This work is supported by the Agency for Science, Technology and Research (A*STAR) of Singapore and an NIH/NHGRI ENCODE grant (1R01HG003521-01).

References

1. ENCODE Project Consortium: The ENCODE (ENCyclopedia Of DNA Elements) Project. Science 306, 636–640 (2004)
2. Iyer, V.R., Horak, C.E., Scafe, C.S., et al.: Genomic binding sites of the yeast cell-cycle transcription factors sbf and mbf. Nature 409, 533–538 (2001)
3. Ren, B., Robert, F., Wyrick, J.J., et al.: Genome wide location and function of dna binding proteins. Science 290, 2306–2309 (2000)
4. Horak, C.E., Mahajan, M.C., Luscombe, N.M., et al.: GATA-1 binding sites mapped in the β-globin locus by using mammalian chip-chip analysis. Proceedings of the National Academy of Sciences 99, 2924–2929 (2002)
5. Weinmann, A.S., Pearlly, S.Y., Oberley, M.J., et al.: Isolating human transcription factor targets by coupling chromatin immunoprecipitation and cpg island microarray analysis. Genes Dev. 16, 235–244 (2002)

6. Weinmann, A.S., Bartley, S.M., Zhang, T., et al.: Use of chromatin immunoprecipitation to clone novel E2F target promoters. Mol. Cell. Biol. 21, 6820–6832 (2001)
7. Hug, B.A., Ahmed, N., Robbins, J.A., Lazar, M.A.: A chromatin immunoprecipitation screen reveals protein kinase cbeta as a direct runx1 target gene. J. Biol.
   Chem. 279(2), 825–830 (2004)
8. Impey, S., McCorkle, S.R., Cha-Molstad, H., et al.: Defining the creb regulona
   genome-wide analysis of transcription factor regulatory regions. Cell 119, 1041–
   1054 (2004)
9. Kim, J., Bhinge, A.A., Morgan, X.C., Iyer, V.R.: Mapping dna-protein interactions in large genomes by sequence tag analysis of genomic enrichment. Nature
   Methods 2(1), 47–53 (2004)
10. Chen, J., Sadowski, I.: Identification of the mismatch repair genes PMS2 and MLH1
    as p53 target genes by using serial analysis of binding elements. Proceedings of the
    National Academy of Sciences 102(13), 4813–4818 (2005)
11. Roh, T.Y., Cuddapah, S., Zhao, K.: Active chromatin domains are defined by
    acetylation islands revealed by genome-wide mapping. Genes Dev. 19(5), 542–552
    (2005)
12. Wei, C.L., Wu, Q., Vega, V.B., et al.: A global map of p53 transcription-factor
    binding sites in the human genome. Cell 124(1), 207–219 (2006)
13. Loh, Y.H., Wu, Q., Chew, J.L., Vega, V.B., et al.: The oct4 and nanog tran-
    scription network regulates pluripotency in mouse embryonic stem cells. Nature
    Genetics 38(4), 431–440 (2006)
14. McAlister, D.: The law of the geometric mean. Proceedings of the Royal Society
    of London 29, 367–376 (1879)
15. Fleming, J., Wallace, J.: How not to lie with statistics: the correct way to summarize
    benchmark results. Communications of the ACM 29, 218–221 (1986)
16. Lin, C.Y., Vega, V.B., Thomsen, J.S., et al.: Whole-genome cartography of estrogen
    receptor alpha binding sites. PLoS Genet 3(6) (June 2007)
17. Shadeo, A., Lam, W.L.: Comprehensive copy number profiles of breast cancer cell
    model genomes. Breast Cancer Res 8(1), R9 (2006)
18. Klinge, C.: Estrogen receptor interaction with estrogen response elements. Nucleic
    Acids Research 29, 2905–2919 (2001)

Appendix: Tables

Table 1. Simulation setups. Five Monte Carlo simulation sets run to assess the
analytical model of random PETn and moPETn clusters formations.

Simulation Set	A	B	C	D	E
Genome length	2Mbp	2Mbp	20Mbp	10Mbp	10Mbp
Number of PETs	300	300	3000	2000	5000
Min. PET length	500bp	700bp	500bp	200bp	300bp
Max. PET length	500bp	700bp	500bp	1000bp	700bp

Table 2. Motif enrichments and ChIP-qPCR validation rate of clusters selected by global thresholding

(A) p53 ChIP-PET clusters

	PET2	PET3	PET4	PET5	PET6	PET7	PET8+
Total clusters	1453	161	66	38	29	13	29
% with motifs	15.97%	59.63%	80.30%	65.79%	89.66%	84.62%	82.76%
ChIP-qPCR success rate	N/A	N/A	100.0%	100.0%	100.0%	100.0%	100.0%

	moPET2	moPET3	moPET4	moPET5	moPET6	moPET7+
Total clusters	1489	140	69	30	26	35
% with motifs	16.25%	67.14%	81.16%	70.00%	88.46%	88.57%
ChIP-qPCR success rate	N/A	100.0%	100.0%	100.0%	100.0%	100.0%

(B) Oct4 ChIP-PET clusters

	PET2	PET3	PET4	PET5	PET6	PET7	PET8+
Total clusters	29453	5556	1540	550	223	102	201
% with motifs	16.74%	24.62%	34.35%	42.36%	52.47%	49.02%	45.77%
ChIP-qPCR success rate	10.00%	9.68%	88.24%	90.48%	100.00%	100.00%	95.00%

	moPET2	moPET3	moPET4	moPET5	moPET6	moPET7+
Total clusters	32739	3734	724	189	93	146
% with motifs	17.57%	27.64%	41.57%	54.50%	70.97%	43.15%
ChIP-qPCR success rate	10.00%	8.82%	95.00%	100.00%	100.00%	100.00%

(C) ER ChIP-PET clusters

	PET2	PET3	PET4	PET5	PET6	PET7	PET8+
Total clusters	5704	930	341	181	124	78	216
% with motifs	40.06%	57.31%	65.69%	70.72%	76.61%	78.21%	83.33%

	moPET2	moPET3	moPET4	moPET5	moPET6	moPET7+
Total clusters	6100	756	281	134	95	208
% with motifs	41.02%	61.90%	64.77%	76.12%	78.95%	85.10%

Table 3. Motif enrichments and ChIP-qPCR validation rate of clusters selected by adaptive thresholding

(A) p53 ChIP-PET clusters

	moPET2	moPET3	moPET4	moPET5	moPET6	moPET7+
Total clusters	0	140	69	30	26	35
% with motifs	N/A	67.14%	81.16%	70.00%	88.46%	88.57%
ChIP-qPCR success rate	N/A	100.0%	100.0%	100.0%	100.0%	100.0%

(B) Oct4 ChIP-PET clusters

	moPET2	moPET3	moPET4	moPET5	moPET6	moPET7+
Total clusters	0	524	717	189	93	146
% with motifs	N/A	36.83%	41.84%	54.50%	70.97%	43.15%
ChIP-qPCR success rate	N/A	16.7%	95.00%	100.0%	100.0%	100.0%

(C) ER ChIP-PET clusters

	moPET2	moPET3	moPET4	moPET5	moPET6	moPET7+
Total clusters	0	552	245	134	95	208
% with motifs	N/A	65.58%	68.57%	76.12%	78.95%	85.10%
ChIP-qPCR success rate	N/A	70.0%	83.3%	100.0%	100.0%	100.0%

Quality of Feature Selection Based on Microarray Gene Expression Data

Henryk Maciejewski

Institute of Computer Engineering, Control and Robotics,
Wroclaw University of Technology,
ul. Wybrzeze Wyspianskiego 27, 50-370 Wroclaw, Poland
Henryk.Maciejewski@pwr.wroc.pl

Abstract. This paper is devoted to the problem of feature selection for class prediction based on results of DNA microarray experiments. A method is presented to objectively compare quality of feature sets obtained using different gene-ranking methods. The quality of feature sets is expressed in terms of predictive performance of classification models built using these features. A comparative study is performed involving means comparison, fold difference and rank-test (Wilcoxon statistic) methods. The study shows that best performance can be obtained using the rank-test approach. It is also shown that the means comparison method can be significantly improved by also taking into account fold-change information. Performance of such mixed methods of feature selection can surpass performance of rank-test methods.

1   Introduction

Genome-wide expression profiling using DNA microarray or similar technologies has become an important tool for research communities in academia and industry. Microarray gene expression studies are designed to obtain insights into yet unknown gene functions/interactions, investigate gene-related disease mechanisms or observe relationships between gene profiles and some factors (such as some risk factor or response to some therapy). Microarray studies motivated development of specific data analysis methods, broadly categorized into *class discovery*, *class comparison* and *class prediction* [12]. Class discovery aims to discover groups of co-regulated genes across the samples tested, or, alternatively, to discover groups of samples similar in terms of their gene expression profile, thus discovering new disease taxonomies [2]. The purpose of *class comparison* studies is to identify genes with most different expression profiles across the classes compared. This identifies groups of genes whose expression is significantly related to the classes and which possibly account for the difference between the classes compared. The purpose of *class prediction* is to build a predictive model for determination of the class membership of samples based on their gene expression profiles. Application areas of this seem very promising not only in research, but in medical diagnosis or prediction of response to treatment. Although US Food and Drug Administration recently approved the first microarray chip to help

M. Bubak et al. (Eds.): ICCS 2008, Part III, LNCS 5103, pp. 140–147, 2008.

doctors administer patient dosages of drugs that are metabolized differently by cytochrome P450 enzyme variant, more wide-spread application of microarrays in clinical or regulatory applications requires that several issues related to accuracy and reproducibility of microarray results as well as analysis of microarray data are resolved.

One step on the way to bringing microarrays to clinical applications was recently made by the Microarray Quality Control (MAQC) Consortium [6], [11] who through a comprehensive experimental study showed that microarrays have grown robust enough to produce data that is reproducible and comparable across different microarray platforms and laboratories. As another conclusion from their research, MAQC recommends using fold-change ratio as a measure of gene ranking for the purpose of identification of differently expressed genes in microarray studies. This recommendation is drawn from the observation that this method of gene selection yields best reproducibility of results across different microarray platforms. MAQC also advices against using means-comparison methods, such as the t-test, for gene ranking. This has opened recent discussion related to validity of gene selection methods. E.g., Klebanov et al. [8] argues that fold-change cannot be regarded superior to the t-test, as it realizes smaller power, and in general, reproducibility of results should not be used as an indication about the adequacy of the gene selection method.

This discussion has motivated the study reported in this paper. Our work addresses another perspective (or criterion) for choosing the *right* method of gene selection. It is proposed to judge the quality of a gene selection method by looking at the information value of genes returned by the method and regarded as features for class prediction. In other words, a gene selection method will be deemed superior if it tends to produce features yielding best predictive performance of sample classifiers. In the following section, an approach is explained to arrive at the quality of features obtained using a given gene selection method. Next, this approach is used to compare three widely used gene selection methods (means-comparison, fold-change and a method based on the statistical rank test). The comparative study is based on the data originally published by Golub et al. [5] and Alon et al. [1]. Finally, it is shown that combining different feature selection methods (e.g., enhancing means-comparison method by also including the fold-change criterion) can result in increased performance of class prediction.

2   Quality of Features from Different Gene Selection Methods

Here we present an approach to express quality of a gene selection method in terms of predictive performance of a classifier using the genes regarded as features. In Sect. 2.1 we discuss the challenges that need to be overcome to build and properly validate performance of a sample classifier based on gene expressions. This should be seen as the motivation for the procedure detailed in Sect. 2.2 for judging the quality of gene selection.

2.1   Classification in High Dimensionality Data

Building a class prediction model based on microarray study data is a challenging task due to very high dimensionality of data obtained and relatively small number of samples available. Microarray studies typically produce data vectors with dimensionality $d \sim 10^3$ to 10^4 (the number of genes observed in one DNA chip), while the number of samples tested is at most $n \sim 10^2$. In other words, microarray studies define an ill-formulated problem of classification, where $d \gg n$, while standard approaches to predictive modeling require that $d \ll n$. This implies that significant dimensionality reduction is required. Another challenge related to the analysis of such data concerns proper estimation of expected predictive performance of the classifier for *new* data. Considering the small number of samples available for model building *and* testing requires that a properly tailored data-reuse approach is used. How these issues will be approached in the procedure in Sect. 2.2 is now developed.

The following notation will be used to represent results of a microarray experiment. Let (x_i, y_i), $i = 1, 2, ..., n$ denote data vectors related to the n samples tested in a microarray study, where $x_i \in R^d$ represents gene expressions from the sample i and $y_i \in C = \{c_1, c_2\}$ denotes the class membership associated with the sample i. For the problem of *class prediction* it is assumed the class membership y_i for each x_i is known prior to analysis. Only the binary classification case is considered here, where $y_i \in \{c_1, c_2\}$; this can be extended to the multi class problem by using ANOVA based metrics for gene ranking (such as the F-statistic).

The problem of class prediction is formally stated in the statistical decision theory [7] as looking for a prediction model $f : R^d \mapsto C$, minimizing the expected prediction error:

$$EPE = E\left[L\left(Y, f\left(X\right)\right)\right] \tag{1}$$

where the *loss function* L, used to penalize misclassification events can be defined as e.g.,

$$L\left(Y, f\left(X\right)\right) = \begin{cases} 1 & \text{for } Y \neq f(X) \\ 0 & \text{for } Y = f(X) \end{cases} . \tag{2}$$

Since in class prediction studies only samples (x_i, y_i), $i = 1, 2, ..., n$ of the random variables X and Y are known, *empirical risk* defined as $\frac{1}{n}\sum_{i=1}^{n} L\left(y_i, f\left(x_i\right)\right)$ is used to estimate the EPE (1). It should be noted that this must be computed based only on the data points *not used* for the purpose of building the model f, and not used in the stage on feature (gene) selection. Considering the small number of data points from a microarray study, EPE can be estimated by repeatedly training and testing the model for different data splits, where a subset of available data is used for feature selection and model building, leaving the remaining (smaller) part for estimation of EPE. This leads to a cross-validation estimate of EPE defined as [7]:

$$CV = \frac{1}{n}\sum_{i=1}^{n} L\left(y_i, f^{-i}\left(x_i\right)\right) \tag{3}$$

where f^{-i} is the classifier fitted to data with the sample x_i removed. This version of cross-validation realizes smaller bias (at the price of higher variance) as compared with the procedures leaving more samples for testing [7].

It should be noted that estimating EPE based on samples used for feature selection leads to overoptimistic estimates of predictive performance of classifiers, as pointed out in [14], and is not an uncommon error in literature.

Another issue concerning class prediction based on microarray data is related to setting the *right dimensionality* of the feature set. Here the well known fact should be considered [7] pertaining to binary classification that in d dimensions $d + 1$ points can be always perfectly separated by a simplest linear classifier. This implies that for microarray data $((x_i, y_i), i = 1, 2, \ldots, n$ where $d \gg n)$, one can always obtain perfect fit of the model to data, providing enough (i.e. $n - 1$) genes are selected. However, such models will not guarantee good predictive performance for new data, as they are prone to *overfitting* meaning small prediction error for training data with high prediction error for test (new) data [12]. This limits the number of genes that should be selected as features for class prediction to no more then the number of data points available in microarray data.

2.2   Quality of a Feature Selection Method

Considering the above mentioned challenges, the following procedure is proposed to arrive at the quality measures attributed to a given feature selection method. In the next chapter three specific feature selection methods are compared using this procedure.

1. Select the value d^* of dimensionality of the feature vector from the range $1..n - 2$.
2. Remove the sample (x_i, y_i) from the original data set (the remaining $n - 1$ samples will be referred to as the training data).
3. Using the training data select d^* genes ranked top by the gene selection procedure considered.
4. Reduce dimensionality of the vectors x by leaving only values of expression of the d^* genes selected (the vectors obtained will be denoted x').
5. Build a classification model denoted f^{-i} by fitting it to the $n - 1$ points x', obtained in the previous step.
6. Compute $e_i = L\left(y_i, f^{-i}(x_i')\right)$.
7. Repeat Steps 2 through 6 for $i = 1, 2, \ldots, n$.
8. Compute $CV_{d^*} = \frac{1}{n} \sum_{i=1}^{n} e_i$ (this estimates the EPE - see (3)).
9. Repeat Steps 1 through 8 for a grid of values d^* spaced evenly in the range $1..n - 2$, using approx. 10 values of d^*.
10. Plot the obtained relationship CV_{d^*} versus d^*.

It is proposed that the quality of a feature selection method be judged by observing the minimum values of CV_{d^*} obtained. Also, comparing the plots obtained in Step 10 for different feature selection methods gives an indication about the quality of features produced by competing methods over a range of different dimensionality models. This approach is used in the sample study shown in the following section.

3 Comparing Quality of Commonly Used Feature Selection Methods

Using the approach proposed in Sect. 2.2, the quality of three different methods commonly used for ranking genes is compared:

1. Gene ranking based on the Wilcoxon statistical rank-test,
2. Gene ranking based on the fold difference (used and recommended by Shi et al. [11]),
3. Gene ranking based on the signal to noise measure (which is an example of direct means comparison methods, such as the t-test).

Feature selection based on the first method requires that the Wilcoxon non-parametric group comparison test is performed independently for every gene. This gives a p-value indicating whether expression of this gene for the groups of samples of class c_1 and c_2 can be considered different. Gene selection using this method returns the set of top d^* (Step 4 in Sect. 2.2) genes ranked by increasing p-value of the test.

Feature selection using the fold difference measure requires that for every gene the ratio of mean expressions from samples of class c_1 and c_2 is computed. More specifically, if for a given gene, the mean value of gene expression from samples of class c_1 and c_2 is denoted μ_1 and μ_2, respectively, then the (log) fold difference measure is defined as:

$$fc = |\log(\mu_1) - \log(\mu_2)| \tag{4}$$

which produces high values if either of the means exceeds the other. Gene selection using this method returns the set of top genes ranked by decreasing value of fc.

Feature selection based on the signal to noise uses the measure defined as:

$$sn = \frac{|\mu_1 - \mu_2|}{\sigma_1 + \sigma_2} \tag{5}$$

where σ_1 and σ_2 are standard deviations of expressions of a fixed gene for the samples of class c_1 and c_2, respectively. Gene selection returns the set of top genes ranked by decreasing value of sn.

Fig. 1 compares the EPE vs. model dimensionality for these three gene selection methods (marked in the plot by 'w', 'sn' and 'fc'). As a classifier we used in this study the multilayer perceptron (MLP) model with one hidden layer. We observe that the minimum value of EPE (i.e., the best predictive performance expected for new, independent samples) is realized for the Wilcoxon method, with 15 genes selected. It can be also observed that for the wide range of different dimensionality models (up to 50 features), Wilcoxon feature selection yields significantly fewer prediction errors then signal to noise or fold difference.

Fig. 1 also includes EPE for a model built using a *combined* method of feature selection (marked in the plot as 'snfc'). This approach includes the following steps:

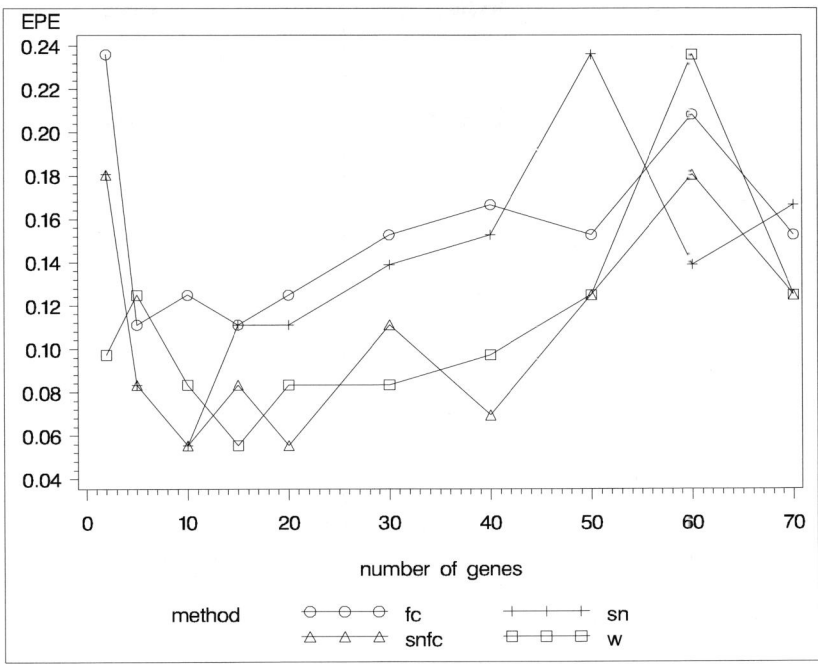

Fig. 1. Expected prediction error of neural network model for different methods of gene selection. (Notation: w=Wilcoxon test, fc=fold difference, sn=signal to noise, snfc=signal to noise with additional fold difference criterion).

1. For each gene, the fc and sn measures are computed according to (4) and (5).
2. Genes are ranked by decreasing values of sn.
3. The required number of top genes is returned, providing a gene realizes at least two-fold difference in expression (i.e., $fc \geq 1$, where in (4) we used the logarithm to the base 2).

The threshold of at least two-fold difference in expression was also used as the feature selection criterion in [11]. Interestingly, features returned from this combined model yield significantly better EPE then features from individual models ns and fc, with minimum values of EPE realized for 10-20 genes. This approach shows similar performance in terms of feature quality to the Wilcoxon method.

The same analysis repeated for a different classification model – logistic regression gives results depicted in Fig. 2.

Basically, the conclusions drawn from Fig. 1 regarding the quality of competing feature selection methods are confirmed: the Wilcoxon method realizes the best predictive performance (for 10-20 features), and the combined method ('snfc') leads to remarkable improvement in feature quality, making this method comparable with the Wilcoxon rank test.

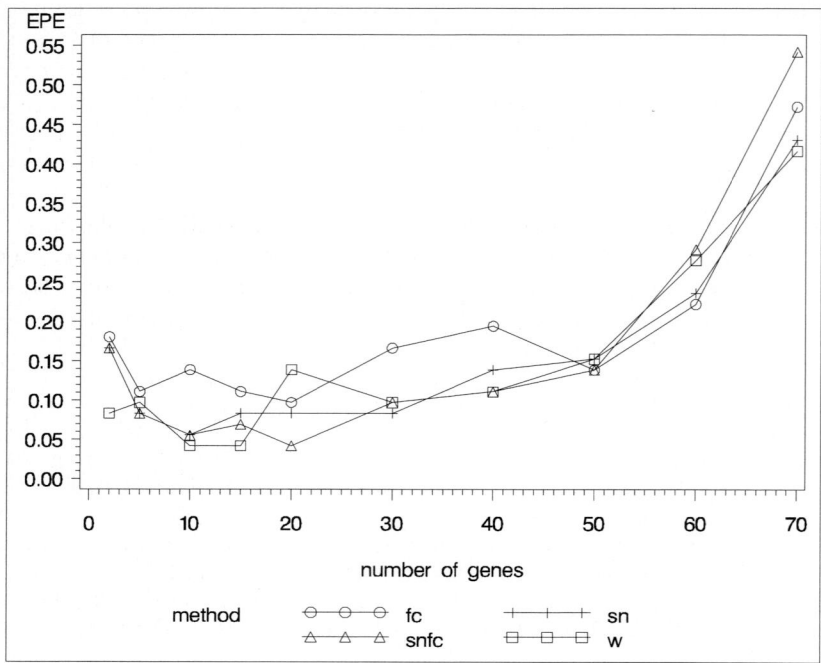

Fig. 2. Expected prediction error of logistic regression model for different methods of gene selection. (Notation: w=Wilcoxon test, fc=fold difference, sn=signal to noise, snfc=signal to noise with additional fold difference criterion).

Figs. 1 and 2 also illustrate model dimensionality related issues: too small a model dimensionality leads to poor prediction performance (due to the model being too simple), while too big a dimensionality leads to overfitting of the model. This compromise should be taken into consideration when setting the *right* dimensionality of a class prediction model.

Similar analysis repeated for the colon data set [1] basically confirms the conclusions drawn from the leukemia study. Again, the Wilcoxon method tends to produce the best predictive performance (the $EPE \approx 0.16$ for 20 genes, using the MLP classifier). Similar results were observed for 10 or 30 features obtained with the combined method.

4   Conclusions

We demonstrated that the quality of gene selection methods can be empirically compared by observing the performance of class prediction models built using features returned by these methods. To obtain a fair picture of the quality, such analysis should not be limited to one pre-fixed number of genes selected, it should rather be made for a representative collection of different dimensionality models, which allows to observe the size of feature vectors yielding good class prediction.

Using this approach, we showed that the Wilcoxon rank test is a superior gene selection method then fold-change or direct means comparison. However, significant improvement can be achieved if different gene selection criteria are used simultaneously. This suggests that feature selection for microarray class prediction probably should comprise information from several different criteria, thus increasing information contents of the feature set. This however requires further research.

Being able to quantitatively rank gene selection methods, as shown in this work, raises another interesting question to what extend the genes selected as best features for sample classification really account for the differences between classes. This open question requires further research in an interdisciplinary team.

References

1. Alon, U., et al.: Broad patterns of gene expression revealed by clustering analysis of tumor and normal colon tissues probed by oligonucleotide arrays. Proc. Natl. Acad. Sci. USA 96, 6745–6750 (1999)
2. Bittner, M., Meltzer, P., Chen, Y.: Molecular classification of cutaneous malignant melanoma by gene expression profiling. Nature 406, 536–540 (2000)
3. Dudoit, S., Shaffer, J., Boldrick, J.: Multiple Hypothesis Testing in Microarray Experiments. UC Berkeley Division of Biostatistics Working Paper Series, Paper 110 (2002)
4. Ewens, W., Grant, G.: Statistical Methods in Bioinformatics. Springer, New York (2001)
5. Golub, T., et al.: Molecular classification of cancer: Class discovery and class prediction by gene expression monitoring. Science 286, 531–537 (1999)
6. Guo, L., et al.: Rat toxicogenomic study reveals analytical consistency across microarray platforms. Nature Biotechnology 24, 1162–1169 (2006)
7. Hastie, T., Tibshirani, R., Friedman, J.: The Elements of Statistical Learning. Data Mining, Inference and Prediction. Springer, New York (2002)
8. Klebanov, L., et al.: Statistical methods and microarray data. Nature Biotechnology 25, 25–26 (2007)
9. Maciejewski, H.: Adaptive selection of feature set dimensionality for classification of DNA microarray samples. In: Computer recognition systems CORES, Springer Advances in Soft Computing, Springer, Heidelberg (2007)
10. Maciejewski, H., Konarski, L.: Building a predictive model from data in high dimensions with application to analysis of microarray experiments. In: DepCoS - RELCOMEX. IEEE Computer Society Press, Los Alamitos (2007)
11. MAQC Consortium [Shi L. et al.]: The MicroArray Quality Control (MAQC) project shows inter- and intraplatform reproducibility of gene expression measurements. Nature Biotechnology 24, 1151–1161 (2006)
12. Markowetz, F., Spang, R.: Molecular diagnosis. Classification, Model Selection and Performance Evaluation, Methods Inf. Med. 44, 438–443 (2005)
13. Polanski, A., Kimmel, M.: Bioinformatics. Springer, Heidelberg (2007)
14. Simon, R., et al.: Pitfalls in the Use of DNA Microarray Data for Diagnostic and Prognostic Classification. Journal of the National Cancer Institute 95, 14–18 (2003)

IMPRECO: A Tool for Improving the Prediction of Protein Complexes

Mario Cannataro, Pietro Hiram Guzzi, and Pierangelo Veltri

Bioinformatics Laboratory, Experimental Medicine Department,
University Magna Graecia, Catanzaro, Italy
{cannataro,hguzzi,veltri}@unicz.it

Abstract. Proteins interact among them and different interactions form a very huge number of possible combinations representable as protein to protein interaction (PPI) networks that are mapped into graph structures. The interest in analyzing PPI networks is related to the possibility of predicting PPI properties, starting from a set of known proteins interacting among each other. For example, predicting the configuration of a subset of nodes in a graph (representing a PPI network), allows to study the generation of protein complexes. Nevertheless, due to the huge number of possible configurations of protein interactions, automatic based computation tools are required. Available prediction tools are able to analyze and predict possible combinations of proteins in a PPI network which have biological meanings. Once obtained, the protein interactions are analyzed with respect to biological meanings representing quality measures. Nevertheless, such tools strictly depend on input configuration and require biological validation. In this paper we propose a new prediction tool based on integration of different prediction results obtained from available tools. The proposed integration approach has been implemented in an on line available tool, IMPRECO standing for IMproving PREdiction of COmplexes. IMPRECO has been tested on publicly available datasets, with satisfiable results.

1 Introduction

The interactions of proteins within a cell are very huge and frequent. They interact composing a very broad network of interactions, also known as *interactome*. If two or more proteins interact for a long time forming a stable association, their interaction is known as *protein complex*. Interactomics study focuses currently: (i) on the determination of all possible interactions and (ii) on the identification of a meaningful subset of interactions. Due to the high number of proteins within a cell, manual analysis of proteins interactions is unfeasible, so the need to investigate interactions with computational methods arises [1]. We focus on interactomics as the study of Protein-Protein Interaction (PPI) as biochemical reaction among proteins, as well as the study of protein complexes.

The most natural way to model PPIs network is by using graphs [2], where proteins are represented as nodes and interactions as edges linking them. The

M. Bubak et al. (Eds.): ICCS 2008, Part III, LNCS 5103, pp. 148–157, 2008.

simplest representation is an undirected graph in which the nodes are labelled with the protein identifiers, while the edges are simple connections (i.e. no labels or directions).

Once the PPI network has been represented as a graph, the biological investigation consists in studying the structural properties of the graph [3]. For example Fig. 1 reports a graph structure where nodes are proteins and edges represent all possible interactions among proteins (nodes). Subgraphs [4], i.e. a subset of nodes and edges, may represent biological relevant and meaningful proteins interactions. For instance, the circled set of nodes and edges connecting them in Fig. 1 is an example of protein interactions representing complexes.

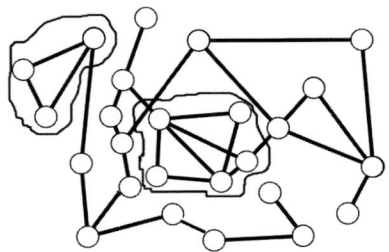

Fig. 1. A graph with dense regions

Consequently, a number of algorithms that predict complexes starting from graphs have been developed [1],[5],[4]. These algorithms, also called complexes prediction tools (or simply predictors), belong to the general class of graph clustering algorithms, where each cluster is defined as a set of nodes of the graph with their connections. Thus, clustering algorithms aim to identify subgraphs. Obviously, the quality of predictors is measured in terms of percentage of complexes biologically meaningful with respect to the meaningless ones. Clustering algorithms take as input a graph representing an interaction network among a set of proteins and an initial configuration (i.e. algorithms parameters such as the number of clusters). While initial configurations mostly depends on clustering algorithms, the initial interaction graph mostly depends on known protein interactions.

Thus, the prediction results are strongly influenced by: (i) the initial configuration of the algorithms and (ii) how valid are the initial protein to protein interactions (i.e., edges in the graph) of the input interaction network (i.e., the graph) [6]. They all apply clustering methodologies based on graph theory. None of them uses biological knowledge while running algorithms to guide the clusters identification or clusters selection.

The idea proposed in this paper (extension of a previous short communication [7]) is to combine different predictor results using an integration algorithm able to gather (partial) results from different predictors, to improve the biological relevance of the protein complexes associated to the output identified clusters. The integration algorithm starts by integrating results (i.e. clusters) obtained by running different available predictors. Three different cases are considered by evaluating the topological relations among clusters coming from the considered predictors: (i) equality: the same clusters are returned by all (or by a significant number of) predictors; (ii) containment: it is possible to identify a containment relation among (a set of) clusters returned by all (or by a significant number of) predictors; (iii) overlap: it is possible to identify an overlap relation among (a set of) clusters returned by all (or by a significant number of) predictors.

It is possible to tune the minimum number of predictors to consider in the three possible cases as well as the kind of investigated relation. The algorithm considers iteratively all the clusters, examines the conditions trying to find possible corresponding selected clusters.

The proposed algorithm works in three phases: i) it firstly parses results coming from different predictors, then (ii) tries to associate them in one of the three possible considered configurations and finally (iii) , it performs the integration phase among clusters. The latter phase is performed by selecting clusters from the set obtained during the second phase. All phases are integrated into an on line available tool called IMPRECO (for IMproving PREdiction of Complexes). IMPRECO manages different formats of results data obtained from different predictors, integrates them and then presents results with evaluation quality measurements. IMPRECO has been tested considering three existing predictors, (MCODE [4], RNSC [5] and MCL [8]), on publicly available datasets showing considerable improvements. Quality measures are obtained by validating the predicted clusters (representing complexes) with respect to experimentally determined protein complexes.

2   The Clustering Integration Algorithm

Let P be a protein interaction network modelled as a graph G, and let PA be a complexes predictor. PA gets G as input and produces a set of subgraphs $C = \{\xi_1...\xi_t\}$, representing clusters, where each cluster ξ_i may be interpreted as a putative protein complex. The proposed integration algorithm receives a set of clustering outputs $CO = \{C_1...C_n\}$ obtained from n different predictors, then it tries to verify the topological relations among clusters and builds three set of clusters one for each of the considered topological relations, i.e. equality, containment and overlap. Finally the three sets are used to build an integrated set of clusters IO, with the aim of improving the quality of results (i.e. in terms of biological meanings) merging the clusters of the three obtained sets. We consider the matching subgraphs problem using only nodes, thus, checking for the correspondence of two subgraphs ξ_1 and ξ_m is reduced to the problem of checking their nodes equality. We now show the three algorithms for equality (Exact Matching procedure), containments (Containment procedure) and overlap (Overlap procedure) relations among clusters.

Exact Matching procedure. Let TD be a dimension threshold value. The equality relation procedure considers clusters that have at most TD nodes. Given a cluster ξ of a clustering output that satisfies this property, the procedure searches for identical clusters in the clustering outputs obtained from the others predictors.

The algorithm stores the corresponding clusters in a list called *Matched*. If at least a minimum number of identical clusters *TM* is found, a representative one is included in a cluster list called *Verify_Equality*, i.e. the list of clusters that satisfies the equality relation. The procedure ends when iteratively all the clusters have been examined. The following pseudo code explains the so far described procedure.

Procedure. ExactMatching (CO, TD, TM)
// CO contains the set of input clusters
// TD is the threshold of dimension
// TM is the minimum number of required clusters
begin
  Matched: List;
  // the list of corresponding clusters
  Verify_Equality: List
  // the list of clusters that verify the equality condition;
    FOR ALL Clusters ξ, $\|\xi\| \leq$ TD,
    // Find the corresponding clusters
    Matched:= FindMatching(ξ, CO);
      IF($\|Matched\|$) $\geq TM$
      $Verify\_Equality$:= $Verify\_Equality + \xi$;
  return $Verify\_Equality$;
  end Exact Matching;

Containment procedure. The Containment procedure considers clusters with more than *TD* nodes. Let ξ a cluster with more than *TD* nodes. The procedure searches in other cluster result sets that ones including ξ. If at least *TM* clusters are found, then one of the found clusters is selected, respectively the smallest or the biggest in terms of nodes depending on an input parameter *IS*. Finally a *Verify_Containment* list of clusters is generated. The procedure ends when iteratively all the clusters have been examined. The following pseudo code explains the so far described procedure.

---

procedure. Containment (CO, TD, TM, IS)
// CO contains the set of selected clusters
// TD is the threshold of dimension
// TM is the minimum number of required clusters
// IS determines the selection of the biggest or smallest cluster
begin
  Matched: List;
  // the list of corresponding clusters
  Let $Verify\_Containment$: List;
  //list of clusters that verify the Containment condition;
    FOR EACH cluster ξ that has dimension higher than TD
    Matched= FindSub(ξ,CO);
      IF($\|Matched\|$) $\geq TM$

```
        IF IS = smallest
           Verify_Containment:= Verify_Containment+ ξ ;
        ELSE
           ξ̃ = max(ξ ∈ Matched);
           Verify_Containment:= Verify_Containment+ ξ̃ ;
   return Verify_Containment
   end Containment;
```

For instance, let us consider a scenario in which three clusters are depicted: (i) includes nodes A, B and C, (ii) includes nodes A, B, C and D, and (iii) includes nodes A, B, C, D and F. Let us suppose that the three clusters come from three different algorithms, that the containment procedure starts with (i), and that TM is set to 2. The nodes of cluster (i) are included in clusters (ii) and (iii), so if $Inserting\_Smallest$ is set to $Smallest$, cluster (i) will be included in $Verify\_Containment$, otherwise if it set to $Biggest$ cluster (iii) is included. $Verify\_Containment$ stores all the clusters that verify the relation. The procedure ends when iteratively all the clusters have been examined.

Overlap procedure. The Overlap procedure considers clusters with more than TD nodes. Let ξ a cluster with more than TD nodes. The procedure searches in other cluster result sets that ones that have an overlap with ξ bigger than a threshold PO. If at least TM clusters are found, then one of them is selected, respectively the smallest or the biggest in terms of nodes depending on an input parameter. Finally a $Verify\_Overlap$ list of clusters is generated. The procedure ends when iteratively all the clusters have been examined. The following pseudo code explains the so far described procedure.

```
procedure. Overlap (CO, TD, TM, II, PO)
// CO contains the set of selected clusters
// TD is the threshold of dimension
// TM is the minimum number of required clusters
// II determines the selection of a cluster
// PO determines the threshold of overlap
begin
   Let Matched : List;
   // The list of corresponding clusters
   Let Verify_Overlap: List;
   //list of clusters that verify the Overlap condition;
   FOR EACH cluster ξ that has dimension higher than TD
      Matched= FindOverlap (ξ,CO,PO);
         IF (‖Matched‖) ≥ TM
            IF II = smallest
               Verify_Overlap:= Verify_Overlap+ ξ ;
            ELSE
               ξ̃ = max(ξ ∈ Matched)
               Verify_Overlap:= Verify_Overlap+ ξ̃ ;
   return Verify_Overlap
   end Overlap;
```

Let us consider the three clusters ((i) A, B and C; (ii) A, B, D and K; and (iii) A, B, F, E and L) that come from three different algorithms and that the overlap procedure starts with (i), and that TM is 2 and PO is 10%. Cluster (i) has an intersection with both (ii) and (iii), and the overlap is higher than PO. Thus, if $Inserting\_Intersected$ is set to $Smallest$, cluster (i) will be included in IO, otherwise if it set to $Biggest$ cluster (iii) is included.

2.1   Integration Algorithm

The whole procedure of integration receives in input a set of clustering outputs CO, then it verifies the three relations by calling the procedures described so far that produce three lists of clusters: $Verify\_Equality$, $Verify\_Containment$ and $Verify\_Overlap$. Each cluster that has been inserted in one list verifies one of the previous relations. Finally, it merges these lists. The following pseudo code explains the so far described procedure.

```
procedure. Integration CO, TD, TM, II, II,PO)
// CO contains the set of selected clusters
// TD is the threshold of dimension
// TM is the minimum number of required clusters
// II and IS determine the selection of a cluster
// PO is the threshold of shared nodes
begin
      Let Verify_Equality: List;
      Let Verify_Containment: List;
    Let Verify_Overlap: List;
      FOR ALL clusters with dimensions lower than TD,
         Verify_Equality = ExactMatching(CO,TD,TM)
      FOR ALL clusters with dimensions bigger than TD,
         Verify_Containment = Containment(CO,IS,TD,TM);
      FOR ALL clusters with dimensions bigger than TD,
         Verify_Overlap = Overlap(CO,II,PO,TD,TM)
      IO = Integrate(Verify_Equality,Verify_Containment,Verify_Overlap);
   end Integration;
```

3   Architecture of IMPRECO

The introduced integration algorithm has been fully implemented in a prototype available through a GUI accessible via a web browser and Java Web Start Technology [1]. The IMPRECO architecture comprises the following components, as depicted in Fig. 2.

Data manager module. It collects the outputs of the different predictors and translates them into a a single formalism known to IMPRECC. Although the formalisms used by the existing algorithms are quite similar, but even small

---

[1] java.sun.com/products/javawebstart/

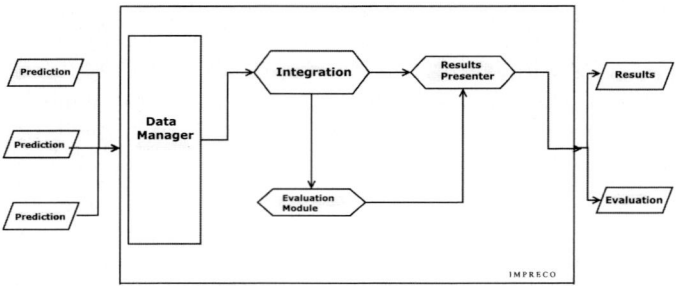

Fig. 2. Architecture of IMPRECO

differences in structure may cause problems in automatic comparison and integration methods. Thus, the data manager must translate such results. Currently, IMPRECO can read the native outputs of MCODE and MCL and can parse the RNSC output format. Users can also specify a clustering in a text file.

Integration Module. It implements the integration strategy. The current version of IMPRECO verifies the three relations in a sequential way. Initially, it builds the set of all clustering outputs starting from data parsed from the data manager. Then it tries to verify the equality relation. Then it finds those that verify the Containment relation, and it finally searches for those that satisfy the Overlap relation.

Evaluation Module. It evaluates the predictions with respect to a reference database, i.e. a catalog of verified complexes . Currently, user can use its own reference databases. Such module compares each predicted complex with those stored in the database and calculates three statistics: sensitivity, positive predictive value and accuracy as defined in [6].

Results presenter. It offers the results through a GUI as represented in Fig. 3. User can also save results in a file for successive processing.

4   Experimental Results

In order to estimate the quality of integration, we used IMPRECO to integrate some clusterings obtained running three existing algorithms over publicly available data. Protein Interaction data of yeast have been downloaded from the website [2]. The website contains data collected from the MIPS database belonging to the catalog of the yeast *Saccharomyces cerevisiae*. Authors collected data of annotated complexes, then they converted them into a graph where each node represents one protein and an edge was created between each pair of polypeptides involved in a common complex. They altered the resulting graph by randomly adding or removing edges. From those graphs we selected a network of 1094 nodes and 14658 edges and we gave it as input to three existing complexes predictors,

---

[2] http://rsat.scmbb.ulb.ac.be/sylvain/clustering_evaluation

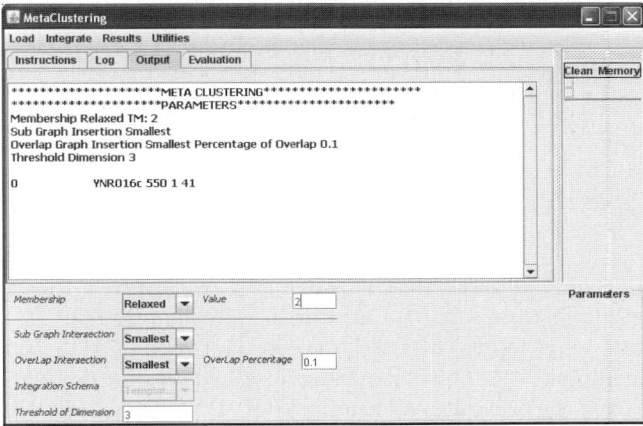

Fig. 3. The GUI of IMPRECO

MCODE, RNSC and MCL, to obtain the proposed clusters. IMPRECO integrated the results inserting in the integrated output those clusters that verified one relation in at least two outputs. For instance, let us consider Fig. 1 that depicts an example of clusters verifying the containment relation that are merged.

Table 1. Example of Considered Clusters

Algorithm	*Cluster*
MCODE-RNSC	YKL193C YDR028C YOR178C YER133W YMR311C
MCL	YKL193C YDR028C YOR178C YER133W YMR311C YER054C YDR028C YOR178C
IMPRECO	YKL193C YDR028C YOR178C YER133W YMR311C YER054C YDR028C YOR178C

We evaluated the performances of both the integrated clustering and the initial clustering. IMPRECO considered each cluster and compared it with the verified complexes. It used as reference the MIPS database catalog [9]. For each complex, it took into account of the matched elements, i.e. the components of a cluster that are presented in the complex. Thus it calculated three statistics as defined in [6]: sensitivity, positive predictive value (PPV) and accuracy(the geometric average of the previous two measures). The first measure represents the ability to detect a complex. The second one estimates the ability to correctly predict a complex. The integrated clustering predicted by IMPRECO resulted in fewer clusters (155) than MCL (165) or RNSC (306) and more than MCODE (73), as shown in Table 2. The integrated set outperforms the other three algorithms in terms of sensitivity (0.89). Conversely, the value of PPV obtained

Table 2. Comparison of IMPRECO vs existing predictions

$Quality Measures$	$MCODE$	MCL	$RNSC$	$IMPRECO$
Number of Clusters	73	165	306	155
Sensitivity	0.34	0.47	0.46	0.89
PPV	0.48	0.69	0.59	0.59
Accuracy	0.40	0.57	0.52	0.70

(0.59) is lower than the best value for MCL (0.70). However, the final accuracy of the integrated clustering (0.70) outperforms all the others.

In our second experiment, we used a second network of with 1034 nodes and 12235 edges available at [3]. This network is obtained by randomly adding and removing edges from the previous. We ran the MCL, RNSC and MCODE algorithms, obtaining respectively 43, 115 and 267 clusters, as summarized in Table 3.

Table 3. Results of experiment 2

$Parameter--Algorithm$	$MCODE$	MCL	$RNSC$	$A1$	$A2$	$A3$	$A4$
$No. of Clusters$	43	115	267	92	92	92	92
$Sensitivity$	0.141	0.488	0.462	0.556	0.207	0.552	0.271
PPV	0.650	0.649	0.594	0.619	0.616	0.584	0.585
$Accuracy$	0.303	0.572	0.524	0.587	0.357	0.567	0.398

The intent of this second experiment is to assess the variation in integration performance when the parameters IS and II are changed. They determine respectively the insertion of the biggest or smallest matching clusters found during the execution of steps 2 and 3. Consequently, these variations do not influence the number of inserted clusters but only their internal structure. To appreciate the impact, we performed four experiments considering the four possible configurations for IS and II: (i) biggest/biggest, identified as $A1$, (ii) smallest/smallest, identified as $A2$, (iii) smallest/biggest, identified as $A3$, and (iv) biggest/smallest, identified as $A4$. Considering the results shown in Table 3, it is evident that configuration $A1$ gives the best results for sensitivity and PPV (0.556 and 0.619) and for accuracy. In contrast, the configuration ($A2$), in which both parameters are set to *Smallest*, gives the worst results. However, we have noticed that this is not a general rule, but must be verified for each dataset.

5   Conclusion

Starting from a protein interaction network, protein complexes can be predicted by the use of clustering algorithms. The combination of different algorithms is a

---

[3] http://rsat.scmbb.ulb.ac.be/sylvain/clustering_evaluation

possible way to improve the prediction performances. This paper addressed this problem proposing a possible strategy of integration. This approach is integrated in an on line available tool: IMPRECO [4]. First experimental results show an improvement with respect to existing predictors. We plan to develop a parallel version of IMPRECO that will execute the data translation and the integration phases as well as the evaluation of results in a parallel way by using a grid infrastructure.

References

1. Sharan, R., Ideker, T., Kelley, B., Shamir, R., Karp, R.: Identification of protein complexes by comparative analysis of yeast and bacterial protein interaction data. J. Comput. Biol. 12(6), 835–846 (2005)
2. Fell, D., Wagner, A.: The small world of metabolism. Nat. Biotechnol. 18(11), 1121–1122 (2000)
3. Lesne, A.: Complex networks: from graph theory to biology. Letters in Mathematical Physics 78(3), 235–262 (2006)
4. Bader, G., Hogue, C.: An automated method for finding molecular complexes in large protein interaction networks. BMC Bioinformatics 4(1), 2 (2003)
5. King, A.D., Przulj, N., Jurisica, I.: Protein complex prediction via cost-based clustering. Bioinformatics 20(17), 3013–3020 (2004)
6. Brohe, S., van Helden, J.: Evaluation of clustering algorithms for protein-protein interaction networks. BMC Bioinformatics 7, 488 (2006)
7. Cannataro, M., Guzzi, P.H., Veltri, P.: A framework for the prediction of protein complexes. In: Abstract in Proceedings of the Bioinformatics Italian Society Conference (BITS 2007) (2007)
8. Enright, A.J., Van Dongen, S., Ouzounis, C.: An efficient algorithm for large-scale detection of protein families. Nucleic Acids Research 30(7), 1575–1584 (2002)
9. Mewes, H.W., Frishman, D., Mayer, K., Mnsterktter, M., Noubibou, O., Pagel, P., Rattei, T., Oesterheld, M.A.R., Stmpflen, V.: Mips: analysis and annotation of proteins from whole genomes in 2005. Nucleic Acids Res. 34(Database issue), D169–D172 (2006)

---

[4] http://bioingegneria.unicz.it/~guzzi

CartoonPlus: A New Scaling Algorithm for Genomics Data

Joanna Jakubowska[1], Ela Hunt[2], and Matthew Chalmers[1]

[1] Department of Computing Science, University of Glasgow, UK
[2] Department of Computer Science, ETH Zurich, Switzerland
asia@dcs.gla.ac.uk, hunt@inf.ethz.ch, matthew@dcs.gla.ac.uk

Abstract. We focus on visualisation techniques used in genome browsers and report on a new technique, CartoonPlus, which improves the visual representation of data. We describe our use of smooth zooming and panning, and a new scaling algorithm and *focus on* options. CartoonPlus allows the users to see data not in original size but scaled, depending on the data type which is interactively chosen by the users. In VisGenome we have chosen genes as the basis for scaling. All genes have the same size and all other data is scaled in relationship to genes. Additionally, objects which are smaller than genes, such as micro array probes or markers, are scaled differently to reflect their partitioning into two categories: objects in a gene region and objects positioned between genes. This results in a significant legibility improvement and should enhance the understanding of genome maps.

Keywords: Genome Visualisation, Visualisation Techniques, Scaling Algorithm, Large Data Sets.

1 Introduction

Medical researchers find it difficult to locate the correct biological information in the large amount of biological data and put it in the right context. Visualisation techniques are of great help to them, as they support data understanding and analysis. We reported our findings from a survey of visualisation techniques used in genome browsers in [8]. We developed a prototype of a new genome browser, VisGenome, which uses the available techniques. VisGenome [9] was designed in cooperation with medical researchers from a hospital. We found that the majority of genome browsers show only a selection of data for one chromosome. This is obvious, because the amount of available information is so large that it is impossible to show all data in one view. Expressionview [4], for example, shows QTLs [1] and micro array probes and no other data. Some of the tools, such as Ensembl [6], show many types of data but use a number of different data views, which make the users disoriented and lost in the tool and data

---

[1] A quantitative trait locus (QTL) is a part of a chromosome which is correlated with a physical characteristic, such as height or disease. Micro array probes are used to test gene activity (expression).

M. Bubak et al. (Eds.): ICCS 2008, Part III, LNCS 5103, pp. 158–167, 2008.

space. Moreover, Ensembl shows as much information as it is possible in one view, instead of offering a view or a panel with additional information. A large number of genome browsers show only a chromosome and do not allow one to see a comparison of two chromosomes from different species. Exceptions include SyntenyVista [7] and Cinteny [15] which show a comparative view of two genomes but are limited with regard to other data, such as micro array probes. On the other hand, SynView [17] visualises multi-species comparative genome data at a higher level of abstraction.

We aim to find a solution which clearly presents all the available information, including all relevant information the biologists wish to see. We aim to find a solution for data analysis which overcomes both representational and cognitive problems.

Here, we describe single and comparative genome representations, see Figure 1 and 2. A single representation is a view which shows data for one chromosome. A comparative representation illustrates relationships between two or more chromosomes.

Our contribution is a scaling algorithm which we call CartoonPlus. Cartoon-Plus allows the users to see data more clearly by choosing one kind of data as basis and scaling other data types in relationship to the basis. The solution does not show data in its natural size but allows one to see relationships between different kinds of data more clearly, especially in a comparative representation.

The paper is organised as follows. Section 2 provides the background about visualisation techniques and their usefulness for medical researchers. Section 3 introduces the visualisation techniques we used in VisGenome and provides details of our new algorithm. We discuss our work in Section 4 and the last section concludes.

2   Related Work

This section examines existing visualisation techniques used in genomics data representation and clarifies why a new scaling algorithm is necessary.

A variety of scientific visualisation techniques are available and could be used for genomics. 2D techniques are very common in gene data visualisation and 3D techniques are rarely used [8]. An exception is [13] which uses a 3D model of the data. In the following we discuss the techniques used in 2D applications.

Fisheye [5] shows detail for an element and its neighborhood, but only an overview for the other elements. It is used in a number of graphical applications, for example for photo corrections, but it is hardly used in biology, with the exception of Wu [12] who used fisheye to show tables representing micro array results. Magic lenses [16] allow the user to transform the data and display extra information, see Zomit [14]. The majority of genome browsers offer scrolling and zooming [1] which are both easy to use. Zooming by buttons is well known and used by the medical researchers. Ensembl [6] uses this kind of zooming. BugView [11] also uses zooming by buttons which makes an impression of smooth zooming. Cartoon scaling is applied to biological data in [7]. The technique

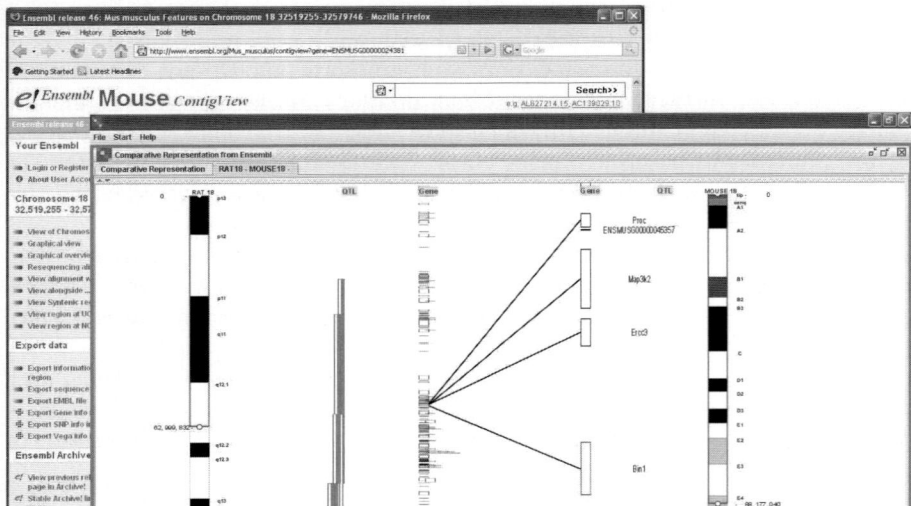

Fig. 1. The comparative representation for the rat chromosome 18 and the mouse chromosome 18. The gene Bin1 in the mouse chromosome 18 is in focus. The background shows additional information from Ensembl for Bin1, activated by clicking on the gene.

deforms the original data and makes it easier to read. SyntenyVista shows all genes in the same size and this makes it clear which genes share a homology link. A true physical representation of genes causes some of them to overlap and the users often cannot precisely see the genes connected by a homology link. This work motivated us to design an improved algorithm for scaling for different kinds of data, and not only for genes. Our new algorithm, CartoonPlus, makes the display of biological data clearer in both single and comparative representations. It makes it easy to see which genes and QTLs share a homology link in a comparative representation and highlights differences and dependencies between different kinds of data in a single representation. Objects that are larger than a basis object form one category. Another category consists of objects smaller than the basis or lying in between basis objects. Those objects contained within a basis object are treated differently than the objects in between.

3   Visualisation Extensions

VisGenome loads QTLs, genes, micro array probes, bands, and markers, and pairs of homologies from Ensembl. It shows single chromosomes or comparisons of two chromosomes from different species. The application uses the visualisation metaphors and algorithms offered by Piccolo [2]. Piccolo puts all zooming and panning functionality and about 140 public methods into one base object class, called PNode. Every node can have a visual characteristic, which makes the overall number of objects smaller than in other techniques which require two objects, an object and an additional object having a visual representation, as in Jazz [2]. A Jazz node has no visual appearance on the screen, and it needs a special object (visual component), which is attached to a certain node in a scene graph and which defines geometry and color attributes. Piccolo supports the same core feature set as Jazz (except for embedded Swing widgets), but it primarily uses compile-time inheritance to extend functionality and Jazz uses run-time composition instead. Piccolo supports hierarchies, transforms, layers, zooming, internal cameras, and region management which automatically redraws the portion of the screen that corresponds to objects that have changed.

In the continuation of the section, we present a new scaling algorithm, CartoonPlus, and then we outline other known visualisation techniques which we implemented.

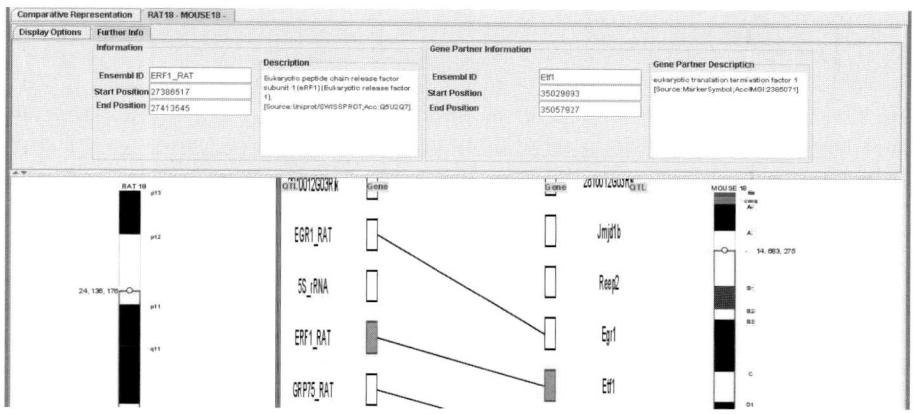

Fig. 2. The comparative representation for the rat chromosome 18 and the mouse chromosome 18. The data is scaled by the scaling algorithm which makes all genes the same size and QTL size depends on genes. Genes ERF1_RAT and Etf1 are linked by a homology line and marked in blue.

Scaling Algorithm. We developed a scaling algorithm for arbitrary genomics data which extends existing solutions. SyntenyVista [7] scaled genes only in a comparative representation. We offer scaling for all data, in both single and comparative representations, see Figure 2 and 3. Previous algorithms were constrained, while the new one scales multiple data types together, with reference to the basis. A user chooses the basis for scaling and then other elements are scaled in relationship to the chosen data type. In the current prototype we chose genes as a basis, so we scale all genes to the same size. An extension of this work is to allow the user to change the basis for scaling interactively. The algorithm looks at other types of data which are smaller or larger than genes, such as markers, micro array probes, or QTLs, and scales them accordingly. We divide all elements smaller than genes into two groups: elements which are in a gene

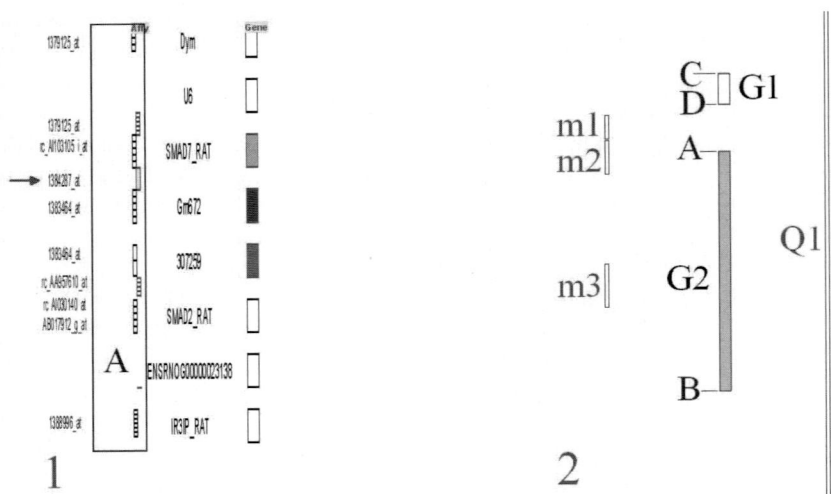

Fig. 3. 1: Single representation for the rat chromosome 18. Three genes (SMAD7_RAT, Gm672, 307259) and one micro array probe set (1384287_at, see arrow) are coloured by different colours, selected by the user interactively. 2: CartoonPlus algorithm (see Figure 4). G2 is a gene which begins at A and ends at B. m1, m2, and m3 are elements smaller than G2 and Q1 is bigger than G2.

region and elements which are in the region between two genes, see Figure 3.1.A. For each type of data holding items smaller than the basis, we create a column holding elements which are situated within the gene boundaries and a second column containing elements which are situated between two genes. For all elements which are in the gene region, we choose the same size for each element, and the same applies to all elements which are in the area between genes. The size of the elements depends on their number in a gene region. This means that if in an area of a gene there is only one marker, it has the same height as the gene, but if there are 10 markers, they together have the same size as the gene (each marker is set to 1/10th of gene height). When an element is on a gene boundary, it is partially in a gene region and partially between two genes, and we situated it in the gene region. We also scale elements like QTLs which are bigger than genes. We look where a QTL begins and ends and we paint it starting at the gene where it begins and ending at the gene where it finishes. The solution allows us to present clearly a homology between genes in a comparative representation and additionally to show relations between micro array probes, markers, genes, and QTLs in two species.

Figure 4 outlines the scaling algorithm. All genes, markers, micro array probes and QTLs are stored in hashtables. The algorithm iterates over all genes (line 2). First we scale markers and micro array probes which are between genes (the previous gene and the current one), see Figure 3.2 object m1 between G1

```
 1 CartoonPlus() {
 2    for(gene in GENEs) {
 3        ResizeAndPaint(gene)
 4        ScaledMarkersBetween = GET_MARKERS_BETWEEN()
 5        for (each marker in ScaledMarkersBetween)
 6            ResizeAndPaint(marker)
 7        ScaledMicroArrayProbesBetween = GET_MICRO_ARRAY_PROBES_BETWEEN
 8        for (each micro_array_probe in ScaledMicorArrayProbesBetween)
 9            ResizeAndPaint(micro_array_probe)
10        ScaledMarkers = GET_MARKERS_IN()
11        for (each marker in ScaledMarkers)
12            ResizeAndPaint(marker)
13        ScaledMicroArrayProbes = GET_MICRO_ARRAY_PROBES_IN()
14        for (each micro_array_probe in ScaledMicroArrayProbes)
15            ResizeAndPaint(micro_array_probe)
16        ScaledQTLs = GET_QTLS_FOR_GENE()
17        for(each QTL from ScaledQTLs)
18            if (QTL.end>D AND QTL.end<=B)
19                ResizeAndPaint(QTL)
20                delete(QTL from ScaledQTLs)
21    }
22 }
23 GET_MARKERS_BETWEEN() {
24    for(marker in MARKERs)
25        if(marker.start>=D AND marker.end<=A)
26            markers.add(marker)
27    return(markers)
28 }
29 GET_MARKERS_IN(){
30    for(marker in MARKERs)
31        if((marker.start<=A AND marker.end>A) OR (marker.start>A))
32            markers.add(marker)
33    return(markers)
34 }
35 GET_QTLS_FOR_GENE(){
36    for(QTL in QTLs)
37        if(QTL.start>D AND QTL.start<=B)
38            QTLs.add(QTL)
39    return(QTLs)
40 }
```

Fig. 4. CartoonPlus algorithm. Hierarchy of object sizes: chromosome $\geq$ QTL$\geq$ gene $\geq$ marker and micro array probe.

and G2. Then we scale markers and micro array probes with a start coordinate before the gene and end coordinate inside the gene or start coordinate inside the gene region, see Figure 3.2 objects m2 and m3, and Figure 4 lines 4-15. Then we place QTLs which begin inside the gene region or in the region between a previous gene and the current gene, see Figure 3.2 object Q1. For each gene we check as well where the end coordinate of a QTL is, and, depending on this, we paint the element. In the pseudo-code we used function ResizeAndPaint which for basis data gives all elements the same size. For small objects, such as m1, m2, or m3, function ResizeAndPaint calculates how many elements are in the gene area or in the area between genes, and divides the area by the number of elements and then the elements are painted in the calculated size. For large elements, ResizeAndPaint calculates the hight of the elements as the beginning of the gene where the QTL starts and end of the gene where it ends. If a QTL begins or ends between genes, the function takes the end of the previous gene or start of the next gene as its coordinates.

Navigation: We offer "overview and detail" views which are manipulated by mouse and keyboard interaction. At the beginning the users see an overview of all chromosomes and can choose the one they would like to see in detailed view. When they see all data for a selected chromosome, the tool gives them the possibility to see an overview of all data, but also details for each part of the data. The users can mark a region which is interesting for them and interact only with the selected part. To make the view clear, instead of presenting all information in one view, we use an info panel which shows additional information for the selected elements on mouse-over (Figure 2).

Marking a Region of Interest: The users can choose a chromosome region of interest (via tab 'Display Options', visible in Figure 2), and manipulate the view only inside the region. This functionality, which marks the region on the chromosome with a red box, is offered by both single and comparative representations. The red box can be moved along the chromosome and its boundaries can be adjusted. The main view shows only the data for the marked region and the users manipulate the data in the selected area. This means that when the user zooms or pans in the main view, all or some of the data from the red square is available. Data outside the coordinates marked by the square is not shown. We found the functionality useful, especially for the users who work with a particular part of a chromosome.

Zooming and Panning: We offer smooth zooming which supports the visual exploration of the chromosome space, based on Piccolo [2]. This provides efficient repainting of the screen, bounds management, event handling and dispatch, picking, animation, layout, and other features. The zooming technique allows the users to keep an area of interest in focus during interaction with the data. Zooming is manipulated by the right mouse button by moving it to the right (zoom in) or to the left (zoom out). Panning uses the left mouse button. Both interactions are easy to use and the users quickly become familiar with them, as confirmed by our study [10].

Focus On: Focus on (Figure 1) makes the focal element large enough, so that its name can be read, moves it to the center of the view, and marks its boundaries in red, which allows the user to see a small part of a viewing history until he changes the region of interest. This means that the user can see which elements he focused on during the session. In a single representation, when the user focuses on an element, all neighbouring elements in the view become proportionally larger in all columns. In a comparative representation, only elements in the chromosome containing the chosen element are changed, and all elements on the other chromosome maintain the same size. This allows the users to see an overview of elements from one chromosome and details for the selected element in the second chromosome. If the user wants all elements in the two columns to be of the same size, he chooses focus elements in both. Then we set the size of all elements to be the same.

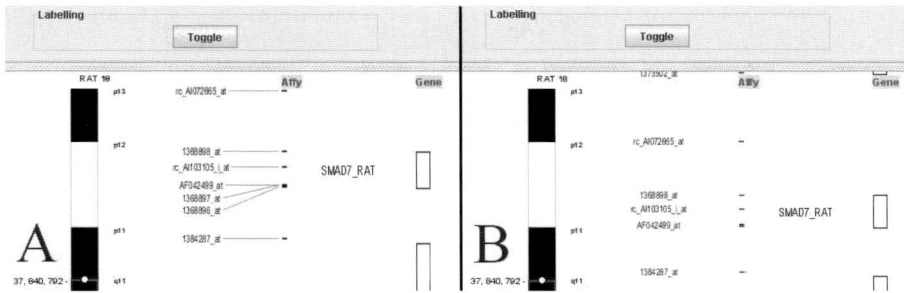

Fig. 5. The single representation for rat chromosome 18. **A** shows all labels and links which connect labels with the elements. **B** shows only a selection of labels shown next to the objects they describe.

Labelling: Because of a large amount of data, there is a problem with labels especially for elements that have the same location. To solve the problem we allow the users to switch between viewing all labels or only a selection. When all labels are visible, they are connected by blue links to the visible elements. When the user moves the mouse close to the element, a link becomes highlighted, which allows the user to localize the element description faster, see Figure 5 A. In selected label view, Figure 5 B, we display only a small subset of labels. If there is enough room, the element name is displayed. For elements with the same coordinates, it is the first element in alphabetic order. We show as next the label for the next element which has enough room for the label.

Additional Information: Many genome browsers place all data into one view, which makes the data difficult to read. We display additional information in an info panel, see Figure 2. In a comparative representation, we show two types of information. We display Ensembl id, coordinates and a description for each element which is pointed to by a mouse. In a comparative representation when the user points to an element from one chromosome which has a homology with an element from the other chromosome, the additional information is displayed for both genes, see Figure 2. Display Options Tab allows the users some data manipulation, like choosing the range of the chromosome region displayed, changing between view with scaled data and unscaled data, or between views with all labels and selected labels. In our solution we do not have to display all information in the main view and this improves usability.

Colours: We use black and white for most data, however, after marking a region of the chromosome, the user can choose color for each of the elements by clicking on the object while pressing Alt. The default colour choice view is displayed and the user can change the colour of the marked element, see Figure 3.1. Additionally, the object boundaries are marked in red during *focus on* and all bands in the chromosomes are coloured by standard colours.

Supporting Data: Ensembl [6] offers data collected from publications and experiments. To help the user contextualise the data, we provide access to

Ensembl by clicking on a feature of interest, which invokes Ensembl in the browser (Figure 1).

Homologies: To support comparative genome analysis, we show chromosomes which have homologies with other chromosomes. Our solution allows the users to identify all homologous chromosomes quickly. When a user looks at all chromosomes in a number of species, and clicks on one, all the homologous chromosomes in other species are highlighted, and facilitate the choice of homology for visual analysis (not shown).

4   Discussion

We examined the visualisation techniques used in genome browsers, and recognised that a number of tools used in biological research implement well known visualisation techniques, but only a few experiment with new techniques. CartoonPlus adds a novel extension to the array of available solutions. It can be used in single and comparative representations. In a single representation, the users can see all data scaled, depending on a chosen basis, which allows them to see clearly which micro array probes and markers are related to a gene. In a comparative representation, the scaling makes homologies between genes clearer.

Among all genome browsers we studied, only SyntenyVista [7] uses a scaling algorithm, however it was used only in a comparative representation and only for genes. The solution we used is novel and it could be useful not only in genomic data but also in different fields of biology and medicine which use one linear scale for many types of objects. We are testing the new technique in an experiment with biological researchers who now use a combination of data from Ensembl and their own lab experiments. We conducted a *user study*, to identify future improvements and assess the usability of our solution and saw that biologists found it useful, especially for scaling small objects (SNPs) [10]. We will next offer the users interactive choice of the basis for the scaling. We want to improve colouring and give the users the option to add colour to a region and not only to a single element.

5   Conclusions

We designed and implemented a new scaling algorithm and combined it with some known visualisation techniques. Our new technique presents the data more clearly, especially in a comparative representation where the users want to see homologies. We believe our visualisation extension improves on the existing tools which try to present as much data as possible or only a predefined subset of data. The combination of scaling, labelling and focus techniques we offer is likely to support an improved understanding of data relationships, as required in biomedical research. In the long term we see significant potential for user control over exactly how and where scaling is done, as in magic lenses [3], although we emphasise the need for a user study to validate this.

References

1. Bederson, B.B., et al.: Pad++: A Zooming Graphical Interface for Exploring Alternate Interface Physics. In: UIST 1994, pp. 17–26. ACM, New York (1994)
2. Bederson, B.B., et al.: Toolkit Design for Interactive Structured Graphics. IEEE Trans. Soft. Eng. 30(8), 535–546 (2004)
3. Bier, E.A., et al.: Toolglass and magic lenses: the see-through interface. SIGGRAPH, 73–80 (1993)
4. Fischer, G., et al.: Expressionview: visualization of quantitative trait loci and gene-expression data in Ensembl. Genome Biol. 4, R477 (2003)
5. Furnas, G.W.: Generalized Fisheye Views. In: CHI, pp. 16–23 (1986)
6. Hubbard, T.J.P., et al.: Ensembl 2007. Nucleic Acids Res. 35(Database issue), D610–D617 (2007)
7. Hunt, E., et al.: The Visual Language of Synteny. OMICS 8(4), 289–305 (2004)
8. Jakubowska, J., et al.: Granularity of genomics data in genome visualisation. TR-2006-221 (2006),
   http://www.dcs.gla.ac.uk/publications/PAPERS/8212/
   AsiaElaMatthewCHI06.pdf
9. Jakubowska, J., et al.: VisGenome: visualisation of single and comparative genome representations. Bioinformatics 23(19), 2641–2642 (2007)
10. Jakubowska, J., et al.: Mixed Paradigm User Study (in preparation)
11. Leader, D.P.: BugView: a browser for comparing genomes. Bioinformatics 20, 129–130 (2004)
12. Min, W., et al.: A fisheye viewer for microarray-based gene expression data. BMC Bioinformatics 7, 452 (2006)
13. Montgomery, S.B., et al.: Sockeye: A 3D Environment for Comparative Genomics. Genome Research 14(5), 956–962 (2004)
14. Pook, S., Vaysseix, G., Barillot, E.: Zomit: biological data visualization and browsing. Bioinformatics 14(9), 807–814 (1998)
15. Sinha, A.U., Meller, J.: Cinteny: flexible analysis and visualization of synteny and genome rearrangement in multiple organisms. BMC Bioinformatics 8, 82 (2007)
16. Stone, M.C., et al.: The movable filter as a user interface tool. In: HCI 1994 Human Factors in Computing Systems, pp. 306–312. ACM Press, New York (1994)
17. Wang, H., et al.: SynView: a GBrowse-compatible approach to visualizing comparative genome data. Bioinformatics 22(18), 2308–2309 (2006)

Automatic Segmentation of Cardiac MRI Using Snakes and Genetic Algorithms

Gustavo Miranda Teixeira, Igor Ramalho Pommeranzembaum,
Bernardo Lino de Oliveira, Marcelo Lobosco, and Rodrigo Weber dos Santos

Departamento de Ciência da Computação, Universidade Federal de Juiz de Fora (UFJF),
Juiz de Fora, Minas Gerais, Brasil
magusbr@gmail.com, tosobaum@yahoo.com.br, belino@acessa.com,
{marcelo.lobosco,rodrigo.weber}@ufjf.edu.br

Abstract. In this work we study and implement techniques for the automatic segmentation of cardiac Magnetic Resonance Images. The methods are based on the active contours algorithm called Snakes, which are adapted and tailored to the specific task of automatic segmentation of the left ventricle of the heart in Magnetic Resonance Images. We propose a new external force to improve the convergence of the Snakes method. In addition, a genetic algorithm is used to find the best set of configuration parameters for the Snakes method. The algorithms are implemented in Java and threads are used to explore data parallelism on shared-memory machines. Tests are performed on 150 short-axis images acquired from two healthy volunteers. Preliminary results suggest the proposed methods are promising and with further development and validation may be used, for instance, for the automatic calculation of cardiac ejection fraction.

Keywords: image processing, snakes, genetic algorithms, cardiac MRI.

1 Introduction

The segmentation of the heart is a common task to be performed in cardiac exams of Magnetic Resonance Imaging (MRI). In one typical exam many images are obtained from different positions of the heart and at different phases of contraction (from systole to diastole). The segmentation is then performed off-line on these images to extract important clinical parameters and information that characterize the function and the anatomy of the heart. For instance, in order to calculate the cardiac ejection fraction or cardiac ejection curves, important parameters that characterize cardiac contraction [1], the medical specialist may need to segment near one hundred of two-dimensional images for a single patient. The cardiac ejection fraction is the relation of the blood cavity volume of the left ventricle during diastole per the volume during systole. For estimating each volume one segments the endocardium in different short-axis images, or slices, of the ventricle (around 10 slices from apex to base) and calculates the areas of the blood cavity in each of these images. The majority of today's commercial software provides segmentation in a semi-automatic way. Therefore, during the segmentation of the cardiac endocardial surface the specialist is

M. Bubak et al. (Eds.): ICCS 2008, Part III, LNCS 5103, pp. 168–177, 2008.

forced to pick around six points in the border between cardiac tissue and the blood cavity of each short-axis image.

The segmentation of the cardiac MRI images is of extreme importance but it is, today, a tedious and an error-prone task. In this work, we investigate the automatic segmentation of the endocardium in cardiac MRI images using the active contour technique named Snakes[2] method. This method is based on a parameterized closed curve. Internal and external forces change the shape of the curve adapting it towards the structure we seek. This curve should be deformed until it minimizes an energy functional determined by the information extracted from the image and by restrictions imposed to the curve, like elasticity and rigidity. In order to achieve better results we have adapted and tailored the Snakes method to the specific task of automatic segmentation of the endocardial surface of the left ventricle of the heart in MRI images. We propose and implement a new external force that further restricts the deformation and overcomes artifacts, such as the papillary muscles. The new method provided better results than those of the traditional Snakes method.

Nevertheless, the results of both implemented methods have shown to be very dependent on their configuration parameters. In this work, we use a Genetic Algorithm (GA) to find the best set of configuration parameters for the automatic segmentation method. The GA is used to solve a minimization problem, where the objective function captures the distance between 25 manually segmented short-axis images and the respective automatic segmented ones. We evaluated the results of the automatic algorithm using 150 manually segmented curves generated by a medical specialist using the best set of configuration parameters found by the GA.

As mentioned before, a single patient exam may consist of over a hundred of cardiac segmentations. Therefore, the computer implementations here described explore the embarrassingly data parallelism of the algorithms using Java threads on a shared-memory machine.

2 Snakes

In this model we consider a Snake as a parameterized curve such as:

$$v(s) = [x(s), y(s)] \; . \tag{1}$$

where x and y are functions which determine the coordinates of the function v, v is the position of the snake and s varies from 0 to 1.

The main idea of the method lies in the minimization of an energy function that involves the Snake curve and features of the image [2, 3]. The energy E associated to the curve is defined in such a way its value is minimum when the curve is near the region of interest, i.e. near the borders of the image:

$$E = \frac{1}{2} \int ((\alpha \cdot |v'(s)|^2 + \beta \cdot |v''(s)|^2) + E_{ext}(v(s))) ds \; . \tag{2}$$

The coefficients α and β represent the elasticity and rigidity of the Snake, respectively, and they define its internal energy. E_{ext} is the energy term related to image characteristics and will be more discussed later. The operators $|v'(s)|^2$ and $|v''(s)|^2$ are the L2 norm of the first and second derivatives of $v(s)$, respectively.

The elasticity coefficient makes the Snake more resistant to traction. The rigidity coefficient makes it more resistant to bending. These two parameters prevent the Snake to become non-continuous or to break during the iteration process of the optimization problem.

The optimization problem above is solved using variational calculus, or being more specific, using the Euler-Lagrange equations[4]. A formal derivation of Eq. (2) gives us:

$$-2\frac{d}{ds}(\alpha\frac{dv}{ds})+2\frac{d^2}{ds^2}(\beta\frac{dv^2}{ds^2})-\|\nabla I\|^2=0 \quad . \tag{3}$$

where ∇I is the classical external energy and represents the image gradient.

The original external energy from the Snake model, E_{ext} is based on the image gradient, ∇I, which points towards the image borders. The main problem of this approach is that it forces the initialization of the Snake to be very close to the region of interest. To overcome this problem we implemented the Gradient Vector Flow (GVF) method [3] with the purpose of increasing the region of influence of the external forces. This is obtained by solving a few steps of the diffusion equation (with a diffusion coefficient equals to γ) to the components of the gradient vector. This procedure spreads the information of ∇I all over the image [3, 5].

2.1 Balloon and Adaptive-Balloon Techniques

There is a particular problem during the segmentation of the endocardial surface: the existence of artifacts that have the same contrast of the object of interest. The presence of these artifacts should be ignored. This is the case of the papillary muscles highlighted in Figure 1.

To overcome this problem we have implemented the Snake method with the so called Balloon Force:

$$F_{Balloon}=k_1.\vec{n} \quad . \tag{4}$$

In this expression k_1 is the constant strength of the force and n is the unitary normal vector to the curve. Depending on the signal of k_1, the force will provide the curve to grow or to shrink. In our implementation it always grows.

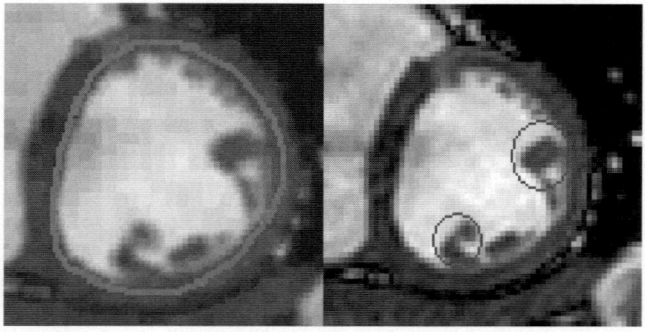

Fig. 1. A correct segmentation (left) avoiding artifacts (right)

However, the balloon force has shown not always to work properly. It was very difficult to adjust the strength parameter. It should be strong enough to overcome the noise and the artifacts but not so strong as to overcome the borders of interest and grow disorderly.

To solve this problem we developed a new force named the Adaptive-Balloon. Its direction is also normal to the curve, but the intensity varies along the Snake and during the iterations of the method. The strength of this force depends on the information of the curve neighborhood.

Since the format of the object to be segmented is convex, we create a force that acts only over those Snake points that have angles between the vector of the GVF and the axial vector at $v(s)$ greater than a predefined value. The axial vector of $v(s)$ connects $v(s)$ to the center of gravity of the image. The equation for the Adaptive-Balloon force is:

$$F_{adapt}(v(s)) = k_2.\vec{n} \text{ , if } \theta > \gamma.$$
$$F_{adapt}(v(s)) = 0, \text{ if } \theta < \gamma. \tag{5}$$

where θ is the angle between the axial vector and the GVF at the Snake point $v(s)$, k_2 is the force strength, $\vec{n}$ is the unitary normal vector and γ is the limit angle previously chosen. Figure 2 shows some examples of the directions of GVF, the axial vector and the angle θ.

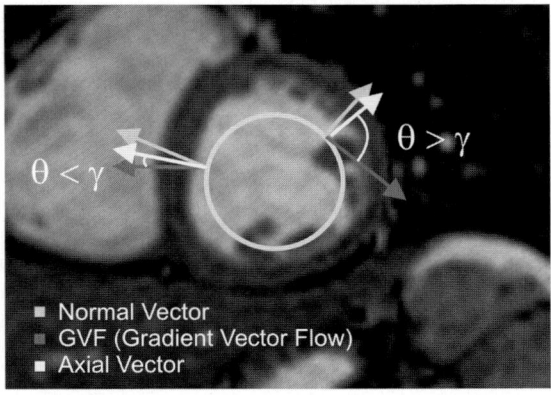

$\theta < \gamma$

$\theta > \gamma$

- Normal Vector
- GVF (Gradient Vector Flow)
- Axial Vector

Fig. 2. The calculation of the Adaptive-Balloon force based on the angle θ between the GVF and the axial vectors

3   Genetic Algorithm

With the Adaptative-Balloon force, the automatic segmentation generated better results than those obtained by the Snake with the classical Balloon force. Nevertheless, the results of both implemented methods have shown to be very dependent on their configuration parameters. One set of parameters may work well in one image, but fail for another. It was not possible to manually find a single set of parameters that produced satisfactory results for all images. Therefore we chose to use a Genetic Algorithm (GA)[6] to find the best set of configuration parameters for the automatic segmentation method. The set of parameters are: The elasticity α; rigidity β; the viscosity γ; the maximum amplitude associated to ∇I, κ; k_1 and k_2, the strength of the Balloon forces as in Eqs. (4) and (5), respectively.

The GA was used to solve a minimization problem, where the objective function captures the distance between 25 manually segmented short-axis images and the respective automatic segmented ones. For each curve we calculate the distance between the curve points and the center of gravity. The center of gravity is taken as the mean of the coordinates of the semi-automatic segmented curve. Therefore, the GA minimizes the error defined by the distance between these two curves:

$$\text{Error} = \frac{\sum_{i=0} \left(Rp_i - Rq_i \right)}{\sum_{i=0} Rp_i} . \tag{6}$$

where Rp_i and Rq_i are distances from the center of gravity to the ith point of the semi-automatic segmented curve and the automatic segmented curve, respectively.

The Genetic Algorithm simulates the evolution of a population of individuals, in our case, the Snake configuration parameters. The individuals, i.e. the Snake configuration parameters, were coded using real numbers. As the population evolves the best adapted individuals survive whereas the others die. The measure of the adaptation level of each individual is called fitness. In our case, the fitness measures the mean distance between the curves obtained automatically and the corresponding manual ones for 25 short-axis images, as described in Eq. (6).

To generate the offspring, the best adapted individuals are selected form the population. The Roulette-Wheel, also known as Fitness Proportionate Selection method [6], was used as the selection algorithm. After the parents are selected, they reproduce, generating the offspring. This process is called crossover. We implemented the Blend Crossover method [7] with a crossover rate of 90%. The offspring may mutate in the process. When this occurs, the new generated individuals are set with new randomly chosen genes within the range of each parameter. The mutation rate was set in a way it affects 10% of the population.

Half of the new individuals are selected to survive regardless of their fitness. The other half is made of individuals with higher fitness than their parents. This way, the individuals with higher fitness survive and we keep a diverse population. The method called elitism [6] was also implemented. It guarantees that the best individual of the population always survives after the selection process. The process described above was repeated two hundred times with a population of one hundred individuals.

4   Implementation

The algorithms developed were implemented in Java [8]. The Java language was chosen due to its rapid prototyping feature, its embedded support for concurrent programming and the wide availability of libraries for numerical methods and image processing. In particular, the software was developed using the ImageJ framework [9]. ImageJ is a public domain software which offers many functionalities to image processing. One of its most interesting features is the possibility to extend its basic functionalities by the development of new plug-ins. That way, new plug-ins were created on ImageJ: two Snake methods, one with the classical Balloon force and another with the new Adaptive-Balloon force; and the Genetic Algorithm.

The implementation of the Snakes methods was based on the finite difference method. Each deformation of the Snake involves a linear system to be solved. The Cholesky method was implemented to solve the linear systems.

A MRI exam generates many images that need to be segmented. Therefore, the implemented algorithms work with sets of images. Each set of images is stored in an internal data structure of ImageJ called ImageStack. The ImageStack is processed in parallel. The parallel implementations explore the embarrassingly data parallelism and were developed for shared-memory machines using Java threads[10]. The implementation of the parallelism is conceptually simple. ImajeJ loads the images from the ImageStack. The plug-in is executed and behaves as a master thread. The master creates new threads, the slaves. Each slave applies the Snakes method to a set of images of the stack and draws the segmentation results directly in the images. Figure 3 ilustrates this process.

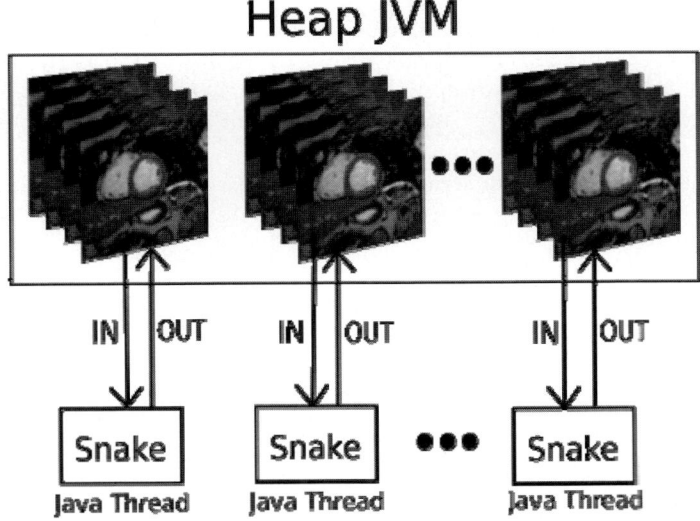

Fig. 3. The parallel implementation

5   Experimental Evaluation

In this section, we present experimental results obtained with a dual Xeon 1.6 GHz, 4 MB of cache, 4 GB of main memory. Each Xeon is a dual core processor. Therefore, all computations were executed using 4 threads. The methods for automatic segmentation were tested using 6 stacks of images obtained from two healthy volunteers. There are 3 stacks per person, each from different short-axis position (apex, mid and base). Each stack has 25 images obtained from different phases of contractions, from diastole to systole and back to diastole. The parallel implementation achieved near linear speedups when using 4 threads. The automatic segmentation of 150 images took less than 5 seconds with 4 threads. To validate the methods the automatic segmentations were compared to manual ones performed by a medical specialist. Figure 4 presents the automatic and manual segmentations for one of the images.

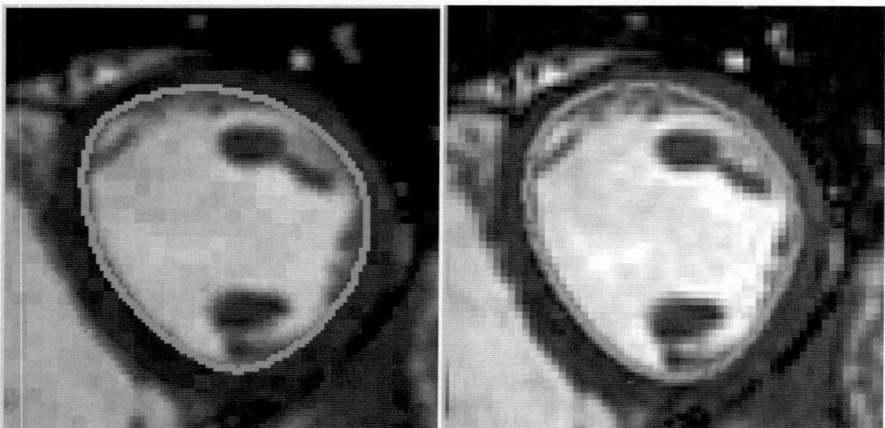

Fig. 4. Manual (left) and automatic segmentation using Snake-Adaptive Balloon method (right)

Table 1. Mean errors using the automatic methods

	Balloon	Adaptive-Ballon
Stack 1	12.2%	10.4%
Stack 2	9.6%	8.3%
Stack 3	9.4%	6.9%
Stack 4	13.4%	11.9%
Stack 5	9.9%	9.1%
Stack 6	9.7%	7.9%
Average	*10.7%*	*9.0%*

5.1   Adaptive-Balloon vs. Balloon Force

As we mentioned before the tuning of parameters for the Snakes methods is a non-trivial task. A set of parameters may achieve good results for one image but not for others. To compare the new developed Adaptive-Balloon force to the classical one, we have first manually tuned the parameters using the Snakes method with the Balloon force for each image. After that, using the same sets of parameters, we executed the Snakes method with the extra Adaptive-Balloon force. Table 1 presents the errors of segmentations obtained with the automatic method compared to the manual segmentations. The error presented in this table is the distance between the two curves, as defined by Eq. (6). The table shows that the new Adaptive-Balloon force technique achieved better results than the classical one.

5.2 Validation of the Method

The Genetic Algorithm optimized the parameters using one of the stacks of 25 images. Figure 5 shows the evolution of the fitness, i.e. the error as defined in Eq. (6), of the best individual (parameter set) per generation. After a few generations we observed that the fitness drops an order of magnitude.

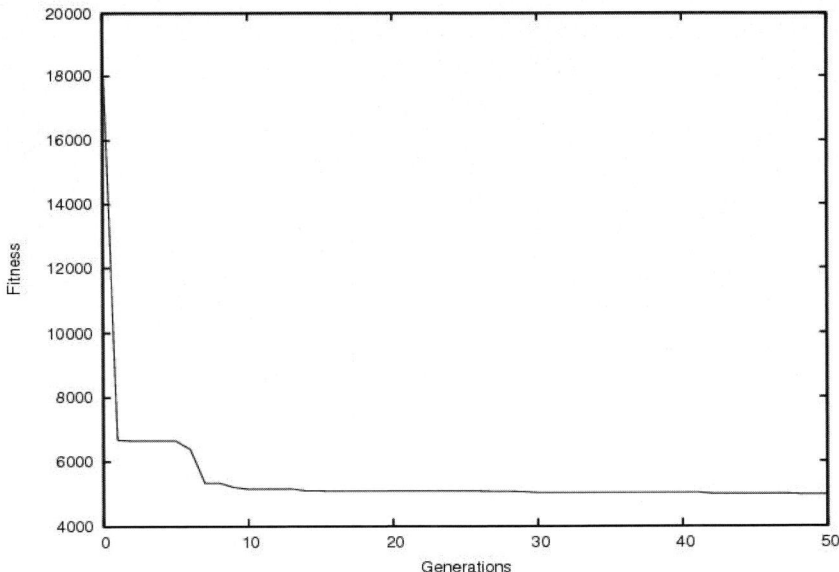

Fig. 5. Fitness value per generations

Table 2. Errors between automatic and manual segmentations

	Mean Error	Standard Deviation
Stack 1	9.8%	3.8%
Stack 2	7.0%	1.8%
Stack 3	6.3%	1.8%
Stack 4	10.6%	3.5%
Stack 5	7.6%	2.7%
Stack 6	7.3%	2.5%
Average	*8.1%*	*2.6%*

The best set of configuration parameters found by the Genetic Algorithm was used to evaluate the effectiveness of our Adaptive-Balloon force technique. Table 2 presents the segmentation errors obtained when applying these configuration parameters to the stacks. Again, the error corresponds to that of Eq. (6). As we can observe, the proposed method obtained satisfactory results, with errors around 8%.

6 Conclusion

The segmentation of the cardiac MRI images is of extreme importance but a tedious and an error-prone task, whereas the automatic segmentation is a challenging task. In this work we proposed, implemented and evaluated a new external force technique, called Adaptive-Balloon, to improve the convergence of the Snakes method, traditionally used to segment images. In addition, a genetic algorithm is used to find the best set of configuration parameters for the Snakes method. The proposed method obtained satisfactory results, with errors around 8%. We developed a parallel implementation to explore the embarrassingly data parallelism. The parallel implementation was very effective, segmenting automatically 150 images in less than 5 seconds. The preliminary results suggest that the methods are promising and with further development and validation they may be used, for instance, for the automatic calculation of cardiac ejection fractions.

Acknowledgments

The authors would like to acknowledge the reviewers for their comments and suggestions that significantly improved the quality of this work. This work was supported by CNPq, FAPEMIG and UFJF. G. M. Teixeira is a scholarship holder of PIBIC/CNPq.

References

1. Kühl, H.P., Schreckenberg, M., Rulands, D., Katoh, M., Schäfer, W., Schummers, G., Bücker, A., Hanrath, P., Franke, A.: High-resolution Transthoracic Real-Time Three-Dimensional Echocardiography: Quantitation of Cardiac Volumes and Function Using Semi-Automatic Border Detection and Comparison with Cardiac Magnetic Resonance Imaging. J. Am. Coll. Cardiol. 43, 2083–2090 (2004)
2. Kass, M., Witkin, A., Terzopoulos, D.: Snakes: Active Contour Models. International Journal of Computer Vision V1(4), 321–331 (1987)
3. Xu, C., Prince, J.L.: Gradient Vector Flow: A New External Force for Snakes. In: IEEE Proc. of the Conference on Computer Vision and Pattern Recognition, pp. 66–71. IEEE Computer Society, Washigton (1997)
4. Forsyth, A.R.: Calculus of Variations. Dover, New York (1960)

5. Xu, C., Prince, J.L.: Generalized Gradient Vector Flow External Forces for Active Contours. Signal Processing 71(2), 131–139 (1998)
6. Eiben, A.E., Smith, J.E.: Introduction to Evolutionary Computing. Springer, Heidelberg (2003)
7. Eshelman, L.J., Schaffer, J.D.: Real-coded Genetic Algorithms and Interval-Schemata. In: Foundations of Genetic Algorithms-2, pp. 187–202. Morgan Kaufman Publishers, San Mateo (1993)
8. Arnold, K., Gosling, J., Holmes, D.: The Java Programming Language. 4th edn. Prentice Hall PTR, Englewood Cliffs (2005)
9. ImageJ, http://rsb.info.nih.gov/ij/
10. Doug, L.: Concurrent Programming in Java: Design Principles and Pattern, 2nd edn. Prentice Hall PTR, Englewood Cliffs (1999)

Computational Tasks in Bronchoscope Navigation During Computer-Assisted Transbronchial Biopsy

Jarosław Bułat[1], Krzysztof Duda[1], Mirosław Socha[1], Paweł Turcza[1], Tomasz Zieliński[2], and Mariusz Duplaga[3]

[1] Department of Measurement and Instrumentation, AGH University of Science and Technology, al. Mickiewicza 30, 30-059 Kraków, Poland
{kwant,kduda,socha,turcza}@agh.edu.pl
[2] Department of Telecommunications, AGH University of Science and Technology, al. Mickiewicza 30, 30-059 Kraków, Poland
tzielin@agh.edu.pl
[3] Collegium Medicum, Jagiellonian University, ul. Skawińska 8, 31-066 Kraków, Poland
mmduplag@cyf-kr.edu.pl

Abstract. The paper presents algorithmic solutions dedicated to computer navigation system which is to assist bronchoscope positioning during transbronchial needle-aspiration biopsy. The navigation exploits principle of on-line registration of real images coming from endoscope camera and virtual ones generated on the base of computed-tomography (CT) data of a patient. When these images are similar an assumption is made that the bronchoscope and virtual camera have approximately the same position and view direction. In the paper the following computational aspects are described: correction of camera lens distortion, fast approximate estimation of endoscope ego-motion, reconstruction of bronchial tree from CT data by means of their segmentation and its centerline calculation, virtual views generation, registration of real and virtual images via maximization of their mutual information and, finally, efficient parallel and network implementation of the navigation system which is under development.

1 Introduction

Virtual bronchoscopy [1] CT-guided approach represents a modern solution to the difficult problem of bronchoscope positioning during medical procedure of transbronchial needle-aspiration biopsy. It makes use of real-time registration of real 2D images (coming from an endoscope) and virtual ones (obtained from virtual camera looking inside a 3D model of bronchial tree, reconstructed from CT patient data by means of segmentation). Usually, the registration of these two-source images is performed using in-the-loop maximization of their: correlation [2] or mutual information [3]. In order to speed-up search for precise virtual camera position, coarse estimation of bronchoscope camera can be performed.

M. Bubak et al. (Eds.): ICCS 2008, Part III, LNCS 5103, pp. 178–187, 2008.

It is usually done from video stream using corresponding points and epipolar geometry [2] or optical flow methods and perspective geometry [4]. In turn, next position of the endoscope camera can be predicted and tracked with Kalman using [5] or Monte Carlo [6] particle filters. Using shape-from-shading technique or triangulation by means of corresponding points it is also possible to extract 3D model of the airways tract from the real-time endoscopic video and try to register it to the 3D model reconstructed from the CT scans. First such attempt has been reported in [7] and further elaborated in [8].

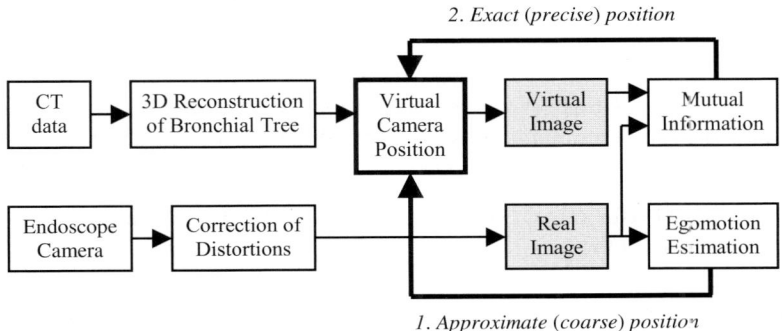

Fig. 1. Block diagram of the system under development

In the paper, some computational and implementation issues concerning the described above bronchoscope navigation scheme are presented. The following aspects are addressed: correction of camera non-linearities, fast approximate estimation of endoscope ego-motion [9], reconstruction of bronchial tree from CT data by means of their segmentation and its centerline calculation [10], virtual views generation, registration of real and virtual images via maximization of their mutual information [11] and, finally, efficient parallel and network implementation of the whole navigation system. The architecture of the system being designed is presented in Fig. 1.

2   Correction of Camera Distortions

The methodology presented in [12], [13] was used for correction of bronchoscope Olympus BF-160 camera lens distortions. As a test image black dotes lying on straight lines have been chosen. The distorted image obtained from bronchoscope camera is presented in Fig. 2a. The applied correction algorithm was based on maximization of the criterion measuering the degree of lines straightness. Using the model of radial distortions, the following polynomial was found:

$$r_c(r) = 4.2009 * 10^{-8}r^4 + 1.5991 * 10^{-10}r^3 + 3.7892 * 10^{-13}r^2 + r \qquad (1)$$

relating the radius r_c in distorted image to the radius r in undistorted image. For better results, unlike [12], [13], in our approach the center of distortions was

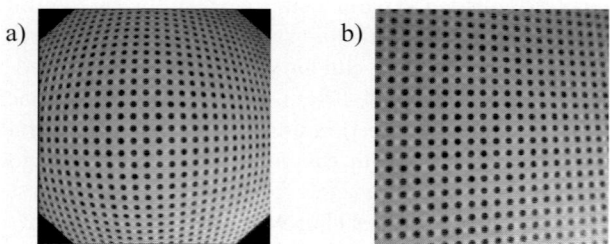

Fig. 2. a) Test image from bronchoscope camera, b) Reconstructed, undistorted image

calculated by means of set of optimizations in the neighborhood of geometric center of the image. The image after correction is presented in Fig. 2b.

3 Segmentation of Bronchial Tree and Calculation of Navigation Path

Image segmentation is the operation of grouping image pixels into separate objects present in a picture. The first step of segmentation algorithm is most often feature extraction and then checking if the specific pixel belongs to the object of interest. In medical CT data segmentation is used to isolate biological structures of interest like whole organs, e.g. bronchial tree, or some interesting, smaller structures like lymphatic nodes or tumors. Segmentation algorithms can be divided into four major groups: *pixel-based*, *region-based*, *edge-based* and *model-based* methods. In our research, the airway tree was segmented with the following steps: data smoothing with 3D gaussian filter, global thresholding and checking 26-connectivity. The exemplary segmentation results are depicted Fig. 3.

Centerline of the segmented bronchial tree is used as a navigation path in: virtual bronchoscopy, planning transbronchial biopsy and guiding bronchoscope's tip during biopsy. Thefore it precise computation is very important. Classification of algorithms for automatic generation of centerline (navigation path) in bronchial tree can be found in [14], [15].

We have proposed a new algorithm based on the distance transform, acting on the segmented bronchial tree, and an original iterative method for path searching [10]. The procedure is equipped with additional heuristic rules that prevent detecting false paths. The algorithm for path detection starts with placing the cube at the beginning of the bronchial tree with sides parallel to CT data coordinates. Fig. 3 shows the position of the cube in bronchial tree during successive steps and values of distance transform on its sides. The transform values on the cube sides are used for setting up the next point of the path. In the case depicted in Fig. 3a, the distance transform shows that the next point of the path should be either in Z or $-Z$ direction. As the direction $-Z$ means going back to previously computed point, the direction Z is chosen. In case depicted in Fig. 3b, from possible $-Z$, Y and $-Y$ the direction Y is used, while the direction $-Y$ is stored and becomes

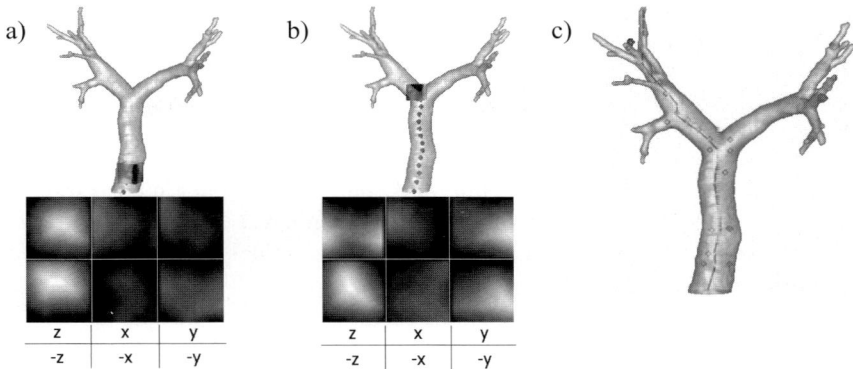

Fig. 3. a), b) Successive steps of computing centerline (navigation path) in bronchial tree (at the bottom values of computed distance transform lying on cube sides are shown); c) Path computed after the first iteration (points connected by the line) are used as starting points for next iterations (points not connected).

starting point for the next iteration (possible branching node) and the direction $-Z$ is neglected for the same reason as previously. Fig. 3c presents result of the first iteration: calculated points of the navigation path that are connected with lines. Consecutive iterations start at points stored as possible branching nodes. The algorithm ends up after checking all branching nodes (what takes 49 iterations in the presented example). Finally, a polynomial of 6-th degree was fitted to calculated points in order to make trajectory smoother.

4   Visualization of Bronchial Tree – Virtual Phantom

The visualization part of the navigation system was developed with Borland C++ Builder and Visualization ToolKit (VTK) cross-platform, open-source library [16]. The Visualization ToolKit makes use of the OpenGL API for 3D graphic card. The surface rendering technique was used for the sake of good performance and quality of generated virtual bronchoscopy images. Surface rendering includes two stages: generation of three-dimensional surface representing bronchial tree walls from CT data and visualization process via a graphic card. Virtual bronchoscopy images (VB) were used for testing motion estimation and navigation algorithms based on image registration.

The process of 3D surface generation consists of: loading a DICOM file with patient's computed tomography (CT) data, cropping CT data to reduce their size and generating isosurface at the level of -500HU by means of marching cubes algorithm [17]. The isosurface on this level goes through the data that represent walls of patient's bronchial tree. The result of computations is the continuous triangular mesh. The triangle strips are created in order to improve rendering performance. For these data a mapper was created to generate OpenGL rendering primitives and actor object for controlling mesh property which is the final step for virtual bronchoscopy image generation. The generated surface can be saved to file for later use.

In order to achieve maximum resemblance with real bronchoscope camera illumination conditions, the virtual light source was set up as follows: it moves along with the camera and its position is the same as camera position. The light is configured as positional (headlight), and the light cone angle corresponds to camera cone angle. To prevent overexposing of nearest surfaces the irregular light intensity along the cone angle was assumed. Light fading attenuation was also used for distance simulation.

5   Fast Estimation of Bronchoscope Egomotion

In order to speed-up egomotion estimation [18] in bronchial environment we use simplified model of geometric relations based on cylindrical shape accompanied by the fixation on a carina [19], what reduces motion's degrees of freedom to four (forward/backward movement, camera rotation, camera tilt in two directions). It is achieved by continuous tracking of the carina (stationary point) illuminated by the camera light source, and by analyzing bronchial wall radial moves relative to fixed point by correlation in polar coordinates [9].

Reverse perspective projection of images before correlations is made by the use of correspondence between z-axis and r-axis derived from the following trigonometric relation (see Fig. 4):

$$\frac{R-r}{z} = tg\varphi = \frac{R}{z-f} \qquad \Longrightarrow \qquad r = R\left(1 - \frac{z}{z-f}\right) \qquad (2)$$

Let us note that R serves only as a scaling factor of the view. In the current research we estimate forward motion, after carina stabilization and camera rotation compensation, as arithmetic mean of directional wall motions. Camera tilt is estimated from geometric mean of these motions.

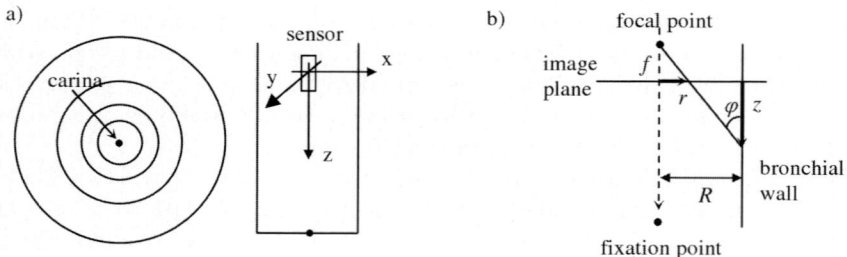

Fig. 4. Applied models: a) segment of bronchial tree (upper perspective projection and x-z cross-section), b) imaging in cylindrical environment with radius R, camera focal length f, radial image axis r and depth from image plane z.

We assessed algorithm accuracy by series of test in virtual cylinders, virtual bronchial trees and on real operational video sequences from transbronchial biopsy [11]. The results of experiments show that accuracy of bronchoscope cumulated motion estimation is within 5% of distance in virtual environments. In Fig. 5a the virtual bronchial tree environment with estimated wall motions is shown. Fig. 5b shows estimation of cumulated forward/backward motion together with imposed forward virtual camera motion. Disturbing factors in this experiment were camera rotation, and x-y plane camera moves.

Fig. 6a shows forward/backward bronchoscope trajectory during real biopsy. This trajectory suggest similarity of frames 7 and 65, being distant in time but close in space, because of the strong backward move followed by the forward move. These frames, shown in Fig. 6b and 6c, confirm this similarity and confirm also satisfactory behavior of our egomotion estimation algorithm.

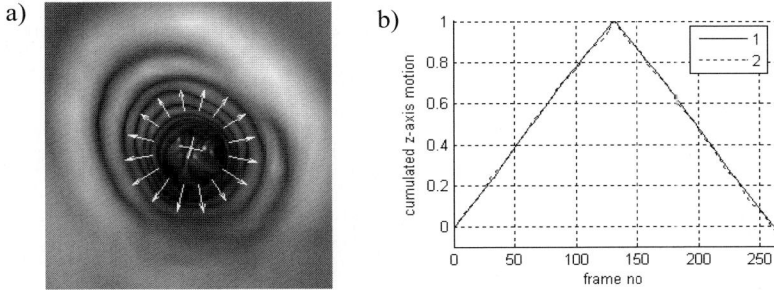

Fig. 5. Example of camera position estimation along z-axis in virtual bronchial-tree phantom: a) Virtual environment with estimated radial wall-move vectors, b) Estimated forward/backward camera trajectories for imposed motion: 1 - camera motion along the path with target fixed on carina, 2 - camera motion with additional camera tilt and rotation and moving target.

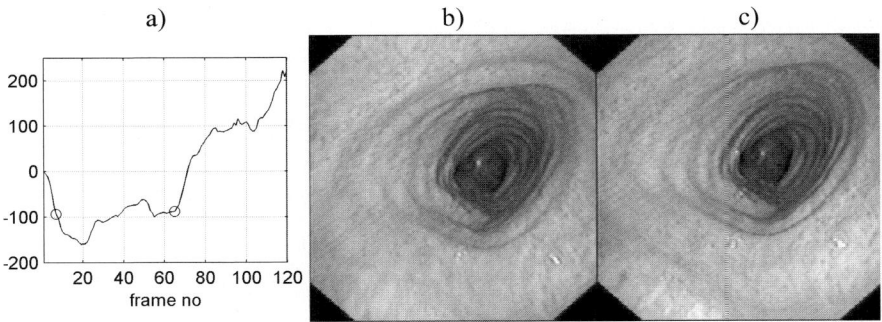

Fig. 6. Estimation of camera position along z-axis from real data: a) estimation result with two frames close in space but distant in time (mark 'o'), b), c) frames 7 and 65

6   Image Registration Using Mutual Information

The information from egomotion algorithm of bronchoscope motions is not sufficient to precise determination of the location of *real* bronchoscope tip in relation
to *virtual* bronchial tree, however, can significantly speed-up navigation process.
Therefore, before successful navigation will be possible, two tasks have to be completed. The first one is to place the *virtual bronchoscope* (the source of virtual
images) in a position corresponding to the real bronchoscope. This is achieved by
adjusting position of virtual bronchoscope in such a way that generated images
are similar as much as possible to images from real bronchoscope.

After setting up the virtual camera starting position, the second task - calibration of egomotion estimation algorithm is performed. Having two images from
real camera at positions z_0 and $z_0 + d$, where z_0 is the starting position and d is
outcome of egomotion estimation algorithm, using appropriate image similarity
measure we try to find such a displacement of the virtual camera position which
makes virtual image as similar to the real one as possible. Egomotion estimation
is used for coarse estimation of virtual camera positition, then image registration
algorithm is used for finer adjustment.

Methods enabling registration of images from the same or different sources
have been extensively developed through the last decades. Numerous papers
were published on this topic [20]. In our approach, in both above described
tasks, mutual information [21] was used as an image similarity measure. It is
based on the concept of joint entropy as given by Shannon for determination of
communication's channel capacity and is defined as follows:

$$I(u, v) = H(u) - H(u \mid v) \tag{3}$$

where $H(u)$ denotes the measure of uncertainty of the value of random variable
u, and $H(u \mid v)$ denotes the same measure but determined with the assumption
that value of random variable v is known. In this way $I(u, v)$ expresses how
much the uncertainty about value of u decreases after getting to know value
v. Correlation between decreasing value of conditional entropy $H(u \mid v)$ and
increasing value of mutual information $I(u \mid v)$ is obvious. Using the Bayesian
theorem: $P(A, B) = P(A \mid B)P(B)$ and the definition of Shannon's entropy

$$H(u) = -\sum_i p_u(i) \log p_u(i), \qquad H(u, v) = -\sum_{i,j} p_{uv}(i, j) \log p_{uv}(i, j), \tag{4}$$

the equation expressing mutual information (MI) may be rewritten into the form

$$I(u, v) = H(u) + H(v) - H(u, v). \tag{5}$$

It includes joint entropy $H(u, v)$, which may be determined on the basis of joint
probability distribution, which in turn can be inferred from the joint histogram
$h(u, v)$ after appropriate normalization.

Exemplary images from real and corresponding virtual camera are presented
in Fig. 7. This figure also shows values of mutual information as a function of virtual camera position. In the experiment, the virtual camera was shifted along the

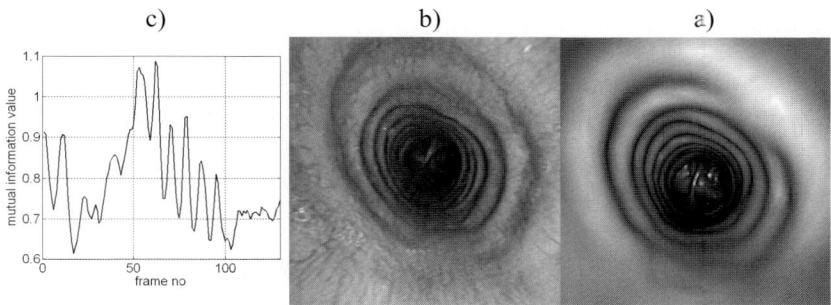

Fig. 7. Example of image registration: a) Mutual information as a function of virtual camera position, b) Real image - frame 71, c) Virtual image corresponding to the real one found by the registration algorithm

computed navigation path lying in the central part of the airways. Observed local maxima of the mutual information curve comes from bronchial tree vertebras.

7 Computation Complexity Analysis

System presented in Fig. 1. have been implemented in Matlab language with the exception of 3D image generation realized by means of hardware supported OpenGL. In spite of using fast matrix calculation and JIT (just-in-time) optimization in the latest version of Matlab, its real-time operation is not possible on a high-end x86 class computer.

Since the video frame rate from bronchoscope is 25 frames per second, we have only 40ms for accomplish one cycle of the proposed navigation algorithm. Execution time of the most important navigation system procedures programmed both in Matlab and C/C++ language is shown in Table 1.

For precise bronchoscope motion estimation it is necessary to perform: one camera correction, one coarse brochoscope motion estimation and on average 25 virtual image generations and mutual information calculations. For that reason one cycle of complete motion estimation needs approximately 5600 ms in Matlab and 620 ms in C/C++. One can see that even for optimized C/C++ version it is not possible to perform the algorithm in real-time.

Table 1. Estimated execution time of the most important navigation system procedures for Pentium 4 3.2 GHz processor (single thread)

Procedure	Matlab time [ms]	C/C++ time [ms]
Camera correction	13	1.4
Coarse motion estimation - egomotion	4600	460
3D image generation (one image)	–	1.4
Mutual information (one image)	46	4.9

There are, however, many possibilities for further execution time optimization. First of all, we can use SSE (Streaming SIMD Extensions) instructions of x86 CPUs instead of plain C/C++. BLAS (Basic Linear Algebra Subprograms) [22] seams to be the most efficient way of using SSE extension. Next, we can take advantage of parallel nature of the presented above navigation algorithm using relatively cheap multicore and multiprocessor systems. Both most demanding procedures (egomotion estimation and mutual information calculation) are highly independent within one cycle of motion estimation. What is more, BLAS library has been recently ported to GPU environment and could take advantage of up to 128 independent processors cores that was introduced in the G80+ generation of the NVIDIA GPUs family with CUDA architecture.

In order to take advantage of massive parallel computation we have chosen MPI (Message Passing Interface) as a inter-process communication framework. MPICH2 (implementation of MPI v2 protocol) provide us a great performance and excellent flexibility. It is available on many hardware and software platforms including Linux and Windows as well as provide network transparency. Using this open source library we are able to split visualization process and computation engine to different computer connected by TCP/IP network. During testing procedure of data (image) and control commands transmitting over MPI we have achieved the following results: up to 1GB/s of messages with data frame between two processes on one computer (four core Xeon), up to 500000 short (128 bytes) control messages per second in the similar setup and up to 95Mbps between two processes on two different computer over 100Mbps Ethernet.

Obtained results have convinced us that it is possible to build described above system working in real-time on the basis of standard PC architecture.

8   Conclusions

In the paper some computational and implementation issues and new solutions for bronchoscope navigation during computer-assisted transbronchial biopsy were presented. In authors opinion the proposed, new, very simple and fast algorithm for real-time estimation of endoscope forward and backward ego-motion is the most crucial in its precise positioning using virtual bronchoscopy CT-based approach. After successful simulation tests real-time implementation of the described navigation system is under development at present. We have already estimated partial algorithms' complexity of this modules and tested some particular hardware solutions: multiprocessor NVIDIA CUDA architecture and MPI-based communication framework which provide high performance parallel environment with minimum development effort.

References

1. Bartz, D.: Virtual Endoscopy in Research and Clinical Practise. Eurographics - Computer Graphics Forum 24(1) (2005)
2. Mori, K., Deguchi, D., Sugiyama, J., et al.: Tracking of a Bronchoscope Using Epipolar Geometry Analysis and Intensity-Based Image Registration of Real and Virtual Endoscopic Images. Medical Image Analysis 6(3) (2002)

3. Sherbondy, A.J., Kiraly, A.P., et al.: Virtual Bronchoscopic Approach for Combining 3D CT and Endoscopic Video. In: Medical Imaging 2000, Proc. SPIE (2000)
4. Helferty, J.P., Higgins, W.E.: Combined Endoscopic Video Tracking and Virtual 3D CT Registration for Surgical Guidance. In: Proc. IEEE ICIP, pp. II-961–964 (2002)
5. Nagao, J., Mori, K., et al.: Fast and Accurate Bronchoscope Tracking Using Image Registration and Motion Prediction. In: Barillot, C., Haynor, D.R., Hellier, P. (eds.) MICCAI 2004. LNCS, vol. 3217, pp. 551–558. Springer, Heidelberg (2004)
6. Deligianni, F., Chung, A., Yang, G.-Z.: Predictive Camera Tracking for Bronchoscope Simulation with CONDensation. In: Duncan, J.S., Gerig, G (eds.) MICCAI 2005. LNCS, vol. 3749, pp. 910–916. Springer, Heidelberg (2005)
7. Bricault, I., Ferretti, G., Cinquin, P.: Registration of Real and CT-Derived Virtual Bronchoscopic Images to Assist Transbronchial Biopsy. IEEE Trans. on Medical Imaging 17(5), 703–714 (1998)
8. Deligianni, F., Chung, A., Yang, G.-Z.: pq-Space Based 2d/3D Registration for Endoscope Tracking. In: MICCAI 2003. LNCS, vol. 2878, pp. 311–318. Springer, Heidelberg (2003)
9. Twardowski, T., Zieliński, T., Duda, K., Socha, M., Duplaga, M.: Fast estimation of broncho-fiberoscope egomotion for CT-guided transbronchial biopsy. In: IEEE Int. Conference on Image Processing ICIP-2006, Atlanta (2006)
10. Duda, K., Duplaga, M.: Automatic generation of a navigation path for virtual bronchoscopy. In: PAK, vol. 5bis, pp. 115–118 (2006) (in polish)
11. Duda, K., Zieliski, T., Socha, M., et al.: Navigation in bronchial tree based on motion estimation and mutual information. In: ICSES, Poland (2006)
12. Vijayan Asari, K., Kumar, S., Radhakrishnan, D.: A New Approach for Nonlinear Distortion Correction in Endoscopic Images Based on Least Squares Estimation. IEEE Trans. on Medical Imaging 18(4), 345–354 (1999)
13. Helferty, J.P., Zhang, C., McLennan, G., Higgins, W.E.: Videoendoscopic Distortion Correction and Its Application to Virtual Guidance of Endoscopy. IEEE Trans. on Medical Imaging 20(7), 605–617 (2001)
14. Kiraly, A.P., Helferty, et al.: Three-Dimensional Path Planning for Virtual Bronchoscopy. IEEE Trans. on Medical Imaging 23(9), 1365–1379 (2004)
15. Bartz, D., Mayer, D., et al.: Hybrid segmentation and exploration of the human lungs. In: IEEE-Visualization-2003 (No. 03CH37496), pp. 177–184 (2003)
16. VTK – The Visualizarion ToolKit, http://www.vtk.org/
17. Lorensen, W.E., Cline, H.E.: Marching cubes: a high resolution 3D surface construction algorithm. Comput. Graph. 21, 163–169 (1987)
18. Tian, T.Y., Tomasi, C., Heeger, D.J.: Comparison of Approaches to Egomotion Computation. In: IEEE Conf. CVPR, pp. 315–320 (1996)
19. Daniilidis, K.: Fixation simplifies 3D Motion Estimation. Computer Vision and Image Understanding 68(2), 158–169 (1997)
20. Zitova, B., Flusser, J.: Image registration methods: a survey. Image and Vision Computing 21, 977–1000 (2003)
21. Pluim, J.P.W., Maintz, J.B.A., Viergever, M.A.: Mutual information based registration of medical images: a survey. IEEE Trans. on Medical Imaging 22, 986–1004 (2003)
22. BLAS (Basic Linear Algebra Subprograms), software library, http://www.netlib.org/blas/

MPEG-7 as a Metadata Standard for Indexing of Surgery Videos in Medical E-Learning

Andrzej A. Kononowicz and Zdzisław Wiśniowski

Department of Bioinformatics and Telemedicine,
Jagiellonian University, Medical College,
ul. Kopernika 7e, 31-034 Krakow, Poland
a.kononowicz@cyfronet.pl

Abstract. The analysis of video recorded surgical procedures is considered to be a useful extension of the medical curriculum. We can foster the development of video-based e-learning courses by working out a unified description method which would facilitate the exchange of these materials between different platforms. Sophisticated metadata enables a broader integration of artificial intelligence techniques into e-learning. The aim of this paper is to present the possibility of combining the MPEG-7 metadata standard with the MeSH classification for indexing of video recordings in medical e-learning. A tool for metadata descriptions of surgical videos in accordance with the MPEG-7 standard is also presented. This tool is part of a larger architecture for the exchange of medical multimedia objects.

Keywords: MPEG-7, e-learning, metadata, medical terminology, learning objects, MeSH.

1 Introduction

E-learning is a contemporary way of learning using multimedia and communication abilities of modern computers and mobile devices (cellular phones, smartphones, palmtops). The main obstacle in the popularization of e-learning is the significant expenditure of time and costs to prepare multimedia materials. We may reduce the problem by building large databases of educational components, which can be used in building more complex courses. Such elements are called learning objects or sharable content objects [1]. On one hand a collective creation and usage of didactic components ensures a larger choice of materials, on the other hand it allows to share the development costs. An example of an initiative promoting e-learning through exchange of learning objects in medicine is the eViP project [2][3]. In this paper we will concentrate on a selected subclass of learning objects which are multimedia learning objects designed to be used in medicine teaching.

It seems to be obvious that the illustration of traditional textual descriptions of procedures by video clips provides the students with broad knowledge. Multimedia learning objects consist of audio/video recordings created during selected surgical procedures. Multimedia databases containing learning objects are a very valuable

M. Bubak et al. (Eds.): ICCS 2008, Part III, LNCS 5103, pp. 188–197, 2008.

source of data for medical educators, however only under the condition that the materials are easily searchable. For that reason, an appropriate description of the content is a crucial part of every multimedia learning object. The description should be easily processed and portable. Metadata designated for medical e-learning should also fulfil some additional requirements specific to the field of medicine – e.g. embedding of existing medical terminologies.

1.1   Usage Scenarios of Multimedia Metadata in Medical E-Learning

Let us consider the case of a student, who is looking for video recordings of minimally invasive surgical procedures. The student enters the keyword "endoscopy" into the search engine. In response, the search engine contacts (for instance by a software agent) a terminology server linked with a medical ontology to look up the definition of the term "endoscopy". It gets the information that endoscopy is a surgical procedure which has several subclasses – e.g. Arthroscopy, Colonoscopy, Gastroscopy or Thoracoscopy. Additionally, the agent contacts also a user preferences database to get the profile of the user (preferred video format, connection bandwidth, accepted languages). Such databases can be integrated with the university's learning management system. The user database may also contain certificates which entitle the student to enter educational database with restricted access. Within the next step, the search agent queries the multimedia databases using the information obtained from the terminology server, web ontologies and the user preferences database. Results obtained from many databases are unified and presented to the user, who selects the adequate material. The list of potential search criteria is large – it can include affected organs or used equipment. Due to a spatial and temporal decomposition of the video file, the transmission can start from the right time point and with the desired organ or equipment highlighted.

Another potential use case example of multimedia metadata is to aid the construction of virtual patients [4]. A medical teacher may use a special authoring tool for building complex patient scenarios (e.g. designated for the OSCE examination). The program analyses the data provided by the teacher and automatically suggests videos and images which could be inserted into the virtual patient. The search algorithms are based on metadata stored in multimedia databases.

1.2   Architecture of a Multimedia E-Learning System in Medicine

From the above described use cases we can derive a theoretical e-learning system architecture exploiting multimedia metadata (Fig.1). Surgery videos are stored in media storage systems and transmitted to users by video streaming servers (e.g. Darwin, Helix or Windows Media Services). Video clips are described by physicians or technical staff trained in indexing of medical multimedia resources using specialized annotation tools. The created metadata is stored in a multimedia database, whereas the medical knowledge is kept on a terminology server (there are already free terminology servers available on the Internet – e.g. UMLS) and as web ontologies (e.g. in OWL format). The user profiles and authorization data are integrated with e-learning platforms by system administrators operating dedicated administrative applications and scripts. Learning objects gathered in the system are used by students (using an interactive learning environment via a standard web browser) or medical educators (using e-learning authoring tools). This paper will concentrate on a selected element of this architecture – the annotation tool for teachers. We will also consider the question what metadata standard should be used.

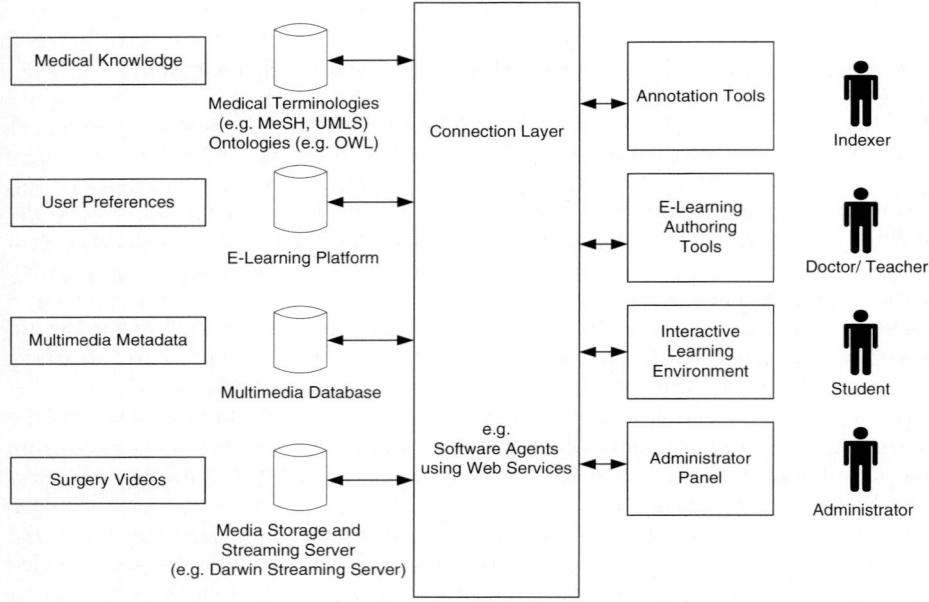

Fig. 1. Architecture of a multimedia e-learning system in medicine

2 Metadata Standards and Terminologies

Much research has already been done in the field of multimedia data description in the past few years. The proposed solutions were based on project specific meta-data, such standards as SMIL, Dublin Core, SMPTE, EBU, TVAnytime or other technologies connected for instance with the Semantic Web initiative (RDF, RDF Schema, OWL, RuleML) [5-8].

2.1 MPEG-7

A significant breakthrough in this field was the release of the MPEG-7 standard in 2001 [9-11]. The official name of the MPEG-7 standard is ISO 15938 Multimedia Description Framework. MPEG-7 is a very flexible specification. It is organized as a collection of tools, which can be used in accordance with the needs of the indexer. The basic building blocks of MPEG-7 are descriptors. Descriptors represent the syntax and semantical meaning of basic description elements (e.g. author's name, media duration, used codecs, textures, audio signal parameters, camera motion descriptions). Descriptor Schemes (DS) consist of related descriptors and smaller description schemes. The syntax of MPEG-7 is based on XML. Descriptors and description schemes are defined in the Description Definition Language (DDL), which is an extension of the XML Schema. DDL allows the user to add new elements to the description. MPEG-7 can be stored and transmitted in textual format or as binary data added to a video stream.

The MPEG-7 metadata descriptors have already been prototypically implemented by several research teams. Tseng et al. [12-13] developed a personalization and summarization system consisting of a MPEG-7 video annotation tool, automatic labeling tools and a summarization and adaptation engine. Caliph and Emir [14] are two applications facilitating semantic descriptions of digital photographs. Tsinaraki at el. [15] proposed a video segmentation tool conform to MPEG-7 for ontology-based semantic indexing. Their tool has been tested in the domain of soccer games. Despite the outstanding possibilities of MPEG-7, there is still lack of applications exploiting this standard in medicine. The authors believe that MPEG-7 as a universal, easily extensible and complex metadata standard is the right choice for the description of learning objects also in the medical domain.

2.2   Terminology – MeSH

The diversity of the natural language hinders the automatic processing of descriptions. Therefore, natural language is often artificially limited to concepts stemming from controlled vocabularies (like classification systems or nomenclatures). Many classification systems and nomenclatures have been created to describe medical knowledge (e.g. ICD, LOINC, Snomed CT, NANDA or MeSH). We have decided to use the MeSH (Medical Subject Headings) thesaurus [16] for the description of medical videos in MPEG-7 standard. This vocabulary has been created by the National Library of Medicine (NLM) with the intention to classify information in the biomedical area. MeSH is successfully used in indexing the MEDLINE database and the NLM catalogues. Different language versions of MeSH (e.g. English, German, French or Polish) already exist. The 2006's version of MeSH contains 23885 descriptors. MeSH descriptors are the building blocks of this classification. The elements are divided into 16 categories (e.g. A: Anatomics, C: General Diagnosis or D: General Drugs and Chemicals), which are divided into further subclasses. Descriptors include a set of semantically related concepts, which consist of one or more terms. A descriptor can have attached attributes, which come from a set of 86 qualifiers (e.g. *abnormalities*, *injuries* or *statistics&numerical data*). Hierarchical (*narrower term/broader term*) and non-hierarchical (*related/see also*) relations exist between the descriptors.

Medical terminologies like MeSH are good starting points for semantic description providing the user with a static knowledge reference. If a more advanced semantic search is needed, additionally, the use of ontologies should be considered. Ontologies encode meanings separately from application code enabling knowledge sharing and support for external reasoning. There exist already many examples of ontologies in the medical domain modeling patient data as well as diagnostic and treatment procedures (e.g. [17]). Jovic at al. [18] explain in their study the construction process of medical ontologies on the example of the heart failure domain. They emphasize the importance of the linkage between ontologies and terminologies. The leading language for expressing ontologies is currently OWL (Web Ontology Language). OWL is usually written in XML/RDF syntax and can be extended by the SWRL rules language. Tsinaraki at el. [19] proposed a framework, called DS-MIRF, for the integration of OWL ontologies with MPEG-7 compliant indexing.

3   MPEG-7 in the Description of Medical Learning Objects

3.1   Video Decomposition

The description of a video file in MPEG-7 may refer not only to the whole clip but also to its fragments. A spatial or temporal decomposition may be distinguished. The first one, expressed by *SpatialDecomposition DS*, allows selecting segments of a picture (e.g. pathological changes, applied medical equipment) and inserting descriptions only of the selected parts. The temporal decomposition (*TemporalDecomposition DS*) enables the partition of the clip into time intervals. It gives the possibility to describe the individual stages of the procedure separately (e.g. preoperative operation, incision, main part of the operation, laying sutures).

3.2   Medical Classifications in MPEG-7

The MPEG-7 standard enables the definition of new classification schemes or importing of the existing ones. The definition of new classifications in MPEG-7 is carried out by the description schemes *ClassificationScheme DS* and *TermDefinition DS*. Concepts derived from declared classifications are placed into the description by

Table 1. Mapping of the MPEG-7 structural annotation classes onto categories of MeSH descriptors

MPEG-7 Structured Annotation	Description of surgery videos	MeSH Categories, Subcategories
Why	Reason for carrying out the procedure. Patient's diagnosis.	C (General Diseases), F3(Mental Disorders)
WhatObject	Names of operated organs or those organs which are visible in the video and are important for students in the opinion of the medical educator.	A (General Anatomy)
WhatAction	Names of performed procedures.	E1-E6 (General Techniques)
How	Medical equipment used in the procedure.	E7 (Equipment and Supplies)
Where	Name of the geographic region in which has the procedure been made.	Z (Geographic Locations)
When	- / No mapping.	-
Who	Patient's characteristic. Patient's profession.	M (Persons)

elements of type *TermUse* and *ControlledTermUse*. This paper focuses on the binding of MeSH terms into the structural description of surgical videos. The structural description in MPEG-7 is represented by the type *StructuredAnnotation*. It may contain any number of *TermUse* instances from seven different categories: actions (element *WhatAction*), persons (*Who*), objects (*WhatObject*), places (*Where*), time (*When*), purpose (*Why*). Each element can enclose a free text description or a reference to a concept from a classification. Table 1 contains the authors' proposal of mapping the MeSH categories onto the MPEG-7 structured annotation types. We also suggest the possible use of the MPEG-7 annotation categories in description of surgical procedures. For instance the element of type *Why* should be used in the description of patient's disease diagnosed, which was the reason for carrying out the operation. We can describe this category by descriptors from the C (*General Diseases*) and F3 (*Mental Disorders*) MeSH subtree.

Example of MPEG-7 code containing a MeSH descriptor.

```
<ClassificationAlias alias="mesh"
     href="http://www.ncbi.nlm.nih.gov/mesh"/>
<!-- ... -->
<TextAnnotation>
    <StructuredAnnotation>
            <WhatAction href=":mesh:D013906">
                    <Name xml:lang="en">Thoracoscopy</Name>
            </WhatAction>
            <WhatObject href=":mesh:D008168">
                    <Name xml:lang="en"> Lung </Name>
            </WhatObject>
    </StructuredAnnotation>
</TextAnnotation>
```

It should be mentioned that the MeSH thesaurus already contains a special qualifier grouping elements for the description of surgical procedures – SU/surg. The qualifier comprises of the following categories: A1-5, A7-10, A13, A14, A17, C1-23, F3. This qualifier could be used theoretically to pick out the MeSH descriptors needed for describing surgical videos. However, in our opinion, the selected set of descriptors is too narrow to fit all concepts useful in the characterization of surgical videos. For instance the qualifier SU/surg does not contain the subcategory E7 – Equipment and Supplies.

4   M7MeDe

As the first element of the proposed architecture of multimedia e-learning systems in medicine, we have decided to implement an annotation tool for surgical videos, which creates descriptions in the MPEG-7 standard using the MeSH classification (Fig.2). The application has been named M7MeDe.

M7MeDe is designed to support the indexation of resources in a medical multimedia library of surgical video recordings. The application enables a temporal

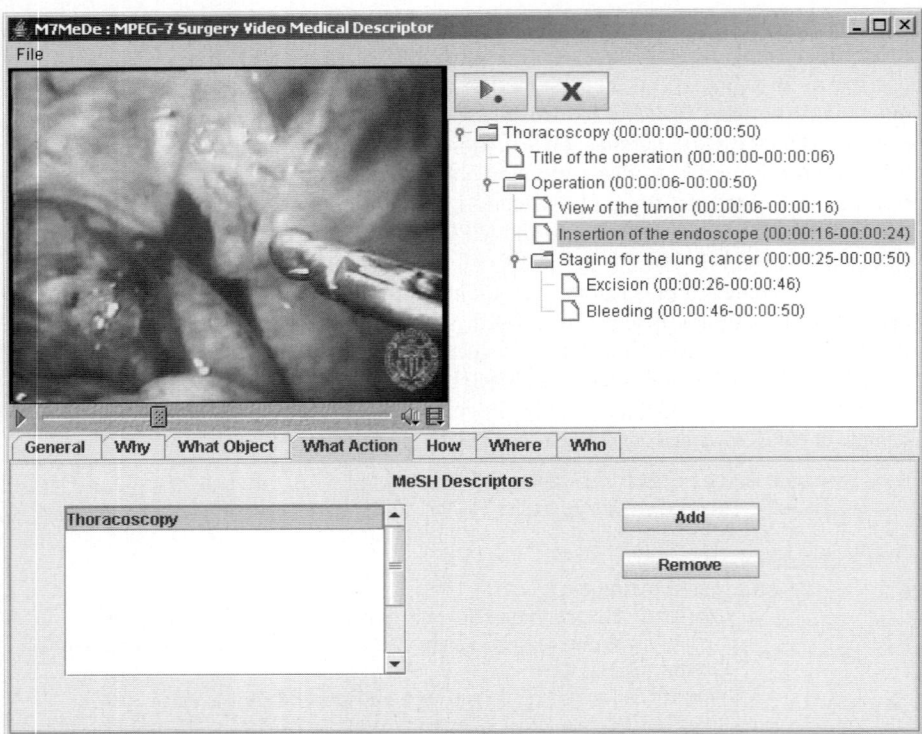

Fig. 2. M7MeDe – an application for describing of multimedia learning objects in the MPEG-7 standard

decomposition of the video, which can be nested in larger segments forming a hierarchical structure. An example of such decomposition carried out with the use of M7MeDe is depicted in Fig 2. The annotated video clip (source of the video [20]) presents a fragment of a thoracoscopy procedure for staging of lung cancer. The first level of video clip description is divided into two parts: the operation's title screen (displaying the operation's title and surgeon's name) and the main operation part. The second part is divided further into three parts: view of the tumor, insertion of endoscope and sampling. The sampling part contains an *excision* and *bleeding* video fragment. We can describe each segment in free text (the *General tab* in Fig 2) or attach keywords from the MeSH thesaurus to a selected MPEG-7 structured annotation category (*Why*, *What Object*, *What Action*, etc). Each annotation category is linked to a subtree in the MeSH-Tree in accordance with the mapping in Table 1. For instance adding a keyword in the *Why* category opens the *General Disease* subtree (Fig.3). Keywords are inherited by subordinated segments (segments which are nested in other segments).

M7MeDe was implemented in Java technology. The MeSH classification was downloaded in the form of XML files, transformed and inserted into a relational

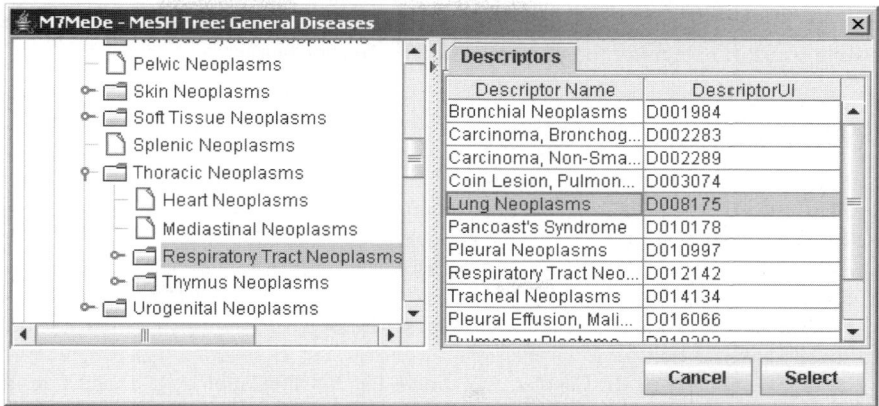

Fig. 3. M7MeDe – Window for selecting of MeSH descriptors in the *General Diseases* category

database. The application uses the Java Media Framework (for video operations) and JAXB (for Java to XML binding).

5   Further Work

The M7MeDe application is still in development stage. Many potential functions are missing (for instance the spatial decomposition of the video or a direct interface to multimedia databases). It is intended to examine the possible relations of medical learning objects described in MPEG-7 to the e-learning standards SCROM and IMS in the future. Further work on this project should also pertain to the remaining elements of the proposed architecture (ontology based search algorithms, placing of multimedia learning objects into learning management systems or construction of authoring tools). Woods et al. [21] showed in their study about indexing of dermatology images, that the use of MeSH alone for indexing finds matching for about one-forth of the terms in their experiment, therefore usage of other classifications and ontologies in the MPEG-7 description beside MeSH will be necessary. For that reason we consider building direct interfaces to the UMLS terminology server and to repositories of web ontologies in the OWL standard.

6   Summary

Well described surgical video recordings are considered to be valuable e-learning materials in medicine teaching. This paper was aimed at discussing the possibilities of using MPEG-7 and MeSH in building multimedia learning objects. The presented application – M7MeDe – allows the division of a video clip into temporal segments and their description with MeSH keywords. The M7MeDe annotation tool is part of a larger architecture which takes advantage of medical metadata. There are many possible ways of extending the presented tool by further features.

References

1. Kononowicz, A.A., Żabińska, M.: Distribution of Learning Objects Based on Agents Technology. Automatyka 9(1-2), 115–126 (2005)
2. eViP – Electronic Virtual Patients Project, http://www.virtualpatients.eu
3. Kononowicz, A.A., Stachoń, A.J., Roterman-Konieczna, I.: Virtual Patient as a Tool for Problem Based-Learning in the Context of the European Project eViP. E-mentor 1(23), 26–30 (2008) (in Polish)
4. Huang, G., Reynolds, R., Candler, C.: Virtual Patient Simulation at U.S. and Canadian Medical Schools. Academic Medicine 82(5), 446–451 (2007)
5. Mrozowski, P., Kononowicz, A.A.: DSS-MEDA – A Web-Based Framework for Video Annotation in Medical E-Learning. Bio-Algorithms and Med-Systems 4(2), 51–56 (2006)
6. Hunter, J., Iannella, R.: The Application of Metadata Standards to Video Indexing. In: Proc. of the 2nd Europ. Conf. on Research and Advanced Technology for Digital Libraries, pp. 135–156. Springer, London (1998)
7. Stamou, G., van Ossenbruggen, J., Pan, J., Schreiber, G.: Multimedia Annotations on the Semantic Web. IEEE MultiMedia 13(1), 86–90 (2006)
8. Carro, S., Scharcanski, J.: A Framework for Medical Visual Information Exchange on the WEB. Comp. Biol. Med. 36, 327–338 (2006)
9. Martínez, J., Koenen, R., Pereira, F.: MPEG-7: The Generic Multimedia Content Description Standard, Part 1. IEEE MultiMedia 9(2), 78–87 (2002)
10. Martínez, J.: MPEG-7: Overview of MPEG-7 Description Tools, Part 2. IEEE MultiMedia 9(3), 83–93 (2002)
11. ISO/MPEG N4242, Text of ISO/IEC Final Draft International Standard 15938-5 Information Technology - Multimedia Content Description Interface – Part 5 Multimedia Description Schemes, MPEG Multimedia Description Schemas Group, Sydney (2001)
12. Lin, C.-Y., Tseng, B.L., Smith, J.R.: VideoAnnEx: IBM MPEG-7 Annotation Tool for Multimedia Indexing and Concept Learning. In: Proc. IEEE Intl. Conf. on Multimedia and Expo (ICME), Baltimore, MD (2003)
13. Tseng, B.L., Lin, C.-Y., Smith, J.R.: Using MPEG-7 and MPEG-21 for Personalizing Video. IEEE MultiMedia 11(1), 42–52 (2004)
14. Lux, M., Klieber, W., Granitzer, M.: Caliph & Emir: Semantics in Multimedia Retrieval and Annotation. In: Proc. of 19th CODATA Conference, Berlin, Germany (2004)
15. Tsinaraki, C., Polydoros, P., Kazasis, F., Christodoulakis, S.: Ontology-based Semantic Indexing for MPEG-7 and TV-Anytime Audiovisual Content. Special Issue of Multimedia Tools and Application Journal on Video Segmentation for Semantic Annotation and Transcoding 26, 299–325 (2005)
16. Nelson, S., Johnston, D., Humphreys, N.: Relationships in Medical Subject Headings. In: Bean, C., Green, R. (eds.) Relationships in the Organization of Knowledge, pp. 171–185. Kluwer Academic Publishers, New York (2001)
17. The National Center for Biomedical Ontology, http://bioontology.org

18. Jovic, A., Prcela, M., Gamberger, D.: Ontologies in Medical Knowledge Presentation. In: Proc. 29th International Conference Information Technology Interfaces. Cavtat, Croatia (2007)
19. Tsinaraki, C., Polydoros, P., Christodoulakis, S.: Integration of OWL Ontologies in MPEG-7 and TV-Anytime Compliant Semantic Indexing. In: Persson, A., Stirna, J. (eds.) CAiSE 2004. LNCS, vol. 3084, pp. 398–413. Springer, Heidelberg (2004)
20. Video Assisted Thoracoscopy, University of Southern California, http://www.cts.usc.edu/videos-mpeg-vidassistthoracoscopy-all.html
21. Woods, J., Sneiderman, C., Hameed, K., Ackerman, M., Hatton, C.: Using UMLS metathesaurus concepts to describe medical images: dermatology vocabulary. Comp. Biol. Med. 36, 89–100 (2006)

Workshop on Tools for Program Development and Analysis in Computational Science

Special Session: Tools for Program Development and Analysis in Computational Science

Jie Tao[1], Arndt Bode[2], Andreas Knuepfer[3], Dieter Kranzlmüller[4],
Roland Wismüller[5], and Jens Volkert[4]

[1] Steinbuch Center for Computing
Karlsruhe Institute of Technology, Germany
[2] Lehrstuhl für Rechnertechnik und Rechnerorganisation
Technische Universität München, Germany
[3] Center for Information Services and High Performance Computing
Technische Universität Dresden, Germany
[4] Institute of Graphics and Parallel Processing
Johannes Kepler University Linz, Austria
[5] Operating Systems and Distributed Systems
University of Siegen, Germany

The use of supercomputing technology, parallel and distributed processing, and sophisticated algorithms is of major importance for computational scientists. Yet, the scientists' goals are to solve their challenging scientific problems, not the software engineering tasks associated with it. For this reason, computational science and engineering must be able to rely on dedicated support from program development and analysis tools.

The primary intention of this workshop is to bring together developers of tools for scientific computing and their potential users. Since its beginning at the first ICCS in 2001, the workshop has encouraged tool developers and users from the scientific and engineering community to exchange their experiences. Tool developers present to users how their tools support scientists and engineers during program development and analysis. Tool users report their experiences employing such tools, especially highlighting the benefits and the improvements possible by doing so.

The workshop covers various research topics, including

- Problem solving environments for specific application domains
- Application building and software construction tools
- Domain-specific analysis tools
- Program visualization and visual programming tools
- On-line monitoring and computational steering tools
- Requirements for (new) tools emerging from the application domain
- Tools for parallel, distributed and network-based computing
- Testing and debugging tools
- Performance analysis and tuning tools
- (Dynamic) Instrumentation and monitoring tools
- Data (re-)partitioning and load-balancing tools

M. Bubak et al. (Eds.): ICCS 2008, Part III, LNCS 5103, pp. 201–202, 2008.
© Springer-Verlag Berlin Heidelberg 2008

– Checkpointing and restart tools
– Tools for resource management, job queuing and accounting

This year a rich number of papers has been submitted to the workshop. From the 25 submissions, nine papers have been selected for a presentation in the conference. These papers give a basic overview and some technical details of the authors' recent research work in the area of software tools, with a focus on optimal libraries, on-line monitoring and steering, performance and pattern analysis systems, automatic code optimization, and systems for supporting application development on the Grid.

The workshop organization team thanks Dr. Martin Schulz, Dr. Josef Weidendorfer, Dr. Karl Fuelinger, and David Kramer for their support in the review process.

BTL++: From Performance Assessment to Optimal Libraries

Laurent Plagne and Frank Hülsemann

EDF R&D, 1 avenue du Général de Gaulle, BP 408, F-92141 Clamart, France
{laurent.plagne,frank.hulsemann}@edf.fr

Abstract. This paper presents the *Benchmark Template Library in C++*, in short **BTL++**, which is a flexible framework to assess the run time of user defined computational kernels. When the same kernel is implemented in several different ways, the collected performance data can be used to automatically construct an interface library that dispatches a function call to the fastest variant available.

The benchmark examples in this article are mostly functions from the dense linear algebra BLAS API. However, **BTL++** can be applied to any kernel that can be called by a function from a C++ main program. Within the same framework, we are able to compare different implementations of the operations to be benchmarked, from libraries such as ATLAS, over procedural solutions in Fortran and C to more recent C++ libraries with a higher level of abstraction. Results of single threaded and multi-threaded computations are included.

1 Introduction

Linear algebra is a field of particular relevance in scientific computing in both, academia and industry. Operations on matrices, dense or sparse, or vectors, large or small, feature in many if not most projects. The prominent position of the topic is reflected in a large number of available implementations.

For a project relying on the BLAS interface on a given target architecture, for example x86, one has the choice between netlib's default implementation [1], ATLAS [2], MKL from Intel [3], ACML from AMD [4] and GotoBLAS [5] to name just a few. Furthermore, there exist other solutions such as Blitz++ [6], MTL [7] or uBLAS [8], which offer the same functionality but use their own interfaces. The obvious question for a user who has to choose among the options is then: "Which option works best in a given computing environment?" In order to answer this question, the user has to solve another problem first: "How to assess and how to compare the performances of the different options?"

This paper describes the *Benchmark Template Library in C++* (**BTL++**) project [9] which is a flexible and user extendible benchmarking framework for computational "actions" (kernels). The extendibility by the user was the dominant design goal of the project. It relies on three main concepts: *implementations, computational actions and performance analysers*. We would like to point out

M. Bubak et al. (Eds.): ICCS 2008, Part III, LNCS 5103, pp. 203–212, 2008.

that by providing realisations of these concepts, a user can interface new libraries, add computational kernels to the benchmarking suite or change the way how or even the type of performance that is measured. Loosely speaking, as long as an implementation of some computational action can be called by a routine from within a C++ program, the chances are high that this implementation can be included in the benchmarking process. The nature of the computational action to be assessed can, of course, be defined by the user. Although BTL++ itself provides different performance analysers that have been sufficient for our purposes, its aim is not to replace dedicated profiling tools like PAPI [10] for instance. On the contrary, the performance counters of PAPI have already been used successfully as performance analysers in the BTL++ framework.

Among the numerous publicly available benchmark suites, BTL++ is related in spirit to the BenchIT [11] project that provides detailed performance evaluations for a fixed set of numerical kernels. While the BenchIT project offers a rich and mature database interface that gathers the performance data for a large variety of architectures, BTL++ is a library that emphasises the extendibility by the user. This generic feature makes BTL++ a very flexible tool with respect to the computational kernels, their implementations and the benchmarking methods.

Originally designed as a flexible benchmark tool, BTL++ now features an optional library generation stage. From the collected measurements, BTL++ can create a new optimal library that routes the user's function calls to the fastest implementation at her disposal. Our performance results underline that the problem size has to be taken into account in the routing process, as the implementation that is fastest for small problems is not necessarily also the fastest on larger ones. Although such a performance assessment/generation sequence is successfully used in projects like ATLAS and FFTW [12], the BTL++ positions itself at a different level since it only generates an **interface** library based on existing implementations. In other words, ATLAS or FFTW are stand alone projects that provide implementations which can be used by BTL++ to generate the optimal interface library on a given machine for a given numerical kernel.

The article is structured as follows. In Sect. 2, the principal building blocks in the implementation of BTL++ are introduced. Section 3 shows benchmark results for both BLAS and non BLAS kernels. In Sect. 4 we use again BLAS examples to demonstrate the BTL++ optimal library generation stage. In Sect. 5 we present our conclusions and discuss some directions for future work.

2   BTL++ Implementation

The development of a Linear Algebra (LA) based code can follow numerous different strategies. It can involve C++ generic LA libraries (uBLAS, MTL, ...), one of the various BLAS implementations (ATLAS, GotoBLAS,...), or being directly hand coded using raw native languages (C/C++, F77, ...). The BTL++ project aims to compare the performance of these strategies. Even though not all of the different strategies take the form of a library in the linker sense, we refer to all implementations for which a BTL++ library interface exists, as BTL++ *libraries*.

The performance of each such BTL++ library is evaluated against a set of computational actions, in our case, LA numerical kernels such as the dot product, the matrix-matrix product and vector expressions like $W = aX + bY + cZ$. Note that the BTL++ kernels are not limited to the operations included in the BLAS API. Also note that a BTL++ library does not have to implement all computational kernels to be included in the comparison. Hence we can take into account specialised implementations like the MiniSSE library [13] for example, which implements only a subset of the BLAS interface. Last, BTL++ provides a set of different performance evaluation methods. This open-source project relies on cooperative collaboration and its design aims to obtain a **maximal legibility and extendibility**. To be more explicit, a user is able to extend the collection of available BTL++ libraries, kernels and performance evaluation methods, as easily as possible. The implementation of BTL++ with object-oriented and generic programming techniques in C++ results in a good modularity and an efficient factorisation of the source code.

The performance evaluation of the axpy operation ($Y = a * X + Y$) using the Blitz++ library will be used as a running example to illustrate the different parts of the BTL++ design.

2.1 BTL++ Libraries Interfaces: The *Library_Interface* Concept

Before measuring the performance of an implementation, one should be able to check its correctness for all considered problem sizes. In order to ease this calculation check procedure, a pivot (or reference) library is chosen to produce the reference results. Since BTL++ is written in C++, the STL has been a natural choice for being the BTL++ reference library. The two vector operands (X and Y) of the axpy operation are initialised with pseudo-random numbers. This initialisation is first performed on the reference STL operands (X_STL and Y_STL). Secondly, these vectors are used for initialising the corresponding Blitz++ operands (X_Blitz and Y_Blitz) via a vector copy. Then, both the STL and Blitz++ libraries perform the axpy operation. A copy operation from a Blitz++ vector to a STL vector and a comparison of two STL vectors are used to check the result.

Obviously, some of the Blitz++ functions that implement the vector copy operations from and to STL vectors could be reused for the implementation of this init/calculate/check procedure applied to another vectorial BTL++ kernel. Moreover, the same functionality has to be implemented for all libraries in the BTL++ library set. In order to give a standardised form for these functions, the BTL++ framework defines the *Library_Interface* concept that consists in the set of constraints on the types and methods that a *Library_Interface model* class must fulfil. See Table 1a) for details. Note that a given *Library_interface model* class can define an incomplete subset of the BTL++ kernels. Of course, the missing kernels cannot be benchmarked for this particular library.

Following our running example, the blitz_interface class modelling the *Library_interface* concept looks like:

Table 1. Sets of types and functions that a user-defined class must provide to model the BTL++ a) *Library_Interface* concept and b) *Action* concepts

a)			b)	
public Types			**public Types**	
RT	Real Type (double or float)		Interface	*Library_Interface* model
GV	Generic Vector Type			
GM	Generic Matrix Type		**methods**	
(static) functions			Ctor(int size)	Ctor with problem size argument
std::string name(void)				
void vector_from_stl(GV &, const SV &)			void initialize(void)	
void vector_to_stl(const GV &,SV &)			void calculate(void)	
void matrix_from_stl(GM &, const SM &)			void check_result(void)	
void matrix_to_stl(const GM &, SM &)			double nb_op_base(void)	
void axpy(RT, const GV &, GV &, int)			**(static) functions**	
+ dot, copy matrix_vector_product, ...			std::string name(void)	

```
template<class real> struct blitz_interface{
  typedef real              RT;
  typedef blitz::Vector<RT>  GV;
  static std::string name( void ){return "Blitz";}
  static void vector_from_stl(GV & B, const std::vector<RT> & B_stl){
    B.resize(B_stl.size()); // Note the () operator for Blitz vectors
    for (int i=0; i<B_stl.size() ; i++) B(i)=B_stl[i];
  }
  static void vector_to_stl(const GV & B, std::vector<RT> & B_stl){
    for (int i=0; i<B_stl.size() ; i++) B_stl[i]=B(i);
  }
  static void axpy(const RT coef, const GV & X, GV & Y, int N){
    Y+=coef*X; // Blitz++ Expression Template !
  }
  ...follows dot, matrix_vector_product,..
};
```

For each library in the BTL++ library set, the definition of the *Library_Interface* concept requires the creation of the corresponding *model* classes, such as ATLAS_interface or F77_interface, for example. To extend the BTL++ library interface collection to another library, one has to define the *Library_Interface* concept *model* class that implements all or a part of the BTL++ kernels. Because of the repetitive nature of the different interfaces, this work is greatly simplified via the inheritance mechanism in C++. We believe that the extendibility of the BTL++ library collection has been confirmed by the successful addition of various sequential [14] and parallel [15] libraries by different users. Now we can make use of this unified interface to implement each element of the BTL++ kernel set.

2.2 *Action* Concept

The BTL++ timing procedure begins with the problem size definition before initialising the reference (STL) and test (e.g. Blitz++) operands. Next, the calculation duration is evaluated with a chosen (user definable) method. Once the calculations have been checked the performance results are stored. From this timing description on can see that all kernels across all libraries can be benchmarked in a similar way. The BTL++ *Action* concept allows the standardisation of these benchmarks by providing a uniform interface to deal with the different kernels. Table 1b) describes this concept. A class modelling the *Action* concept is implemented for each BTL++ kernel. The previously introduced *Library_Interface* concept allows us to implement only one *Action* template *model* class per BTL++ kernel. For example, the `axpy` operation is handled by the following `action_axpy<>` template class:

```
template<class LIB_INTERFACE> class action_axpy {
public :
    ...
  static  std::string name( void ){return "axpy";}
  double nb_op_base( void ){return 2.0*_size;}
  void calculate( void ) { LIB_INTERFACE::axpy(_coef,X,Y,_size);}
private :
  typename STL_INTERFACE::gene_vector  X_stl,Y_stl;
  typename LIB_INTERFACE::gene_vector  X,Y;
};
```

All the other BTL++ kernels are handled by their corresponding template *Action* class (action_dot, action_copy, . . .). For example, one could execute the following code:

```
action_axpy< blitz_interface<double> > a(1000);
a.calculate();     // Compute axpy
a.check_result(); // Compare with STL reference result
```

3   The Benchmark Results

In this chapter we present some performance comparisons obtained with BTL++ and discuss various ways of using the collected information. Due to the extendible design of BTL++, the effort needed to specify which computational actions to compare is rather small, so that, once a new library has become available, a user can easily decide which kernels to test. The results of a benchmark run are currently stored in a file the name of which identifies the library, the action and the floating point type used in the computations. This rather rudimentary storage solution for the benchmarking results is most likely to be replaced by a lightweight data base in the future. In any case, once the performance data has been collected, it is straightforward to create graphical representations or tables of comparison as the examples in this article show.

The observed performance results for a matrix-matrix multiplication in Table 2 illustrate clearly the performance advantage of optimised BLAS libraries, such as ATLAS or GotoBLAS, over straightforward, but un-tuned implementations such as the FORTRAN77 example. The results for Blitz−+ and uBLAS reiterate the point that alternative interfaces to the same computational action can be accommodated. The gap between the tuned and the un-tuned implementations is independent of the programming language and the programming paradigm used. All un-tuned, or shall we say straightforward, implementations, be it in C, FORTRAN77 or C++ (procedural or object-oriented), suffer from the same performance problem as they do not take the memory hierarchy of the hardware into account.

Table 2. Performances of Matrix-Matrix product ($A \times B$) and $X = \alpha Y + \beta Z$ operations in MFlop/s. Small matrices are understood to have up to 333 rows, while large matrices have more than 333 rows. For vectors, the small/large size threshold is set to 10^5 elements. These definitions were found to be adequate on a 1.6 GHz PentiumM processor with 1MB L2 cache. The performance values are the algebraic mean over all measurement points in the respective category. The GNU Compiler Collection version 3.3.5 provided the C, C++ and Fortran compilers, the optimisation level was "-O3".

$A \times B$	small matrices	large matrices	$X = \alpha Y + \beta Z$	small vectors	large vectors
ATLAS 3-7-24	**1228**	**1409**	C	**797**	134
Goto baniasp-r1.15	1171	1333	STL	795	134
MKL 9.0	1111	1192	Blitz++ 0.9	691	135
C	758	208	blocked ET	673	**200**
Blitz++ 0.9	685	203	uBLAS (Boost 1.32)	635	135
STL	670	184			
f77	646	270			
uBLAS (Boost 1.32)	617	180			

For more information we refer to the web page of the project [9]. In particular, the web site provides detailed information on the interfaces to the different libraries and how the computational actions were implemented when the standard BLAS call was not available.

In addition to identifying which implementation works best in a given computing environment, BTL++, like any benchmark, can be used to compare different environments, such as different compilers or different machines. However, this aspect of building up a knowledge base over time is not integrated into the BTL++ installation. As indicated earlier, we plan to change the data output and storage part of our benchmarking framework, which will make the building up of and the information retrieval from the performance data base much easier.

To illustrate the point that the benchmark kernels are not limited to BLAS operations, and to show that libraries do not always offer the best performance, we present results for the vector operation $X = \alpha Y + \beta Z$. The results in Table 2 and Fig.1a) show that for large vectors, the authors' blocked expression template performs better than all the other options by on average 48% [16].

The results in Fig. 1b) demonstrate that BTL++ is not limited to mono-threaded computational kernels. The multithreaded computations were carried

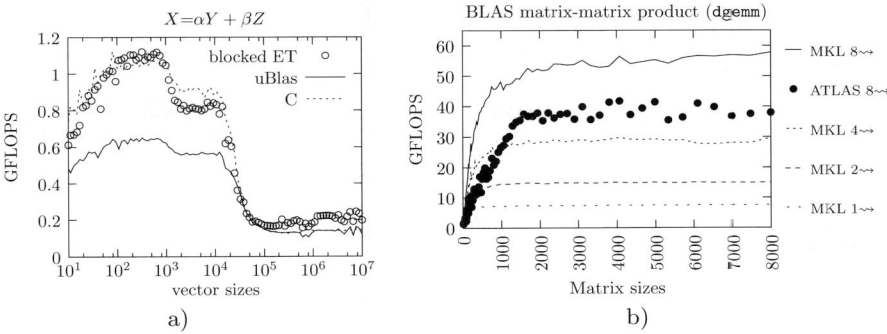

Fig. 1. a) Performance data in GFlop/s for the $X = \alpha Y + \beta Z$ operation. b) Multi-threaded performance results in GFlop/s for the dgemm routine on eight processor cores (2 sockets, 4 cores per socket).

out on a dual processor/quad core Intel Xeon with 2.83GHz with Intel MKL version 10 and ATLAS version 3.8.0.

4   Generation of the Optimal BLAS Library

In this chapter, we show how the so-called BTL++ hybrid library, which is the *optimal* BLAS interface library on a given machine, is constructed from previously collected performance data.

The principle of the BTL++ BLAS hybrid implementation is very simple. For each considered subroutine f of the BLAS API and for each library L that implements f, the BTL++ tool suite provides a set of performance measurements $\{\mathrm{perf}(\mathtt{f}, L, s)\}$ at different problem sizes s. From these results the BTL++ project selects two libraries L_1 and L_2 as well as a threshold size t such that the performance sum $S(L_1, L_2, t)$ is maximised:

$$S(L_1, L_2, t) = \sum_{s \leq t} \mathrm{perf}(\mathtt{f}, L_1, s) + \sum_{s > t} \mathrm{perf}(\mathtt{f}, L_2, s) \ . \tag{1}$$

The result of the optimisation step is an automatically generated C file named `Hybrid.c` which contains the switches for all BLAS subroutine implementations. For the PentiumM target, the generated call for the daxpy operation in `Hybrid.c` reads as follows:

```
void cblas_daxpy(const int N, const double alpha,
                 const double *X, const int incX,
                 double *Y, const int incY){
  if (N<18738){
    MiniSSE1_cblas_daxpy(N,alpha,X,incX,Y,incY);
  }
  else{
```

```
    ATLAS_cblas_daxpy(N,alpha,X,incX,Y,incY);
  }
}
```

Again for the PentiumM at 1.6GHz, the result of the optimisation process, i.e. the choice of libraries and the threshold values, for different BLAS operations is given in Table 3.

Table 3. BTL++ interface automatically generated for the PentiumM (1.6GHz) target

BLAS routine	Best Library for small problem sizes	Small/large threshold	Best Library for large problem sizes
daxpy	MiniSSE	18738	ATLAS
dcopy	Netlib	86974	ATLAS
ddot	MiniSSE	18738	ATLAS
dgemm	ATLAS		ATLAS
dgemv	MKL		MKL

Note that a switch between two libraries is only generated when needed. In other words, when one particular implementation offers the best performance across the board, no switch statement is generated and hence, no run time overhead occurs.

4.1   BLAS Implementation Wrappers

The main difficulty in building the hybrid BLAS is to cope with the differences in the BLAS installations. For example, MKL is distributed as a dynamic library (libmkl.so) implementing the C BLAS interface (c_blas_xxx), while ATLAS automatically builds a static library (libatlas.a) implementing the same interface. Last example, the Goto library is built dynamically (libgoto.so) for numerous computer targets but only implements the BLAS F77 interface that has to be used through the static netlib CBLAS wrapper (libcblas.a).

The main idea in the BTL++ approach is to wrap each library, whether static or not, that contributes to the optimal BLAS implementation into a dynamic library and to use the dlopen()/dlsym() functions to load these dynamic wrappers when needed.[1]

To illustrate that the generated library interface does indeed offer the best performance possible, we show the results for the daxpy and dcopy operations in Fig. 2.

---

[1] In order to avoid infinite recursions at link time, the proposed solution relies on the -Bsymbolic option of the gnu ld linker, which binds the static library symbols to the intermediate dynamic wrappers. The portability of this solution is an open issue.

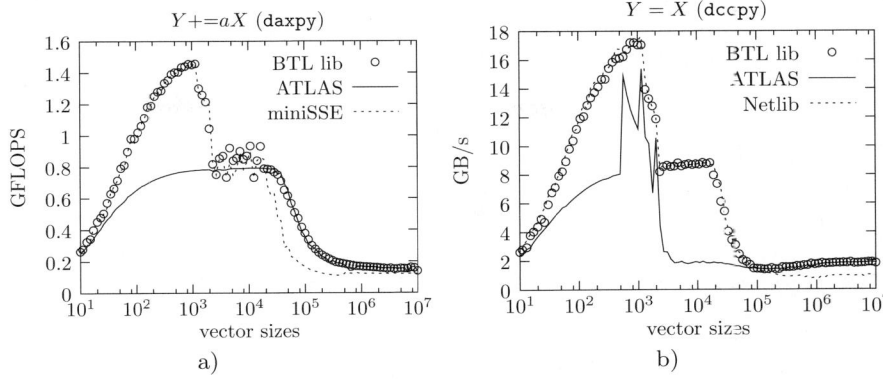

Fig. 2. Performance data for the `daxpy` (a) and `dcopy` (b) operations using the generated, optimal BLAS interface (BTL lib). The threshold values and the best implementations available are determined automatically.

5   Conclusion and Outlook

In this article, we have described the benchmarking project BTL++, which helps a user to assess and compare the performance of user definable, computational actions in her given computing environment of hard- and software. The paramount design goal of the BTL++ project was flexibility, which allows a user to adapt or extend the three parts (libraries, computational kernels and performance analysers) of the framework to her needs with little effort. The only requirement the approach currently imposes is that the user defined parts can be called from a C++ main program.

Several extensions, ranging from interfacing new libraries over adding new computational kernels to adapting the approach to parallel computing, have been carried out successfully by different users. The fact that these extension efforts were possible without interaction with the BTL++ developers indicates that the design goal of user extendibility has been achieved.

We have shown how the results of the benchmarking exercise of the BLAS operations can be used to create a new library automatically that for each function and in each problem size range calls the best implementation available. Numerical results confirm that the generated interface library does indeed obtain the same performance as the best implementation identified in the benchmarking comparison.

Concerning the evolution of the project, we have identified several areas of possible improvement. In the area of parallel computing, we intend to build on the work described in [15] to enlarge the scope of our benchmarking tool to distributed and shared memory programs.

The current implementation of the generation of the optimal interface library has been developed for Linux/GNU x86-platforms. The question concerning the portability of this implementation remains open.

Acknowledgements. The authors are grateful to their colleagues Christian Caremoli, Ivan Dutka-Malen, Eric Fayolle and Antoine Yessayan for their valuable help and expertise.

References

1. Netlib: BLAS web page, `http://www.netlib.org/blas`
2. Whaley, R.C., Petitet, A.: Minimizing development and maintenance costs in supporting persistently optimized BLAS. Software: Practice and Experience 35(2), 101–121 (2005)
3. Intel: MKL web page can be found from, `http://www.intel.com`
4. AMD: ACML web page, `http://developer.amd.com/acml.jsp`
5. Goto, K., van de Geijn, R.A.: Anatomy of a High-Performance Matrix Multiplication. ACM Transactions on Mathematical Software 34(3) (September 2007)
6. Veldhuizen, T.L.: Arrays in blitz++. In: Caromel, D., Oldehoeft, R.R., Tholburn, M. (eds.) ISCOPE 1998. LNCS, vol. 1505, pp. 223–230. Springer, Heidelberg (1998)
7. Siek, J.G., Lumsdaine, A.: The matrix template library: Generic components for high-performance scientific computing. Computing in Science and Engineering 1(6), 70–78 (1999)
8. Walter, J., Koch, M.: uBLAS web page, `http://www.boost.org/libs/numeric/ublas`
9. Plagne, L.: BTL web page, `http://projects.opencascade.org/btl`
10. Browne, S., Dongarra, J., Garner, N., Ho, G., Mucci, P.: A Portable Programming Interface for Performance Evaluation on Modern Processors. The International Journal of High Performance Computing Applications 14(3), 189–204 (2000)
11. Schöne, R., Juckeland, G., Nagel, W.E., Pflüger, S., Wloch, R.: Performance comparison and optimization: Case studies using BenchIT. In: Joubert, G.R., Nagel, W.E., Peters, F.J., Plata, O.G., Tirado, P., Zapata, E.L. (eds.) Parallel Computing: Current & Future Issues of High-End Computing. Proceedings of the International Conference ParCo 2005, vol. 33, pp. 877–884. Central Institute for Applied Mathematics, Jülich, Germany (2006)
12. Frigo, M., Johnson, S.G.: The design and implementation of FFTW3. Proceedings of the IEEE 93(2), 216–231 (2005)
13. Berghen, F.V.: miniSSEL1BLAS web page, `http://www.applied-mathematics.net`
14. Petzold, O.: tvmet web page, `http://tvmet.sourceforge.net`
15. Mello, U., Khabibrakhmanov, I.: On the reusability and numeric efficiency of C++ packages in scientific computing (2003), `http://citeseer.ist.psu.edu/634047.html`
16. Plagne, L., Hülsemann, F.: Improving large vector operations with C++ expression template and ATLAS. In: 6th intl. workshop on Multiparadigm Programming with Object-Oriented Languages (MPOOL 2007) (July 2007)

DaStGen—A Data Structure Generator for Parallel C++ HPC Software

Hans–Joachim Bungartz, Wolfgang Eckhardt, Miriam Mehl,
and Tobias Weinzierl

Technische Universität München, Boltzmannstr. 3, 85748 Garching, Germany
weinzier@in.tum.de
http://www5.in.tum.de

Abstract. Simulation codes often suffer from high memory requirements. This holds in particular if they are memory-bounded, and, with
multicore systems coming up, the problem will become even worse as
more and more cores have to share the memory connections. To optimise data structures with respect to memory manually is error-prone
and cumbersome.

  This paper presents the tool DaStGen translating classes declared in
C++ syntax and augmented by new keywords into plain C++ code.
The tool automates the record optimisation, as it analyses the potential
range of each attribute, and as the user can restrict this range further.
Herefrom, the generated code stores multiple attributes within one single primitive type. Furthermore, the tool derives user-defined MPI data
types for each class. Using the tool reduces any algorithm's memory footprint, it speeds up memory-bounded applications such as CFD codes, and
it hides technical details of MPI applications from the programmer.

1   Introduction

Writing software for the Computational Sciences is a task becoming more and
more challenging: For several years, performance has been the only metric measuring the quality of software. Today, scientists insist on fast code that, in addition, is extendable, portable, maintainable and is delivered in time keeping step
with the hardware's and algorithms' development. Furthermore, the software is
to be embedded seamlessly into the whole simulation pipeline [10].

  Many approaches have been developed to tackle these challenges: Problem
solving environments [9] address the software's integration into the application
landscape. Sophisticated language libraries such as expression templates [12] and
well-engineered, standardised application programming interfaces such as MPI
allow for an abstraction from hardware and optimisation details. Domain specific languages enable scientists and engineers to concentrate on modelling, and
profiling tools and simulators [13] detect bottlenecks during the development.
All these examples are chosen arbitrarily, and the list is to be continued.

  In this paper, we focus on the question whether the object modelling facilities
of C++ are satisfying for scientific software. Several observations can be made:

M. Bubak et al. (Eds.): ICCS 2008, Part III, LNCS 5103, pp. 213–222, 2008.

- The range of a variable's value frequently is well-known in Scientific Computing. C++ cannot exploit this knowledge which results in a waste of memory:
  Two integers with range from 0 to 500, e.g., typically consume eight bytes
  although the whole information could be stored within two bytes.
- Many algorithms don't have to store all their record's attributes all the
  time. Often, some attributes are not needed for example in-between two
  iterations. In analogy to databases, we either call an attribute persistent or
  non-persistent if we want to be able to discard it. With C++, there is no way
  to define attributes that are sometimes not to be stored within containers
  depending on the object's life cycle.
- Mapping C++ classes onto MPI data types is an error-prone task as the
  language has no built-in support for MPI and as developers often are experts
  in Mathematics, Physics or Engineering, e.g., and not that familiar with MPI
  interna.

During the development of a new CFD code called Peano [1], these issues came
into play. They even influence considerably any realisation discussion up to now:
CFD codes often are memory-bounded [4]. High memory requirements for objects that are processed each iteration are thus annoying. As we implement a
domain decomposition [2], each developer has to map his records onto MPI data
types. All the people involved have a strong mathematical and CFD background.
To force them to think about memory layout, techniques saving individual bits
and MPI specifics keeps people from their actual work and, sometimes, the resulting code does not meet high quality standards.

Thus, we extend the C++ programming language with new modelling directives and provide a code generator acting as preprocessor and mapping these
extensions onto standard C++. The generator is called DaStGen (data structure
generator). We implemented it as plain, lightweight, stand-alone tool in Java, i.e.
it does not require any third-party components and is easy to extend and tailor.
The parser front-end is realised with SableCC [3]. Using DaStGen reduces the
amount of memory needed by the application, reduces the development time,
and reduces the number of bugs typically introduced by programmers. In this
paper, we present the generator and the language extension, as we believe that
many C++ applications besides CFD codes can also benefit from the techniques
implemented.

Hereby, we turn our attention to the primitive built-in C++ types apart from
floating point numbers, and we discuss how to reduce the amount of memory
required for them. We have not found a similar, as simple to use and lightweight
tool for this issue so far. Also, we touch on the subject inheritance, but we
do not take into account all the problems introduced by pointers and dynamic
data structures. The idea to generate MPI data types automatically is not new.
Actually, it has been discussed several times in more detail including dynamic
data structures and pointers (see the tools MPIECC [11] or C++2MPI [5], e.g.).
Nevertheless, our restricted but much simpler solution fits perfectly into the
concept of a precompiler reducing memory requirements, our solution allows for

exchange of a subset of an object's attributes, and the resulting code comes along without requiring for additional libraries and tools.

The remainder is organised as follows: First, we present the extension of the C++ class declaration directives, and we specify the mapping onto plain C++ (Sec. 2). Second, we give three small application examples extracted from the CFD code: classes for a solver for the Poisson equation, classes for an adaptive Cartesian grid management, and, finally, a combination of the two into a solver for the Poisson equation on adaptive Cartesian grids (Sec. 3). Third, we show the reduction of memory requirements, a source code extract proving the amount of manual work saved, as well as some preliminary runtime results (Sec. 4). Some concluding remarks in Sec. 5 close the discussion.

2   Specification

In DaStGen, the user defines classes' attributes in C++ syntax. These attributes have to have a built-in type. Arrays are supported. The generator converts the specification into C++ code where the attributes are hidden and setter and getter operations are provided. Hereby, DaStGen also supports preprocessor macros and compile-time constants. The user just has to declare the latter ones with the keyword `Constant:` before they are used. Afterwards, one can for example define an attribute `double a[D]`, where D is a symbol defined when the generated code is compiled.

2.1   Boolean Attributes and Bit Fields

If an attribute's type is boolean, C++ realises the attribute inefficiently in terms of memory, as the compiler maps the type onto a primitive data type such as byte [6]. The attributes of the instances of the class in Fig. 1, e.g., are mapped to a sequence of $2 + D$ primitives. If a boolean is stored in one byte, this results in a memory need of $2 + D$ bytes for information that could be encoded in $2 + D$ bits. DaStGen tracks all the booleans and arrays of booleans in a class, and stores the resulting information into one single primitive per class — the target bit field.

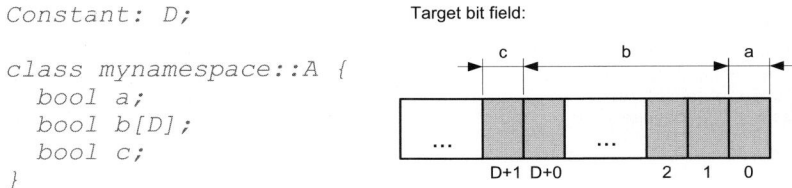

Fig. 1. Very simple class declaration (left) and corresponding target bit field storing all the $D + 2$ bit values (right). This target bit field would be used if all the attributes were marked as packed.

The type of the bit field is declared via `Packed-Type:` and it typically is a byte or an integer.

In the example in Fig. 1, the target bit field's first bit represents the value of attribute a, the bits 1 up to D represent b, and the $(D+2)$th bit holds c. The generated getter and setter extract and modify information within the bit field. They are implemented via fast bit-wise operations. Whether DaStGen packs data into the target bit field or maps it directly on a standard C++ attribute is controlled via the keyword `packed` for each attribute, i.e. the mapping is applied only if the keyword `packed` is written before the type of an attribute. The code in Fig. 1 lacks this keyword. Thus, it would be mapped 1:1 onto a C++ class.

2.2   Enumerations and Integers with Restricted Range

As enumerations are mapped onto numbers by any C++ compiler internally, the idea is obvious to take them into account for the bit packing, too. Otherwise, each enumeration with k variants needs at least one byte, although one could encode the information in $\lceil \log_2 k \rceil$ bits. Hence, DaStGen reserves a fitting number of bits within the target bit field for each enumeration type and stores the information there, if the enumeration attribute is marked by `packed`.

A similar reasoning holds for integers where the range is known a priori. DaStGen allows the programmer to append `from x to y` to any integer attribute marked with `packed` and, afterwards, uses only the actual number of bits required to store the value. This works for an arbitrary number of integers.

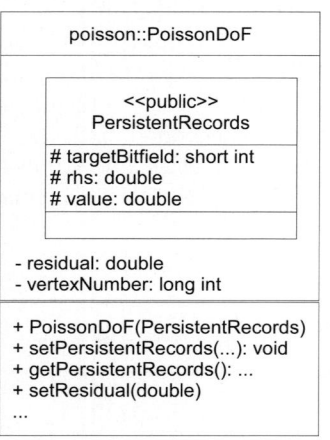
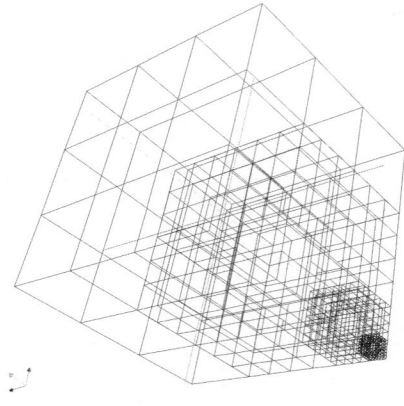

Fig. 2. The persistent attributes are stored within an embedded type of their own (left). Furthermore, all the packed attributes of type boolean, array of bit and integer augmented by a range are stored within one bit field. Right-hand side: Adaptive grid resolving a singularity for the Poisson equation.

2.3   Persistent Attributes

Characteristically for Scientific Computing, often not all the attributes of a class are valid throughout the complete object life cycle. For our PDE solver, e.g., we had temporary variables such as the local residual being important during an iteration but useless in-between two iterations. If one stores objects within a container (an array, e.g.), the size of the object determines the memory needed as the container does not know which attributes have semantics and which do not and, thus, are not to be stored. Memory is wasted.

Hence, we introduce the keyword pair `persistent` and `discard`. For the generated C++ class, DaStGen creates an embedded subclass holding only attributes marked as `persistent` (see Fig. 2). Depending on the programmer's needs, one can thus store the complete object or only this embedded type. Getters, setters and a constructor to read and modify the embedded type are created automatically.

2.4   Attributes Sent Via MPI

MPI offers the powerful mechanism of user-defined data types making MPI work with C++ objects. Nevertheless, to write the mapping between C++ classes and the MPI data types is error-prone, and the code has to be updated every time the data structure is modified. This task becomes cumbersome if several `#ifdef` require for different mappings to be written for different compile environments.

DaStGen automatically generates user-defined MPI data types for all `#ifdef` combinations. Similar to Sec. 2.3, users often want to communicate only a subset of the attributes in order to downsize the MPI messages. Thus, DaStGen introduces an additional keyword `parallelize` (or the equivalent `parallelise`) to mark attributes to be taken into account for communication via MPI.

Each generated class then has a public static attribute `Datatype` of type `MPI_Datatype`. The attribute can be passed to MPI opperations like any built-in MPI type. Thus, the generated code does neither require any additional libraries or files, nor does DaStGen pose any restrictions on the type of MPI operations. The user can send and receive both (subsets of) objects and arrays of them. A serialisation of data connected by pointers is beyond our scope.

2.5   Inheritance and Hierarchy Flattening

The additional attribute markers of DaStGen allow for a significant reduction of object size by packing information into bit fields and distinguishing between attributes that are persistent and those that are not. Furthermore, DaStGen simplifies the communication via MPI. To facilitate the modelling of data structures, we find it useful to support inheritance. Yet, if the inheritance is mapped directly to C++ inheritance, this implies additional memory requirements to hold the inheritance relationship, and it keeps DaStGen from optimising the memory consumptions aggressively.

If a class holds one bit and one of several subclasses holds another bit, these two bits cannot be packed into one target bit field with standard C++ mechanisms. Although a workaround might be possible, we decided to introduce a flattening operator into DaStGen: Every time DaStGen is passed a (multiple) inheritance with the keyword Extends:, the involved classes are merged and the hierarchy is removed. This works for multiple levels of inheritance, too. Afterwards, the merged class is processed as a standard class modelled via DaStGen. The inheritance structure is thus lost in the generated code, but the memory consumption can be reduced. Furthermore, the generation of MPI datatypes works straightforward.

2.6   Name Transformations, MPI Aspects and C++ Interna

Some additional features such as a plugin mechanism to add aspects [7] to the generated code or to validate class and attribute names concerning coding conventions complete the generator. For the application examples, e.g., we provide a plugin adding send() and receive() operations to the classes, and adding #pragma pack directives to tune the memory layout to minimal memory requirements.

3   Application Example

The following three examples extracted from our CFD code Peano [1] demonstrate the DaStGen extensions in action. In the end, we will use them to measure the gain in memory efficiency in Sec. 4.

3.1   A Solver for the Poisson Equation

For a simple solver for the Poisson equation arising from our CFD code's pressure computation, we modelled the unknowns of the PDE as follows:

```
Packed-Type: short int;
class poisson::PoissonDoF {
  enum DoFState { UNDEF, INNER, DIRICHLET };
  persistent packed DoFState state;
  persistent double rhs;
  persistent parallelise double value;
  discard parallelise double residual;
  discard long int vertexNumber;
};
```

Throughout the solver iteration, the local residual is needed for computations and the degree of freedom's number of is required by the visualisation interface. Both values are not to be stored in between iterations. For parallelisation, the solver implements a domain decomposition. Hereby, only the unknown's actual value and the residual have to be interchanged as all the other values can be determined locally.

3.2   An Adaptive Grid Management

Besides the solver, we implemented a grid management for adaptive Cartesian grids [8]. Here, the basic element is the grid vertex holding all local structural information such as refinement properties, refinement status and the information if the vertex is a hanging node. Information on the vertex's position and refinement level is only stored or communicated in the debug mode. Thus, debugging becomes easier.

```
Constant: DIMENSIONS;
class grid::Vertex {
  enum RefinementControl {
    UNREFINED,REFINED,REF_TRIGGERED,REFINING,COARS_TRIGGERED,COARSENING
  };
  persistent packed bool isPersistentVertex;
  persistent parallelise packed RefinementControl refinementControl;
  persistent parallelise packed bool refinementData[DIMENSIONS];
  #ifdef Debug
  persistent parallelise packed int level;
  persistent double x[DIMENSIONS];
  #endif
};
```

3.3   An Adaptive Poisson Solver

Finally, we combined the code for solving the Poisson equation with the adaptive grid management. The class fitting to this combination has to hold the attributes of both the Poisson solver's and the adaptive grid management's class. To avoid code duplication, we introduced a new vertex class inheriting from the two classes. Hereby, PoissonDoF.def and Vertex.def are the file names the examples from above are written to.

```
Extends: PoissonDoF.def;
Extends: Vertex.def;
class poisson::PoissonVertex {};
```

4   Results

This section evaluates the influence of the DaStGen features on our CFD code. The experiments were conducted on a Pentium 4, 3.4 GHz processor with 2 GByte RAM. They yield three interesting observations: First, due to packing the bits and splitting up the data into persistent and not persistent attributes, the amount of memory required per object is reduced by a factor of two or more (Table 1). The memory reduction effect especially benefits from the inheritance flattening: The size of PoissonVertex is smaller than Vertex's size added PoissonDoF's size.

Second, there's no runtime breakdown to be observed although the getter and setter for the packed records now incorporate several bit shift operations

Table 1. Number of bytes required with different features from Sec. 2 enabled. d represents the constant DIMENSIONS. Memory alignment is switched off.

	poisson PoissonDoF	grid Vertex d=2	poisson PoissonVertex d=2	grid Vertex d=3	poisson PoissonVertex d=3
debug mode	36	19	55	20	56
1:1 mapping	36	7	43	8	44
persistent	20	7	27	8	28
packed	34	2	34	2	34
persistent and packed	18	2	18	2	18
memory improvement	2	3.5	2, 38	4	2, 44

(Fig. 3). Further investigations are necessary, but the result suggests that the amount of additional work to be done is compensated by the speedup due to the reduction of data to be transferred by the memory bus. This does not hold if one artificially increases the computational work to be done per instance, i.e. this property only holds for memory-bounded algorithms.

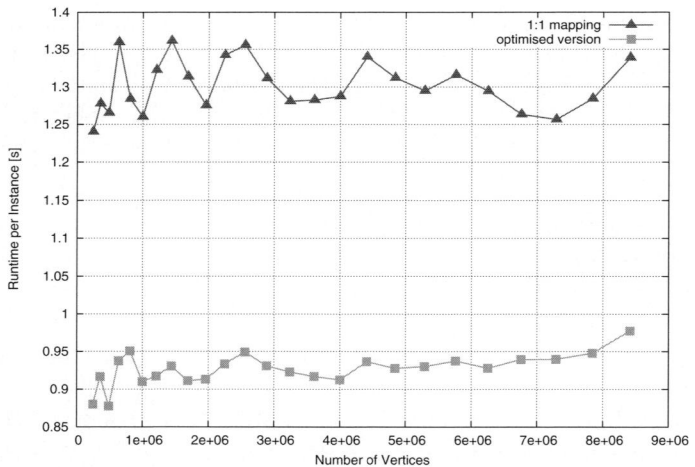

Fig. 3. Runtime per object instance for 1:1 mapping and the version generated by DaStGen. Measurements result from the two-dimensional case.

Finally, the source code fragment in Fig. 4 shows how many technical details for the mapping of C++ classes onto MPI data types can be hidden from the developer due to the usage of the generator.

```
const int Attributes = 3;
MPI_Datatype subtypes[Attributes] =
  {MPI_DOUBLE,MPI_DOUBLE,MPI_UB};
int blocklen[Attributes] =
  {1,1,1};
MPI_Aint     disp[Attributes];
PoissonDoF dummy[2];
MPI_Aint base;
MPI_Address(
  &(dummy[0]),
  &base);
MPI_Address(
  &(dummy[0]._persistent_records._value),
  &disp[0]);
MPI_Address(
  &(dummy[0]._residual),
  &disp[1] );
MPI_Address(
  &(dummy[1]._persistent_records._value),
  &disp[2] );
for (int i=0; i<Attributes; i++)
  disp[i] -= base;
MPI_Type_struct(
  Attributes, blocklen, disp,
  subtypes, &PoissonDoF::Datatype );
MPI_Type_commit(
  &PoissonDoF::Datatype );
```

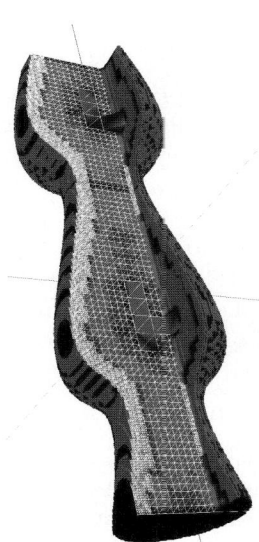

Fig. 4. Source code extract from the generated MPI data type declaration (left). Typical grid for our CFD experiments (right).

5   Concluding Remarks

DaStGen is a small but powerful tool to map augmented C++ classes to a memory efficient realisation supporting MPI in plain C++. The resulting code requires less memory than a standard implementation, and it runs faster when the application is memory-bounded. Although writing the efficient implementation manually is not of great difficulty for an experienced programmer, using DaStGen is of great value: It reduces the coding time significantly. it reduces the number of bugs in pure technical parts of the code, it allows the programmer to focus on algorithms and semantics instead of the realisation, and it enhances the readability and reusability of the model.

With all the multicore systems coming up, we believe that the performance impact of the records' size will become more and more important. Throughout the discussion of the record size, we have concentrated on integers and enumerations. Yet, the bigger part of a simulation's memory footprint typically stems from floating point numbers. How to apply a packing to these is not clear a priori. Furthermore, it is questionable whether the performace would benefit, as no equivalent for the fast bit-wise operations is available, as many architectures handle floats completely different compared to integers (they are not held in the L1-cache, e.g.), and as floating point operations last long. Although DaStGen's valuable keyword **persistent** is independent of this reasoning, further research on record size reduction might be profitable.

Right now, getting rid of technical details enables faster rapid application prototyping, and, due to simple modelling facilities, people are more willing to evaluate different algorithmic approaches based upon different record layouts. From a technical point of view, the parallel extension of any application's part becomes simpler. Hence, the tool shows how scientific code—especially memory-bounded code—benefits from application and domain specific language extensions allowing for sophisticated modelling, efficient memory usage and aspect oriented programming.

Acknowledgements. This work has partially been funded by DFG's research unit FOR493, the DFG project HA 1517/25-1/2 and TUM's International Graduate School of Science and Engineering (IGSSE) as part of the Excellence Initiative of the German Federal and State Governments.

References

1. Brenk, M., Bungartz, H.–J., Mehl, M., Muntean, I.L., Neckel, T., Weinzierl, T.: Numerical Simulation of Particle Transport in a Drift Ratchet. SIAM Journal of Scientific Computing (2008) (accepted)
2. Bungartz, H.–J., Mehl, M., Weinzierl, T.: A Parallel Adaptive Cartesian PDE Solver Using Space–Filling Curves. In: Nagel, W.E., Walter, W.V., Lehner, W. (eds.) Euro-Par 2006. LNCS, vol. 4128, pp. 1064–1074. Springer, Heidelberg (2006)
3. Gagnon, E., Hendren, L.J.: SableCC, an Object-Oriented Compiler Framework. In: TOOLS 1998: Proceedings of the Technology of Object-Oriented Languages and Systems, pp. 140–154. IEEE Computer Society, Los Alamitos (1998)
4. Gropp, W., Kaushik, D., Keyes, D., Smith, B.: Towards Realistic Performance Bounds for Implicit CFD Codes. In: Proceedings of Parallel CFD 1999, pp. 241–248. Elsevier, Amsterdam (1999)
5. Hillson, R., Iglewski, M.: C++2MPI: A Software Tool for Automatically Generating MPI Datatypes from C++ Classes. In: Proceedings of the International Conference on Parallel Computing in Electrical Engineering, pp. 13–17. IEEE Computer Society, Los Alamitos (2000)
6. Hyde, R.: Write Great Code - vol. 2: Thinking Low-Level. In: Writing High-Level. No Starch Press (2004)
7. Kiczales, G., Lamping, J., Menhdhekar, A., Maeda, C., Lopes, C., Loingtier, J.-M., Irwin, J.: Aspect-Oriented Programming. In: Aksit, M., Matsuoka, S. (eds.) ECOOP 1997. LNCS, vol. 1241, pp. 220–242. Springer, Heidelberg (1997)
8. Mehl, M., Weinzierl, T., Zenger, C.: A cache-oblivious self-adaptive full multigrid method. Numerical Linear Algebra With Applications 13(2-3), 275–291 (2006)
9. Parker, S.G., Weinstein, D.M., Johnson, C.R.: The SCIRun computational steering software system. Modern Software Tools in Scientific Computing, 1–40 (1997)
10. President's Information Technology Advisory Committee: Computational Science: Ensuring America's Competitiveness (2005)
11. Renault, É.: Extended MPICC to Generate MPI Derived Datatypes from C Datatypes Automatically. In: Cappello, F., Herault, T., Dongarra, J. (eds.) PVM/MPI 2007. LNCS, vol. 4757, pp. 307–314. Springer, Heidelberg (2007)
12. Veldhuizen, T.: Expression Templates. C++ Report 7(5), 26–31 (1995)
13. Weidendorfer, J., Kowarschik, M., Trinitis, C.: A Tool Suite for Simulation Based Analysis of Memory Access Behavior. In: Bubak, M., van Albada, G.D., Sloot, P.M.A., Dongarra, J. (eds.) ICCS 2004. LNCS, vol. 3038, pp. 440–447. Springer, Heidelberg (2004)

RMOST: A Shared Memory Model for Online Steering*

Daniel Lorenz[1,2], Peter Buchholz[1], Christian Uebing[3], Wolfgang Walkowiak[1], and Roland Wismüller[2]

[1] Experimental Particle Physics, University of Siegen, Germany
[2] Operating Systems and Distributed Systems, University of Siegen, Germany
[3] Center for Information and Media Technology, University of Siegen, Germany
daniel.lorenz@uni-siegen.de

Abstract. Online steering means to visualize the current state of an application which includes application data and/or performance data, and to modify data in the application. Thus, in online steering the application as well as the steering tool must concurrently access and modify the same data at run time. In this paper a new model for online steering is presented which models the mechanism of online steering as access to a distributed shared memory. The integrity requirements of the steered application are analyzed. The integrity can be ensured through an appropriate consistency model. Finally, the online steering system RMOST is presented which is based on the distributed shared memory model and can be used to steer Grid jobs from the High Energy Physics experiment ATLAS.

1   Introduction

In recent times, scientific simulations increased both in complexity and in the amount of data they produce. Often, the simulations run on batch systems in clusters or computational Grids and do not support interactivity during the runtime of the simulation. Online steering of an application enables the visualization of intermediate results, performance data, or other application data, and the invocation of actions, e.g. modification of a parameters by the user at runtime of the job. The user can interactively explore parameter realms, debug his program, or optimize performance. Beause the user sees results earlier, he can evaluate results earlier and react before the job has finished. Thus, online steering accelerates scientific research and saves ressources.

In this work application means the steered program. A steering tool is distinguished from a steering system. A steering tool is the interface to the user which visualizes data and offers the user the possibility to enter commands, e.g. modifications of a parameter. A steering system comprises all extentions to the application, external components, specialized steering tools, and extentions to offline visualization tools to enable steering. For example, a steering system can

---

* This work is partly funded by the *Bundesministerium für Bildung und Forschung* (BMBF) as part of the German e-Science Initiative (Contract 01AK802E, HEP-CG).

M. Bubak et al. (Eds.): ICCS 2008, Part III, LNCS 5103, pp. 223–232, 2008.

comprise steering libraries on the application side, special processes which are needed for the communication or automated decision making, and libraries on the steering tool side to extend an existing visualization.

Until now, various steering systems have been developed [1,2,3,4,5,6,7] for supporting the scientist with interactive control over his simulation. Existing systems provide means to retrieve data from the application and invoke actions in the remote application. Typically, the application is instrumented with calls to a steering library to enable the sending of data to a remote visualizer, or to apply commands from the user. The steering system manages the data transport to a customized user interface. One of the reasons why steering systems are not used is the required effort to instrument a legacy application for steering.

In this paper another approach is used, which views steering similar to distributed shared memory (DSM). Any steering is basically the change of state of an application and a state change corresponds to a change in memory. Thus, steering can be modeled as a case for DSM because all steering actions can be reduced to memory access operations. This approach simplifies the application of steering systems to existing software and improves the efficiency of steering.

If a data object is modified in a running application without any synchronization with the execution of the applications, severe errors may occur.

In these cases the integrity of the data in the application is broken. To protect the integrity, rules are needed which define the order of access operations on the shared data. The necessary rules define a consistency model. Though various steering tools exists, until now no consistency model for online steering exists. In most cases the integrity problem is not addressed or left to the user.

Based on the DSM-based steering model, the new online steering system RMOST (Result Monitoring and Online Steering Tool) [8,9] was developed. The advantages for the user are the ease of use of a DSM-like approach and the build-in consistency guarantees.

2 Formalism for the DSM Based Model for Online Steering

In online steering, the application and the steering tool access both the same data. If the steering tool and the application run in the same address space, it is a trivial task. But if visualization and simulation have different address spaces, e.g. if they are located on different machines, a mechanism to access the remote data is needed. Thus, online steering can be modeled as DSM. The advantages of a DSM model are that the complexity of distributed data is hidden from the user of the steering system, and it looks like accessing only local data for the steering tool and the application. The steering system completely handles the communication and it supports the programmer with the consistency guarantees to maintain data integrity.

In the DSM based model of online steering, two kinds of processes exists with different roles and properties. Firstly, n application processes $p_1, ..., p_n$ exist. The application may synchronize $p_1, ..., p_n$ with any mechanism, e.g. MPI, or shared memory. However, the synchronization within the application is out of scope of this work. Secondly, m steering processes $p_{n+1}, ...p_{n+m}$ exist, each representing a steerer in a collaborative environment. The data objects which can be visualized or steered reside in the distributed shared memory. Each data object o has a home location $H(o) \in p_1, ..., p_n$ which is one of the application processes. The steering processes are not chosen for home locations, because the steerers may detach and thus causing the home location to be inaccessible.

Three kinds of memory operations exist: read operations r, write operations w, and synchronization operations s. Read and write operations are denoted as $o(p, x, v)$ where $o \in \{w, r\}$ specifies the operation type, p is the process that perform the operation, x is the memory location, and v is the value that is written or read. Synchronization operations are denoted as $s(p, x)$. A process p_i is viewed as a sequence of memory operations $S_i = \{o_1, o_2, ...\}$ with $o_i \in \{w, r, s\}$. A process *sees* a write operation w if a current read operation would return the value written by w. A write operation w is *visible* to a process p if p can see w.

Each application process p_i is associated with a logical clock T_i, which indicates the progress of the process. T_i in incremented when p_i release an update to the distributed shared memory, or when p_1 sees an update of another process. Thus, synchronization operations imply clock incrementations. An *epoch* is the interval between two consecutive clock incrementations. The furthest common logical time $T_{min} = \min(T_1, ..., T_n)$ is the minimum time of all application processes. The furthest time $T_{max} = \max(T_1, ..., T_n)$ is the maximum time of all application processes.

3   Data Integrity

Data integrity is an important prerequisite to obtain correct results from the application. This means a steering system should ensure that the displayed data is consistent in itself, and any modifications must preserve the integrity of the data within the application. In this section, the effects that might affect the integrity are analyzed which lead to two integrity conditions. The first one is the *inner-process* condition, and the second one is the *inter-process* condition.

3.1   The Inner-Process Condition

The inner-process condition requires that the data in the application must not be modified externally during certain operation intervals, and that the write operations of the application to shared objects become only visible if the data is in a well-defined state. For example, assume one formula is computed where one variable x appears at different places in the formula. The result can only be correct if the value of x stays the same during the whole computation. Another case could be a numerical n-body simulation. While it is allowed to modify parameters between each simulated time step, the value should stay the same inside each simulated time step.

Also the modifications of the application to shared data should become visible only at well defined places. Imagine several properties of different input objects are computed. If the object is visible and displayed after the computation of the first few properties while the other properties stem from another input, the displayed result is propably incorrect and can be misleading. Thus, to preserve the inner-process condition, changes of the application must only become visible at well defined points, and changes by the steerer must only be applied by the application at well-defined *synchronization* points. Typically, one epoch is bounded by two synchronization points, which implies that synchronization points match the incrementations of the logical clock.

3.2   The Inter-Process Condition

The inter-process condition considers the different progress of different processes. Firstly, it requires that write operations of the steering processes must be seen in all processes at the same time step. Secondly, values of displayed data objects must stem from the same epoch.

For example, suppose a parallel simulation iterates over several time steps and each process computes a part of the overall result. If changing a boundary parameter, one would like to change this parameter at all processes in the same epoch. If a steerer changes the value of the parameter in the DSM, the system must ensure that the modification is viewed by all processes at the same epoch.

Another case occurs if a steerer wants to display a distributed object which is modified by several processes, and each process computes a part of the whole object. The steerer must only see the writes of all processes up to T_{min} to retrieve an internally consistent data set. Writes of processes that have proceeded further ahead must not be visible to the steerer to provide a well-defined display of intermediate results.

4   Consistency Models

To ensure the integrity of the data in online steering, each process must view access operations to the shared memory according to certain rules. For each given set of access operations, a consistency model is defined through the possible orders in which each process is allowed to see the memory accesses [10]. Thus, a consistency model can be used to maintain data integrity. In this section, consistency models are evaluated which fulfill the requirements for data integrity in online steering. One consistency model will not satisfy all cases, because not all data objects require both integrity requirements analyzed in Sec. 3. Some data objects have no integrity conditions and can be treated completely asynchronous, some data objects have only the inner-process condition, and some data objects require both conditions. Thus, different consistency models are appropriate to each of these cases. The case that data objects have only the inter-process condition is not considered, because the inter-process condition implies the existence of epoches. The transition points between two epoches define the synchronization points where values may be read or modified.

4.1   Consistency for the Inner-Process Condition

The inner-process condition allows the application and distribution of updates only at special synchronization points. The desired existence of special synchronization points leads to a consistency model which is similar to weak consistency [11]. Two possibilities exist which behave different in the following case: Let p_1 be an application process and let p_2 be a steerer process that viewed the accesses $w(p_1, x, 1)$, and $s(p_1)$. Now, p_2 executes $w(p_2, x, 2)$, and $r(p_2, x, ?)$ before it sees another $s(p_1)$. Which value should $r(p_2, x, ?)$ return?

1. $r(p_2, x, 1)$ returns the current value of the application. The newly written value is not seen until the next synchronization operation. This model delays the execution of the write operation after the next synchronization operation, thus it is called *delayed weak consistency*. This consistency model displays always a consistent set of values from the application, but it has the effect that a read operation at the steerer may not return the value written by the previous write operation. This behavior can be interpreted as display of results. An advantage of this model is that it does not require a sequential order of the synchronization points.
2. $r(p_2, x, 2)$ returns the value recently written by the same process. This leads to weak consistency with the modification that updates are applied exactly at the next synchronization operation, instead of *latest* at the next synchronization operation. In this case a read operation of the steerer may return a value that is not consistent with the results from the application. The displayed data equals the value the application sees when it enters the next epoch. It can be interpreted as display of the configuration.

Interestingly, the sequential consistency [12] is too strong for the inner-process condition. In most DSM systems, the usage of relaxed consistency models is driven by the better performance of the relaxed models compared to strong consistency models, but the programmer wants his program to behave like sequential consistency [12] would be used. In the case of online steering, strong consistency would not provide the desired behavior. In the example shown above both cases violate the rules of sequential consistency. With delayed weak consistency a read does not return the value of the most recent write, and with weak consistency p_1 and p_2 view write operations in different orders.

If synchronization operations are not global but only for one or a few data objects, release like consistency models [13] can be derived. But this reduces the advantage of a simple instrumentation, because it requires more detailed information about which data is updated at each synchronization point.

4.2   Consistency for Both Conditions

In this case it must be ensured that the steerers retrieve all values from the same epoch, and all application processes apply all modifications at the same epoch. At every given time, each epoch can be assigned to one of the following three groups:

- The *past* are those epoches T that are finished by all application processes: $T < T_{min}$.
- The *future* are those epoches T that are not yet entered by any application process: $T > T_{max}$.
- The *presence* are the epoches T that do neither belong to the past nor to the future: $T \in [T_{min}, T_{max}]$.

Each write operation w is tagged with a time stamp $T(w)$. Write operations of an application process p will be tagged with the timestamp of the process $T(w) = T(p)$. Write operations of a steering processes will be tagged with $T_{max} + 1$. Thus, each data object has a schedule of values assigned to it. A read operation of data object x by a steering process will always return the most recent value v from the past. Read operations of an application process p at time step $T(p)$ will always return the most recent value v from the viewpoint of the process.

This consistency model is called *schedule consistency*. Steerers can only write to the future and read from the past. It has the effect, that modifications are not seen immediately, but after a delay which depends on the length of the presence. The delayed weak consistency is a special case of the schedule consistency with the presence comprising only one epoch. Formally, this effect is caused by an reordering of write and read operations in the steerer processes. Writes that occur before a read in program order may be seen later than the read.

4.3   Consistency with No Integrity Conditions

Beside parameters or results which probably have the inner-process or inter-process condition, data objects with a producer-consumer access pattern exists which require none of the integrity conditions. These data object have one producer, which is the only process writing to this data object, and one or more consumer processes who read this data object. For example, processor load or other monitoring data has neither the inner-process nor the inter-process condition. For those data the update intervals or delays implied by the weak or schedule consistency may be inappropriate. These data objects are independent from other data objects by definition, thus Pipelined RAM consistency [14] should be sufficient. Pipelined RAM consistency ensures that all processes view the writes of a process p in the order they are executed by p.

5   Implementation in RMOST

RMOST (Result Monitoring and Online Steering Tool)[1] [8,9] is an online steering system for Grid Jobs of the High Energy Physics (HEP) experiment ATLAS [15]. It consists of an application independent implementation of the presented DSM approach for online steering, and a thin integration layer into the ATLAS software. Through the DSM-based approach it is possible to enable steering of Grid

---

[1] RMOST can be downloaded from http://hep.physik.uni-siegen.de/grid/rmost

Jobs in the ATLAS experiment without modification of the source code. Currently, only sequential applications with one steerer are supported. Its architecture consists of four main layers:

1. The communication layer realizes a communication channel between the application and the steering tool. The Grid communication channel of RMOST is described in [16].
2. The data consistency layer implements a DSM system with the consistency models described in Sec. 4.
3. The data processing layer is a place holder for any data processing performed by the steering system like filtering, or automated evaluation.
4. The data access layer provides tools for data access. For example, in RMOST a preloaded library replaces standard library calls in order to observe file accesses. Another (not yet implemented) possibility is to monitor method calls by modifying a classes' virtual table.

5.1   Data Consistency Layer

The data consistency layer provides a framework for several consistency protocols implementing different consistency models. The framework consists of the registry, the manager, and an interface for consistency protocols.

The registry contains for every steerable data object its name, the used protocol, and the access methods of the local copy. If several processes register data with same name, these objects are considered as local copies of the same value. The manager handles all communication in a separate thread. Asynchronous messages are immediately forwarded to the appropriate protocol, while synchronized message types are buffered until the next synchronization operation.

Consistency protocols can send messages via the manager and are called on every synchronization point, when a message is received for it, or if a data object is accessed which uses this protocol. Currently, for delayed weak consistency, pipelined RAM consistency, and blockwise delayed weak consistency both an invalidate and an update protocol are implemented. As example, the update protocol for the weak and for the delayed weak consistency are explained:

If a process uses the update protocol for the weak consistency and a write occurs, it sets a modification flag for this data object. At the next synchronization point, the process sends updates of all data objects whose modification flags have been set. If a process receives an update, it is buffered until the next synchronization point. At this point the new value is applied. If an update was received and the local modification flag of this data object is set, too, the value from the steering process has priority. For weak consistency the synchronization points must be ordered sequentially. Thus, a process must obtain a synchronization lock before it can execute a synchronization point.

If the application uses delayed weak consistency, it performs the same actions as with weak consistency except that it does not obtain the synchronization lock. If the steering tool performs a write operation, the new value is buffered, but not yet applied locally. At the next synchronization point an update is send to the application. If the application receives an update, it applies it at its next synchronization point. Afterwards, it sends an acknowledgment back to the steering tool. The steering tool applies the new value at the next synchronization point after receiving the acknowledgment.

6 Application of RMOST

The ATLAS [15] experiment is performed at the Large Hadron Collider (LHC) at CERN. Beside many others, the most prominent goal of the ATLAS experiment is to find the Higgs particle which is responsible for the masses of particles.

The experiment software framework Athena [17] was created for the computation of the data and is commonly used in the HEP community. The processed data consists of collision events which can be computed independently. In general, the desired results are statistics over several thousands events.

The Athena framework [17] provides different components which can be plugged together by the user through a so called job options file. Furthermore, Athena can be extended with customized components contained in a shared library. The different components can be categorized into several basic classes. The two important classes for the implementation of RMOST are algorithms and services. The core of an Athena job is a list of algorithms which are executed for each event. Services provide functionality to other components.

The ROOT toolkit [18] is commonly used for offline visualization of physical results. It provides an interface to extend ROOT with new classes which are located in a shared library and loaded dynamically. Modifications and recompilations of the ROOT toolkit and the Athena framework to enable steering are hardly accepted by the HEP community. Thus, for the integration of RMOST in the Athena framework a new algorithm RM_Spy was developed which can be applied to the Grid job by editing the job options file. RM_Spy enables the steering of the job execution, monitoring of intermediate results in the output files, and modification of the job options file. The steering API is encapsulated by a new Athena service RM_SteeringSvc. Thus, steering of Athena jobs is enabled without modification of the source code of existing components. Other components can be extended with customized steering features by using the RM_SteeringSvc.

Steering can be made available to ROOT by dynamically loading an extra library with interface classes for ROOT to RMOST. It allows to modify steerable parameters, or view progress information from the job. Through preloading of the RMOST file access library, the steering system intercepts file accesses and fetches or updates the according parts of the file. Thus, the existing offline visualization in ROOT can be used for online monitoring of intermediate results and steering without modifications of the source code.

7 Related Work

Until now, no general DSM-based model for steering exists. However, some steering tools provide tools to support the user to maintain the integrity of the data.

Closest to this work is the Pathfinder [5] steering system. Steering actions and the program's execution are both viewed in terms of atomic transactions. They address the issue to consistently apply steering actions to a parallel message passing program. A steering action is consistent if it is applied in a consistent snapshot of the parallel program. An algorithm is presented which detects inconsistent steering actions. The mayor issue is to define points in a parallel application where steering actions can be consistently applied. As result a global ordering of all transactions exists, which leads to sequential consistency.

CUMULVS [2,19] is a steering tool which allows to make checkpoints of a parallel program. An algorithm is presented to capture distributed data objects consistently by stopping processes that have already processed ahead until all processes reached an equal progress. While this algorithm is similar to the presented schedule consistency, CUMULVS has no DSM-based model for steering.

EPSN [20] requires a description of the structure of the application. For each steerable data object, areas are defined where the data object may be read or changed. The source code of the application must be instrumented with markers to the abstract structure. VASE [6] follows similar principles. The integrity problem is brought to an abstract level which can simplify the problem for the user. However, the decision where a data object may be accessed without disrupting integrity stays with the user. Both have no DSM-based approach for steering.

In RealityGrid [1] a client/server based steering system was developed. The steering library only informs the application on events which must be handled by the user. The user may use predefined library calls to react on events, but a DSM like mechanism does not exist. The steering actions are performed in a single steering library call to reduce the effort of instrumentation. Because events are processed in a single function, by default, weak consistency is implicitly realized.

8   Conclusions and Future Work

The data integrity of an application can be destroyed through online steering. Two major conditions that ensure data integrity are identified, the inner-process condition and the inter-process condition. Online steering is viewed as a access to distributed shared memory. The integrity of the data can be maintained if the steering systems provide certain consistency guarantees. This allows to easily apply steering to existing legacy codes. In the case of the ATLAS experiment it was possible to enable offline legacy codes for online steering without changing existing codes by just adding components to the framework. Furthermore, offline visualization tools could be used for online visualization.

The necessary conditions may vary between different objects of the same application, thus a steering system should support a number of consistency models. First measurements show, that invalidate protocols effectively avoid overload on the network, which happens in stream-based steering tools. On the other hand, update protocols are faster for small amounts of data. We are currently working on an automatic selection between update or invalidate protocols that dynamically adapts to the environment and optimizes the performance.

References

1. Jha, S., et al.: A computational steering API for scientific Grid applications: Design, implementation and lessons. In: Workshop on Grid Application Programming Interfaces (September 2004)
2. Geist, G.A., et al.: CUMULVS: Providing fault-tolerance, visualization and steering of parallel applications. Int. J. of High Performance Computing Applications 11(3), 224–236 (1997)
3. Vetter, J.S., et al.: High performance computational steering of physical simulations. In: Proc. of the 11th Int. Symp. on Parallel Processing, pp. 128–134. IEEE, Los Alamitos (1999)
4. Ribler, R.L., et al.: The Autopilot performance-directed adaptive control system. Future Generation Computer Systems 18(1), 175–187 (2001)
5. Hart, D., et al.: Consistency considerations in the interactive steering of computations. Int. J. of Parallel and Distributed Syst. and Networks 2(3), 171–179 (1999)
6. Brunner, J.D., et al.: VASE: the visualization and application steering environment. In: Proc. 1993 ACM/IEEE conference on Supercomputing, pp. 560–569 (1993)
7. Brodlie, K., et al.: Visualization in grid computing environments. In: Proc. of IEEE Visualization 2004, pp. 155–162 (2004)
8. Lorenz, D., et al.: Online steering of HEP Grid applications. In: Proc. Cracow Grid Workshop 2006, Cracow, Poland, Academic Computer Centre CYFRONET AGH, pp. 191–198 (2007)
9. RMOST webpage, http://www.hep.physik.uni-siegen.de/grid/rmost
10. Steinke, R.C., Nutt, G.J.: A unified theory of shared memory consistency. J. of the ACM 51, 800–849 (2004)
11. Dubois, M., et al.: Memory access buffering in multiprocessors. In: ISCA 1986: Proc. of the 13th annual int. symp. on Computer architecture, pp. 434–442. IEEE Computer Society Press, Los Alamitos (1986)
12. Lamport, L.: How to make a multiprocessor computer that correctly executes multiprocess programs. IEEE Trans. Comp. C-28(9), 690–691 (1979)
13. Gharachorloo, K., et al.: Memory consistency and event ordering in scalable shared-memory multiprocessors. In: 25 Years ISCA: Retrospectives and Reprints, pp. 376–387 (1998)
14. Lipton, R.J., Sandberg, J.S.: PRAM: A scaleable shared memory. Technical Report CS-TR-180-88, Princeton University (September 1988)
15. The ATLAS Experiment, http://atlasexperiment.org
16. Lorenz, D., et al.: Secure connections for computational steering of Grid jobs. In: Proc. 16th Euromicro Int. Conf. on Parallel, Distributed and network-based Processing, Toulouse, France, pp. 209–217. IEEE, Los Alamitos (2008)
17. Athena Developer Guide, http://atlas.web.cern.ch/Atlas/GROUPS/SOFTWARE/OO/z0_obsolete/architecture/General/Tech.Doc/Manual/AthenaDeveloperGuide.pdf
18. Brun, R., et al.: ROOT - an object oriented data analysis framework. In: Proc. AIHENP 1996 Workshop. Number A 389 in Nuclear Instruments and Methods in Physics research, pp. 81–86 (1997) (1996)
19. Papadopoulos, P.M., et al.: CUMULVS: Extending a generic steering and visualization middleware for application fault-tolerance. In: Proc. 31st Hawaii Int. Conf. on System Sciences (HICSS-31) (January 1998)
20. Esnard, A., Dussere, M., Coulaud, O.: A time-coherent model for the steering of parallel simulations. In: Danelutto, M., Vanneschi, M., Laforenza, D. (eds.) Euro-Par 2004. LNCS, vol. 3149, pp. 90–97. Springer, Heidelberg (2004)

A Semantic-Oriented Platform for Performance Monitoring of Distributed Java Applications

Włodzimierz Funika, Piotr Godowski, and Piotr Pęgiel

Institute of Computer Science, AGH, al. Mickiewicza 30, 30-059 Kraków, Poland
funika@agh.edu.pl, {flash,pegiel}@student.agh.edu.pl
Phone: (+48 12) 617 44 66; Fax: (+48 12) 633 80 54

Abstract. In this paper we present an approach to semantic performance analysis in on–line monitoring systems. We have designed a novel monitoring system which uses ontological description for all concepts exploited in the distributed systems monitoring. We introduce a complete implementation of a robust system with semantics, which is not biased to any kind of the underlying "physical" monitoring system, giving the end user the power of intelligent monitoring features like automatic metrics selection and collaborative work.

1 Introduction

The design of a distributed application is in many cases a challenge to the developer ([1,2,3,4]). On the one hand, there are the limitations and performance issues of distributed programming platforms. So one of the most important tasks is to increase the performance and reliability of distributed applications. On the other hand, the developer must assure that the application manages and uses distributed resources efficiently. Therefore, understanding application's behaviour through performance analysis and visualization is crucial. It is especially true now, when many distributed systems exploit the SOAP protocol, where functionality of the program is implemented as Web Services. The monitoring of data flow between components could be very helpful for the user to discover performance problems with a system.

The biggest problem when using performance tools (especially, these working "on-line") is their complexity. Thus many users benefit from often less complex but easier to use tools. So a very important task is to ease user's interactions with the monitoring system, moreover, to turn these interactions into a kind of collaboration activities with the system, which involve other users. Certainly, "simple" should not imply "limited functionality" related to performance evaluation. Nowadays, more and more developed software use software agents which *guide* the user step-by-step. Such agents usually use a semantic description of software's features and through the analysis of user's behaviour provide suggestions what to do to achieve a desired result.

A similar approach can be used in tools used for performance monitoring in the distributed environment. The first steps have been done (AutoPilot, PerfOnto),

M. Bubak et al. (Eds.): ICCS 2008, Part III, LNCS 5103, pp. 233–242, 2008.

but their authors aimed at developing their own architecture for semantic description (mostly based on feedback from metrics) from scratch, omitting the already existing solutions (like OWL/RDF). The Semantic Web paradigm has introduced the concept of semantic description of resources (OWL/RDF, DAML) and services (mostly Web Services – OWL-S, DAML-S). We can leverage from existing standards to develop a performance monitoring tool using some knowledge which describes performance metrics, and this is the primary goal of our paper.

The rest of this paper is organized as follows: Section 2 discusses a motivation and system use cases. Related work is discussed in Section 3, in Section 4 we present our proposed ontology and system architecture for on-line monitoring system with semantics, followed by Summary and Future work in Section 5.

2   System Use Cases

In this paper we are presenting a semantic-oriented monitoring infrastructure called *SemMon*. The architecture of the tool fits into the OMIS model [10] and is capable to co-operate with available monitoring systems (like J-OCM[8], JMX[1]).

The complexity and heterogeneity of the technologies necessitates to introduce semantics into the distributed computing monitoring because; the large amount of hardware, software and network environments makes monitoring a challenging task. Semantics enables the system to automatically process data without supervision or customized processing for specific areas, enables "understanding" what is really monitored, which in turn reduces the time the user spends on manually searching for issues and shortens the system learning curve. Having a semantic description and taxonomy of the monitored elements and their contexts, the system is "smart" enough to guide its user throughout the whole monitoring/analysis process. The user can focus on its main task: to find performance issues within limited time, based on the system guidance coming from historic analysis and being able to add their own measurements when needed.

Semantics in the monitoring architecture should exploit as much as possible from existing solutions, libraries and tools, with special attention paid to Open Source software and solutions developed in European projects (like GOM [9] developed in the K-Wf Grid project [7]). The following general use cases show the usability of the designed system.

The *user* should be able to:
- monitor the performance of a Java application running under control of a *physical* monitoring system
- use the system in an automatic way with a set of metrics which are meaningful for the user and a desired result
- get information about metrics that should be called in a next step.

The *system administrator* should be able to:
- create, destroy, and insert a semantic description of available metrics and elements of the monitored system

---

[1] Stands for *Java Management Extensions*.

- provide new metrics in a *physical* monitoring system, and describe them in semantic way
- manage historical performance data.

The work should be based on a portable (preferably XML) client–server protocol both for sending and receiving requests about semantic descriptions. Developing a system as a Web Service (with semantic description provided) leads to a universal solution, which fits into the Semantic Web paradigm. The developed system should be designed in such a way that it should work with any existing "native" grid-enabled monitoring system. The system should be able to integrate with existing ontologies describing resources and performance measurements which should be a great benefit for system administrators. The designed system should be able to be extended with sensors and metrics strongly related to the structure of the monitored application to point the actual and the most accurate source of the data.

3   Related Work

In this section we will concentrate on those available monitoring systems where semantics or flexible monitoring architecture are introduced.

Gemini [5] is a Grid monitoring framework that fulfills a gap between resources monitoring components and monitoring services clients. It performs measurements using a set of loadable modules called sensors which retrieve monitoring data on its own or by using external applications for this purpose. Although the Gemini framework is powerful in its flexibility of adding new *sensors*, it does not use any kind of semantics for selecting performance metrics to run and analyse or for providing any guidance to the user.

Autopilot [11] has been developed within the Grid Application Development Software (GrADS) Project [6] and is responsible for adaptive control of distributed applications. Autopilot's architecture comprises performance sensors and a decision control unit using fuzzy logic to analyse received data from sensors and preparing messages to actuators. Autopilot is the very first example of exploiting some kind of semantics usage, or rather *fuzzy logic* usage to help with monitoring and adaptation actions.

PerfOnto [12] is a new approach to performance analysis, data sharing and tools integration in Grids that is based on ontology. PerfOnto is an OWL ontology describing experiment-related and resource-related *concepts*. The experiment-related concept describes experiments and their associated performance data on applications. The prototype PerfOnto system is able to search data in an ontological (i.e. using a knowledge base) manner, e.g. to find a code region executed on a particular node with a metric exceeding a threshold value, thus giving a hint to the site scheduler to migrate a job to another node. PerfOnto gives a rich description of performance data, but does not provide any automation for using it. Whereas using much of PerfOnto's taxonomy and retaining the main idea of describing resources in form of ontology, we were able to significantly extend it and provide adaptation algorithms.

4   Overview of the SemMon System

The visualization of monitoring data in a "user friendly" form is one of the most key features provided by any performance monitoring system. Due to the great amount of gathered information, proper presentation and interpretation of observation results becomes a very complicated task. So steering the visualization of monitoring data involving making decisions what, when, in what form, under which circumstances should be presented to the user is a challenge.

A high level architecture overview of SemMon is introduced in Fig. 1.

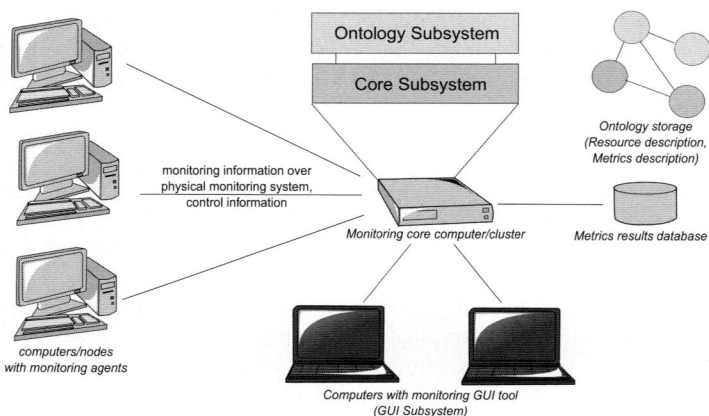

Fig. 1. System architecture as distributed environment. Monitoring core computers include *Core Subsystem* and *Ontology Subsystem*.

The heart of the model are *computers* that provide a primary system functionality like processing an ontology with Resource Capabilities and Metrics or storing monitoring data.

To support *knowledge persistency* a database is required. This functionality is implemented in the *Ontology subsystem*. Another part of this node is support for a "physical" monitoring system. This subsystem has to provide a functionality for registering monitoring agents as well as for processing monitoring data.

The second part of the system contains computers with monitoring agents. Agents expose monitored resources to our monitoring system. All of them will register to the Core subsystem, afterwards Core is able to introspect possible resources that are exposed. Agents are programs on the *nodes/computers* that access the "physical" monitoring system, e.g. JMX, JOCM.

The last part of the system are GUI clients connected to the Core subsystem. GUI is an environment for *collaborative work* – the users share metric ranks between different GUI instances in order to help other users in proper decision making. In the following we focus on a description of the components of the SemMon monitoring system.

4.1   Ontology Subsystem

The Ontology subsystem is the heart of the whole system. The key aspect to understand here is the *ontology* term. An ontology is an explicit specification of a conceptualization. In such an ontology, definitions associate the names of entities (e.g., classes, relations, functions, or other objects) with a human readable text describing what the names are meant to denote, and formal axioms that constrain the interpretation and well formed use of these terms, formally specified with the OWL language. The Ontology Web Language (OWL) is intended to be used when the information needs to be processed in an automatic way by applications, as opposed to situations where the content only needs to be presented to humans. OWL has powerful facilities for expressing meaning and semantics and thus OWL goes beyond all other similar languages in its capability to represent a machine interpretable content on the Web.

The Ontology subsystem contains methods for parsing, automatic interpretation, searching, creating, and, finally, saving and sharing ontology data. The Ontology subsystem brings a unique feature to the designed system: the capability of *interpreting what is monitored* both for system users and (what is even more important) for the system itself. Using the knowledge deployed in the underlying ontology data, the system is aware what is monitored and what should be monitored in a next step within the monitored application's lifetime. Every single type of resource accessible to the monitoring system is described in the OWL ontology and reflects a natural computing resources hierarchy. Part of the description or even the whole of it can be updated.

Resources in question are: *Resource classes* (like Node, CPU, JVM), *Resource instances* (i.e. OWL instances of resources available in the underlying monitoring system, like `CPU_i386_node2_cluster1`) and the *measurable attributes* for the resource instances. Each Resource class defines which measurable attributes are available for its instances. A measurable attribute, called in this paper *ResourceCapability*, might be both an atomic attribute (like `LoadAvg1Min`) or an OWL superclass for a set of *ResourceCapabilities*. This way a natural hierarchy of capabilities can be constructed. A special property `hasResourceCapability` is a glue between Resources and *ResourceCapabilities*. Any type of Resource can contain any number of Resource Capabilities. Fig. 2 shows the Resources ontology class hierarchy while Fig. 3 presents a fragment of the ResourceCapabilities ontology class hierarchy.

An ontology describes metric concepts like the OWL classes or individuals describing metrics available to be executed by the user. Metrics ontology reflects the metrics hierarchy (i.e. from the most generic metric to the most specific one) in order to provide a rich description for ontology reasoners. Metrics can be *simple*, i.e. the metric is able to measure only one attribute or *custom*, which means that the metric can be applied to as many capabilities as required and it is even possible to provide custom implementations for metrics (*user-defined metrics*).

A metrics ontology is designed from a flat list of all available metrics to be considered by the monitoring system. However, having only a flat list without a hierarchy (specialization) introduced, it is impossible to provide any powerful

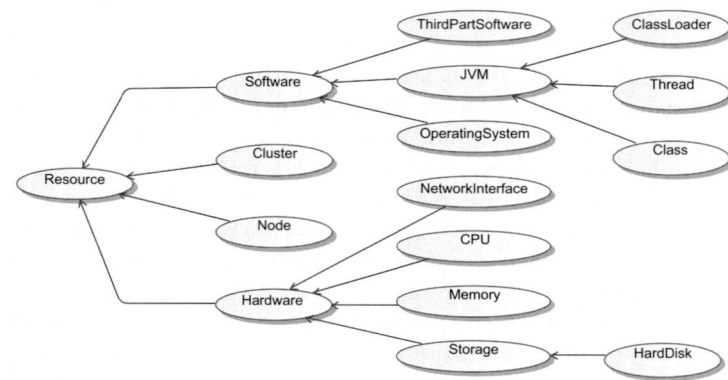

Fig. 2. Resources ontology diagram

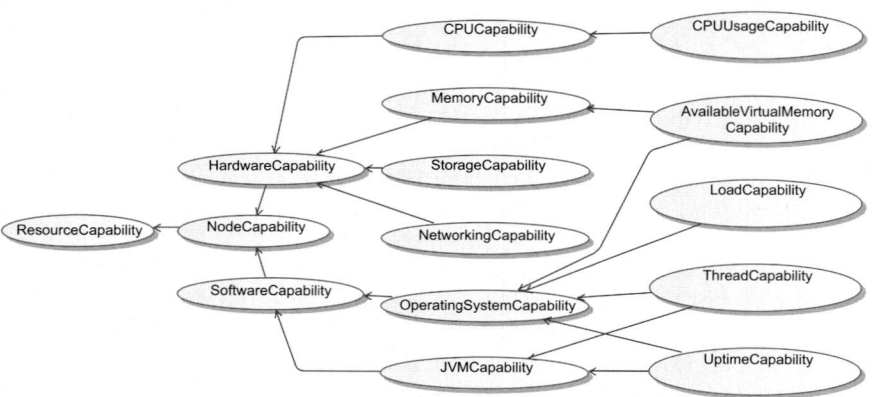

Fig. 3. ResourceCapabilities ontology diagram

reasoning process. This is because no "generic-specific" or "is related to" relationships are provided. Looking at a flat list of all possible metrics, the next step is to find out which of them are *generic* and which are *specific*. Such relationships can be expressed in an ontology as the `rdfs:subClassOf` property. A sample superclass metric might be `SoftwareMetric` with its specific subclass `JVMThreadCPUTimeMetric`. As a result, metrics form a tree which can be used for a reasoning process.

A special metric property `monitors` is a glue between the Metrics ontology and the Resources and ResourcesCapabilities ontology. Property `monitors` has a domain in the `AbstractMetric` class (and its subclasses) and a range in the `ResourceCapability` classes. Because the cardinality of this property is not limited, any type of `AbstractMetric` is able to monitor any number of capabilities. This means that the total number of measurements available in the system does not equal to the number of subclasses and individuals of the `AbstractMetric` class, but is a sum of cardinalities of `monitors` properties in the metrics ontology. Metric property `hasCustomImplementationClass` is used to inform the system that the metric is a *custom metric*, i.e. has its own implementation. This property points to the fully qualified Java class name implementing the Custom Metric interface. Custom Metric has its own implementation rules that is exactly returned as a measurement process, which is explained as follows. Since Custom Metric can access the Core public API, and the Resources registry, it can request any number of capabilities' values from the underlying monitoring system. The only contract that Custom Metric must meet is to return a single number each time it is requested for.

It should be noted that the described Ontology subsystem should be able to coexist with any already existing ontology for resources description and matching.

4.2   Core Subsystem

The Core subsystem is responsible for connecting to the underlying monitoring system's initialization (using its protocol adapter mechanism), deploying, initializing and executing metrics (including user-defined metrics), providing an interface to the Ontology subsystem and last but not least, exposing a public (remote) interface for GUI clients to connect to. Core also manages GUI clients subscribed to the list of connected resources, running metrics, running metric values and alarms (i.e. conditional action metrics notifications). The Core subsystem comprises three components – *Adapter, Resource Registry*, and *Remote interface for GUI*. The *Adapter* component follows the commonly used Adapter structural design pattern and is used for "translation" of all Core requests into the requests specific to the underlying monitoring system (JMX, J-OCM, OCM-G, etc.). Due to the major differences and interface incompatibilities of a wide range of monitoring systems available on the market, a common interface called *Protocol Adapter* is designed. *Resource Registry* is a service that leverages both Core and Protocol Adapter. Resource Registry holds (with Protocol Adapter) all the resource instances found as visible in the underlying monitoring system and maps them into Core identifiers. Protocol Adapter resolves incompatibility issues between different physical monitoring systems. User-defined metrics have full access to the public Core API. Therefore a user-defined metric (implementing the *Custom Metric* interface) can introspect Resource Registry and with a Protocol Adapter implementation is capable to send a specific query to the underlying monitoring system. This feature is useful when the underlying monitoring system has some specific features, not covered by the generic Protocol Adapter interface.

Remote interface for GUI allows remote GUI clients to connect to SemMon to enable collaborative work and provides:

– notifications for: newly attached and detached *monitoring systems*, started and stopped *measurements* on the Core subsystem, and, finally, notifications for measurement values
– interface for alarms – Alarms are *conditional action metric* notifications. When some *action metric* is running on the Core subsystem and its value ex-

ceeds a declared *threshold action value*, all the *unconditional action metrics* that are declared in the underlying ontology are sent as notifications to all the subscribed GUI users. The user is enabled to take an action to resolve the alarm (e.g. to start a new metric from within a list of metrics suggested by the system).

4.3   Sample Use of SemMon

The below real-life example shows a few of the key SemMon features. A SemMon system user is monitoring a complex distributed application with critical problems relating to unstable memory usage over the application life time occurring only on a single node of the cluster. The monitored application is a WebService-enabled Java server, deployed across a cluster of processing nodes placed behind a restrictive firewall with a load balancer. The usual behaviour on the correctly deployed system in question is to consume 50% of the CPU time for each node and create no more than 100 threads per JVM instance.

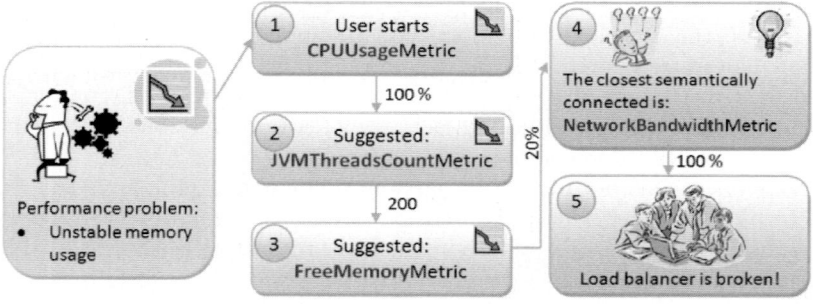

Fig. 4. Sample analysis with SemMon

At first, the user decides to monitor a CPU usage on the heaviest loaded node (see step *1* in Fig. 4). When a CPU burst lasts for at least 3 minutes, SemMon deducts from the metrics ontology a "critical" situation, and calculates the next most probable metric to start. Since CPU load is semantically connected with the number of JVM live threads, the adequate metric is suggested and the user follows this guidance (*2*). Again, the number of threads is irrationally high (over 200 threads), so a new alarm is raised and SemMon "reasons" that since the CPU load and number of threads were already monitored, it would be reasonable to monitor memory usage (*3*). Since the memory usage is at 80% level of the available virtual memory, a new alarm is raised. This moment is the key in the described monitoring scenario: since the observed CPU load, memory usage, and JVM threads count are extremely high, the algorithm selects the *Network Bandwidth* metric to run (*4*). A motivation for doing this is that this metric had been frequently selected by other system users in the past and it is semantically connected with memory usage, CPU load, and JVM threads (there is a possibility that application threads are processing data incoming from the network). SemMon has calculated the best matching metric and the user is able to see that almost the whole available network bandwidth is consumed by the incoming traffic. This suggests that either the cluster system is overloaded (which is not the case since we observe high load only on the certain node), or the load balancer is broken. The user checks the network bandwidth on the rest of the cluster nodes and does not observe any significant incoming traffic. This leads to the conclusion that the problem lies not in the monitored application, but in the load balancing component (5).

Please note that the path followed by the user comprises both hardware (low level) and software (high level) metrics. It shows how flexible a reasoning process might be when the knowledge stored in SemMon holds possibly a full description of the environment. It is also possible to track down performance issues not only in the monitored application, but also in its environment.

5   Summary and Future Work

The main objective of this paper was to present the design and implementation of a robust and flexible semantics-oriented monitoring system, SemMon. It seems to be one of the first complete approaches to the joint "worlds" of on-line distributed monitoring and Semantic Web.

The *SemMon* system extensively uses ontology for semantic description of all concepts used in. It is as much flexible as it can be, starting from picking up automatic ontology changes, through automatic metric selection assistance, collaborative users' knowledge leveraging, user-defined metrics, finally, to the extensible and clear visualisation options.

There are still places for improvements. There is a unresolved problem with performing part of the computations on the clients to improve system scalability by reducing the size of performance data sent to a central database. An important task is to explore algorithms for reasoning in the ontology frameworks. Although there are some improvements in the query algorithms, they are just based on additional caching layer rather than optimizing algorithms.

Acknowledgements. The research is partially supported by the EU IST 0004265 CoreGRID and 031857 `int.eu.grid` projects with the related SPUB-M grant.

References

1. Gerndt, M., Wismüller, R., Balaton, Z., Gombás, G., Kacsuk, P., Németh, Zs., Podhorszki, N., Truong, H.-L., Fahringer, T., Bubak, M., Laure, E., Margalef, T.: Performance Tools for the Grid: State of the Art and Future. APART-2 Working Group, Research Report Series, Lehrstuhl für Rechnertechnik und Rechnerorganisation (LRR-TUM) Technische Universität Muenchen, vol. 30. Shaker Verlag (2004) ISBN 3-8322-2413-0
2. Podhorszki, N., Kacsuk, P.: Presentation and Analysis of Grid Performance Data. In: Kosch, H., Böszörményi, L., Hellwagner, H. (eds.) Euro-Par 2003. LNCS, vol. 2790, pp. 119–126. Springer, Heidelberg (2003)

3. Reed, D.A., Ribler, R.L.: Performance Analysis and Visualization. In: Foster, I., Kesselman, C. (eds.) Computational Grids: State of the Art and Future Directions in High-Performance Distributed Computing, pp. 367–393. Morgan-Kaufman Publishers, San Francisco (1998)
4. Wismüller, R., Bubak, M., Funika, W.: High-level application-specific performance analysis using the G-PM tool. Future Generation Comp. Syst. 24(2), 121–132 (2008)
5. Balis, B., Bubak, M., Labno, B.: GEMINI: Generic Monitoring Infrastructure for Grid Resources and Applications. In: Bubak, M., Unger, S. (eds.). Proc. Cracow Grid Workshop 2006. The Knowledge-based Workflow System for Grid Applications, pp. 60–73. ACC Cyfronet AGH, Poland (2007)
6. Berman, F., Chien, A., Cooper, K., Dongarra, J., Foster, I., Johnsson, L., Gannon, D., Kennedy, K., Kesselman, C., Reed, D., Torczon, L., Wolski, R.: The GrAds project: Software support for high-level grid application development. Technical Report Rice COMPTR00-355, Rice University (2000)
7. Bubak, M., Fahringer, T., Hluchy, L., Hoheisel, A., Kitowski, J., Unger, S., Viano, G., Votis, K., K-WfGrid Consortium: K-Wf Grid Knowledge based Workflow system for Grid Applications. In: Proc. Cracow Grid Workshop 2004, p.39. ACC CYFRONET AGH, Poland (2005) ISBN 83-915141-4-5
8. Funika, W., Bubak, M., Smętek, M., Wismüller, R.: An OMIS-based Approach to Monitoring Distributed Java Applications. In: Kwong, Y.C. (ed.) Annual Review of Scalable Computing, ch.1, vol. 6, pp. 1–29. World Scientific Publishing Co. and Singapore University Press, Singapore (2004)
9. Krawczyk, K., Slota, R., Majewska, M., Kryza, B., Kitowski, J.: Grid Organization Memory for Knowledge Management for Grid Environment. In: Proc. Cracow Grid Workshop 2004, pp.109-115. ACC CYFRONET AGH, Krakow, Poland (2005) ISBN 83-915141-4-5
10. Ludwig, T., Wismüller, R., Sunderam, V., Bode, A.: OMIS – On-line Monitoring Interface Specification (Version 2.0). LRR-TUM Research Report Series, vol. 9. Shaker Verlag, Aachen (1997)
11. Ribler, R.L., Vetter, J.S., Simitci, H., Reed, D.A.: Autopilot: Adaptive Control of Distributed Applications. In: Proc. 7th IEEE High-Performance Distributed Computing Conference (1998)
12. Truong, H.-L., Dustdar, S., Fahringer, T.: Performance metrics and ontologies for Grid workflows. Future Generation Comp. Syst. 23(6), 760–772 (2007)

A Tool for Building Collaborative Applications by Invocation of Grid Operations

Maciej Malawski[1], Tomasz Bartyński[2], and Marian Bubak[1,2]

[1] Institute of Computer Science, AGH, Mickiewicza 30, 30-059 Kraków, Poland
[2] Academic Computer Centre CYFRONET AGH, Nawojki 11, 30-950
Kraków, Poland
{malawski,bubak}@agh.edu.pl, t.bartynski@cyfronet.pl

Abstract. The motivation for this work is the need for providing tools which facilitate building scientific applications that are developed and executed on various Grid systems, implemented with different technologies. As a solution to this problem, we have developed the Grid Operation Invoker (GOI) which offers object-oriented method invocation semantics for interacting with computational services accessible with diverse middleware frameworks. GOI forms the core of the ViroLab virtual laboratory engine and it is used to invoke operations from within experiments described using a scripting notation. In this paper, after outlining the features of GOI, we describe how it is enhanced with a mechanism of so-called *local gems* which allows adding high-level support for middleware technologies based on the batch job-processing model, e.g. EGEE LCG/gLite. As a result, we demonstrate how a molecular dynamics program called NAMD, deployed on EGEE, was integrated with the ViroLab virtual laboratory.

Keywords: tools, application building, Grid computing, virtual laboratory, EGEE, DEISA.

1 Introduction

Grid infrastructures have been considered the most appropriate platform for computational science for many years [1]. Mainstream projects providing such infrastructures in Europe include EGEE [2] and DEISA [3], but there are also other initiatives using various middleware frameworks, often based on a service-oriented architecture or component models. Building applications that can utilize these infrastructures remains a challenging task for programmers, due to the relatively low-level interfaces to computing resources, often limited to simple batch job submission. Therefore, research in the field of providing tools for the development of such programs is of great importance.

Such a challenge is faced by the *Virtual Laboratory* [4], which is developed in the scope of the ViroLab project. The experiments in this virtual laboratory are high-level applications which orchestrate many computational tasks run on the Grid. The notation used for specifying experiment plans uses the Ruby scripting language. This approach allows specifying arbitrary complex experiments in a

M. Bubak et al. (Eds.): ICCS 2008, Part III, LNCS 5103, pp. 243–252, 2008.

modern object-oriented dynamic language, thus giving the programmer full control and flexibility in experiment design. Scripts, being written in a full-fledged programming language, can define experiment logic using a rich set of control structures and also perform some computations locally.

To access the underlying Grid resources, a dedicated module of the virtual laboratory, called the Grid Operation Invoker (GOI) [5], has been developed. It applies an object-oriented model with remote procedure call semantics to dispatch computation in a uniform manner using diverse middleware technologies. Web Services and MOCCA [6] components were supported at the first development stage. GOI introduces multiple levels of abstractions, called *Grid Objects*, which allow users to interact with various middleware systems.

The main goal of the research presented in this paper was to extend GOI to support middleware technologies which are based on the job processing model, which is the case with EGEE and DEISA. Such infrastructures provide scientists with computational power, storage and a wide range of scientific applications. However, their resources are accessed with tools dedicated for one specific middleware package, which enables submitting jobs or sequences of jobs. In order to solve a scientific problem, it is often required to combine results produced by a set of these tools, as well as by local applications. This procedure is time-consuming and can be performed only by skilful users. Research can be facilitated by integrating all local tools, Web Services and Grid jobs into a single experiment which uses a uniform and simple notation to describe all steps of a scientific process and automate it entirely.

This paper is organized as follows: Section 2 gives an overview of the related work on providing access to Grid middleware systems. Subsequently, in Section 3, we introduce the main concepts of the Grid Operation Invoker and then, in Section 4 – its role in the virtual laboratory. Subsequently, in Sections 5 and 6, a detailed description of enhancements which were provided to add support for job-based middleware systems on the example of LCG/gLite (EGEE) is presented. In Section 7 we report on experiments which were performed in the virtual laboratory exploiting this new tool. The final section includes a summary and a brief presentation of future work.

2   Related Work

Numerous software frameworks have been developed to provide high-level access to Grid services using heterogeneous middleware systems. The Grid Application Toolkit (GAT) [7], currently evolving into the Simple API for Grid Applications (SAGA), provides a language-neutral API to basic Grid use cases, such as operations on files, monitoring events, resources, jobs, information exchange, error handling and security. However, it does not introduce an object-oriented API to invoke applications. A similar approach has been undertaken by the authors of the Grid Services Base Library (GSBL) [8], but it is still limited to such operations as job submission and file transfers.

Another high-level approach is implemented in NetSolve/GridSolve [9], which is an RPC-based system where a client delegates the execution of an operation to a selected server providing input parameters. The server executes the appropriate service and returns output parameters or error status to the client. Since GridSolve requires installation of specific servers, its usage on such infrastructures as EGEE is not straightforward.

Portal-based systems, like GridPortlets and OGCE [10], also provide the means of accessing multiple middleware technologies. These solutions are usually dependent on a specific portal technology (e.g. Java portlets), although recently, in VINE project, there have been efforts to extend their usability to more general applications.

Of note are systems used for migrating so-called *legacy code* applications to Grid or to Grid Services. Examples of such systems include LGF [11] which wraps legacy code as Globus 4 services on a fine-grained level, or GEMLCA [12], which offers a more coarse-grained approach. However, they are limited to a single middleware suite, such as Globus 4.

Other platforms which aim to facilitate the usage of Grids by scientific applications include workflow systems, such as K-Wf Grid [13], which manages workflows on multiple levels of abstraction; Kepler [14], which allows integrating multiple actor models, and Taverna [15], successfully applied to many life-science applications. The main drawback of workflow systems, in comparison to the scripting approach, is the limited expressiveness of graphical notations when applied to more complex experiments.

3   Grid Operation Invoker: Abstractions over the Grid

The Grid Operation Invoker is designed as a module of the Virtual laboratory engine which is responsible for communication with diverse underlying middleware technologies. Fig. 1 illustrates the Grid Object hierarchy. The main reason behind introducing this hierarchy and its associated layers of abstraction was that the complexity of the heterogeneous, distributed environment should be hidden from end users. Developers of an application should not be concerned with manually interfacing all underlying middleware technologies – they should instead be focused on the problem they are solving.

Each *Grid Object Class* is an abstract entity which defines a set of *Grid Operations*. These operations are invoked from the script, while the actual computation is performed on a remote machine. Each Grid object class may have multiple *Implementations* with different middleware technologies representing the same functionality. Each of the implementations may have multiple *Instances*, possibly running on different resources, thus with different levels of performance. Grid object instances of a specific class may use a variety of middleware suites and therefore must be interfaced using their specific protocols. Moreover, Grid objects may have various properties, such as stateless or stateful interaction mode, synchronous or asynchronous operation invocation or being private or shared between experiments runs and users. Developers are not concerned about finding the optimal instance and interfacing with it; however, they must be aware of each Grid object's properties. For instance, they must know whether a Grid object they are using preserves state between invocations of operations.

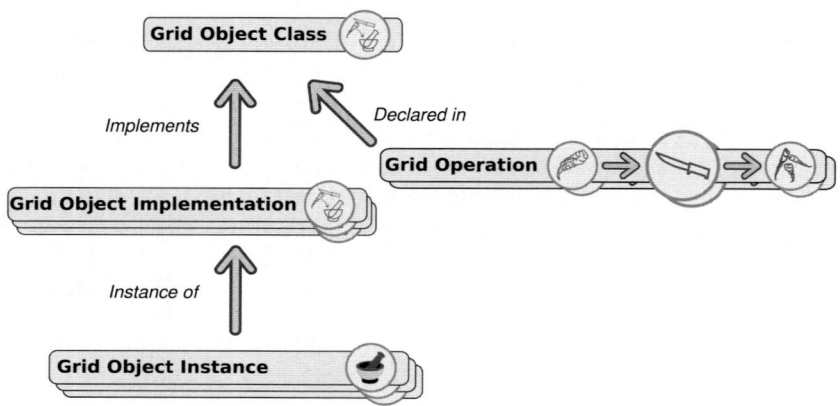

Fig. 1. Three levels of the abstraction over the Grid environment

```
require 'cyfronet/gridspace/goi/core/g_obj'

begin
    drs = GObj.create('org.virolab.DrugRankingSystem2')
    mutations = 'P1M I2L S3T P4Q E6G T7C V8T P9L V10N K11F V35T T39K'.split(' ')
    res = drs.drs('ANRS', 'reverse_transcriptase', mutations)
    puts res
end
```

Fig. 2. A sample ViroLab experiment invoking the drug ranking Web service using the Grid Object library

A sample script demonstrating the invocation of the Decision Support System (DSS) which suggests a drug ranking for a patient with a specific set of HIV mutations is shown in Fig 2. `GObj` is a factory for creating Grid objects representing the DSS Web service. Upon instantiation, the operations of a Grid object can be invoked directly, as seen in the next line of code. Please note that by using Ruby string operations, such as `split()`, simple conversions are possible (this would be nontrivial in the case of graphical workflow systems and would often require specific converter or adapter services).

4   Architecture of Grid Operation Invoker

The Grid Operation Invoker is a JRuby implementation of the library that provides a uniform interface to multiple middleware technologies. It supports abstraction over the heterogeneous environment described in Section 3.

Fig. 3 shows how GOI is positioned in the context of other modules of the virtual laboratory. GOI is a part of GSEngine, which is the main execution server for experiments, with an embedded JRuby interpreter. Descriptions of technical information of Grid Objects are stored in the external Grid Resource

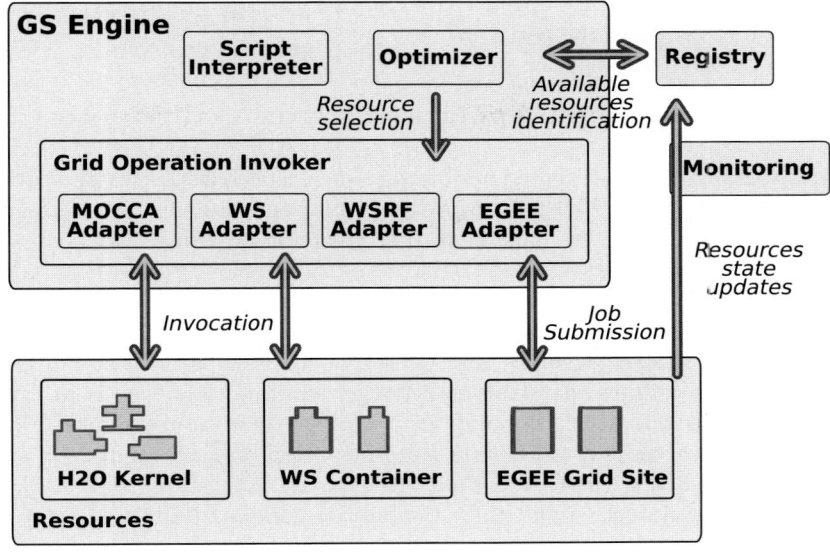

Fig. 3. Grid Operation Invoker in the context of the GSEngine

Registry service and the Optimizer module is responsible for selection of optimal instances if more than one instance is available for a specific object. The Grid Object Invoker has a modular architecture, which allows plugging in adapters for different technologies. Web Services and MOCCA components were supported from the beginning, while adapters for job-based middleware technologies such as EGEE and DEISA, are the subject of this research and are described in detail in the following sections.

5   Local Gems

Grid objects which represent application-specific functionality are often referred to as *gems*, by analogy to RubyGems [16] – a standard for distributing Ruby libraries. Examples of *ViroLab gems* are such services as the Drug Resistance Service [17] or the RegaDB HIV sequence alignment and subtyping tools [18], all wrapped as Grid Objects.

In addition to the Grid Objects corresponding to remote computations, *Local gems* are introduced as a way of representing local computation as a Grid object. From the application developer's point of view local gems are another computational technology and are accessed via the same uniform interface as other technologies. Local gems enable one to download the source of a Ruby class, evaluate the source at runtime and execute it locally. They facilitate sharing single classes that provide functionality usable for a scientific community, which is, however, too lightweight to be exposed as e.g. a Web service. Local gems are registered in the Grid Resource Registry as a *Grid Object Instance* and their source code is stored in the registry.

6   Representation of Applications Executed as Jobs

In order to integrate job-oriented middleware such as EGEE LCG/gLite with the Grid Operation Invoker, an object-oriented representative of a job is required. We require a Grid object implementation (technology adapter) which would delegate the invocation of its operations to the submission of jobs using specific Grid middleware, and return a result upon successful completion of a job.

In contrast to the already implemented adapter classes, capable of producing client-side *Grid Object Instance* representatives of Web Service and MOCCA middleware, it is not possible to implement a generic factory for representatives of jobs. Web Service and MOCCA components are, by their nature, object-oriented and are contacted using a well-defined interface. Representatives of *Grid Object Instances* published with these technologies are actually stubs (proxies) which provide the same interface and therefore can be generated automatically. Job-oriented middleware enables us to execute command-line applications which do not provide a remote API. Functionality provided by an application is organized in a set of methods, and is determined on the basis of command-line input parameters. As a consequence of this fact, the application has to be wrapped with a special class that exposes its functionality as Grid object methods.

```
require 'cyfronet/gridspace/goi/utils/lcg/edg_u_i_wrapper'
require 'cyfronet/gridspace/goi/utils/lcg/job_spec'

class NamdWrapper

 def molecule_simulate_submit(jobName, inputs, outputs, nodeNumber=1)
   @jobSpec = JobSpec.new
   @jobSpec.executable='/bin/bash'
   @jobSpec.arguments=('$VO_VOCE_SW_DIR/NAMD_2.6/namd.run #{inputs[0]}')
   if nodeNumber > 1
      @jobSpec.add_property('JobType', '"MPICH"')
      @jobSpec.add_property('NodeNumber', nodeNumber.to_s)
   end
   @jobSpec.stdoutput= jobName + '.out'
   @jobSpec.stderr= jobName + '.err'
   @jobSpec.add_to_output_sandbox(@jobSpec.stdoutput)
   @jobSpec.add_to_output_sandbox(@jobSpec.stderr)
   inputs.each{ |input|   @jobSpec.add_to_input_sandbox(input) }
   outputs.each{ |output| @jobSpec.add_to_output_sandbox(output) }
   return @jobSpec
 end
 ...
```

Fig. 4. Local class wrapping an EGEE *NAMD* job: the _submit() method

Each wrapper class should be application-specific. It is common practice in various legacy-code wrapping systems to define a special descriptor language (e.g. XML-based) to specify the mapping between object operations and specific command-line parameters or program execution. In our case, since Ruby is used as the implementation language, it is natural to also use Ruby for specification of this mapping. Therefore, we have decided that wrapper classes would be local gems, able to prepare inputs, submit a job, manage it and retrieve results. Such local gems, in turn, can use a lower-level Ruby API to interact with middleware-specific job management operations.

As an example of how such a local gem is built, let us consider a *NamdWrapper* class which enables us to use the molecular dynamics NAMD [19] application (Fig. 4), which has been installed on a Cyfronet EGEE site. The wrapper uses two classes provided by the Grid Operation Invoker: *EdgUIWrapper*, that allows using the EDG User Interface, and *JobSpec*, which generates a JDL file for a given job. The EdgUIWrapper class provides a Ruby API to LCG/gLite middleware by wrapping the command-line user interface.

```
...
  def molecule_simulate_get_output
    out = edg_job_get_output(@edgJobId)
    xscFile = out.get_file('alanin.xsc')
    return xscFile
  end
```

Fig. 5. Local class wrapping an EGEE *NAMD* job: the _get_output() method

The wrapper class must implement two methods for each method of the *Grid Object Instance* representative. For instance, representative of the NAMD application provides a *molecule_simulate()* method and the wrapper class must implement *molecule_simulate_submit()* and *molecule_simulate_get_output* methods. The former method (Fig. 4) generates a JDL file for a job that performs the requested computation and performs data conversions if required. The latter method (Fig. 5) retrieves the job's output and converts it to Ruby objects.

7   Sample Applications

As stated earlier, the Grid Operation Invoker was implemented and integrated with the ViroLab Virtual Laboratory. In addition to previously developed adapters for Web Services and MOCCA technologies, support for local gems is now available. Support for EGEE LCG/gLite middleware was implemented using local gems. In a similar way, gems from RegaDB for HIV genotyping [18] were integrated, using WTS [20] services.

A sample application made accessible in the Virtual laboratory is the NAMD molecular dynamics package. The source code of a sample experiment that performs a molecular dynamics simulation for an alanine amino acid in a water environment is presented in Fig. 6. At the beginning of the script, the developer requires a *GObj* class that provides a uniform interface to create *Grid Object Instance* representatives. Subsequently, a representative of the NAMD application is created. Properties of the job are defined (job name, input and output files and, finally, the number of nodes used for computation). The next line invokes the *molecule_simulate()* method which automatically generates a JDL file for the job, submits the job, monitors its status and retrieves results upon successful completion or informs the user about an error.

```
require 'cyfronet/gridspace/goi/core/g_obj'

namd = GObj.create('cyfronet.gridspace.gem.Namd')

jobName = 'namd'
inputs = ['alanin.namd', 'alanin.params', 'alanin.psf', 'alanin.pdb']
outputs = ['alanin.coor', 'alanin.vel', 'alanin.xsc']
nodeNumber = 4

namd.molecule_simulate(jobName, inputs, outputs, nodeNumber)
```

Fig. 6. Application submitting a job that executes *NAMD* on the EGEE

This experiment may be further extended by adding invocations of the locally installed VMD (Visual Molecular Dynamics) [21] toolkit, which is a molecular visualization program, in order to display the obtained results.

GOI is also used to run other experiments in the Virtual laboratory, including a full experiment called "from genotype to drug resistance". Data mining experiments can also be built using the Weka [22] toolkit wrapped as MOCCA components. At present, more services are being added as new gems; among them the Web services from the European Bioinformatics Institute [23].

8   Summary and Future Work

In this paper we have described the Grid Operation Invoker as a tool facilitating application building, used by scientific experiment developers who are familiar with simple script programming. To hide the complexity of details which are usually required to deal with Grid middleware technologies, we have introduced the Grid Object abstraction, which represents any type of computational resource. Support for Web services and MOCCA components as sample technologies was present from the earliest version of GOI. In this paper we have described how GOI was extended to support Grid objects representing local processing by *local gems* and how this mechanism can be used to add support for middleware systems based on the batch job submission processing model. By introducing simple wrapper classes implemented in Ruby, we can add an object-oriented interface to various applications executed on EGEE using LCG/gLite middleware. This new technology was tested using the NAMD molecular dynamics package and integrated into the ViroLab Virtual Laboratory.

Further work involves analysis, design and implementation of an introspection mechanism that will enable interactive execution of experiments. Moreover, we will work on a module enabling us to plug in diverse security mechanisms, such as GSI [24] and Shibboleth [25]. Adding support for more middleware technologies, such as UNICORE [26], AHE [27] and WSRF [28], is also under development. Thereafter we would like to integrate more *Grid Object Instances* to be able to build more complex experiments.

Acknowledgments. This work has been made possible through the support of the European Commission ViroLab Project [29] Grant 027446. This research is also partly funded by the EU IST CoreGRID Project and Polish SPUB-M grants. The authors would like to express their gratitude to Mariusz Sterzel for his hints on integrating NAMD with our system and deployment on EGEE, and to Tomasz Gubala, Eryk Ciepiela and Piotr Nowakowski for their comments and input.

References

1. Foster, I., Kesselman, K.: Scaling System-Level Science: Scientific Exploration and IT Implications. Computer 39(11), 31–39 (2006)
2. EGEE Homepage (2006), http://public.eu-egee.org/
3. DEISA project (1999), http://deisa.org
4. The ViroLab Virtual Laboratory Website (2008), http://virolab.cyfronet.pl/
5. Bartynski, T., Malawski, M., Gubala, T., Bubak, M.: Universal Grid Client: Grid Operation Invoker. In: PPAM. LNCS. Springer, Heidelberg (to appear, 2007)
6. Malawski, M., Bubak, M., Placek, M., Kurzyniec, D., Sunderam, V.: Experiments with distributed component computing across grid boundaries. In: Proceedings of HPC-GECO/COMPFRAME Workshop in Conjunction with HPDC 2006, pp. 109–116 (2006)
7. Grid Application Toolkit (2004), http://www.gridlab.org/WorkPackages/wp-1/
8. Bazinet, A.L., Myers, D.S., Fuetsch, J., Cummings, M.P.: Grid Services Base Library: A high-level, procedural application programming interface for writing Globus-based Grid services. Future Generation Comp. Syst. 23(3), 517–522 (2007)
9. NetSolve/GridSolve (2006), http://icl.cs.utk.edu/netsolve/
10. Zhang, C., Kelley, I., Allen, G.: Grid portal solutions: a comparison of GridPortlets and OGCE: Research Articles. Concurr. Comput. Pract. Exper. 19(12), 1739–1748 (2007)
11. Bubak, M., Baliś, B., Sterna, K., Bemben, A.: Efficient and Reliable Execution of Legacy Codes Exposed as Services. In: Shi, Y., van Albada, G.D., Dongarra, J., Sloot, P.M.A. (eds.) ICCS 2007. LNCS, vol. 4487, pp. 390–397. Springer, Heidelberg (2007)
12. Delaitre, T., Kiss, T., Goyeneche, A., Terstyánszky, G., Winter, S.C., Kacsuk, P.: GEMLCA: Running Legacy Code Applications as Grid Services. J. Grid Comput. 3(1-2), 75–90 (2005)
13. Neubauer, F., Hoheisel, A., Geiler, J.: Workflow-based Grid applications. Future Generation Comp. Syst. 22(1-2), 6–15 (2006)
14. Altintas, I., Jaeger, E., Lin, K., Ludaescher, B., Memon, A.: A Web Service Composition and Deployment Framework for Scientific Workflows. In: ICWS. IEEE Computer Society, Los Alamitos (2004)
15. Oinn, T.M., Greenwood., R.M., Addis, M., Alpdemir, M.N., Ferris, J., Glover, K., Goble, C.A., Goderis, A., Hull, D., Marvin, D., Li, P., Lord, P.W., Pocock, M.R., Senger, M., Stevens, R., Wipat, A., Wroe, C.: Taverna: lessons in creating a workflow environment forthe life sciences. Concurrency and Computation: Practice and Experience 18(10), 1067–1100 (2006)

16. Thomas, D., Fowler, C., Hunt, A.: Programming Ruby – The Pragmatic Programmer's Guide, 2nd edn. The Pragmatic Programmers (2004)
17. Sloot, P.M.A., Tirado-Ramos, A., Altintas, I., Bubak, M., Boucher, C.: From Molecule to Man: Decision Support in Individualized E-Health. Computer 39(11), 40–46 (2006)
18. de Oliveira, T., Deforche, K., Cassol, S., Salminen, M., Paraskevis, D., Seebregts, C., Snoeck, J., Janse van Rensburg, E., Wensing, A.M.J., van de Vijver, D.A., Boucher, C.A., Camacho, R., Vandamme, A.-M.: An automated genotyping system for analysis of HIV-1 and other microbial sequences. Bioinformatics (2005)
19. Phillips, J.C., Braun, R., Wang, W., Gumbart, J., Tajkhorshid, E., Villa, E., Chipot, C., Skeel, R.D., Kalé, L., Schulten, K.: Scalable molecular dynamics with NAMD. J. Comput. Chem. 26(16), 1781–1802 (2005)
20. Witty Services (2007), http://wts.sf.net
21. Humphrey, W., Dalke, A., Schulten, K.: VMD – Visual Molecular Dynamics. Journal of Molecular Graphics 14, 33–38 (1996)
22. Witten, I.H., Frank, E.: Data Mining: Practical machine learning tools and techniques. Morgan Kaufmann, San Francisco (2005)
23. Labarga, A., Valentin, F., Anderson, M., Lopez, R.: Web Services at the European Bioinformatics Institute. Nucl. Acids Res. (2007)
24. Foster, I.T., Kesselman, C., Tsudik, G., Tuecke, S.: A Security Architecture for Computational Grids. In: ACM Conference on Computer and Communications Security, pp. 83–92 (1998)
25. Shibboleth (2008), http://shibboleth.internet2.edu/
26. Unicore (2008), http://www.unicore.org
27. Application Hosting Environemt (2007), http://www.realitygrid.org/AHE/
28. The WS-Resource Framework (2007), http://www.globus.org/toolkit/
29. ViroLab (2006), http://virolab.org

Using MPI Communication Patterns to Guide Source Code Transformations

Robert Preissl[1,2], Martin Schulz[1], Dieter Kranzlmüller[2],
Bronis R. de Supinski[1], and Daniel J. Quinlan[1]

[1] CASC, Lawrence Livermore National Laboratory, USA*
{preissl2,schulzm,bronis,dquinlan}@llnl.gov
[2] GUP, Johannes Kepler University Linz, Austria/Europe
{rpreissl,dk}@gup.jku.at

Abstract. Optimizing the performance of HPC software requires a high-level understanding of communication patterns as well as their relation to source code structures. We describe an algorithm to detect communication patterns in parallel traces and show how these patterns can guide static code analysis. First, we detect patterns that identify potential bottlenecks in MPI communication traces. Next, we associate the patterns with the corresponding nodes in an abstract syntaxtree using the ROSE compiler framework. Finally we perform static analysis on the annotated control flow and system dependence graphs to guide transformations such as code motion or the automatic introduction of MPI collectives.

1 Introduction

With today's increasingly complex and highly scalable parallel systems, we require approaches that combine static (compile time) and dynamic (run time) information to capture sufficient information to optimize their applications. Similar techniques already support feedback guided compilation; however, these approaches limit the kind of data they collect and the optimizations that they apply (often restricted to simple localized decisions like inlining or code layout).

In this paper, we apply this principle to the analysis and optimization of global communication behavior, which is a primary source of inefficiency in parallel machines [2]. We use a suffix tree algorithm to detect repeating patterns in communication traces and map the patterns to static data structures. For the latter part we rely on ROSE, a comprehensive and flexible toolkit for the generation of source-to-source translators. We present early experiences on static source transformations exploiting the additional runtime information.

Sec. 2 gives an overview of the approach and shows the most important steps of the process. In Sec. 3 we formally define the repeating structures in strings and then compare the runtime behavior and memory usage of two different

---

* Part of this work was performed under the auspices of the U.S. Department of Energy by Lawrence Livermore National Laboratory under Contract DE-AC52-07NA27344. (LLNL-CONF-400356).

M. Bubak et al. (Eds.): ICCS 2008, Part III, LNCS 5103, pp. 253–260, 2008.

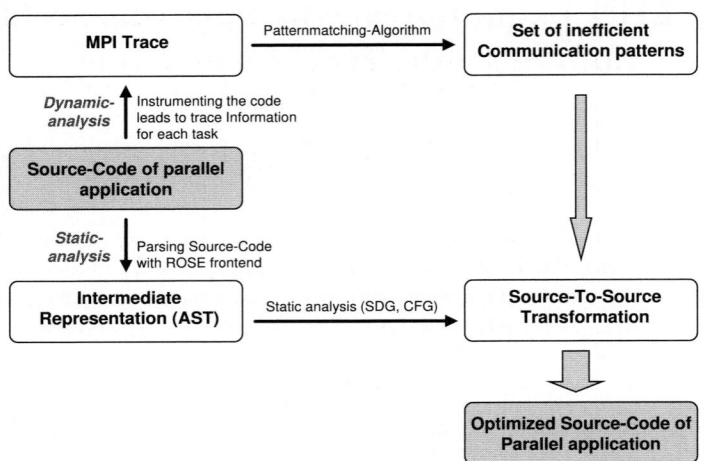

Fig. 1. Flow diagram of the main approach

algorithms that use the structures to find MPI communication patterns. Sec. 4 gives an overview of the ROSE compiler infrastructure that enables us to perform static analysis of the associated regions in the parallel code and to carry out source-to-source transformations. Sec. 5 provides a concrete example of a successfully transformed parallel code that achieves better performance by overlapping communication and computation. In the final sections we discuss aspects of future research and present our conclusions.

2   Combing Static and Dynamic Analysis

The analysis of MPI communication patterns has traditionally been used for high-level understanding and error detection [3,4,5]. In this work we extend the use of patterns to static analysis and source code transformations for performance optimization. We detect repetitive patterns of inefficient communication at runtime (e.g., poorly implemented broadcast operations) and use this information to optimize these heavily used structures in existing applications statically by replacing them with more efficient equivalent operations.

Fig. 1 gives an overview of our approach: we instrument the target MPI application, generate an MPI trace of the program executed under a given set of parameters, and then use pattern matching to isolate recurring inefficient communication structures. We also generate an abstract syntax tree (AST) of the application and perform a static analyses to extract control and data flow. We then map the detected patterns onto this information and use these annotations to guide potential source-to-source transformations.

This process comes with four major challenges:

1. Detecting MPI operations that would make recurring patterns interesting
2. Finding and filtering the patterns that involve these operations

3. Determining if the optimizations applicable to the patterns can be applied safely to the AST (possibly through specialization)
4. Transforming the related source code fragments

3   Detecting Patterns in MPI Communication Traces

MPI communication patterns are repeating communication structures in an MPI event graph [6]. First each MPI event is encoded into a 32 bit integer, such that each task's trace information is represented by an array of integer variables. For each task, we compute all repetitive and maximal repetitive sequences [7].

Definition 1. *A maximal pair of a string S of length n is a triple $(p1, p2, l)$, such that*

$$
\begin{aligned}
S[p_1, p_1 + l - 1] &= S[p_2, p_2 + l - 1], \ but \\
S[p_1 - 1] &\neq S[p_2 - 1] \ \ and \ \ S[p_1 + l] \neq S[p_2 + l]
\end{aligned} \tag{1}
$$

where p_1, p_2 denote the starting positions of the two substrings and l gives their lengths. A maximal repeat is a string represented by such a triple.

We use the maximal repeats, which are a subset of all repeated sequences, to select a start sequence. Starting from a maximal repeat, we identify global communication patterns that they contain. Using maximal repeats is crucial for finding repeating communication patterns efficiently. We have currently implemented two ways to detect repetitive structures in MPI traces: a naive algorithm based on convolution and a more advanced technique using suffix trees.

Our naive convolution method aligns the left end of the pattern P with the left end of the string S and then compares the characters of P and S left to right until either two unequal characters are found or until P is exhausted. It then shifts P to the right one place and repeats the comparisons process. If n is the length of P and m is the length of S, this approach makes $\Theta(nm)$ [7] comparisons in the worst case. Since there is possibly a huge number of MPI events in the trace, this method take significant runtime.

Our more advanced compressed suffix tree approach borrows ideas from computational biology for finding repeating gene structures in huge DNA sequences [7]. The suffix tree for a string S with length m is a rooted directed tree with exactly m leaves and whose labels correspond to substrings of S. For any leaf i, the concatenation of the edge-labels on the path from the root to leaf i exactly spells out the suffix of S starting at position i.

Definition 2. *An internal node v of the suffix tree for string S is called left diverse if at least two leaves with suffix position i and j in $v's$ subtree have different left characters, that is, $S[i - 1] \neq S[j - 1]$*

Theorem 1 (D. Gusfield). *The substring α labeling the path to an internal node v of the suffix tree for string S is a maximal repeat iff v is left diverse.*

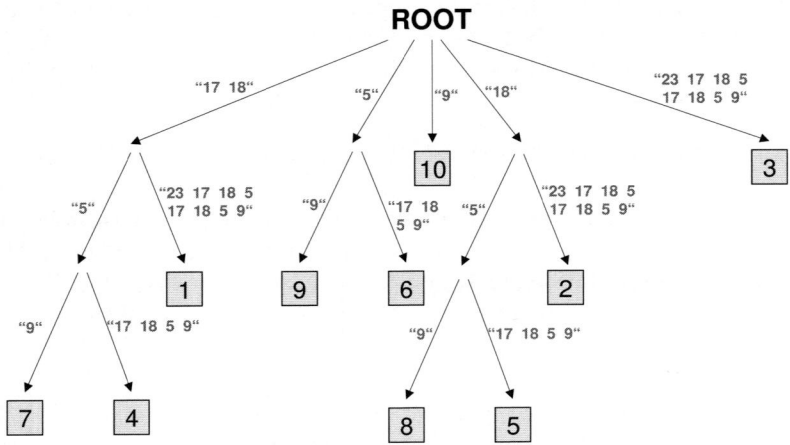

Fig. 2. Suffix for a string representing MPI events of a trace

D. Gusfield describes algorithms (e.g., Ukkonen's algorithm) for constructing such trees with all its suffix links for a string S with length m in $\Theta(m)$ time [7]. However, these algorithms are not space efficient, particularly with large alphabets, as with MPI traces. Thus, we use an alternative approach. The compressed suffix tree [8] for a string S with length m of an alphabet Σ occupies $\Theta(m \, log|\Sigma|)$ bits. The final time requirement for creating the tree is $\Theta(m \, log|m| \, log|\Sigma|)$, being reasonably close to the best current theoretical result [9].

Fig. 2 shows a suffix tree of the string $S =$ "17, 18, 23, 17, 18, 5, 17, 18, 5, 9", in which each integer represents an MPI event. We can easily see from the suffix tree that $\langle 17, 18 \rangle$ with starting positions (1, 4 and 7) and $\langle 17, 18, 5 \rangle$ with starting positions (4 and 7) are maximal repeats. In addition to these (maximal) repeats, S has the repeat $\langle 18, 5 \rangle$ with starting positions (5 and 8).

We must choose a repeat on a specific task from which to start the pattern matching process. Since each task typically has many maximal repeats, the time required to compute a communication pattern for all maximal repeats is generally prohibitive. Hence, we first filter start-repeats and then compute only communication patterns that involve these filtered events. This filter step is either guided by special seed events, e.g. events in the trace with a huge difference between start- and end-time, or by focusing on selected MPI operations.

Fig. 3 shows an example MPI trace in which the events in bold highlight a repeating set of MPI operations that our pattern-detection algorithm finds. This pattern is potentially equivalent to a broadcast operation using a tree-based communication structure. In this example, our static analysis must verify that the pattern is a broadcast operation and, if so, that we can replace the associated code segments with an equivalent, but probably more efficient version in the form of a native MPI collective. This is promising because it will enable programmers, that do not have a broad knowledge of MPI to apply more efficient functions defined by the MPI standard in without having to use them explicitly.

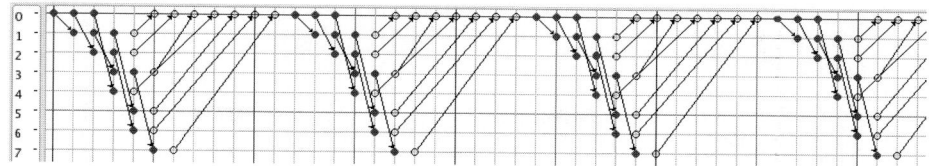

Fig. 3. A detected pattern representing a broadcast operation in an MPI trace

4   Static Analysis with ROSE

A communication pattern extracted from the runtime trace only points to a potential bottleneck. Further, the pattern is initially only valid for the particular input set used during the application run. We must verify the pattern occurs across all control flows and investigate pattern specific global data flow constraints (e.g., in the case of a suspected broadcasts that the same data is communicated in all messages). The required information to achieve this goal is encapsulated in the System-Dependence-Graph (SDG) [10] and the Control-Flow-Graph (CFG).

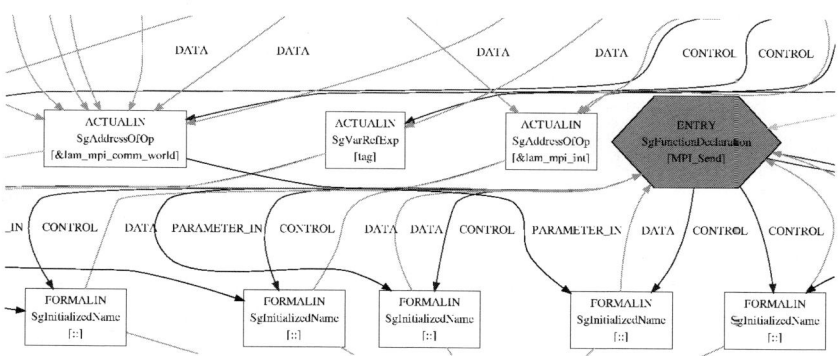

Fig. 4. Excerpt of the System-Dependence-Graph (SDG) of a parallel program

We use the ROSE open-source compiler infrastructure to generate both of these graphs. ROSE is a tool kit to generate custom source-to-source translators. It provides mechanisms to translate input source code into an intermediate representation (AST) [11], libraries to traverse and manipulate the information stored in the AST, as well as mechanisms to transform the changed AST information back into valid source code. The representation within the AST as well as the supporting data structures is powerful enough to readily exploit knowledge of the architecture, parallel communication characteristics, and cache architecture in the specification of the transformations [12].

Fig. 4 illustrates a small excerpt of an SDG generated by ROSE providing both data and control dependency information. The specific example shows the nodes

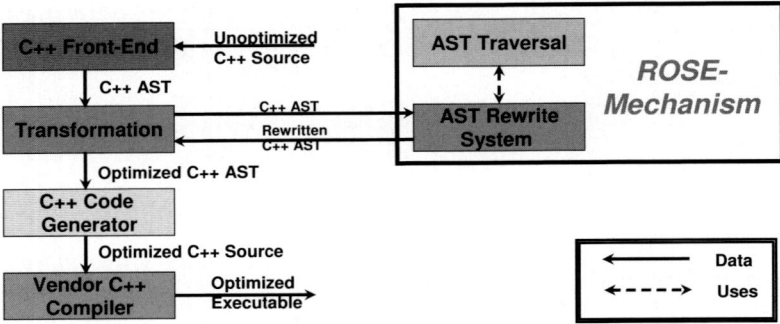

Fig. 5. The ROSE compiler infrastructure

and data-dependence egdes that represent flow of data between statements or expressions as well as control-dependence edges that represent control conditions on which the execution of a statement or expression depends, in this case around an MPI_Send function in a parallel application.

The flow diagram in Fig. 5 reflects the complete approach, transforming unoptimized C++ code based on user defined abstractions into highly optimized code. In our approach we have defined specific transformations to MPI code structures based on dynamic analysis results in the form of inefficient MPI communication patterns.

5   Examples and Early Results

The following simple example illustrates how we can use our techniques to replace blocking with non-blocking communication to overlap communication and computation for better overall performance. Fig. 6 shows the structure of the corresponding "Late-Sender" pattern: the receiving task wastes useful time in waiting for a message to be sent by another task. We detect this pattern from our communication trace and then apply a source code transformation using the ROSE

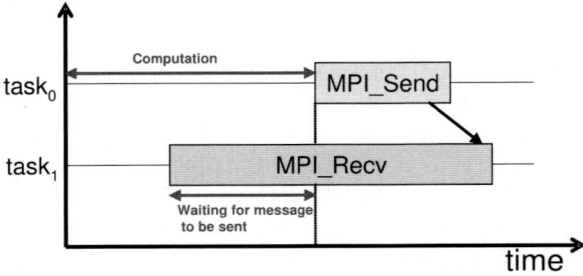

Fig. 6. Late-Sender pattern

```
Data x, y                                    MPI_Request req
for ( nr_repeats ) {                         Data x, y
    if( task_0 ) {                           for ( nr_repeats ) {
        Do_Computation ( x )                     if( task_0 ) {
        MPI_Send ( task_1, x )                       Do_Computation ( x )
    }                                                MPI_Send ( task_1 , x )
    if( task_1 ) {              Source-Code           }
        MPI_Recv( task_0, x )   Transformation       if( task_1 ) {
    }                             ─────────▶              MPI_Irecv( task_0 , x, &req )
    if( task_1 ) {                                   }
        Do_Computation ( y )                     if( task_1 ) {
    }                                                Do_Computation ( y )
    if( task_1 ) {                                }
        Do_Computation ( x )                     if( task_1 ) {
    }                                                MPI_Wait ( &req )
}                                                    Do_Computation ( x )
                                                 }
                                             }
```

Fig. 7. Source code transformation: before (left), after (right)

compiler framework that introduces non-blocking communication. as shown in the pseudo-code in Fig. 7. Our transformation replaces the blocking MPI_Recv with its non-blocking MPI_Irecv version and adds the matching MPI_Wait so as to provide the largest possible communication/computation overlap allowed by data dependencies on the receive buffer.

6   Conclusion and Future Work

This paper presents our work to automate the analysis and optimization of parallel scientific applications by combining dynamic runtime information in the form of communication patterns with static analysis and code transformation. We use the dynamic information to identify the Send- and Receive Events that are part of inefficient communication patterns and are good targets for source code optimizations.

In our ongoing research, we are extending the library of inefficient MPI communication patterns that we optimize beyond the communication/computation overlap transformation described here. In particular, we will introduce static and dynamic analysis as well as the corresponding transformation engine for automatically adding collective operations. Further, we will provide several other novel methods for selecting and filtering MPI seed events that will help us find more complex patterns that lead to communication bottlenecks.

References

1. The List of Worlds 500 Fastest Supercomputers, www.top500.org
2. Bassetti, F., Davis, K., Quinlan, D.: Improving Scalability with Loop Transformations and Message Aggregation in Parallel Object-Oriented Frameworks for Scientific Computing. In: Computing, Information, and Communications Division, Los Alamos, NM, USA (1998)

3. Kranzlmüller, D., Grabner, S., Volkert, J.: Event Graph Visualization for Debugging Large Applications. In: Proc. of SIGMETRICS Symposium on Parallel and Distributed Tools, SPDT Philadelphia, PA, USA, pp. 108–117 (1996)
4. Kranzlmüller, D.: Communication Pattern Analysis in Parallel and Distributed Programs. In: Proc. of the 20th IASTED International Multi-Conference Applied Informatics (AI 2002), International Symposia on Software Engineering, Databases, and Applications, International Association of Science and Technology for Development (IASTED). ACTA Press, Innsbruck (2002)
5. Knüpfer, A., Kranzlmüller, D., Nagel, W.E.: Detection of Collective MPI Operation Patterns. In: Kranzlmüller, D., Kacsuk, P., Dongarra, J. (eds.) EuroPVM/MPI 2004. LNCS, vol. 3241, pp. 259–267. Springer, Heidelberg (2004)
6. Kranzlmüller, D.: Event Graph Analysis for Debugging Massively Parallel Programs. PhD thesis, GUP Linz, Johannes Kepler University Linz, Austria (2000)
7. Gusfield, D.: Algorithms on Strings, Trees, and Sequences. Computer Science and Computational Biology (1997)
8. Dixit, K., Gerlach, W., Maekinen, Vaelimaeki, N.: Engineering a Compressed Suffix Tree Implementation. Department of Computer Science, Series of Publications C, Report C-2006-37, University of Helsinki, Finland (2006)
9. Hon, W.-K., Sadakane, K., Sung, W.-K.: Breaking a time-and-space barrier in constructing full-text indices. In: FOCS 2003: Proceedings of the 44th Annual IEEE Symposium on Foundations of Computer science, Washington, DC, USA, p. 251 (2003)
10. Horwitz, S., Reps, T., Binkley, D.: Interprocedural slicing using dependence graphs. ACM Transactions on Programming Languages and Systems, p. 26 (1990)
11. Schordan, M., Quinlan, D.: A source-To-Source Architecture for User-Defined Optimizations. Lawrence Livermore National Laboratory, USA (2003)
12. Quinlan, D.: Compiler Support for Object-Oriented Frameworks. Lawrence Livermore National Laboratory, USA (1999)
13. Panas, T., Quinlan, D., Vuduc, R.: Tool Support for Inspecting the Code Quality of HPC Architectures. Lawrence Livermore National Laboratory, USA (2007)

Detection and Analysis of Iterative Behavior in Parallel Applications

Karl Fürlinger and Shirley Moore

Innovative Computing Laboratory,
Department of Electrical Engineering and Computer Science,
University of Tennessee
{karl,shirley}@eecs.utk.edu

Abstract. Many applications exhibit iterative and phase based behavior. We present an approach to detect and analyze iteration phases in applications by recording the control flow graph of the application and analyzing it for loops that represent iterations. Phases are then manually marked and performance profiles are captured in alignment with the iterations. By analyzing how profiles change between capture points, differences in execution behavior between iterations can be uncovered.

Keyword: Phase detection, control flow graph, continuous profiling.

1 Introduction

Many applications exhibit iterative and phase based behavior. Typical examples are the time steps in a simulation and iteration until convergence in a linear solver. With respect to performance analysis, phase knowledge can be exploited in several ways. First, repetitive phases offer the opportunity to restrict data collection to a representative subset of program execution. This is especially beneficial when tracing is used due to the large amounts of performance data and the challenges involved with capturing, storing, and analyzing it. Conversely, it can be interesting to see how the iterations differ and change over time to expose effects such as cache pollution, operating system jitter and other sources that can cause fluctuations in execution time of otherwise identical iterations.

In this paper we present an approach to detection and analysis of phases in threaded scientific applications. Our approach assists in the detection of the phases based on the control flow graph of the application if the developer is not already familiar with the code's structure. To analyze phase-based performance data we modified an existing profiling tool for OpenMP applications. Based on markups in the code that denote the start and end of phases, the profiling data is dumped to a file during the execution of the application (and not only at the end of the program run) and can be correlated to the application phases.

The rest of this paper is organized as follows. Section 2 describes the technique we used to assist the developer in detecting iterative application phases. In Sect. 3 we describe the analysis of performance data based on phases using the existing profiling tool called ompP. In Sect. 4 we describe an example of applying our

M. Bubak et al. (Eds.): ICCS 2008, Part III, LNCS 5103, pp. 261–267, 2008.

technique to a benchmark applications, in Sect. 5 we describe related work and conclude in Sect. 6.

2   Iterative Phase Detection

Our approach to identify iterative phases in threaded applications is based on the monitoring and analysis of the control flow graph of the application. For this, we extended our profiling tool ompP.

ompP [1] is a profiling tool for OpenMP applications that supports the instrumentation and analysis of OpenMP constructs. For sequential and MPI applications it can also be used for profiling on the function level and the phase detection described here is similarly applicable. ompP keeps profiling data and records a call graph of an application on a per-thread basis and reports the (merged) callgraph in the profiling report.

Unfortunately, the callgraph of an application (recording caller–callee relationships and also the nesting of OpenMP regions) does not contain enough information to reconstruct the control flow graph. However, a full trace of function execution is not necessary either. It is sufficient that for each callgraph node a record is kept that lists all predecessor nodes and how often the predecessors have been executed. A predecessor node is either the parent node in the callgraph or a sibling node on the same level. A child node is not considered a predecessor node because the parent–child relationship is already covered by the callgraph representation. An example of this is shown in Fig. 1. The callgraph (lower part of Fig. 1) shows all possible predecessor nodes of node A in the CFG. They are the siblings B and C, and the parent node P. The numbers next to the nodes in Fig. 1 indicate the predecessor nodes and counts after one iteration of the outer loop (left hand side) and at the end of the program execution (right hand side), respectively.

```
P() {                              A() {          C() {
   for(i=1; i<5; i++ ) {              X();            Z();
      A();                            Y();         }
      B();                         }
      C();
   }
}
```

```
                 predecessor                          predecessor
                 list                                 list
   P                              P
   +-A            (P:1)             +-A                (P:1,C:4)
   |   +-X        (A:1)             |   +-X            (A:5)
   |   +-Y        (X:1)             |   +-Y            (X:5)
   +-B            (A:1)             +-B                (A:5)
   +-C            (B:1)             +-C                (B:5)
       +-Z        (C:1)                 +-Z            (C:5)
```

Fig. 1. Illustration of the data collection process to reconstruct the control flow graph

Implementing this scheme in ompP was straightforward. ompP already keeps a pointer to the *current* node of the callgraph (for each thread) and this scheme is extended by keeping a *previous* node pointer as indicated above. Again this information is kept on a per-thread basis, since each thread can have its own independent callgraph as well as flow of control.

The previous pointer always lags the current pointer one transition. Prior to a parent → child transition, the current pointer points to the parent while the previous pointer either points to the parent's parent or to a child of the parent. The latter case happens when in the previous step a child was entered and exited. In the first case, after the parent → child transition the current pointer points to the child and the previous pointer points to the parent. In the latter case the current pointer is similarly updated, while the prior pointer remains unchanged. This ensures that the previous nodes of siblings are correctly handled.

With current and previous pointers in place, upon entering a node, information about the previous node is added to the list of previous nodes with an execution count of 1, or, if the node is already present in the predecessor list, its count is incremented.

The data generated by ompP's control flow analysis can be displayed in two forms. The first form visualizes the control flow of the whole application, the second is a layer-by-layer approach. The full CFG is useful for smaller applications, but for larger codes it can quickly become too large to comprehend and cause problems for automatic layout mechanisms. An example of an application's full control flow is shown in Fig. 2 along with the corresponding (pseudo-) source code.

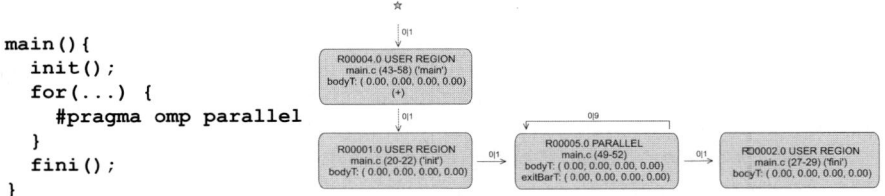

Fig. 2. An example for a full control flow display of an application

Rounded boxes represent source code regions. That is, regions corresponding to OpenMP constructs, user-defined regions or automatically instrumented functions. Solid horizontal edges represent the control flow. An edge label like $i|n$ is to be interpreted as thread i has executed that edge n times. Instead of drawing each thread's control flow separately, threads with similar behavior are grouped together. For example the edge label $0-3|5$ means that threads $0, 1, 2,$ and 3 executed that edge 5 times. This greatly reduces the complexity of the control flow graph and makes it easier to understand.

Based on the control flow graph, the user has to manually mark the start and end of iterative phases. To mark the start of the phase the user adds the directive **phase start**, to mark the end **phase end**.

3   Iterative Phase Analysis

The phase based performance data analysis implemented in ompP works by capturing profiling snapshots that are aligned with the start and end of program phases. Instead of dumping a profiling report only at the end of the program execution, the reports are aligned with the phases and the change between capture points can be correlated to the activity in the phase. This technique is a modification of the incremental profiling approach described in [2] where profiles are captured in regular intervals such as 1 second.

The following performance data items can be extracted from phase-aligned profiles and displayed to the user in the form of 2D graphs.

Overheads. ompP classifies wait states in the execution of the OpenMP application into four overhead classes: synchronization, limited parallelism, thread management and work imbalance. Instead of reporting overall, aggregated overhead statistics, ompP's phase analysis allows the correlation of overheads that occur in each iteration. This type of data can be displayed as two-dimensional graphs, where the x-axis correlates to execution time and the y-axis displays overheads in terms of percentage of execution time lost. The overheads can be displayed both for the whole application or for each parallel region separately. An example is show in Fig. 5.

Execution Time. The amount of time a program spends executing a certain function or OpenMP construct can be displayed over time. Again, this display shows line graphs where the x-axis represents (wall clock) execution time of the whole application while the y-axis shows the execution time of a particular function or construct. In most cases it is most useful to plot the execution time sum over all threads, while it is also possible to plot a particular thread's time, the minimum, maximum or average of times.

Execution Count. Similar to the execution time display, this view shows when certain constructs or functions got executed, but instead of showing the execution time spent, the number of invocations or executions is displayed in this case.

Hardware Counters. ompP is able to capture hardware performance counters through PAPI [3]. Users selects a counter they want to measure and ompP records this counter on a per-thread and per-region basis. Hardware counter data can best be visualized in the form of heatmaps, where the x-axis displays the time and the y-axis corresponds to the thread id. Tiles display the normalized counter values with a color gradient or gray scale coding. An example is show in Fig. 4.

4   Example

In this example we apply the phase detection and analysis technique to a benchmark from the OpenMP version (3.2) of the NAS parallel benchmark suite. All experiments have been conducted on an four processor AMD Opteron based

SMP system. The application we chose to demonstrate our technique is the CG application which implements the conjugate gradient technique.

The CG code performs several iterations of an inverse power method to find the smallest eigenvalue of a sparse, symmetric, positive definite matrix. For each iteration a linear system $Ax = y$ is solved with the conjugate gradient method.

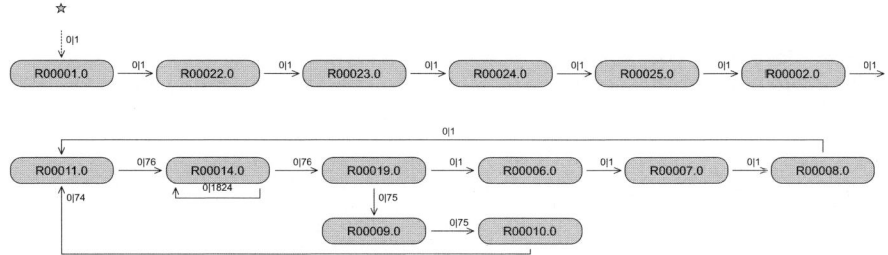

Fig. 3. (Partial) control flow graph of the CG application (size C)

Fig. 3 shows the control flow graph of the CG application (size C). To save space, only the region identification numbers Rxxxx are shown in this example, in reality the control flow nodes show important information about the region such as region type, location (file name and line number) and execution statistics in addition. Evidently the control flow graph shows an iteration that is executed 76 times where one iteration takes another path than the others. This is the outer iteration of the conjugate gradient solver which is executed 75 times in the main iteration and once for initialization.

Using this information (and the region location information) it is easy to identify the iterative phase in the source code. We marked the start and end of each iteration with an **phase start** directive and each end the ends with a **phase end** directive. Using directives (compiler pragmas in C/C++ and special style comments in FORTRAN) similar to OpenMP directives has the advantage that the normal (non-performance analysis) build process is not interrupted while the directives are translated into calls that cause ompP to capture profiles when performance analysis is done and ompP's compiler wrapper script translates the directives into calls implemented by ompP's monitoring library.

Fig. 4 shows the overheads over time display of for the iterations of the CG application with problem size C. Evidently, the only significant overhead identified by ompP is imbalance overhead and the overhead does not change much from iteration to iteration with the exception of two peaks. The most likely reason for these two peaks is operating system jitter, since the iterations are otherwise identical in this example.

Fig. 5 shows the heatmap display of the CG application with four threads. The measured counter is PAPI_FP_OPS. In order to visually compare values, absolute values are converted into rates. The first column of tiles corresponds to the initialization part of the code which features a relatively small number of floating

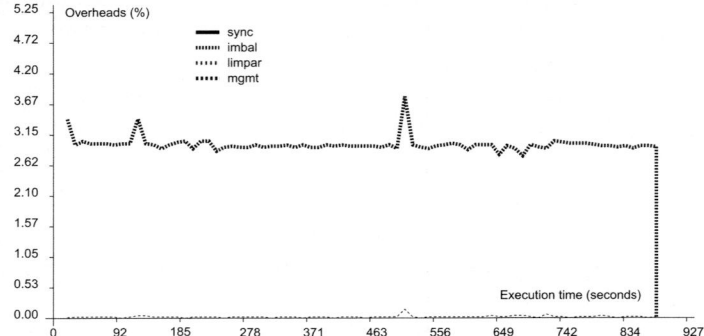

Fig. 4. Overheads of the iterations of the CG application. X-axis is wallclock execution time in seconds, while th y-axis represents the percentage of execution time lost due to overheads.

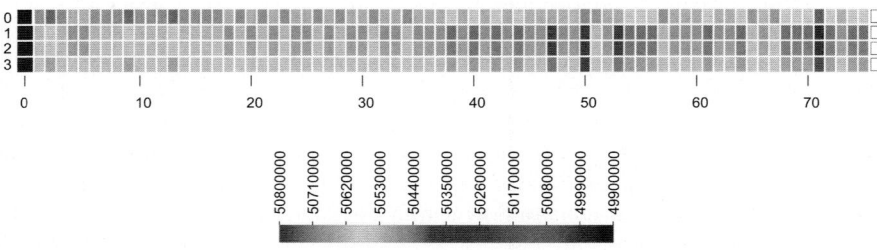

Fig. 5. Performance counter heatmap of the CG application. X-axis is phase or iteration number, the y-axis corresponds to the thread ID.

point operations, the other iterations are of about equal size but show some difference in floating point rate of execution.

5   Related Work

Control flow graphs are an important topic in the area of code analysis, generation, and optimization. In that context, CFGs are usually constructed based on a compiler's intermediate representation (IR) and are defined as directed multi-graphs with nodes being basic blocks (single entry, single exit) and nodes representing branches that a program execution *may* take (multithreading is hence not directly an issue). The difference to the CFGs in our work is primarily twofold. First, the nodes in our graphs are generally not basic blocks but larger regions of code containing whole functions. Secondly, the nodes in our graphs record transitions that have actually happened during the execution and do also contain a count that shows how often the transition occurred.

Detection of phases in parallel programs has previously been applied primarily in the context of message passing applications. The approach of Casas-Guix et al.

[4] works by analyzing the autocorrelation of message passing activity in the application, while our approach works directly by analyzing the control flow graph of the application.

6   Conclusion

We have presented an approach for detecting and analyzing the iterative and phase-based behavior in threaded applications. The approach works by recording the control flow graph of the application and analyzing it for loops that represent iterations. This help of the control flow graph is necessary and useful if the person optimizing the code is not the code developer and does not have intimate knowledge.

With identified phase boundaries, the user marks the start and end of phases using directives. We have extended a profiling tool to support the capturing of profiles aligned with phases. In analyzing how profiles change between capture points, differences in execution behavior between iterations can be uncovered.

References

1. Fürlinger, K., Gerndt, M.: ompP: A profiling tool for OpenMP. In: Mueller, M.S., et al. (eds.) IWOMP 2005 and IWOMP 2006. LNCS, vol. 4315. Springer, Heidelberg (2008)
2. Fürlinger, K., Dongarra, J.: On Using Incremental Profiling for the Performance Analysis of Shared Memory Parallel Applications. In: Kermarrec, A.-M., Bougé, L., Priol, T. (eds.) Euro-Par 2007. LNCS, vol. 4641, pp. 62–71. Springer, Heidelberg (2007)
3. Browne, S., Dongarra, J., Garner, N., Ho, G., Mucci, P.J.: A portable programming interface for performance evaluation on modern processors. Int. J. High Perform. Comput. Appl. 14, 189–204 (2000)
4. Casas-Guix, M., Badia, R.M., Labarta, J.: Automatic phase detection of MPI applications. In: Proceedings of the 14th Conference on Parallel Computing (ParCo 2007), Aachen and Juelich, Germany (2007)

Guided Prefetching Based on Runtime Access Patterns*

Jie Tao[1], Georges Kneip[2], and Wolfgang Karl[2]

[1] Steinbuch Center for Computing
Forschungszentrum Karlsruhe
Karlsruhe Institute of Technology, Germany
jie.tao@iwr.fzk.de
[2] Institut für Technische Informatik
Universität Karlsruhe (TH)
Karlsruhe Institute of Technology, Germany

Abstract. Cache prefetching is a basic technique for removing cache misses and the resulting access penalty. This work proposes a kind of guided prefetching which uses the access pattern of an application to prohibit from loading data which are not required. The access pattern is achieved with a data analyzer capable of discovering the affinity and regularity of data accesses. Initial results depict a performance improvement of up to 20%.

1 Introduction

During the last years, the gap between processor and memory speed has considerably widened. This indicates a significant performance degradation for applications with a large number of cache misses, because an access to the main memory takes hundreds of cycles while an access in the cache needs only several ones. According to [8], the SPEC2000 benchmarks running on a modern, high-performance microprocessors spend over half of the time stalling for loads that miss in the last level cache.

Cache misses can be caused by several reasons, for example, accessing a memory word at the first time, cache size smaller than that of the working set, or data evicted from the cache due to mapping conflict but used again. The resulted cache miss is correspondingly called compulsory miss, capacity miss, and conflict miss.

The most efficient way to reduce compulsory miss is prefetching [2]. This technique attempts to load data into the cache before it is requested. A key issue with this technique is to avoid prefetching data that is not required in a short time. Such inefficient prefetching may cause more cache misses because the prefetched data can directly, or indirectly, evict frequently reused data out of the cache. Moreover, prefetching consumes memory bandwidth. This additional

---

* This work was conducted as Dr. Tao was a staff member of the Institut für Technische Informatik.

M. Bubak et al. (Eds.): ICCS 2008, Part III, LNCS 5103, pp. 268–275, 2008.

memory traffic may degrade performance of multiprocessor systems, especially the emerging multi-core processors, where the main memory is shared across all on-chip processors.

In this work, we use a data analyzer to detect the affinity and regularity of data accesses and then apply this information to guide prefetching. The base of this analyzer is a memory reference trace achieved by instrumenting the assembly code during the compiling process. Based on this trace, the analyzer runs optimized analysis algorithms to discover repeated access chains and access strides. The former is a group of accesses that repeatedly occur together but target on different memory locations. The latter is a constant distance between accesses to successive elements of a data array. It is clear that both access chain and stride depict which data is requested.

These findings can be used to guide hardware prefetching. However, within this work we use them to perform software prefetching for the reason of easy implementation. We implemented a source-to-source compiler that takes an original program as input, and creates a new version of the same code but with prefetching instructions inserted. These instructions are formulated based on the output of the data analyzer, which shows both the access pattern and their occurrence in the source code. We studied this approach with several sample programs. Initial results depict a performance improvement of up to 20%.

The remainder of this paper is organized as following. Section 2 briefly describes existing research work in the field of cache prefetching. In Section 3 the analysis tool, together with the instrumentor for access trace, is introduced. This is followed by a detailed description of the precompiler in Section 4. In Section 5 first experimental results are presented. The paper concludes in Section 6 with a short summary and several future directions.

2   Related Work

Earlier techniques for cache prefetching are simple, where a prefetch is always issued to the next cache block [13] or several consecutive blocks [7]. This approach is still used by processor-associated prefetching, e.g. Pentium 4, for a straightforward implementation. However, due to its inefficiency current research work focuses on prefetching with access pattern like stride and linked memory references. Hardware-based prefetches often use a specific hardware component to trace repeated execution of a particular memory instruction and detect thereby reference strides, while software approaches usually rely on a compiler to analyze array references in program loops.

Baer and Chen [1] deployed a hardware function unit with the basic idea of keeping track of access patterns in a Reference Prediction Table (RPT). This RPT stores the last known strides. Prefetches are issued when the strides between the last three memory addresses of a memory reference instruction are the same.

Another hardware-based prefetching applies the Markov predictor [6] that remembers past sequences of cache misses. When a miss is found which matches a miss in the sequence, prefetches for the subsequent misses in the sequence are issued. The advantage of this approach is to be able to prefetch any sequence of memory references as long as it has been observed once.

The converse approach is software-based prefetching, in which access patterns are usually discovered by compilers. Luk and Mowry [9] use a compiler to detect linked memory references. Since addresses are not known at compile time, the compiler observes data structures containing an element that points to some other or the same data structure. Inagaki et al. [4] also uses profiling to guide compilers to perform stride prefetching. They proposed a profiling algorithm capable of detecting both inter- and intra-iteration stride patterns. This is a kind of partial interpretation technique and only dynamic compiler can perform such profiling. The profile information is gathered during the compiling process. The algorithm is evaluated with a Java just-in-time compiler and experimental results show an up to 25.1% speedup with standard benchmarks.

In summary, for detecting access patterns hardware approaches can take advantage of the runtime information but do not know the references in the future. In addition, they need specific support of hardware components and can hence not be commonly applied. Software prefetching is general and more accurate, however, the accuracy can only be achieved by compiler techniques in combination with runtime profiling.

We deploy a straightforward approach to achieve the access patterns. This is a separate analysis tool which has not to be integrated into modern sophisticated compilers. Since the analysis is based on memory references performed at the runtime, the accuracy of the access pattern can be guaranteed. More importantly, the tool can find all strides associated either with arrays or linked memory references. Additionally, this analysis tool delivers access sequences that are a number of references targeting different addresses but frequently issued successively. Clearly, this information also specifies the prefetch targets.

3   Pattern Acquisition

As mentioned, the information for our guided prefetching is directly acquired from the runtime references. For this, we developed a pattern analysis tool which is based on a code instrumentor.

3.1   Code Instrumentation for Access Trace

The instrumentor is an altered version of an existing one called *Doctor*. *Doctor* is originally developed as a part of Augmint [11], a multiprocessor simulation toolkit for Intel x86 architectures. It is designed to augment assembly codes with instrumentation instructions that generate memory access events. For every memory reference, *Doctor* inserts code to pass the accessed address, its size, and the issuing process to the simulation subsystem of Augmint.

We modified *Doctor* in order to achieve an individual instrumentor independent of the simulation environment. The main contribution is to remove from *Doctor* the function calls to Augmint. For example, *Doctor* relies on Augmint to acquire the process/thread identifier of the observed memory access. Within the modified version, we use the *pthread* library to obtain this parameter. Hence, the new instrumentor is capable of generating memory access traces for multithreading programs based on the *pthread* library, like the OpenMP applications.

This instrumentation results in the recording of each memory reference at runtime. Besides the access address and the thread ID, we also store the position in the source code for each access. This information is needed to map the accesses to the source code and further to correctly insert prefetching instructions in it. For a realistic application, the trace file could be rather large. It is possible to use specific strategy, such as lossy tracing [10], to reduce the size of the trace file, however, for the accuracy of the analysis results, needed for the guided prefetching, a full trace is essential. In this case, we store the access records in a binary format and additionally develop a tool for transforming the binaries to the ASCII form if needed. This scheme also reduces the runtime overhead introduced by the instrumentation.

3.2   Affinity Analysis

For acquiring access patterns we developed an analysis tool [14] which currently detects both access chain and access stride. The former is a group of accesses that repeatedly occur together but target on different memory locations. An example is the successive access to the same component of different instances of a *struct*. It is clear that this information directly shows the next requested data which is actually the prefetching target. The latter is the stride between accesses to the elements of an array. Similarly, this information also indicates which data is next needed.

For detecting address chains the analysis tool deploys Teiresias [12], an algorithm often used for pattern recognition in Bioinformatics. For pattern discovery, the algorithm first performs a *scan/extension* phase for generating small patterns of pre-defined length and occurrence. It then combines those small patterns, with the feature that the prefix of one pattern is the suffix of the other, into larger ones. For example, from pattern DFCAPT and APTSE pattern DFCAPTSE is generated.

For detecting access strides the analyzer applies an algorithm similar to that described in [5]. This algorithm uses a search window to record the difference between each two references. References with the same difference are combined to form an access stride in the form of $< start\_address, stride, repeating\_times >$. For example, a stride $< 200, 100, 50 >$ indicates that starting with the address 200, memory locations with an address difference of 100, e.g. 300, 400, 500 etc., are also requested, and this pattern repeats for 50 times.

The detected access patterns are stored in an XML file. Together with each access chain or stride, the position in the source code is also recorded. This allows us to track the patterns in the program source and hence to find the location for inserting prefetching codes.

4   Automatic Prefetching

With the access pattern in hand, we now need a precompiler to generate an optimized source program with prefetching inserted. This is actually the task of a parser. We could directly deploy an existing parser of any compilers and build the prefetching on top of it, however, theses parsers are usually of the complexity that is not necessary for this work. In this case, we developed a simple parser specifically for the purpose of prefetching.

The key technique with the parser is to determine what and where to prefetch. For the former the access chains and strides clearly give the answer. However, the access patterns are provided in virtual addresses, while the parser works with variables. Therefore, an additional component was implemented for the goal of transforming the addresses in the pattern to the corresponding data structures in the program.

Previously, we relied on specific macros to register variables and generate a mapping table at the runtime, but currently we obtain the mapping table automatically by extracting the static data structure from the debugging information and instrumenting the *malloc* calls for dynamic ones.

Based on the variable names, the parser can build the prefetching instructions like $prefetch(p)$, where p is the variable to prefetch. Nevertheless, for arrays and linked references such as the *struct* data structure in C/C++ programs, it must additionally examine the corresponding code line to acquire other information about e.g. dimensions of an array.

The other important issue with prefetching is to decide the location where the prefetching instruction is inserted. This location directly determines the prefetching efficiency. For example, if the data is loaded into the cache too earlier, it can be evicted from the cache before it is used. On the other hand, if the prefetching is not issued early enough, the data is potentially not in cache as the processor requests it.

However, it is a quite tedious work to accurately compute the prefetching location. For simplicity, in this initial work we put the prefetching instruction several iterations before the associated. Users are asked to specify a prefetch distance, i.e. the expected number of iterations.

Knowing where to prefetch, the parser now has to identify the individual code line in the program. For this, the source code is scanned and key words that represent the start or end of a code line, like *if*, *for*, and *;*, are searched. Comments and directives for parallelization are removed before the scanning process for efficiency, but inserted into the resulted program again after this process.

Overall, based on the parser and other components, an optimized version of an application is generated with each prefetching instruction corresponding to a detected access pattern. This new version can be traditionally compiled and executed on the target architecture supporting software prefetching.

5   Initial Experimental Results

We used several applications to examine the effectiveness of this approach. In the following, we show two representative results, one achieved with the matrix multiplication code and the other with a realistic application.

The matrix multiplication program is used to examine how the prefetch distance influence the overall performance of an application. As mentioned in the previous section, we simply place the prefetching instruction several iterations, specified by the user, before the data is requested.

For this small code, our pattern analyzer found two strides, but no access chains. Theses strides correspond to both input matrices, A and B, one with an access stride of 1 (A) and the other of n (B) where n is the length of a matrix row.

Both matrices are prefetched based on the observed strides. We run the program on a Pentium 4 machine using different prefetch distances and the execution time is measured.

Figure 1 shows the experimental results, where the x-axis presents the various prefetch distance, while the y-axis depicts the improvement which is calculated with the execution time of the original code divided by that needed for running the code version with prefetching.

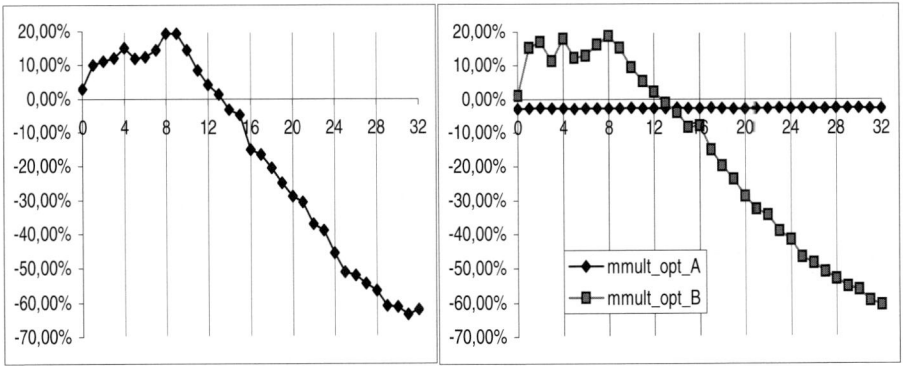

Fig. 1. Performance change with the prefetch distance (left: prefetching both A and B, right: prefetching A or B)

Observing the left diagram in the figure, it can be seen that the performance improvement arises up to a prefetch distance of 8 where the maximum of 20% is reached, and then decreases. This indicates that with a prefetch distance of smaller than 8 the data perhaps have not been loaded to the cache. For a larger prefetch distance, however, some prefetched data have been potentially evicted from the cache as the processor requests them. Hence, the best prefetch distance for this code is 8.

For a more detailed insight into the prefetching impact, we additionally measured the execution time with prefetching only one matrix, either A or B. The result is depicted in the right diagram of Figure 1.

As expected, matrix B performs well with prefetching and is therefore the contributor of the performance achieved with the overall program. However, prefetching A shows a performance lost of about 3%. It can also be observed that the prefetch distance does not influence the prefetching efficiency. This is caused by the fact that Pentium 4 performs hardware prefetching that preloads the successive data block of the accessed one. In this case, the requested data are prefetched by the hardware, indicating that our software prefetching is not necessary, but introduces overhead. For matrix B with a large access stride, nevertheless, the hardware prefetching is not efficient. In this case, our guided prefetching shows its power.

Now we observe a realistic application to evaluate the feasibility of this approach. The chosen application implements a 2D Discrete Wavelet Transform (DWT) algorithm, which is usually applied for image and video compression. This algorithm mainly contains two 1D DWT in both directions: horizontal filtering processes the rows followed by vertical filtering processing the columns. Our pattern analyzer detected strides with the input and out images in both filtering functions. The application is executed on an AMD and a XEON processor individually. Figure 2 depicts the experimental results, where the x-axis presents the image size while the y-axis shows the improvement in execution time to the original code.

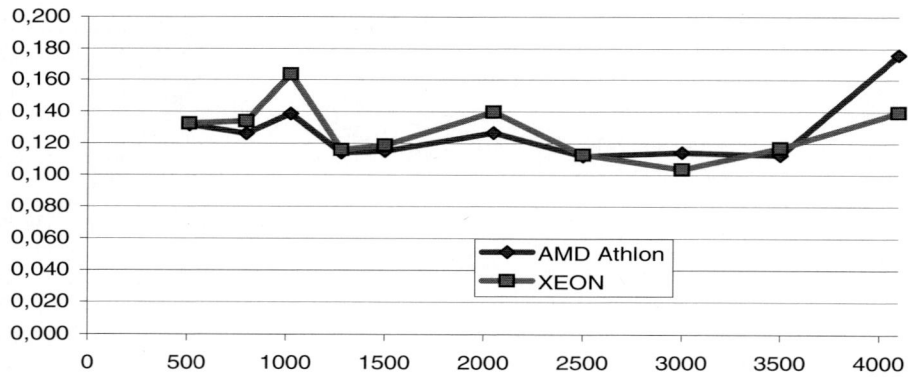

Fig. 2. Performance of the DWT program on different architectures

As can be seen, we achieved peak performance with each image size of a power of two. This is because the DWT algorithm has data locality problem with theses image sizes [3]. Hence, the optimization, specifically for improving the cache performance, shows its effectiveness. Overall, we achieved a performance gain of up to 18% with this application.

6   Conclusions

In this paper, we proposed a kind of guided prefetching for tackling cache problems. The prefetching is based on access patterns acquired by a self-developed analysis tool. Initial experimental results show a speedup of up to 20% in execution time.

However, in this prototypical implementation we have not deployed heuristics for computing the accurate prefetching position. This will be done in the next step of this research work. In addition, we intend to discover other access patterns that could be used for choosing the prefetch targets.

References

1. Baer, J.-L., Chen, T.-F.: Effective Hardware-Based Data Prefetching for High-Performance Processors. IEEE Transactions on Computers 44(5), 609–623 (1995)
2. Berg, S.G.: Cache prefetching. Technical Report UW-CSE 02-02-04, Department of Computer Science & Engineering, University of Washington (February 2004)
3. Chaver, D., Tenllado, C., Pinuel, L., Prieto, M., Tirado, F.: 2-D Wavelet Transform Enhancement on General-Purpose Microprocessors: Memory Hierarchy and SIMD Parallelism Exploitation. In: Proc. Int. Conf. on the High Performance Computing (December 2002)
4. Inagaki, T., et al.: Stride Prefetching by Dynamically Inspecting Objects. In: Proceedings of the ACM SIGPLAN Conference on Programming Language Design and Implementation, pp. 269–277 (June 2003)
5. Mohan, T., et al.: Identifying and Exploiting Spatial Regularity in Data Memory References. In: Supercomputing 2003 (November 2003)
6. Joseph, D., Grunwald, D.: Prefetching using markov predictors. IEEE Transactions on Computers 48(2), 121–133 (1999)
7. Jouppi, N.P.: Improving Direc-mapped Cache Performance by the Addition of a Small Fully-associative Cache and Prefetch Buffers. In: Proceedings of the 17th Annual International Symposium on Computer Architectures, May 1990, pp. 364–373 (1990)
8. Lin, W., Reinhardt, S., Burger, D.: Reducing DRAM Latencies with an Integrated Memory Hierarchy Design. In: Proceedings of the 7th International symposium on High-Performance Computer Architecture, January 2001, pp. 301–312 (2001)
9. Luk, C., Mowry, T.C.: Compiler-Based Prefetching for Recursive Data Structures. In: Proceedings of the Seventh International Conference on Architectural Support for Programming Languages and Operating Systems, October 1996, pp. 222–233 (1996)
10. Marathe, J., Mueller, F., de Supinski, B.: A Hybrid Hardware/Software Approach to Efficiently Determine Cache Coherence Bottlenecks. In: Proceedings of the International Conference on Supercomputing, June 2005, pp. 21–30 (2005)
11. Nguyen, A.-T., Michael, M., Sharma, A., Torrellas, J.: The augmint multiprocessor simulation toolkit for intel x86 architectures. In: Proceedings of 1996 International Conference on Computer Design (October 1996)
12. Rigoutsos, I., Floratos, A.: Combinatorial Pattern Discovery in Biological Sequences: the TEIRESIAS Algorithm. Bioinformatics 14(1), 55–67 (1998)
13. Smith, J.E.: Decoubled Access/Execute Computer Architectures. In: Proceedings of the 9th Annual International Symposium on Computer Architectures, July 1982, pp. 112–119 (1982)
14. Tao, J., Schloissnig, S., Karl, W.: Analysis of the Spatial and Temporal Locality in Data Accesses. In: Alexandrov, V.N., van Albada, G.D., Sloot, P.M.A., Dongarra, J. (eds.) ICCS 2006. LNCS, vol. 3992, pp. 502–509. Springer, Heidelberg (2006)

Performance Tool Workflows

Wyatt Spear, Allen Malony, Alan Morris, and Sameer Shende

Performance Research Laboratory
Department of Computer and Information Science
University of Oregon, Eugene OR 97403, USA
{wspear,malony,amorris,sameer}@cs.uoregon.edu

Abstract. Using the Eclipse platform we have provided a centralized resource and unified user interface for the encapsulation of existing command-line based performance analysis tools. In this paper we describe the user-definable tool workflow system provided by this performance framework. We discuss the framework's implementation and the rationale for its design. A use case featuring the TAU performance analysis system demonstrates the utility of the workflow system with respect to conventional performance analysis procedures.

1 Introduction

Performance analysis is an important component of software development, especially in high performance and parallel computing. With the proliferation of multicore systems and the growing reliance upon parallel computing in science and industry, the number of programmers who need to analyze and optimize application performance as a matter of course is likely to increase.

Collecting performance data can be a complicated and time consuming undertaking. Depending on the performance metrics one intends to collect and the tools employed the steps may include source level instrumentation, compilation with performance tool specific compilers or options, execution with performance tool specific options or composed with a data collection tool, data collection, storage, analysis, format conversion and visualization. It often requires knowledge of several distinct tools to effectively perform even a subset of these tasks. This must be accomplished and any technical hurdles must be overcome before the true goal of performance analysis, actually using collected performance data to improve the efficiency of an application, can be pursued.

Because performance analysis tools are usually command-line based, multi-step performance analysis procedures are generally either done by hand or performed with scripts. Such scripts are invariably specific to the tools being used and sometimes even the application being analyzed. In any case, managing performance tool inter-operation is left to the end user. Expertise in the use of individual tools and the collective use of multiple tools must be developed individually, or obtained via documentation rather than any easily deployable programmatic means.

In more conventional venues of software development the Integrated Development Environment has found favor for its productivity enhancing features. Few

M. Bubak et al. (Eds.): ICCS 2008, Part III, LNCS 5103, pp. 276–285, 2008.

IDEs offer significant support for the development of high performance applications. Where IDEs do offer performance analysis solutions, they are typically unique to the IDE in question. If other tools are required the user must return to the command line or resort to manual data manipulation.

The performance analysis framework for Eclipse attempts to address these difficulties by providing an extensible, modular, general system for defining performance analysis workflows. Workflow systems have been shown to increase efficiency and ease of use for complex computational activities composed of discrete steps [6]. With the growing complexity of and need for performance analysis, the benefits of workflow techniques seem quite applicable. The expertise required to perform a given performance analysis task or series of tasks can be encapsulated in a workflow definition for easy distribution and deployment. So long as the necessary tools are available on the system, the user need only select the desired workflow and set any necessary starting parameters.

The performance analysis framework we have developed offers a modular, extensible solution to the problem of performance analysis in IDEs by encapsulating existing command line based tools. In addition to the basic requirement of offering performance analysis tool functionality, it allows tools to be linked together in a workflow of performance analysis steps. The components comprising this system and the steps that led to its development are described below.

2   Eclipse

The Eclipse integrated development environment [3] began as a platform for the development of Java applications. In recent years its functionality has expanded to support multiple languages and programming paradigms. The C/C++ Development Tools (CDT) [2] and Photran [14] projects provide functionality for C/C++ and Fortran development, respectively. The Parallel Tools Platform (PTP) [7] project extends the capabilities of the CDT and Photran by offering parallel development, launch and debugging. These tools make Eclipse a promising means of enhancing traditional command line tools with IDE based development techniques in high performance computing. However, the requirements of high performance computing extend beyond the productivity-improving features of a standard IDE.

The tools available for performance analysis and the breadth of features that they cover make development of new tools specifically for a given IDE a difficult task. This is further complicated by the fact that most parallel application developers are already accustomed to some set of tools and development procedures available on the command line. Those who take the first steps toward IDE based parallel application development may find the productivity gains provided by the IDE significantly curtailed if they are faced with unfamiliar or incomplete tools. The solution to this problem is to wrap the IDE around existing tools, fully exposing their capabilities in the GUI. By making performance tools available within Eclipse we not only simplify their usage, often the tools incorporated into the environment are made more usable than in their initial command-line based incarnations.

The performance analysis framework required a set of general, modular components that could be combined to allow any command-line based performance analysis tool to be incorporated. The movement of other projects toward supporting parallel application development on the Eclipse platform and the platform's modularity and extensibility makes it an ideal focus for this work.

3   TAU

The performance analysis framework was initially created to add support for the TAU (Tuning and Analysis Utilities) [8][10]. Like many other performance analysis systems TAU is primarily oriented toward usage on the command line. It supports numerous platforms and parallel programming paradigms including MPI and OpenMP. Its data collection options include tracing, profiling, call-path profiling, hardware counter data collection and inter-operation with various other performance analysis systems.

The performance data produced by TAU can be broadly classified as either trace output or profile output. Depending on TAU's configuration one or both types of data may be generated in any of several formats. TAU provides several performance analysis tools, including ParaProf [1], a scalable graphical profile analysis tool. It also provides support for database management of performance profiles via the PerfDMF [4] database system.

TAU's flexibility necessitates some complexity in deployment. A new configuration must be created for each combination of performance analysis options to be used. In most cases the application to be analyzed must be instrumented with TAU API calls and linked with the appropriate TAU library.

The use of TAU is greatly simplified by the use of the Performance Database Toolkit (PDT) [5] and TAU's compiler wrapper scripts. PDT enables automatic parsing and instrumentation of a project's source code, using selective instrumentation if necessary. TAU's compiler wrapper scripts automatically include the necessary invocations of PDT and modifications to the compilation command to include TAU libraries. Collecting performance data can be as simple as replacing the default compiler command with the TAU compiler script in a project's makefile and specifying the desired compiler and environmental options. However, once the raw performance data has been collected there are still numerous paths to be taken in manipulation and analysis of that data.

4   The TAU Plug-In

The most fundamental way to use TAU in conjunction with Eclipse is to replace the default compilers used by Eclipse's build system with the TAU compiler scripts. Although this can be done manually, the numerous options and features available to TAU at compile time must still be known to the user to be invoked. Furthermore the process of adjusting analysis parameters and rebuilding the application when different performance analysis data are desired still necessitates manual string editing. Manually collecting performance data with TAU in Eclipse

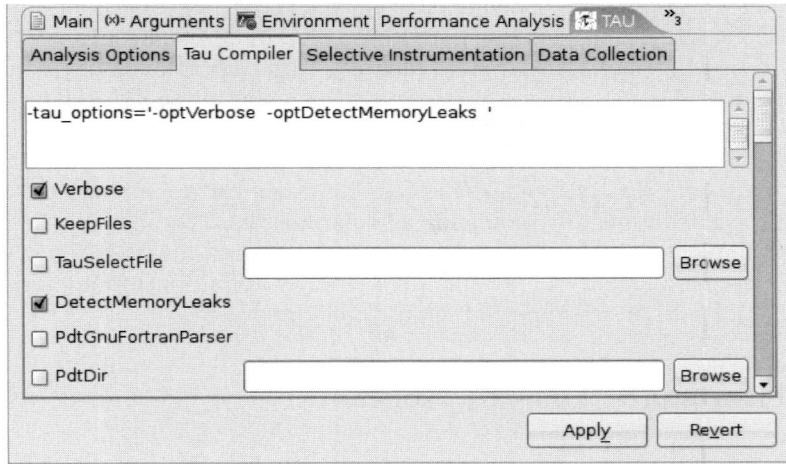

Fig. 1. TAU Compiler Option Selection

is no more difficult than typical command-line based usage. The same generally applies to other performance tools that one might attempt to deploy in Eclipse. There is clear room for improvement.

The core of the TAU plug-in for Eclipse [9] allows the user to select a TAU configuration graphically. The visual exposure and online documentation of TAU's performance data collection options and the easy selection of those options within the Eclipse environment (Fig. 1) makes the plug-in a significant improvement over manual inclusion of TAU in the build configuration.

The TAU plug-in also provides options for automatically storing generated profile data and viewing that data in the ParaProf profile analysis tool. In effect, it provides a single, static workflow using a pre-set series of tools

5   The Performance Framework

The development of the TAU plug-in and subsequent refactorings and optimizations resulted in a modular set of routines for the invocation of tools at various stages of the Eclipse build and launch procedures. It also resulted in the creation of a dynamic UI generator which can read user specified option definitions and create a corresponding set of options in Eclipse's user interface. Once defined, these options are automatically applied to the relevant build or launch steps.

Though these components had been directed toward the use of TAU in Eclipse, many of their capabilities are just as easily applied to other performance analysis tools. The TAU specific elements of the core plug-ins amounted to the strings defining the compiler and option names and the logic specific to TAU configuration and option selection. Rather than require other tool developers to develop their own Eclipse plug-ins, we determined to extend the TAU plug-ins to support essentially arbitrary series of commands which could be assembled for any given tool without significant knowledge of Eclipse plug-in development.

Converting the plug-in to a system capable of dynamic command specification was not a trivial undertaking. The parameters of a selected tool workflow needed to be associated with data structures that could efficiently propagate through each stage of performance data generation and analysis. Furthermore, the internal representations of tool specific configurations had to support essentially arbitrary combinations of commands and parameters. Fortunately, the highly modular implementation of the performance framework makes it fairly simple to add new functionality or extend the workflow definition format should an analysis tool with an unsupported interface be discovered.

6   Performance Analysis Workflow

There are two essential goals of the workflow definition format. The first is to grant flexibility to tool developers to integrate their tools into Eclipse without the need to modify their tools or to devote a great deal of time to learning Eclipse plug-in development. The second is to grant performance tool users the ability to easily configure and deploy project-specific performance analysis capabilities within Eclipse with a relatively small investment of time and effort.

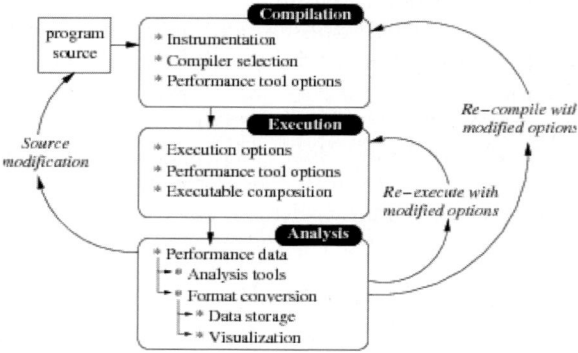

Fig. 2. A General Performance Analysis Workflow

The performance workflows are defined in an XML document stored outside of the Eclipse workspace. This document may contain an arbitrary number of individual workflow definitions. Workflow definitions are broken down into compilation, execution and analysis phases (Fig. 2). A single workflow may consist of several tools. By composing performance data collection and analysis tools it is possible to produce an instrumented program, execute that program, collect the performance data and display that data in the chosen performance data viewer with a single mouse click.

6.1   Compilation

The compilation phase definition specifies compilers and compiler arguments to be used by the Eclipse build system. This is primarily useful for performance tools that require recompilation of the program for instrumentation. Typically this section will include specification of a compiler, often a compiler wrapper script associated with the performance tool, and any relevant arguments. The arguments specified in the tool definition will be combined with any arguments provided by the build configuration selected for the build process.

Workflows that define a compilation phase require the user to select a build configuration rather than an existing executable file in Eclipse's launch configuration. The selected build configuration us used as the basis for the construction of the new executable. The user interface adjusts for the selection of an executable or a build configuration depending on the requirements of the selected workflow.

The default behavior of the performance plug-ins, when recompilation is defined in the workflow, is to compile and execute the program and then run any specified analysis tools once data generation is complete. However the execution phase may be disabled, for example if the user intends to run the compiled program outside of Eclipse.

6.2   Execution

The execution phase accommodates the use of tools that prepend the project's compiled executable. This element was intended to support performance tools such as perfsuite's psrun [13]. However, it has also proven useful for composing applications for more general purposes. The ability to define a tool configuration that automatically runs a compiled application with mpirun, for example, has proven useful on systems where the PTP is unavailable. The composition of multiple applications in the launch phase is supported. A typical use of this capability is to initiate a parallel launch of an application using mpirun composed with an analysis tool.

Unlike most of the other workflow components, the modifications to the standard launch system for the execution phase to support arbitrary tools were almost entirely independent of the work done to support TAU. TAU itself does not make use of composed executable launching. Additionally, unlike modifying the build configuration system, manually arranging a composed executable launch in Eclipse is not a straightforward process.

6.3   Analysis

The analysis phase is initiated after execution. It consists of a series of commands and their arguments to be run in sequence. Information on the output of the workflow's performance tool is used to pass performance data to the specified post-processing, data-management or visualization applications. Additionally, the output of one application can be directed as an argument to another. The flexibility of the tool definitions available in the analysis phase allow a great deal of creativity in its use by tool developers and users.

It is difficult to dynamically capture the output of the executed program for use in a given analysis-phase application. In some cases the names of generated performance files can not be predicted. Presently it is necessary specify the output to be accessed in the tool definition on a per-application basis. Eventually it should be possible to specify this dynamically from within the launch configuration.

7   Use Cases

The large number of performance tools and the varied capabilities of each tool presents a vast number of combinations which can be difficult to wade through for those who simply want to collect performance data and optimize their software without focusing on the technical details of performance analysis. Even when the procedure is straightforward, undertaking a multi-step performance analysis workflow manually can be time consuming. The use cases presented here provide some insight into the utility of performance analysis workflows.

7.1   TAU

A common use case for the TAU performance analysis system calls for generation of trace and profile data for a given application. The trace data is then merged, converted into a trace format associated with a particular trace viewer and analyzed in that trace viewer. The profile data is stored in a database and viewed in ParaProf.

The individual steps are relatively simple once one becomes familiar with the various tools involved. However, even with expertise in the various tools, the procedure is still time consuming. To get from compilation to data visualization as rapidly as possible, it is necessary to observe each of the individual steps and initiate the next as soon as the previous is complete.

A workflow defined in the performance analysis framework makes this procedure much more straightforward. After selecting the relevant TAU workflow (Fig. 3) it is still necessary to select a TAU configuration and specify any compilation or execution specific options. However the user interface provided for these options require only introductory knowledge of TAU's capabilities.

With the relevant options selected the performance workflow may be launched. The application will be recompiled with TAU's compiler wrapper scripts, automatically generating instrumented source and linking it with the relevant libraries. It bears note that for a tool lacking automatic instrumentation capabilities instrumentation may still need to be performed manually.

The TAU-instrumented executable will be launched immediately with any parameters specified in the Eclipse launch configuration, in addition to those provided by the TAU configuration interface. Depending on the type of TAU configuration specified trace and/or profile data may be produced along with any output normally generated by the program being analyzed.

By default, profile data will be uploaded to a user-specified PerfDMF database if one is available. The ParaProf profile analysis tool will be launched on the profile data stored within the database. The profile files are deleted if the database

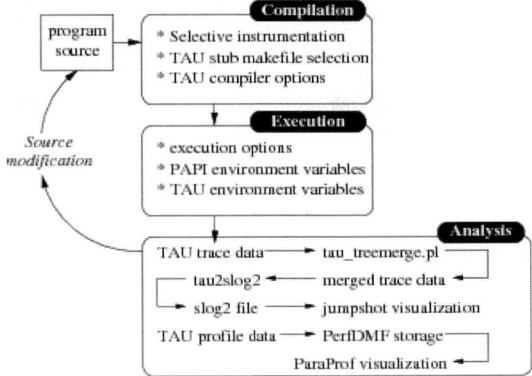

Fig. 3. A Workflow for a Common TAU Use Case

upload is successful, otherwise they are placed in a local directory identified by the launch configuration name and timestamp. The logic for these operations is coded directly into the performance framework. This behavior is left from the original TAU specific plug-in. However a similar series of analysis steps could easily be specified using the workflow definition system.

Trace data manipulation was not supported directly in the TAU specific iteration of the plug-in. In this workflow trace management consists of the following steps. First tau_treemerge.pl is called to merge tracefiles from multiple processes. tau2slog2 is then called on the merged trace output to produce a .slog2 file. Finally, to display the trace Jumpshot is called on the .slog2 file.

Once a launch configuration is configured with the desired workflow and the relevant performance collection and application options are specified it will persist within the Eclipse workspace. Changes to the source code or the launch parameters can be tested immediately with minimal oversight by the user.

7.2   Valgrind

A relatively simple workflow using a tool quite different from TAU involves the use of the memory analysis tool valgrind [12]. A valgrind workflow requires only the specification of the valgrind executable in the execution phase, along with any desired options.

When this workflow is selected in the launch configuration no other modifications are necessary. The executable of the selected project will be composed with valgrind and any options defined in the workflow. The launch will then proceed normally. The output from valgrind will be displayed in Eclipse's console view unless different output behavior for valgrind is specified in the workflow.

8   Future Work

As of this writing, some of the performance framework's more advanced capabilities are still somewhat limited to supporting TAU. One priority is extending the selective instrumentation capabilities of the system to support the various selective instrumentation schemes of alternative performance analysis tools. A related goal is to support the selection tool options usable by multiple tools, such as PAPI hardware counters [11], for tools other than TAU using a single interface.

Currently the analysis tools are only invoked after an execution phase has completed. Although this covers the most typical use cases, it may be useful to initiate an analysis workflow on performance data already present on the filesystem. Support for analysis-only operations is a priority, though the interface for this feature is not likely to be an extension of the launch configuration.

Presently the workflow definition format only allows for linear workflows using some combination of compilation, execution and analysis commands. Soon, the modular nature of the workflow components will allow the addition of some logical elements to the workflow system. For example, rather than terminate at the end of the analysis phase, data collected in analysis could be evaluated and the compilation and execution phases repeated with modified parameters until a certain set of conditions is met. This will be useful for performance analysis procedures that require iterative data collection. The initial implementation of this capability will likely be to facilitate performance scalability testing.

Ultimately we hope to provide a means of visually defining and modifying tool workflows within Eclipse. This will remove the need for users or developers to interact with the tool definition XML file and make the full capabilities of the workflow system more accessible and obvious.

9   Conclusion

By providing a general framework for performance analysis in a popular IDE we hope to simplify the process of performance analysis just as IDEs have assisted with simplifying other aspects of the software development cycle. Ideally this work can benefit not only existing software developers, but will also be of use to newcomers to high performance software engineering who may be more accustomed to IDE-based software development.

Perhaps more importantly, we have created a means for expert users to encapsulate their expertise in a programmatic way. A given series of performance analysis operations can be defined as a workflow by a tool developer or advanced user. This can then be made available to others who desire the final output of the potentially complex performance analysis procedure, but have no desire to engage in the manual, multi-step process every time they need to collect the data from their application.

References

1. Bell, R., Malony, A.D., Shende, S.: A Portable, Extensible, and Scalable Tool for Parallel Performance Profile Analysis. In: Proc. EUROPAR 2003 conference. LNCS, vol. 2790, pp. 17–26. Springer, Heidelberg (2003)
2. CDT - C/C++ Development Tools, http://www.eclipse.org/cdt

3. Eclipse, http://www.eclipse.org
4. Huck, K., Malony, A.D., Bell, R., Li, L., Morris, A.: PerfDMF: Design and implementation of a parallel performance data management framework. In: Proc. International Conference on Parallel Processing (ICPP 2005). IEEE Computer Society. Los Alamitos (2005)
5. Lindlan, K.A., Cuny, J., Malony, A.D., Shende, S., Mohr, B., Rivenburgh, R., Rasmussen, C.: A Tool Framework for Static and Dynamic Analysis of Object-Oriented Software with Templates. In: Proceedings of SC 2000: High Performance Networking and Computing Conference, Dallas (November 2000)
6. Oinn, T., et al.: Taverna/myGrid: aligning a workflow system with the life sciences community. In: Taylor, I.J., Deelman, E., Gannon, D.B., Shields, M. (eds.) Workflows for e-Science: scientific workflows for Grids, pp. 300–319. Springer, Guildford (2007)
7. PTP - Parallel Tools Platform, http://www.eclipse.org/ptp
8. Shende, S., Malony, A.D.: The TAU Parallel Performance System. International Journal of High Performance Computing Applications, ACTS Collection Special Issue (2005)
9. Spear, W., et al.: Integrating TAU With Eclipse: A Performance Analysis System in an Integrated Development Environment. In: Gerndt, M., Kranzlmüller, D. (eds.) HPCC 2006. LNCS, vol. 4208, pp. 230–239. Springer, Heidelberg (2006)
10. TAU - Tuning and Analysis Utilities,
    http://www.cs.uoregon.edu/research/tau/home.php
11. Moore, S., Cronk, D., Wolf, F., Purkayastha, A., Teller, P., Araiza, R., Aguilera, M., Nava, J.: Performance Profiling and Analysis of DoD Applications using PAPI and TAU. In: Proceedings of DoD HPCMP UGC 2005. IEEE, Nashville, TN (2005)
12. Nethercote, N., Seward, J.: Valgrind: A Framework for Heavyweight Dynamic Binary Instrumentation. In: Proceedings of ACM SIGPLAN 2007 Conference on Programming Language Design and Implementation (PLDI 2007), San Diego, California, USA (June 2007)
13. Kufrin, R.:PerfSuite: An Accessible, Open Source Performance Analysis Environment for Linux. In: 6th International Conference on Linux Clusters: The HPC Revolution 2005. Chapel Hill, NC (April 2005)
14. Photran - Fortran Development Tools, http://www.eclipse.org/photran/

Workshop on Software Engineering for Large-Scale Computing

Workshop on Software Engineering for Large Scale Computing (SELSC)

Daniel Rodríguez[1] and Roberto Ruiz[2]

[1] Computer Science Department, The University of Alcalá
Ctra. Barcelona km. 33.6. 28871 Alcalá de Henares (Madrid), Spain
`daniel.rodriguezg@uah.es`
[2] Computer Science Department, Pablo de Olavide University
Ctra. Utrera km. 1, 41013 Sevilla, Spain
`robertoruiz@upo.es`

1 Motivation

Computational Science techniques are increasingly being applied in both research institutions and industry to for example, financial applications, bioinformatics applications, physical applications, data mining applications based on grid or clusters etc. As with other disciplines there are specific issues when planning and developing such applications. For example, those applications are generally distributed and based upon Grid environments or clusters with specific management and planning issues, design, testing, etc.

This workshop was created with the intention of analysing software engineering issues specific to these types of applications such as adaptation of processes for, planning, management, verification and validation, testing, quality measurement, etc.

Also, there is also a motivation for further research on the other direction, i.e., the application of computational intelligence techniques to software engineering issues as a result of the creation of large metrics databases collected from software projects. Examples of computational techniques applied to Software Engineering include genetic algorithms, Bayesian networks, system dynamics, visualization, search based software engineering etc. Many emergent issues are subject to research efforts combining both computational techniques and Software Engineering.

2 Topics of Interest and Objectives

Therefore, topics of interest of this workshop included:

- Software processes for computational science applications
- Analysis and development of computational applications
- Testing of computational applications
- Computational intelligence techniques applied to software engineering
- Mining software engineering repositories
- Management issues for handling large amount of data
- Execution of data mining algorithms on grid and cluster environments
- Etc.

M. Bubak et al. (Eds.): ICCS 2008, Part III, LNCS 5103, pp. 289–290, 2008.

And the objectives included:

- To bring together researchers and practitioners from computational science and software engineers backgrounds
- To steer discussion and debate on various aspects and issues related to software engineering processes when applied to computational problems
- How to apply software engineering techniques to computational intelligence problems
- To open research directions that are deemed essential by the researchers in the field and industry

The 9 selected papers out of the 19 submitted cover these topics from both points of view, i.e., software engineering issues for large scale applications and using computational science for helping with software engineering issues.

3 Program Committee and Acknowledgements

The organisers of this workshop would like to thank the organisers of main ICCS conference for their support, specially, Dr G D Dick van Albada. Also thanks to all members of the programme committee for all their effort reviewing and improving the submitted papers:

- Javier Dolado – Basque Country University, Spain
- Mark Harman - King's College London, UK
- Rachel Harrison - Stratton Edge Consulting, UK
- Andreas Jedlitschka - Fraunhofer IESE, Germany
- Taghi M. Khoshgoftaar - Florida Atlantic University, USA
- Andres A. Folleco - Florida Atlantic University, USA
- Dietmar Pfahl - University of Calgary, Canada
- Marek Reformat - University of Alberta, Canada
- Daniel Rodríguez - Univ. of Alcalá, Spain
- Mercedes Ruiz - Univ. of Cádiz, Spain
- Roberto Ruiz - Univ. of Seville, Spain
- Manoranjan Satpathy - General Motors, India
- Florian Urmetzer - University of Reading, UK
- Christian Weihrauch - University of Reading, UK
- Andrea Weise - University of Reading, UK

Finally, this workshop was also supported by grant number CCG07-UAH-TIC-1588, jointly supported by the University of Alcalá and the autonomic community of Madrid (*Comunidad Autónoma de Madrid*).

Modeling Input Space for Testing Scientific Computational Software: A Case Study

Sergiy A. Vilkomir[1], W. Thomas Swain[1], Jesse H. Poore[1], and Kevin T. Clarno[2]

[1] Software Quality Research Laboratory,
Department of Electrical Engineering and Computer Science, University of Tennessee,
Knoxville, TN 37996, USA
{vilkomir,swain,poore}@eecs.utk.edu
[2] Reactor Analysis Group, Nuclear Science and Technology Division,
Oak Ridge National Laboratory, Oak Ridge, TN 37923, USA
clarnokt@ornl.gov

Abstract. An application of a method of test case generation for scientific computational software is presented. NEWTRNX, neutron transport software being developed at Oak Ridge National Laboratory, is treated as a case study. A model of dependencies between input parameters of NEWTRNX is created. Results of NEWTRNX model analysis and test case generation are evaluated.

1 Introduction

Testing scientific computational software has been the subject of research for many years (see, for example, [3, 4, 6, 7]). Because of the increase in size and complexity of such software, automation of scientific software testing is very important. A part of this task is test case selection from a large input space.

For test automation, model-based approaches are most effective. In particular, we use Markov chain models to support automated statistical testing [13, 14]. The method and tools were initially applied to systems where sequences of discrete stimuli cause software responses and changes in the state of use. However, the behavior of computational software is more typically a function of a large multi-parameter static input space rather than sequences of discrete stimuli. Often all input parameters are entered as a batch, and then the software computes results with no further interaction with users.

Although other testing methods (combinatorial testing [2, 5], etc.) are applicable to this problem, they are not as directly supportive of automated testing as the method based on directed graphs and Markov chains. We considered this problem in [12] where a method based on dependency relations encoded into Markov chain models was described. When mutual dependencies limit the set of valid input combinations, the method captures only the valid combinations. In this paper, we consider an application of this method to scientific software[1]. The case study is NEWTRNX [1] – a neutron transport simulation developed at Oak Ridge National Laboratory (ORNL)[2].

---

[1] This research is supported by the University of Tennessee Computational Science Initiative in collaboration with the Computing and Computational Sciences Directorate of Oak Ridge National Laboratory.

[2] Research sponsored by the Laboratory Directed Research and Development Program of Oak Ridge National Laboratory (ORNL), managed by UT-Battelle, LLC for the U. S. Department of Energy under Contract No. DE-AC05-00OR22725.

M. Bubak et al. (Eds.): ICCS 2008, Part III, LNCS 5103, pp. 291–300, 2008.
© Springer-Verlag Berlin Heidelberg 2008

For NEWTRNX, as in most multiphysics simulations, the large number of input parameters and values makes manually selecting an arguably sufficient set of test cases very challenging. As a first step towards test automation, a model of dependencies among NEWTRNX input parameters is created. Then special tools can be used to generate test cases automatically from the model. In our test automation, we used JUMBL [10] – the J Usage Model Builder Library developed in the Software Quality Research Laboratory (SQRL) at the University of Tennessee.

Sec. 2 presents a brief review of the method for modeling the input space. In Sec. 3, we describe the neutron transport software in very general terms. In Sec. 4, we apply our method to this specific case to create a model of dependencies among NEWTRNX input parameters. Sec. 5 contains results of model analysis and test case generation for the model from Sec. 4, using the JUMBL. Conclusions are discussed in Sect. 6.

2 Modeling Input Space for Software Test Case Generation

Modeling an input space for the purpose of test case generation was presented in [11] for independent input parameters and in [12] for the situations when dependencies between input parameters exist. The basis of the method is representation of the input space as a directed graph, where nodes represent input parameters and incoming arcs represent specific values of the parameters. Any path through the graph (a sequence of arcs) corresponds to a test case, so the graph can be used as a model for test case generation.

From a usage point of view, there is some probability (relative frequency) that the parameter will have a particular value. We associate these probabilities with corresponding arcs of the graph, creating a probability distribution over the exit arcs of each node. This creates a model which satisfies the probability law of Markov chains, allowing analysis of the testing process via well-known mathematical techniques and software.

Input space modeling can be illustrated with the following small example. Consider a system with three input parameters i_1, i_2, and i_3, which can take the following values:

- $i_1 \in \{1, 2, 7, 9\}$
- $i_2 \in \{2, 5, 8\}$
- $i_3 \in \{3, 6, 9\}$

If these parameters are independent, then all combinations are possible. One model of the input space for this situation is shown in Fig. 1.

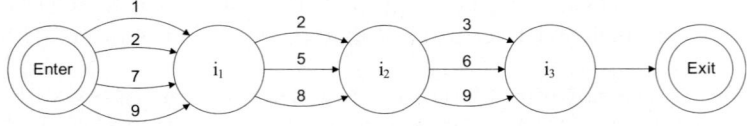

Fig. 1. Model for independent parameters

When dependencies among parameters exist, we modify the model by splitting nodes and merging nodes [12]. Splitting nodes is used for a dependency between two parameters. For example, consider the following dependency between i_2 and i_3: $P_1(i_2) \Rightarrow i_3=6$, where $P_1(i_2)$ is the characteristic predicate for set $\{2, 8\}$. In other words, if i_2 takes value 2 or value 8, then i_3 shall take only the value 6. To encode this dependency, we split i_2 node into to nodes: (i_2, P_1) and (i_2, P_2), as shown in Fig. 2, where P_2 is the characteristic predicate for set $\{5\}$. The model in Fig. 2 now satisfies the Markov probability law. The model can then be analyzed as a Markov chain and used for test case generation.

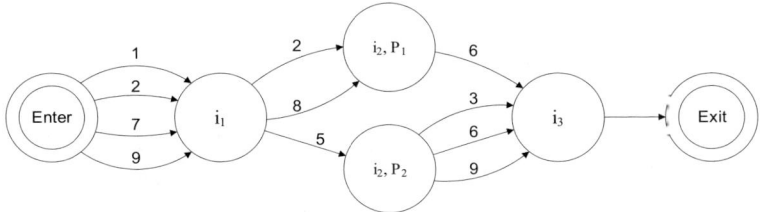

Fig. 2. Model of dependency between i_2 and i_3

Merging nodes is used for dependencies among several parameters. For example, suppose the pair (i_2, i_3) is dependent on $i1$:

- $P_3(i_1) \Rightarrow (i_2, i_3) \in \{(5, 9), (8, 3), (8, 9)\}$, where $P_3(i_1)$ is the characteristic predicate for set $\{2, 9\}$.
- $P_4(i_1) \Rightarrow (i_2, i_3) \in \{(2, 3), (2, 6), (2, 9), (5, 3), (5, 6), (8, 6)\}$, where $P_4(i_1)$ is the characteristic predicate for set $\{1, 7\}$.

This dependency can be modeled in three steps (Fig. 3):

- Creating a new derived parameter (i_2, i_3) by merging parameters i_2 and i_3,
- Splitting parameter i_1 according to predicates P_3 and P_4,
- Establishing arcs between i_1 and (i_2, i_3) based on the possible values of (i_2, i_3).

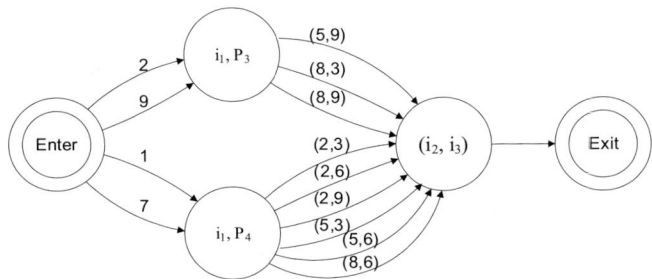

Fig. 3. Model of the dependency between i_1 and (i_2, i_3)

Detailed information about this method, including the application of the method to different types of dependencies, results of test case generation, and more examples can be found in [12]. The purpose of the current paper is to apply this approach to an existing computational science software application.

3   A Case Study: Neutron Transport Software (NEWTRNX)

The NEWTRNX transport solver [1] was developed at ORNL to provide proof-of-principle software for high-fidelity modeling of nuclear fission energy systems. The primary function of the solver is to provide the heat generation distribution of neutrons.

Realistic nuclear reactor simulation is important because it requires an accurate understanding of the interactions of multi-scale, multi-physics phenomena through complex models that may only be tractable with leadership-class computing facilities. The distribution of neutrons within the nuclear vessel is described by the seven-dimensional isotropic source-driven or forward eigenvalue neutral-particle Boltzmann transport equation (equations 1-3) [8], where the fundamental unknown, ψ, is the time-dependent space-velocity distribution of neutrons in the system.

$$\frac{1}{v}\frac{\partial \psi}{\partial t}+\overline{\Omega}\bullet\overline{\nabla}\psi+\sigma_{t}\psi =\frac{1}{4\pi}\left\{ \begin{array}{ll} \displaystyle\int_{0}^{\infty}dE'\sum_{p=0}^{P}\sigma_{s}^{p}\sum_{q=-p}^{p}\phi^{pq}Y_{pq}+Q & isotropic \\[2em] \displaystyle\int_{0}^{\infty}dE'\sum_{p=0}^{P}\sigma_{s}^{p}\sum_{q=-p}^{p}\phi^{pq}Y_{pq}+\left[\frac{1}{k}\right]\int_{0}^{\infty}dE'\,\sigma_{f}\phi^{00} & forward \end{array} \right\} \quad (1)$$

$$\phi^{pq}=\int_{0}^{4\pi}d\Omega'\;\psi\left(\overline{r},E',\overline{\Omega}',t\right)Y_{pq}^{*}\left(\overline{\Omega}'\right) \qquad (2)$$

$\overline{r}=space\,(x,y,z)$

$\left(E,\overline{\Omega}\right)=velocity\,(energy\,of\,the\,neutron,direction\,of\,travel)$

$\psi\left(\overline{r},E,\overline{\Omega},t\right)="angular\;flux"$

$\phi^{pq}\left(\overline{r},E',t\right)=pq^{th}\,angular\,moment\,of\,the\,"angular\;flux"$

$\sigma_{t}\left(\overline{r},E,t\right)=total\,cross\,section$

$\sigma_{s}^{p}\left(\overline{r},E'\rightarrow E,t\right)=p^{th}\,harmonic\,moment\,of\,the\,scattering\,cross\,section$ (3)

$\sigma_{f}\left(\overline{r},E'\rightarrow E,t\right)=energy\,distribution\,of\,fission\,cross\,section$

$Y_{pq}\left(\overline{\Omega}\right)=pq^{th}\,spherical\,harmonic\,moment$

$v\left(E\right)=speed\,of\,the\,neutron$

$Q\left(\overline{r},E,\overline{\Omega},t\right)=external\,source\,of\,neutrons$

$k=largest\,eigenvalue\,(real\,and\,positive)$

Solving for the space-velocity distribution for even simple geometries requires a large number of degrees of freedom (dof) for an accurate solution (traditionally 10^9 dof per time-step for small single-processor calculations). To solve the equation, the direction of neutron travel is discretized into a set of uncoupled discrete-ordinate directions for a given neutron energy. Similarly, the energy (speed) of the neutron utilizes a piecewise-constant ("multigroup") discretization. This leads to a multi-level iterative solver where "outer" iterations solve for the coupling of the energy terms for all space and angular moments and "inner" iterations solve for the coupling of all directions for a given energy "group" [8].

There are various methods to solve the inner and outer iterations, as well as possible communications strategies. Many of these options are dependent on other options. For example, there are many ways to choose the discrete-ordinate directions, such as a 3d level-symmetric or a 2d/1d product quadrature set [8]. The type of quadrature set and number of directions constrain the number of angular moments that can be accurately computed. This leads to a large and inter-dependent input space that must be regularly tested for robustness and accuracy. This testing task necessitates automation.

4   Modeling Dependencies Among Input Parameters of NEWTRNX

One important aspect of NEWTRNX is the ability to specify different problem definitions and various mathematical methods for solving each defined problem. It allows the user to choose various input options by setting values of the following input parameters:

- mode – type of the problem (isotropic or forward)
- nSpherical – number of harmonics moments used to represent the anisotropic scattering source
- quadrature – type of quadrature set utilized (level-symmetric or product)
- nSn – level-symmetric quadrature set order (even numbers)
- nAziProd – azimuthal (2d) portion of the product quadrature set
- npolar – polar (1d) portion of the product quadrature set (3d is the product of nAziProd and npolar)

The parameters mode and nSpherical can be assigned their values independently, but there are dependencies between quadrature and other parameters:

- If quadrature equals symmetric, then parameter nSn is used and nAziProd and npolar are not used.
- If quadrature equals product, then parameters npolar and nAziProd are used but nSn is not used.

To reflect this dependency, we split parameter quadrature into two nodes, as described in Sec. 3.1. The structure of the dependency is presented in Fig. 4 (values of the parameters are not shown).

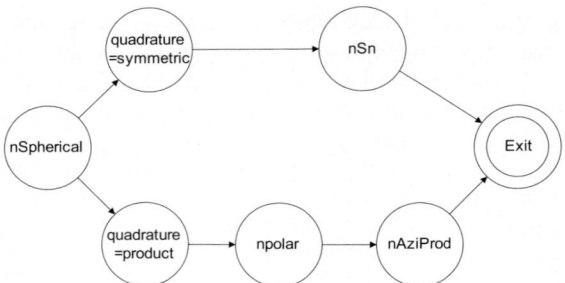

Fig. 4. Structure of the dependency among quadrature and nSn, nAziProd, npolar

The dependency among nSpherical, nSn, nAziProd, and npolar is shown in Table 1. The left column contains all possible values of nSpherical, and the other columns contain corresponding values of nSn, nAziProd, and npolar for every value of nSpherical.

Table 1. Dependency among nSpherical and nSn, nAziProd and npolar

nSpherical	nSn	nSpherical	npolar	nAziProd
{0, 1}	{2, 4, 6, 8, 10, 12, 14, 16}	{0, 1}	{1, 2, 3, 4, 5, 6}	{1, 2, 3, 4, 5,..., 12}
{2}	{6, 8, 10, 12, 14, 16}	{2, 3}	{2, 3, 4, 5, 6}	{2, 3, 4, 5,..., 12}
{3}	{10, 12, 14, 16}			
{4}	{14, 16}	{4, 5}	{3, 4, 5, 6}	{3, 4, 5,..., 12}
{5}	-			

Five different groups of values of nSpherical correspond to only four groups of values of nSn and to three groups of values of nAziProd and npolar. To reflect this, we split the node for nSpherical into five different nodes (Fig. 5). Then for each of these five nodes, we reflect the dependency on quadrature by splitting the node for quadrature into two nodes.

The model in Fig. 5 shows all dependencies between pairs of NEWTRNX input parameters and can be used for test case generation. Note that the model contains only necessary states (five for nSpherical, four for nSn, and three for nAziProd and npolar).

5 Test Case Generation for NEWTRNX

The model in Fig. 5 is a Markov chain model of valid input parameter combinations. We use the JUMBL [10] for the following tasks:

- Analysis of the Markov chain model.
- Test case generation based on the model.

To define a model for input to the JUMBL, the model is described in The Model Language (TML [9]). First the JUMBL is used to check logical consistency of the model and then to produce a model analysis report. General information includes numbers of nodes, arcs, stimuli, etc., as shown in Fig. 6 for the NEWTRNX model. Detailed information includes long run model statistics for nodes, arcs, and stimuli. Of particular interest are statistics for stimuli (Fig. 7) because they directly describe the long run use of different input values represented by the model.

Here "occupancy" is the number of occurrences of a given value (stimulus) divided by the total number of stimuli occurrences. "Mean occurrence" is the average number of times the specific value occurs in a single test case.

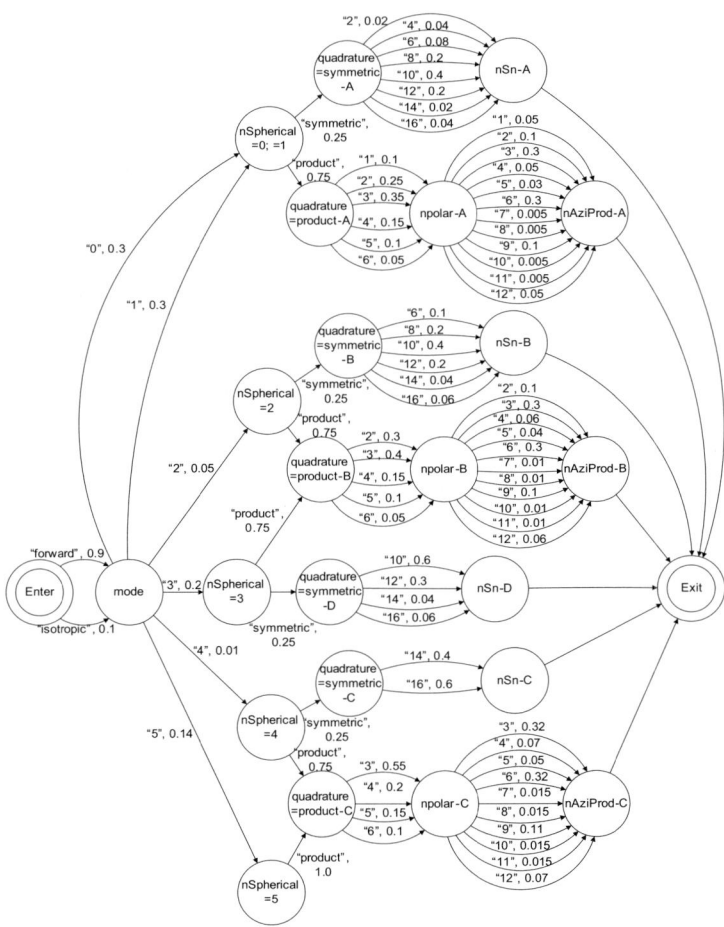

Fig. 5. Dependencies between input parameters of NEWTRNX

Node Count	25 nodes
Arc Count	92 arcs
Stimulus Count	37 stimuli
Expected Test Case Length	5.785 events
Test Case Length Variance	2.534 events
Transition Matrix Density (Nonzeros)	52.8E-3 (33 nonzeros)
Undirected Graph Cyclomatic Number	9

Fig. 6. NEWTRNX model statistics (fragment from JUMBL output)

Stimulus	Occupancy	Mean Occurrence (visits per case)
Exit	0.172860847	1
mode=forward	0.155574762	0.9
mode=isotropic	17.2860847E-3	0.1
nAziProd=1	3.88936906E-3	22.5E-3
nAziProd=10	1.09550562E-3	6.3375E-3
nAziProd=11	1.09550562E-3	6.3375E-3
nAziProd=12	7.61884183E-3	44.075E-3
nAziProd=2	11.019879E-3	63.75E-3

Fig. 7. NEWTRNX stimulus statistics (fragment from JUMBL output)

Table 2. NEWTRNX weighted test cases (fragment)

N	Probability	mode	nSpherical	quadrature	nSn	npolar	nAziProd
1	0.027	forward	3	symmetric	10	-	-
2	0.027	forward	0	symmetric	10	-	-
3	0.027	forward	1	symmetric	10	-	-
4	0.022	forward	5	product	-	3	6
5	0.022	forward	5	product	-	3	3
6	0.021	forward	0	product	-	3	6
7	0.021	forward	0	product	-	3	3
8	0.021	forward	1	product	-	3	6
9	0.021	forward	1	product	-	3	3
10	0.016	forward	3	product	-	3	6

Various types of test cases can be generated from the model including model coverage tests, random tests, and weighted tests. Coverage tests are generated as the minimal set of test cases that cover every arc in the model. Thus, to cover all arcs in the NEWTRNX model, 53 test cases were generated. Random test cases are generated according to the probabilities on the arcs. Weighted test cases are those generated in order of decreasing probability. For random and weighted tests, the number of test cases can be specified. A separate file is created for every test case. The ten highest probability test cases for the NEWTRNX model are shown in Table 2, with their individual probabilities of occurrence. Statistics for the 53 coverage tests and the ten most likely tests are provided in separate test analysis reports (Fig. 8).

Random test cases can be generated for reliability estimation. The "optimum" reliability represented by a particular set of test cases can be computed prior to test execution by assuming that all test cases will be successful. These values can be used during test planning for estimation of the required number of test cases. When testing is completed and the number of failures is known, operational reliability estimates are included in a test analysis report.

Node Count	25 nodes
Arc Count	92 arcs
Stimulus Count	37 stimuli
Test Cases Recorded	10 cases
Nodes Generated	19 nodes / 25 nodes (0.76)
Arcs Generated	25 arcs / 92 arcs (0.27173913)
Stimuli Generated	12 stimuli / 37 stimuli (0.324324324)
Nodes Executed	0 nodes / 25 nodes (0)
Arcs Executed	0 arcs / 92 arcs (0)
Stimuli Executed	0 stimuli / 37 stimuli (0)

Node Count	25 nodes
Arc Count	92 arcs
Stimulus Count	37 stimuli
Test Cases Recorded	53 cases
Nodes Generated	25 nodes / 25 nodes (1)
Arcs Generated	92 arcs / 92 arcs (1)
Stimuli Generated	37 stimuli / 37 stimuli (1)
Nodes Executed	0 nodes / 25 nodes (0)
Arcs Executed	0 arcs / 92 arcs (0)
Stimuli Executed	0 stimuli / 37 stimuli (0)

a) Weighted test cases   b) Coverage test cases

Fig. 8. NEWTRNX model test case statistics (fragment from JUMBL output)

6 Conclusions

Specification of all valid test cases from a large input space can be a challenging task, especially when there are dependencies among input parameters. We have presented a method for solving this problem and demonstrated its practical application on the neutron transport software tool NEWTRNX.

The selection of test cases is performed in two steps. First, a Markov chain model of the input space is created, reflecting dependencies among input parameters. Second, the JUMBL library of software tools is used for model analysis and test case generation. Results for NEWTRNX test planning are provided. The case study shows the applicability of model-based statistical testing for testing large scientific computational software systems. The next phase of this effort will investigate methods for automating both test execution and results checking.

References

1. Clarno, K., de Almeida, V., d'Azevedo, E., de Oliveira, C., Hamilton, S.: GNES-R: Global Nuclear Energy Simulator for Reactors Task 1: High-Fidelity Neutron Transport. In: Proceedings of PHYSOR–2006, American Nuclear Society Topical Meeting on Reactor Physics: Advances in Nuclear Analysis and Simulation, Vancouver, Canada (2006)

2. Cohen, D.M., Dalal, S.R., Fredman, M.L., Patton, G.C.: The AETG system: An approach to testing based on combinatorial design. IEEE Transactions on Software Engineering 23(7), 437–444 (1997)
3. Cox, M.G., Harris, P.M.: Design and use of reference data sets for testing scientific software. Analytica Chimica Acta 380(2), 339–351 (1999)
4. Einarsson, B. (ed.): Accuracy and Reliability in Scientific Computing. SIAM, Philadelphia (2005)
5. Grindal, M., Offutt, J., Andler, S.F.: Combination testing strategies: A survey. Software Testing, Verification, and Reliability 15(3), 167–199 (2005)
6. Hatton, L.: The T experiments: errors in scientific software. IEEE Computational Science and Engineering 4(2), 27–38 (1997)
7. Howden, W.: Validation of scientific programs. Comput. Surv. 14(2), 193–227 (1982)
8. Lewis, E., Miller Jr., W.F.: Computational Methods of Neutron Transport. ANS (1993)
9. Prowell, S.: TML: A description language for Markov chain usage models. Information and Software Technology 42(12), 835–844 (2000)
10. Prowell, S.: JUMBL: A Tool for Model-Based Statistical Testing. In: Proceedings of the 36th Annual Hawaii International Conference on System Sciences (HICSS 2003), Big Island, HI, USA (2003)
11. Swain, W.T., Scott, S.L.: Model-Based Statistical Testing of a Cluster Utility. In: Sunderam, V.S., van Albada, G.D., Sloot, P.M.A., Dongarra, J. (eds.) ICCS 2005. LNCS, vol. 3514, pp. 443–450. Springer, Heidelberg (2005)
12. Vilkomir, S.A., Swain, W.T, Poore, J.H.: Combinatorial test case selection with Markovian usage models. In: Proceedings of the 5th International Conference on Information Technology: New Generations (ITNG 2008), Las Vegas, Nevada, USA (2008)
13. Walton, G., Poore, J.H., Trammell, C.: Statistical Testing of Software Based on a Usage Model. Software: Practice and Experience 25(1), 97–108 (1995)
14. Whittaker, J., Poore, J.H.: Markov Analysis of Software Specifications. ACM Transactions on Software Engineering and Methodology 2(1), 93–106 (1993)

Executable Platform Independent Models
for Data Intensive Applications[*]

Grzegorz Falda[1], Piotr Habela[1], Krzysztof Kaczmarski[2], Krzysztof Stencel[3],
and Kazimierz Subieta[1]

[1] Polish-Japanese Institute of Information Technology, Warsaw, Poland
[2] Warsaw University of Technology, Warsaw, Poland
[3] Institute of Informatics Warsaw University, Warsaw, Poland
{gfalda,habela,stencel,subieta}@pjwstk.edu.pl,
k.kaczmarski@mini.pw.edu.pl

Abstract. In this paper we investigate the capabilities and shortcomings of
UML 2 in the field of executable modelling, with special focus on database applications. As envisioned in the MDA initiative, this sets the challenges like
achieving platform-independence of the base model, complete specification of
behaviour in an adequate way and making the model an unambiguous and suitable input for model compilers. Additionally, we need to provide sufficiently
powerful and abstract means of creating expressions over data, so that our platform-independent language is not inferior in that matter compared to database
query languages known from DBMSs.

Keywords: Executable Models, Platform Independent Modelling, Database
Modelling, Database Query Languages, Model Driven Architecture, UML,
OCL.

1 Introduction

The main expectation behind model-driven development based on platform-
independent models (PIMs) is earning more value in the process of model development. As far as the idea of Model Driven Architecture (MDA) [3] is considered, this
leads to the following requirements:

- Ability to accompany the modelling solution with highly automated means of producing executable code.
- Platform-independence, which isolates the investment made on producing a model,
  against the future changes of the target deployment platform (for example from
  particular solution of data persistence – relational or object database).
- Representing behaviour inside the model in a form that offers a level of abstraction
  higher than the one anticipated at the target platform code (otherwise the expressiveness of the model would be questionable and producing efficient code would
  be problematic).

---

[*] Supported by the EC 6-th FP, Project VIDE, IST 033606 STP.

M. Bubak et al. (Eds.): ICCS 2008, Part III, LNCS 5103, pp. 301–310, 2008.

- Modelling notions offered to the developer based on existing technologies and standards. This is important from the point of view of the cost of adoption, as well as because the reuse of existing modelling tools is desirable.
- All details (e.g. type information) essential for model compilers need to be easily available in the model. For example, if one platform-specific programming language represents updates of local variables and persistent data uniformly, but on the other platform the processing may differ, model should make the distinction easily available. In other words, while in the concrete syntax offered to a modeller transparency and uniformity are desirable, the amount of information actually recorded in the model needs to take into account processing requirements of various target platforms and respective model compilers.

In this paper we describe a solution for creating and executing platform independent models based on UML [4] in the field of data intensive applications, being able to process large data sets using a higher-level query language. In the rest of this section, we discuss model execution. In the second section, we describe a language created for purposes of PIM specification, while in the third section we briefly present the IDE for this kind of models.

1.1 Model Execution

The technology of executable models is originated from the Shlaer-Mellor method [6]. According to [2], Executable UML is a profile of UML allowing the developer to define the behaviour of a single subject matter in a sufficient detail that it can be executed and tested. An executable platform-independent UML model describes only the data and the behaviour, without making any platform specific decisions. It does not specify organization of data, even an implementation involving classes and objects is not necessarily required. An Executable UML model does not define neither system distribution nor the number and allocation of separate threads. Each platform calls for a different model compiler according to specific organization of hardware and software. This covers also the operating system(s) (if any), the programming language(s) of the platform and solutions concerning data persistence. Further platform specific issues are the approaches to optimization, transaction framing, fault-tolerance and more.

We consider the ability to execute the model to be valuable for the following three main purposes:

- **Software construction.** Complete platform-independent models can be used to produce software for the target platform. That's where the increased effort on creating models is expected to pay with the savings on coding activities at target platform. The degree to which this step can be automated depends, among other factors, on the complexity of platform-specific technical details.
- **Testing and debugging.** Assuming that the model compilers under consideration constitute a mature technology (i.e. construction of platform specific artifacts is a more or less automated step and is not error prone itself), the burden of assuring that software would work according to the developers' intent is shifted to the PIM level. Hence the need of easily available model execution engine (not necessarily the same as the target deployment platform assumed), which could be invoked

from the development environment. In this case, the additional challenge (apart from just running the model) is allowing its stepwise execution and tracking.

- **Prototyping and validating with final users.** In cases when moving from PIM to the target code involves significant effort (which – given the complexity and multi-tiered nature of the technologies used for building typical business applications – may be a quite usual situation), it becomes essential to validate models somehow before proceeding to platform-specific work. Involvement of the prospective users is desirable; however, the model itself can be not comprehensible enough. Hence, we assume the need for at least rudimentary means of prototyping the user interface that could cooperate with executable PIMs.

The above considerations on model execution lead us towards a language that, at least from the designer/programmer point of view, resembles typical database programming languages. This means, that although its area of application lies somewhere between traditional modelling tools and programming languages, the level of precision and skills necessary brings the requirement for the developers of programming background.

1.2   A Textual Language for the PIM

Here an important question arises: why do we need a textual language at all? Textual language is necessary to describe precise behaviour of models since visual syntax operates on higher level of abstraction and usually leaves many details unspecified. Visual programming/modelling is therefore used only for a limited set of programming constructs leaving possibility for programmers/modellers to switch to textual syntax any time and complete programming of details.

There is also no CASE tool currently available that is able to create complete UML actions and activities models. Most of them model only selected features of UML standard. This selection is arbitrary and different in each tool making them very often incompatible.

Efficient processing of data structures in typical business applications requires means of specifying expressions (queries) with similar power to SQL, OQL or XMLQuery. Again, a natural choice for this is to adapt an existing OMG standard – namely, the OCL (Object Constraint Language). The language is extensible, prepared to be used with UML, and the role of "query language" is among the applications anticipated by its authors.

Concluding, our textual VIDE language defines a textual syntax that is human readable and similar to other programming languages, supporting necessary UML behavioural constructs accompanied with OCL expressions and queries (in this part using the standard OCL's syntax) but used exclusively on PIM level.

2   A Programming Language Built on UML and OCL

As stated in OCL 2.0 [5] specification, the main reason for introducing this language in UML models is to be able to describe expressions:

"Object Constraint Language (OCL) is a formal language used to describe expressions on UML models. (…) OCL expressions can be used to specify operations/actions that, when executed, do alter the state of the system. (…) UML modellers can also use OCL to specify queries on the UML model, which are completely programming language independent."

We then might expect, that OCL makes UML a tool not only for modelling but also capable of creating platform independent executable models, with behaviour specified by OCL and UML. In this section we shall analyze the integration of these two standards and try to conclude whether they can be used for real-life business database applications development.

2.1 Responsibilities of the Standards

It is important to note that not all features of UML and OCL are useful from VIDE PIM language's point of view. If we analyze OCL 2.0 and UML 2.1 standards, we can observe that their functionality overlaps in many places, for example: we have *ReadStructuralFeatureAction* in UML and similar *PropertyCallExp* in OCL, *CallOperationAction* in UML and corresponding *OperationCallExp* in OCL, etc. To avoid functional and semantic inconsistencies, we divided responsibilities between these two languages following a simple rule: OCL is used as a query and expression language to access data, while UML is used to update data and to cover other imperative constructs. Here we describe the responsibilities of both standards in our solution:

- UML – defines data structures, defines program behaviour in the domain of imperative constructs (loops, conditionals, program blocks and exceptions), updates values and objects in the system;
- OCL – defines queries over data objects described by UML models and stored in the system and represents operation calls.

2.2 Problems of Integration

The described integration exposed various problems, however. We must admit that even though OCL 2.0 is advertised as a "language that may be used in UML models in any place where an expression may be used", it is not fulfilling this expectation. Below we enumerate just a few of the most important problems:

- OCL expression cannot access local variables defined for example in method bodies;
- Although there is a conversion specified from UML types to OCL types, there is no explicit definition of the opposite conversion. It is then formally impossible to consume OCL expression results in UML actions and other UML constructs;
- There is an important problem of correct and common interpretation of collection types. In UML, a collection is represented by multiple values. OCL defined dedicated collection types, which are containers for stored values. When OCL expression is accessing UML multiple value, it is converted to appropriate OCL collection instance. On the other side also OCL expression (or query) may return multiple values, which are packed in a collection type. However, from UML's point of view, OCL collection is just a single value. There is no reverse mapping from OCL collections to UML multiple value variables. Because of that, standard UML cannot treat OCL collections properly and cannot handle them for example in Expansion Regions;

- OCL allows a programmer to define custom types (for example for collections) but does not define how they should be stored in UML models;
- OCL expressions cannot refer to any of UML actions. This may be unimportant if OCL is only used simply as expression and query language, but may become critical if one would like to use do meta-programming and analyze UML behavioural models during query execution.

2.3   Limitations of Textual Programs Represented in UML Models

It was surprising to notice that UML cannot cover, in a reasonably straightforward way, all variety of constructs that are handy for textual programmers. In fact, some constructs become so popular through their presence in mainstream programming languages, that their straightforward support in UML seems highly desirable. There are three types of these limitations.

First one is connected to textual syntax shortcuts. For example programmers often use shortcuts for assignments or incrementing operations (like +=, *=, ++, etc). There are no corresponding notions in UML. When such constructs are expanded to complete statements and then translated to appropriate UML elements, we loose information about the original syntax, which is important if we want to be able to switch back to a textual view of the model.

Second one appeared in constructs like loops and switch statements. UML contains only one element for all kinds of loops (although with appropriate flags to distinguish for example between do…while and while...do). Similarly, the switch statement is modelled with elements which are also used for if...then...else statement. Also different types of comments fall into this category. Again, we loose information on the original syntax in the textual code.

Last but not least, there is a potentially useful information about the current state of textual view, which also should be stored in the model to be able to present exactly the same view to the programmer when moving back from model view.

Above limitations are not really any kinds of flaws of the standard, but rather points, which must be taken into consideration when designing a tool for high-level programming based on UML. Thanks to extensibility mechanisms and model annotations, these programming requirements may be fulfilled quite simply.

2.4   UML Class Diagram as an Object Database Schema

Adopting class diagrams to represent object database schema requires several additional explanations. From object oriented database point of view we need at least names for all objects (names for classes are not enough) and starting point for queries (so called set of root objects). Root objects constitute main part of the schema, which may be compared to a main module or an entry point for procedural programs. Fig 1. presents an example schema adopted to needs of object database.

There are five classes: *Person, Author, Editor, Book* and *SoldItem*. They are defined using regular UML notions. From database point of view we need to give also names to objects, that is instances of these classes. To achieve this goal we introduce a container singleton object, which will be a main module of our system: *Bookstore-System*. It contains properties named authors, editors, books and soldItems which in fact model a place, where our root object will appear. *BookstoreSystem* is in the same time context and starting point for all queries.

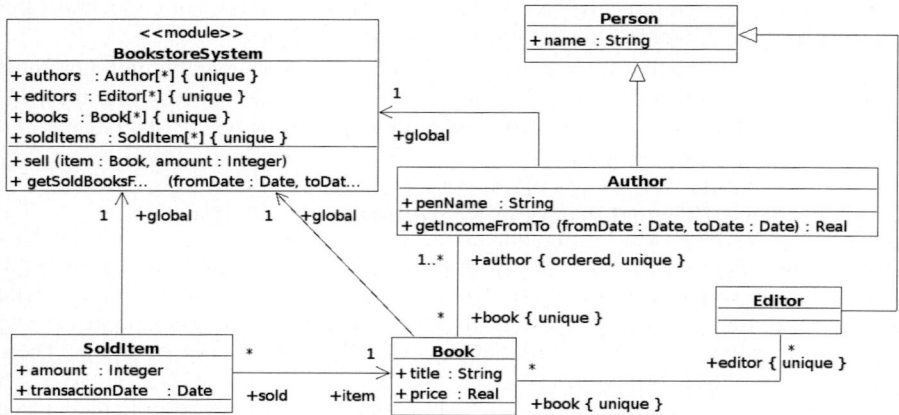

Fig. 1. Example UML schema adapted to database needs

Another observation which we made during implementation of UML execution engine is lack of global variables which may be important for database programs. We simply should be able to write a query starting from root objects in any possible context in the schema. Unfortunatelly even when *Author* class instances are embedded as properties of *BookstoreSystem* (as authors collection) procedure getIncomeFromTo(from:Date, to:Date):Real cannot access for example books collection. To make is possible we added explicitly associations from all objects to *BookstoreSystem* object, which is the global context for the schema.

2.5 Means of Achieving Semantically Complete Model Compiler Input

In order to achieve functionally complete and semantically consistent models, we have chosen not to extend UML's meta model, willing to be able to analyze VIDE models in any UML 2.0 capable CASE tool. Thus, all the constructs which appear in the VIDE code repository are standard meta model compliant. There are, however, three types of modifications that had to be done to achieve semantic completeness and to overcome the fact some constructs are inexpressible in the standard way:

1. Adding more information to UML constructs. This was implemented using EMF annotations facility. All information required for modelling tools and for easing the translation from textual code to models and vice versa is provided this way (like annotation letting us to distinguish := and += when transformed to Add*ValueAction element) .
2. Limiting model details to the level acceptable by VIDE programs and expressible by VIDE programmers. This was done simply by omitting UML meta model constructs that are not relevant to VIDE.
3. Extending the semantics of OCL-UML integration. This was done by clear explanation of new or updated behaviour of already existing UML constructs, which, similarly to UML and OCL standards, needed to be communicated to all involved

developers of the toolset element; particularly – to model compiler developers, and may be subject to future standardization process.

3   Model Editor and Compiler (IDE)

3.1   Model Creation and Compilation Process

A user of the VIDE editor first creates the data model of the application. The data model has the form of a UML class diagram. Then, the user generates stubs of the methods declared by this diagram. These stubs can be filled by means of one of the code editors: either textual, visual or other domain specific wizards. The method bodies are then type checked, compiled and eventually put into the UML model repository besides the elements implied by the class diagrams.

If the user wants to execute and test the model, he/she will have to provide a connection to an execution platform (in our case ODRA database server). Model compilation generates code for specific platform (in our case ODRA) directly from the UML repository. The generated code is then run on the ODRA server thus installing all the objects of the compiled model. The user can now write and run ad-hoc code (e.g. method calls) in order to test methods or the sole schema. The whole process is presented on the semi-formal Fig. 2. Three actors are depicted on this picture: the user, the VIDE editor and the ODRA database. Dashed arrows denote derivation dependencies between documents (diagrams, code and repository content). Solid labelled arrows represent authoring or usage relationships between actors and documents. For example, the solid arrow, which connects the user with the class diagram, means that the user *creates* this diagram.

3.2   ODRA Platform and Platform Specific Parameters

The experimental DBMS ODRA (Object Database for Rapid Application Development) [1] is the model execution platform in our project. ODRA is built according to the Stack-Based Architecture (SBA) which has been presented in a number of papers (e.g. [7]), but currently the best description of SBA can be found in [8]. ODRA DBMS perfectly suits the needs of a model execution platform because of language SBQL, programming abstractions and adaptive semi-strong type checker.

This platform implies that persistent objects are similar to volatile ones and are generally indistinguishable. ODRA assumes that all local variables are only temporal, while properties of classes marked as *module* are persistent. ODRA model compiler translates typical object manipulating statements to ODRA language, which operates on dedicated object store. Other compilers, for example for Java platform, could translate them to code working on JDO or other JEE solution.

ODRA also assumes bidirectional references, exceptions throwing and catching and procedure driven program execution. All these platform dependent features are introduced by ODRA model compiler.

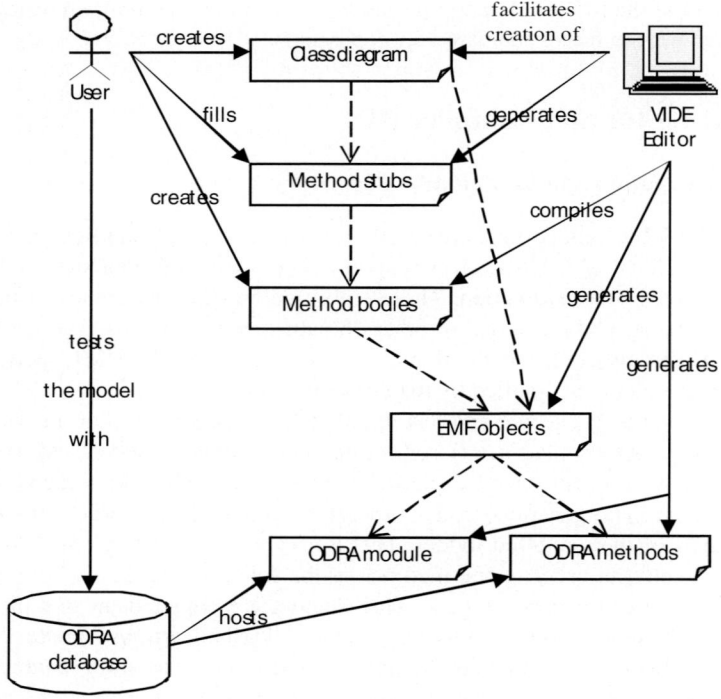

Fig. 2. ODRA IDE with integrated model compiler

4   Example Model and Its Execution

In this chapter we present a continuation of an example application shown in Fig 1. The structure of objects is already defined there. Now, we add some very high level behaviour inside procedure getIncomeFromTo. The languge is a minimal textual representation for UML Actions and Activities supplied with OCL expressions. Its code may look as follows:

```
context Bookstore::Author.getIncomeFromTo body {

    authorsBooks : Bag[0..*]( Book );
    soldItemsFromTo : Bag[0..*]( SoldItem );

    authorsBooks insert global.books->
      select( b | b.author.name->includes(self.name) );

    soldItemsFromTo insert global.soldItems->select(
      (transactionDate >= fromDate) and
        (transactionDate <= toDate) );

    return soldItemsFromTo->select( i | global.books->
      includes(i.item) )->collect( i |
        i.amount * i.item.price * 0.05) -> sum();

}
```

This simple procedure was divided into three queries but obviously can be much more compressed. This rather declarative code is transformed automatically to UML PIM model, which is partially shown in Fig. 3. You may see an activity with all behavioural UML constructs inside. They are then transformed to code acceptable by ODRA database, sent there and executed. A user may analyse results, call procedures or execute queries on a resulting object database, which is implementing all structures and behaviours from UML model.

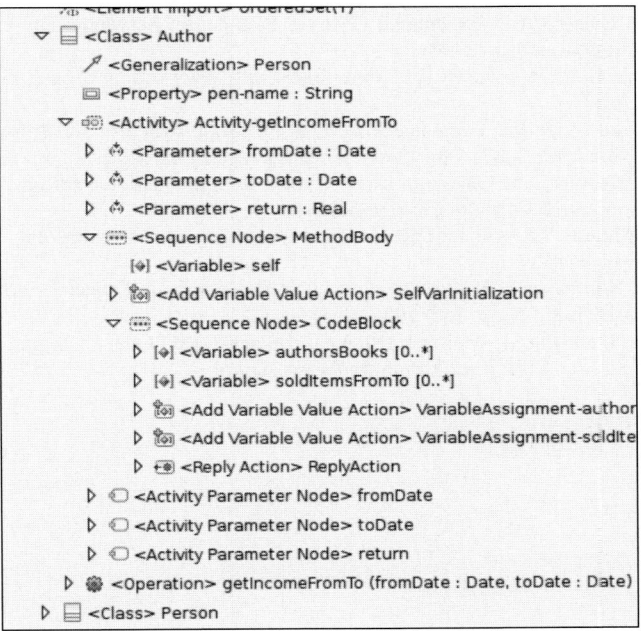

Fig. 3. A part of PIM repository for BookstoreSystem example

5   Conclusions

In this paper we investigated possibilities of using UML for executable modelling in the field of database applications. Due to its popularity, rich data model and the presence of modelling and model transformation tools, UML seems a promising mean of development of platform independent executable models. Applying that approach to applications based on a database brings a significant challenge for the development of model compiler solutions. Investigating the model compilers based on a fully object oriented system eases the development of model execution engine and will allow to find out, how much of the model compiler functionality is related to data model differences.

Apart from that, this area of application requires powerful constructs for building expressions. This brings the need of adapting and applying OCL as a query language for Executable UML PIMs. However, due to the shortcomings of existing versions of those specifications, achieving the above outlined functionality needs to involve research and updates to future versions of the standard.

References

1. Lentner, M., Subieta, K.: ODRA: A Next Generation Object-Oriented Environment for Rapid Database Application Development. In: Ioannidis, Y., Novikov, B., Rachev, B. (eds.) ADBIS 2007. LNCS, vol. 4690, pp. 130–140. Springer, Heidelberg (2007)
2. Mellor, S.J., Balcer, M.J.: Executable UML: A Foundation for Model-Driven Architecture. Addison Wesley, Reading (2002)
3. OMG: MDA Guide Version 1.0.1 (June 2003), http://www.omg.org/cgi-bin/doc?omg/03-06-01
4. OMG: Unified Modeling Language Specification (Superstructure and Infrastructure) Version 2.1.2 (November 2007), http://www.omg.org/spec/UML/2.1.2
5. OMG: Object Constraint Language OMG Available Specification Version 2.0 (May 2006), http://www.omg.org/cgi-bin/doc?omg/06-05-01
6. Shlaer, S., Mellor, S.J.: Object-Oriented Systems Analysis: Modelling the World in Data. Yourdon Press (1988)
7. Subieta, K., Kambayashi, Y., Leszczylowski, J.: Procedures in Object-Oriented Query Languages. In: VLDB 1995, pp. 182–193 (1995)
8. Subieta, K.: Stack-Based Approach (SBA) and Stack-Based Query Language (SBQL), Description of SBA and SBQL (2007), http://www.sbql.pl

OCL as the Query Language for UML Model Execution[*]

Piotr Habela[3], Krzysztof Kaczmarski[3,4], Krzysztof Stencel[2,3], and Kazimierz Subieta[1,3]

[1] Institute of Computer Sciences of the Polish Academy of Sciences, Warsaw, Poland
[2] Institute of Informatics, Warsaw University, Warsaw, Poland
[3] Polish-Japanese Institute of Information Technology, Warsaw, Poland
[4] Faculty of Mathematics and Information Science, Warsaw University of Technology, Warsaw, Poland
{habela,stencel,subieta}@pjwstk.edu.pl,
k.kaczmarski@mini.pw.edu.pl

Abstract. Despite the specification of OCL mentions "query language" as one of its possible applications, there are rather few efforts in that direction. However, the problem becomes central where applying MDA to data intensive application modelling is considered. Recently added UML elements of Actions and Structured Activities make it possible to represent a level of detail similar to the one of common programming languages, but data processing requires adequate querying capability as well. As the OMG specification of the UML family, the Object Constraint Language becomes the most obvious candidate to serve this purpose. In this paper we research this role of OCL. Especially, we address the issues of seamless integration with UML metamodel and the useful features of query languages that are missing from OCL.

Keywords: OCL, query language, MDA, UML.

1 Introduction

The approach of model-driven software development and the MDA initiative in particular, sketch the vision of the next big step in raising the level of abstraction and flexibility of programming tools. While any method which focuses on modelling activities can be considered "model-driven", the key expectation behind MDA is achieving a productivity gain through the automating software construction based on models. This results in a significant shift of expectations regarding modelling constructs – from being merely a semi-formal mean for outlining and communicating project ideas, to machine-readable specification demanding precise semantics. Thus MDA creates a spectrum of model applications. In this paper we focus on one of them, namely executable models. If models are to be executable, precise semantics is inevitable. Furthermore, executable models could blur the distinction between modelling and programming, since they would facilitate automatic production of executable code.

---

[*] Supported by the EC 6-th FP, Project VIDE, IST 033606 STP.

M. Bubak et al. (Eds.): ICCS 2008, Part III, LNCS 5103, pp. 311–320, 2008.

In our approach to this problem we strive to combine available standards in order to ease the adoption of the developed solution. If most things to be learned by a prospective user are relatively well-known (which should concern OO standards), the user will be more eager to accept and employ such a solution. In our opinion UML Structures unit seems to be rich and versatile enough to be considered as a foundation for a data model used in the platform-independent development. A number of semantic details need to be clarified to achieve that aim though. To make the model complete, the means of imperative programming need to be available at the PIM level. To raise the intuitiveness and productivity compared to the mainstream platform-specific technologies, the statements and queries should be integrated into a single language in a truly seamless way. We will also provide an execution engine for PIMs as a reference implementation. It is essential as a modelling tool component serving for platform-neutral model validation.

Within the VIDE project [14], as a part of visual development toolset we have developed VIDE-L, which stands for VIDE language. VIDE-L is a textual language for PIM level representation of behaviour in terms of UML Actions and Activities with the Object Constraint Language (OCL) as an expression/query language. Most applications developed nowadays are database applications in the sense that their functionality highly depends on persistent and shared business data. On the PIM level we have a chance to avoid the impedance mismatch and provide a query language seamlessly integrated with the host language. We have chosen to use OCL as this query language, since its specification claims that it is suitable for this purpose. In this paper we will describe our reference implementation of OCL as a database query language and a textual language for UML Actions and Activities as the host language. This reference implementation of VIDE-L is based on the Stack-Based Approach to query languages [8–10]. We use its query language (SBQL) as the assembly language underlying OCL executions. OCL queries are mapped onto SBQL queries which are eventually executed. SBA/SBQL is a well-elaborated execution framework with very general data store model, strong typing [11] and query optimisation [12],[13]. This solves the issue of type checking and optimisation of OCL queries. Those features also prove that the idea of the language is realistic in terms of the requirements the code generated from it needs to meet at the execution platforms. As a small off-topic remark we note that SBA has also a powerful updateable view mechanism, which makes OCL views stand just behind the corner.

The rest of the paper is organized as follows. In Sec. 2 ODRA DBMS is presented. Sec. 3 and Sec. 4 describe the problems solved during the implementation of OCL. Sec. 5 gives some examples of using OCL in method bodies, ad-hoc queries and programs. Sec. 6 outlines open details and some of standards' refinements needed. Sec. 7 shows related works while Sec. 8 concludes.

2   ODRA DBMS

ODRA (Object Database for Rapid Application development) is a database management system which implements SBA and OCL on top of it. Users can post queries either in SBQL or OCL. SBQL is treated as the assembly language, so OCL queries are compiled to and executed as SBQL queries. If the client query is formulated in OCL, it will be mapped onto SBQL during parsing. OCL-SBQL mapper is a component of the ODRA parser. Since SBQL is very powerful, there is no limitation on query languages which can be implemented this way. Soon, we plan to add to ODRA mapper components for other query languages: XQuery and RDQL. ODRA implements type checking and query optimization. Mapping any query language onto SBQL makes it possible to employ its optimization and strong typing capabilities. After mapping onto SBQL the query is type checked, and optimized by rewriting (also view expansion if possible) and exploitation of indices. Then it gets compiled to bytecode. The compiled query is then executed by the bytecode interpreter.

3 OCL Grammar Disambiguation

The main problem with the implementation of OCL was implied by its ambiguous syntax. The specification [3] contains the whole chapter titled *Concrete syntax*. Unfortunately this syntax is (probably intentionally) ambiguous. Let us take a look at the following obvious example fragments of rules taken from the specification:

```
OclExpressionCS ::= PropertyCallExpCS
OclExpressionCS ::= VariableExpCS
...
PropertyCallExpCS ::= ModelPropertyCallExpCS
...
ModelPropertyCallExpCS ::= AttributeCallExpCS
...
AttributeCallExpCS ::= simpleNameCS isMarkedPreCS?
...
VariableExpCS ::= simpleNameCS
```

If the parser has to analyze an `OclExpressionCS` and sees an identifier token (i.e. `simpleNameCS` in the terminology of OCL) on the input, it cannot make a choice which rule to reduce. Among the rules shown above it could be `VariableExpCS` or `PropertyCallExpCS`. However, there are a lot of other possibilities in the grammar for syntactical analysis of an identifier. . It is a well known fact, that the OCL grammar as defined in the language specification is ambiguous [15]. The specification uses contextual information, which is not available during a purely syntax based analysis (such as parsing).

Since ODRA DBMS is written in Java, we have chosen Cup from many available LALR(1) parser generators. When run for the first time on the OCL grammar, Cup reported 104 shift/reduce conflicts and 164 reduce/reduce conflicts. Moreover, 23 rules were never reduced, mainly due to conflicts. We strived to disambiguate this grammar and eventually succeeded. Most of the branches of `OclExpressionCS` had to be deleted or factored upwards to the rule for `OclExpressionCS`. We faced the last mile problem. 20% of the effort was devoted to the first 80% conflicts, but 80% of the effort was devoted to the last 20% conflicts. The less the number of conflicts was, the harder was to remove of remaining conflicts. We managed to retain the weirdest part of the syntax, i.e. Smalltalk-like prefix and infix method calls:

```
OperationCallExpCS ::=
           OclExpressionCS simpleNameCS OclExpressionCS
...
OperationCallExpCS ::= simpleNameCS OclExpressionCS
```

The question is why the specification [3] calls it a concrete syntax, if it is not suitable for parsing? This rather defines an abstract syntax. The so called 'disambiguating rules' require the parser to consult some meta information in a database. This is not the way how generated LALR(1) parser make their choices.

4   Improvements Towards Data Intensive Operations

First of all, we have to agree with [18] that OCL's syntax is not the best for database programmers. It is not intuitive, too complicated and too elaborate. However, for the reasons mentioned in the introduction we follow the standard.

We are also conscious that database community would welcome modification of several operations. One of them would be a Cartesian product operator, which is too limited in its abilities. Another candidate for extension would be introduction of a transitive closure operation, which is as important for modern databases as recursion for programming languages. Its lack, even if justified in OCL, certainly limits queries possible to be expressed.

For basic consistency with database systems we added some features, which are not modifying the language syntax and would be obvious for any programmer. OCL defines only two aggregation functions: size and sum. Using them we can compute the number of employees of a department or the total salary in a department, e.g.:

```
Dept->allInstances()->select(name='toys').employs
  ->size()

Dept->allInstances()
  ->select(name='toys').employs.salary->sum()
```

We added min, max and avg aggregation, so that more statistics can be computed:

```
Dept->allInstances()
  ->select(name='toys').employs.salary->avg()

Dept->allInstances()
  ->select(name='toys').employs.salary->min()

Dept->allInstances()
  ->select(name='toys').employs.salary->max()
```

The rest of the implementation of OCL was relatively easy, since SBQL (the query language we mapped OCL onto) is quite powerful compared to OCL. There were almost no problems in finding this mapping.

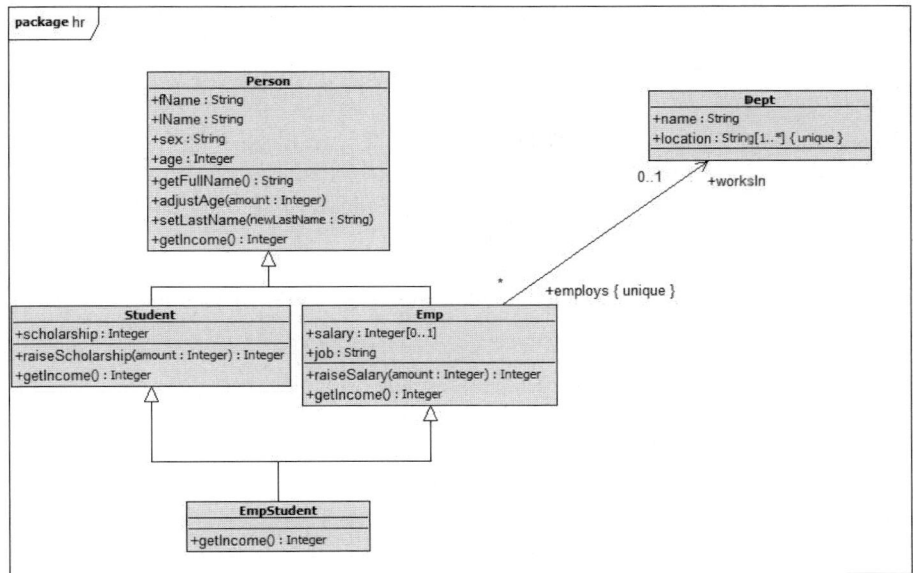

Fig. 1. Example database schema

5 Examples

In this section we will show some examples of the code written in VIDE-L (OCL is seamlessly embedded in VIDE-L). We use the database schema from Fig. 1. The schema also contains multiple inheritance which is smoothly implemented in ODRA. VIDE-L with OCL will be used in method bodies and ad-hoc queries.

Let us start from definitions of some methods. Of course we have to augment the OCL `context` phrase, since it was not originally intended to define arbitrary method bodies. We will use the keyword `body` to indicate that a particular `context` phrase introduced a method body. Here are the methods of the class Person (apart from the method *getIncome* which will be discussed later):

```
context Person::getFullName() : String
   body { return fName + ' ' + lName; }

context Person::adjustAge(amount : Integer)
   body { age += amount; }

context Person::setLastName(newLastName : String)
   body { lName := newLastName; }
```

The method *raiseScholarship* is rather straightforward:

```
context Student::raiseScholarship(amount : Integer) :
Integer  body {      scholarship += amount;
                     return scholarship; }
```

The method *raiseSalary* is more complex since we have check if a subobject *salary* exists, because it is optional. If it does not exist, it must be created and inserted by appropriate UML Action represented in textual syntax. Note the smooth integration of the OCL query and VIDE-L.

```
context Emp::raiseSalary(amount : Integer) : Integer
  body {     if (self.salary->size() = 0)
               self.salary insert amount;
             else self.salary := amount;
             return self.salary; }
```

The *getIncome* method is overridden in all classes in the hierarchy of class *Person*:

```
context Person::getIncome() : Integer
  body { return 0; }

context Student::getIncome() : Integer
  body { return scholarship; }

context Emp:: getIncome() : Integer
  body { return
      if salary->size() = 0 then 0 else salary endif; }

context EmpStudent:: getIncome() : Integer
  body {
    return scholarship
      + if salary->size() = 0 then 0 else salary endif;}
```

These methods and multiple inheritance are properly handled by the ODRA run-time, so to compute the total income of all persons in the database we issue the query:

```
Person->allInstances()->collect(getIncome())->sum()
```

We can also find the average salary for each department by means of the following query which reminds a dependent join:

```
Dept->allInstances()->collect(d |
  Tuple { dept = d,
          totalSal = d.employs.getSalary()->avg() } )
```

Now we will show some examples of ad-hoc imperative statements which can be executed against such a database. The first one assigns minimal salary in a department to all employees of this department who do not have established salaries yet:

```
Emp->allInstances()->select(salary->size()=0) foreach
  { e | e.raiseSalary(e.worksIn.getSalary()->min()); }
```

The next one moves all employees from the Toys department to the Research department. We will show two ways of finding all subject employees. The first one starts from the Toy department (*Dept*) and collects all its employees:

```
Dept->allInstances()->select(name='Toys')
  ->collect(employees) foreach { e |
      e unlink worksIn;
      e link worksIn to Dept->allInstances()
        ->select(name = 'Research'); }
```

The second one starts from employees (*Emp*) and selects those who work in the toy department. The body of the loop is the same.

```
Emp->allInstances()->select(worksIn.name = 'Toys')
foreach { e |
   e unlink worksIn;
   e link worksIn to
     Dept->allInstances()->select(name = 'Research'); }
```

The last example program gives 10% raise to all students which are also employees of departments located in Warsaw. Note that the selection of departments by location is slightly more complex because it is a multi-valued attribute.

```
EmpStudent->allInstances()
   ->select(worksIn.location->exists(l | l = 'Warsaw') )
foreach { es | es.raiseSalary(es.getSalary * 0.1)); }
```

As we can see even complex tasks can be solved by relatively simple OCL queries. Furthermore, imperative constructs of VIDE-L embed OCL queries in a very natural way. VIDE-L has statements which handle collections (like the **foreach** statement) returned by OCL queries. Each expression in this language is a query and vice versa. The impedance mismatch is mostly eliminated this way.

6   Improving the Integration with the Imperative Part

Despite the obvious benefit of reusing popular specification (that is, OCL) for the purpose of a model-level query language, a number of issues arise resulting from the fact this purpose was not fully foreseen at the time OCL was designed. Some of the problems can be removed by updating UML and OCL specifications to fully integrate them and to reduce redundancy between UML Actions and OCL expressions. To this extent the postulates would include:

- Completing the UML expressions metamodel part with the means of accessing local variables defined by UML – e.g. inside method bodies.
- Reducing the number of UML actions by those that overlap with OCL (various "read" actions dealing with: properties, variables, extents, links, and *self* variable).
- Unifying a type system between OCL and UML to assure bidirectional interoperability (so that not only OCL can read any UML-defined features, but also that OCL expression results can be consumed by UML actions and activities).

To illustrate, how the pragmatic features of the language are dependent on unifying the types between UML and OCL, consider the case of tuple results. The example below illustrates, how the style of coding changes (for a pure query method), depending whether the tuple results are allowed for UML methods.

The example (see Fig. 2 for its schema) assumes producing a nested data structure retrieved from objects of several different classes. This may be needed for constructing report or e.g. for feeding a GUI forms. The first version (*getOrderDetailsObjects* operation) uses objects being created inside a method. The second version (*getOrderDetailsTuples*) attempts to take full advantage of the OCL, and hence is implemented by a single OCL expression.

Although, due to the way UML class diagrams can describe nested structures, the both approaches are similarly complex in terms of their static model, the difference of method behaviour code is significant.

Consider the first variant that assumes that Tuple types are supported only inside OCL expressions and (accordingly to straightforward understanding of UML Actions validity constraints) each assignment deals with a single value.

```
ShopModule.getOrderDetailsObjects(in cName : String) :
OrderDetailsClass [0..*] {
  oList : Bag [0..*] (OrderDetailsClass);
  order->select(customer.name=cName) foreach { o |
    oDetail : OrderDetails =
      OrderDetailsClass create { id := o.ID;
                                 custName := o.customer.name;
                                 comments := o.comments };
    o.items foreach oDetail.item insert
      ItemInfoClass create { prodName := product.name;
                             prodQuantity := quantity};
    oList insert oDetail; }
  return oList;}
```

Now we can compare it with the variant in which tuples are allowed to be used in operation result declarations.

```
ShopModule.getOrderDetailsTuples(in cName : String) :
OrderDetailsTuple [0..*] {
  return order->select(customer.name=cName)
    ->collect(o | Tuple {    id = o.ID,
      custName = o.customer.name, comments = o.comments,
      item = o.item->collect( Tuple { prodName =
          product.name, prodQuantity = quantity}) }); }
```

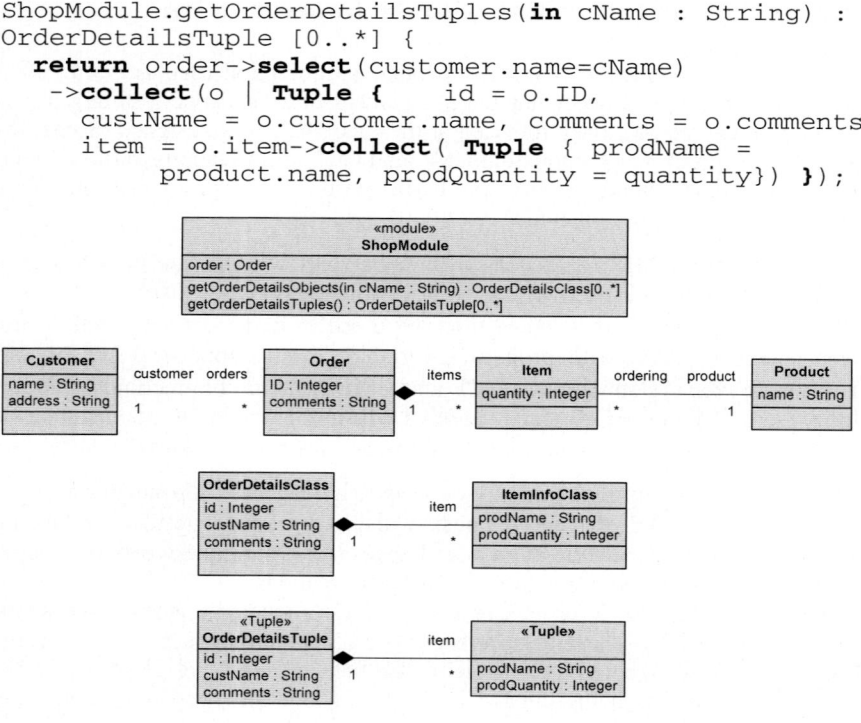

Fig. 2. Exemplary schema involving tuple results

As can be seen, allowing the use of Tuple types in method signatures can spare us a number of statements. In this particular example, two variable declarations, two foreach loops (realized by UML's *ExpansionRegion* construct), two object creation actions and several respective assignment actions are avoided.

7 Related Work

Although our VIDE-L seems to be the first application of OCL as a database query language, there were other OCL based tools which should be mentioned. Most of them implemented OCL in version 1.4 or 1.5. We know of only two that are compliant with the latest specification. Dresden OCL Toolkit [16] provides an OCL 2.0 parser and interpreter but is a metamodel based solution. It provides an API to define and execute constraints of UML models in version 1.5. It is not intended to be a database query engine, thus it does not provide any kind of optimizations and it does not define any database specific operators. Another implementation of OCL is MDT-OCL (Eclipse Model Development Tools) plug-in [17]. It provides a single metamodel integrated from UML 2.1 and OCL 2.0 plus OCL parser and interpreter. Its main purpose is to evaluate UML model constraints, thus to work on M2 meta-level, but may also be used to query data stored as a model instance. It is probably one of the best OCL based tools but still cannot be accepted as a database solution.

8 Conclusions

In this paper we described the implementation of OCL as a database query language. We presented the problems which were solved during this work (especially with ambiguous grammar) and augments which have been added to OCL so that it could be called a query language. Our implementation efforts have a wider purpose. OCL was embedded into a high level programming language VIDE-L. In our opinion this embedding is perfectly smooth and allows formulating even complex queries and program quite compactly. This creates a good starting point for further research which aims at development of modeling tools which follow MDA philosophy of creating machine-readable executable specifications to be executable on PIM level. Since these tools will use standards like OCL and UML Actions and Activities, it can be easier adopted by the community of developers and modellers.

References

1. Object Management Group: Unified Modeling Language: Superstructure version 2.1.1 (February 2007), http://www.omg.org/cgi-bin/doc?formal/2007-02-05
2. Mellor, S.J., Scott, K., Uhl, A., Weise, D.: MDA Distilled: Principles of Model-Driven Architecture. Addison-Wesley, Reading (2004)
3. Object Management Group: Object Constraint Language version 2.0 (May 2006), http://www.omg.org/cgi-bin/doc?formal/2006-05-01
4. Hailpern, B., Tarr, P.: Model-driven development: The good, the bad, and the ugly. IBM Systems Journal: Model-Driven Software Development 45(3) (2006)

5. Ambler, S.W.: A Roadmap for Agile MDA. Ambysoft (2007), http://www.agilemodeling. com/essays/agileMDA.htm
6. Thomas, D.A.: MDA: Revenge of the Modelers or UML Utopia? IEEE Software 21(3), 15–17 (2004)
7. Warmer, J., Kleppe, A.: Object Constraint Language, The: Getting Your Models Ready for MDA. Addison-Wesley, Reading (2003)
8. Subieta, K.: Stack-Based Approach (SBA) and Stack-Based Query Language (SBQL) (2008),http://www.sbql.pl
9. Subieta, K., Kambayashi, Y., Leszczyłowski, J.: Procedures in Object-Oriented Query Languages. In: Proc. VLDB Conf., pp. 182–193. Morgan Kaufmann, San Francisco (1995)
10. Subieta, K.: Theory and Construction of Object-Oriented Query Languages. Editors of the Polish-Japanese Institute of Information Technology (2004) (in polish)
11. Lentner, M., Stencel, K., Subieta, K.: Semi-strong Static Type Checking of Object-Oriented Query Languages. In: SOFSEM, pp. 399-408 (2006)
12. Płodzień, J., Kraken, A.: Object Query Optimization through Detecting Independent Sub-queries. Inf. Syst. 25(8), 467–490 (2000)
13. Płodzień, J., Subieta, K.: Query Optimization through Removing Dead Subqueries. In: Caplinskas, A., Eder, J. (eds.) ADBIS 2001. LNCS, vol. 2151, pp. 27–40. Springer, Heidelberg (2001)
14. Visualize all moDel drivEn programming, http://www.vide-ist.eu/
15. Akehurst, D., Patrascoiu, O.: OCL 2.0 - Implementing the Standard for Multiple Metamodels. In: Proceedings of the Workshop OCL 2.0 - Industry Standard or Scientific Playground? November 2, Electronic Notes in Theoretical Computer Science, vol. 102, pp. 21–41 (2004)
16. Dresden OCL Toolkit, http://dresden-ocl.sourceforge.net
17. Model Development Tools OCL, Eclipse Foundation, http://wiki.eclipse.org/MDT
18. Vaziri, M., Jackson, D.: Some Shortcomings of OCL, the Object Constraint Language of UML. In: Proceedings of the Technology of Object-Oriented Languages and Systems (TOOLS 34'00), pp. 555–562. IEEE Computer Society, Los Alamitos (2000)

Managing Groups of Files in a Rule Oriented Data Management System (iRODS)

Andrea Weise[1], Mike Wan[2], Wayne Schroeder[2], and Adil Hasan[3]

[1] Centre for Advanced Computing and Emerging Technologies (ACET), University of Reading, UK
a.weise@reading.ac.uk
[2] San Diego Supercomputing Center (SDSC), University of California, San Diego, USA
[3] Science and Technology Facilities Council, Rutherford Appleton Laboratory, UK

Abstract. The iRODS system, created by the San Diego Supercomputing Centre, is a rule oriented data management system that allows the user to create sets of rules to define how the data is to be managed. Each rule corresponds to a particular action or operation (such as checksumming a file) and the system is flexible enough to allow the user to create new rules for new types of operations. The iRODS system can interface to any storage system (provided an iRODS driver is built for that system) and relies on its' metadata catalogue to provide a virtual file-system that can handle files of any size and type.

However, some storage systems (such as tape systems) do not handle small files efficiently and prefer small files to be packaged up (or "bundled") into larger units. We have developed a system that can bundle small data files of any type into larger units - mounted collections. The system can create collection families and contains its' own extensible metadata, including metadata on which family the collection belongs to. The mounted collection system can work standalone and is being incorporated into the iRODS system to enhance the systems flexibility to handle small files.

In this paper we describe the motivation for creating a mounted collection system, its' architecture and how it has been incorporated into the iRODS system. We describe different technologies used to create the mounted collection system and provide some performance numbers.

1 Introduction

The SDSC Storage Resource Broker (SRB) is a data grid management system, which is able to unite and manage storage media of many kinds on heterogeneous systems across the network and, as a result, to make the storage infrastructure appear transparent for the end user. The software supports data grids, digital libraries, and persistent archives [1]. The success of the Storage Resource Broker has shown the demand for such data management systems. Currently petabytes of data are stored in various SRB systems, such as the Biomedical Information

M. Bubak et al. (Eds.): ICCS 2008, Part III, LNCS 5103, pp. 321–330, 2008.

Research Network, UK eScience (e-minerals/e-materials project) or Taiwan National Digital Archives Program. The SRB is a complex system with built-in abilities which are not easily changeable. Therefore, one of the main goals for developing a new data management system is to provide the user with a system which is easier to administrate and is flexible enough to change with different existing environments.

The Rule Oriented Data System (iRODS), also developed by the San Diego Supercomputing Center (SDSC), is a state of the art open source client-server middleware to manage huge amount of data in a grid environment. It supports in addition to the SRB features real-time sensor data collection and large scale data analysis [2].

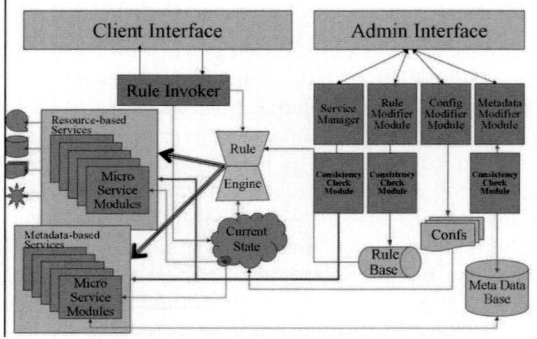

Fig. 1. iRODS Architecture

One of the main difference between iRODS and SRB is that within iRODS the data is managed automatically by rules [3]. Fig. 1 [2] displays the iRODS architecture with its' main modules. The client connects to the server and the client request will always go through the rule engine. Another subject which distinguish the iRODS from SRB are the levels of virtualisation which are workflow, management policy, service and rule virtualisation [2]. This is also provided through the administration interface with its' modules.

With the introduction of rules, the user is able the manage their own data in almost any way. Additionally, the implementation of services to manage or process the data, such as data conversion can be easily archived by the end user. The above mentioned rules are operations which are executed on the server side [2]. Each iRODS server has its' own rule engine. Fig. 2 [2] shows a flow diagram, which gives a basic overview about the way the rule engine works. The rule engine acts as interpreter of the rules [2] and is one of the core modules of iRODS. Rules are composed of the actual event, conditions, action sets and recovery sets [2]. The rule format can be seen as follows.

```
actionDef|condition|workflow-chain|recovery-chain
```

Rules consist of micro services. Micro services are C-functions [2], which can be provided by anyone to organise and structure the data. This above format put in action could look like the following example.

$$\texttt{HAAW(*A,*B)||HAAW-Bundle(*A,*B)|nop}$$

The name of the rule is `HAAW`. For this rule no conditions are attached. The micro service which is executed is called `HAAW-Bundle`. Two parameters (`A` and `B`) are passed to the function. Further, there is no recovery function provided. The sign "|" servers as a separator [2]. There are different types of rules namely, rules which can be applied immediately, rules where the application can be deferred and rules that could be applied periodically.

Due to the rule and micro service structure the iRODS becomes a highly configurable and flexible data management system, a system which is able to respond to many kinds of different requirements. Those requirements can be located in the data preprocessing or postprocessing. In terms of managing files within iRODS rules enable the user to automatically archive or organise files. To achieve that desired automatism a certain framework is needed, which will allow to plug-in any archive system.

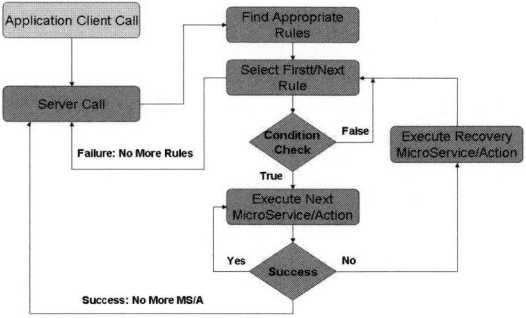

Fig. 2. Rule Engine Flow Diagram

2   Mounted Collection - Concept

2.1   Motivation

Nowadays almost every system has the desire to perform as fast as possible. While working with SRB a common problem was faced when archiving a large number of small files [4]. Small files typically make very inefficient use of mass storage capabilities which can have deleterious effects on the system performance [4]. This problem becomes very clear when tape systems are involved. The operations required to successfully write a file to a tape are time consuming, e.g. seek operation to find the right position [4].

To overcome the above listed performance problems within SRB the container concept was introduced. However, this concept of containers has certain shortcomings. The way containers are handled within SRB are not very efficient. Once the data is deleted from the container the space is not released and can not be reused. Therefore, the size of the container does not decrease. Further, the container concept does not leave much space for different archiving systems and are unfortunately tied to SRB. It is not possible to extract and therefore handle a container outside of the SRB system. To extract individual files from a container the metadata held within SRB is necessary. Consequently, the need emerged to develop a flexible system, which will be able to accept different kinds of mounted collections or archiving systems and which will overcome the weak points of the container concept.

2.2   Requirements

Out of the aforementioned reasons the following requirements for the mounted collections are stated. The mounted collection should

- be able to store all types of files and be a single file only
- be able to add, replace (update) and delete files to the mounted collection
- be able to pre-size the mounted collection
- be able to create mounted collection families
- hold checksums of files inside the mounted collection
- be as compact as possible (or compressible)
- be accessible/ usable inside and outside of iRODS
- be searchable to gain information about the files in the mounted collection without actually unpacking the mounted collection
- provide the ability to add limited metadata
- reflect the original folder structure
- should hold sufficient metadata to manage the mounted collection files in different environments

The possibility to take a whole mounted collection out of the iRODS system and still be able to access the information which are held in the mounted collection is a useful feature. This will allow to move huge amount of data efficiently between different systems.

In addition of the basic functionality a few further issues were taken into consideration during the development. The operations in connection with mounted collection have to be reasonable fast. Further, the overhead of the mounted collection structure itself should be as small as possible concerning the needed memory. To be able to use the mounted collections outside of the iRODS environment efficiently, flexible metadata capabilities are very useful. The possibility to compress individual files within the mounted collection will make the system more desirable. More, the mounted collection should support hierarchies and should be platform independent with a long term support.

Before integrating the mounted collection into iRODS a standalone mounted collection application was developed. By doing so, the performance of the mounted collection operation could be determined. Furthermore, through the application the possibility is provided to handle the collection outside of iRODS.

There are many concepts on the market to organise data and information in a structured way. A very efficient way is offered by databases. During the development databases were examined regarding speed, interface possibilities, database file structure, database operation possibilities and open source based code.

Finally, three database engines were chosen - SQLite version 3.3.8., Berkeley Database version 4.4.20 and Apache Derby version 10.2.2.0.

SQLite is a small C library that implements a self-contained, embeddable, zero-configuration SQL database engine [5]. This and the fact that the database file only consist of one single file made this engine an ideal candidate. As for the Berkeley DB, the developer promise a high performance and embeddable engine [6]. Derby is a light database engine using the SQL standard [7]. To be able to give an appropriate statement concerning the mentioned criteria all database engines were tested.

3   Finding the Right System

First of all, a standalone application was developed, which is currently a command line tool that performs the required mounted collection operations. On one hand, this application allows the user to handle the collections outside iRODS. On the other hand, the application is used for performance tests of the individual collections operations to determine the most suitable database engine.

The mounted collection can hold any kind of data (files). With each data certain metadata are associated, e.g. folder structure or file size, which will enable other potential applications to manage the collections with its' content successfully. The metadata can be extended according to the users request. For each file a checksum is computed using the Message Digest 5 (MD5) hash algorithm. This will detect any tampering with or corruption of the mounted collection. Files can be added, extracted, updated. The collection content can be listed in two different ways, short and long versions.

The test environment consists of functions, which are later embedded into the iRODS system. Each mounted collection operation is presented by a function. These procedures perform basic operation, e.g. open a collection, add file to a collection or delete file from a collection. For each function a certain version for each database engine has been implemented under the restriction to keep the code as similar as possible.

For testing purposes 1,000 to 100,000 files were created with different sizes (1KB - 10 MB). Then the time was measured for the mounted collection operations. Fig. 3 depicts the time measurement for all three database engine, executing the insert and extract function.

The Berkeley database is faster than the SQLite engine. The results for the Derby engine are not acceptable for the needed purpose. Currently Java programs are slower in performance than C programs. First of all, the Java code is

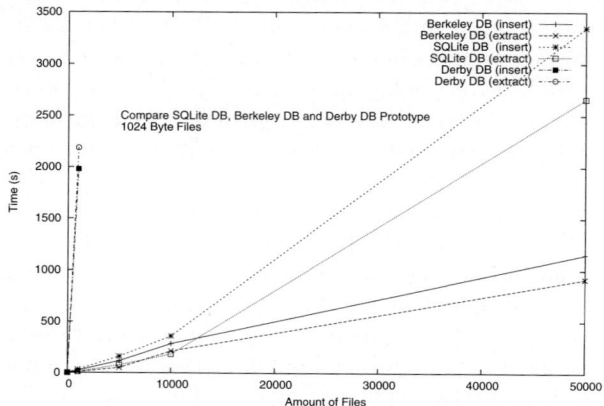

Fig. 3. Time to Store 1 KByte Files

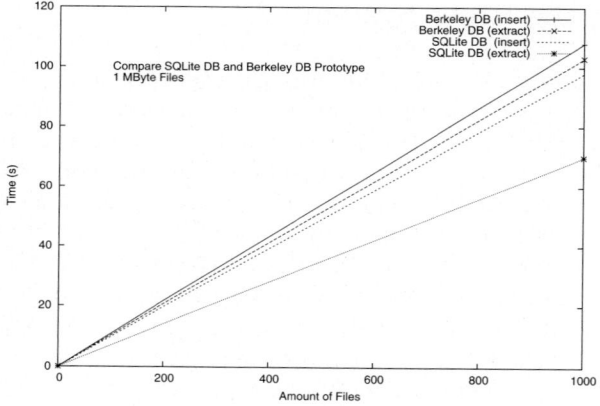

Fig. 4. Time to Store 1 MByte Files

run in a virtual machine. This causes a significant delay in the start up progress, since everything is being loaded [8]. By increasing the size of the files the SQLite database engine becomes faster compared to the Berkeley engine (Fig. 4)

Concerning the needed memory, the SQLite and Berkeley engines are very similar. The SQLite database consist only of one database file (Berkeley creates one file for each table). Due to the fact, that the SQLite database engine perfoms faster with growing file size and the SQLite API is easier to incorporate into C code, the SQLite database is the most suitable for a mounted collection prototype.

4   Mounted Collections within iRODS

By integrating the mounted collection into iRODS two things need to be achieved. Firstly, the collection forming process has to be executable manually and secondly, through rules. To accomplish that it is important to understand the concept of the iRODS framework.

4.1   iRODS Framework

The client-server architecture of the iRODS separates the software in two major parts. The icommands is one possibility to present the client side. Each basic file operations such as list or make a directory, are presented by an individual commands, e.g. `ils`, `imkdir` or `iput`, to submit a file to the iRODS. According to the given task (command), on the client side a certain amount of preprocessing is executed, e.g. mapping file name to the iRODS environment.

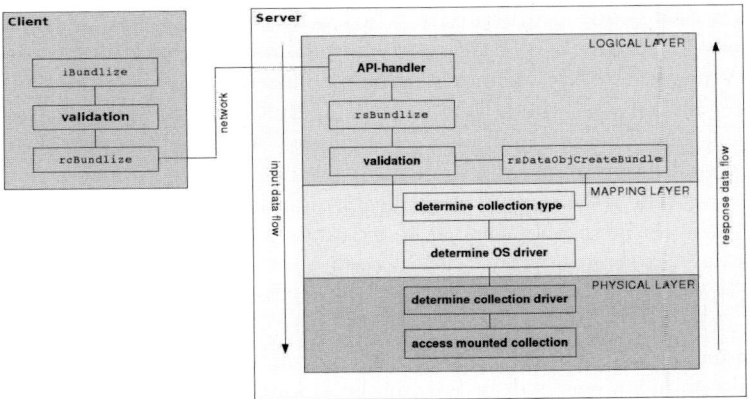

Fig. 5. iRODS Mounted Collection Framework

The iRODS can be divided in three basic layers on the server side, as shown in Fig. 5. The user requests are evaluated and processed in the logical layer. The logical layer queries the iRODS catalogue (ICAT). The ICAT (a database) holds all the metadata and information needed to manage the data within iRODS. To be able to map the logical name to the right physical location the mapping layer serves as glue layer. The actual access to the physical medium is done in the physical layer. On the basis of this layer frame work the iRODS becomes highly scalable. By providing the correct driver in the physical layer the system can adjust to many different environments. Based on that concept, the mounted collection framework was developed.

4.2   Mounted Collection Framework

General Concept. Mounted collections are only managed on the server side. That means, a file can not be transfered directly from the client into a collection. The user has to submit a file first to the iRODS and then the file can be transfered from there to a collection. This goes conform with rule oriented idea of iRODS. Once a file is relocated to the mounted collection, the actual file and its' metadata will be deleted from the iRODS environment. The collection itself is able to hold certain metadata, which are sufficient to reintegrate the files back into the iRODS.

A new icommand is introduced - `ibundle`. This command accepts a directory. This directory will be bundlized on the server side. It is not possible to submit a single file directly to a collection using an icommand. Further, the user has the possibility to specify the name of the new mounted collection. The iRODS itself contains "special" files. Those files are not visible to the user and can not be accessed directly by the user. The location of those files are determined by the iRODS system. Mounted collections are defined as special files they will be stored under their own collection hierarchy, which will be determined by the management system. This way, all the collection files will be hold in one location and the user can not accidentally tamper with the collection files. Mounted Collection files are registered with the ICAT, indicating that this file is a mounted collection/archive file. The user will also has a choice of different collection types. If the corresponding drivers are provided any archive or collection type is possible.

Retrieving a file from a mounted collection directly to the user (client side) is also not possible. First, the user has to extract the file within the iRODS. From there the file can then be transfered to the user.

Furthermore, the user will be able to segment the collection file. This is called mounted collection families. The user can specify a maximum size of a single collection member. If this size is exceeded a new member of this family is created. The family management is done within the mounted collection itself.

Framework. A basic overview of the mounted collection framework embedded into the iRODS framework can be seen in Fig. 5. On the client side the icommand `ibundle` is provided. The submitted user information, e.g. collection name or collection type, are evaluated on the client side. Those formated information are put into a structure, which will be transfered to the server side. From the iRODS framework an application interface (API) handler is provided, which will connect the client API call with its' opponent call on the server side. On the server side the new API `rsBundlize` was developed. The client counterpart is `rcBundlize`. In the logical layer the information provided by the the client are processed (validation), e.g. is the collection really existing? If the mounted collection is not available it will be created. In order to do that, another API was developed - `rsDataObjCreateBundle`. This will create the mounted collection and register this as a special file at the ICAT. To meet the conditions of the existing iRODS framework the new API's follow the concept of dividing the processing into 3 basic layers. That means, procedures for the logical layer, which involves all ICAT handling, are provided. Further, glue functions are developed to be able to map the information to the physical layer. The physical layer has also mapping routines. First of all, the mapping to the correct collection type driver has to be done (mapping layer). After this, the mapping to corresponding driver for the current operating system in use is needed (physical layer). Each mounted collection/archive system may use a different descriptor to access their own collection/archive. This creates the need to allow any descriptor needed. The demanded flexibility is offered through the mapping system.

The glue function and the function to access the media can also be seen as API calls. Through this level of abstraction any other function from the level above is able to use the API calls in the lower level. This leads to the possibility to extended existing functionality to increase the functionality of the system. For example the icommand ils, which is used to list directory contents, is extended to list the mounted collection content as well.

The new mounted collections framework structure is modular. That means, each of the new functions can be accessed easily from the core functions of the iRODS system. This makes the framework accessible for the rule engine. With the provided functions microservices can easily be created and therefore, rules to automatically make use of the mounted collection can be established.

5 Future Work

The current prototype fulfils the requested conditions. But the SQLite database has a deflating limitation. Currently this engine supports terabyte-sized databases and gigabyte-sized strings and blobs [5]. To avoid any of those limitations a collection will be developed, which will be a hybrid out of the SQLite engine and 7-Zip. 7-Zip is a open source file archiver with a high compression ratio [9]. The combination of a collection with a high compression rate and certain amount of useable metadata, which will be fast searchable, will make the mounted collections even more interesting. With the incorporation of 7-Zip the limitation in size will only depend on the hardware.

Security is an important, but complicated issue. First of all, the mounted collections, since handled outside of iRODS needs more security features such as password protection and encryption possibilities. The user also only wants to authenticate once at the system. There are different systems available on the market. Kerberos is such a service [10]. Another identity management possibility is provided by Shibboleth [11]. In future both authentication management system could be embedded within iRODS.

6 Conclusion

In this paper the iRODS framework was briefly described and its' difference to the Storage Resource Broker are highlighted. A certain emphasis was placed on rules, which define the rule oriented data system. The need for mounted collections or archives within such systems where pointed out, which lead to the presentation of the philosophy behind the new mounted collection concept. To reach a high flexibility in handling the collection a standalone application was developed. The new collection is based on a database engine. The standalone application was used to determine the best database engine through performance tests. Some results are presented within this paper. Further, major development steps which led to the mounted collection framework are outlined. The paper shows how the iRODS framework the mounted collection framework defines and how the collection was successfully incorporated into the rule oriented data management system. With the collection framework embedded into the iRODS this management system becomes even more flexible and powerful compared to other existing systems. Furthermore, the made enhancement of iRODS by the described collection framework makes the system more attractive for the public. iRODSs' rule oriented nature is supported in any way by the here presented contribution.

Acknowledgements

The authors would like to thank The National Science Foundation (NSF) [USA], The National Archives and Records Administration (NARA) [USA] and The Science and Technology Facility Council [UK] who support this research financially. Further, the authors would like to thank Prof. V.Alexandrov (ACET) for his support while developing this software.

References

1. Rajasekar, A., Wan, M., Moore, R., Schroeder, W., Kremenek, G., Jagatheesan, A., Cowart, C., Zhu, B., Chen, S.-Y., Olschanowsky, R.: Storage resource broker - managing distributed data in a grid. Technical report, San Diego Supercomputer Center (SDSC), University of California
2. About irods (September 19, 2007), http://irods.sdsc.edu/index.php/Main_Page
3. Rajasekar, A., Wan, M., Moore, R., Schroeder, W.: A prototype rule-base distributed data management system. Technical report, Paris, France (May 2006)
4. Strong, B., Corney, D., Berrisford, P., Folkes, T., Moreton-Smith, C., Kleese-Van-Dam, K.: Key lessons in the efficient archive of small files to the cclrc mss using srb. Technical report, IEEE (2005)
5. About sqlite (September 13, 2007), http://www.sqlite.org
6. Why oracle berkeley db? (September 13, 2007), http://www.oracle.com/database/berkeley-db/index.html
7. What is apache derby? (September 13, 2007), http://db.apache.org/derby/
8. Lewis, J.P., Neumann, U.: Performance of java versus c++ (accessed on December 3, 2007), http://www.idiom.com/~zilla/Computer/javaCbenchmark.html
9. 7-zip (September 17, 2007), http://www.7-zip.org/
10. Clifford Neumann, B., Tso, T.: Kerberos: An authentication service for computer networks. Technical report, Institute of Electricaland Electronics Engineers (September 1994)
11. Shibboleth, http://shibboleth.internet2.edu/

Towards Large Scale Semantic Annotation Built on MapReduce Architecture*

Michal Laclavík, Martin Šeleng, and Ladislav Hluchý

Institute of Informatics, Slovak Academy of Sciences,
Dúbravská cesta 9, Bratislava, 845 07
laclavik.ui@savba.sk

Abstract. Automated annotation of the web documents is a key challenge of the Semantic Web effort. Web documents are structured but their structure is understandable only for a human that is the major problem of the Semantic Web. Semantic Web can be exploited only if metadata understood by a computer reach critical mass. Semantic metadata can be created manually, using automated annotation or tagging tools. Automated semantic annotation tools with the best results are built on different machine learning algorithms requiring training sets. Another approach is to use pattern based semantic annotation solutions built on NLP, information retrieval or information extraction methods. Most of developed methods are tested and evaluated on hundreds of documents which cannot prove its real usage on large scale data such as web or email communication in enterprise or community environment. In this paper we present how a pattern based annotation tool can benefit from Google's MapReduce architecture to process large amount of text data.

Keywords: semantic annotation, information extraction, metadata, MapReduce.

1 Introduction

Automated annotation tools can provide semantic metadata for semantic web as well as for knowledge management [4] or other enterprise applications [11].

Pattern based automatic or semi-automatic solutions for semantic annotation or tagging are usually based on NLP, information retrieval or information extraction fields or minimally method algorithms common in the mentioned fields are applied.

Information Extraction - IE [1] is closed to semantic annotation or tagging by Named Entity recognition – NE defined by series of MUC conferences.

Semi automatic annotation approaches can be divided into two groups with regards to produced results [1]:

- identification of concept instances from the ontology in the text
- automatic population of ontologies with instances in the text

One of pattern based solutions for semi-automatic annotation is Ontea [2] [3] that uses regular expression patterns to detect or create instances in ontology. In our previous

* This work is supported by projects NAZOU SPVV 1025/2004, Commius FP7-213876, SEMCO-WS APVV-0391-06, VEGA 2/7098/27.

M. Bubak et al. (Eds.): ICCS 2008, Part III, LNCS 5103, pp. 331–338, 2008.

works [2] [3] we compared Ontea with other annotation methods and we conducted experiments to demonstrate its success rate above 60% that is comparable to well known annotation methods with easier applicability on concrete domain specific application due to relatively simple method built on regular expressions. This is another reason behind our decision to port Ontea into MapReduce architecture. We believe other well known semantic annotation or IE solutions such as C-PANKOW, KIM, GATE or different wrappers can be ported into MapReduce architecture. For survey on semantic annotation please see [4] [5] [14].

To our best knowledge the only semantic annotation solution which runs on distributed architecture is SemTag [6]. It uses the Seeker [6] information retrieval platform to support annotation tasks. SemTag annotates web pages using Stanford TAP ontology [7]. However, SemTag is able to identify but not create new instances in the ontology. Moreover, its results as well as TAP ontology are not available on the web for a longer period of time.

In our previous work we ported semantic annotation into Grid [3] with good results but with no easy and direct implementation and results integration. Thus we have focused on different parallel and distributed architectures.

Google's MapReduce [8] architecture seems to be a good choice for several reasons:

- Information processing tasks can benefit from parallel and distributed architecture with simply programming of Map and Reduce methods
- Architecture can process Terabytes of data on PC clusters with handling failures
- Most of information retrieval and information extraction tasks can be ported into MapReduce architecture, similar to pattern based annotation algorithms. E.g. distributed grep using regular expressions, one of basic examples for MapReduce, is similar to Ontea pattern approach using regular expressions as well.

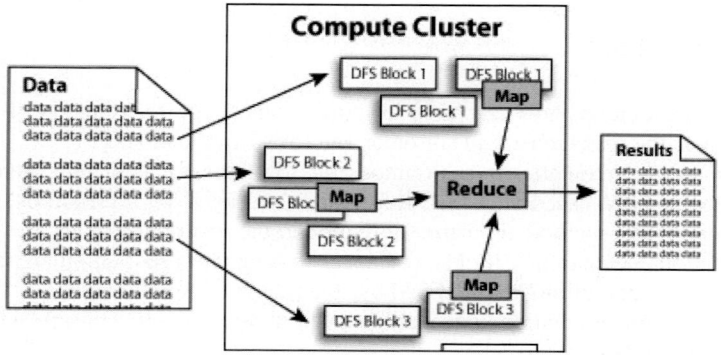

Fig. 1. MapReduce Architecture figure (source: Hadoop website)

On Figure 1 we can see main components of the MapReduce architecture: Map and Reduce methods, data in distributed file system (DFS), inputs and outputs. Several replicas of data are created on different nodes, when data are copied to DFS. Map tasks are executed on the nodes where data are available. Results of Map tasks are key value pairs which are reduced to results produced by Reduce method. All developer need to do is implement Map and Reduce method and architecture will take care of distribution, execution of tasks as well as fault tolerance. For more details on MapReduce please see [8].

Two open source implementation of MapReduce are available:

- Hadoop [9], developed as Apache project with relation to Lucene and Nuch information retrieval systems, implemented in Java. Hadoop is well tested on many nodes. Yahoo! is currently running Hadoop on 10,000 nodes [15] in production environment [16].
- Phoenix [10], developed at Stanford University, implemented in C++.

In this paper we discuss work in progress - porting of pattern based semantic annotation solution Ontea into MapReduce architecture and its Hadoop implementation. We provide preliminary results on 8 nodes Hadoop cluster on email documents.

2 Ontea

The method used in Ontea [2] [3] is comparable particularly with methods such as those used in GATE, C-PANKOW, KIM, or SemTag. It process texts or documents of an application domain that is described by a domain ontological model and uses regular expressions to identify relations between text and a semantic model. In addition to having pattern implementation over regular expressions, created Ontea's architecture allows simply implementation of other methods based on patterns such as wrapers, solutions using document structure, language patterns similar to GATE, C-PANKOW and many others. Ontea [17] is being created as an Open source project under Sourceforge.net.

2.1 Ontea Scenarios and Results Examples

Current Ontea implementation can be executed in 3 different scenarios:

- *Ontea*: searching relevant individuals in knowledge base (KB) according to generic patterns
- *Ontea creation*: creating new individuals of objects found in text
- *Ontea IR*: Similar as previous with the feedback of information retrieval methods and tools (e.g. Lucene) to get relevance computed above word occurrence and decide weather to create instance or not.

Table 1. Examples of Instances and Patterns

#	Text	Instance	Patterns – regular expressions
1	Apple, Inc.	Company: Apple	*Company:* ([A-Za-z0-9_+)[,]+(Inc|Ltd)
2	Mountain View, CA 94043	Settlement: Mountain View	*Settlement:* ([A-Z][a-z]–[ ]*[A-Za-z]*)[ ]+[A-Z]{2}[ ]*[0-9]{5}
3	laclavik.ui@savba.sk	Email: laclavik.ui@savba.sk	*Email:* [-_.a-z0-9]+@[-_.a-zA-Z0-9]+\.[a-z]{2,8}
4	Mr. Michal Laclavik	Person: Michal Laclavik	*Person:* (Mr.|Mrs.|Dr.) ([A-Z][a-z]+ [A-Z][a-z]+)

New application scenarios can be created by combination of Result Transformers, which is discussed in next chapter.

2.2 Ontea Architecture

The fundamental building elements of the tool are the following java interfaces and extended and implemented objects:

- *ontea.core.Pattern*: interface for adaptation for different pattern search. Currently implemented pattern search uses regular expressions *PatternRegExp*.
- *onetea.core.Result*: a class representing annotation results by means object instance of defined type/class. Its extensions are different types of instances depending on implementation in ontology (Jena, Sesame) or as value and type pairs.
- *ontea.transform.ResultTransformer*: interface that after implementation contains different types of transformations among annotation results. Thus it can transforms set of results and include in transformation various scenarios of annotation such as relevance, result lemmatization, transformation of found value/type pairs (Table 1) into OWL instances in sesame or Jena API implementation. It is used to transform type value pairs into different type value pairs represented e.g. by URI or lemmatized text value. It can be also used to eliminate irrelevant annotation results.

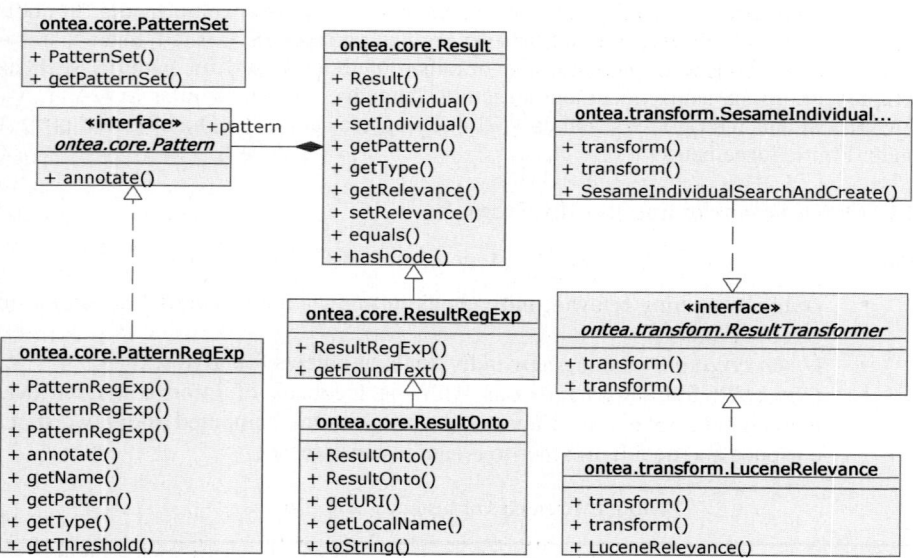

Fig. 2. Basic classes of Ontea platform

On the Figure 2 you can see *Result* class, *Pattern* and *ResultTransformer* interfaces. Such design allows extending Ontea for different patterns implementations or for the integrations of existing pattern annotation solutions. Also it is possible to implement various result transformations by implementing *ResultTransformer*, which can be used also as inputs and outputs between Map tasks in MapReduce architecture.

2.3 Integration of Ontea with External Tools

Ontea tool can be easily integrated with external tools. Some tools can be integrated by implementation of result transformers and other need to be integrated directly.

- *MapReduce*: Large scale semantic annotation using MapReduce Architecture – is main topic of this article. Integration with Hadoop requires implementation of Map and Reduce methods as described in next chapter.
- *Language Identification*: In order to use correct regexes or other patterns, often we need to identify language of use. For this reason it is convenient to integrate Ontea with language detection tool. We have tested Ontea with Nalit [11]. Nalit is able to identify Slovak and English texts as well as others if trained.

As already mentioned some integration can be done by implementing Result transformers:

- *Lemmatization:* When concrete text is extracted as representation of an individual, often we need to lemmatize found text to found or create correct instance. For example capital of Slovakia can be identified in different morphological forms: *Bratislava, Bratislave, Bratislavu*, or *Bratislovou* and by lemmatization we can identify it always as individual *Bratislava*. We have tested Ontea with Slovak lemmatizer Morphonary [12]. It is also possible to use lemmatizers or stemmers from Snowball project [18], where java code can be generated.
- *Relevance Identification*: When new instance is being created or found, it is important to decide on instance relevance. This can be solved using information retrieval methods and tools such as Lucene [19]. When connecting with Lucene, Ontea asks for percentage of occurrence of matched regular expression pattern to detected element represented by word on used document set. Document set need to be indexed by Lucene. Example can be *Google, Inc.* matched by pattern for company search: \\s+([-A-Za-z0-9][ ]*[A-Za-z0-9]*),[ ]*Inc[.\\s]+", where relevance is computed as "Google, Inc." occurrence divided by "Google" occurrence. Use of Lucene is related to *Ontea IR* scenario and *LuceneRelevance* implementation of *ResultTransformer* interface. Similarly, other relevance algorithms such as cosine measure can be implemented. This was used for example in SemTag [6].
- *OWL Instance Transformation*: Sesame, Jena: Transformation of found key – value pairs into RDFS or OWL instances in Sesame or Jena API. With this integration, Ontea is able to find existing instances in knowledge base if existing and create new once if no instance found in DB. Ontea also use inference to found appropriate instance. For example if Ontea process sentence "Slovakia is in Europe." using pattern for location detection *(in\near)* +(\\p{Lu}\\p{L}+) following type value pair is detected *Location: Europe*. If we have Location ontology with Subclasses as Continents, Settlements, Countries or Cities and Europe is already present as instance of continent, Ontea can detect existing Europe instance in knowledge base using inference.

3   Ontea Ported into Hadoop

For porting Ontea or any semantic annotation solution it is important to understand results of annotations as well as how they can correspond to key/value pairs - outputs of Map and Reduce methods to be implemented in MapReduce architecture. In table 1 we show a simple example of Ontea possible annotation results such as settlements, company names, persons or email addresses. Used regular expressions are simplified to be more readable and understandable.

In the Map method, input is a text line which is processed by Ontea's regex patterns and outputs are key value pairs:

- Key: string starting with detected instance type and continue with instance value similar to *instance* row in table 1. This can be extended to return also instance properties e.g. address, phone or email as properties of company.
- Value: File name with detection of instance. It can be extended with position in file e.g. line number and character line position if needed.

Basic building blocks of Ontea are the following java classes and interfaces described earlier, which can be extended. Here we describe them in scope of MapReduce architecture:

- *ontea.core.Pattern*: interface for adaptation of pattern based searching in text. Main Pattern method *Pattern.annotate()* runs inside of Map method in MapReduce implementation.
- *onetea.core.Result*: a class which represents the result of annotation – an ontology instance. It is based on the type and value pairs as in table 1, *instance* column. Ontology results extension contains also URI of ontology individual created or found in ontology. Results are transformed into text keys as output of Map method in MapReduce implementation.
- *ontea.transform.ResultTransformer*: interface which transform results of annotation. Transformers are used in Map or Reduce methods in MapReduce implementation to transform individuals into OWL file or eliminate some results using Ontea IR scenario.

3.1   Ontea Running on Hadoop MapReduce Cluster

We wrapped up Ontea functionality into Hadoop MapReduce library. We tested it on Enron email corpus [20] containing of 88MB of data and our personal email containing of 770MB of data. We run same annotation patterns on both email data sets, on single machine as well as 8 node Hadoop cluster. We have used *Intel(R) Core(TM)2 CPU 2.40GHz* with *2GB RAM* hardware on all machines.

As you can see from Table 2, the performance increased 12 times on 16 CPUs in case of large data set. In case of smaller data set it was only twice faster then on single machine and MapReduce overhead is much more visible. In the table 2 we present only 2 concrete runs on 2 different datasets, but in reality we have executed several runs on these datasets and computational time was very similar so we can conclude that times presented in table 2 are very close to average.

Table 2. Performance and execution results

Description	Enron corpus (88MB)	Personal email (770MB)
Time on single machine	2min, 5sec	3hours, 37mins, 4sec
Time on 8 nodes hadoop cluster	1min, 6sec	18mins, 4sec
Performance increased	1.9 times	12 times
Launched map tasks	45	187
Launched reduce tasks	1	1
Data-local map tasks	44	186
Map input records	2,205,910	10,656,904
Map output records	23,571	37,571
Map input bytes	88,171,505	770,924,437
Map output bytes	1,257,795	1,959,363
Combine input records	23,571	37,571
Combine output records	10,214	3,511
Reduce input groups	7,445	861
Reduce input records	10,214	3,511
Reduce output records	7,445	861

In our tests we run only one Map method implementation and one Reduce method implementation. We would like to implement also passing Map results to another Map method as an input and thus fully exploit potential of *ResultTransformers* in Ontea architecture. However, we believe that this new tests does not change – decrees performance of semantic annotation on MapReduce architecture.

4 Conclusion and Future Work

In this paper we discussed briefly how pattern based semantic annotation could benefit from MapReduce architecture to process a large collection of data. We demonstrated how Ontea pattern solution could be ported to implement basic Map and Reduce methods. Furthermore we provided preliminary results on 8 node Hadoop cluster. As we can see from preliminary results, performance on large datasets is very reasonable on Hadoop. MapReduce architecture is scalable to thousands machines. We believe semantic annotation can be successful only if able to annotate or tag large collections of documents.

In our future work we would like to test MapReduce also on several Map tasks in a row and publish implemented code under Ontea.sourceforrge.net project. We also want to use MapReduce architecture to solve concrete application domains such as geographical location identification of web pages and large scale email processing to improve automated email management and semantic searching.

References

1. Cunningham, H.: Information Extraction, Automatic. Encyclopedia of Language and Linguistics, 2nd edn. (2005)
2. Laclavik, M., Seleng, M., Gatial, E., Balogh, Z., Hluchy, L.: Ontology based Text Annotation OnTeA; Information Modelling and Knowledge Bases XVIII. Frontiers in AI, vol. 154, pp. 311–315. IOS Press, Amsterdam (2007)
3. Laclavik, M., Ciglan, M., Seleng, M., Hluchy, L.: Ontea: Empowering Automatic Semantic Annotation in Grid. In: Wyrzykowski, R., et al. (eds.) PPAM 2007, LNCS, vol. 4967, Springer, Heidelberg (2008)
4. Uren, V., Cimiano, P., Iria, J., Handschuh, S., Vargas-Vera, M., Motta, E., Ciravegna, F.: Semantic annotation for knowledge management: Requirements and a survey of the state of the art. Journal of Web Semantics 4(1), 14–28 (2005)
5. Reeve, L., Han, H.: Survey of semantic annotation platforms. In: SAC 2005: Proceedings of the 2005 ACM symposium on Applied computing, pp. 1634–1638. ACM Press, New York (2005)
6. Dill, S., Eiron, N., et al.: A Case for Automated Large-Scale Semantic Annotation. Journal of Web Semantics (2003)
7. Guha, R., McCool R.: Tap: Towards a web of data, http://tap.stanford.edu/
8. Dean, J., Ghemawat, S.: MapReduce: Simplified Data Processing on Large Clusters, Google, Inc. OSDI 2004, San Francisco, CA (2004)
9. Lucene-hadoop Wiki, HadoopMapReduce (2008), http://wiki.apache.org/lucene-hadoop/HadoopMapReduce
10. The Phoenix system for MapReduce programming (2008), http://csl.stanford.edu/~christos/sw/phoenix/.
11. Laclavik, M., Seleng, M., Hluchy, L.: ACoMA: Network Enterprise Interoperability and Collaboration using E-mail Communication. In: Expanding the Knowledge Economy: Issues, Applications, Case Studies. IOS Press, Amsterdam (2007)
12. Vojtek, P., Bieliková, M.: Comparing Natural Language Identification Methods based on Markov Processes. In: Slovko - International Seminar on Computer Treatment of Slavic and East European Languages, Bratislava (2007)
13. Krajči, S., Novotný, R.: Lemmatization of Slovak words by a tool Morphonary. In: TAOPIK (2), Vydavateľstvo STU, pp. 115–118 (2007) ISBN 978-80-227-2716-7
14. Corcho, O.: Ontology-based document annotation: trends and open research problems. International Journal of Metadata, Semantics and Ontologies 1(1), 47–57 (2006)
15. Open Source Distributed Computing: Yahoo's Hadoop Support, Developer Network blog (2007), http://developer.yahoo.net/blog/archives/2007/07/yahoo-hadoop.html
16. Yahoo! Launches World's Largest Hadoop Production Application, Yahoo! Developer Network (2008), http://developer.yahoo.com/blogs/hadoop/2008/02/yahoo-worlds-largest-production-hadoop.html
17. Ontea: Pattern based Semantic Annotation Platform, SourceForge.net project (2008), http://ontea.sourceforge.net/
18. Snowball Project (2008), http://snowball.tartarus.org/
19. Apache Lucene project (2008), http://lucene.apache.org/
20. Klimt B., Yang Y.: Introducing the Enron Corpus. In: CEAS, 2004 (2008), http://www.ceas.cc/papers-2004/168.pdf, http://www.cs.cmu.edu/~enron/

Managing Large Volumes of Distributed Scientific Data

Steven Johnston, Hans Fangohr, and Simon J. Cox

Southampton Regional e-Science Centre,
University of Southampton, United Kingdom
sjj698@zepler.org, {h.fangohr,s.j.cox}@soton.ac.uk

Abstract. The ability to store large volumes of data is increasing faster than processing power. Some existing data management methods often result in data loss, inaccessibility or repetition of scientific simulations. We propose a framework which promotes collaboration and simplifies data management. We propose an implementation independent framework to promote collaboration and data management across a distributed environment. The framework features are demonstrated using a .NET Framework implementation called the Storage and Processing Framework.

Keywords: File Object Model, Data management, database.

1   Introduction

It is generally accepted that processing power doubles every 18 – 24 months and data storage density every 12 – 18 months [1]. The result is that the cost per Gibibyte (GiB) to the end user is falling which indicates that the ratio of processing power to data storage will decrease. This opens up new opportunities to consider more efficient methods of data management.

Databases for example [2], provide a low level data management repository capable of storing and manipulating data. They are often considered too rigid for dynamic data and too complex for non-technical users.

Many users utilise proprietary or custom applications as an alternative to generic databases [3] [4]. These are usually designed for a particular task and tend to be more user friendly but often less flexible. For example the BioSimGrid project [5] [6] is establishing a worldwide repository for simulation results using Grid [7] technologies to distribute and manage the large volumes of data.

Filesystems are ideal for managing data but support for storing and searching the metadata is minimal [8].

Users prefer to keep data in a format with which they are familiar. File systems are capable of dealing with large volumes of data and databases are suitable for managing metadata about the data. There is a need to leverage the benefits of both advanced databases and robust filesystems to provide a simple and easy to use data management solution for scientific users.

Based on these challenges, we propose a method to assist with the collaboration and management of data, particularly scientific data. The proposed method

M. Bubak et al. (Eds.): ICCS 2008, Part III, LNCS 5103, pp. 339–348, 2008.

aims to transparently bring advanced database features to the inexperienced database user.

We propose a method of associating code with data files by treating them as objects. This provides users with additional functions or operations on a file, without having to customise code to process the file. We fully describe the proposed method and demonstrate a specific implementation of the proposal.

2   File Object Method

Most operating systems are capable of associating an application to a specific data file type, *e.g.* files with the extension `txt` are often associated with a text editor. This enables the user to open the file by 'double-clicking' an icon and relying on the operating system to open the appropriate application, capable of reading the file.

This mechanism is achieved either by associating the file extension, `txt` with an application or by associating the Multipurpose Internet Mail Extensions (MIME) [9] type with an application. Alternatively it is possible to inspect the content of the file to determine its type using a file signature [10].

When a user loads an application associated with a data file, the application can be thought of as a library of functions or methods which are capable of operating on that file. For example a `txt` file is opened using an editor which has features to count the words or change the data encoding.

We propose an infrastructure to extend this common functionality to support custom 'applications', in the form of user defined code. Instead of a data file having an associated application we propose that the data file has associated functions and methods which originate from user code.

We propose treating files like programming objects in an infrastructure called the File Object Method (FOM). The FOM associates code with files providing users with the ability to execute these routines as methods on file objects. The aim of the FOM is to extend the usefulness of flatfiles using object oriented programming techniques. This can then be used by hybrid database systems, as well as by users who manage data using flatfiles.

Extending files to appear as objects ensures users execute the correct methods on the correct file types, thus removing the responsibility for users to ensure they have the correct file format.

There are two ways with which a user can associate methods to files in the FOM. The first is to simply associate code with a single file *i.e.* its path and filename. The second is to associate a `type` with each method or set of methods, and then associate this with file types. This means that users can deposit a file into the FOM, and without any intervention be able to list and execute methods that are associated with that file type, using the data they have just deposited. In the text file example, when a user adds a new text file to the file system they will be able to see that there is a method to count the words in that file.

3   Implementation

We demonstrate the FOM using a prototype, called the Storage and Processing Framework (SPF) which demonstrates all the FOM features in a secure and distributed environment.

The SPF is implemented using the .NET framework and is based on web services. The underlying infrastructure supports a secure and distributed file system upon which we demonstrate the FOM features using two examples. The examples are written in different .NET languages and used to demonstrate the multi-language support of the SPF.

The SPF has three key components: the storage service, the storage manager and the client layer.

The storage service manages the data, mapping it to a physical resource. The SPF can have many storage services each controlling a single resource. All the storage services are controlled by a single storage manager. The storage manager is the point of entry for the client layer which exposes all the SPF features to the end users. As each storage service has to register with the storage manager, the user layer can locate any data in the SPF.

3.1   Storage Layer

The key objective of the storage layer is to provide a mechanism for accessing files on a given machine via a web service. By implementing the storage layer it is possible to show how this FOM model can be used to perform calculations in a distributed environment. This storage layer is not intended to be a substitute for a distributed file system and is merely a testing platform for the SPF.

The storage layer is responsible for making the files transparently accessible to the client layer, regardless of location. The storage layer is built up with many storage services and a single storage manager, both of which are web services.

3.2   Storage Service

Each machine has one storage service which manages the data stored on that resource. The storage service is responsible for taking files and storing them on the storage space provided by a resource. The service then responds to requests for files and information about files. The storage service instance has two key components: i) the storage API and ii) the DLL manager.

The storage API maps the SPF file requests to local files and is responsible for invoking the DLL manager. When a file is deposited, it is stored in the local file system and the name and file type are registered with the local SQL database.

The users .NET code is compiled into an assembly called a Dynamically Linked Library (DLL). The DLL manager stores all the user's code and maps it to files and file types. When a file is selected, either programmatically or through the user interface, the DLL manager provides information about the user's code associated with that file and what methods are available. When a user executes a method the DLL manager locates the code and executes the constructor using the local file as a parameter.

The DLL manager caches the results returned by an SPF method. These results are stored in the DLL SQL database and are used to speed up SPF execution. The DLL cache stores the last modified time of all SPF files. When an SPF method is invoked the timestamps are compared. If an SPF file has been altered the SPF cache is flushed.

3.3   Storage Manager

Each storage service has to register itself with the storage manager which is responsible for receiving file requests from the client layer and returning a list of storage web services that store the requested data file.

There are many storage services: one per machine and one storage manager in the SPF. The FOM architecture can support more than one storage manager to allow users to have more than a single point of entry. This can assist with load balancing although the SPF implementation utilises a single storage manager.

The storage manager's function is to provide a point of entry for the client service. When a client requests a file it first asks the storage manager to return the file providing the location of the client web service which requires the file. The storage manager is a lightweight index of the files stored in the SPF. If a requested file is unknown the storage manager polls all the known storage services requesting the file. The location of files are cached in the storage manager and the client layer.

3.4   File Objects

The storage layer identifies files using a SPF Uniform Resource Indicator (URI). Each storage service has its own root folder where all the SPF data files are located. The SPF URI is relative to the root folder. Thus each file does not need to have the same location on each storage service.

The client layer treats a replicated file as a single file object. When a user requests data from the file, the most appropriate storage service is selected. This is based on Central Processing Unit (CPU) utilisation but can be substituted for other machine parameters *e.g.* memory or storage requirements. If the file replications are not synchronised, the client layer flags the file as dirty. If the user chooses to ignore this then the file with the most recent time stamp is used. When the client layer marks a file as dirty it notifies all the out-of-date replicas of the storage service where the most recent data is stored. The replica storage services then synchronises the data files.

The client layer can query all known storage services for a list of files and directories contained within a specific SPF folder.

3.5   Client Layer

All the client features are accessible through the client Application Programming Interface (API). This interface provides a .NET library upon which all the client layers are built.

The client API must have a point of entry into the SPF. This is accomplished by supplying the location of one storage manager web service. Once this is established, the API can query the SPF for files, names, locations and associated user code. If the API does not know of the storage service where a data file is located, it first asks the storage manager to supply a list of valid storage services. The locations of files are cached in the API to speed up frequent access. The API can then query one of the storage services to retrieve information about the associated code.

The client API provides methods to retrieve data files as well as query the associated user code. This provides a list of associated user classes which can in turn provide a list of associated user methods.

Once the user's method has been selected the API can call the storage service to invoke the associated code on the selected SPF file. The results are then transferred back to the API layer where users are free to manipulate the data. The results of a user method invocation are returned as .NET objects which the users can then use programmatically in future code.

The Graphical User Interface (GUI) provides a visual representation of the underlying API capabilities and is intended as an example application of the SPF. All the features exposed in the GUI can be programmatically utilised by the user.

4 Example Scenario

To demonstrate the FOM capabilities we provide two examples.

A C# class which provides metadata about a specific file. The `Advanced FileInfo` class takes a file name as a constructor parameter and provides three methods to return information about the selected file, using the following methods.

The `contact` method returns a string indicating whom the user can contact for further information about the class. The `getFullLocalPath` method returns the full path and file name of the selected file. This is useful in a distributed environment as the SPF file name masks the real location of the data files. The `getLastAccessTime` method returns a `DateTime` object showing when the file was last accessed.

The second example, written in Visual Basic, the `WordCount` class takes a filename as a constructor parameter and provides the following methods to return information about the selected file:

The `countWords` method returns the number of words in the selected text file. The `countLines` method returns the number of lines in the selected text file. The `countCharacters` method returns the number of characters in the selected text file

Both .NET classes are compiled into a .NET library (DLL) which are then imported into the SPF so they become available to all users.

4.1 Selecting an SPF Method

The SPF windows GUI shown in Fig. 1 is divided into five regions. The lower region displays the properties of the currently selected object.

The upper four regions are used to find and select methods associated with data files. For example, to select the `GetLastAccessedTime` method on the `readme.txt` file, the user's four steps are shown in Fig. 1.

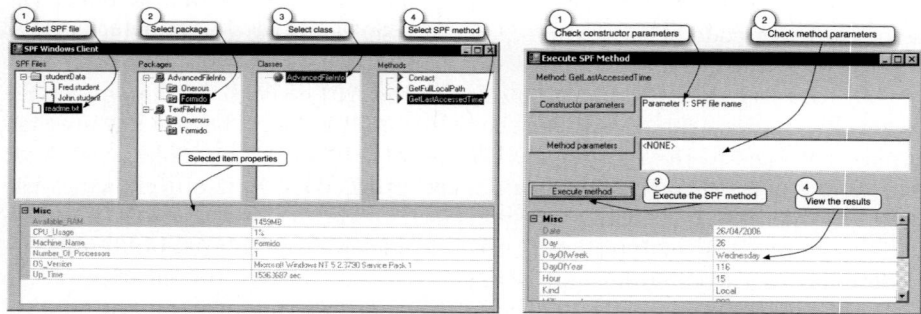

Fig. 1. The SPF client GUI (*Left*). The GUI to execute an SPF method *Right*.

Step 1: When the user interface loads, the client layer connects to a storage manager web service. The client then requests that all storage services return a list of all files and folders stored within the root SPF directory. These files are then collated and displayed in the *SPF file* tree view of the user interface. In this example the user has selected the `readme.txt` file.

Step 2: When a file is selected in the *SPF file* tree, the *packages* tree is then populated. This provides a list of all the packages (collections of classes) which are associated with the selected file. In this example the `readme.txt` file has two associated packages, the `AdvancedFileInfo` and the `TextInfo` package. Not all storage services have the capability to run all user code as the DLL may not exist on a particular resource. The client layer will automatically select a storage service to execute a method base on the machine's load. In the GUI a user can see which machines are capable of running particular user code. In this example the user has selected the `AdvancedFileInfo` package.

Step 3: When a package is selected in the *packages* tree, the *classes* tree is populated. This is a list of all the classes contained within the selected package. In this example there is just one class, `AdvancedFileInfo`.

Step 4: When a class is selected in the *classes* tree, the *methods* tree is populated. When the `AdvancedFileInfo` class is selected the three methods are displayed. The user can then select a method and execute it.

4.2   Executing an SPF Method

Once a user has selected the SPF method to execute, it can then be invoked and the results returned. When a user selects a method in the *SPF Windows client* the *Execute SPF method* window appears, as shown in Fig. 1.

In the GUI interface, users can supply primitive parameters using the constructor parameters option. In this example the class only has one constructor which takes the file name, thus the default options are used.

The default behaviour of the SPF is to execute a method without any parameters. Users can provide parameters by selecting the method parameter option.

Once a user has set the optional constructor and method parameters it can be invoked using the execute method button. This will cause the storage service where the file is located to request the DLL Manager to create an instance of the class and invoke the selected method. The results from this method are then serialised and passed back to the client layer using web services. The client layer returns the results object to the calling interface, in this example the GUI.

The results of the method are returned as an object, displayed the properties of the object in the results view as show in Fig. 1. It is expected that a user will take the return object and utilise it in a calling application. The GUI is intended to demonstrate the capabilities of the underlying API.

5   File Object Method Features

All the storage services run on the host machine's file system and the structure of the files remain unchanged. Simulation and experimental data can be written directly into the storage service using the user's preferred method.

User codes must be compiled into a class library (DLL) which is common practice in the .NET framework.

When a class is selected the associated files and file types are displayed. To associate additional files with the currently selected class, users can add full SPF file names. Users can associate the selected class with any file of a particular type. In this example entering `*.txt` will associate the `AdvancedFileInfo` class with all text files.

Directories and files are treated the same in the SPF, thus users can associate code on a one-to-one basis with directories.

The example shown in section 4 demonstrates how users can discover methods using the client API. All the methods associated with a data file can be listed, queried and executed.

When an SPF method is executed the results are returned as .NET objects. These objects allow the user to programmatically integrate the SPF into existing applications.

Users cannot directly return SPF method results as file objects. If the SPF method creates a file when executed this will then appear in the same directory as the data file. This mechanism can be used to retrieve data from SPF methods as file objects.

5.1   Load Balancing

The load balancing is managed in the API layer. When an SPF method is called the API looks at all the storage services and selects the one with the lowest CPU load. The rules which determine how a machine is selected are stored in a single class. Further SPF optimisations can include additional rules to manage the load balancing. For example, rules which take the available RAM, network speed and available storage can be added to the load balancing rules.

5.2   Security

All the SPF user accounts are created and managed by the Windows operating system and Active direcotry (AD). The SPF security is managed at the operating system level ensuring that the data and hosting machines remain robust against malicious users. When data is deposited it is marked as read-only for all users and read-write for the depositing user. Other users can be given permission to alter data files using the OS user permission settings. Users can change the file permissions to suit their task using the methods currently used.

All the code executed in the SPF framework runs as a restricted user. The SPF, by default only supports managed code. This reduces the users ability to execute malicious code on any of the SPF storage services.

Users can see all the methods available on any data file, the users code is restricted by the OS user account under which it is executed.

Users have the ability to turn on transport security features in the SPF. For example the Message Transmission Optimisation Mechanism (MTOM) data encryption can be used to encrypt data sent to the storage services. The .NET framework supports Simple Object Access Protocol (SOAP) extensions to encrypt web service data.

In addition the option to run reliable message transacted messages is available using the web service standards.

5.3   Method Results Cache

When an SPF method is executed the results are requested from a storage service. Each storage service is completely autonomous and is responsible for invoking the DLL Manager. This provides a good opportunity to cache previously computed results.

Every time an SPF method is executed the calling parameters and the results are stored by the DLL manager. If the method is executed with the same parameters the cached results are returned.

The cache is only valid for a single storage service to ensure that data is kept consistent. If the data in a file is changed all the cached method results are removed.

6   Discussion

The ability of the FOM to support many programming languages adds complexity to the project. Wrapping or integrating different programming languages can often be impossible or prone to errors. It is for this reason that the FOM will never support *all* programming languages.

To ensure that the FOM is non-intrusive and of benefit to users we aim to support as many of the key programming languages as possible. It is important that users are free to write FOM methods in their programming language of preference. This is enhanced with the use of the .NET Framework which supports many programming languages.

The issue of code quality will always be in dispute as it is not possible to check all the code submitted. This feature remains in the FOM specification as the aim is to provide some assurances, i) user's code cannot compromise the host system, ii) user's code cannot corrupt the data stored in the FOM.

The FOM data security comes from the underlying Operating System (OS), providing a user has permissions to change data the FOM permits the operation.

The FOM does not control the names that users provide for methods and classes, hence naming clashes are expected. This could be overcome using a namespace similar to that used in Extensible Markup Language (XML) or providing an internal FOM name. The end result has to ensure that the FOM is capable of dealing with methods of the same name from different users.

When users submit code it is copied to a single storage service. Multiple submissions are required if the user wishes to make the code available on many machines. The client API supports methods to copy user code to other machines. Currently this copies the user's compiled library to a different machine. It is possible to automatically copy all users' code to all storage services but this does not deal with code dependencies. It is for this reason that code exists only on the machine where it is deposited.

The SPF results cache is only flushed when a data file has been altered. Since an SPF method may return large volumes of data the cache can quickly grow. Currently the cache is limited by available storage space. It would be beneficial to change this cache so that only methods which are time consuming are cached. There needs to be a mechanism to limit the size of the cache and rules dictating which cached results are removed first.

7   Summary

We have proposed a concept, called FOM, where files are treated as objects, exposing users' code as methods on these file objects. The aim is to enhance the user's data management experience without drastic changes to the user's existing workflow.

The key objectives of the FOM are to preserve the user's data format, and the ability to reuse existing code such that the FOM is non-intrusive to the user.

The FOM provides operations which enable users to deposit and retrieve data into the repository. Once the data is stored in the FOM, users have the ability to submit and execute code on the data.

The FOM objectives and operations are discussed along with an implementation independent description of the FOM. The FOM provides a series of features which are then described in the context of the SPF implementation.

We demonstrate all the FOM features using a fully-functional prototype called the SPF. This prototype is implemented using the .NET framework and integrates data across a distributed environment.

We demonstrate the SPF capabilities using two user examples written in different programming languages. The SPF provides the ability to execute methods on remote machines and returns the results as .NET objects. The capabilities of the SPF underlying client API are demonstrated with the use of a GUI.

Using this interface we show how users can locate data and view its associated methods. The GUI can invoke SPF methods and display the results.

Future works involves securing user code, leveraging existing distributed file systems and interoperability across heterogeneous systems.

References

1. Moore, G.: Cramming more components onto integrated circuits. Electronics 38(8), 114–117 (1965)
2. Brown, E.: An overview of sql server 2005 beta 2 for the database administrator. Journal (July 2004)
3. Benson, D.A., Karsch-Mizrachi, I., Lipman, D.J., Ostell, J., Wheeler, D.L.: Genbank: update. Nucleic Acids Research 32 (September 2004)
4. Berman, H., Westbrook, J., Feng, Z., Gilliland, G., Bhat, T., Weissig, H., Shindyalov, I., Bourne, P.: The protein data bank. Nucleic Acids Research 28 (2000)
5. Ng, M.H., Johnston, S., Wu, B., Murdock, S.E., Tai, K., Fangohr, H., Cox, S.J., Essex, J.W., Sansom, M.S.P., Jeffreys, P.: Biosimgrid: grid-enabled biomolecular simulation data storage and analysis. Future Generation Computer Systems (2006) doi:10.1016/j.future.2005.10.005
6. Tai, K., Murdock, S., Wu, B., Ng, M.H., Johnston, S., Fangohr, H., Cox, S.J., Jeffreys, P., Essex, J.W., Sansom, M.S.P.: Biosimgrid: towards a worldwide repository for biomolecular simulations. Organic & Biomolecular Chemistry 2, 3219–3221 (2004)
7. Foster, I., Kesselman, C., Tuecke, S.: The anatomy of the grid: Enabling scalablevirtual organisations. International Journal on Supercomputer Applications (2001)
8. Apple: Spotlight overview. Developer Connection (April 2005)
9. Freed, N., Borenstein, N.: Multipurpose internet mail extensions. The Internet Engineering Task Force, RFC-2045 (November 1996) (Online; accessed February 20, 2006), www.ietf.org/rfc/rfc2045.txt
10. Sammes, A.J., Jenkinson, B.: Forensic Computing: A Practitioner's Guide. Springer, London (2007)

Discovering Knowledge in a Large Organization through Support Vector Machines

J.A. Gutiérrez de Mesa and L. Bengochea Martínez

Computer Science Department, The University of Alcalá
Ctra. Barcelona km. 33.6. 28871 Alcalá de Henares (Madrid), Spain
{jagutierrez,luis.bengochea}@uah.es

Abstract. Much of the information used by an organization is collected in the form of manuals, regulations, news etc. These are grouped into controlled documentary collections, which are normally digitized and accessible via a content management system. However, obtaining new knowledge from collected documents in an organization requires not only sound search and retrieval of information tools, but also the techniques to establish relationships, discover patterns and provide overall descriptions of the entire contents of the collection. This article explores the nature of knowledge and the role that occupy the documentary collections as a source of obtaining him knowledge. It also describes the collection of documents will be used along the exposure of this study and the techniques of processing information in order to obtain the desired results. This paper describes the use of computational methods, support vector machines in particular, in a large organisation for document classification.

Keywords: Stemming, Indexation, Support Vector Machines, Documentation and Knowledge management.

1 Introduction

Knowledge Management (KM) as discipline has acquired importance in recent years. The number of scientific articles devoted to this discipline has increased in recent years. One of the main features of knowledge management is its heavy reliance on related disciplines such as information retrieval, data mining, databases and content management systems (CMS). It can even become the standard technology for the implementation of programs of knowledge management [1].

2 Document Collection

In this work, we used a collection of articles published by a Spanish newspaper between July 1, 2004 and June 30, 2006 (two years). The collection consisted of a total of 2,067 documents. The sum of all the words in all documents of the collection was 883,425. The number different character was 104 and the total number of characters used in all documents was 5,441,472. Most of the documents have a length between 350 and 500 words (83%).

M. Bubak et al. (Eds.): ICCS 2008, Part III, LNCS 5103, pp. 349–357, 2008.

2.1 Modelling Documents

Regardless of who is elected one level or another in the choice of terms, in all cases are going to get vectors with many dimensions. In our case, the dimensionality of space vector is given by the number of different words that are used in each and every one of the documents that are part of the collection. Of the 883,425 words contained in the collection, 37,402 are different, which is the vocabulary V of the collection of 37,402 words.

$$W = \sum_{i=1}^{n} w_j = 883,425; \tag{1}$$

One way to reduce the dimensionality is to delete words that do not add any meaning to the text (empty words) and another way is to group words that have the same root in a single lexical (Stemming) such that the total number of different words is reduced. These two processes are described below.

2.2 Normalization Process

In order to be able to run the algorithms, the first step is to transform the documents into plain text and extract the vectors that represent each of the documents. Then, there is a standardization step to facilitate the extraction of measures, such as frequencies and being the normalization the most common operation [2].

2.3 Selection of the Vocabulary

Since the documents will be classified according to their textual content, it is possible to discard all those terms that do not provide relevant information for this purpose. Human languages include many words that are only used to articulate phrases, but do not add any meaning to the text. Also, those words have with very high frequencies. Other words less frequent but more useful for the text to discriminate on the basis of their content.

The set of words that can be regarded as irrelevant or empty for a given language, in our case Spanish, is a priori comprised of the following categories: common adjectives, articles, adverbs, prepositions, conjunctions. Interjections, pronouns, auxiliary verbs (e.g., be, can, do, etc.) and modal verbs (e.g., power, hold, sing, etc.).

Once the list of empty words is built and after their removal, we have the following values: total number of empty words is 503,198; total number no empty words: 380,227 and number of different not empty words: 36,352.

While the elimination of empty words considerably reduces the size of the text, another technique is to remove those words does not reach a certain threshold to reduce the dimensionality of space vector [3]. This technique is based on the idea that when a term appears very infrequently in a collection of documents, their discrimination capabilities is virtually zero, so it can be ruled out at the time of building the model to represent the documents [4]. With this threshold, and after the stemming the process described in the next section, the size of vocabulary will increase from $V=13,256$ to be $V = 6,768$, i.e. get a reduction of the dimensionality of nearly only 50% by removing words that appear only once, twice or three times in the set of all documents that form the collection.

3 Stemming

The basis of a lemmatizer consists of a finite state machine that tries to represent changes in a certain suffix stem. Each suffix involves a series of rules that express how a suffix has been incorporated into the stemming. Since, there can be many variations and exceptions for the same suffix, the PLC can sometimes be quite complex. From these bases, are developing various algorithms stemming for years, such as those based on the probability that a word belongs to the class defined for a stem [5].

Almost all lemmatizers are built upon the foundation of the work by Lovin [6] in 1968 and variants such as those described by Dawson [7], Porter [8] and Chris D. Dave [9]. We have also built a Lemmatizer to apply to documents from the collection object of our study, based on the works of Porter and other more specific to the Spanish language [10].

3.1 Vector Construction

With a very large number of elements that are zero, the following the nomenclature is used to represent every element of the non-zero vector: *{Position: Value}*, where *Position* is an ordinal representing the position it occupies in the lexeme vocabulary, and *Value* is the measure of the contribution that lexeme in the full meaning of the document, D_i. Therefore, a document is represented by the vector:

$$D_i = \{w_1 : f_1, w_2 : f_2, \dots \quad w_{ni} : f_{ni},\} \tag{2}$$

The metrics to be used is *TF x IDF* and vectors will be standardized ($|D_i| = 1$) so that the values of f_j will be given as:

$$f_j = \frac{TF(w_j, D_i)\log(\dfrac{|D_i|}{DF(w_j)})}{\sqrt{\sum_j \left[TF(w_j, D_i)\log(\dfrac{|D_i|}{DF(w_j)})\right]^2}} \tag{3}$$

Where $TF(p_j, D_i)$ represents the frequency with which appears lexeme that took the position p_i in the document D_i; $DF(p_j)$ is the frequency of that same lexeme in the entire collection. Applying the formula to documents, get vectors as shown in Fig. 1 and who will be that we use from this point forward.

4   Classification of Documents

In our study, we use the thesaurus Eurovoc [11] for selecting the categories to which documents may be assigned. This choice was justified by the need to have a package that covers all possible areas addressed in the documents. Moreover, it has been developed by experts following strict criteria.

Prior to the construction and implementation of our own classifier based on the technique known as Support Vector Machine (SVM) [12], which is the one that obtained better results classifying documents [13] , we will give a brief description of some of the most commonly used methods for grading.

In all cases it is building a model by automatic learning from a set of documents previously tagged by an expert. The model thus constructed will be able to deduct the class to which should be given every new document unknown to be present. This type is called supervised learning, because it gives the system the list of categories to which they belong all documents of a collection. A system of unsupervised learning, which builds a model able to infer the existence of clusters of documents, and hence "discover" a class structure that is not known in advance. The action taken by this type of system will call the "grouping of documents", and will be treated in the following point, to differentiate it from the classification of documents "or" text categorization "study in this point.

4.1   Support Vector Machines

Recent studies [14], [15], [16], [17] show that Support Vector Machines (SMV) are the preferred method for text classification. Unlike other methods, SVM can work efficiently with thousand of dimensions whereas in other classifiers, when there is a large number of attributes with little discrimination power, attributes need to be discarded by some preprocessing filters affecting their performance [18]. However, despite their high accuracy documented in numerous publications, and perhaps because of their complexity, SVMs have failed to completely replace simpler methods of automatic classification such as Naïve-Bayes [19].

SVM is based on the concept of minimizing risk structural which is found in the vector space which is represented as vectors documents, hyperplane separating those who belong to two different categories, and also do so with the greatest possible margin of separation. The position in space, occupy any new document, the class to determine who should be allocated. It is therefore a classifier binary and to build a multi classifier must be calculated so as hyperplans classes there.

To carry out the classification of the collection, we are going to use the program package SVM light [20] that allow us to employ algorithms SVM learning, with different parameters and kernel functions, to suit the nature of our problem. However, how to use through orders or commands in text mode has lifted us to develop a GUI to implement the programs. In the preparation phase are formed vectors that are going to represent the set of documents or evidence of learning, properly labelled according to their membership in the class for which we construct the grid. Through panel "svm_learn" (Fig. 1) allows the user to execute the learning module with the options you choose. To carry out this operation is necessary and at least one file of learning. The outcome of this panel will be a file with the model for classifying built.

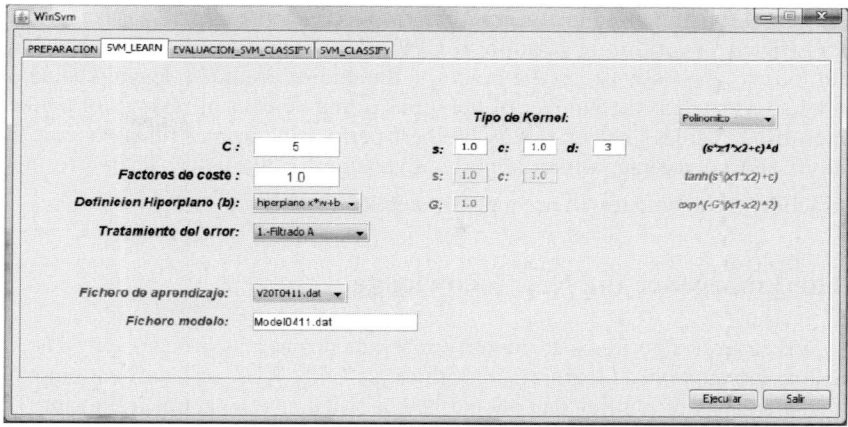

Fig. 1. Panel WinSvm for obtaining model in the training phase

5   Experimental Results

To calculate the optimal values to apply when constructing models for qualifying, we chose the three categories that belong to a larger number of documents on joint training. These are: "04 Life policy" "08 International relations" and "28 Social Affairs". In order to have two sets of classified documents manually, one for the training phase and the other to check the behaviour of the model built, the initial set of 104 documents have been divided into two sets of 52 papers each. It has been tested with the four types of kernel function possible and the results are shown in Table 1.

Table 1. Classification with 52 papers training

Model construction				Test over 52			
Category	Kernel	Iterant.	Kern Evl.	%Success	#Failures	Precision	Recall
04	Lineal	17	1556	69,23	16	72,22	54,17
04	Polynomic	19	3044	71,15	15	73,68	58,33
04	Sigmoid	9	2490	46,15	28	46,15	100,00
04	RBF	20	3099	73,08	14	75,00	62,50
08	Lineal	19	1666	71,15	15	60,00	35,29
08	Polynomic	23	3264	71,15	15	75,00	17,65
08	Sigmoid	16	2870	67,31	17	-	-
08	RBF	26	3429	67,31	17	50,00	5,88
28	Lineal	19	1666	71,15	15	-	-
28	Polynomic	26	3429	73,08	14	100,00	6,67
28	Sigmoid	17	2934	71,15	15	-	-
28	RBF	22	3209	71,15	15	-	-

The kernel function used to construct the model of learning, in all cases is $\text{SVM}^{\text{light}}$, a very efficient algorithm in relation to CPU time. Therefore, the only criterion we should look to choose between the modes is the rate of accuracy, leaving aside other considerations such as the number of iterations or the number of reviews of the kernel function used. Table 1 shows that is the best-performing kernel functions are linear functions and polynomic, with a slightly advantage for the latter. Therefore, for the construction of the binder will use a polynomic kernel $(\vec{x}_1 * \vec{x}_2 + 1)^3$.

5   The Emergence the New Knowledge

There are several algorithms to identify relevant phrases in a document. The most interesting are supervised learning algorithms, as C4.5, KEA or GenEx attempting to document as a set of phrases that should be classified as relevant or irrelevant. To do so, it must provide before a set of documents belonging to a body similar to those discussed and whose relevant phrases are known in advance. From this set and through a process of training builds a table of discretization of the characteristics associated with the terms deemed relevant documents joint training. Once the system is trained, the process of automatic identification of the terms will be considered the metadata value of a new document. It consists of the following: after a period of normalization of the text obtained a first relationship sintagmes candidates, discarding those that do not meet a number of conditions (that is not its length between a maximum and a minimum preset, which begin or end with empty words, which do not reach a minimum frequency of occurrence, etc.). . It was also put to candidates for a phase Stemming with the aim of considering only the roots of words and thus increase the value of their frequencies. Then, a discreet rate of each term, based on the following values:

- Relative frequency of occurrence of S phrase in the text in relation to the overall control (TF × IDF), as measured:

$$TF \times IDF = \frac{frecuency(S, D)}{size(D)} \times (-\log_2 \frac{1 + df(S)}{N}) \tag{4}$$

  where frequency *(S, D)* is the number of times the phrase appeared in the paper *S D* size *(D)* is the number of words that has the document, df *(S)* indicates the number of documents corpus overall contain the term *S* (adds 1 to avoid log 0) and *N* is the total number of documents in the overall corpus).

- Distance, in the words from the beginning of the text until the first appearance of the words. The result is a number between 0 and 1 which represents the portion of the document that precedes the first appearance of a word:

$$distance = \frac{first\ time\ to\ see\ a\ sintagme}{total\ words\ of\ the\ document} \tag{5}$$

- Frequency with which he has already been considered as relevant among the objects of all control. This measure is known as frequency *k* and the underlying idea is that a word candidate is more likely to be relevant if it has been found as other relevant documents corpus training.

For the construction program, we started KEA system, developed in the "Digital Libraries and Machine Learning Labs" [21] at the University of Waikato and is distributed under the GNU Public License.

5.1 Clustering

The grouping a collection of papers has historically been perceived by the researchers as a discovery tool and to help reduce redundancy and demand cognitive [22]. A system of grouping should have the ability to assign each new document to the group most appropriate and should therefore be able to solve three problems: how to create groups, how to identify the relationships between the groups and how to keep the group system.

The most interesting approach in the form of documents and added that in addition, has the advantage of providing direct labels of the groups, is to extract the most relevant phrases for each document in the collection and use as a criterion for grouping. The relevant phrases are good descriptors of the topics covered in a document and therefore help build subspaces small, but representative of space full of documents. The method used to form aggregates is to sort the relevant phrases by the number of documents that share, from highest to lowest. The first group of documents on this list, will form the nucleus of the first added, and the term will be shared by the label of this aggregate. Then he goes through the list of documents added and are appended documents with which it shares other relevant phrases. When this process is completed recursive, passed to the next term of the ordered list, and so complete.

Table 2. Aggregates from syntagms of long> = 1 in category 28

Social Issues	
Descriptor Eurovoc	**Aggregates**
2811.- Movimientos migratorios	Ley de Extranjería
2816.- Demografía y población	-
2821.- Marco social	alto el fuego política antiterrorista
2826.- Vida social	matrimonio homosexual Juan Pablo II
2831.- Cultura y religión	EE UU Benedicto XVI Bin Laden Conferencia Episcopal
2836.- Protección social	Ceuta y Melilla accidentes de tráfico
2841.- Sanidad	Severo Ochoa
2846.- Urbanismo y construcción	plan de choque

During the process of forming aggregates in each category, you get the average length of the documents that form, expressed as the average number of days between July 1, 2004 and the date of publication of document (column "Days (*Dias*)" on the Table 2).

6 Conclusions

In this work, we presented the development of a bespoke computational science application that is going to be used by a large organization to classify documents. To do so, on the one hand, we presented the latest developments in the techniques of automatic classification and clustering textual documents. On the other hand, we showed how to build and validated models using a medium size collection of documents text in Spanish to perform measurements and results that were not previously available. Results show that it is possible to generate patterns of searching documents from the collection, exclusively using automatic learning techniques based on statistical methods, without having to implement other techniques of natural language processing.

References

1. Pérez-Montoro, M.: Sistemas de gestión de contenidos en la gestión del conocimiento. Textos universitaris de biblioteconomia i documentació. 14. Facultat de Biblioteconomia i Documentació. Universitat de Barcelona (2005)
2. Baeza-Yates, R., Ribeiro-Neto, B.: Modern Information Retrieval. ACM Press, Addison-Wesley, New York (1999)
3. Yang, Y., Pedersen, J.: Intelligent information retrieval. Intelligent Systems and Their Applications, IEEE 14, 4 (1999)
4. Luhn, H.P.: The automatic creation of literature abstracts. IBM Journal of Research and Development 2, 159–165 (1958)
5. Allan, J., Kumaran, G.: Details on Stemming in the Language Modeling Framework. Center for Intelligent Information Retrieval. Department of Computer Science. University of Massachusetts Amherst. Technical Report No. IR289 (2001)
6. Lovins, J.B.: Development of a Stemming Algorithm. Mechanical translation and computational linguistics 11, 22–31 (1968)
7. Dawson, J.: Suffix removal and word conflation. Bulletin of the Association for Literary & Linguistic Computing, 33–46 (1974)
8. Porter, M.F.: An algorithm for suffix stripping. Originally published in Program 14(3), 130–137 (1980)
9. Paice, C.D.: Another stemmer. ACM SIGIR Forum archive 24, 56–61 (1980)
10. Figuerola, C.G., Gómez, R., López, E.: Stemming and n-grams in Spanish: An evaluation of their impact on IR. Journal of Information Science 26, 461–467
11. Eurovoc: Tesauro Eurovoc. Presentación alfabética permutada. Edición 4.2 - Lengua española. Comunidades Europeas (2006) ISSN 1725-426
12. Joachims, T.: Leaning to classify text using SVM Methods Theory and Algorithms. Kluwer Academic Publishers, Dordrecht (2001)
13. Sebastiani, F.: Classification of text, automatic. In: Brown, K. (ed.) The Encyclopedia of Language and Linguistics, vol. 14. Elsevier Science Publishers, Amsterdam (2006)

14. Flach, P.A.: On the state of the art in machine learning: A personal review. Artificial Intelligence 131, 199–222 (2001)
15. Yang, Y., Zhang, J., Kisiel, B.: A scalability analysis of classifiers in text categorization. In: Proceedings of the 26th annual international ACM SIGIR conference on Research and development in information retrieval. ACM Press, New York (2003)
16. Lewis, D.D., Yang, Y.R., Tony, G., Li, F.: RCV1: A New Benchmark Collection for Text Categorization Research. The Journal of Machine Learning Research 5 (2004)
17. Lai, C.-C.: An empirical study of three machine learning methods for spam filtering. Knowledge-Based Systems 20, 249–254 (2007)
18. Gabrilovich, E., Markovitch, S.: Text categorization with many redundant features: using aggressive feature selection to make SVMs competitive with C4.5. In: Proceedings of the twenty-first international conference on Machine learning, ICML 2004. ACM Press, New York (2004)
19. Gayo Avello, D.: BlindLight - Una nueva técnica para procesamiento de texto no estructurado mediante vectores de n-gramas de longitud variable con aplicación a diversas tareas de tratamiento de lenguaje natural". Univ. Oviedo. Dpto. de Informática (2005)
20. Joachims, T.: Leaning to classify text using SVM Methods Theory and Algorithms. Kluwer Academic Publishers, Dordrecht (2001)
21. Frenk, et al.: Domain-specific keyphrase extraction. In: Proc. Sixteenth Int. Joint Conference on Artificial Intelligence, pp. 668–673. Morgan Kaufmann, San Francisco (1999)
22. Roussinov, D., Chen, H.: A Scalable Self-organizing Map Algorithm for Textual Classification: Neural Network Approach to Thesaurus Generation. Communication and Cognition 15(1-2), 81–112 (1998)

An Event-Based Approach to Reducing Coupling in Large-Scale Applications

Bartosz Kowalewski[1], Marian Bubak[1,2], and Bartosz Baliś[1]

[1] Institute of Computer Science, AGH, Krakow, Poland
[2] Academic Computer Centre CYFRONET AGH, Krakow, Poland
kowalewski.bartosz@gmail.com, {bubak,balis}@agh.edu.pl

Abstract. Large-scale distributed applications tend to become more and more complex and hard to develop, and execute. Approaches used when building such systems have to be revised, in order to decrease coupling within the code and increase productivity during the development process. Especially event-based programming applied to Web services should gain much attention. In this paper, we present an experimental publish/subscribe infrastructure, which introduces robust event-based mechanisms to be used with Web services-enabled applications. The concept of this solution is built upon the extensibility and configurability principles. We show that performance gap between traditional distributed event-based technologies and the Web also services-based approach is not necessarily as significant as most people tend to think.

Keywords: Large-scale computing, distributed computing, decoupling, Web services, event infrastructure, publish/subscribe, WS-Notification.

1 Introduction

The main characteristic of an event-based design is its striving to decrease coupling in a system [5]. Coupling, usually contrasted with cohesion, is the degree of association between modules, components, subsystems, etc. [6] Over the past three decades a lot of attention has been paid to this concept, both in scientific and commercial computing. Besides many other measures, the level of coupling started to be treated as quality metrics for software design. It has been observed that tight coupling can lead to problems in all phases of application's life cycle. It can make introducing changes in particular parts much complicated and force developers working on distinct modules to frequently synchronize their activities. Tight coupling can also make it impossible to test components separately. On the other hand, loose coupling is usually achieved at the cost of performance: "Loose coupling intentionally sacrifices interface optimization to achieve flexible interoperability among systems that are disparate in technology, location, performance and availability" [11].

Cain and McCrindle state that unmanaged coupling is an indicator of potential productivity bottlenecks [1]. They accentuate the impact of a flawed

M. Bubak et al. (Eds.): ICCS 2008, Part III, LNCS 5103, pp. 358–367, 2008.

architecture on the number of developers that can work in parallel. They conclude with the statement that improperly coupled software causes people to be improperly coupled.

The event-based programming paradigm is built upon the concept of an event object, an entity that represents a situation, an occurrence of interest to third parties. This approach reduces the overall complexity of a system, providing much flexibility when coding, testing and maintaining the application. Unfortunately, it can lead to an increase in complexity of the parts' internals, making it extremely difficult to understand their operation without analyzing the rest of the system.

Event-based approach is becoming more important in contemporary software systems, as the volume of real-time data that enterprises must process and manage continues to increase. The most advanced event-based approach – Complex Event Processing (CEP), which is predicted to become mainstream in the nearest future, emphasizes the act of processing multiple events, aiming at identifying the meaningful ones and discovering complex events. CEP introduces new areas where events could be applicable and defines additional challenges to be faced [14]. Complex Event Processing is even predicted to be heavily used in Enterprise Service Buses.

Many real world examples could also be observed in the Grid environment. The Flood Forecasting Simulation Cascade (FFSC) [9], developed as a part of the K-Wf Grid project, is a loosely coupled large-scale application addressing the complex problem of flood prediction. FFSC takes advantage of the Service-Oriented Architecture (SOA) to handle real-time forecasting. In such application scenarios Web services-based event-driven mechanisms seem a natural approach, providing a flexible and efficient means to handle high data rates.

The publish/subscribe pattern, decoupling subscribers from publishers, has become a critical part of many system architectures [3] and is increasingly being used in a Web services context. The Web services, which are currently the preferred standards-based way to realize Service-Oriented Architecture (SOA), are emerging as the next generation platform for large-scale distributed applications. While the standard enables applications to communicate over Internet protocols, it is still lacking built-in mechanisms supporting the publish/subscribe model. As a response to the need of standardization two partially overlapping specifications, offering a foundation for event-driven architectures built using Web services, were proposed: Web Services Notification (WSN) [13] and Web Services Eventing [12]. Both of these families of specifications and related white papers define standard interoperable protocols through which Web services can exchange event objects. The single greatest drawback of the event-based approach applied to Web services is the delegation of validation of message payload against some contract to the application logic. The WSDL documents based on WS-Notification or WS-Eventing standards will not provide definition for the format of message payload.

Running event-based applications using the Web services technology still remains a challenging problem. There is no single widely accepted standard, providing interoperability between any Web services-based publish/subscribe applications. Another important issue is that the Web Services Notification and Web Services Eventing specifications do not address all of the problems which arise when using publish/subscribe communication with Web services and are considered to be rough drafts, not solid, stable specifications.

Currently none of the publish/subscribe infrastructures based on the above-mentioned specifications provide features that would make them successfully compete with the traditional approaches, like Java Message Service (JMS) or CORBA Notification Service. The performance of these frameworks is far from being satisfactory. What is more, these solutions fail to provide straightforward ways to be employed into an application without the need to understand various complicated mechanisms used. The main assumption is that the designer and the developer should be unaware of the complexity of the infrastructure.

In this paper, we describe our approach to development of an efficient publish/subscribe infrastructure compliant with the Web Services Notification specification (WSN-PSI). The main advantages of this solution are its extensibility and configurability which make it possible to adjust the solution to meet various requirements. While providing these features, WSN-PSI still puts emphasis on performance, showing that the gap between traditional distributed event-based technologies and Web services-based solutions does not have to be enormous.

2   Background

A number of projects is being developed, that aim in providing event-based infrastructures build upon Web services. Those solutions implement WS-Eventing, WS-Notification, or both of these specifications. Hopefully, a new composite standard, WS-EventNotification, was announced and is planned to be available in 2008 [4]. It is intended to solve the interoperability problems between the two specifications currently used.

The WS-Messanger [10] is an implementation of both the Web Services Notification and Web Services Eventing specifications. This solution attempts to support mediation between these two incompatible standards.

Apache Muse [7] provides the functionality defined in Web Services Notification. It also implements other WS-* specifications – Web Services Resource Framework and Web Services Distributed Management.

ServiceMix [8] is an Open Source Enterprise Service Bus (ESB) with support for WS-Notification. This functionality is provided as a binding component and cannot run separately.

A new emerging framework for notification in Grid systems is presented in [2]. This content-based distributed notification service is based on WS-Notification. Unfortunately, the project is still in its initial phase.

Despite the number of Web services-based publish/subscribe infrastructure providers, none of them offers a high degree of extensibility and configurability. What is more, those solutions do not meet the performance requirements of contemporary distributed systems and cannot compete with the traditional event-based solutions. Furthermore, most of the WS-Notification implementations mentioned above are inherent parts of larger, more complex solutions, and cannot run separately. One is unable to only use the publish/subscribe modules.

3   Concept of WSN-PSI

3.1   Requirements

The basic and obvious assumption that the solution is to be founded on the concept of Service-Oriented Architecture and Web services implicates numerous requirements, like service contract or reusability. These obvious requirements will not be enlisted within the objectives below. The following paragraphs present features that would define a noteworthy Web services-based publish/subscribe solution.

Among the most significant functional requirements we can mention:

- well-defined message exchanges – all of the communication scenarios associated with actions other than notification exchange should be precisely defined, enabling full compatibility between different implementations or versions of the infrastructure,
- dynamic reconfiguration of the environment – endpoints taking part in communication need have no knowledge of other endpoints prior to registration or subscription, nodes can be created and destroyed without having negative impact on infrastructure operation,
- messaging brokers – it should be possible to decouple producers and consumers by placing intermediary services (brokers) between them,
- topics – a topic (subject) should be attached to every message, enabling clients to separate independent message exchanges,
- filtering – it should be possible to limit number of messages sent to a consumer using filter constraints.

Non-functional requirements concerning our publish/subscribe solution are as follows:

- standards-based approach – will provide optimal solution adoption and reduce efforts involved in building and maintaining applications founded on this infrastructure,
- performance – reduce complexity to get as fast and efficient infrastructure as it is possible with contemporary Web services technologies,
- usability – the solution has to be easy to employ in order to produce a specialized messaging infrastructure, minimizing the effort required to switch from using the standard request/response pattern to the event-based approach,
- extensibility – should provide for change while minimizing impact to existing infrastructure fragments,
- configurability – the constituent parts of the solution should be configurable and exchangeable,
- scalability – the infrastructure should be capable of dynamic messaging endpoint allocation and environment reconfiguration.

3.2   Design

We decided to build the infrastructure on the basis of the already mentioned WS-Notification (WSN) specification to ensure better adoption of the solution. WSN is a natural successor of the Open Grid Services Infrastructure (OGSI) Notification standard, which makes it the approved event-based approach for the Grid environment. Moreover, WS-Notification provides many advanced features, mechanisms and application scenarios. Last but not least, using WS-Notification lets us observe which areas are sill covered inadequately by the available standards and need to be improved.

The overall architecture of the system is founded on entities defined in WS-Notification. WSN-PSI implements a large part of WS-BaseNotification and WS-BrokeredNotification, and a basic subset of concepts defined in WS-Topics. Worth mentioning is the fact that WSN-PSI does not restrict selection of cooperating nodes to some predefined architectures and does not require endpoints taking part in communication to have knowledge of other endpoints prior to subscription or registration. Because of this approach, WSN-PSI provides really flexible and dynamically reconfigurable environment. Also by providing support for intermediary services WSN-PSI empowers the process of decoupling.

The whole infrastructure is built with the assumption that the only way to make the system usable is to provide a high level of extensibility and configurability. WSN-PSI provides a collection of configurable building blocks, enabling one to construct a specialized publish/subscribe solution. They were designed with care for loose coupling, high cohesion, and the fundamental object-oriented design principle – the single responsibility principle. Fig. 1 presents the approach used to implement WSN services.

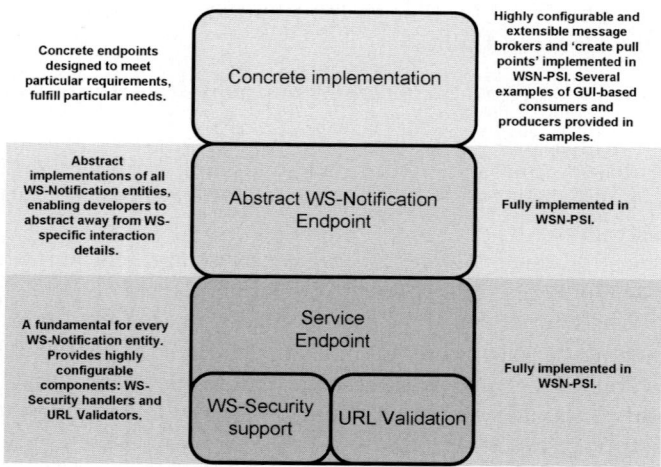

Fig. 1. Design of a WS-Notification-based entity

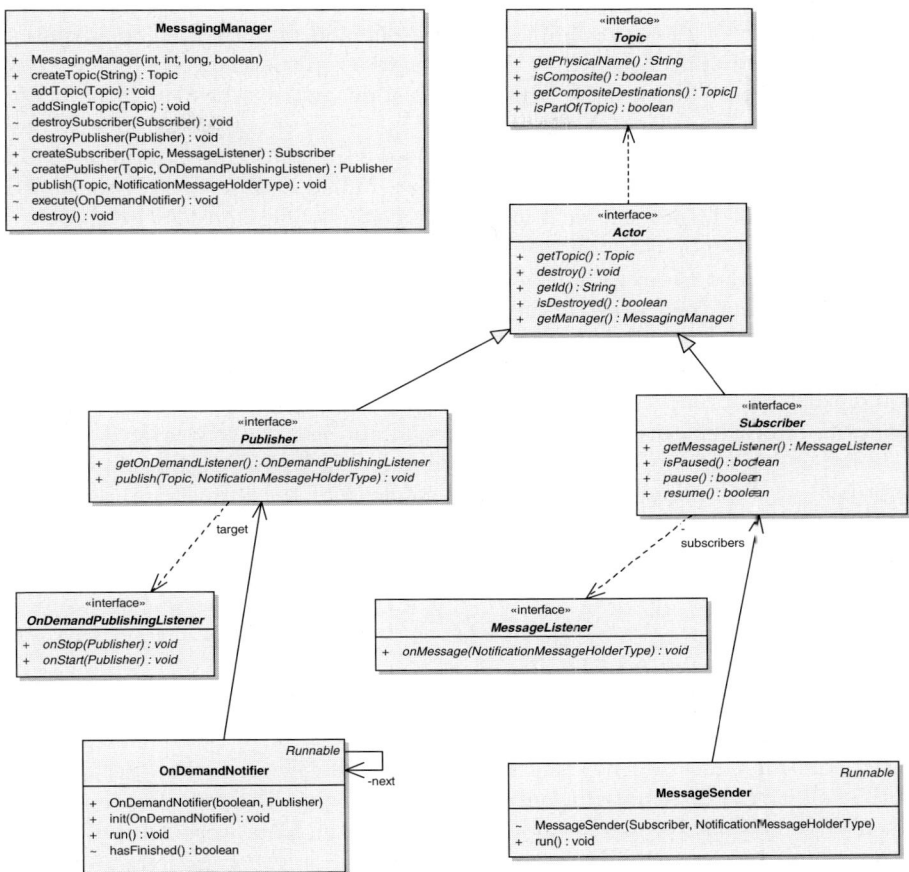

Fig. 2. WSN-PSI: a simplified class diagram of the internal brokering system

The full-featured and highly configurable notification broker is a good illustration of the application of this model. Mechanisms dedicated to broker customization include, among others, the ability to select the method of event notification dispatching inside the broker. One can currently choose between JMS-based dissemination and an efficient internal dispatching technique. Fig 2 provides a simplified class diagram, presenting the design of this internal brokering system incorporated into WSN-PSI. This module defines only two types of actors: *Publishers* and *Subscribers*. They can be created and are managed by a centralized *Messaging Manager*, build upon a thread pool dedicated to executing long tasks. Every *Subscriber* is provided a *Message Listener*. Each time a notification message is routed to the *Subscriber*, a task wrapped into a *Message Sender* object is sent to the *Messaging Manager*, which plays the role of a task scheduler. Similarly, every demand-based *Publisher* is provided an *On-Demand*

Publishing Listener. Publishers utilize *On-Demand Notifiers* to wrap tasks associated with pausing and resuming WSN on-demand publishers. The advantage of the described approach is that the messages are dispatched locally, no external messaging provider is used. This reduces additional overhead associated with transferring the messages. What is more, this broker directly uses build-in Java thread pooling mechanisms which also causes an increase in performance. One significant drawback of this approach is that, unlike in JMS-based broker, it is impossible to plug into the local messaging system using non-WSN clients.

The design is not the only element that makes the solution highly usable and efficient. Also the selection of technologies has a significant impact on the characteristics of the infrastructure. WSN-PSI is founded on many widely accepted standards and libraries. As it was stated when specifying requirements, one of the main principles employed when designing the infrastructure was to use standardized approaches whenever possible. Adoption of well-known standards makes the infrastructure easier to understand and use, reducing efforts involved in building and maintaining applications founded on WSN-PSI. Standards-based approach also simplifies the process of evolution of the infrastructure and increases portability. To ensure required performance level, technologies had to be picked carefully. For example XFire, which proved to be really efficient, was chosen as the SOAP framework.

4   Performance Analysis

The approach described here attempts to show that performance of Web services-based solutions does not necessarily have to be much worse than performance of the traditional distributed event-based technologies. To prove this hypothesis a series of tests was prepared and executed. All the tests were run on a single PC machine and were designed to present the same single scenario implemented using different technologies. This scenario defined two endpoints, a master and a slave, communicating through a broker. Messages were sent from the master to the broker, further to the slave node and back to the master through the intermediary service. Each time round trip time (RTT) was evaluated and the time required to send a message from one node through the broker to the other endpoint was calculated. This scenario was repeated 1000 times using messages with 1024 characters long payloads.

The tests cover two configurations of WSN-PSI infrastructure and the most significant WS-Notification implementations currently available. The first WSN-PSI notification broker version was based on internal message dispatching mechanisms, while the second one used dispatching mechanisms implemented on the basis of JMS. The ServiceMix ESB was run with the Servicemix-wsn2005 binding component deployed. The WS-Notification implementation provided by the Apache Software Foundation, Apache Muse, had to be modified to be suitable for the test scenario. Muse does not implement the notification broker functionality and an internal bridge between notification consumer and producer had to be added. To have the possibility to compare Web services-based solutions and the distributed event-based technologies a sample Java Message Service (JMS) specification implementation provided by JBoss Application Server was also taken into account. Tab. 1 presents results of the experiments.

Table 1. Average delivery time

Technology/solution	Average delivery time (ms)
Java Message Service (JMS)	11.8
WSN-PSI (internal)	16.1
WSN-PSI (JMS)	26.4
ServiceMix	28.5
Apache Muse	42.2
WS-Messenger	106.9

The experimental data shows that it is possible to create a Web services-based solution that will successfully compete with JMS. The results also suggest that most of the infrastructures currently available tend to underperform and could easily be improved. On the other hand, one should take into account that using the same payload sizes for the messages being disseminated in different tests does not make the overall size of the messages equal. Various strategies to building notification messages, possibly incorporating WS-Addressing support, may have considerable impact on the empirical results.

Additionally, we conclude that the performance gap between traditional event-based approaches and the Web services-based solutions grows considerably with the number of advanced filtering and Quality of Service (QoS) mechanisms employed.

5   Possible Applications

WSN-PSI is designed to cover a wide spectrum of scenarios. It provides a high level of scalability, being 100% compliant with the WS-Notification specification. This causes the topology not to be restricted by any means, enabling developers or users to have full freedom to build their own specialized configuration. WSN-PSI also uses a layered architecture with many interchangeable modules, easily configurable using XML files.

The intermediary service is fully implemented and could be used with minimum changes in the configuration files. The project also provides many stubs and samples written in Java programming language. These could straightforwardly be used to incorporate Web Services Notification functionality into an existing large-scale application. All of these services could be deployed using the embedded Jetty WebServer. This significantly simplifies the process of packaging and running the application, and could be used when the standard, Web ARchive-based, approach to deployment is undesirable.

WSN-PSI could also be used in scenarios requiring support for security. The infrastructure uses WS-Security, which provides basic security mechanisms for Web Services Notification message exchanges. Unfortunately, the mechanisms employed when handling WS-Security data cause a serious decrease in performance.

Worth mentioning is the fact that the Java programming language was chosen for developing the experimental solution, to provide sufficient level of portability and to be able to take advantage of existing implementations of Web services-based technologies. Java is the only programming language used in WSN-PSI. Development of sample infrastructures for other languages is considered to be out of scope of this work and may be added in the future phases of the project. Also no cross-language clients are provided. Fortunately, the Web Services Notification compliant WSDL files shipped with WSN-PSI could by used to generate clients in any programming language.

6   Conclusions and Future Work

In this paper we presented WSN-PSI: an efficient publish/subscribe infrastructure that could be successfully used to decrease coupling in large-scale Web services-based applications. The solution was designed to meet tough performance requirements and make Web services-based technologies an option when choosing communication means for a highly decoupled system.

One considerable drawback of using the event-based approach with Web services, which was identified during early development, is the incompleteness of Quality of Service (QoS) mechanisms available. QoS comprises various aspects of messaging, including reliable delivery, order, security, duplicate elimination and many more. The family of second-generation Web services documents (WS-*) provides WS-Security, WS-ReliableMessaging and WS-Reliability specifications. Unfortunately, these standards are immature and do not cover all of the desirable mechanisms. Moreover, both WS-Notification and WS-Eventing leave many areas unstandardized, increasing the risk of incompatibility between various infrastructures. Absence of some advanced mechanisms, like support for event object routing, may also cause the more complex configurations to malfunction.

Future plans for the WSN-PSI project include, amongst all the other goals, experimenting with various technologies that could be applicable to the infrastructure. It also seems reasonable to experiment with some specifications from the huge family of second-generation Web services documents, other than WS-Notification and WS-Security already used in WSN-PSI. Furthermore, the project lacks more complex tests in a real grid environment.

Acknowledgments. This work was supported by EU projects ViroLab IST-027446 and CoreGRID IST-004265 with the related Polish grants SPUB-M.

References

1. Cain, J., McCrindle, R.: An Investigation into the Effects of Code Coupling on Team Dynamics and Productivity. In: IEEE 26th Annual International Computer Software and Applications Conference (COMPSAC) (2002),
http://www.blunder1.demon.co.uk/research/COMPSAC2002.pdf
2. Quiroz, A., Parashar, M.: A Framework for Distributed Content-based Web Services Notification in Grid Systems. Future Generation Computer Systems 24, 452–459 (2008)
3. Carzaniga, A., Di Nitto, E., Rosenblum, D., Wolf, A.: Issues in Supporting Event-based Architectural Styles. In: ICSE 1999 Workshop on Engineering Distributed Objects (EDO 1999) (1999),
http://www-serl.cs.colorado.edu/~carzanig/papers/isaw3.pdf
4. Hewlett Packard Corporation, IBM Corporation, Intel Corporation, and Microsoft Corporation. Toward converging Web service standards for resources, events, and management, http://download.boulder.ibm.com/ibmdl/pub/software/dw/webservices/Harmonization_Roadmap.pdf
5. Eugster, P., Felber, P., Guerraoui, R., Kermarrec, A.-M.: The Many Faces of Publish/Subscribe. ACM Computing Surveys, http://www.irisa.fr/paris/Biblio/Papers/Kermarrec/EugFelGueKer03ACMSurvey.pdf
6. Faison, T.: Event-Based Programming: Taking Events to the Limit, ch. 1. Apress (2006)
7. The Apache Software Foundation. Apache Muse, http://ws.apache.org/muse/
8. The Apache Software Foundation. ServiceMix – WSN 2005 (2005),
http://incubator.apache.org/servicemix/servicemix-wsn2005.html
9. Habala, O., Mališka, M., Hluchý, L.: Service-based Flood Forecasting Simulation Cascade in K-Wf Grid. In: K-WfGrid - The Knowledge-based Workflow System for Grid Applications, Proceedings of CGW 2006, vol. II (2006)
10. Huang, Y., Slominski, A., Herath, C., Gannon, D.: WS-Messenger: A Web Services-based Messaging System for Service-Oriented Grid Computing,
http://www.cs.indiana.edu/~yihuan/research/HuangY-WSMessenger.pdf
11. Kaye, D.: Loosely Coupled: The Missing Pieces of Web Services, ch. 10. RDS Associates (2003)
12. Microsoft, IBM, TIBCO, Bea Systems, and Computer Associates. Web Services Eventing, http://www.w3.org/Submission/WS-Eventing/
13. OASIS. Web Services Notification,
http://www.oasis-open.org/committees/tc_home.php?wg_abbrev=wsn
14. Wu, E., Diao, Y., Rizvi, S.: High-Performance Complex Event Processing over Streams. In: 2006 ACM SIGMOD International Conference on Management of Data (2006),
http://avid.cs.umass.edu/sase/uploads/pubs/sase-sigmod2006.pdf

Exploring Cohesion, Flexibility, Communication Overhead and Distribution for Web Services Interfaces in Computational Science

Miguel-Angel Sicilia and Daniel Rodríguez

Computer Science Department, The University of Alcalá
Ctra. Barcelona km. 33.6. 28871 Alcalá de Henares (Madrid), Spain
{msicilia,daniel.rodriguezg}@uah.es

Abstract. Computational science studies often rely on the availability of large datasets through the Web. Web services (WS) provide a convenient way for making those datasets available, since they rely on a standard and widely available technology. However, there are many ways to devise a Web service interface for a given dataset, and the resulting interfaces vary in their properties related to cohesion, distribution, flexibility and communication overhead, among other parameters. This paper explores these attributes and provides some directions on how they can be measured. Concretely, a well-known cohesion metric is explored as a way to characterize the possible kinds of Web service interface designs. This is discussed for a concrete distributed context of service-oriented architectures.

Keywords: Computational science, Web services, cohesion, performance, distribution, flexibility.

1 Introduction

Computational science requires the use of large amounts of data for the construction of models and simulation analyses in order to solve scientific problems. In recent years, vast amounts of data have been made available through the Web as Web services (WS) provide a way to obtain pieces of these kinds of information through standard Web technology, and many on-line databases currently use WS interfaces [8]. Most of these interfaces provide access to schemas containing the scientific data through a collection of services that clients must call to obtain the data.

It is known that current the current way of implementing Web services based on SOAP[1] messaging over the HTTP[2] protocol incorporates some additional overheads [7]. Depending on the design of the Web services interfaces, client applications would require to issue a larger or smaller number of Web service invocations, thus incurring in more or less overhead respectively. An extreme case occurs when a single Web service retrieves all the data in a particular dataset, which reduces to the minimum the

---

[1] http://www.w3.org/TR/soap12-part0/
[2] http://www.w3.org/Protocols/

M. Bubak et al. (Eds.): ICCS 2008, Part III, LNCS 5103, pp. 368–375, 2008.

number of remote method calls but clearly compromises flexibility to build applications on top of it. In the opposite extreme, a design in which the data in each entity in the data model is provided by a separate Web service provides maximum flexibility. In addition, some Web service interface designs delegate complex query processing to the server while others only retrieve data, and processing needs to be done in the client side. Depending on the computing power provided by the server, one or other option could be desirable. In consequence, Web service interface design critically impacts the flexibility, performance and distribution properties of a system. These trade-offs can be analysed with metrics defined for structured or Object Oriented (OO) systems. For example, Perepletchikov et al [6] have analysed the design of 3 different designs for a Service Oriented Architecture (SOA) with traditional metrics such as Lines of Code and Cyclomatic Complexity [2, 4] and the set of OO metrics defined by Chidamber and Kemerer (C&H) [1]. As authors state, most OO metrics need to be adapted for Web services, mainly assuming that a class is a service or set of services (business process).

Designers facing the analysis or design of Web service interfaces could find beneficial the availability of metrics, indicators or guidelines for measuring designs flexibility, degree of distribution and communication overheads. Computational science applications are typically characterized by relatively stable data schemas but with large volumes of data, which implies that communication overhead is an issue for this kind of applications. Also, since new research directions would impose additional requirements, designers need to be able to adapt systems to unpredicted new uses, and thus, *flexibility* is also a requirement.

This paper attempts to delineate some possible indicators that could be used to measure how a concrete Web service design affects these variables. It explores basic design issues and the potential use of cohesion as an indicator of some properties of the quality of the design.

The rest of this paper is structured as follows. Section 2 discusses the some metrics and the reformulation of a metric for Web services. Next, we present a small case study in Section 3. Finally, Section 4 concludes the paper and points to future research directions.

2   Communications Overhead, Flexibility and Web Services Design

The Web service architecture defined by the W3C enables applications to communicate over the Internet, in other words, Web services allow applications to access software components through standard Web technology. The use of Web services introduces, however, additional overheads in communications as a trade-off for the increase in flexibility they provide. In particular, due to the usage of XML, not only requests and replies are larger when compared to traditional Web interactions [5] but also the need for parsing the XML code in the requests adds additional server overhead.

Here we are concerned with computational science applications, in which (i) large amounts of data need to be transferred through the net, and (ii) the interfaces need to be flexible to serve the needs of different applications and research needs. Web service interfaces that wrap datasets can be devised with different styles, and this affects performance due to communication overheads. In general, finer grain service calls produce additional latency due to the increase in the number of verbose SOAP messages and established connections needed. In contrast, coarse-grained interfaces will reduce network latency but are less flexible for applications that require only some specific pieces of whole information. In addition, some Web services perform complex queries against the schema that should be performed at the client side if they were not available.

These tradeoffs could be subjected to measure in an attempt to develop a metric that combines flexibility, performance and distribution when designing Web services interfaces. Since the amount of services offered for the same schema will ultimately depend on the parts of the data model that are read, *cohesion* can be used as a candidate property to develop the measures sought. In the area of OO systems, one of the metrics defined by Chidamber and Kemerer (C&K) [1] is the *Lack of Cohesion in Methods (LCOM)*. It is as a quality metric of the cohesiveness of a class by measuring the number of method pairs that do not have common instance attributes. More formally, LCOM measures the extent to which methods reference the class instance data. Let us consider a class C_1 with n methods $M_1, M_2, ..., M_n$. and let $\{I_j\}$ = set of instance variables used by method M_i. There are n such sets $\{I_1\},...,\{I_n\}$. Let $P = \{(I_i, I_j) \mid I_i \cap I_j = \varnothing\}$, and $Q = \{(I_i, I_j) \mid I_i \cap I_j \neq \varnothing\}$. If all n sets $\{I_1\},...., \{I_n\}$ are $\varnothing$ then let $P = \varnothing$.

$$LCOM = \begin{cases} |P| - |Q| & if \ |P| > |Q| \\ 0 & otherwise \end{cases}$$

For example, consider a class C with 3 methods (M_1, M_2, M_3) and Let $\{I_1\}$ = {a, b, c, d, e}, $\{I_2\}$ = {a, b, e}, and $\{I_3\}$ = {x, y, z}. $\{I_1\} \cap \{I_2\}$ is nonempty but $\{I_1\} \cap \{I_3\}$ and $\{I_2\} \cap \{I_3\}$ are null sets. LCOM is the number of number of null intersections – number of nonempty intersections), which in this case is 1 A high value of *LCOM* implies that there is a lack of cohesion, i.e., low similarity between the methods of a class and as a result, the class can be composed of unrelated objects. High cohesiveness of methods within a class is desirable, since classes cannot be divided and promotes the *encapsulation*. Low cohesiveness increases complexity, thereby increasing the likelihood of errors during the development process.

There are other reformulations of cohesion that can be found in the literature since the original *LCOM* C&K metric has been criticized. For example, Henderson-Sellers [3] comments that two classes can have a *LCOM=0* while one has more common attributes than the other. Also, there is no maximum value so it is difficult to interpret the values. As a result Henderson-Sellers [3] defined a modification, *LCOM-HS*, as follows. Let us consider a set of methods $\{M_I\}$ (I=1,...,m) assessing a set of attributes $\{A_j\}$ (j=1,...,a) and the number of methods that access each attribute $\mu(A_j)$, then define *LCOM-HS* is defined as:

$$LCOM - HS = \frac{\left(\frac{1}{a}\sum_{j=1}^{a}\mu(A_j)\right) - m}{1 - m}$$

The LCOM-HS range is between 0 and 2. If all methods access all attributes, then $\Sigma \ \mu(A_j)$ = ma, so that *LCOM-HS = 0*, which indicates perfect cohesion. If each method access only 1 attribute then $\Sigma\,\mu(A_j) = a$ and *LCOM-HS* = *1*, which indicates already a high level of lack of cohesion. Therefore, for OO systems values near 0 are preferred for this metric, where most methods refer to most instance variables.

The same measure can be extrapolated to the case of Web services, if we consider methods to be *Web Services*, and attributes to be *Entities* in the data model. Therefore, the reformulated the Henderson-Sellers *LCOM* metric for Web services, let us call this metric *LCOM-WS*, can be defined as:

$$
LCOM - WS = \begin{cases} \dfrac{\left(\dfrac{1}{|E|}\sum_{j=1}^{|E|}a(e_j)\right)-|S|}{1-|S|} & if\,|S| > 1 \\ 0 & if\,|S| = 0 \end{cases}
$$

When all information is retrieved in a single service, the denominator is 0, so a separate definition is provided for it. Obviously, a single service retrieving all the data is fully cohesive according to the idea behind LCOM, but it lacks flexibility; the whole set of information, which many applications possibly do not need (at least at a give instant of time), is retrieved for all calls all the time. Same as with the original metric, a case of minimum cohesion occurs when every entity has just one associated Web service to retrieve the data from it, including intermediate tables. In this case, the result of the formula is equals to 1, yielding a high lack of cohesion value. Regarding processing in that case, the server acts as a simple processor of queries on single entities, and no join or other kind of expensive computations are carried out.

The key here is to explore why low cohesion leads to more efficient interfaces in terms of the ratio of data transferred to total communication overload. They, however, tend to be less flexible when obtaining concrete data elements. When cohesion is increased, so is the network overload as more data than needed is transferred (interfaces are of coarser granularity).

3   Case Study

For the sake of contrasting designs, we will consider a data schema ξ that for practical purposes will be considered a relational data schema. The schema is considered to be formed by a group of entities, $E=\{e_1,..., e_n\}$, some of which are related by referential integrity constraints $e_j \rightarrow e_k$. Then, a maximum cohesiveness (i.e., low Lack of Cohesion) in the Web services interface will be achieved when the entire data set is retrieved in a single Web service, or horizontal portions of the dataset (subsets of the tuples in a table, for example) which take one piece of information for each of the entities. Then, here communication overheads is minimal (but maximal latency, i.e., time to transfer the whole dataset or most of it), except for the case that horizontal slices are taken, e.g when the Web services are instructed to retrieve tuples in slice of say, 30 rows per call. Obviously, the case of minimal communication overhead approximately corresponds to download or bulk export functionality. In this case, C&K LCOM is 0, which means high cohesion.

As a way to analyze the impact of Web services design in flexibility, performance and distribution in computational science applications, we will study a concrete, relatively simple application, using the Multilocus Sequence Typing (MLST) database[3] containing genetic information of different types of organisms. Figure 1 provides a summary of one of its databases in the form of a relational schema[4]. There are some attributes that occur several times depending on the kind of organism or information represented, e.g. `allelic_profiles`. Also, the table locus is actually a set of tables, but these issues do not affect our current analysis, since we can consider the tables to be conceptual entities accessed by the WS interface.

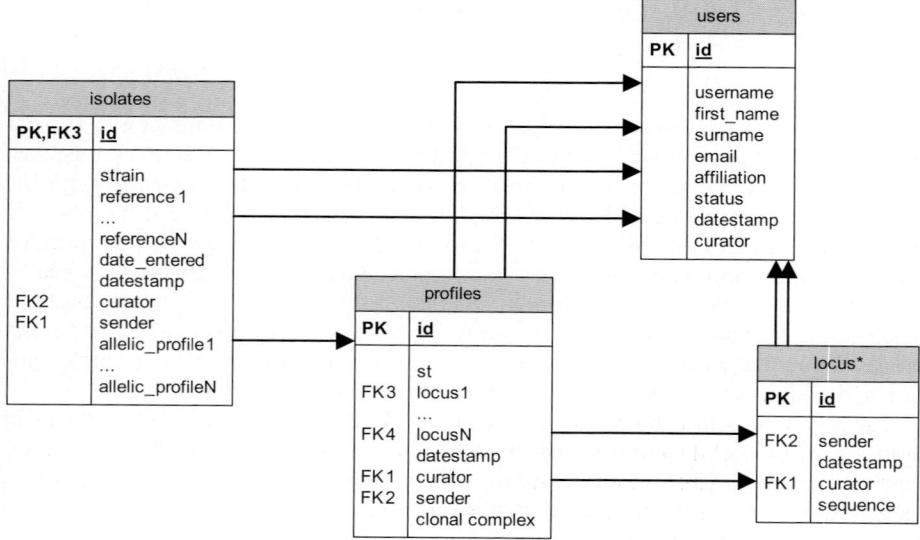

Fig. 1. Database Schema for the MLST database

Table 1 provides the counts and attributes of the Web services interface provided by MLST at the time of this writing. Services marked with (*) have been excluded from the analysis since they are actually providers of URI-based ways of retrieving data, which is a mechanism not homogeneous with the rest of them. The column "attributes retrieved" identifies which attributes are the outputs of Web services. It should be noted that some of the services do not actually retrieve actual information, but information on the database schema or on the number of records (rows) contained in a table, these are marked as such in the "support iteration" column, since their intent is that of providing information for calling other services. Also, the column "processing" indicates which services are considered to retrieve information from several entities, thus involving a join or other kind of processing.

---

[3] http://pubmlst.org/

[4] The data model is an abstraction on the documentation of the MLST database for the purpose of analysis, we do not claim it is actually the underlying data schema of the service.

Table 1. Web Services interfaces provided by MLTS

Web Service	Entities read	Attributes retrieved	Multiplicity	Support iteration	Processing
getIsolate	Isolates	all	1		
getIsolateCount	Isolates	none	1	yes	
getIsolateFields	Isolates	none	N	yes	
getRelatedIsolateIdsByProfile	isolates, profiles	id	N		join
getRelatedIsolateIdsByST	isolates	id	N		logical
isolateQuery	isolates	id	N		logical
Blast	locus	id	N		complex
getAlleleCount	locus	none	1	yes	
getAlleleSequence	locus	sequence	1		
getAlleleSequences	locus	sequence	No		
getAlleleSequencesURL (*)					
getForwardConsensus	locus	sequence	1		complex
getLocusLength	locus	none	1	yes	
getReverseConsensus	locus	sequence	1		complex
locusBlast	locus	id	N		complex
locusQuery	locus	id	N		complex
getClonalComplex	profiles	clonal complex	1		
getClonalComplexes	profiles	clonal complexes	N		
getProfile	profiles	Several	1		
getProfileCount	profiles	None	1	Yes	
getProfileListURL (*)					
getRelatedProfilesByST	profiles	All	N		
getRelatedProfilesByProfile	profiles	All	N		
getRelatedSTsByProfile	profiles	St	N		
getRelatedSTsByST	profiles	St	N		
getSTs	profiles	St	N		

If we apply the LCOM-WS and LCOM-CK metrics described above, the information can be contrasted with the design of the WS interface as showed in Table 2. Case (iii) in the Table is the counting for the interfaces analyzed in Table 1, and cases (i) and (ii) are the two "extreme" cases commented before. The contrast of cases (i) and (ii) shows that the former is measured as highly cohesive, but it provides the minimum flexibility since all the data is transferred in a single invocation. In contrast, the flexibility in case (ii) is maximum, in the sense that applications can obtain the concrete information blocks required down to the granularity of single entities (this can be seen considering the getIsolate, getAlleleSequence and getProfile services). For case (ii), cohesion is low, but cases of non-flexible interface might also yield intermediate cohesion values. For example, two Web services that obtain two unrelated parts of a data model would yield a *LCOM-WS* of 0.33. The first consequence then is that the studied *cohesion metrics cannot be used to compare any arbitrary WS interface design when flexibility is a requirement.* Flexibility is a property of design that can be inspected from the specification of WS. Table 2 shows also how the two extreme cases coincide in that they rely all the processing of data to the client, but they are opposites in communication overhead and flexibility. It should be noted that some of the services in Table 1 provide server-side processing of considerable

Table 2. Results of the Analysis

Property	Case (i): A single WS	Case (ii): A WS per entity	Case (iii): Actual interface	Case (iii): Actual interface + one more query WS	Case (iii): Actual interface + five more query WS
LCOM-WS	0	1	0.77	0.74	0.63
LCOM-CK	0	6	274	270	254
Flexibility	Minimum	maximum	maximum[5]	=	=
Communications overhead	Minimum	maximum	intermediate	Eventually slightly increased	Eventually slightly increased
Distribution	all client side	all client side	6 out of 24	7 out of 25	11 out of 29

complexity, e.g. the `blast` service provides sequence matching through heuristics with no linear computational complexity in the worst case.

Then, for interfaces with maximum flexibility, the increase in cohesion comes from the addition of services that require joining information from different entities or performing complex queries, which entails delegating workload to the server side. This idea could be used to explore metrics of distribution in Web service interfaces, which are relevant for the design of solutions to complex applications. It is important to consider that data servers on the Web could be facades for parallel computing or Grid systems, so that delegating to the server is the preferred way of developing the application. In that direction, a possible additional metric for distribution could be derived from the WMC (*Weighted Methods per Class*) metric proposed by C&K [1]. Fig 2 depicts this abstract relationship.

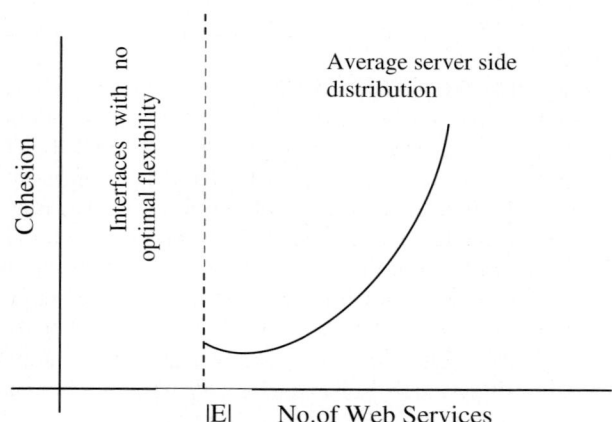

Fig 2. Relationship between no. of WS, cohesion and distribution flexibility

---

[5] We do not consider the fact that some of the attributes in the users entity can not be retrieved with the list of services presented, since flexibility here deals with not getting undesired data only.

4 Conclusions and Future Work

The comparison of potential designs for WS interfaces to computational science data-sets reveals the importance of considering at least cohesion, flexibility, communication overload and distribution. Using the data entities as units, maximum flexibility is achieved in these interfaces by methods retrieving information for each of the entities, and measures of lack of cohesion go smaller if additional query services are provided. These services are often used in WS interfaces to distribute part of the processing to the server side, since they could be executed at the client side after getting the data. The data gathered for the case study points out that flexible interfaces will yield high values of the LCOM metric, and the provision of additional query services make these figures lower. However, less flexible interfaces would yield better cohesion figures but loosing entity-based flexibility.

Future work will deal with measuring the actual impact of additional communication overload in these interfaces as well as defining new metrics that combine trade-offs between the different properties studied in this paper.

Acknowledgements. This research was supported by grant number CCG07-UAH-TIC-1588, jointly supported by the University of Alcalá and the autonomic community of Madrid (*Comunidad Autónoma de Madrid*).

References

1. Chidamber, S.R., Kemerer, C.F.: A metrics suite for object-oriented design IEEE Transactions on Software Engineering 20(6), 476–493 (1994)
2. Fenton, N.E., Pfleeger, S.L.: Software Metrics: a Rigorous & Practical Approach, International Thompson Press (1997)
3. Henderson-Sellers, B.: Object-Oriented Metrics Measures of Complexity. Prentice Hall, Upper Saddle River (1996)
4. McCabe, T.J., Watson, A.H.: Software Complexity. Journal of Defense Software Engineering 7(12), 5–9 (1994)
5. Oh, S., Fox, G.C.: Optimizing Web Service messaging performance in mobile computing. Future Generation Computer Systems 23(4), 623–632 (2007)
6. Perepletchikov, M., Ryan, C., Frampton, K.: Comparing the Impact of Service-Oriented and Object-Oriented Paradigms on the Structural Properties of Software. In: Meersman, R., Tari, Z., Herrero, P. (eds.) OTM-WS 2005. LNCS, vol. 3762, pp. 431–441. Springer, Heidelberg (2005)
7. Tian, M., Voigt, T., Naumowicz, T., Ritter, H., Schiller, J.: Performance considerations for mobile WS. Computer Communications 27(11), 1097–1105 (2004)
8. Tiwari, A., Sekhar, A.K.T.: Workflow based framework for life science informatics. Computational Biology and Chemistry 31(5-6), 305–319 (2007)

Workshop on Collaborative and Cooperative Environments

Collaborative and Cooperative Environments

Christoph Anthes[1], Vassil Alexandrov[2], Dieter Kranzlmüller[1],
Jens Volkert[1], and Gerhard Widmer[3]

[1] GUP, Institute of Graphics and Parallel Processing
Johannes Kepler University, Altenbergerstraße 69, A-4040 Linz, Austria
[2] Centre for Advanced Computing and Emerging Technologies,
The University of Reading, Reading, RG6 6AY, United Kingdom
[3] CP, Institute of Computational Perception
Johannes Kepler University, Altenbergerstraße 69, A-4040 Linz, Austria

1 Motivation

Technological advances in high-speed networking and computational grids do not only transform the methods applied to everyday science, but also the collaboration and cooperation between scientists at almost arbitrary locations around the world. The additional provision of multi-sensory, immersive Virtual Reality interfaces as tools to improve the collaboration between groups of human users is another hot topic in this research domain, which will most likely increase the potential benefits of these distributed research communities. The vision to facilitate large scale, complex simulations, which may be steered through natural and intuitive interfaces, is both intriguing and of high scientific interest.

This workshop at the ICCS 2008 in Cracow comprises the best submitted papers concerning the application and usage of collaborative and cooperative environments, as well as the technologies supporting them in the scientific and industrial context. The workshop on collaborative and cooperative environments has offered the possibility to discuss the different approaches in this domain, to show the latest results, products, or research prototypes to potential users, and to establish connections between developers and users of associated technologies. The attendants were asked to present and discuss the following technologies:

- Collaborative and cooperative tools and environments
- Development of associated parallel and distributed computing solutions
- Integration of networking and grid computing technology
- Provision of multi-sensory, natural and intuitive interfaces
- Immersive Virtual and Augmented Reality approaches
- Requirement studies for future collaboration tools
- Evaluation of existing collaboration environments and practical experiences

Each of the submitted papers has been reviewed by at least three international referees in this domain. The highest ranking contributions are presented here.

M. Bubak et al. (Eds.): ICCS 2008, Part III, LNCS 5103, pp. 379–380, 2008.
© Springer-Verlag Berlin Heidelberg 2008

2   Overview of Contributions

The paper by Choiński et al., *Multi-Agent System for Collaboration in Hybrid Control*, investigates an ontology-based multi-agent system augmented with web engineering for the for validation of hybrid control of biotechnological plant. In order to acquire knowledge and configure the multi-agent system, the data gathered from the web is reduced, filtered and validated.

The problem of geographically distributed software engineering is addressed by Penichet et al. in their contribution *Design and Evaluation of a Service Oriented Architecture-based Application to Support the Collaborative Edition of UML Class Diagrams*. The authors present a collaborative CASE tool called CE4WEB to support the edition of UML diagrams using the Service Oriented Architecture paradigm.

Another contribution from the domain of software engineering, *g-Eclipse - a contextualised framework for Grid users, Grid resource providers and Grid application developers*, by Kornmayer et al. introduces an eco-system to access Grid infrastructures with support for contextualised user roles. The abstraction layer of the g-Eclipse system, its integration in the Eclipse framework and the main use cases are presented.

Zuzek et al. propose a model to support the process of setting up a Virtual Organization (VO) in their paper *Formal Model for Contract Negotiation in Knowledge-based Virtual Organizations*. They discuss in detail the formal model underlying the process of contract negotiation and how the ontological description of domains related to given VO supports this process.

In *An Approach for Enriching Information for Supporting Collaborative e-Work*, Anya et al. combine latent semantic analysis, domain task modelling and conceptual learning to enrich information in order to support collaborative e-Work. They illustrate their approach using the prototypical e-Workbench system.

The contribution by Dunk et al., *Dynamic Virtual Environments Using Really Simple Syndication*, investigates the use of really simple syndication (RSS) to dynamically change virtual environments (VE). Instead of simulating weather conditions in training scenarios, actual weather conditions can be incorporated, improving the scenario and immersion. This weather data is gathered by incorporating an up-to-date RSS feed in the VE.

Jamieson et al. present another contribution from the Virtual Reality domain. In *Immersive Co-operative Psychological Virtual Environments (ICPVE)* they discuss their approach to develop a variety of different applications from the field of psychology, using a single framework.

In the paper *Environment for collaborative development and execution of virtual laboratory applications* by Funika et al., a user interface system is introduced which enables collaboration between developers and users, to improve experiments and introduce more refinements to the research conducted within the ViroLab Virtual Laboratory.

Multi-Agent System for Collaboration in Hybrid Control

Dariusz Choiński, Witold Nocoń, and Mieczyslaw Metzger

Faculty of Automatic Control, Electronics and Computer Science,
Silesian University of Technology,
ul. Akademicka 16, 44-100 Gliwice, Poland
{dariusz.choinski,witold.nocon,mieczyslaw.metzger}@polsl.pl

Abstract. Research and design in novel branches of technology usually require experts with different skills to cooperate on the same project. Participation of external experts is crucial for the best possible results of any significant experimentations and research. Successive iterations in these tasks need flexible validation. An ontology-based Multi-Agent System augmented with Web engineering is proposed for validation of hybrid control of biotechnological plant. For acquiring knowledge through Web, the proposed system carries out reduction, selection and validation of data and of its structures and then reconfiguration of multi agent hybrid control system.

Keywords: MAS, ontology, hybrid control, validation, scalability and reconfigurability.

1 Introduction

Research and design in novel branches of technology usually require experts with different skills to cooperate on the same project. For example, designing state-of-the-art biotechnological processes requires not only a deep biological and chemical understanding of the phenomena employed, but also an integration of the process with control and computer systems that are used to monitor and supervise the process. Hence, collaboration between biotechnological engineers or scientists and control or computer engineers is crucial.

Design and development, as well as operation and control of modern industrial processes are characterised by no common understanding and terminology related to these tasks (see e.g. [1]). On the other hand, research and design quite often require continuous iterations (changes) during experimentations or design phase. An additional, remote expert can be helpful in both design and operating control with an application of Internet (see e.g. [2]). In such activities, better understanding between humans and information systems within the net is expected. These requirements can be fulfilled using Semantic Web (see e.g. [3]). The fundamentals of such kind of understanding are based on ontologies, with appropriate rules (see e.g. [4]). Ontologies are simple for discrete systems because of being based on discrete logic, however, for continuous processes with infinite number of states when the problem is more difficult, creating ontology may be possible in some cases [4], [5]. Synthesis of ontology for physical processes is similar to synthesis of phenomenological models when object-based description is used.

M. Bubak et al. (Eds.): ICCS 2008, Part III, LNCS 5103, pp. 381–388, 2008.

Using object-based description with decomposition to subsystems and functions, it is possible to treat subsystems as hybrid automata [6], [7] and to apply flexible control of hybrid systems [8], [9]. The major advantage of a hybrid system deals with its particular proprieties when event-driven part (hybrid automaton subsystem) and time-driven part (continuous state equations subsystem) can be corrected separately without disturbing the whole system. Hybrid control facilitates recognition of discrete states required by technology. If it is possible to define those states and technology rules, than binding of those states by ontology is also possible. In this case a Multi Agent System (MAS) iterative improvement of hybrid control is applicable (the MAS notion is well-established in the scientific domain – see [10] as an example of general description and [11] of an industrial applications). In this paper such kind of hybrid control for a biotechnological process is proposed, developed and evaluated.

The proposed hybrid control system compiles techniques discussed above. Description of industrial equipment and instrumentation as well as automatic control deals with hierarchical description – even for small plants it has a very complicated structure which is difficult to survey. On the other hand, browsing of database is based on relation models. That is why, in most cases, these tasks are incoherent. Our contribution includes: a) a mechanism that reflects dynamical hierarchical structure in a relational database, b) possibility for browsing the process data by the remote expert, with the hierarchical structure of the process reproduced from the relational database. It can be also noticed, that for acquiring knowledge through Web, the proposed system carries out reduction, selection and validation of data and its structures. On the basis of such information, it is possible to take appropriate operating decisions (for example choosing a sequence transition to a new process state). Such operations should be treated as a particular kind of scalability and reconfigurability of a controlled process under consideration because, according to external demands, the process can change its structure, number of states and configuration of agents.

The paper is organised as follows. Sect. 2 provides a brief description of a multi agent system used for control. Sect. 3 provides formalisms for ontology-based MAS. The proposed implementation is presented in sect. 4. Concluding remarks are presented in sect. 5.

2   Hybrid-Hierarchical Control of Biotechnological Process

The biotechnological pilot-plant designed and operated at the Faculty of Automatic Control, Electronics and Computer Science serves as a platform for investigations regarding activated sludge process in aquatic environment. Depending on the needs of the researching team, the structure of the biological process involved may be changed. For example, the plant may be operated as a continuous or sequencing activated sludge process, the later involving cyclic utilization of the biological reactor for reaction and settling phases, the former involving continuous sedimentation of activated sludge in the settler with recycle of the thickened sludge back to the reactor.

Those possibilities of different modes of process operation make it necessary to automatically move the process state into a different region. Therefore, the considered object model is described as a state machine augmented with differential equations, namely as a hybrid automaton. Because of the complexity of a biotechnological process model, it is convenient to split the model into smaller operating entities that are associated with a particular, locally valid subsystem.

The necessity of state transitions, partitioning of the process model and the complexity of the whole controlled process makes the application of MAS for control purposes justified. Fig. 1 illustrates this problem, where Ω denotes a set of objects states. For example, the object state may be changed from continuous control (Ω_i) to discrete control (Ω_{i+1}) for changing biomass concentration and identification of OTR (oxygen transfer rate) and return to continuous control (Ω_{i+2}) of object with additional function based on calculated OTR coefficient for better dissolved oxygen control.

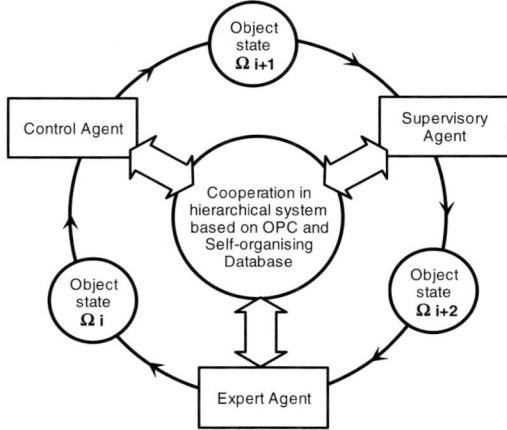

Fig. 1. An example [2] of object states transition in hybrid hierarchical and multi-agent control system

Structure of the control system is hierarchical, and consists of the following hierarchically dependent agents:

- Control agent – implementing all the closed-loop and open-loop control algorithms, hence this part of the control system is time-driven, e.g., the measurements are read and controls are transmitted to the plant in fixed time intervals regardless of the process behaviour.
- Supervisory agent – responsible for general supervision of the process performance and basic automatic compensation against process fluctuations
- Expert agent – provides remote expert knowledge in case of some off-nominal situations than can not be dealt with by the supervisory or control agent.

Because the presented control system has been designed hierarchically and the control scheme is hybrid in nature, the whole system may be considered as agent-ready, that is, application of MAS should not require any additional adaptation of the system. This structure will also enable the control system to be easily adapted in case of the system's expansions, modification and scaling.

3   Design of a Multi-Agent Control System

Formalism of the technological projects, the DCS (Distributed Control System) architecture and software is based on, is commonly realized by CAD software compatible with the IEC 61346 standard. This standard specifies rules for structuring and reference designations. The structure defines dependencies and relations between objects that are considered in the process of designing, construction, realising, operation, exploitation and disposal. A set of mutually connected objects is a system. Information about objects and about the system may be chosen based on different aspects. Therefore, the structure of the system and of the particular objects may be described in many different ways. The standard provides three examples of structures: function-oriented, location-oriented and product-oriented. The particular structures are organized hierarchically and should specify the information regarding the system, content of the particular documents and composition of reference designations.

The functions hierarchy tree enables engineers of different specializations to work on the project using a consistent naming convention, hence reducing organizational effort. In addition, the CAD software used for the reference designations creation, enables automatic update of this functions hierarchy tree in case of the system's expansions, modification and scaling. The presented hierarchy is also used to specify the OPC structure of the information needed for the implementation of control algorithms. OPC technology is selected because it is versatile and commonly used in process automation systems. Based on component ontology, different subsystems of the automated process are distinguished.

Apart from the hierarchy of functions, synthesis of control algorithms for the automated process requires additional information (Fig. 2). First, knowledge about the process behaviour, both static and dynamic must be considered in order for a correct control algorithm to be synthesised. Such knowledge is usually represented by phenomenological models existing as a set of ordinary or partial differential equations extended with conditions for changing the process state by certain transitions. Hence, as was discussed in sect. 2, the system is modelled as a hybrid automaton. The taxonomy of functions based on phenomenological models, together with the particular component ontology, serves as a basis for the needed functions to be implemented.

Additionally, knowledge about the process and control equipment boundary conditions and constraints must be taken into account. Control algorithms must not be created regardless of those constraints and conditions. Therefore, once a particular subsystem is distinguished, a deterministic finite state automaton is created that takes the boundary conditions and constraints into account.

Both the taxonomy-based functions and the knowledge about constraints serves as a basis for a hybrid control system realization. One should note, that because the hierarchy of functions based on the IEC 61346 standard [12] is automatically updated in case of system expansions or modifications (realized by the CAD software), update of the control system software and/or the control system architecture is facilitated. Easy scaling of the control system is therefore enabled.

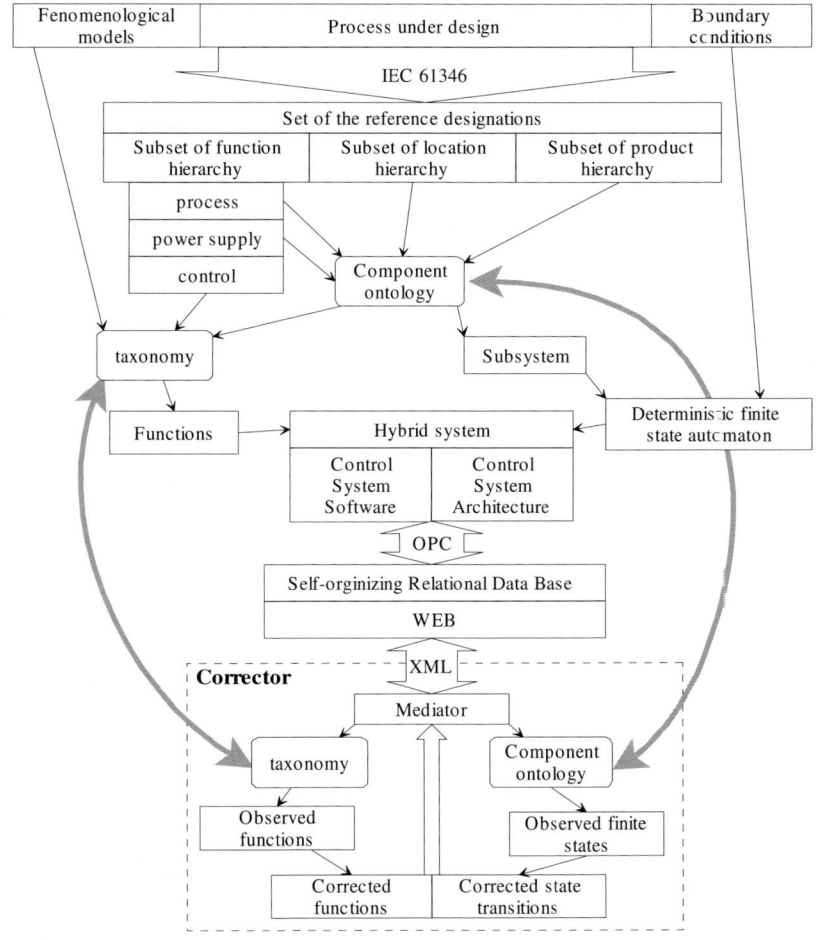

Fig. 2. System architecture

4   Implementation

Implementation of control system architecture, algorithms and software for a biotechnological process requires collaboration of different specialists. Once a preliminary control system architecture is created, distant experts my be employed to correct this system in a proper way. Fig. 3 presents a general idea of a remote corrector taking part in designing and testing of the control system.

The Corrector application enables reading of the hierarchical structure of items in database, selection of those that are interesting for the Corrector, monitoring of current and historical data and the realization of control algorithms that utilize this data. Mechanisms for establishing connection to the database, sending the defined user configuration and sending values of items with write-to-OPC-server request into the

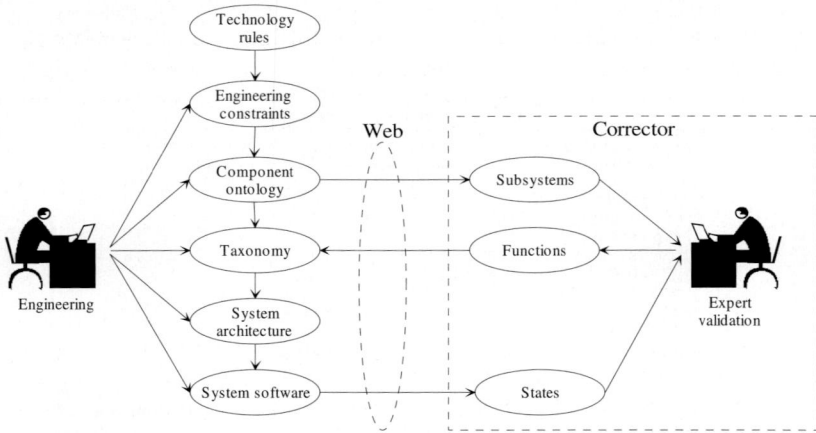

Fig. 3. Corrector realization in the Web environment

database, are also implemented. Visualization and coupling of dynamically loaded clips to the measurements is realized as well.

The most crucial element of the Corrector is a Mediator [13] which, by gathering knowledge about the plant and by gathering data, ensures limitation of basic errors resulting from a bad understanding of structure and states of the system (refer to Fig. 3).

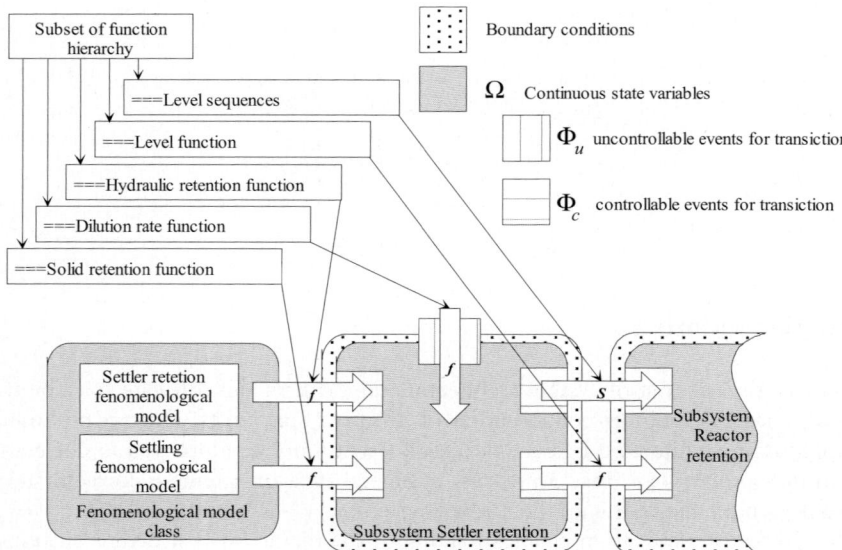

Fig. 4. Taxonomy for model and functions

The taxonomy is a set of tools that assure encapsulation and polymorphism of phenomenological models (Fig. 4). Encapsulation should ensure division of model properties into the private part that is enclosed in the subsystem parameter range, and into the public part that defines interactions between other subsystems. At the same time, formalized methods exist that enable modifications of model properties. Consecutive derived classes correspond to models. Methods and properties of the base class are virtual. Private properties are only modified by methods. The base class, designed as an abstract class, will enable a polymorphism-based design of controls utilizing models represented by objects. This enables the assignment of the base class address to the derived model of a particular object.

In order to avoid the need for creation of a virtual constructor of the derived class, the base class possesses a cloning method that clones the derived-class object. The abstract base class may for example represent a class of models that are based on the mass balance and reaction rate. The parameters of this type of models are encapsulated so that only the allowed procedures (methods) may change those parameters. Every consecutive type of models (a class) is created using the inheritance mechanism, hence by creating child classes. On the other hand, an object of a particular derived-class is created as a pointer (an address) to the base class (polymorphism). Because of that, an extended class of models may by used in the distributed hierarchical system. The major reason for such description is the possibility of utilizing programming tools that enable creation of distributed, object-oriented applications, for example the EN 61499 [14] compatible CAD software.

5   Concluding Remarks

The biotechnological pilot-plant designed and operated continuously for three years at the Faculty of Automatic Control, Electronics and Computer Science serves as a platform for investigations regarding activated sludge processes in aquatic environment. Depending on the particular set of control algorithms applied, the structure of the biological process involved may be changed. For example, the plant may be operated as a continuous or sequencing activated sludge process, the later involving cyclic utilization of the biological reactor for reaction and settling phases, the former involving continuous sedimentation of activated sludge in the settler with recycle of the thickened sludge back to the reactor.

Such kind of operations are difficult tasks. Hence, a participation of external experts in process operations as well as in equipment corrections is crucial for successful maintenance of the biological processes and for the process improvement. The proposed system evidently enables these tasks. The proposed system has been beneficially functioning for two years. Therefore, future work will include adaptation of this system to other types of continuous processes.

Acknowledgements. This work was supported by the Polish Ministry of Science and Higher Education, using funds for 2006-2008, under grant no. N514 006 31/1739.

References

1. Marquardt, W., Nagl, M.: Workflow and information centered support of design processes—the IMPROVE perspective. Computers and Chemical Engineering 29, 65–82 (2004)
2. Choiński, D., Nocoń, W., Metzger, M.: Multi-Agent System for Hierarchical Control with Selforganising Database. In: Nguyen, N.T., Grzech, A., Howlett, R.J., Jain, L.C. (eds.) KES-AMSTA 2007. LNCS (LNAI), vol. 4496, pp. 655–664. Springer, Heidelberg (2007)
3. Berners-Lee, T., Hendler, J.: Publishing on the semantic web - The coming Internet revolution will profoundly affect scientific information. Nature 410(6832), 1023–1024 (2001)
4. Borst, P., Akkermans, H., Top, J.: Engineering ontologies. Int. J. Human – Computer Studies 46, 365–406 (1997)
5. Morbach, J., Yang1, A., Marquardt, W.: OntoCAPE—A large-scale ontology for chemical process engineering. Engineering Applications of Artificial Intelligence 20, 147–161 (2007)
6. Leduc, R.J., Lawford, M., Dai, P.: Hierarchical Interface-Based Supervisory Control of a Flexible Manufacturing System. IEEE Transactions on Control Systems Technology 14(4), 654–668 (2006)
7. Lynch, N., Segala, R., Vaandrager, F.: Hybrid I/O Automata. Information and Computation 185, 105–157 (2003)
8. Cassandras, C.G., Pepyne, D.L., Wardi, Y.: Optimal control of a class of hybrid systems. IEEE Transactions on Automatic Control 46(3), 398–415 (2001)
9. Schaft van der, A.J., Schumacher, J.M.: Compositionality issues in discrete, continuous, and hybrid systems. International Journal of Robust and Nonlinear Control 11(5), 399–539 (2001)
10. Wooldridge, M., Jennings, N.R.: Intelligent agents: theory and practice. The Knowledge Engineering Practice 10(2), 115–152 (1995)
11. Marik, V., McFarlane, D.: Industrial Adoption of Agent-Based Technologies. IEEE Intelligent Systems, 27–35 (January/February 2005)
12. International Electrical Commision: IEC 61346-1, Industrial Systems, installations and equipment and industrial products - Structuring principles and reference designations, 1st edn. (1996)
13. Wiederhold, G.: Mediators in the Architecture of Future Information Systems. IEEE Computer 25, 38–49 (1992)
14. IEC Technical Committee TC65/WG6, IEC61499 Industrial Process Measurement and Control – Specification IEC Draft (2000)
15. Kraska, J.: An Internet-Based Mobile Control Algorithms. Master Thesis, Institute of Automatic Control, Silesian University of Technology (2006) (in polish)

Design and Evaluation of a Service Oriented Architecture-Based Application to Support the Collaborative Edition of UML Class Diagrams

Victor M.R. Penichet, Jose A. Gallud, Ricardo Tesoriero, and Maria Lozano

LoUISE Research Group – Computer Science Research Institute (I3A), Castilla-La Mancha University, 02071 Albacete, Spain
{vpenichet,jgallud,mlozano,ricardo}@dsi.uclm.es

Abstract. Developers in modern and geographically distributed software companies need to collaborate because most of them are part of big development teams involved in large projects. CASE tools are commonly used to model software applications. However, most of them are not really prepared to support collaboration in the sense that they do not provide developers with a real collaborative environment. In this paper, we show the design and the implementation of a cooperative CASE tool called CE4WEB to support the edition of UML diagrams using the Service Oriented Architecture (SOA) paradigm. The tool described in this paper demonstrates the successful use of groupware tools and Software Engineering techniques. An evaluation method is also presented to show the validity of the proposal.

Keywords: groupware, SOA, collaborative CASE tool.

1 Introduction

Nowadays, software development tends to be decentralized; online collaboration among different development teams is something that software companies wish to incorporate in their development environments.

Positive effects of the globalization phenomenon are, among others, the possibility of configuring distributed development teams belonging to different countries that collaborate in the same project. SourceForge or Linux are good examples of the distributed software engineering [8].

CSCW research field helps people to work better by means of computers. Software development is one of the scenarios where groupware may improve the way programmers work together providing collaborative tools and methods [7].

In this paper, the design and the implementation of CE4WEB (Cooperative UML Editor for the Web) is presented. CE4WEB is a cooperative CASE tool that supports the edition of UML [12] diagrams. This tool demonstrates the emerging use of groupware tools and techniques in software engineering.

The structure of the paper is as follows: Section 2 describes how collaboration is supported by the common CASE tools. Section 3 describes the design and the implementation of our proposal. Section 4 shows how the tool works in real cases. Section 5

M. Bubak et al. (Eds.): ICCS 2008, Part III, LNCS 5103, pp. 389–398, 2008.

describes the evaluation method we have performed to validate the tool and Section 6 shows the metrics applied and the results obtained from the experiment. Finally, some conclusions and future works are presented in Section 7.

2 Collaborative CASE Tools: An Overview

Traditional CASE tools provide some collaborative functionality; therefore we can consider them as collaborative tools. However, from the CSCW research field point of view, most common CASE tools do not provide a real cooperative scenario for software development. In this section, we describe the way in which these tools support collaboration in the modeling process of software applications with UML.

Modelling provides the first step to improve workgroup, though market demands more powerful tools to support UML modelling by combining the functions of traditional UML applications with the advantages of real time collaboration [9].

The evolution of UML modelling tools covers these stages: standalone tools, repository-based model sharing, Web-based model sharing, Real-time model sharing and collaboration. The main motivation of our work is the lack of UML editors specially designed to support collaboration. Many authors identify this lack as one of the future needs of CASE tools [1, 2, 5, 13].

Rational Rose supports some kind of collaboration by allowing the parallel development of a model by dividing it in versionable units, sharing models among different teams and a complete integration with common version control tools. The Teamwork release of Visual Paradigm for UML offers a similar support for version controlling. Microsoft Visio offers collaboration support by means of SharePoint Portal Server to publish and share diagrams among the different users. EclipseUML is integrated inside the Eclipse framework. Team's members can access to the artefacts from different stages of the development at any time. Enterprise Architect for UML is one of the most powerful tools from the team support point of view. This environment offers the usual version control system and other additional functions to maintain all members of the team informed and connected with the project. PoseidonUML and Konesa are well-known environments that offer an advanced collaboration support for development teams. They supply advanced functions as real-time collaboration, versioning control, central repository, news board, etc.

All these tools have common features: support for UML diagrams, reverse engineering, code generation, version controlling, and so on, but collaboration support is poor or incomplete. CO2DE [11] is the tool we have used as reference. It is a cooperative class diagram editor where users can work simultaneously in the same diagram keeping a version list.

Some authors establish that the main collaborative features that are missed in these known tools [3] are the following:

- Collaboration: Although PoseidonUML and Konesa support collaboration, most of them do not provide real-time collaboration.
- Perception: The tools revised do not support user perception of the work developed by other members of the team.

- Communication: Users communication is not supported.
- Workgroup memory: There is no support for storing the information that the workgroup produces when different people are working in the same project.

3 Designing and Implementing the Collaborative Editor

In this section, we describe the design and implementation of our cooperative editor called CE4WEB. The description is focused on the cooperative aspects of the system.

As a result of the former analysis of the most common UML tools, we have defined the following list of the most important requirements that a cooperative CASE tool should include:

- Allow the geographical distribution of users working in the same diagram.
- Support real time cooperation (synchronous working).
- Support informal and formal communication.
- Store and share a collective memory.
- Provide a way to let users know the work developed by others (awareness).

The first requirement made us to select the Web platform, as some authors recommended [4, 6]. In this way, we can support the distribution of users and their need of synchronous and asynchronous collaboration. The second requirement implies the use of a concurrency control system which includes coordination services and data consistency. Regarding the third requirement, we decided to include the following units: an alarm and notifications system, an asynchronous communication system (electronic mail), a comment and annotations system and a synchronous communication system (chat). The fourth requirement implies the use of a repository or a group memory space implemented by a versioning system. The tool also provides a log file to store all users' actions. And the last requirement can be reached by providing a list of participants, a friendly interface where the changes introduced by any user are immediately reflected on the interface. We have also decided to apply the telepointer technique used in CO2DE [11].

The system should also accomplish a series of non functional requirements such as portability, easiness, low response time, availability, and generality (not only for UML diagrams).

The main components of the system are the following. An UML Editor that allows the creation and modification of UML class diagrams in a cooperative way, a component responsible for the synchronous cooperative process of the UML Editor (SCS), a reduced version of a control system that supports maintenance and management processes (Versioning Control System, SCV), the user management system responsible for controlling and maintaining users and workgroups (SGU), and a system to control the access to the Database containing both the application data and the information generated the by workgroups (CAD).

The general view of the system architecture containing the components described above is shown on Fig. 1.

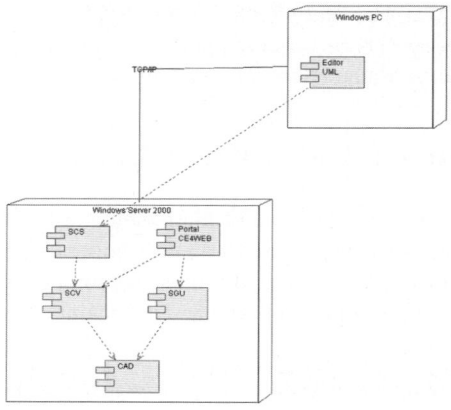

Fig. 1. System architecture

4   Collaborative Editing of UML Diagrams: CE4WEB at Work

We have followed the typical schema of three-layer architecture in the implementation of the system. The layout layer is implemented by the UML Editor and the Web
Application. The business layer is implemented by the Users Management System,
the Versioning Control System and the Synchronous Collaboration System. The data
layer is implemented by the Data Access Component and the database.

The UML Editor allows a real-time interaction among multiple users to create or
modify an UML class diagram. The editor offers two main behaviours: as a UML
editor, it offers a limited functionality compared to Rational Rose or MS Visio; as a
collaborative tool, it offers a shared blackboard, a list panel of connected users, a list
panel of actions, the position of the users' pointers, a computer log for actions and
changes, and a chat for communicating.

A group of users that want to work together in the same project have to log on the
system. The *User and Group System* offer users the possibility of organizing themselves in groups sharing resources and a common space. This part of the system is
responsible for group and security management.

When users log on the system, different options are offered to them. Users can access a previous project or create a new one. Users can see the information about the
group they belong. An interesting point here is the possibility for a user to join an
active session and cooperate with other users in the same group or starting a new
session.

The *Versioning Control System* is responsible for offering a shared workgroup
memory. The information and artifacts are organized in the following way: a project
is composed by a set of diagrams. Each diagram has associated a set of versions. A
version is created in a session. A new version is generated when the diagram is modified. Fig. 3 shows a versioning tree, where each node represents a different version of
a diagram and the number represents the order in which they were created.

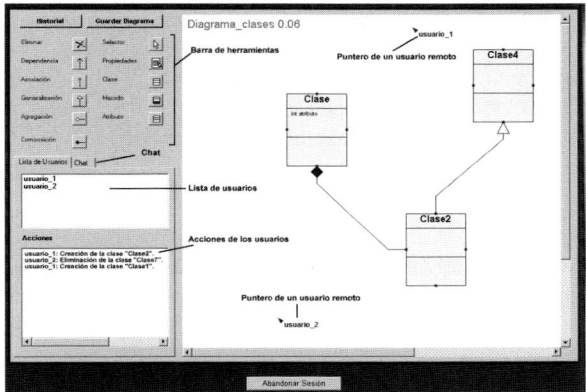

Fig. 2. UML editor's interface

A user can see the information of a specific version by selecting one of the nodes. The information of a version is shown in Fig. 4. This figure has several tabs containing different functionality; reviewing the status of the diagram, showing the list of users' actions, showing a list of the chat's posts, and the users' annotations.

The *Synchronous Collaboration System* supports the communication, coordination and information sharing processes that allow users to work synchronously when they are editing an UML diagram. This component guarantees shared data integrity and is responsible for propagating users' actions efficiently.

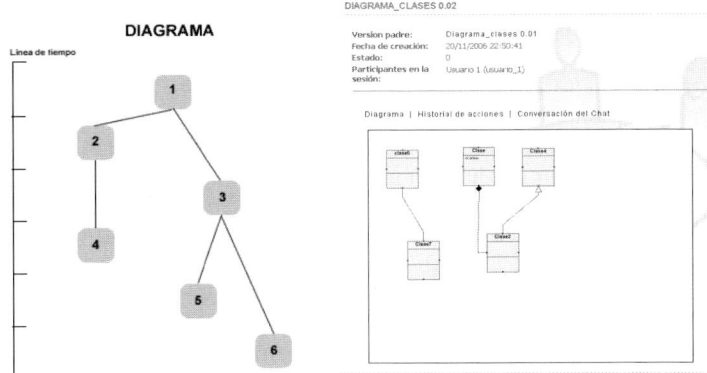

Fig. 3. Information page about diagram versions (right) and version tree (left)

The communication between the Web client and the server is implemented by means of Web services. In this way, the UML collaborative editor takes advantage of the SOA paradigm.

The CE4WEB has been developed with C# using a special library called *Netron Graph*. The client application does not need any *ActiveX* or *COM* component but it uses a technology called Fusion that allows embedding *WinForm* controls in a HTML page. This technology is similar to *Java Applets* or *JavaBeans*. These technologies can be used to build smart Web clients. In this case, the UML editor is encapsulated in a DLL file called *ControlUML.dll* and it is stored in the server, together with the other application components. The assembled is embedded in the Web page *Editor.aspx* using the *<object>* HTML tag.

```
<OBJECT id="EditorUML" standby="Loading UML Editor..."
classid="http:ControlUML.dll#ControlUML.Control_UML"
VIEWASTEXT></OBJECT>
```

The *classid* parameter indicates: the name of the assembled: *ControlUML.dll* and the class implementing WinForm: *ControlUML.Control_*UML. When the *Editor.aspx* is invoked, the assembled travels together with the HTML code to the client browser and is executed in the client side.

5   Evaluating the Collaboration in the CE4WEB Tool

Several interesting quality aspects of the CE4WEB tool could be evaluated. We may divide the system in different layers and select some parameters to be measured in each layer. If we consider architecture, we should evaluate the tool performance by monitoring the CPU and network utilization among other parameters. On the other hand, we might consider collaboration aspects only and focus the evaluation on how the CE4WEB tool helps people to work together. General users' satisfaction should also be measured as an important quality dimension.

First of all, we had to define the objective of our evaluation (hardware, software, and architecture or usability aspects) and then we had to determine the evaluation method including the specification of the different experiments to be carried out and the metrics we should use in those experiments.

CE4WEB was designed to allow software designers to work in groups; therefore our main interest was to know if the tool actually helped people to work together in the same project, modelling the same diagram. Thus, assessing the collaborative issues of the CE4WEB CASE tool was the main goal of our evaluation.

The next step is to find out how to evaluate groupware users' satisfaction. It must be noted that user's satisfaction and group's satisfaction cannot be obtained in the same way, although it is possible to adapt some traditional HCI evaluation methods to test groupware systems.

Taking all these considerations into account, we describe the evaluation method we defined in order to assess the main aspects of groupware systems, as is cooperation, coordination, information sharing and collaboration together with the spatial and temporal dimensions.

Groupware main features were assessed during the process of modelling a unique medium-sized UML diagram. This artefact was concretely the class diagram corresponding to a booking system for sport facilities and other associated materials (rackets, balls, etc). The diagram has nine classes, nine relationships and a set of methods and attributes associated to each class. Users can introduce more classes and relationships if needed.

Regarding people involved in the experiment, we decided to test the CE4WEB tool using groups of 2, 3, 4, and 5 people. Testing groupware tools take an extra time because users have to learn previously the different tasks and the very tool as well. In order to reproduce a realistic scenario, we defined different roles on each experiment. Two roles were defined: Depicting classes, and depicting relationships.

Other tasks, as introducing methods, attributes and comments were left out of the roles and could be completed by whoever remaining idle, once completed the classes and the relationships.

After describing important aspects of the evaluation, it is time to present the evaluation method, which covers the following phases. Warming-up phase: Before beginning the experiment, each user learns the basic functions of the system in order to avoid measuring the user's experience instead of the degree of collaboration supported by the tool. In this phase users employed main functionalities of the CE4WEB tool. These functions were the same as the ones to be used during the experiment. Lab 1: Synchronous modeling of the UML diagram by a group of two people. Lab 2: Synchronous modeling of the UML diagram by a group of three people. Lab 3: Synchronous modeling of the UML diagram by a group of four people. Lab 4: Synchronous modeling of the UML diagram by a group of five people.

As it can be noted, all the experiments were carried out at the same time, which is known as synchronous cooperation. Regarding the spatial dimension, each user was in his/her office and all the experiments were performed on the Internet. It is important to remark this fact, because most of the known collaborative CASE tools only work in an intranet because of the chosen technology.

One important consideration we took into account when performing the experiment was the fact that users could only communicate by means of the functionality provided by the tool. This requirement was important in order to evaluate the quality of the groupware features supported by the tool itself without any external help.

Asynchronous experiment was not considered in this occasion, because conditions of synchronous collaboration are much harder and interesting than the asynchronous ones in order to assess the CE4WEB tool.

Next section presents the metrics employed and the outcomes of the experiment.

6 Presentation of Metrics and Results

The evaluation process started with the Warm-up phase. Each user was guided by a personal instructor who taught him/her the basic functionality of the CE4WEB tool. A correct instruction of users takes over 30 minutes and also some extra time for users to learn by themselves. To perform the different experiments, users received a list of 15 tasks that should be completed in 10 minutes.

Before presenting the results of the experiment, we will talk about the selected metrics to measure the quality degree of the cooperation in CE4WEB.

Users involved in the different experiments should write down the next values: Number of tasks completed in the defined period; Number and description of errors introduced, if any; and a list of suggestions and improvements and Fill in the satisfaction questionnaire (10 questions).

All users completed all tasks in the predefined time. Additionally, the more users were involved in the experiment the faster tasks were performed. This is an interesting aspect when it is needed to point out some benefits of groupware systems. In this case, we are talking about obtaining a higher throughput when the number of users that are working together is higher.

Some errors were detected during the execution of the experiments. The most important one is related to the need of using a communication mechanism from the very beginning, just when users log into the system. The CE4WEB tool offers a Chat tool but it is only available once users have entered to the drawing area to start the creation of the model.

The satisfaction questionnaire was defined using the recommendations of the ISO/IEC 9126-4, part 4 about Quality in Use [10]. The formula to measure users' satisfaction can be shown here:

$$Sq = \sum_{l=1}^{L} \frac{l}{L} \cdot \frac{x_l}{N} \tag{1}$$

Where Sq is the satisfaction level of question q, L is the amount of possible answers (in this case 5), x_l is the amount of people that answered l in question q, and N is amount of people that filled in the questionnaire.

The questionnaire was composed by ten questions evaluating whether the system is capable of performing the task, whether the system works properly, whether the tasks have been performed on time, there have been problems to perform tasks, the problems found could be solved, the cooperative task has been a richer experience than if it would have been done in an individual way, the communication among the members of the group was correct, the tool provides the sufficient cooperative functions needed to perform tasks in a cooperative way, the tool allows information exchange among the different users and whether the tasks have been performed synchronously.

The users' satisfaction degree is expressed in 5 levels. The satisfaction degree mapping is the following: 1-absolutely disagree, 2-disagree, 3-indifferent, 4-agree, 5-absolutely agree.

This questionnaire was filled in by 3 university professors and 3 PhD students related to the collaborative research field. The results obtained applying this expression to the questionnaire that the users filled in can be seen in Fig. 4.

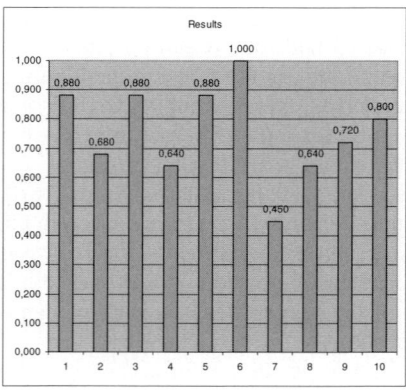

Fig. 4. Users' satisfaction results (values for each question)

Only in question number 6 users were completely agree with the system. This question was "The collaborative task has been a richer experience than if it would have been done in an individual way". On the other hand, we can note the very low value in question 7. This question was "The communication among the members of the group was correct" and this was due to the lack of suitable tools inside the tool to enable communication from the very beginning. This lack of communication systems before starting the drawing process made users to use additional communication tools to bypass some complicated tasks. A complementary view of the former results can be found in Table 1. In this case we can see the results of users' satisfaction according to each participant.

Table 1. Users' satisfaction by participant

Participant	1	2	3	4	5	Mean	Std. Dev.	Max	Min
Satisfaction	3.8	3.9	3.9	3.9	3.7	3.84	0.08	3.9	3.7

Analyzing Table 1 we can conclude that users are satisfied enough with the collaboration quality offered by the CE4WEB tool as the values obtained are near 4 that means "Agree". As it can be noted, the standard deviation is very low, what shows the coincident opinion among the different users.

7   Conclusions and Future Work

The wide use of broadband communications technology propitiates the growing of collaborative systems. This is particularly useful in the field of Software Engineering where developers and software companies need modern collaborative tools in order to face complex projects in a globalized world. Many experts have noted that most of the problems appearing in software projects are due to the lack of communication, collaboration and coordination among the members of the team.

This paper presents a cooperative editor to support the process of modelling UML diagrams called CE4WEB. This CASE tool supports the main CSCW features: coordination, communication and information sharing. This tool has been developed as a SOA application in order to cover the real needs demanding by development teams when they are physically distributed in different countries.

The tool has been evaluated using a special evaluation method defined to measure the quality of the collaboration, communication, coordination and information sharing provided by the tool. The evaluation outcomes demonstrate the validity of the implemented tool and point out the need of this kind of collaborative tools.

The paper shows also the effort put in the definition of the evaluation method. However it is possible to make a recurrent question about the validity of the method applied and additional tools should be introduced to prove that formally. This is part of the future work.

References

1. Altmann, J., Weinreich, R.: An Environment for Cooperative Software Development. Realization and Implications. In: Thirty-First Annual Hawaii International Conference on System Sciences, vol. 1, pp. 27–37 (1998)
2. Amin Farshchian, B.: A Framework for Supporting Shared Interaction in Distributed Product Development Projects. Department of Computer and Information Science. Norwegian University of Science and Technology (Norway) (2001), Online proceedings http://csgsc.idi.ntnu.no/2001/?page=proceedings
3. Borges, M., Araújo, R.M., Dias, M.: A Framework for the Classification of Computer Supported Collaborative Design Approaches. In: III CYTED-RITOS International Workshop in Groupware – CRIWG 1997, El Escorial (España), pp 91–100 (1997)
4. Borges, M.: Designing collaboration through a web-based groupware infrastructure. International Journal of Computer Applications in Technology 19, 175–183 (2004)
5. Cook, C., Churcher, N.: An Extensible Framework for Collaborative Software Engineering. Software Visualization Group, Department of Computer Science, University of Canterbury (Nueva Zelanda) (2003)
6. Goguen, J.A., Lin, K., Gea, M., Cañas, J.: Web-based Support for Cooperative Software Engineering. In: Proceedings International Symposium on Multimedia Software Engineering, Taipai (Taiwan), pp. 167–191(25) (2000)
7. Grudin, J.: CSCW: History and Focus. University of California. IEEE Computer, 27, 5, 19–26 (1994)
8. Herbsleb, J., Mockus, A., Finholt, T., Grinter, R.: An Empirical Study of Global Software Development: Distance and Speed. In: Proceedings of the 23rd International Conference on Software Engineering, ICSE 2001, pp 81–90 (2001)
9. Hurwitz Report: Collaborative UML Development. CanyonBlue Incorporated (2001)
10. ISO/IEC TR 9126-4:2004. Software engineering.Product quality.Part 4: Quality in use metrics. JTC 1/SC 7; ISO Standards. ICS: 35.080. Stage date: 2004-03-17
11. Meire, A.: Suporte à Edição Cooperativa de Diagramas utilizando Versões. Universidade Federal do Rio de Janeiro (2003)
12. Object Management Group. UML Superstructure Specification, v2.0 (2005)
13. Saeki, M.: Communication, Collaboration and Cooperation in Software Development - How Should We Support Group Work in Software Development? In: Proceedings of the 2001 Asia Pacific Software Engineering Conference, APSEC 2001, pp 12–20 (2001)

g-Eclipse – A Contextualised Framework for Grid Users, Grid Resource Providers and Grid Application Developers

Harald Kornmayer[1], Mathias Stümpert[2], Harald Gjermundrød[3], and Paweł Wolniewicz[4]

[1] NEC Laboratories Europe, IT Division, Rathausallee 10, 53757 St. Augustin, Germany
harald.kornmayer@it.neclab.eu
[2] Forschungszentrum Karlsruhe, Institut für wissenschaftliches Rechnen, Postbox 3640, 76021 Karlsruhe, Germany
[3] University of Cyprus, PO Box 20537, 75 Kallipoleos Str. 1678 Nicosia, Cyprus
[4] Poznan Supercomputing and Networking Center, 61-704 Poznan, ul. Noskowskiego 10, Poland

Abstract. As the future pervasive and ubiquitous computing environment will be composed of resources from local computing, Grid, SOA and Web infrastructures, the complexity of this distributed system will increase significantly. At the same time the end user wants easy and simple access to his computing environment while he receives more responsibilities i.e. for the composition and the management of the system. In order to perform his daily work, the end user needs a general workbench toolset which supports customisation and contextualisation for the user. The g-Eclipse framework offers an eco-system to access Grid infrastructures with support for contextualised user roles. Currently, the g-Eclipse framework includes contextualised perspectives for Grid end users, Grid resource provider and Grid application developers. The abstraction layer of the g-Eclipse system, its integration in the Eclipse framework and the main use cases are presented.

1 Introduction

Wide-scale distributed infrastructures, called Grids [1], emerged in the recent years to enable the sharing of geographically distributed, heterogeneous computing, storage and network resources. These infrastructures are connected by Grid middlewares which offer the basic services to interact with the underlying Grid infrastructure, but each middleware system follows a slightly different approach, although there is a trend towards interoperable, service-oriented implementations of Grid-services.

Grid providers can build "Virtual Organisations" on top of these infrastructures, which users can belong to, in order to solve complex problems. The benefit of such a general infrastructure for scientific and commercial applications was demonstrated by many Grid projects. But many of the potential users restrain

M. Bubak et al. (Eds.): ICCS 2008, Part III, LNCS 5103, pp. 399–408, 2008.

themselves from using the Grid because of the inherent complexity of using and interacting with Grid technologies. Furthermore, the Grid user is often limited to the "end user" role, but a user may also take the roles of providing resources, applications and/or services. A user of Grid resources does not cover only one role in only one context as in desktop computing, but many roles are aggregated in a single Grid user depending on the tasks he wants/has to fulfill [2]. Therefore, user-friendly and intuitive user interfaces are needed to make the look-and-feel of Grid infrastructures similar to that of existing computer desktop systems. Additionally, these frameworks must provide support for contextualisation and customisation for the Grid user to support his daily work.

The g-Eclipse project [3] currently develops such a framework for different Grid roles and contexts. As it relies on the Open Source framework Eclipse [4], the g-Eclipse project delivers extensions for the Eclipse workbench to integrate Grid resources and to support the different Grid roles. The future g-Eclipse framework aims to devise a middleware independent, integrated workbench toolset to enable contextualisation for different Grid user roles.

Results of this work are presented in this paper which is structured as follows: The different roles and contexts are described in Sec. 2. A short overview of the Eclipse framework is given in Sec. 3. In Sec. 4 the g-Eclipse architecture, the abstraction layer for the Grid resources and its integration into the Eclipse framework are presented. Sec. 5 describes a number of tasks that a user, operator or developer can perform using the g-Eclipse framework. Sec. 6 concludes.

2   Grid Roles and Contexts

Grid users act on service-oriented infrastructures in different roles and contexts. An analysis of these roles and contexts build the base for the design and development of a contextualised Grid workbench.

2.1   Grid Roles

The following main roles of Grid users have been identified within the g-Eclipse project.

Grid application users want to interact with Grid resources in the same manner as with local resources to perform their daily work. I.e. they start applications and monitor the progress of the submitted jobs. The access to their distributed data (i.e. opening, copying, moving, renaming, visualising, ...) is another important action they need to perform.

Grid operators manage the distributed Grid resources. This includes the configuration of the infrastructure and services as well as the monitoring, testing and benchmarking of Grid resources.

Grid application developers develop applications for Grid infrastructures including compiling, debugging and deployment of applications and services. They want the freedom to develop their applications with their preferred programming language.

The interaction with Grid resources should not differ from the interaction with local resources for these three Grid user roles.

2.2   Grid Contexts

Even with different roles, a Grid user can interact with Grid infrastructures in different contexts. The following contexts had big impact on the design of the g-Eclipse framework.

Virtual Organisations (VO): As Grids are seen as a tool for distributed collaborations, users organise themselves and their resources into Virtual Organisations (VO) in order to improve scalability and manageability while addressing security issues. Of course, a single user can belong to multiple VOs. A resource provider allocates his resources on a VO basis instead of a per actor basis to improve scalability. This means, that resources are supplied to VOs based on some service level agreements and not to individual actors. A VO can be seen as a virtual brace around the distributed resources and services which were connected to collaboratively solve a task. Membership in an Virtual Organisation is managed by some kind of membership service. Based on different access roles in a VO and a given membership of the users, the access to the distributed resources is managed. This collaborative aspect of Grid infrastructures is seen as very essential for the design of the g-Eclipse framework.

Projects: In his daily work, a user organise his tasks and subtasks in projects and folders. The Eclipse framework follows this approach by organising files and configurations within a *project* in the workbench. The project is a placeholder for all possible interactions which an actor initiates in a specific context. This approach would be helpful for Grid users too, as not all tasks can be organised in a common Grid workspace. I.e. an Grid user can have multiple research projects that are using resources from different VOs at the same time. It was envisioned that the information belonging to one research project should be organised into a Grid project which includes a description of the VO assigned for this project. All resources and artefacts of a Grid project (data files, job descriptions, information about submitted jobs, . . .) should be collected within the scope of one project.

3   Eclipse as Underlying Platform

To reach the goal of a reliable, contextualised framework for Grid users, the g-Eclipse team decided to reuse the Eclipse platform [4]. The Eclipse framework was designed as an open platform for a wide range of tools and the initial contribution was an integrated development environment for Java. The central point of the Eclipse architecture and framework is its plug-in architecture, a component-based software architecture that leads to a clear and modular design [5]. In the Eclipse world, every plug-in amends the functionality of other plug-ins. This is achieved by the underlying OSGi [6] framework that defines the dependencies between the different plug-ins, and how and when additional plug-ins are loaded. In addition, the Eclipse framework relies on the mechanisms of extension points and extensions. An extension point is a definition of how to enhance existing functionality. This way of building software components leads to an extensible architecture with well-defined interfaces.

The graphical front-end of Eclipse is organised with the concepts of views, editors, wizards, preference pages, etc. These components provide the basic functionality to integrate new GUI elements into the framework. These basic elements are grouped in so called perspectives. Perspectives determine the visible actions and views within the workbench, but go well beyond this by providing mechanisms for task oriented interaction with resources in the Eclipse Platform. Users can rearrange their workbench and therefore customise it to their needs and habits with the help of these components.

4   g-Eclipse Architecture

The overall architecture of the g-Eclipse framework is shown in Fig. 1. Based on the Eclipse framework, g-Eclipse provides a core Grid model [7] for the integration including abstract implementations. The user interface (UI) components of the g-Eclipse framework will reuse the components of the Eclipse framework (i.e. Views, Editors, Wizards, ...). Middleware specific implementations can be build on top of the g-Eclipse core model and will reuse the UI components of the g-Eclipse framework. This architecture enables the g-Eclipse framework to be middleware independent and to deliver two middleware specific implementations for gLite [8] and GRIA [9] up to now.

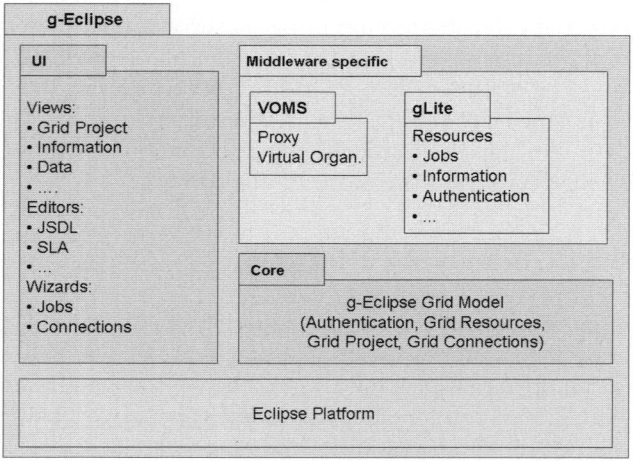

Fig. 1. The g-Eclipse Architecture

4.1 Integration with Eclipse

The integration of UI elements in the Eclipse workbench relies on the interfaces *IAdaptable* and *IResource* of the Eclipse framework. The g-Eclipse interface *IGridElement* helps to integrate the g-Eclipse Grid model in the Eclipse framework by providing the method *getResource*. All other interfaces of the g-Eclipse Grid model are based on this interface. The Grid element and its subclasses offer further methods to integrate both local entities and remote Grid elements.

4.2 The Abstraction Layer of the g-Eclipse Grid Model

The abstraction layer of the Grid model contains a multitude of Java interfaces that define the basic functionalities of the various Grid model elements. Fig. 2

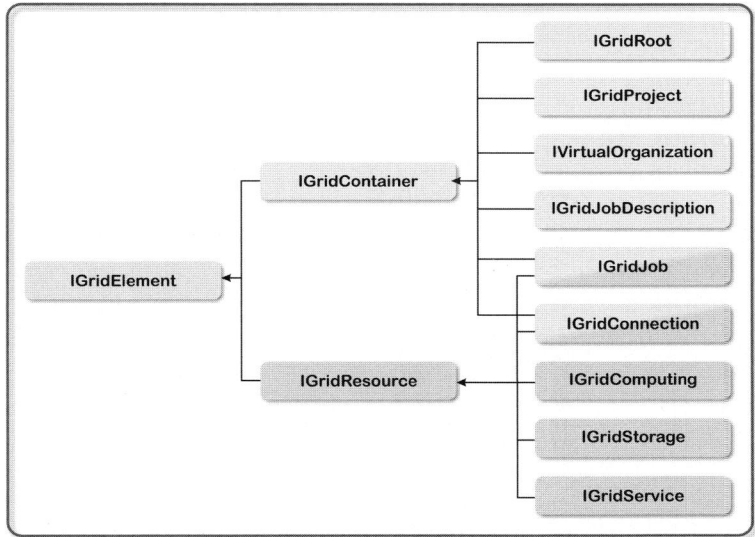

Fig. 2. The Interface of the g-Eclipse Core Grid Model

shows a simplified outline of the inheritance tree of these interfaces. As this layer only contains interfaces, multiple inheritance is allowed. This is used to map the structure of local and Grid elements to the model as well as mapping the relations between the different elements. Basically, the leaves of this inheritance tree are thought to be implemented either internally by the model or externally by middleware specific implementations.

4.3 Grid Project

The Grid project shape the base for the integration of g-Eclipse in the Eclipse environment. As described in Sec. 2, a Grid project is the fundamental context for a user to interact with a Grid infrastructure. A Grid project is a direct child of the Grid workspace root and holds all information necessary to access the Grid. Within g-Eclipse, a Grid project is always connected with one Virtual Organisation. Furthermore, a Grid project follows some structure and consists of the following standard folders (see Fig. 3):

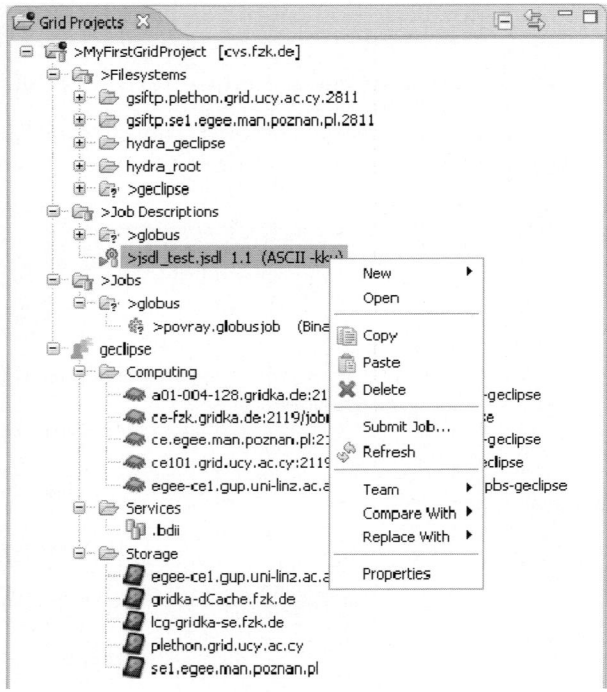

Fig. 3. The g-Eclipse Grid project view

Filesystems holds the connections to local and remote filesystems and enables the seamless access to users' data.

Job description holds the definition of Grid jobs created by the user and enables the user to start computation jobs on Grid infrastructures.

Jobs holds instances of Grid jobs, which are submitted Job descriptions. These instances will be updated frequently to get the actual status.

Virtual Organisation holds the computing and storage resources and available services of the VO of this project. This folder is purely virtual as all information about its entries will be collected from the VO declaration or from VO specific information systems.

By defining a clear correlation between a Grid project and a Virtual Organisation the g-Eclipse project nature and the corresponding Grid project view is essential for the contextualisation of the user workbench. Therefore the Grid project view will be part of the different perspectives for the different Grid user roles and their implementations as Eclipse perspectives.

5   Contextualised Grid User Perspectives

The role based contextualisation of users within the g-Eclipse framework relies on the Eclipse perspective pattern. The current g-Eclipse perspectives consist of a set of GUI elements provided by plug-ins and which combined provide an abstraction of the Grid and assist the actor in the role based Grid interaction.

5.1   Grid Application User

The goal of the user perspective is to simplify the procedure of accessing Grid resources. By using this perspective a Grid application user is able to create a Grid project including remote Grid resources, manage its data, utilise an editor to specify the job particulars and manage the execution of them on a Grid infrastructure.

The predefined user perspective for Grid application users is shown in Fig. 4. This contextualised Grid workbench perspective consists of the following components:

- The Grid project view is part of all contextualised and predefined g-Eclipse perspectives and shown on the top-left of Fig. 4. This view enables the access to Grid data, Grid job descriptions, Grid jobs and to the virtual resources of the assigned VO of the project.
- More details about the Grid resources are available from the Grid information view (see bottom-left). The shown Grid information view supports the GLUE information schema [10].
- The central part of the workbench is the editor area, where data content is shown. In Fig. 4 the JSDL multipage editor is shown as an example of integrated tools of the g-Eclipse framework. This editor supports the Grid application user in the definition of Grid jobs by supporting the JSDL schema [11].
- On the right, the properties view of the Eclipse framework shows the properties of the selected Grid element in the Grid project view.
- The Authentication Token view (bottom right) is part of the security infrastructure of the g-Eclipse framework and supports the Grid user with details about his current security tokens.
- The Jobs view, shown in the lower part of Fig. 4 shows the status of current and past Grid jobs.
- The Connection view (not shown) enables the user to create, move, copy and delete directories and files on local or remote storage devices.
- The Job Description wizard and Job Submission wizard (not shown) supports the user in the process of creating job description and submitting it to the Grid.

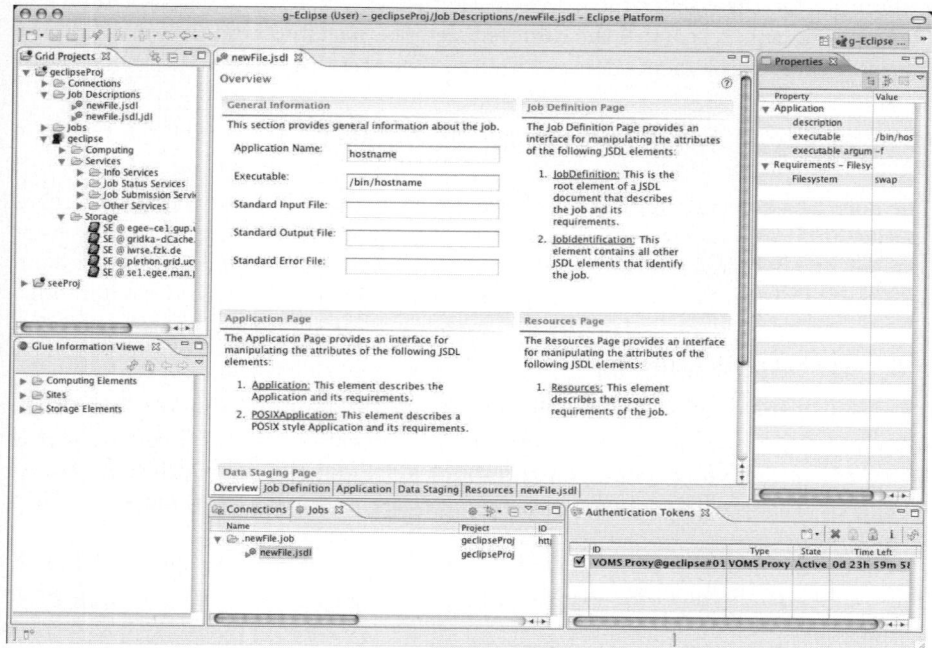

Fig. 4. The g-Eclipse Grid Application User Perspective

5.2   Grid Operator

Grid users can manage and test existing Grid infrastructures by using the g-Eclipse operator perspective. Additionally to the Grid project view and the Grid information view, the operator perspective consists of the following predefined components:

- The Batch service editor supports the operator of a Grid computing resource to manage the available batch system with the help of a graphical editor. The details of the selected batch resource are shown in the properties view.
- The Batch Job view enables the mapping from Grid Jobs to local jobs in the batch queue. With the help of this view, resources providers can organise the execution of jobs in the local batch system.
- The Terminal view enables the operator to connect himself directly to a computing or storage resource.

5.3   Grid Application Developer

Support for the development of Grid applications is provided by the g-Eclipse developer perspective. Nowadays, Grid applications are legacy applications and therefore the migration of these application must be supported by the developers perspective. Applications are developed as separate projects in the local workspace. With the help of the g-Eclipse perspective and its wizards these applications can be compiled and debugged on remote resources, deployed on selected Grid resources. For the development of more complex Grid workflows, a dedicated workflow editor is currently being integrated in the g-Eclipse framework.

5.4   Customisation

The above mentioned perspectives offer only predefined configurations for the proposed Grid user roles. As g-Eclipse is build on top of the Eclipse framework it supports the customisation of these perspectives. The layout of the perspectives is persistent between sessions. Users can customise each of the proposed perspectives to their needs and preferences. Furthermore, new Grid roles can be supported by defining a corresponding predefined g-Eclipse perspective.

6   Conclusions

The g-Eclipse project provides an extensible framework to enable various Grid actors to access Grid resources in an intuitive and easy way. g-Eclipse relies on the Eclipse framework and provides a Grid model to seamlessly integrate Grid resources into the Eclipse framework. Based on this middleware independent Grid model, the g-Eclipse framework can be extended to support different Grid middleware systems. The concepts of Virtual Organisations and Grid projects are important to contextualise the work of Grid users.

The g-Eclipse framework offers a set of tools which are composed to so called g-Eclipse perspectives to support different Grid actors. Currently, three actors – Grid application users, Grid resource operators and Grid application developers – have been identified and are supported with predefined contextualised Grid workbenches. These predefined workbenches can be customised by the Grid users to adjust the workbench to their needs and preferences. Grid actors can easily switch to different workspaces depending on the planned interaction with the Grid infrastructure.

The g-Eclipse framework has been available for potential users since end of 2007. The framework will be tested with an application from the airplane industry when support for the GRIA middleware is completed. The interest in the g-Eclipse framework is increasing and more users are expected to use the g-Eclipse framework for their applications. Reports on user experience with the g-Eclipse framework will be addressed in the future.

The g-Eclipse Grid model and its integration in the Eclipse perspectives framework offer a framework that can be contextualised and customised to the need of future Grid users. The architecture of g-Eclipse enables further mashups between Grid GUI components, SOA GUI components and GUI for Web2.0 by defining personalised and contextualised user workbenches and environments.

Acknowledgments. This work was partly supported by the EU project g-Eclipse (#FP6-2005-IST-034327). We would also like to thank the member institutions of the g-Eclipse consortium and all the project members. Furthermore, g-Eclipse is supported by the Eclipse foundation.

References

1. Foster, I., Kesselman, C.: The Grid: Blueprint for a New Computing Infrastructure. In: Concepts and Architecture, ch. 4, pp. 37–64. Elsevier, Amsterdam (2004)
2. Schilit, B., Adams, N., Want, R.: Context-aware computing applications. In: Proceedings of IEEE Workshop on Mobile Computing Systems and Applications, Santa Cruz, California, December 1994, pp. 85–90. IEEE Computer Society Press, Los Alamitos (1994)
3. The g-Eclipse Project, http://www.geclipse.eu
4. The Eclipse Project, http://www.eclipse.org
5. Gamma, E., Beck, K.: Contributing to Eclipse. Principles, Patterns, and Plugins. Addison-Wesley Longman, Amsterdam (2003)
6. OSGi Alliance, http://www.osgi.org
7. g-Eclipse architecture II, Public project deliverable D1.5 of the g-Eclipse project, http://www.geclipse.eu/fileadmin/Documents/Deliverables/D1.5.pdf
8. Hemmer, F., Laure, E., Barroso Lopez, M., Di Meglio, A., Fisher, S., Guy, L., Kunst, P., Prelz, F.: Middleware for the Next Generation Grid Infrastructure. In: Proceedings of CHEP 2004, Intelaken, Switzerland (2004), http://glite.web.cern.ch/glite
9. Surridge, M., Taylor,S., De Roure, D., Zaluska, E.: Experiences with GRIA - Industrial Applications on a Web Services Grid. In: Proceedings of the First International Conference on e-Science and Grid Computing pp. 98–105 (2005), http://www.gria.org
10. Andreozzi, S., Burke, S., Field, L., Fisher, S., Kønya, B., Mambelli, M., Schopf, J.M., Viljoen, M., Wilson, A.: GLUE Schema Specification, version 1.2, Final Specification (December 3, 2005)
11. Anjomshoaa, A., Brisard, F., Drescher, M., Fellows, D., Ly, A., McGough, S., Pulsipher, D., Savva, A.: JSDL Specification, Version 1.0 (November 7, 2005)

Formal Model for Contract Negotiation in Knowledge-Based Virtual Organizations

Mikołaj Zuzek[1], Marek Talik[1], Tomasz Świerczyński[1], Cezary Wiśniewski[1], Bartosz Kryza[2], Łukasz Dutka[2], and Jacek Kitowski[1,2]

[1] Institute of Computer Science, AGH-UST, Krakow, Poland
[2] Academic Computer Centre CYFRONET AGH, Krakow, Poland
{bkryza,dutka,kito}@agh.edu.pl

Abstract. In this paper we propose a formal model which supports the process of setting up a Virtual Organization by means of allowing contract negotiation between parties pariticipating in such VO. The negotiated contract is used to configure the necessary Grid middleware components for the purpose of contract enforcement during the lifetime of the Virtual Organization. We present a brief overview of the framework and discuss in detail the formal model underlying the process of contract negotiation and how the ontological description of the domains related to given VO supports this process.

Keywords: Virtual Organization, Contract Negotiation, Ontology, Grid

1 Introduction

As Grid systems are being moved from academic and research facilities to more challenging business and commercial applications, such issues as control of resource sharing become of crucial importance. In order to manage and share resources within the Grid the idea of Virtual Organizations emerged, which enables sharing only subsets of resources among partners of such Virtual Organization within a potentially larger Grid setting. In order to support creation and management of such Virtual Organization, Grid middleware must support several aspects such as security, resource sharing policy definition and enforcement, resource discovery and usage limited according to the VO policy and other. We propose a semantic based approach, implemented in the form of a framework, called FiVO (Framework for intelligent Virtual Organization) that supports creation and management of dynamic Virtual Organizations with special focus on authorization of access to resources based on ontologies [1,2]. In this paper we focus on its contract negotiation component and in particular on the formal model which allows to control and verify the negotiations process and its result. The contract, described by a special ontology, provides all information necessary for configuration of Virtual Organization in a Grid system, by automatically translating proper contract statements to configuration options of such systems as for example VOMS (Virtual Organization Management System) [3] or PERMIS [4]. Contracts also allow for specification of non-functional parameters of the envisioned

M. Bubak et al. (Eds.): ICCS 2008, Part III, LNCS 5103, pp. 409–418, 2008.

VO collaborations, especially SLA's (Service Level Agreements) in order to provide for the Grid monitoring layer necessary data for the contract enforcement [5]. Our work is being evaluated within the EU-IST project Gredia [6], on two commercial applications. First is related to inter-banking solution for automatic credit-scoring of bank users credit requests. The second one is a media application oriented on providing a collaborative environment for nomadic journalists.

2   Related Work

The idea of supporting Virtual Organizations with formal model of contract negotiation process is not present in existing works. However, some attempts were already made in order to deal with supporting Virtual Organizations with contract based agreements on how to share the resources of the entities participating in some VO. The authors of [7] describe requirements for automating the contract management in a VO. They identify 3 kinds of contracts in a VO: business contract, ICT contract and ASP (Application Service Provider) contract. In [8] an attempt was made to formalize a definition of contract based multi-agent Virtual Organization. The authors define 4 key properties of VOs: Autonomy, Heterogeneity, Dynamism and Structure. They use terminology from agent-based systems, e.g. they refer to the VO itself as an agent. The contract is defined as a set of commitments, goals and agents in some context. The paper introduces a formal definition of a hierarchical VO with a set of agents (which can be VOs themselves), policies, goals and commitments. The VO is then a set of bilateral contracts between agents in a VO, and can be more easily defined in a distributed setting. For example for 3 partners and 2 contracts $A \leftrightarrow B$ and $B \leftrightarrow C$, A and C don't event need to know about each other. Another example of contract based VO's is presented in [10]. Authors present web-Pilarcos J2EE based agent framework for managing contract based Virtual Organizations. The contract itself is an object (J2EE EntityBean) and can be in several states such as In-negotiation, Terminated etc. The proposed solution is not based on ontologies, and the metadata reasoning is mentioned briefly. The proposed architecture has many different components - which might make it hard for integration with custom systems - should rather provide a more unified interface based on easily adaptable standards. The paper discusses the basic requirements for a VO contract such as modeling of service behaviour, communication services and some non-functional properties such as QoS. In [9] authors present an event based protocol for decision making in Virtual Organizations, for multi-agent systems. The authors introduce a voting protocol based on RONR (Robert's Rules of Order). Thorough discussion of requirements necessary for a general VO management system can be found in [11].

3   Contract Negotiation Approach in Gredia

The GREDIA project aims at the development of a Grid application platform, providing high level support for implementation of Grid business applications concerned with users mobility. This platform is generic in order to combine both

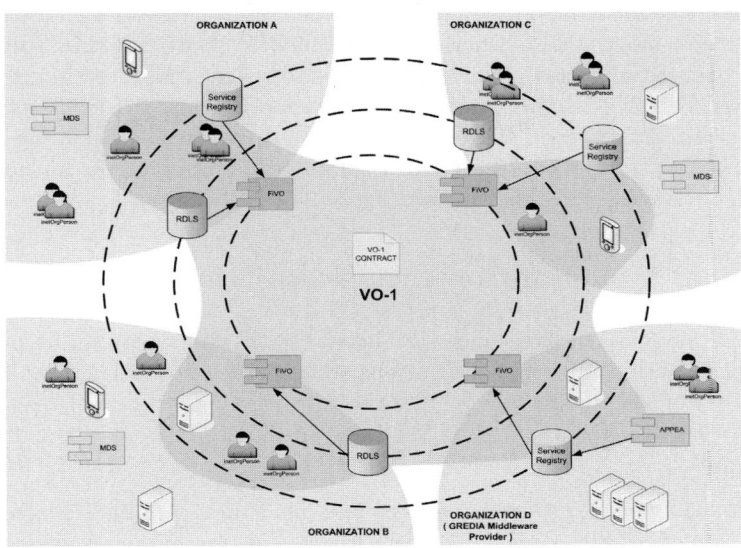

Fig. 1. FiVO overall vision in a distributed Gredia environment

existing and arising Grid middleware, and facilitates the provision of business services, which mainly require access and sharing of large quantities of distributed annotated numerical and multimedia content. One of the main GREDIA features is its focus on mobile users to exploit Grid technologies in a seamless way by enabling mobile access and sharing of distributed content. The potential results of the platform are being validated through two pilot applications, including media and banking. Fig. 1 presents example deployment of the FiVO framework in a distributed Gredia environment. Four organizations are sharing their resources within the VO-1. FiVO component is deployed within each organization and is responsible for storing semantic descriptions of its contents (i.e. resources provided to other organizations). These descriptions can include such aspects of organization as its structure and business logic described in proper ontology as well as hardware, data and service resources available and provide for sharing with other organizations including security and QoS rules. FiVO instances deployed in particular environment can connect through a peer-to-peer overlay, based on Grid Organizational Memory (GOM) knowledge base [12], in order to form a VO and collaborate on emerging tasks. The main feature of FiVO is the support for contract negotiation and management, which enables coordinated establishment of agreement among partners who want to create a new Virtual Organization. In order to enable organizations to define an unambigous agreement on how to share their resources in the form of ontology, the contract ontology provides all the necessary generic concepts, covering such aspects as the notion of Virtual Organization, security and authorization rights, Quality

of Service in the form of Service Level Agreement statements as well as the negotiation process itself. These generic concepts have to be extended by the domain
specific concepts of a particular VO in order to reflect its actual intented goal.
The contract itself is simply a set of ontology individuals. Abstract statements
from the contract are used to configure the VO middleware, including security
and monitoring infrastructure, in order to enforce proper access authorization
and QoS, as described by the contract.

4   Formal Model of Contract Negotiation

The formal model of contract negotiations in FiVO framework will be presented
by the introduction of several definitions. In general we can say, that the contract is a set of statements which state the rules of cooperation between parties,
referred to here as agents (A) and the rules specifying how their resources (X)
can be shared.

Definition 1. *Let $R : A \to 2^X$ be resource ownership function which assigns
the sets of resources to agents who own them, with the assumption that agents
do not own the same resources at the same time, i.e.:*

$$\forall a_1, a_2 \in A : a_1 \neq a_2 \Rightarrow R(a_1) \cap R(a_2) = \emptyset \tag{1}$$

Definition 2. *Atom, p, is a logical expression concerning one agent and one of
his resources. The set of all possible atoms can be defined as:*

$$P = \{p = (a, x, z, d) : a \in A \wedge x \in R(a) \wedge$$
$$d = \{(u, v) : u \in \Omega \wedge v \in Val(u)\}\} \tag{2}$$

*where a is the agent, x is the resource owned by the agent, z represents an action
that is supposed to be taken by the agent and d contains the parameters of the
action (Ω is the set of possible attributes and Val maps the parameters to their
possible values).*

Definition 3. *Statement, s, is a set of atoms in the form of a logical sentence. It
represents agents' requirement or commitment depending on whether the statement is still under negotiations or is already accepted. All atoms of a given
statement must relate to one agent. The set of all possible statements can be
defined as:*

$$S = \{s = (p_1, p_2, ..., p_n) : n \in N_+ \wedge p_i = (a, x_i, z_i, d_i), i = 1, 2, ..., n\} \tag{3}$$

In order to allow separation of the negotiation process for the whole VO into
partial negotiations on subsets of resources that will be included in the overall
contract, we introduce the concept of a negotiation table.

Definition 4. *The negotiation table is a couple (O, C) of disjunctive sets of
statements, where O includes opened statements which are still being negotiatied,*

and C contains already accepted statements (closed). The set of all possible tables can be defined as:

$$\Phi = \overset{\circ}{\Phi} \cup \hat{\Phi} = \{(O, C) \in 2^S \times 2^S : O \cap C = \emptyset\} \tag{4}$$

We say that negotiation table is closed when it does not contain any opened statements, otherwise it is opened. Thus we can define the sets of opened and closed negotiation tables as:

$$\overset{\circ}{\Phi} = \{(O, C) \in \Phi : O \neq \emptyset\} \tag{5}$$

$$\hat{\Phi} = \{(O, C) \in \Phi : O = \emptyset\} \tag{6}$$

Since all negotiations take place on at least one negotiation table we can define a negotiation state as a subset of the possible negotiations set, i.e.:

Definition 5. *The negotiation state is any set of negotiation tables, i.e. $q \subset \Phi$*

The changes in the negotiation tables can only be achieved by means of sending proper messages. These messages define who states what about which resources:

Definition 6. *By message, m, we mean a quintuple defined as:*

$$(a, I, Y, U, R) \in M = A \times M_I \times M_Y \times M_U \times 2^{\hat{\Phi}} \tag{7}$$

where a represents the agent who sends the message, I is the set of pairs table-statement, which specifies which statements should be added to which negotiation table (M_I), thus:

$$M_I = \{I \subset \overset{\circ}{\Phi} \times S : ((O, C), s) \in I \Rightarrow s \notin O \cap C\} \tag{8}$$

The above definition disallows adding statements which are already in a given negotiation table. Y is the set of statements accepted by the agent (only statements already added to the given negotiation table can be accepted), i.e:

$$M_Y = \{Y \subset \overset{\circ}{\Phi} \times S : ((O, C), s) \in Y \Rightarrow s \in O\} \tag{9}$$

U specifies the set of statements updated by the current message and it contains triples, which contain the set on which the statement is updated, statement already existing in this table and the new statement which does not yet exist in this table:

$$M_U = \{U \subset \overset{\circ}{\Phi} \times S \times S : ((O, C), s_1, s_2) \in U \Rightarrow s_1 \in O \wedge s_2 \notin O \cup C\} \tag{10}$$

The statement from the set U can only modify parameters (d) of existing statements, thus the following holds:

$$((O, C), (a_1, x_1, z_1, d_1), (a_2, x_2, z_2, d_2)) \in U \Rightarrow a_1 = a_2 \wedge x_1 = x_2 \wedge z_1 = z_2 \tag{11}$$

Finally the set R is the set of tables rejected by the agent. It cannot contain already closed negotiation tables, as well as it cannot contain the sets mentioned by statements in I, Y and U, thus we have:

$$((O,C),s) \in I \vee ((O,C),s) \in Y \vee ((O,C),s_1,s_2) \in U \Rightarrow (O,C) \notin R \quad (12)$$

Since agent cannot accept statements which are not related to him, we have:

$$(a,I,Y,U,R) \in M \wedge ((O,C),(\tilde{a},\tilde{x},\tilde{z},\tilde{d})) \in Y \Rightarrow a = \tilde{a} \quad (13)$$

Now, we can use these definitions to verify that a message is valid with respect to the negotiation state:

Definition 7. *Let $V : 2^{\Phi} \to M$ be a mapping, such that:*

$$m = (a,I,Y,U,R) \in V(q) \Leftrightarrow$$
$$[((O,C),s) \in Y \vee ((O,C),s_1,s_2) \in U \vee (O,C) \in C] \Rightarrow (O,C) \in q \quad (14)$$

In a given negotiations state, as correct messages from the set $V(q)$, we mean those, which accept or modify statements from the negotiation tables which are actually contained within that negotiation state. Invalid messages, are those, which relate to statements not present in any of the negotiation tables for the given state.

Definition 8. *Let $Q : \{(q,m) \in 2^{\Phi} \times M : m \in V(q)\} \to 2^{\Phi}$ be a transfer function which assigns to every negotiation state and message the consecutive state. Assuming $m = (a,I,Y,U,R)$ we get:*

$$Q(q,m) = q \cup (\bigcup_{((O,C),s) \in I}(O \cup \{s\}, C)) \setminus (\bigcup_{((O,C),s) \in Y}(O,C)) \cup$$
$$(\bigcup_{((O,C),s) \in I}(O \setminus \{s\}, C \cup \{s\})) \setminus (\bigcup_{((O,C),s) \in Y}(O,C)) \cup$$
$$(\bigcup_{((O,C),s_1,s_2) \in U}(O \cup \{s_2\} \setminus \{s_1\}, C)) \setminus (\bigcup_{((O,C),s_1,s_2) \in U}(O,C)) \setminus C \quad (15)$$

The message modifies the negotiation state according to these rules:

- statements of the form $((O,C),s)$ from set I create new statements s in the negotiation table (O,C), so the negotiation table changes to $(O \cup \{s\}, C)$
- statements of the form $((O,C),s)$ from set Y change the statement to close, i.e. it moves them from the set O to set C: $(O \setminus \{s\}, C \cup \{s\}$
- statements of the form $((O,C),s_1,s_2)$ from the set U, cause replacement of the statement s_1 with statement s_2 within the table (O,C), thus: $(O \cup \{s_2\} \setminus \{s_1\}, C)$
- negotiation tables from the set C are removed from the input state

Each valid message sent to a negotiation table, modifies the state of the negotiation process. Thus we can define a mapping which gives us the negotiation state from the sequence of messages:

Definition 9. *Let us define a mapping* $\varphi : D_\varphi \rightarrow 2^\Phi$, *such that:*

$$\emptyset \in D_\varphi \tag{16}$$

$$\varphi(\emptyset) = \emptyset \tag{17}$$

$$\mathop{\forall}_{i=0,1,2,\ldots,n+1} m_i \in M \wedge \Psi = (m_0, \ldots, m_n) \wedge \tilde{\Psi} = (m_0, \ldots, m_{n+1}) \wedge \Psi \in D_\varphi$$
$$\Rightarrow [m_{n+1} \in V(\varphi(\Psi)) \Rightarrow \tilde{\Psi} \in D_\varphi \wedge \varphi(\tilde{\Psi}) = Q(\varphi(\Psi), m_{n+1})] \tag{18}$$

The above mapping maps sequences of messages into negotiation states. An empty sequence is assigned an empty negotiation state. If a given negotiation sequence Ψ belongs to the domain of state function (which means it represents correct negotiation process) then the value of function $\varphi(\Psi)$ is the state of negotiations after applying all messages from this function.

Definition 10. *Negotiations is any finite sequence of messages which belongs to the domain of state function* φ.

We say that negotiations completed successfully when the last state contains at least one closed negotiation table and does not contain any opened negotiation tables. Thus finally we can define a contract:

Definition 11. *A contract is a set of statements contained in the final negotiations state. i.e.:*

$$K = \mathop{\cup}_{(O,C) \in \varphi(\Psi)} C \tag{19}$$

The definitions presented above specify formal way of negotiating a contract for a VO, therefore VO can be understood as a practical realization of the above defined contract. This allows to verify that the negotiations were fair and that they completed successfully, i.e. all parties were finally satisfied. Another step is to use the definition of a VO goal, defined as a set of requirements for the contract to automatically verify, whether all initial requirements for the VO were satisfied. In order to support dynamic VOs, where the conditions can change during the operation of VO, the contract, as defined above can be ammended, by starting the negotiations again with the current state of the contract as the initial state.

5   From Contract Model to VO Ontology

The formal model provides means to implement a negotiation framework which can allow parties to define the rules of cooperation within a given Virtual Organization. After the contract is successfuly negotiated, the framework should ensure the rules agreed upon are obeyed. According to the presented formal model a sample statement from the EasyLoan banking Virtual Organization looks like below:

$s = StCreditCalculation =$
$\{p = (a='\#Marco', \; x='\#CreditCalculationService', \; z='\#ProvidesService',$
$\quad d=\{('\#TimeToComplete', \; '15minutes'), \; ('\#accessRole','\#BankClerk')\})\}$

These statements are encoded in the Web Ontology Language with respect to proper contract ontology, and the above statement in this ontology is rendered as:

```
<j.0:VirtualOrganization rdf:ID="EasyLoan">
 <j.0:name rdf:datatype="...">
        EasyLoan Application VO</j.0:name>
 <j.0:administeredBy rdf:resource="1#Marco"/>
 <j.3:hasContract>
  <j.3:Contract rdf:ID="EasyLoanContract">
   <j.3:hasStatement>
    <j.3:Statement rdf:ID="StCreditCalculation">
     <j.3:hasAtom>
      <j.3:Atom rdf:ID="AtomCreditCalculationService">
       <j.3:hasActor rdf:ID="#Marco"/>
       <j.3:hasResource>
        <j.2:Service rdf:ID="CreditCalculationService">
         <j.0:isOwnedBy rdf:resource="#HappyBank"/>
         <j.0:belongsTo rdf:resource="#EasyLoan"/>
        </j.2:Service>
       </j.3:hasResource>
       <j.3:hasAction>
        <j.3:Action rdf:ID="ProvidesService"/>
       </j.3:hasAction>
       <j.3:hasParameter>
         <j.3:Parameter>
          <j.0:hasQoSAttribute>
           ---- j.5:QoSAttribute - #j.6:TimeToComplete
          </j.0:hasQoSAttribute>
          <j.6:hasValue>
           ---- { "15",  #owlTime::unitMinute }
          </j.6:hasValue>
         </j.3:Parameter>
        </j.3:hasParameter>
        <j.3:hasParameter>
         <j.3:Parameter>
          <j.0:hasAuthorizationRestriction>
           ---- j.5:accessRole
          </j.0:hasAuthorizationRestriction>
          <j.6:hasValue>
          ---- { "#BankClerk" }
          </j.6:hasValue>
         </j.3:Parameter>
```

```
    </j.3:hasParameter>
    ---- more parameters ...
   </j.3:Atom>
  </j.3:hasAtom>
   </j.3:Statement>
  </j.3:hasStatement>
 </j.3:Contract>
 </j.3:hasContract>
</j.0:VirtualOrganization>
```

In Fig. 2, we can see sample contract ontology visualization for EasyLoan Virtual Organization of the banking scenario.

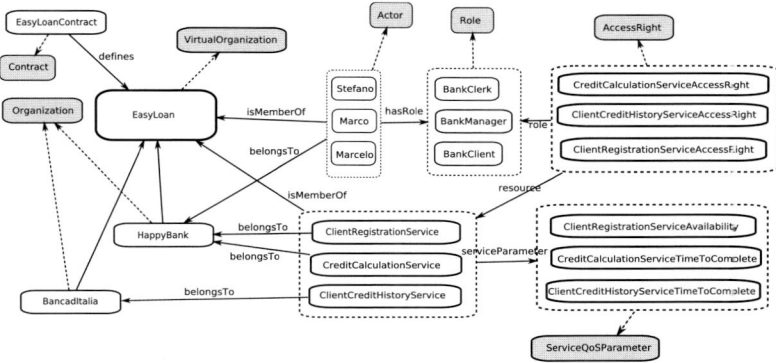

Fig. 2. Graphical representation of ontological contract for Banking application

This contract is used by FiVO to configure Grid middleware services such as VOMS, PERMIS or MDS in order to actually deploy the Virtual Organization in the Grid environment. Further enforcement of the contract statements is performed automatically by these services.

6   Conclusions and Future Work

In this paper we have presented the vision and architecture of the FiVO (Framework for Intelligent Virtual Organizations) which enables contract negotiation and management for Grid-based Virtual Organizations in a business setting. We believe that such functionality will foster the adoption of Grid and Virtual Organizations in commercial applications by simiplifying the process of Virtual Organization inception and management of agreements on how the resources of each participating organizations should be shared among partners of a VO. The framework is currently being implemented and the future work will include, development of ontologies required for contract definition and evaluation of the framework on pilot applications. The formal model will be extended with a equivalence relation by means of the type of resources and allowing OR operator between atoms in a statement.

Acknowledgements. The authors want to acknowledge the support of the EU Gredia Project (IST-FP6-034363) and AGH University of Science and Technology grants 11.11.120.777 and 500-08.

References

1. Kryza, B., Dutka, L., Slota, R., Kitowski, J.: Supporting knowledge-based dynamic virtual organizations with contracts. In: Proc. of eChallenges 2007 Conference and Exhibition, The Hague, Netherlands, October 24-26, 2007, pp. 937–945 (2007)
2. Kryza, B., Dutka, L., Slota, R., Pieczykolan, J., Kitowski, J.: Gvosf: Grid virtual organization semantic framework. In: Bubak, M., Turala, M., Wiatr, K. (eds.) Proc. of Cracow Grid Workshop 2006 (CGW 2006), pp. 104–110. ACK-Cyfronet AGH, Krakow (2007)
3. Alfieri, R., Cecchini, R., Ciaschini, V.: Voms, an authorization system for virtual organizations. In: Fernández Rivera, F., Bubak, M., Gómez Tato, A., Doallo, R. (eds.) Across Grids 2003. LNCS, vol. 2970, pp. 33–40. Springer, Heidelberg (2004)
4. Chadwick, D.W., Otenko, A.: The permis x.509 role based privilege management infrastructure. Future Generation Comp. Syst. 19, 277–289 (2003)
5. Alipio, P., Neves, J., Carvalho, P.: An ontology for network services. Computing and Informatics 26(5), 543–561 (2007)
6. Gredia Consortium: Gredia project website, http://www.gredia.eu
7. Shelbourn, M., Hassan, T., Carter, C.: 3.1. Legal and contractual framework for the VO. In: Camarinha-Matos, L.M., Afsarmanesh, H., Ollus, M. (eds.) Virtual Organizations Systems and practices, pp. 167–176. Springer, Heidelberg (2005)
8. Udupi, Y.B., Singh, M.P.: Contract enactment in virtual organizations: A commitment-based approach. In: Proc. of the Twenty-First National Conference on Artificial Intelligence and the Eighteenth Innovative Applications of Artificial Intelligence Conference, Boston, Massachusetts, USA, July 16-20, 2006. AAAI Press, Menlo Park (2006)
9. Pitt, J.V., Kamara, L., Sergot, M.J., Artikis, A.: Formalization of a voting protocol for virtual organizations. In: 4rd International Joint Conference on Autonomous Agents and Multiagent Systems (AAMAS 2005), Utrecht, The Netherlands, July 25-29, 2005, pp. 373–380. ACM, New York (2005)
10. Metso, J., Kutvonen, L.: Managing virtual organizations with contracts. In: Workshop on Contract Architectures and Languages (CoALa 2005) (2005)
11. Ratti, R., Camarinha-Matos, L.M., et al.: Specification of vo creation support tools. Technical report, Ecolead Consortium (2006)
12. Kryza, B., Slota, R., Majewska, M., Pieczykolan, J., Kitowski, J.: Grid Organizational Memory - Provision of a High-level Grid Abstraction Layer Supported by Ontology Alignment. Future Generation Computer Systems (FGCS) 23(3) (March 2007)

An Approach for Enriching Information for Supporting Collaborative e-Work

Obinna Anya, Atulya Nagar, and Hissam Tawfik

Intelligent and Distributed Systems Lab, Deanery of Business and Computer Sciences,
Liverpool Hope University, Liverpool, United Kingdom L16 9JD
{05008721,nagara,tawfikh}@hope.ac.uk

Abstract. As a result of the high level of knowledge required in collaborative e-Work as well as its changing work contexts, e-Work support systems need to provide not only information in the form of documents and articles, but also expert-level explanations in the form of supporting literature and references to theories and related cases, to justify retrieved information and offer cognitive support to e-Work. In this paper, we present a novel approach for enriching information for supporting collaborative e-Work, which combines latent semantic analysis, domain task modelling and conceptual learning. We illustrate the potential of our approach using our e-Workbench system. e-Workbench is a prototype system for adaptive collaborative e-Work.

Keywords: Collaborative e-Work, information enrichment, concept-based knowledge acquisition, cognitive support.

1 Introduction

Collaborative e-Work is highly knowledge-intensive, and involves experts, often, with different knowledge backgrounds and from different work organisations who share knowledge in order to arrive at an optimal decision or problem-solving strategy [2], [9]. Such work activity requires both documented and experience-based knowledge. As a result, collaborative e-Work support systems need to provide not only information in the form of documents and articles, but also expert-level explanations in the form of supporting literature and references to theories and related cases to justify retrieved information and offer cognitive support to augment workers' ideas.

Different techniques have been proposed in advanced information processing, the semantic Web and knowledge management, towards information enrichment. In machine learning, natural language processing and information retrieval, various techniques have been combined in efforts to identify features for indexing information resources. A number of these methods employ shallow statistical inferences that do not typically result in knowledge-rich representations. Techniques in natural language processing have been explored but their huge reliance on grammar makes them less attractive, especially in domains where problem-solving methodologies have not recorded using strict grammatical structure [3]. Within the Semantic Web community, the use of metadata annotations, such as tagging [4] has proved immensely popular as a technique for enriching information resources, but its employment of domain experts makes it less cost-effective.

M. Bubak et al. (Eds.): ICCS 2008, Part III, LNCS 5103, pp. 419–428, 2008.

This paper presents a novel approach for enriching information for supporting collaborative e-Work, which combines latent semantic analysis (LSA), domain task modelling (DTM) and conceptual learning. The focus is to enable the work environment to acquire sufficient knowledge of a work domain in terms of key terms and concepts within the domain and their relationships, (previous) cases, as well as possible tasks and task goals in order to provide expert-level explanations to justify retrieved information resources and offer cognitive support to augment workers' ideas. We illustrate the potential of our approach using e-Workbench [5].

2   Related Work

A number of researches have been carried in areas, such as advanced information processing, computer supported cooperative work and group decision making, knowledge management and the semantic Web, towards information enrichment for various purposes. Feng et al. [1] proposed a model of an information space, consisting of knowledge and document subspaces, to enable the acquisition of knowledge from digital libraries. The authors noted that two shortcomings in the effort towards satisfying man's information needs to support decision making are inadequate strategic level cognition support and inadequate knowledge sharing facilities. Ackerman [8] presented the Answer Garden, which is aimed to achieve knowledge acquisition in two intertwined ways – by making relevant information retrievable and by making people with knowledge accessible. The issue of common understanding in collaborative decision making is addressed in [7] through argumentative discourse and collaboration. Evangelou et al. [6] presented an approach for supporting knowledge-based collaborative decision making, which aims at developing knowledge management services for the capturing of organisational knowledge in order to augment teamwork, and thus enhance decision making quality.

An underlying goal of most of these efforts is to analyse information and/or integrate multiple knowledge resources so as to derive a semantic relationship among the basic concepts. The novelty in our approach lies in the ability to employ an integrated approach aimed to enable a work support system to acquire sufficient knowledge of a work domain in order to cognitively provide expert-level explanations to justify the use of an information resource.

3   e-Workbench Overview

We developed e-Workbench in [5], a prototype intelligent system aimed to equip future collaborative workspaces to adapt to work, and creatively support problem solving and decision making by leveraging on distributed knowledge resources in order to proactively augment workers' capabilities. The ultimate goal is to enable the work environment to become not only a workspace, but also 'a co-worker' and 'a collaborator' as a result of its knowledge of work and creative participation in problem solving and decision making.

We use a semi-automated approach in enabling e-Workbench to acquire knowledge of work and appropriately understand the users' domain of work. This involves:

(1) 'Training' e-Workbench to learn about key concepts and threads of ideal problem solving strategies within the domain of work. (2) Generating knowledge element models (KEMs) of the domain based on a DTM as well as cases and tasks that constitute best problem solving strategies. With the knowledge acquired during training and the KEMs generated, e-Workbench is able to retrieve appropriate information to assist in decision making and provide justifications for e-workers' views. Fig. 1 depicts an overview of the e-Workbench approach.

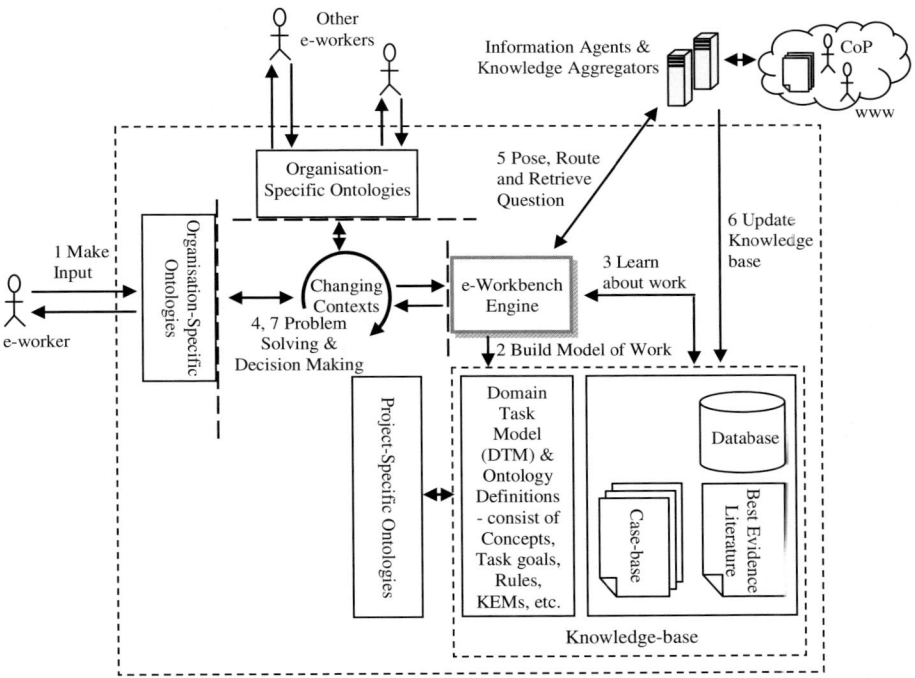

Fig. 1. The e-Workbench Approach

4   Information Enrichment Approach

Our approach to information enrichment for supporting collaborative e-Work involves the following processes:

- Posing a question to a knowledge expert or an information retrieval agent such as a search engine, and obtaining retrieved results.
- Selecting the most appropriate result(s) based on domain expert knowledge.
- Using LSA to infer deeper meanings from the selected documents.
- Using DTM, conceptual learning and project-specific ontologies to construct concept structures, relevant cases and KEMs based on work context tree (WCT) in order to acquire contextual knowledge about work.

- Mapping the knowledge extracted to the meanings inferred from the selected documents in order to conceptually enrich retrieved information.

During collaborative problem solving and decision making, information is usually retrieved when a user poses a question to a knowledge expert or a search engine in order to resolve an exception. Our experience in developing e-Workbench, however, shows that this process of enriching the retrieved information cannot be completely automated particularly with regard to e-Work because of its rapidly changing work contexts and the likely use of loosely coupled teams from multiple domains of expertise. As a result, we use a semi-automatic approach, which is described as follows, to achieve this.

4.1 Query Analysis

A search query is manually elaborated by a domain expert in order to accentuate relevant terms and domain concepts that relate to the query. The elaboration process involves (1) extracting representative terms, which are likely to contain domain knowledge and which actually contain the expected results from a given query, and (2) incorporating relevant terms that may not appear in the query, but could be used to capture key concepts and domain rules within the domain. The extracted and incorporated terms are used to build an accentuated matrix of key terms/concepts used during the LSA-based indexing process (see sec. 4.4). During the elaboration process, terms with different names across different sub-domains, e.g. *civil engineering*, *building site* and *estate development*, which semantically refer to the same object, are identified. We refer to such terms as synonyms. On the other hand, terms, which we refer to as polysemys that have different shades of meaning across sub-domains, are equally identified. Our elaboration approach enables us to solve the problem of the apparent inability of Web users to formulate effective search queries that accurately reflect their information needs [10], which thus poses the challenge of finding the optimum search word size of a query that will give the most effective search result. Our technique is to identify the key terms within a search query, or that are related to a search query, their synonyms and polysemys, accentuate those terms within the query, and relate them to the concept and context of work for effective knowledge support. The accentuated terms from the search query may also be used as subsequent queries. Based on our conceptual query analysis, each term is related to a domain concept, and relevant documents selected from the retrieved query results.

4.2 Conceptual Learning

Using a conceptual learning mechanism [11], e-Workbench is trained on the domain of work. The goal of conceptual learning is to enable the system to build a "knowledge space" for understanding relevant concepts, principles and facts within the identified domain(s), and a "conceptual space" for inferring relationships among them. This forms the semantic memory [12] of e-Workbench, which allows it to build the cognitive capabilities required for accurate perception of concepts (and objects) within the domain of work. In intelligent systems, perception is accomplished by the ability of a system to recognise and analyse features of concepts (and objects), which it encounters (and interacts with). For a given search query, we identify two levels of concepts: the high level composite concept and the low level primitive concept. Using simple mapping rules, we denote these generically, and using an example from the domain of civil construction, as:

<composite concept>::=<composite concept> | <primitive concept>
<primitive concept>::=<feature> | ... | <feature>
<civil engineering>::=<building site>|<road construction>
<building site>::=<engineer>|<hoarding>|<concrete>|<scaffolding>|< crane>
<hoarding> ::= <is-a-fence>|<is-temporary>|
                  <screens-off-a-building-site>|<serves-to-discourage-theft>

where ::= denotes a classification or mapping rule, and | denotes a semantic operator.

A feature is a psychological representation of properties of the world that can be processed independently of other properties and that are relevant to a task, such as categorisation, in collaborative e-Work. They are identified by their functional role in cognitive processing, such as the act of including (or excluding) an entity as a member of a category [13]. We represent a feature as a 2-tuple:

feature (V, T)

where V is a set of properties of a feature and T, a set of types for corresponding properties. $V \neq \phi$; $T \neq \phi$.

The next stage of our conceptual learning technique consists of building the episodic memory [12] of e-Workbench. Episodic memory is used for storing events (within collaborative e-Work) having features of a particular time and space, and for storing episodes and traces of learning – about skills, problem solving strategies and processes, decision making procedures, contextual use of information resources – that occur within relevant cases through case-based reasoning. Episodic memory is associative in nature and content-addressable [12]. As a result, it could be used to appropriately hold episodic information about a case in collaborative e-work by storing tasks based on concept as well as context, storing logical associations between tasks, specifying appropriate action plans and storing dynamic references to information resources. In e-Workbench, this memory is populated with cases, which are, in turn, composed of conceptually categorised tasks. A case and a task are both represented as a 3-tuple entity as follows:

case (P, A, S)
task (C, R, E)

where P is the problem (or case) description, which includes the initial problem state, relevant concepts and their feature values, a description of problem execution context and possible actors (people and virtual roles), one or more tasks and their (sub)goals as well as associated action plans or behaviour streams required to achieve those goals; A is an action plan or behaviour stream whose execution transforms the problem from the problem state to a goal state; and S is the expected result when A is applied to P. C is the case under which the task is valid or, at least, applicable, R is the set of resources required to perform the task, and E is the set of operations of corresponding resources.

4.3 Domain Task Modelling

The domain concepts related to the given search query are manually analysed using WCT [5]. The goal is to filter out, from the DTM, possible motives of a worker's actions and ideas, and correlate them conceptually based work goal. To achieve this, we analyse the given e-Work project in terms of the domain(s), existing knowledge, given terms of reference, previous cases and possible tasks and task goals. The work context tree enables us to generate semantically rich service descriptions used to meaningfully encapsulate ideas and knowledge resources within an e-Work task structure. We refer to these semantic rich service descriptions as KEMs. We use a KEM to refer to concept knowledge, which could be described by an information resource (i.e., an entity that has identity, for example, given by a URI) that is capable of supplying coherent statements of facts, empirical results, reasoned decisions or ideas (i.e., data, information, processes), which can be applied to justify the use of retrieved information in collaborative problem solving and decision making, to corroborate or refute a worker's view or to build new knowledge.

Fig. 2 shows a WCT. The root, KW, of the top-down tree is the given e-Work project or problem of interest. The root node contains three items: the domain ontology, D, which provides domain permissible procedures, rules and conceptual information relevant to KW, existing knowledge, K and work goal, G. K comprises theories, stories, hypotheses, philosophies, assertions, rules, metaphors and initial work input, in the form of terms of references, relevant to KW. G is the expected result of work. The next level consists of a set of nodes that describes cases within the KW. Each case node contains two items: the work, KW (as defined in the root node) and the case context, C_C. C_C comprises goals, motives, conditions and information that pertain to the case. The third level consists of nodes that describe tasks in the KW. Each task node consists of three items: the next upper level case node, C_x $(1 \le x \le n)$, the task context, C_T and the task goal, O. The fourth level (the leaves) consists of the KEMs. A KEM has four items: the next upper level task, T_x $(1 \le x \le n)$, a knowledge descriptor, S, the role, R, the effect, E and N, the referred cognitive node(s). The knowledge descriptor provides metadata descriptions about the KEM. R is the action performed or knowledge supplied by KEM, while E is the expected change brought about by R in T_x. N refers to the node(s) in a knowledge network that possibly has (or have) the resource (information, service or human expertise) required to perform T_x, or augment the process of performing T_x or taking a decision towards performing T_x. N could be Web resources, denoted by a URL, non-Web resources, e.g. a book, human agents, such as an expert in the given domain of work or non-human agents, such as knowledge repositories, Web sites, content and referral databases, avatars, and "webbots" [14] that have additional information with which to support retrieved information. The three nodes provide three cognitive planes, with which to analyse work at the domain, conceptual and task levels.

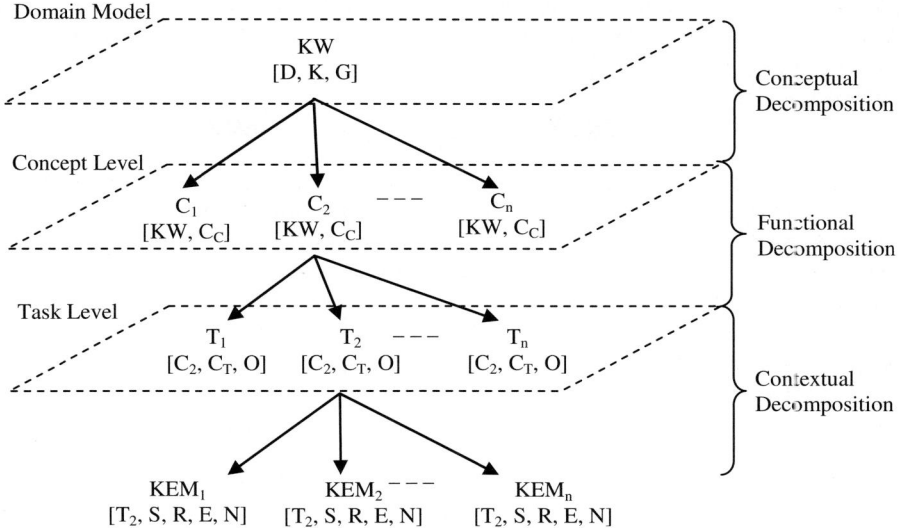

Fig. 2. Work Context Tree for Generating Knowledge Element Models

4.4 Latent Semantic Analysis

We use LSA [15] to infer deeper meanings and relations from the words, phrases and sentences in an elaborated search query, and associate those meanings to the documents retrieved. Using LSA, we aim to enable the system to predict what terms "actually" are implied by a query or apply to a retrieved document. Table 1 shows four documents D1, D2, D3 and D4, which are the selected results from the query: *importance of hoarding materials*. Key terms from the elaborated search query are shown on the first row of the table. The key concept being considered is *building construction*, and is represented in the table by p. The context of work, q, is described by the phrase "*a building construction site*".

From the table, the user would have considered documents 2, 3 and 4 relevant (column R). Document 2 contains words within the context of work, while document 3 contains a key word, but relates to neither the underlying concept nor the context of work. Document 1, though not considered relevant by the user, matches the query because it relates to the underlying concept, while document 4 matches the query because it relates to both the concept and context of work.

Table 1. Sample Term by Document Matrix

	hoarding	building	material	board	fence	p	q	R	M
D1	*	*				*			*
D2			*	*	*		*	*	
D3	*							*	
D4	*	*	*			*	*	*	*

LSA uses the technique of singular value decomposition in which a rectangular matrix, A of terms by documents, m x n, is decomposed into a product of three other matrices from which the original matrix can be approximated by linear combination, as denoted by eqn. 1:

$$A_{(mxn)} = T_{O(mxm)} \, S_{O(mxn)} \, D_{O(nxn)}{}' . \qquad (1)$$

such that T_O represents the term matrix, $D_O{}'$ represents the document matrix, and S_O, a diagonal matrix containing singular values arranged in descending order. Our goal is to capture the significance of every term in the query to the underlying concept of work. As shown in fig. 3, the g highest singular values identify the g most importance concepts in T_O, which is represented by a $T_{(mxg)}$ matrix. The weights in $S_{(gxg)}$ reflect the importance of concepts in $T_{(mxg)}$. Multiplying $T_{(mxg)}$ by $S_{(gxg)}$ results in the accentuation of the entries (concepts) in $T_{(mxg)}$.

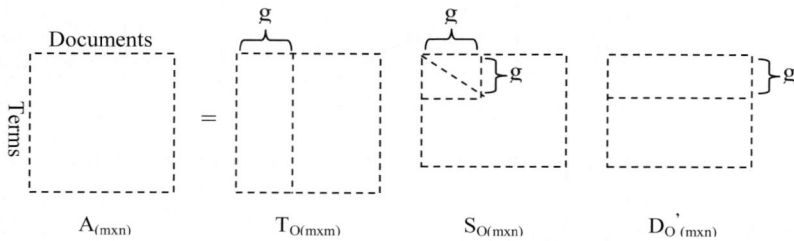

Fig. 3. Singular Value Decomposition of a Term by Document Matrix

4.5   Enriching Retrieved Information

Fig. 4 shows our model for mapping extracted concepts to acquired knowledge in order to enrich retrieved information for supporting collaborative e-Work. The model consists of a knowledge space (KS), an information space (IS) and a cognition support area (CS). A retrieved information resource is analysed (see sec. 4.1) and input into the knowledge space, and KEMs are generated from it based on fig. 2. KS contains the DTM for the given e-Work project, ontology definitions and work goal as well as concept models, cases and possible tasks and task goals. In e-Workbench, ontologies are defined based on the combined ontologies and knowledge structure of all organisations participating in the given e-Work project as well as the DTM of the e-Work project (see fig. 1). Within the IS, users and agents search for relevant documents using the generated KEMs and key terms as search guide. From the generated KEMs, a KEM is selected. Work context information is applied to it; and best evidence literature information, used to enrich it to provide justifications for retrieved resources. Concepts and deeper meanings are extracted from the retrieved documents (see sec. 4.2 – sec. 4.4) to provide explanations and enhance their information base. Finally cognition support (expert-level knowledge) is provided in the CS in the form of rich information (i.e. enhanced retrieved document + supporting explanations and justifications.

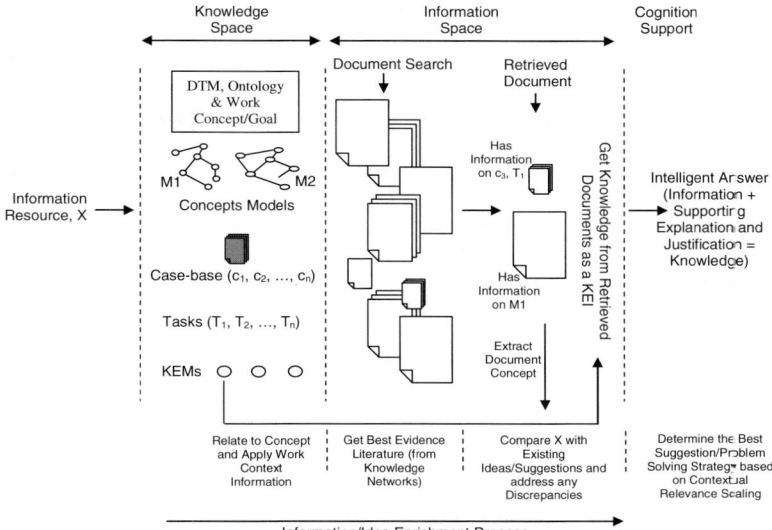

Fig. 4. Information Enrichment Framework

5   Conclusion and Future Work

This paper presents an approach for enriching information for supporting collaborative e-Work, which combines LSA, DTM and conceptual learning. The focus is to enable the work environment to acquire knowledge of the domain of work in terms of key terms and concepts within the domain and their relationships to previous cases, as well as possible tasks and task goals in order to provide expert-level explanations to justify retrieved information resources, and offer cognitive support to augment workers' ideas. Our future work will focus on developing mechanisms to keep track of the changing work contexts in collaborative e-Work so as to ensure that enriched information resources are effectively used to cognitively support decision making and problem solving.

References

1. Feng, L., Jeusfeld, M., Hoppenbrouwers, J.: Beyond Information Searching and Browsing: Acquiring Knowledge from Digital Libraries. INFOLAB Technical Report ITRS-008, Tilburg University, The Netherands (February 2001)
2. Nof, S.: Design of Effective e-Work: Review of Models, Tools, and Emerging Challenges. Production Planning and Control 14(8), 681–703 (2003)
3. Asiimwe, S., Craw, S., Wiratunga, Taylor, B.: Automatically Acquiring Structured Case Representations: Smart Way. In: Ellis et al. (eds.) Applications & Innovations in Intelligent Systems XV, pp. 45–58. Springer, London (2007)

4. Tanasescu, V., Streibel, O.: Extreme Tagging: Emergent Semantics through the Tagging of Tags. In: Proc. of the Int.l Workshop on Emergent Semantics & Ontology Evolution, ISWC/ASWC 2007, Busan, S Korea (2007)
5. Anya, O., Nagar, A., Tawfik, H.: A Conceptual Design of an Adaptive and Collaborative e-Work Environment. In: Proc. of the 1st Asian Modelling Symposium, Asia 2007, Thailand, March 27-30 (2007)
6. Evangelou, C., Karacapilidis, N., Tzagarakis, M.: On the Development of Knowledge Management Services for Collaborative Decision Making. Journal of Computers 1(6), 19–28 (2006)
7. Karacapilidis, N., Papadias, D.: Computer-Supported Argumentation and Collaborative Decision Making: The Hermes System. Information Systems 26(4), 259-277(19) (2001)
8. Ackerman, M.: Augmenting the Organizational Memory: A Field Study of Answer Garden. In: Proc of the ACM Conf on Computer Supported Cooperative Work (CSCW 1994), October 22-26,1994, pp. 243–252. ACM, Chapel Hill, NC (1994)
9. Experts Group: New Collaborative Working Environments 2020, EUROPEAN COMMISSION Information Society Directorate-General, Report of the Experts Group on Collaboration @ Work, Brussels (February 2006)
10. Smyth, B.: Adventure in Personalised Web Search. In: proc of 27th SGAI Int'l Conf on Innovative Techniques and Applications of AI, Cambridge, December 12-14 (2007)
11. Wiig, E., Wiig, K.: On Conceptual Learning, Knowledge Research Institute, Inc. Working Paper 1999-1 (1999)
12. Ramamurthy, U., D'Mello, S., Franklin, S.: Modified Sparse Distributed Memory as Transient Episodic Memory for Cognitive Software Agents. In: IEEE Int'l Conf on Systems, Man and Cybernetics, vol. 6, pp. 5858–5863 (October 2004)
13. Rogosky, B., Goldstone, R.: Adaptation of Perceptual and Semantic Features. In: Carlson, L., van der Zee (eds.) Functional Features in Language and space: Insights from perception, categorisation and development, pp. 257–273. Oxford University Press, Oxford (2005)
14. Carley, K.: Smart agents and organizations of the future. In: Lievrouw, L., Livingstone, S. (eds.) Handbook of new media, pp. 206–220. Sage, London (2002)
15. Deerwester, S., Dumais, S., Furnas, G., Landauer, T., Harshman, R.: Indexing by latent semantic analysis. Journal of the American Society for Information Science 41(6), 391–407 (1990)

Dynamic Virtual Environments Using Really Simple Syndication

Andrew Dunk, Ronan Jamieson, and Vassil Alexandrov

Centre for Advanced Computing and Emerging Technologies
The University of Reading
a.dunk@reading.ac.uk
http://www.acet.reading.ac.uk

Abstract. This paper investigates the use of really simple syndication (RSS) to dynamically change virtual environments. The case study presented here uses meteorological data downloaded from the Internet in the form of an RSS feed, this data is used to simulate current weather patterns in a virtual environment. The downloaded data is aggregated and interpreted in conjunction with a configuration file, used to associate relevant weather information to the rendering engine. The engine is able to animate a wide range of basic weather patterns. Virtual reality is a way of immersing a user into a different environment, the amount of immersion the user experiences is important. Collaborative virtual reality will benefit from this work by gaining a simple way to incorporate up-to-date RSS feed data into any environment scenario. Instead of simulating weather conditions in training scenarios, actual weather conditions can be incorporated, improving the scenario and immersion.

Keywords: Virtual Reality, Weather, RSS, XML, OpenSG.

1 Introduction

Virtual Reality is a way of showing information to users in such a way that it can be interacted with, as you would expect to interact with the real world. However it is not limited to this, it also allows users to view fictional items or objects that would not be possible to visualise easily in the real world. The most popular method for achieving a virtual reality is to use three dimensional visualisations for the users to view. The senses of touch (haptics), sound, smell and even taste have been experimented with as well.

A CAVE [4] is a fully immersive 3D environment able to display stereo imagery that can span 4 walls, the floor and ceiling, giving the user the impression that they are physically in the virtual environment being displayed. Presence is a measure of how accepting a user is to a virtual environment, or how detached they are from the real world. By bringing real world information into the virtual one like the current weather, or information that users are accustomed to, like a mobile phone telephone book, helps users to accept the virtual environment, increasing their presence.

M. Bubak et al. (Eds.): ICCS 2008, Part III, LNCS 5103, pp. 429–437, 2008.

This work is a continuation of the Virtual Weather [5] project, which aimed to create a virtual reality environment to simulate some basic weather types over a loaded model. This work will concentrate and generalise the area of downloading and parsing data from the Internet, by implementing a suitable Extensible Markup Language (XML) parser to be used for both configuring the application and parsing any RSS data downloaded from the Internet. This generalisation enables relationships of any RSS data to variables of a rendering engine. Including but not limiting the application to downloading different weather sources and rendering them into a virtual environment. Fig.1 shows the result of an RSS weather feed downloaded from the Internet and its representation rendered in a CAVE.

Collaborative virtual environments (CVE), scenario training, product searching, and other real world data related applications can benefit from this work by gaining a simple way to incorporate up-to-date RSS feed data into their systems. The ability to dynamically change environment variables from RSS feeds enables a vast amount of information to be incorporated into a CVE, increasing immersion, productivity, and quality of the system.

Another goal for the project is platform independance, to allow the application to run on multiple architectures.

Fig. 1. Downloaded RSS weather data representation rendered in the CAVE

2   Related Work and Background

There is a large amount of work in rendering weather and animating weather conditions, including photo-realistic weather animations in real time. Modern day computer games are becoming involved in using more realistic weather simulations, quite recently Second Life [7] has just started improving its environment by integrating a new photo realistic weather rendering system [6] from Windward Mark Interactive [1] called Nimble™and WindLight™which use physics based algorithms for rendering volumetric clouds and calculating the effects the atmosphere has on a lighting model. The Second Life game environment has simulated weather patterns that change during play but the weather information is generated for the game environment and does not relate to the real world weather. This must not be confused with the 3D Weather Data Visualization in Second Life [11] which is an in game 3D map of the United States of America with up-to-date weather data of America.

Architects use similar technologies to view their designs in the real world, enabling them to see casting shadows and other aesthetics that may be difficult to visualise normally. These systems can simulate the effects of the sun during any time of the day, but do not gather weather information from the Internet.

There are a few games that support real world weather conditions, these include some of the more recent flight simulators like the Microsoft Flight Simulator [9], and Black & White [10] a real time strategy game. Both these games use dedicated weather information that is specifically formed to work with the application and is downloaded from a dedicated server. In Black & White you are able to simulate a weather location by typing in your zip or postal code into the game, and Microsoft Flight Simulator simulates up-to-date weather of the location you are flying in.

A mobile phone application named Mobile Weather [3] uses Yahoo Weather [12] RSS feeds to display location based weather information to the user on their mobile phone but does not render the data in any way, it displays the data in an easy to read format.

3   Overview of Approach

This work differs from other work by having the ability to use any RSS data source, linking select information with a rendering engine. This section describes the integral sections of the work and how they combine together with a rendering engine. The use case describes this projects use in rendering a weather simulation. Fig.2 shows a basic representation of the systems structure.

3.1   RSS Downloader

All underlying networking used to connect to the Internet and download RSS feeds is achieved through the use of sockets and a very basic set of HTTP 1.1 protocol [8] commands. Connections to a web server can also be made via a proxy with or without basic authentication, the proxy server settings are set in the configuration file for the application. Sensitive data such as proxy authentication can be left blank in the configuration file, the application will ask for these details at runtime.

When downloading an RSS feed, first a connection request is made to the specified server, on success a page request of a known RSS feed is sent and the response is stored into memory, the RSS data is retrieved from this response and passed to the XML parser.

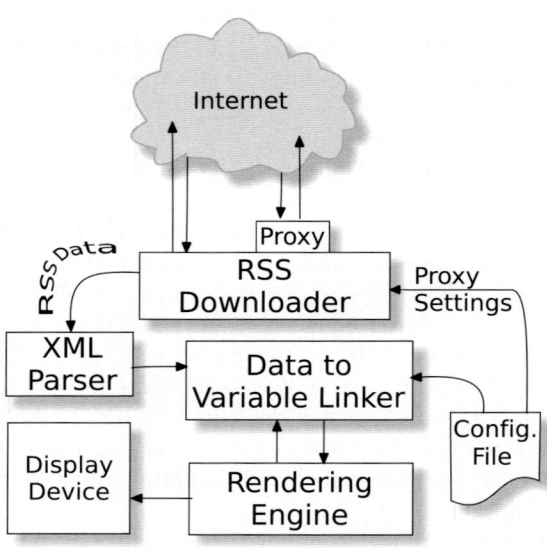

Fig. 2. A System Overview

3.2   XML Parser

RSS feeds are written using XML, therefore a very small, fast and effective, open-source, cross platform C++ XML parser [2] is used within the application to both parse the downloaded RSS feeds, and the configuration files that help link the RSS feed data to relevant variables in the rendering engine as well as general configuration options for the application itself.

The XML parser loads a full XML file into memory, this can either be from physical file or memory location, it parses the data and generates a tree structure representing the XML file. Once parsed successfully its possible to easily explore the tree and retrieve any data required. Modifications to the tree are also possible. Memory management is totally transparent through the use of smart pointers so there is no need to call `new, delete, malloc or free,` decreasing the possibility of memory leaks and increasing its efficiency.

3.3   Data to Variable Linker

On request from the rendering engine the data to variable linker will use the RSS configuration file information to request the required RSS data from the XML parser for a given variable name and return the data to the rendering engine.

3.4   RSS Configuration File

A new configuration file is created for every RSS feed downloaded from the Internet unless it already exists. The generated configuration file lists all the variables found in the RSS feed as a comment, to be used as a reference when editing the configuration file. This does not necessarily include all possible variables as the feed may change over time. For accuracy it is advised to still read about the the the particular RSS feed you are attempting to use from its origin. The file needs to be edited to create links between the RSS data and the rendering engine, once this is done it dose not need editing unless the RSS feed changes.

The configuration files are implemented in this way so that its possible to easily change the data being used within the system, it also gives the added advantage of quickly and easily adapting the system to new or changed RSS feeds. An example of the RSS configuration file is given below.

```
<!-- XML Config file RSS feed links -->
<?xml version="1.0" encoding="UTF-8" standalone="yes" ?>
<feed>
        <url></url>
        <date></date>
</feed>
<link name="">  <!-- name to request from rendering engine -->
        <rssData><!-- RSS data to find -->
                <type></type>
                <name></name>
        </rssData>

        <!-- return rules: return, equal, notequal, grater, less -->
        <return>
                <value rule="" data=""></value>
                <default></default>
        </return>
</link>
```

Each link has a name, this would be the name requested from the rendering engine. The rssData values are used to find the RSS data required, type refers to an attribute or node, and name refers to the type name in the RSS feed. It is possible to set up basic rules when returning data, these are **return** which simply returns the RSS data value, **equal**, **grater**, and **less**, which compare the RSS data to some constant value, and will return the RSS data value or a given return value if the rule is true. If the rule fails or is false a default value is returned instead.

4   Use Cases

Accurate weather simulation and prediction can be a complicated and very compute intensive process, by using pre processed weather data that gives the predicted weather, this task of compiling the metrological data into useful information is skipped. There are may sites on the Internet that use RSS feeds for weather prediction information, Yahoo Weather being one of them. This section describes the use of a weather RSS feed in conjunction with the described work and a weather rendering engine to produce a virtual environment that's able to display the current weather of a specified geographical location around the world.

4.1   Rendering Engine

The rendering engine is able to animate various weather effects including the sun, clouds, rain, and the effect of the wind. The engine has been written in a modular form and is now not limited to these weather types, additional modules can be added to increase the diversity of weather patterns that the engine can simulate. Each module of the rendering engine is able to take parameters associated with their type of weather, the rendering engine currently has nine parameters that are used when rendering all weather simulations and are mentioned in the following sections. All the module parameters are linked with the RSS configuration files mentioned above, and the rendering engine requests the required data from the data to variable linker during runtime.

Rain and Clouds. The rain and cloud objects are able to load a 3D model to be used for each rain drop or cloud rendered in the environment. They are effected by the wind speed and direction, during animation the objects are pushed in the direction of the wind. The objects are randomly scattered over the environment and in the case of clouds are morphed in size to give the appearance of different cloud shapes. The volume of rain and cloud objects rendered related to the parameters passed to the rendering engine and can render from zero to `MAX_RAIN` or `MAX_CLOUD` number of objects which is defined at compile time of the application. Transitions between different amounts of rain and clouds are performed gradually over time to give a more realistic weather changing effect.

Sky Background. The entire environment is enclosed in a coloured dome with makes the background sky. The sky colour can be change to any value and will gradually fade to that value from it's current colour over time. During cloudy rainy weather a greyer sky is perceived, where as hot weather will produce a brighter shade of blue.

During sunrise and sunset hours the sky will change accordingly, during the night the sky will darken.

The Sun. The sun uses the current time of day and places itself accurately in the sky according to sunrise and sunset times. The sun will light all of the models in the environment according to its position in the sky. It is also possible to speed up time and see a whole day's cycle.

Fog. Fog was implemented in a novel fashion; a plane is placed in front of the user, and as the user moves the plane moves with them. Different transparencies are set depending on the weather visibility. If the plane is set to be completely transparent, it shows the scene as normal and is not noticed. A semi-translucent plane shows a partially foggy scene and an opaque plane results in a very foggy scene. Fog was implemented this way for simplicity, at a slight sacrifice of aesthetic look. Transitions between visibility levels are changed over time to give the effect of fog either emerging or dispersing. Fig.3 shows an example of the simple fog effect.

Fig. 3. An example of a scene with and without fog present

Wind. The wind object does not render any graphics but manipulates the rain and cloud objects. The wind speed and direction are converted into a translation that is applied to both the clouds and rain. The wind will randomly deviate slightly from its true speed and direction to more closely resemble wind in the real world as it is never constantly blowing in exactly the same direction or speed all the time.

Time. The time object has two main roles. It either returns the current time of day converted into minutes, or a custom time interval can be set to increase or decrease the speed of time. The time object is used mostly by the sun but can also be used to check for updated RSS feeds after an elapsed time period.

Configuration File. There are many weather feeds available on the Internet today and each is specific to it's source, therefore a system needed to be developed that would allow easy configuration of this received data so it could be easily converted into useful information for the rendering engine.

The applications configuration file allows you to map selected values from a specified RSS feed to the weather parameters needed by the rendering engine. It also stores default values for parameters that cannot be mapped, details of the location the RSS weather data is associated with, proxy details and authentication information if needed, a specific model can be loaded with the weather simulation, as well as custom models for the rain and clouds. The functionality for adding custom files for the rain and cloud models was added for diverse effects like being able to render a scene with it literally raining cats and dogs, or to use a superior cloud model than the one provided. An example of the application configuration file is given below.

```
<!-- XML Config file for mangoXML Weather Simulation -->
<?xml version="1.0" encoding="UTF-8" standalone="yes" ?>
<configuration>
        <proxy host="" port="">
                <authentication username="" password="" />
        </proxy>
        <location name="">
                <rss host="" file="" />
                <model file="" />
        </location>
        <defaults>
                <model file="" />
                <rain  file="" />
                <cloud file="" />
        </defaults>
</configuration>
```

If a proxy isn't used then these details should be removed from the configuration file so that the application knows not to attempt to connect via proxy. If any attributes are left blank, they will be requested during runtime of the application, giving the option of not needing to save usernames and passwords into the configuration file.

5 Conclusion

This work demonstrates a novel way of using RSS data from the Internet enabling an easier way to integrate it into an application or virtual environment. The system is completely configurable and very easy to use. With the added ability of being able to connect to the Internet via proxy services using basic authentication methods.

Many applications will benefit from this work as they gain a simple way to incorporate up-to-date RSS feed data into their systems. The work give the ability to dynamically change environment variables from RSS feeds, and enables a vast amount of information to be incorporated into their systems, increasing productivity, and quality.

The case study has been successfully implemented, and the addition of an XML parser and variable linker has increased its usability, giving it the ability to use many different weather sources. The application is also cross platform compatible and able to run on multiple architectures.

6 Future Work

Further tests from other RSS sources need to be carried out to show that similar results can be achieved with different RSS feeds. The application still has much more to offer, further development in the XML configuration files could include a more generalised and intelligent way to integrate the RSS feed data to the rendering engine so that the configuration files are easier to configure. Incorporating all the defined variables of the rendering engine into the configuration files will allow further changes to the rendering engine without recompiling the source code. Implementing a callback structure for the RSS data linker would allow seemless updates to single variables rather than requesting an update for all variables from the engine. The rendering engine can have many graphic improvements including volumetric clouds and fog. Weather after effects like wet shiny surfaces during and after rainfall. Inclusion of further weather types, snow and hail could easily be produced from a derived rain class. The inclusion of audio would increase the realism of the weather types. The work should also be expanded into other areas of use to enable an easy way of intergrating RSS data into applications, opening them to larger resources of data.

References

1. Windward Mark Interactive, `http://www.windwardmark.net`
2. Frank, I., Vanden Berghen: C++ xml parser,
   `http://www.applied-mathematics.net/tools/xmlParser.html`
3. Brukakis, Dimitri: Mobile weather for s60 smartphones,
   `http://www.ubahnstation.net/projects/mweather`
4. Karelitz, Demiralp, D.B., Cagatay, S.Z., Jackson, C.D., Laidlaw, Cave, D.H.: fish-tank virtual-reality displays: A qualitative and quantitative comparison. IEEE Transactions on Visualization and Computer Graphics 12(3), 323–330 (2006)
5. Dunk, A.: Virtual weather. SCARP (2006)
6. Linden Labs. Linden lab shines new light on second life with acquisition of technology from windward mark interactive (May 2007),
   `http://lindenlab.com/press/releases/05_21_07`
7. Riley, D.: Better clouds, wind coming to second life (May 2007), `http://www.techcrunch.com/2007/05/21/better-clouds-wind-coming-to-second-life/`
8. Jeffrey, C., Henrik, M., Nielsen, F., Masinter, L., Leach, P.J., Berners-Lee Roy, T., Fielding, T., Gettys, J.: Hypertext transfer protocol – http/1.1. (June 1999),
   `http://www.w3.org/Protocols/HTTP/1.1/rfc2616.pdf`
9. Microsoft Game Studios. Microsoft flight simulator,
   `http://www.fsinsider.com/Pages/default.aspx`
10. Lionhead Studios. Black & white, `http://www.lionhead.com/bw/about.html`
11. Weber, A.: 3D Weather data visualization in second life (October 2006),
    `http://www.secondlifeinsider.com/2006/10/28/3d-weather-data-visualization-in-second-life/`
12. Yahoo! Weather Weather prediction and RSS feeds. `http://weather.yahoo.com`

Immersive Co-operative Psychological Virtual Environments (ICPVE)

Ronan Jamieson, Adrian Haffegee, and Vassil Alexandrov

Advanced Computing and Emerging Technologies Centre,
The School of Systems Engineering, University of Reading,
Reading, RG6 6AY, United Kingdom
r.jamieson@reading.ac.uk

Abstract. Virtual Reality (VR) has been used in a variety of forms
to assist in the treatment of a wide range of psychological illness. VR
can also fulfil the need that psychologists have for safe environments in
which to conduct experiments. Currently the main barrier against using
this technology is the complexity in developing applications. This paper
presents two different co-operative psychological applications which have
been developed using a single framework. These applications require different
levels of co-operation between the users and clients, ranging from
full psychologist involvement to their minimal intervention. This paper
will also discuss our approach to developing these different environments
and our experiences to date in utilising these environments.

1   Introduction

Psychologists are constantly seeking new methods and technologies for treating
a range of psychological illness. They also have a need to created realistic
controllable environments in which to conduct research into the different psychological
aspects of human behaviour. One such technology that can fulfil both
these requirements is VR, which provides users with the means to create 3D
computer generated virtual environments (VE's). These virtual environments
can be viewed in a variety of ways using different display technology from non-immersive
systems (e.g. PC desktops) to semi-immersive (e.g. head mounted
displays, HMD's[1]) right to fully-immersive systems (e.g. CAVE[2]). A fully immersive
system allows us to be bodily immersed within the virtual environment,
this assists in creating a feeling of presence. As mentioned by Slater "presence
means that the user constructs a mental spatial model out of virtual stimuli
and the perception of the self in the virtual environment"[3]. The advantage
that a fully-immersive system gives the therapist/psychologist is, providing the
client/user with a increased feeling of presence. This will help lead to a more
successful treatment outcome for the client, as the environment will be more involving.
Our applications will be focusing on utilising a fully immersive system
for these reasons.

There are many different options on what constitutes a co-operative virtual
environment, and some of these will be discussed in more detail in the next section.
In terms of what is considered a co-operative virtual environment for this

M. Bubak et al. (Eds.): ICCS 2008, Part III, LNCS 5103, pp. 438–445, 2008.

paper the following approach has been taken which has had to also consider the terms from a physiological point of view as well. From a psychological viewpoint different types of environments require varying levels of co-operation between client/user and therapist/psychologist. Therefore we have used the term in the following context. A full co-operative environments requires continual feedback from the client to enable the therapist to modify the environment. This modification is required to enable the client to overcome their phobias/fears and bring about a positive treatment outcome. Where as an automatic co-operative environment reduces the needs on the physiologist to continually control the environment as once the user completes a set task then the system can automatically move onto the next defined task, this allows the physiologist to focus on analysing the actions of the user. Also all actions of the user (e.g. head movement, length of time in a particular position/area) is recorded by the system for further off-line analysis. It should be noted that co-operation should not be confused with collaboration, in terms of VE these are two distinct different approaches.

The following section examines the related work concerning using VR for treating physiological illness and as an experiential tool. It also outlines current research into what is considered a co-operative virtual environments. Section three provides an overview of how we have used our framework to create our different environments and discuss some of the specific requirements that each environment has. This is then followed by an examination of our two application areas that were chosen. Finally our conclusions are presented and future work if required is outlined.

2   Related Work

When investigating previous work carried out in this area it becomes apparent that there are two main topics that needed to be considered. The first being the use of VR in the treatment of psychological illness. The second being, what is considered co-operation when using VEs.

The main focus for using VR with treating clients psychological problems has focused on using HMD's to created the virtual environment in which clients are treated. Using this techology the most successfully treated illness have been situational based phobias. This is due to the fact the VR is a powerful medium which allows for the creation of realistic environments. Situational phobias are related to a persistent fear of a particular stimulus and therefore leads to an avoidance of that stimulus. Acrophobia is an example of this type of phobia, it is concerned with the fear of heights. Traditional therapy uses in-vivo techniques to treat this, where as VR has been found to be as if not more successful than this technique[4]. With the added benefits of the being able conduct the treatment in the privacy and safety of the therapists office. Also complex and potentially costly situations are possible to recreate virtually. Another area that VR is currently being investigated as a potential therapeutic tool is Post Traumatic Stress Disorder (PTSD)[5].

When implementing a co-operative psychological VE it is important to understand what interactions will be required and approaches others have taken. In an traditional VE when dealing with users, who share the same VE and are manipulating the same objects, there are some clear distinctions to be made in the way that they work together. In terms of VE there is a difference between co-operation and collaboration. According to Broll, co-operative implies joint editing of shared objects, while collaborative additionally allows truly concurrent editing[6] . Where as Margery defines co-operation as a situation where two or more people interact on the same object in a concurrent but co-operative way[7]. As our interactions involve both the client/user and therapist/psychologist our version of co-operation is very similar but different, depending on what outcome is required. In our case both parties are jointly engaged in the environment and through their actions (conscious and sub-conscious) act on objects with in the scenario which effects the overall outcome of the session.

3   System Overview

Our two different environments have been developed using a single framework, VieGen[8], which is used for the creation and control of Virtual Interactive Environments. It aims to provide users with limited technical ability the power to create their own virtual environments without the need of learning a complex programming language. It also has a API for more advanced users.

A system overview of our approach can be seen in figure 1. This approach allows us to take advantage of the modular nature of VieGen.

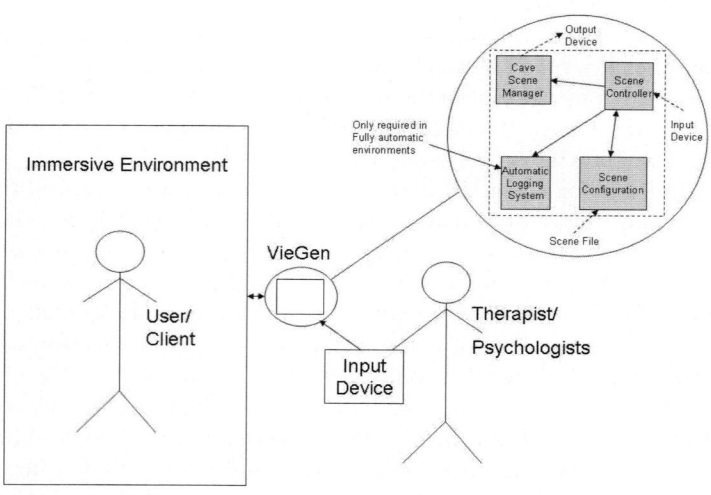

Immersive Co-operative Virtual Environments – System Overview

The following features of the framework are used in our applications:

3.1 Scene Configuration

The Scene Configuration module is responsible for loading the scene file. The scene file is a text file written either in plain text or XML. It contains a description of the objects we require to create the environment, also the position and scale of these objects. We can also assign attributes to objects if we require (e.g. a car object can be dynamic and thus will be able to move around the scene either on a random or pre-set path).

3.2 Logging

The Logging Module allows the psychologist to automatically capture data from the experimental sessions. This data can be of various different parameters and date/time stamped. An example of these parameters could be the position the users head was in at a particular location within the environment and for how long they remained in a particular pose. It is a fully configurable module, which allows the psychologist to decide which parameters they require capturing before the commencing of the session.

3.3 Cave Scene Manager (CSM)

The Cave Scene Manager Module is used for the internal representation of a user in the VE and manages the basic hierarchy. It provides a generic abstraction layer to the underlying VR hardware allowing system independence. By utilising the tracked users head position and field of view, it manages which objects would be observed and thus require rendering. It controls how this should be achieved and maps the resulting display onto the VR hardware.

3.4 Scene Controller

The Scene Controller Module takes care of the maintenance and control of the active environment. It handles the position and movements of any dynamic objects. It looks after the the logical hierarchies between objects, and uses these for grouping their behaviour.

4 Application Areas

Creating applications that fulfil the needs of psychologists and therapists has traditionally being an expensive and complex process, involving large amounts of a developers time and effort. This has not always delivered what the end users required. But by using our approach we aim to assist the end users in developing co-operative environments that fulfil their needs, by allowing them to quickly and easily build these environments themselves.

Two specific applications have been develop which address different psychological needs with varying levels of co-operation. The first application is concerned with addressing the requirements of creating a fully co-operative environment for the treatment of situational phobias. Where as the second application is used to create a automatic co-operative experimental environment for studying aspect of human behaviour. These have been both developed using the same framework, using modules of the VieGen framework that fulfil their requirements.

4.1   Fully Co-operative Virtual Environment

As mentioned previously VR has been used for the successful treatment of situational phobias. One such phobia, that has very serious consequences for the clients who suffer from it is, Post Traumatic Stress Disorder (PTSD). This is usually caused by the client being placed in a scenario that causes a traumatic event to occur to that person. This could be a violent assault, terrorist attacks, military combat amongst others. We have based our application on a military combat environment. This application was required to be fully co-operative due to the fact that the client must engage with their therapist. They share the environment in terms of the client and therapist are communicating continually about the environment and situations that occur within the session. They discuss the feelings/emotions that gets invoked during the session. This co-operation must take place to allow the client to move on emotionality/psychologically.

The therapist uses an input devices with contains a GUI, which allows them to control the environment variables (e.g. time of day, audio, movement of people/vehicle). Based on the clients case notes, the therapist will have a record of the traumatic event that lead to the clients PTSD. From this information the therapist can increase the intensity of the environment (e.g. more explosions, insurgents appearing) and guide the client to a successful outcome. This outcome will involve a reduction in the clients PTSD symptoms. The current application is based on a simple environment that allows the client to navigate through an scenario that resembles a downtown market in a middle eastern country (see Figure 2(a)), there are also people and Vehicles within the environment (e.g. soldiers, Humvee). Other features include the ability to add extra audio effects (e.g. gunfire, explosions, sirens), we can also add and remove insurgents (see Figure 2(b)) from the environment.

The environment has been tested on a number of volunteers. Due to the serious nature of the illness, it was not possible to intially test the environment with clients suffering from PTSD. The volunteers time was used to test the immersion of the environment versus a HMD version. Also tested was the interaction with the different elements within the environment (e.g. insurgents being added and remove from the scene). The audio elements are vital to the feeling of presence with the environment and various combinations where tested with the volunteers feeding back on the most realistic audio effects.

Fig. 1. (a) Market Place (b) Insurgent Appearing

4.2   Automatic Co-operative Virtual Environment

The ability to create controllable, measurable and realistic environments is a requirement for psychologists that seek to study different aspect of human behaviour. The developed application was required to be a fully automatic co-operative environment, the user would complete the task set and the system would move onto the following task, thereby the system and user are automatically co-operating with the psychologist. This feature allowed the psychologist to observe and analyse the users behaviour and reactions. The application developed was concerned with customer shopping behaviour, with the aim being to carry out a psychological analysis of whether different layouts of products, both in positioning and in shelf dressing around them, would have any effect on the choice of users' selected products.

Two series of trials of the environment were carried out with a range of different users. The first being a Search Test, this test was to record the movements and time taken for a user attempting to locate a product in a VE. The user started next to one of a number of different shelf layouts. They were shown one or two virtual products when they then had to locate on the shelf. By recording how the user moved with the environment, including their head positions and movements, it was possible to reconstruct the way they acted within the test. The purpose of this test was to discover if there was any relationship between searching and the layouts used. An example of this trail can be seen in Figure 3(a).

The second trail was a Primed Selection Test, this test required the user to move along a fixed path within the environment. Along the path were ten categories of products. The first time along the path the subjects were asked to select specific items from a list, and the second time they were free to choose themselves. The hypothesis being investigated, was whether priming the users with the sample product would cause them to select, or at least notice those items. An example of this trail can be seen in Figure 3(b).

These trials were conducted with 16 users, who completed both tasks. The main points of note that came from these trails were that the effects of the differnet layouts did have an noticeable effect on the users ability to find particular products and that if a user was primed about a certain product they could find it quicker, and were more likely to choose the primed product when given a free choice.

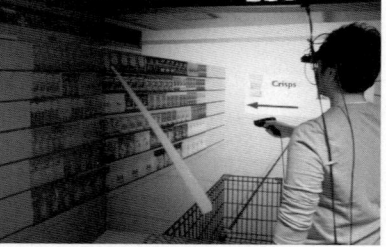

Fig. 2. (a) Search Trail (b) Primed Selection trail

5 Conclusion and Future Work

This paper has described the successful development of two different co-operative psychological environments for immersive VR systems. Each application has incorporated the required levels of co-operation that is needed. This ranged from fully co-operative to automatic co-operative environments. These applications are currently at different stages of implementation, the automatic co-operative environment (i.e. customer shopping behaviour) has been utilised in successfully testing various different hypothesis on a wide range of different consumers, where as the fully co-operative environment (i.e. PTSD) is awaiting user trails which should commence shortly.

Future work will include expanding the range and complexity of the consumer trails in the automatic environment. We will also endeavour to increase the cross section of consumers that are to be tested. Automatic environments that incorporate tracking the eye position of the consumer rather than head position will be investigated as this should increase the accuracy of the gaze data captured. Increasing the realism of the fully co-operative environment by the use of haptic feedback will be researched. The range of audio clips for both environments will be expanded which also should increase the feeling of presence within each application. Expanding the range of PTSD environments will be investigated, to try and discover if there is a need for this approach within other professions which are exposed to potentially stressful scenarios (e.g. fire-fighters, police). Research will also be carried out into the possibilities and advantages of enabling the environments to be networked.

References

1. Sutherland, I.: A head-mounted three-dimensional display. In: Proceedings of Fall Joint Computer Conference, vol. 33, pp. 757–764 (1968)
2. Cruz-Neria, C., et al.: The CAVE: audio visual experience automatic environment. Communication of the ACM 35(6), 64–72 (1992)
3. Slater, M., et al.: Immersion, Presence and Performance in Virtual Environments: An Experiment using Tri-Dimensional Chess. ACM Virtual Reality Software and Technology (1996)

4. Schuemie, M.J., et al.: Treatment of Acrophobia in Virtual Reality: a Pilot Study. In: Broeckx, F., Pauwels, L. (eds.) Conference Proceedings Eurcmedia, Antwerp, Belgium, May 8-10, pp. 271–275 (2000)
5. Rizzo, A.A., et al.: User-Centered Design Driven Development of a VR Therapy Application for Iraq War Combat-Related Post Traumatic Stress Disorder. In: Proceedings of the 2006 International Conference on Disability, Virtual Reality and Associated Technology, pp. 113–122 (2006)
6. Broll, W.: Interacting in distributed collaborative virtual environments. In: VRAIS, Los Alamitos, pp. 148–155 (March 1995)
7. Margery, D., et al.: A general framework for cooperative manipulation in virtual environments. In: Gervautz, M., Hildebrand, A., Schmalstieg, D. (eds.) Virtual Environments, Eurographics, pp. 169–178. Springer, Heidelberg (1999)
8. Haffegee, A., et al.: Creation and Control of Interactive Virtual Environments. In: Alexandrov, V.N., van Albada, G.D., Sloot, P.M.A., Dongarra, J. (eds.) ICCS 2006. LNCS, vol. 3992, pp. 595–602. Springer, Heidelberg (2006)

Environment for Collaborative Development and Execution of Virtual Laboratory Applications

Włodzimierz Funika[1], Daniel Harężlak[2], Dariusz Król[2],
and Marian Bubak[1,2]

[1] Institute of Computer Science AGH, al. Mickiewicza 30, 30-059 Kraków, Poland
[2] ACC CYFRONET AGH, ul. Nawojki 11, 30-950 Kraków, Poland
{funika,bubak}@agh.edu.pl, d.harezlak@cyf-kr.edu.pl,
dkrol@student.agh.edu.pl

Abstract. This paper presents the solutions for a user interface environment which have been developed within the ViroLab Virtual Laboratory to enable developers and medical researchers to develop and execute experiments. Experiments require support in form of a script editor easily extendible by a number of additional functions related to the functionality of the Virtual Laboratory, like sharing experiments, and an experiment management mechanism which enables the experiments to be executed with a number of facilities allowing, e.g. tracking the execution, logging errors. Moreover, two user groups, developers and users, need to collaborate to improve experiments and introduce more refinements to the research conducted, which imposes many requirements on the environment which becomes the main collaboration platform for the researchers. The work identifies these requirements in the domain of medical sciences and proposes solutions for efficient and convenient collaboration.

Keywords: virtual laboratory, experiment plan, user interface, collaboration environment, distributed development.

1   Introduction

Working in research projects which involve building applications in new fields of science often demands implementing reliable multi-purpose graphical user interfaces. One of the main tasks of the EU IST ViroLab project [1], [2] is to provide an easy way for different specialists – computer scientists as well as medical researchers – to access distributed resources with sharing and processing clinical data coming from medical practitioners.

The way of accessing data or using computational resources should be simplified so the end users do not need to know about the underlying infrastructure. Groups of researchers working on a project should use the facilities provided by the computer domain, as they would use laboratory equipment during conducting an experiment. ViroLab Virtual Laboratory defines *experiment* as a "process that combines together data with a set of activities that act on that

M. Bubak et al. (Eds.): ICCS 2008, Part III, LNCS 5103, pp. 446–455, 2008.

data in order to yield experiment results" [3]. Furthermore, it is required that the software would constitute a virtual laboratory in which scientists from different geographical locations may collaborate and take part in an experiment. This involves preparing a presentation layer which is well-tailored for specific needs of scientific collaboration environments.

One of the primary functionalities the presentation layer must support is to provide facilities for the developer to edit an experiment plan with any enhancements which allow to share experiments (store/retrieve), access data and services (browse/fetch), test experiment plans Grid (track execution/report errors). Whenever an experiment plan is considered mature enough to be submitted for a real execution, some other presentation functionality is needed for the user, which assumes that via a standard Web browser they can launch the experiment plan and get a feedback in form of execution results, moreover, observe whether the experiment is running well or any errors occur. Challenges to be addressed are easing the development process, bridging a gap between the developer and the user, the reusability of components [4] and providing a common collaboration space, a typical issue of distributed development in various domains (e.g. [5]).

This paper presents an approach how to address the needs of two different groups of users, the ones who plan experiments and those who run them and how to meet their demands which turn into collaboration issues. A novel approach is presented to cover the development on the Grid and the interaction with the Grid problems.

The rest of the paper is organized as follows: we give an overview of related work in Section 2, then we are coming to a solution we have adopted for the research, in Section 3. The implementation status presented in Section 6 is followed by concluding remarks and future plans.

2   Related Work

A graphical user interface is built according to the specific requirements of a given project. In the area of grid programming, many efforts were undertaken to satisfy the needs of end users with respect to graphical user interfaces, especially if we are speaking of application composition. Many solutions refer to the workflow approach of representing the internal component dependencies of the applications.

In the K-WF Grid project [6], the User Interface is available through the Grid-Sphere portal [7]. A task is defined through the Grid Workflow User Interface (GWUI) which is a Java applet. The user starts a workflow definition by providing a text sentence which describes the problem. At the next stage this sentence is analyzed and a proper context based on it is selected. GWUI is underlied by GWES – Grid Workflow Execution Service. All conflicts from within this layer are automatically delegated into GWUI – after that the user is able to see a description of the problem.

P-Grade project [8] uses a special designed graphical language – *Grapnel*. It is used for defining relations between resources. This language hides calls to low level communication functions; it uses predefined topology templates for communication between resources (pipe, ring, etc.). A workflow is defined graphically. The use of the Grapnel language resulted in the emergence of a Grapnel language-oriented Editor (GRED) which features automated positioning of ports and flow controls elements and support of grid portals. While this approach is very user-friendly for beginners, experienced users want to have more control on what and how is designed, e.g., with text editors.

Taverna project [9] represents another approach to building application workflows problem, mainly used in bioinformatic domain. User Interface is represented by a standalone application (called Taverna Workbench) within which the user composes a workflow from prepared blocks, e.g. services. No technical knowledge about workflows is required, thus it can be used by the scientist without programmistic skills.

The `myExperiment` project [10] is closely connected to Taverna. Its main goal is to provide a so called collaborative space for the Taverna project community via web portal. It can be used to present (as a screenshot exported from Taverna Workbench), share or collect feedback from users about a workflow.

However, these solutions do not decouple the processes of building and executing the applications, which very often induces additional overhead for end-users who also need to be experts in workflow construction and constrains the environment just to executing workflows without the capabilities that are required by scientific communities (e.g. sharing knowledge, managing results, etc.).

Recently, the area of web development has gained a lot of attention thanks to the new technologies often related to as *Web 2.0 technologies*. There are many examples of applying the techniques that come with the new trend of web development in e-Science [11]. The capabilities of the new approach allow to create collaboration environments based on well-known web standards which are already successfully used by social communities all over the world, such as MySpace [12] or Facebook [13]. The content in such environments is provided and assessed by the users themselves, which makes it very dynamic and rich. An important goal would be to create such environments applicable to the science domain, where knowledge is provided and shared in communities of scientists, using standard web tools (web browsers, web services).

3   Requirements for UI and a Solution in Virolab

Our approach to providing user interfaces reflects the adopted assumptions on classifying the ViroLab users' community into two groups with respect to user interfaces.

The first user group of the ViroLab Virtual Laboratory are *experiment developers* who combine their domain knowledge and technical skills to plan and develop new experiments. Therefore providing tools that support experiment development process is crucial. Another aspect that should be kept in mind is collaboration. Mechanisms for sharing knowledge and experience should be as important as development tools. These features are provided by a dedicated

Integrated Development Environment (IDE), we call ViroLab *Experiment Planning Environment* (EPE). The idea behind EPE is to gather the facilities supporting the above activities into a single integrated environment.

A challenge is to prepare and integrate a set of tools that will be on the one hand powerful enough to satisfy an advanced user and on the other hand it should be user friendly, so that users who are not familiar with applied technologies could use it. In either situation EPE should decrease the complexity level of the experiment management and publishing process.

The functionality of the environment is concentrated around two main parts that address the above mentioned issues:

- a dedicated editor for experiment development that provides syntax highlighting and code auto-completion,
- the Experiment Repository client that enables sharing experiments between developers and publish experiments to the scientific community.

Moreover, due to the distributed nature of computation in the Virolab Virtual Laboratory a tool that provides information about available computational resources, which are stored in the *Grid Resource Registry*, is needed. Another feature that facilitates experiment planning is access to the data model which is available in an ontology form. Also an *Outline* view that shows each object from the source code, e.g., variables or methods definitions, is available.

¿From the collaboration point of view, EPE provides a mechanism for exchanging information between experiment developers and scientists. It is also possible to share a single experiment among a group of developers who are geographically distributed. It is especially important when developers from different institutes want to collaborate on a single project.

The second user group of the Virtual Laboratory involves the *experiment users* who need to manipulate the queue of run experiments and trace the execution of experiment plans. This group concentrates mostly around clinicians and doctors of specific medical domains, who are not assumed to have complex knowledge about scientific computing. Therefore developing a generic interface for experiment management is a challenge. Our analysis of user requirements has resulted in the identification of four main purposes regarding the functionality of the *Experiment Management Interface* (EMI).

The first task is to *locate a desired experiment* by browsing through various experiment repositories and then through experiment versions. The repositories are populated by developers in the process of creating experiments with EPE. Different versions deliver different functionalities with respect to the same experiment plan (e.g. enhancements like mail notification or use of a service with different algorithm). Another task would be to *manage the experiment execution* and to *monitor its status*. This part also includes interacting with the experiment by providing data when needed. The data requests are planned by the experiment developer during the implementation phase. The third task is to *gather, store* and *share results*. This involves building a collaboration space for the users executing the experiments, which enables sharing the data returned by an experiment and the knowledge which comes with the results. Finally, it is desirable to enable *providing feedback on the quality* of an experiment to its developers. And last but not least is the user-friendliness of the interface, which refers to both accessibility and use.

While the described solution delivers two separate graphical user interfaces, specific for each user group, there are possible common fields of cooperation between these two user groups. It can comprise reporting problems of the correctness of experiment plans, identifying new requirements, enhancements, etc. This issue imposes a need in a *collaboration space* between EMI and EPE users. This role is played by an experiment repository, which is used by EPE to store and manage versions of the developed experiments and by EMI to list and present details of the available experiments. The solution under discussion allows to use more than a single repository, one per specific field of interest.

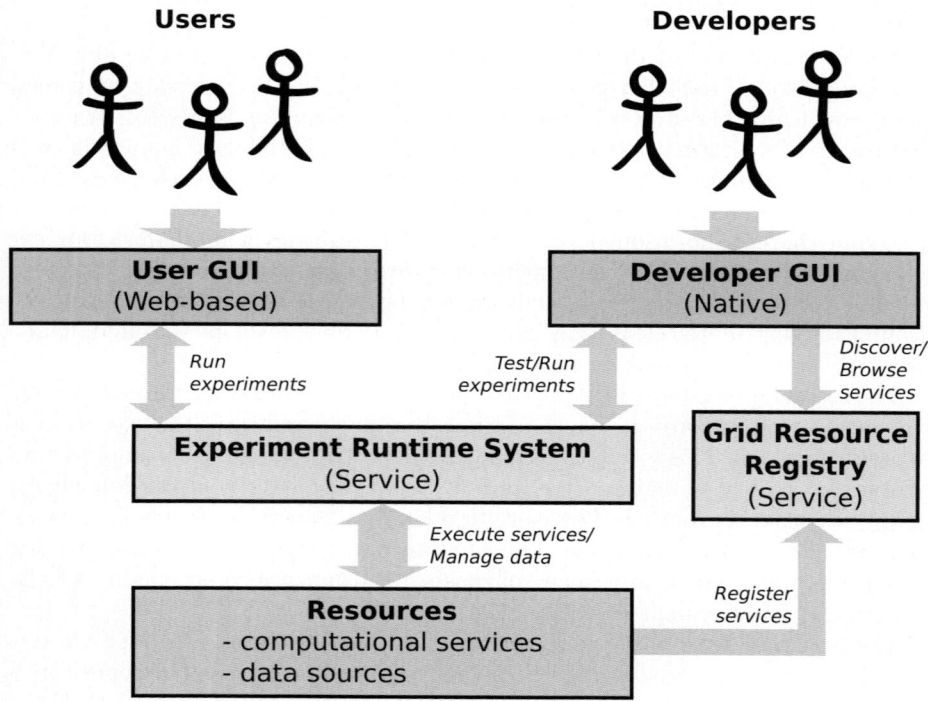

Fig. 1. General concept of user interfaces in ViroLab

In Fig. 1 an overview of User Interfaces with the underlying system structure is presented. On the left, experiment-user side, an easily accessible solution using well-known standards is required. To accomplish this, the interface is implemented as a lightweight web application. The concept of thin client enables users to manage their experiments through even mobile devices.

4   Experiment Planning Environment

As mentioned above, the main goal of EPE is to simplify the experiment development process by providing an extendible collection of dedicated tools. In order to fulfill this requirement, EPE is based on the Eclipse Rich Client Platform (Eclipse RCP) [14]. The core of the platform includes a component called Equinox which is an implementation of the OSGi core framework specification [15]. From the user's viewpoint this means that each part of the environment is treated separately and can be loaded during runtime on demand, which is known as the *lazy initialization* technique.

As mentioned, we do not want to limit the environment to a fixed set of tools. Therefore another feature of the EPE architecture should be extendibility. From the very beginning Eclipse RCP was assumed to be a platform that should be extended easily by new features. Therefore, a special kind of component, called *plugin*, that provides some functionality along with information about itself which is processed by the platform, is used. Thus it is possible to develop new plugins and add them to the environment or publish them and allow other experiment developers to make use of them.

However, plugins make the environment too fine grained, thus it becomes difficult to manage EPE that may contain dozens of plugins. To overcome this problem, the *feature* was defined as a group of plugins that are logically linked and address a certain developer activity. This approach facilitates organizing the internal structure of EPE and extending its functionality.

The current version of the EPE contains the following elements (see Fig. 2):

- *Workbench* that is the central point of the application that manages the layout of each part of the EPE. From the workbench one can choose a *view* or an *editor* to open.
- *Script editor* which supports the experiment development process. It aims to provide such functions as: syntax coloring and code assistance.
- *Experiment Repository client* that supports experiment sharing operations. Dedicated wizards enable to export experiments to or import them from the Experiment Repository.
- *Connector* to the execution service. After setting up the experiment interpreter (e.g. GSEngine) one can start running an experiment with one click. Different versions of the interpreter can be applied along with different experiments.

5   Experiment Management Interface

To support interactions with the user EMI provides a set of related visual components that allow to perform dedicated actions (e.g. run experiment) via a web browser. In Fig. 3 several dependencies between visual components and server-side client libraries are presented. The user is operating in the client layer which is available through a web browser. This allows for platform-independent interactions between the users and the experiment. Each of the visual components is

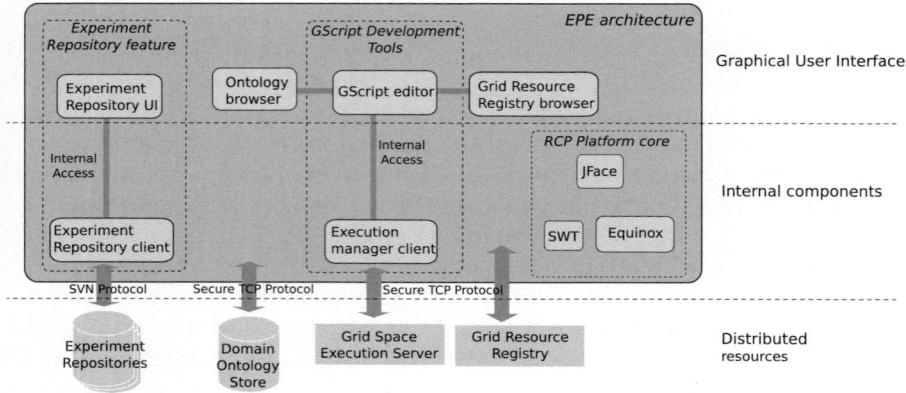

Fig. 2. EPE components and their dependencies

supported by a relevant client library which communicates with the underlying resource. The basic functionality of the EMI interface includes three components:

- *Repository Browser Component* – This component is responsible for presenting available experiments to the user. It is possible to connect to several repositories at a time and browse their contents. The repositories are implemented as SVN code stores. To provide coherence between many repositories a uniform directory structure of the experiments had been established. Access to such repositories is possible only through dedicated clients which keep the content according to the agreed standard.
- *Execution Manager Component* – To execute experiments and monitor their execution status, including handling user data requests, this component was implemented. It may execute more than one experiment at a time and present the actual state which can currently be either *running, input awaiting* or *finished*. On the server-side, a client library is responsible to forward experiment execution requests to the Grid Space Server [16]. Between the client and the server an independent communication protocol is used based on the secure TCP layer.
- *Result Management Component* – This component manages experiment results. They may be assessed and either removed or saved in an external Result Store for future use. One of the most important features of this component is the ability to share the results with other researchers. This follows the idea of delivering a collaboration space for groups of scientists.

Another feature supported by the Repository Browser Component is the feedback mechanism. This lets users executing an experiment contact a developer of this particular expeirment and report on its quality. This ensures a user-developer loop to be a part of the solution's lifecycle and improves the cooperation between the two groups.

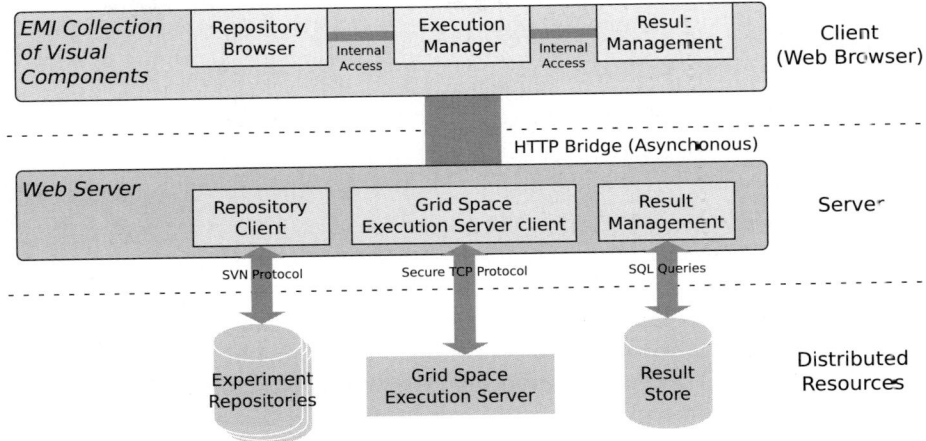

Fig. 3. EMI visual components and their dependencies to server-side clients and underlying resources

The communication model between the client and the server layers is asynchronous. This is a consequence of using Web 2.0 mechanisms such as AJAX. While in the standard *HTTP request-response* model all the contents of a web page including each of the components would have to be refreshed, using new mechanisms allows for much lower communication overhead between the client browser and the server, because only the content that needs updating is refreshed. This also decouples the requests coming from individual visual components and allows for convenient and less centralized web programming.

Another advantage of using the new web programming techniques, namely GWT [17] is the possibility of communication between individual visual components on the client side without the server's participation. This allows for fast content reloads in the client's web browser (as far as interactions with the server are not needed).

6   Current Usage of the Environment

The prototypes of both Experiment Planning Environment and Experiment Management Interface are integrated with other components of the ViroLab Virtual Laboratory [3]. Now they are being improved according to the reported end-user requirements. Both user groups successfully exploit and test the interfaces, thus cooperating on new scientific experiments. It is also freely available from [1], where one may provide feedback or notify about a useful feature that could be included. The EMI web interface is also available and accessible through the project page [3].

Both environments have already been used to implement and run a series of experiments involving simple tests, data access and virology, namely *Nucleotide sequence*, *HIV subtype* or *Sequence alignment* experiments. A full list of experiments can be found at [1]. The implementation already includes using different middleware technologies and infrastructures as well as interaction with the end users running the experiments.

7   Summary and Future Work

In the paper an approach to delivering a presentation layer for Grid users and developers is presented. The layer aims at fulfilling the collaboration requirements of those groups by providing on the one hand an integrated Grid development environment (EPE) and on the other hand a web-based management interface (EMI). The environments enable tight collaboration among and between those groups which is seen as a novelty in bridging the gap between Grid developers and users. One of main goals of this solution is to reach out beyond the scope of the ViroLab and provide a generic approach to using the Grid. Present prototypes already work and provide most of the functionality described in this paper. Many experiments have already been developed and successfully executed which acknowledged the feasibility of the solution. Subsequent improvements are applied to the system and include the following.

The main goal of the development layer is to provide the user with facilities that support experiment development process. Due to the fact that experiment is planned with a special script language (GScript) it is crucial to provide an extendible and helpful *script code editor*. This layer can also involve an integrated preprocessor, semantic code inspection, code autocompletion and an ontology browser. Within this layer the user can access the ViroLab Grid Environment and execute a script code (created in the script code editor) and trace the execution.

The prototypes of EPE and EMI are being enhanced to meet further requirements requested by target users. Specifically, for the EPE environment, additional administrative plugins are required and a more advanced autocompletion mechanism for convenient experiment planning. As for the EMI, a more user-friendly interface with a better feedback support and result management system is going to be implemented. The interactivity issues of the web client are also addressed, for this purpose Web 2.0 tools such as AJAX are used in the current implementation course. Further plans include implementing an advanced mechanism for requesting user data that supports building complex and dynamic forms. Another extension will let different formats of results to be presented to the users for convenient data analysis.

Acknowledgments. This work is partly funded by the European Commission under the ViroLab IST-027446 and the IST-2002-004265 Network of Excellence CoreGRID projects, as well as, the related Polish grant SPUB-M.

References

1. Virolab Virtual Laboratory Home Page, http://www.virolab.org
2. Sloot, P.M., Tirado-Ramos, A., Altintas, I., Bubak, M., Boucher, C.: From Molecule to Man: Decision Support in Individualized E-Health. Computer 39(11), 40–46 (2006)
3. Gubala, T., Balis, B., Malawski, M., Kasztelnik, M., Nowakowski, P., Assel, M., Harezlak, D., Bartynski, T., Kocot, J., Ciepiela, E., Krol, D., Wach, J., Pelczar, M., Funika, W., Bubak, M.: Virolab virtual laboratory. In: Proceedings of Cracow Grid Workshop 2007, pp. 35–40. ACC CYFRONET AGH (2008)
4. Goble, C., Roure, D.D.: Grid 3.0: Services, semantics and society. In: Proceedings of Cracow Grid Workshop 2007, pp. 10–11. ACC CYFRONET AGH (2008)
5. Anthes, C., Volkert, J.: A toolbox supporting collaboration in networked virtual environments. In: Sunderam, V.S., van Albada, G.D., Sloot, P.M.A., Dongarra, J. (eds.) ICCS 2005. LNCS, vol. 3516, pp. 383–390. Springer, Heidelberg (2005)
6. Knowledge-based Workflow System for Grid Applications, http://www.kwfgrid.net
7. Gridsphere Project Home Page, http://www.gridsphere.org
8. P-grade Project Home Page, http://www.lpds.sztaki.hu/pgrade/main.php?m=1
9. Oinn, T., Addis, M., Ferris, J., Marvin, D., Senger, M., Greenwood, M., Carver, T., Glover, K., Pocock6, M.R., Wipat, A., Li, P.: Taverna: a tool for the composition and enactment of bioinformatics workflows. Bioinformatics 20(17), 3045–3054 (2004)
10. myExperiment Project Home Page, http://www.myexperiment.org
11. Fox, G.C., Guha, R., McMullen, D.F., Mustacoglu, A.F., Pierce, M.E., Topcu, A.E., Wild, D.J.: Web 2.0 for grids and e-science. In: INGRID 2007 - Instrumenting the Grid, 2nd International Workshop on Distributed Cooperative Laboratories - S.Margherita Ligure Portofino. (2007)
12. MySpace.com: An international site that offers email, a forum, communities, videos and blog space (2008)
13. Facebook Utility Home Page, http://www.facebook.com
14. Eclipsepedia RCP Description, http://wiki.eclipse.org/Rich_Client_Platform
15. OSGI Specifications Home Page, http://www2.osgi.org/Specifications/HomePage
16. Ciepiela, E., Kocot, J., Gubala, T., Malawski, M., Kasztelnik, M., Bubak, M.: Gridspace engine of the virolab virtual laboratory. In: Proceedings of Cracow Grid Workshop 2007, pp. 53–58. ACC CYFRONET AGH (2008)
17. Google Web Toolkit – GWT, http://code.google.com/webtoolkit

Workshop on Applications of Workflows in Computational Science

International Workshop on Applications of Workflows in Computational Science (AWCS 08)

Adam Belloum[1], Zhiming Zhao[1], and Marian Bubak[1,2]

[1] Informatics Institute, University of Amsterdam, Amsterdam, the Netherlands
{adam,zhiming,bubak}@science.uva.nl
[2] Institute of Computer Science AGH, Krakow, Poland
{bubak}@agh.edu.pl
http://staff.science.uva.nl/~zhiming/workshop/awcs/2008/

Abstract. The goal of the Workshop on Applications of Workflows in Computational Sciences (WACS) is to provide a forum for sharing knowledge and experience on developing workflow applications, and to highlight important requirements for developing workflow systems. This short paper gives an overview on scientific workflow management systems and their application in e-Science, and introduces the topics of the Workshop. Several research focuses on utilizing scientific workflow in enhancing e-Science applications and on developing workflow management systems will be enumerated.

Keywords: scientific workflow, e-Science, application integration, data management, resource discovery, frameworks, security.

1 Aims and Scope

In many scientific domains, workflows become an important mechanism to prototype and perform complex experiments and to achieve scientific discoveries. A workflow system allows scientists to wrap and integrate legacy applications via an intuitive interface, to schedule the computing of different tasks at an abstract level, and to automate the processes in data processing. In this way, domain scientists can effectively utilize available resources and focus on the logic of the experiments instead of low level technical details.

During the passed years, a large number of workflow systems have been developed, e.g., DAGMan[1] and Pegasus[2] focus on managing massive computing tasks for processing distributed data, Taverna[3] integrates distributed web services based components, and Kepler[4] provides flexible GUI to prototype workflows and execute them. Decoupling workflow engine from the GUI and implementing the engine as a service allow better interoperability among workflow management systems, a good example for this interoperability has been achieved by invoking the WS-VLAM engine[5] from the Taverna workbench as a sub-workflow in a Taverna workflow.

M. Bubak et al. (Eds.): ICCS 2008, Part III, LNCS 5103, pp. 459–462, 2008.

Most of the SWMS are historically driven by applications in specific domains, e.g., bio informatics, high energy physics, and astronomical observations. Investigating the common characteristics in these domain specific systems and implement them as part of a generic framework emerges as an important need for e-Science applications. A number of research projects such MyExperiment[6] in the UK and VL-e in the Netherlands[7] aim to realize a Grid enabled generic framework where scientists from different domains can share their knowledge and resources, and perform domain specific research. Sharing knowledge and resource requires more interoperability among the major workflow management systems. More sophisticated solution are needed to achieve a seamless integration among workflow, approaches like the workflow bus[8] developed in the context of VL-e present a potential solution to the interoperability problem. Different requirements for supporting domain specific applications are important driving force for the development of workflow systems. It is therefore an important motivation for us organize such workflow to collect use cases from different application domains and understand specific requirements.

An overview of available scientific workflow systems and their application in e-Science is given in[9], while the challenges in this area are examined in[10]. Recently, workflow systems are build exploiting achievements in semantics and ontologies[11], [12], [13]. The goal of the Workshop on applications of workflows in computational sciences (WACS) is to provide a forum for sharing knowledge and experience on developing workflow applications, and highlight important requirements for system development.

2   Workshop Content

The Workshop addresses a number of issues related to the development of workflow applications.

Data management. Jablonski et al. present an infrastructure called DaltOn for managing the complexity in processing and integrating scientific data. DaltOn provides a data centered vision, namely data logistics, on modeling processes of collecting, storing, transporting and processing data in scientific experiments. The DaltOn aims at decoupling the data processing issues, e.g., syntactic and semantic conversions, from the actual workflow steps, and promoting the reuse of experiment logics.

Resource discovery. Huang discusses a resource discovery in distributed environment. A P2P environment called Virgo is used as the test bed. A multi-level virtual group architecture is used to manage the complexity of the system.

Security. Yau, et al. discuss the security and privacy issues scientific workflows. A trusted computing based scheme is proposed for selecting trusted resource providers, for protecting confidentiality and integrity of job information, and for auditing data for process provenance.

Usability. Buckingham et al. present a web accessible scientific workflow system called GPFlow. A model for collection processing based on key aggregation and slicing which guarantee the processing integrity and facilitates automatic association of inputs. The GPFlow workflow defined on a single value may be lifted to operate on a collection of values with no change required to the workflow. Combing this workflow framework with Grid infrastructure will be the next step to achieve.

Frameworks. Jakimovski et al. present new framework for the Gridification of Genetic Algorithms. The framework enables easy implementation of Genetic Algorithms and also enables researchers easy and stable usage of the Grid for their deployment. The design of the framework was based on principles that make it very open and extensible.

Wibisono et al. describe ongoing efforts for designing a framework for performing parameter sweep experiments. The paper discusses the requirements gathered from use cases in various scientific domains indicate that interactivity is needed but not fully supported by most of existing frameworks designed to support parameter sweep applications. Preliminary design of a framework that would support interactivity is presented.

Application use cases. Fernandez-Quiruelas et al. describe how a well known climate model (CAM) can take advantage of the Grid computing power. In this work usability and robustness are the major requirements because the potential end-users have little background in computer systems and should not be bothered with the complexity of the underlying infrastructure. This requires involve managing a complex workflow involving long-term jobs and data management in a user-transparent way.

3   Summary and Outlooks

Several related workshops have been organized during passed years, e.g., WSES[1], SWF[2], and SWBES[3] The development of workflow management systems faces two important challenging issues. On one hand, the domain specific experiments from different applications require customized solutions in workflows for particular problems; on the other hand, to enable knowledge transfer and information sharing between different domains, a generic workflow solution is also demanded. A successful workflow system no only means it has mature conceptual design and engineering but more importantly it can be effectively enhance real life applications. The usability of a workflow system is essential to introduce the system to different domains scientists: not only suitable interface for composing and executing workflow, but also a set of user oriented tools for viewing, moving and processing data and for the provenance of the workflow and reproducing the

---

[1] International Workshop on Workflow Systems in e-Science,
http://staff.science.uva.nl/~zhiming/workshop/wses/
[2] International Workshop on Scientific WorkFlows,
http://www.cs.wayne.edu/~shiyong/swf/
[3] International Workshop on Scientific Workflow and Business workflow standards in e-Science, http://staff.science.uva.nl/~adam/workshops/e-science2007/cfp-swbes-2007.htm

results. Sharing the knowledge in meaningful workflows is becoming be an important requirement for e-Science framework. It is necessary to integrate different technologies, e.g., semantic based annotation and searching, and collaborative working facilitates, with scientific workflow system.

Acknowledgement

We would like to express our great appreciation for all PC members, referees and colleagues who have supported AWCS 08.

References

1. DAGMan. Directed acyclic graph manager (2005),
   http://www.cs.wisc.edu/condor/dagman/
2. Gil, Y., Deelman, E., Blythe, J., Kesselman, C., Tangmunarunkit, H.: Artificial intelligence and grids: Workflow planning and beyond. IEEE Intelligent Systems 19(1), 26–33 (2004)
3. Oinn, T., Addis, M., Ferris, J., Marvin, D., Senger, M., Greenwood, M., Carver, T., Glover, K., Pocock, M.R., Wipat, A., Li, P.: Taverna: A tool for the composition and enactment of bioinformatics workflows. Bioinformatics Journal, online (June 16, 2004)
4. Altintas, I., Berkley, C., Jaeger, E., Jones, M., Ludäscher, B., Mock, S.: Kepler: An extensible system for design and execution of scientific workflows. In: SSDBM, pp. 423–424 (2004)
5. Wibisono, A., Belloum, A., Inda, M., Roos, M., Breit, T., Hetrzberger, L.O., Korkhov, V., Vasunin, D.: Vlam-g: Interactive dataflow driven engine for grid-enabled resources. Scientific Programming 15(3), 173–188 (2007)
6. Goble, C.A., De Roure, D.C.: myexperiment: social networking for workflow-using e-scientists. In: WORKS 2007: Proceedings of the 2nd workshop on Workflows in support of large-scale science, pp. 1–2. ACM, New York (2007)
7. VL-e. Virtual laboratory for e-science (2005), http://www.vl-e.nl/
8. Zhao, Z., Booms, S., Belloum, A., de Laat, C., Hertzberger, B.: Vle-wfbus: a scientific workflow bus for multi e-science domains. In: Proceedings of the 2nd IEEE International conference on e-Science and Grid computing, Amsterdam, the Netherlands, December 4- 6, 2006, pp. 11–19. IEEE Computer Society Press, Los Alamitos (2006)
9. Taylor, I., Deelman, E., Gannon, D., Shields, M. (eds.): Workflows for e-Science. Springer, Heidelberg (2007)
10. Gil, Y., et al.: Examining the Challenges of Scientific Workflows. IEEE Computer 40(12), 24–32 (2007)
11. Bubak, M., Unger, S. (eds): K-WfGrid - The Knowledge-based Workflow System for Grid Applications. In: Proceedings of CGW 2006, vol. II, http://www.cyfronet.krakow.pl/cgw06/, ISBN 978-83-915141-8-4
12. Lee, S., Wang, T.D., Hashmi, N., Cummings, M.P.: Bio-STEER: A Semantic Web Workflow Tool for Grid Computing in the Life Sciences. Future Generation Computer Systems 23(3), 497–509 (2007)
13. Truong, H.-L., Dustdar, S., Fahringer, T.: Performance Metrics and Ontologies for Grid Workflows. Future Generation Computer Systems, vol 23(6), 760–772 (2007)

Framework for Workflow Gridication of Genetic Algorithms in Java

Boro Jakimovski, Darko Cerepnalkoski, and Goran Velinov

University Sts. Cyril and Methodius,
Faculty of Natural Sciences and Mathematics,
Institute of Informatics,
Arhimedova bb, 1000 Skopje, Macedonia
{boroj,darko,goranv}@ii.edu.mk

Abstract. In this paper we present new Java framework for Gridification of Genetic Algorithms. The framework enables easy implementation of Genetic Algorithms and also enables researchers easy and stable usage of the Grid for their deployment. The design of the framework was based on principles that make it very open and extensible. The Grid components use pure Java implementation of Grid job submission and retrieval for the Glite grid middleware by using Web Services (WS). The framework was tested on the SEEGRID testbed. Using this framework we have developed a pilot application for optimizing data warehousing VIS problem.

1 Introduction

Evolutionary algorithms (EA) are a computational model inspired by the natural process of evolution. They have been successfully used for solving complex optimization problems. Genetic algorithms (GA), a subclass of EA, search for potential solution by encoding the data into a chromosome-like structure. The search is done over a set of chromosomes (population) with repetitive application of recombination, mutation and selection operators until certain condition is reached. One repetition is called a generation.

Usually the search for a solution using GA is a long and computationally intensive process. Fortunately the GA is easily parallelized using data partitioning of the population among different processes. This kind of parallelism ensures close to linear speedup, and sometimes super-linear speedup. Over the past years many variants of parallelization techniques are exploited for parallelization of GA [1][2].

Computational Grids [3] represent a technology that enables ultimate computing power at the fingertips of users. Today, Grids are evolving in their usability and diversity. New technologies and standards are used for improving their capabilities.

Gridified genetic algorithms have been efficiently used in the past years for solving different problems [4][5]. The Grid architecture resources are very suitable for GA since the parallel GA algorithms use highly independent data parallelism. Gridification and effective utilization of the powerful Grid resources represent a great challenge. New programming models need to be adopted for implementation of such

M. Bubak et al. (Eds.): ICCS 2008, Part III, LNCS 5103, pp. 463–470, 2008.

parallel algorithms. In this paper we present the Java GA Grid Framework that will make this utilization easier and more efficient.

The rest of this paper is organized as follows. In Section 2, we present the architecture of the Java Grid framework for GA (JGFGA). Section 3 describes the Grid components of the framework. Real usage of the framework is presented in Section 4 as we present the pilot application GROW. Finally, Section 5 concludes this paper and gives future development issues.

2   Java Framework for GA

In this section we will present the architecture of the new Java Grid framework for Genetic Algorithms responsible for implementation of GA. First we will start with the architecture design issues and later progress towards presentation of the framework components.

2.1   Designing the Framework

Various GA frameworks have been developed in the past years, implemented in different programming languages. The reason for developing new GA framework was not for introducing new model for parallelizing GA, but to enable easier implementation, better portability and grid execution of parallel GA. The framework is implemented in Java because of its platform independence and good OO properties.

The framework design is founded on the following concepts: *modularity* – the framework should be defined as collection of base components that enable modular composition when designing a solution; *extensibility* – one of the main aspects influencing the development of the framework is to provide base components that enable easy extension of the framework functionality, i.e. new algorithms or new chromosome type with new evaluation function can easily implement this by extending classes or implementing interfaces that are part of the framework; *flexibility* – of the way the framework is used either by simply using already defined classes or by implementing extensions, or by choosing parallel or serial execution and other similar aspects; and *adaptability* – a framework should be as adaptable as possible because the process of implementation of GA can be divided into several phases, so the framework should enable researchers to easily implement new or adapt existing solutions by changing phase implementation.

2.2   Framework Components

The Framework organization can be divided in two parts: GA implementation and GA gridification. The GA implementation part of the framework consists of components that give easy and custom implementation of GA optimizations. On the other hand the GA gridification components enable workflow grid parallelization of the execution of the implemented GA optimizations. We continue with more detailed description of both parts.

The framework is organized in three main packages: gridapp.grid, gridapp.ga and gridapp.util. The focus in this section will be the *gridapp.ga* package containing the core GA classes. Some of the classes are abstract classes, intended for further implementation and specialization. Other classes are normal classes that implement some aspects of the GA execution, which are not designed to be extended. Most common implementations of the abstract classes can be found in *gridapp.ga.impl* package. The package *gridadd.util* contains utility classes used by many different classes in the framework. The last package *gridapp.grid* will be discussed in the next section. We continue with the presentation of the *gridapp.ga* and *gridapp.ga.impl* class design and available classes.

2.3 GA Classes

Fig. 1a presents the design of the core classes that implement or allow the implementation of GA in the framework. We will briefly describe this design.

The abstract class *Gene<T>* represents one gene. Every real implementation of a gene needs to inherit this class and implement needed methods. The reason why *Gene* class is defined as generic class of type *T* is because one gene is an array of alleles. *T* is the type of one allele. Available implementations of the *Gene<T>* class that can be found in the *gridapp.ga.impl* package are shown in Fig. 1b.

The abstract class *Chromosome<T extends Gene>* represents one chromosome. It can be seen that chromosomes are derived using one *Gene* type, and the main purpose of the *Chromosome* class is to act as a container of *Genes* and implement structure for gene organization. The only subclass included in the framework that implements the *Chromosome* class is the *ArrayChromosome<T>* class that organizes the genes into an array. If this is not sufficient the users can implement different structures for the chromosomes. The *Chromosome* class has several methods for accessing and manipulation of its genes. Mainly these are methods for accessing the genes using indexes, manipulation of the structure such as cloning or copying parts of the gene which later are used for recombination or mutation.

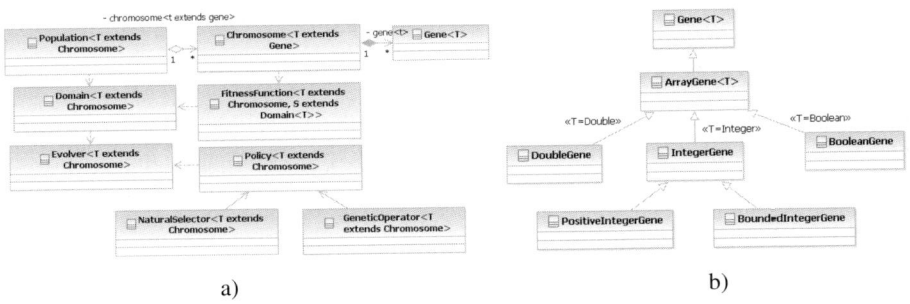

a)   b)

Fig. 1. a) Core Genetic Algorithm class design and connection; b) Gene<T> class hierarchy

The class *Population<T extends Chromosome>* represents single population of chromosomes. This class is not abstract since it is only used as a container of chromosomes, with standard methods for data access and manipulation.

The class *Domain<T extends Chromosome>* plays key role in GA execution. This class represents a structure that holds additional information on the chromosomes and genes needed for their interpretation. As shown previously, genes and chromosomes carry only raw data organized in certain data structure. The *Domain* class adds order and meaning to this data, usually by storing additional data specific for the problem. The class is not abstract, since simple problems that do not need additional information can step over this class.

The class *FitnessFunction<T extends Chromosome, S extends Domain<T>>* is the class that implements the fitness function evaluation for a certain chromosome type and domain over that chromosome type. The *Domain* is critical since it holds the main information for interpretation of the chromosomes. The *Domain* class also holds a link towards its *FitnessFunction* and *Population*, and hence represents a description of the whole problem we are solving. This makes it one of the key ingredients to the class *Evolver* that implements the actual process of evolution of the population.

Another class that is used by the *Evolver* is the *Policy<T extends Chromosome>* class. The *Policy* class specifies the rules the *Evolver* will use to implement the optimization. In another words the *Policy* is a "program" that *Evolver* will follow. The policy uses the subclasses of the abstract classes *NaturalSelector* and *GeneticOperator* to specify which operators will be used to create new chromosomes, and which selectors will be used to select the new chromosomes.

The classes *NaturalSelector* and *GeneticOperator* are inherited in many classes from the *gridapp.ga.impl* package. Some of them are: *RouletteWheelSelector*, *TournamentSelector*, *LinearRankSelector*, *EliteSelector*, *RandomSelector*, *CrossoverOperator* and *MutationOperator*.

3   Gridification of the GA Framework

In this section we will present the Grid components of the Grid Java framework for GA. Gridification of GA optimizations can be divided in two aspects. The first aspect is the choice of parallel GA technique. The second aspect is concerned with the underlying Grid connectivity with the Glite grid middleware.

3.1   Parallel Genetic Algorithms

The Parallel Genetic Algorithms (PGAs) are extensions of the single population GA. The well-known advantage of PGAs is their ability to perform speciation, a process by which different subpopulations evolve in diverse directions simultaneously. They have been shown to speed up the search process and to attain higher quality solutions on complex design problems [6][7].

There are three major classes of PGA: Master-slave, Cellular and Island. Master-slave parallelization uses single population of chromosomes and parallelizes only the chromosome evaluation part of the optimization. This makes it suitable for usage where the parallel environments have shared memory. Cellular PGA also consists of single chromosome population, but the computation can be spatially structured. This is mostly suitable for massively parallel systems, consisting of large number of processing elements organized in a topology, which is followed by the PGA. Most widely used and most sophisticated PGA is the Island PGA, or in other words Multi-population PGA. This approach enables parallel nearly-independent execution of populations. The only connection between the populations is occasional migration of chromosomes. This PGA is suitable for message passing parallelism environments.

The nature of the Grid makes it best suited for Island PGA for achieving high performance parallelism. The available cluster resources can be used using MPI or similar parallelization mechanism. This approach is extended in the Grid-enable Hierarchical PGA (HPGA)[8] which uses two level of population distribution. The first level makes several independent islands distributed over several clusters. On each cluster several parallel jobs are started which take part of the island (sub-island) on which they run the GA. After several iterations the sup-islands are rejoined and mutation is done. This process is repeated. Another positive aspect of using several clusters in parallel is having bigger population and separation into islands might increase diversity and speed up convergence.

3.2 Grid Workflow Genetic Algorithms

The JGFGA implements the PGA by using workflow execution. A grid workflow is a directed acyclic graph (DAG), where nodes are individual jobs, and the vertices are inter-job communication and dependences. The workflow inter-job communication is implemented by input and output files per job. More precisely the first job outputs data into files, and after the job terminates the files are transferred as input files to the second job. gLite WMS service takes care of the job scheduling and file transfers.

The JGFGA enables implementation of PGA by menas of four classes (jobs): *Breeders, Migrator, Creator* and *Collector,* all members of the *gridapp.grid* package. Breeders take as input a domain file and policy file. The domain file is simply Java serialized Domain object, while policy file is a java serialized Policy object. Having a population and a policy the Breader calls the Evolver and iterates several generations. The resulting population is again serialized into a domain file. Migrators on the other hand take as input several domain files, execute the inter-population migrations and output one resulting domain. The Creator and Collector classes enable easy creation of new random populations and collect several populations into one.

Having this four classes, currently we have implemented the class *JobGraph<T extends Chromosome>* that enables automatic workflow generation. Generated workflows contain several iterations of Breeder and Migrator jobs where each node is mapped to a separate Grid job. One iteration of breeding is called an epoch. An example of a sample workflow is shown on Fig. 2. The same class enables users to generate JDL (Job Description Language) files specifying the workflow for gLite grid middleware. These files are later submitted using the Grid submission tools. The parameters that can be given to JobGraph in order to model the Grid execution are: number of islands, number of epochs and migration width. Additional parameters that specify the population and policy are number of iteration per epoch and size of a single population. Further parameters that define Grid execution characteristics are the Retry counts which make the workflow more resilient when some job hits problematic Grid site and fails to execute.

The scheduling and execution of the generated workflows are not controlled by the framework. The Grid mapping decisions are done by the gLite WMS service. We plan in later developments to introduce additional properties for defining constraints in the JobGraph class that will enable guided mapping of jobs to Grid resources.

It can easily be seen that JGFGA has superior flexibility than HPGA. Most important disadvantage is the restrictions of HPGA for inter island material exchange.

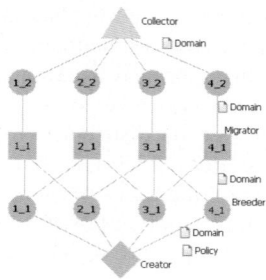

Fig. 2. Sample Grid Workflow GA. The circles are breeders and squares are Migrators.

3.3 Java Grid Framework Components

The Grid components of the framework that enable usage of the gLite Grid testbeds are based on the Workload Management System (WMS) and Logging and Bookeeping (LB) Web Services (WS). The framework is bundled with several libraries that support the WMSWS and LBWS connectivity. Additional functionality of the framework is Grid authentication by using VOMS-proxy-init. This makes the framework completely independent from the Glite UI installation and enables users to use it from any available platform.

In order for the framework to work, the user only needs to provide his/hers valid certificate, and put it in *.globus* directory in the his/hers home. For convenience we decided that it is best to provide the certificate in *.pkcs12* format, even though the framework will work if the certificate is in *.pem* format. Additionally the user needs to specify the directory with the CA certificates and vomses configuration and voms certificates. Best place for this directories are *.globus/certificate*, *.globus/vomses*, *.globus/vomsdir*.

For simplicity reasons the Grid components are organized in a single class called *GridServices*. The *GridServices* class offers the following methods: *buildProxy, isProxyValid, jobListMatch, jobSubmit, dagJobSubmit, getJobStatus, getJobOutput* and *jobPurge*. The constructor of the class requires for a *Properties* object that specify the Grid configuration parameters: *tmpDir, userGlobusDir, proxyFile, vomsesDir, vomsDir, caDir, delegationId, WMProxyURL* and *LBProxyURL*. If some of the parameters are not supplied the default values are assumed.

4 Implementation of Pilot Application

The pilot application that was successfully implemented using the Java Grid Framework for GA is the GRid Optimization for data Warehousing (GROW). The application problem area is VIS optimization of Data Warehouses. We choose this application as it was addressed in our previous research on GA optimizations [9][10].

The performance of the system of relational data warehouses depends of several factors and the problem of its optimization is very complex. The main elements of a system for data warehouse optimization are: definition of solution space, evaluation function and the choice of optimization method. The solution space includes factors relevant for data warehouse. Previous research has shown performance and quality aspects of different approaches towards solving this problem. Some of them use genetic algorithms for the search for optimal result [10].

The focus on this paper is not to develop new optimization algorithm for the VIS problem, but to implement the problem using the Java Grid Framework for GA.

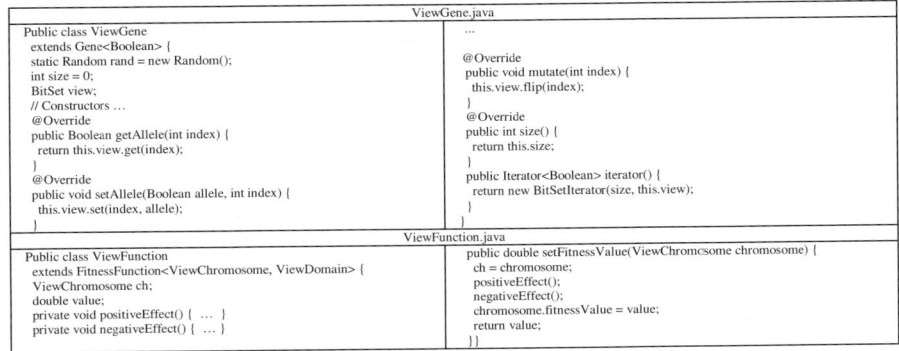

Fig. 3. Implementation of GROW GA

The implementation of GROW follows the path of every GA implementation in JGFGA. It starts with the implementation of *Gene* and *Chromosome* and later continues with the implementation of the *FitnessFunction* class for the problem in conjunction with the *Domain* class. All classes need to be implemented as extensions of classes from the framework mentioned above.

The GROW application implements *ViewGene* using *BitSet* structure and overriding the required methods as shown in Fig. 3. The *ViewChromosome* is extension of the *ArrayChromosome* class and uses *ViewGene* objects. The evaluation *ViewFunction* uses the implemented *ViewChromosome* and *ViewGene* classes and is implements by defining positive and negative functions (Fig. 3). At the end the *Domain* is implemented to hold the *Population*, *ViewFunction* and additional information. In order to enable easier gridification prior to the implementation of the applications frontend the GROW GA was packed into a single jar archive.

The developed pilot application for data warehousing is currently running on the SEE-GRID testbed. We are currently using the application for our future research in the field of GA for Data Warehousing.

5 Conclusion and Future Work

In this paper we presented the new Java Grid Framework for GA. The framework represents a powerful tool in the hands of researchers, enabling easy creation of Gridified GA optimizations.

Our current research is oriented on testing different kinds of approaches for Grid paralellizations of GA, mainly focused on what kind of models of workflow parallelization (based on previously defined parameters) can be most effective when using the Grid infrastructure.

Further development of the JGFGA is oriented towards implementation of grid data management and job wrapping per island in order to overcome the problem with job failure. The job wrapping can be easily facilitated since the input and output files of each job are in the same format (Domain files). Hence in failed nodes we can skip the epoch iteration and just copy the input files as output files. This approach will decrease the convergence of the optimization process and might produce worse final results, but will increase Grid job success rate which is the biggest problem with gridified workflows nowadays.

References

1. Nowostawski, M., Poli, R.: Parallel Genetic Algorithm Taxonomy. In: Proceedings of the Third International conference on knowledge-based intelligent information engineering systems (KES 1999), Adelaide, pp. 88–92 (1999)
2. Cantu-Paz, E.: A Survey of Parallel Genetic Algorithms. Calculateurs Paralleles, Reseaux et Systems Repartis 10(2), 141–171 (1998)
3. Foster, I., Kesselman, C. (eds.): The Grid: Blueprint for a New Computing Infrastructure. Morgan Kaufmann Publishers, San Francisco (1999)
4. Herrera, J., Huedo, E., Montero, R.S., Llorente, I.M.: A Grid-Oriented Genetic Algorithm. In: Sloot, P.M.A., Hoekstra, A.G., Priol, T., Reinefeld, A., Bubak, M. (eds.) EGC 2005. LNCS, vol. 3470, pp. 315–322. Springer, Heidelberg (2005)
5. Imade, H., Morishita, R., Ono, I., Ono, N., Okamoto, M.: A Grid-Oriented Genetic Algorithm Framework for Bioinformatics. New Generation Computing 22(2), 177–186 (2004)
6. Cui, J., Fogarty, T.C., Gammack, J.G.: Searching databases using parallel genetic algorithms on a transputer computing surface. Future Generation Computer Systems 9(1), 33–40 (1993)
7. Sena, G.A., Megherbi, D., Isern, G.: Implementation of a parallel genetic algorithm on a cluster of workstations: travelling salesman problem, a case study. Future Generation Computer Systems 17(4), 477–488 (2001)
8. Lim, D., Ong, Y.-S., Jin, Y., Sendhoff, B., Lee, B.-S.: Efficient Hierarchical Parallel Genetic Algorithms using Grid computing. Future Generation Computer Systems 23(4), 658–670 (2007)
9. Velinov, G., Kon-Popovska, M.: Solving View and Index Selection Problem Using Genetic Algorithm. In: Proc. Second Balkan Conference in Informatics, BCI, pp. 180–192 (2005)
10. Velinov, G., Gligoroski, D., Kon-Popovska, M.: Hybrid greedy and genetic algorithms for optimization of relational data warehouses. In: Proc. of the 25th IASTED International Multi-Conference: artificial intelligence and applications, pp. 470–475 (2007)

Complex Workflow Management of the CAM Global Climate Model on the GRID

V. Fernández-Quiruelas, J. Fernández, A.S. Cofiño, C. Baeza, F. García-Torre,
R.M. San Martín, R. Abarca, and J.M. Gutiérrez

University of Cantabria, Spain. SENAMHI, Perú. UDEC, Chile.
(On behalf of the EELA team)
valvanuz.fernandez@gestion.unican.es
http://www.meteo.unican.es

Abstract. Recent trends in climate modeling find in GRID computing a powerful way to achieve results by sharing computing and data
distributed resources. In particular, ensemble prediction is based on the
generation of multiple simulations from perturbed model conditions to
sample the existing uncertainties. In this work, we present a GRID application consisting of a state-of-the-art climate model (CAM) [1]. The
main goal of the application is providing a user-friendly platform to
run ensemble-based predictions on the GRID. This requires managing
a complex workflow involving long-term jobs and data management in a
user-transparent way. In doing so, we identified the weaknesses of current
GRID middleware tools and developed a robust workflow by merging the
optimal existing applications with an underlying self-developed workflow.

Keywords: GRID computing, workflow, long term jobs, climate models,
CAM model, El Niño phenomenon, GRID-CAM application.

1 Introduction

GRID technologies emerged in the 90's as a way to share computer resources
and other scientific equipment across geographically distributed locations in a
user-transparent way [2]. By sharing computer resources it is meant not only
to share their storage capacity, but also the computer power, which would be
used to run applications. The user transparency relies on what is referred to as
"middleware", a software layer between the applications and the GRID infrastructure. A number of research and commercial projects have developed different
middleware solutions and applications (e.g. the EGEE project [3] is the reference in GRID development in Europe). New applications ported to the GRID
demand new services which are not always available in the existing middleware.
In this paper, we present a new paradigmatic example on the area of numerical
climate simulation which demands solutions in terms of, e.g., job duration and
workflow management.

The EU-funded project EELA (E-Infraestructure shared between Europe and
Latin America) aims at bringing the e-Infrastructures of Latin American countries to the level of those of Europe, identifying and promoting a sustainable

M. Bubak et al. (Eds.): ICCS 2008, Part III, LNCS 5103, pp. 471–480, 2008.

framework for e-Science [4]. Among other tasks, EELA aims at identifying new applications to be ported to the GRID. The present paper describes the new developments achieved as a result of porting a climate application to the GRID under the EELA framework with the goal of analysing el Niño phenomenon, which is a key factor for Latin-American (LA) climate prediction. El Niño has a special interest due to its direct effect in the Pacific coast of South America and, in particular, in Peru and Chile (EELA LA partners).

We selected a Global Circulation Model (GCM; see Section 2) as the first application to be ported to the GRID, since any further simulation or analysis step would require a global simulation as starting point. The particular features of the GCM (experiments lasting beyond proxy certificates lifetime, control of jobs, etc) are described in Section 3. Using the existing middleware solutions (Section 4) we designed a new application developing extra middleware to run the GCM in the GRID with a specific workflow, solving most of the problems encountered.

2   Climate Modeling and GRID Computing

Climate models are complicated computer programs which require large amounts of CPU power. Most of them are parallelized. However, the GRID cannot make the most of this kind of parallelism, since the latency across geographically distributed computers would render the program completely inefficient.

Apart from computer parallelism, climate science is recently making use of a large number of simulations, referred to as "ensemble", of the same phenomenon in order to assess the uncertainty inherent to the simulation [5,6]. Ensembles of simulations with varying parameters are also used for sensitivity experiments and many other applications. Each simulation in an ensemble is independent of the others and can be run asynchronously. This kind of parametric jobs is well suited for the GRID, since each simulation can be carried out in different nodes and the results are made available as a uniform data set in the Logical File Catalogue (LFC; see Section 4 below) [7], ready to be analyzed.

Unlike volunteer computing projects, such as climateprediction.net [8], where the GCM needs to be simplified and most of the results thrown away to avoid the overloading of the volunteer hosts, the GRID allows running a full state-of-the-art model and store the regular output information.

A GCM poses specific problems to the GRID (see Section 3), which cannot be solved by the existing general solutions to easily port legacy applications to the GRID. Solutions such as GEMLCA [9] use the application to be ported as a black box and, thus, cannot monitorize intermediate states of the simulation or manage the delivery of completed output files to the catalog.

2.1   Climate Model Used

Dynamical climate models are mathematical models that numerically solve the nonlinear equations governing the atmosphere on a global lattice with horizontal resolutions ranging from 50km to 300km, depending on the application. These models require a set of initial conditions (values of climate variables – wind, pressure, temperature, etc, – on the lattice points at the starting time) to propagate the solution forward in time.

In order to analyze the atmospheric part of the global climate system, we selected the CAM model (Community Atmosphere Model), which is the latest in a series of atmosphere GCMs developed at NCAR for the weather and climate research communities [1]. The model can be run either in parallel (using MPI) or as a single process. The single-process version has been deployed and run in the EELA testbed with T42 resolution: 128 (longitude) × 64 (latitude) and 27 vertical levels, i.e. 221184 points per time step. The model produces 32 3-D and 56 2-D variables over the lattice. Therefore it is expensive in CPU-time and storage capacity. The simulation of a year takes approximately 48 CPU hours (i.e. 100 years would take 7 CPU months) and produces 197 MB per time step (i.e. more than 720 GB per century). We are interested on simulating the climate during 1.5 years to study El Niño phenomenon. The application we designed aims to perform sensitivity experiments by running an ensemble of simulations with varying parameters (related to the sea surface temperature).

3 Requirements and Workflow Management

It is currently uncommon the use of GRID computing to run long-term jobs, due to the high rate of job failure and the CPU-time limitations for the jobs on the local management system (typically only jobs lasting less than 48 hours are allowed). These problems become critical for long simulations such as those performed with climate models and other similar Earth Science applications. Thus, unlike many other applications ported to GRID, earth science applications need to make use of advanced techniques in workflow management. In particular, the climate application described in this paper has the following requirements:

1. Failure aware: Due to the nature of GRID there are several reasons which may cause job failures in the testbed, including heterogeneity of resources, CPU-time limited queues, etc.
2. Checkpointing for restart: The complexity of the climate model runs may require jobs to be restarted in a different working nodes due, for instance, to the excessive duration of the job.
3. Monitoring: Since the climate simulations last for a long time, we need to be aware of the simulation status once it has been sent to the testbed: whether the model is running or not, which time step is being calculated, which files have been uploaded to Storage Elements [10], which is the last restarting point, etc.
4. Data and Metadata storage: The goal of our application is the generation of output information that can be easily accessed by users, so data and metadata should be stored in an appropriate form.

The above requirements made necessary the development of a goal-oriented workflow manager in order to run the experiments and analyze the results with a minimum of human intervention. Therefore, we developed the application GRID-CAM which is a "GRID workflow management tool for simulating climate with CAM".

3.1   The GRID-CAM Application

In this section we briefly introduce and define the different components involved in a typical climate simulation. We define an EXPERIMENT as an ensemble of simulations (parametric jobs) designed to answer some scientific question (a single execution is the simplest experiment); each of these executions is called a REALIZATION and requires a set of input data to run the model in the prescribed simulation period (typically one year). A particular type of experiments are those related to climate sensitivity studies. In this case the different sets of input data are obtained from a single one including certain user-defined perturbations to form the ensemble (perturbed initial or boundary conditions, etc.).

The lowest level component of our application is a JOB. This component matches with a standard GRID job and cannot be related one to one with a realization since realizations cannot be guaranteed to finish in a single job. In general, a realization requires several jobs to complete, each one restarted from the previous one. As the job is running, the model generates information (files and metadata) that has to be available from every other component of the GRID: restart files (for failure recovery), current simulation time step, number of restarts, job id (for monitoring purposes), statistical information, output data, etc. Hereinafter, all the data and metadata generated by the models will be referred to as OUTPUT INFORMATION.

Therefore, numerical climate simulation on the GRID requires the management of a complex workflow formed by experiments composed of realizations split across jobs. This workflow is not trivially managed by the currently available GRID middleware, so new features are necessary for a proper execution of climate simulations.

4   Middleware Used in GRID-CAM

The *gLite* middleware is an integrated set of components designed to enable resource sharing in GRID [11]. The core components of the gLite architecture are the following:

- User Interface (UI): It is the access point to the GRID.
- Computer Element (CE): A set of computing resources localized at a site (i.e. a cluster, a computing farm).
- Worker node (WN): The cluster nodes where the jobs are run.
- Storage Element (SE): Separate service dedicated to store files.

The *Logical file catalog(LFC)* [7] is a secure GRID catalog containing logical to physical file mappings. The primary function of the LFC is to provide central registration of data files distributed amongst the various Storage Elements [10].

On the other hand, *AMGA* [12] is the gLite Metadata Catalogue, and we just use it as a classical GRID-enabled database where we store all the status and metadata information we need.

We also used *GridWay* [13], which is a GRID meta-scheduler that gives a scheduling framework similar to that found on local Resource Management systems, supporting resource accounting, fault detection and recovery and the definition of state-of-the-art scheduling policies. Compared with the LCG workload management it is much faster and easy to use [14].

Besides the previous existing middleware products, some GRID developments were necessary in order to deploy the climate application and to develop the appropriate workflow elements. These new components are described in the following sections.

4.1   The Grid Enabling Layer (GEL)

Climate models are mature applications with thousands of lines of code, which need to be ported to the GRID introducing small modifications to the code to perform system calls to specific applications which are in charge of interacting with the GRID on behalf of the climate model. To this aim, we developed a new software layer, referred to as GRID Enabling Layer (GEL), which provides the model with the ability to interact with the GRID. The slightly modified source code of the model plus its GEL conform a fully featured GRID application. Since climate models are developed by external institutions, this approach is the best suited to keep up with the most recent updates with the least effort, since only the small modifications to interact with the GEL need to be introduced at key points of any new release.

The GEL provides the following capabilities:

- Realization monitoring: Since our simulations last for a long time, we need to know their status once they have been sent to the testbed: If the model is preparing the WN or running, which step of time is calculating, which files has uploaded to SE-LFC, which is the last restart pointer, etc. This is analyzed in detail en the next section.
- Management of restart: Each time CAM dumps a new restart file, the GEL uploads the restart files to the nearest SE and register them in the LFC. It also publishes the restart field associated to this experiment in the AMGA database. This way, if the job fails and the realization is rescheduled to another WN, it will continue calculating from this time step.
- Data and Metadata management: In order to store all the output and restart information generated by the model, we need that the metadata and files are permanently registered in a place accessible from any component of the GRID (AMGA and LFC-SE).

The above issues were solved by introducing Fortran system calls at 4 specific points of the CAM source code. These calls execute the GEL scripts which carry out the previously mentioned tasks. The GEL consists of a series of scripts written in higher level languages (Shell, Perl and Python) allowing for a faster development process and an easier interface with the middleware.

4.2   Workflow Design Using GridWay and AMGA

In order to manage the workflow needed to build an unattended application we used GridWay for scheduling the jobs with re-schedule-on-failure capabilities and AMGA for monitoring.

GridWay Configuration. After considering several job managers, we found that GridWay meta-scheduler was the one that best fulfilled our requirements, since GridWay is able to detect job failures for any of the problems mentioned in Section 3, and it is able to re-schedule the failed jobs to another CE. Moreover, once the re-scheduled job starts to run in the WN, a component developed within the GRID-CAM application queries the AMGA database to find the latest restart files for this realization in order to continue the simulation started for the previous job. We have also adopted an additional monitoring feature provided by GridWay. For debugging purposes, while the job is running in the WN, a monitor script (running also in the WN) checks the status of the job. This monitor can copy the output and error files of our job to the UI with a given frequency. In this way, from the UI we monitorize the exact status of each of our realizations. Additionally, key information for the application workflow is stored in AMGA (Section 4.2).

When an ensemble of simulations is sent to the GRID, each realization of the ensemble is converted to a GridWay job that is sent to the scheduler. When GridWay receives the jobs, it searches the CE better suited to our application needs and chooses the best among them. To do so, it uses a powerful scheduling policy that takes into account the user requested requirements (memory, CPU, etc.) and an heuristic scheduling based on the jobs sent in the past. For instance, if all jobs sent to a CE failed, GridWay will not send jobs to that site again.

The components and flow of our workflow design are shown in Fig. 1.

Finally, in order to manage the issue of the expiration of the proxy, which affects every long lasting job, we used the *myproxy* credential management system as a provisional solution that is able to extend the authenticated time to one week. More research is required to deploy longer term unattended climate simulations.

Monitoring with AMGA. The AMGA database has two different tasks in the application. On one hand, it is used to store the information generated by the experiments executed in the GRID. On the other hand, it is used for monitoring purposes, storing all the status information about each of the simulations as metadata information. The tables and relationships used by GRID-CAM are shown in Fig. 2. Some of them are also relevant to the workflow, as described below:

- EXPERIMENT: When preparing the experiment, this table is filled with the perturbation type used (multiplicative, random, etc), the number of realizations and a description and dates of start and end.

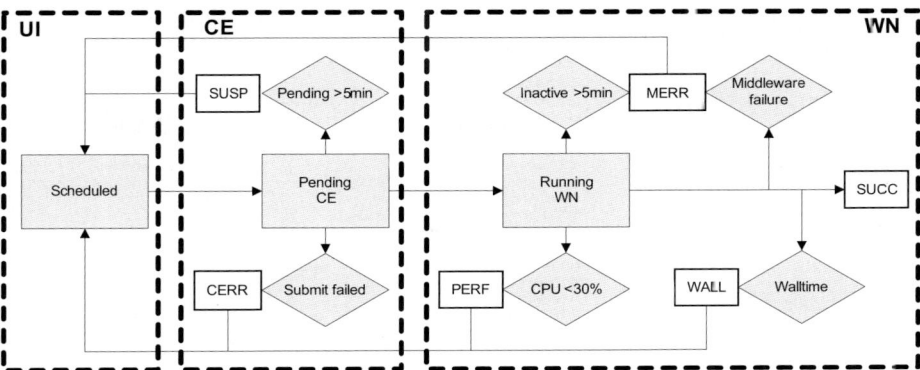

Fig. 1. Components and flow diagram of GRID-CAM. See Section 5 for details of the error signals.

- REALIZATION: Each realization can be executed in many different nodes. This table keeps track of current time step, restart files, id cf the current job executing the realization, etc.
- JOB: This table is used to keep track of the different jobs used in an experiment. It stores the timing information, the WN and the realization it contributed to. Most of this information is stored for statistical purposes.
- OUTPUTFILE: Each realization generates a number of files as it runs. This table stores metadata and access information for the files stored in the catalog. This speeds up the data discovery process.

5   Experimental Results

In order to test the GRID-CAM application, we ran a simple experiment consisting of 100 realizations simulating the climate on El Niño region during a period of two years; to this aim, we used different initial conditions as input for the realizations (perturbed sea surface temperatures). The GRID-CAM workflow used part of the resources from the EELA testbed and was executed within an arbitrary week. Therefore, the results reported here are just a particular illustration of the application's performance and cannot be considered for testbed comparison or benchmarking purposes.

In order to make our experiment as realistic as possible and to observe the efficiency of the workflow manager, we used the full list of sites from the EELA project. Some of the sites used for this experiment are located in Latin American countries. This made even more likely the occurrence of errors in the workflow due to network latency problems (this is one of the EELA challenges).

The experiment lasted one week and at the end of this period 89 realizations concluded successfully, 7 were still running and 4 crashed without finishing. Regarding the workflow, we obtained the following results:

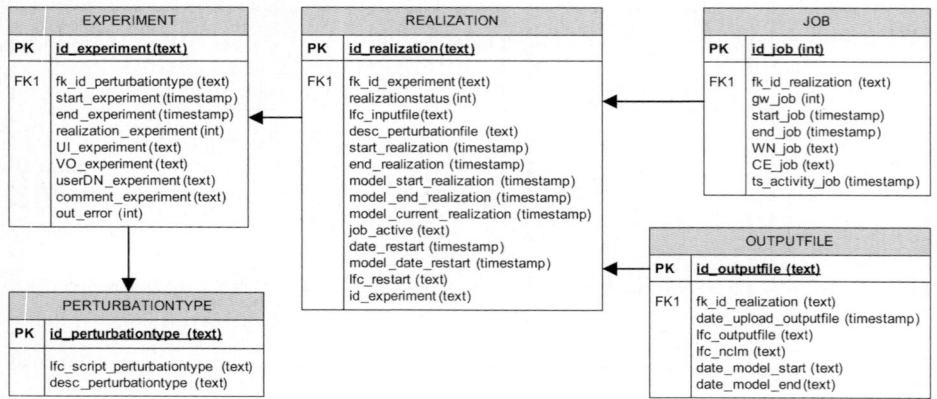

Fig. 2. Structure of the AMGA database used to store metadata and status information for the GRID-CAM application

- GridWay needed to run 1080 Globus jobs to complete the 100 realizations, from which:
  28% SUCC or WALL: Finished OK (reaching the end of the simulation or exhausting the walltime allowed by the local queue)
  31% CERR: Failed in the CE. These failures were due to misconfigured CE.
  17% SUSP: Suspension timeout. GridWay is configured to kill a Globus job if the job is waiting on the CE queue more than 5 minutes.
  8% PERF: Killed by our monitor in the WN because the CPU time dedicated to the job in the WN was lower than a 30%.
  16% MERR: Killed by our monitor because the GEL experienced problems contacting the GRID middleware (SE, CE or AMGA). Most of these jobs were run in Latin America and the main cause of failure was a network outage.
- The experiment generated 300GB of output data replicated in 2 different SE in Europe. Metadata of this data was also successfully published in the AMGA database for later use.
- Our workflow failed to manage 4 realizations. After analysing the output, we discovered that the errors were due to middleware errors that we did not manage. We have fixed the application to solve this problem.

The amount of CERR errors was caused by the misconfiguration of 2 sites where the jobs failed systematically. The rate of SUSP is also higher than expected. During the experiment there were some sites that did not accept jobs because their queues were collapsed.

6   Conclusions

We presented a successful port of a state-of-the-art global climate model (CAM) to run unattended on the GRID. The port consisted of two main components:

The Grid Enabling Layer to allow CAM interface with the GRID middleware and a failure-aware workflow built on the GridWay meta-scheduler. The application was tested in a realistic experiment.

The main conclusion of this test is that, although many problems (inherent to the GRID [15]) arose during the execution of the experiment, the GRID-CAM workflow was able to restart the simulations in most of them, allowing to finally obtain nearly 90% of successful complex realizations suitable for a statistical study of the problem at hand. Part of the identified errors have been already corrected, thus the performance of the workflow is expected to increase after a few more tests.

Without the workflow developed (and considering only the unrealistic –nearly useless– case of climate runs running for less than 1 simulated year) the success rate would drop to a 28%.

GRID-CAM is able to run an ensemble of indefinitely long climate simulations split in jobs of any duration (as imposed by the local queues at each site) in an unattended and user-transparent way. This application is general enough to support a wide range of the experiments currently being run in the climate science.

Acknowledgments. This work has been partial funded by the EELA project under the 6th Framework Program of the European Commission (contract no. 026409). J. F. is supported by the Spanish Ministry of Education and Science through the Juan de la Cierva program.

References

1. Collins, W.D., Rasch, P.J., Boville, B.A., Hack, J.J., McCaa, J.R., Williamson, D.L., Kiehl, J.T., Briegleb, B., Bitz, C., Lin, S.J., Zhang, M., Dai, Y.: Description of the NCAR Comunity Atmospheric Model (CAM 3.0). Technical Report NCAR/TN-464+STR, National Center for Atmospheric Research (2004), http://www.ccsm.ucar.edu/models/atm-cam/docs/description/description.pdf
2. Foster, I., Kesselman, C.: The grid. Blueprint for a new computing infrastructure. Morgan Kaufmann Publishers, San Francisco (1999)
3. Enabling Grids for E-sciencE (EGEE), http://www.eu-egee.org
4. E-infrastructure shared between Europe and Latin America (EELA), http://www.eu-eela.org
5. Palmer, T.N.: The economic value of ensemble forecasts as a tool for risk assessment: From days to decades. Quart. J. Royal Meteor. Soc. 128, 747–774 (2002)
6. Hagedorn, R., Doblas–Reyes, F.J., Palmer, T.: The rationale behind the success of multi–model ensembles in seasonal forecasting - I. Basic concept. Tellus 57A, 219–233 (2005)
7. Official Documentation for LFC and DPM, https://twiki.cern.ch/twiki/bin/view/LCG/DataManagementDocumentation
8. Allen, M.: Do it yourself climate prediction. Nature 401, 642 (1999)
9. Delaitre, T., Kiss, T., Goyeneche, A., Terstyanszky, G., Winter, S., Kacsuk, P.: GEMLCA: Running legacy code applications as Grid services. Journal of Grid Computing 3, 75–90 (2005)

10. Delgado, A., Méndez, P., Donno, F., Sciabá, A., Campana, S., Santinelli, R.: LCG-2 user guide (2004), http://edms.cern.ch
11. Delgado, A., Méndez, P., Donno, F., Sciabá, A., Burke, S., Campana, S., Santinelli, R.: gLite 3 user guide (2007),
    https://edms.cern.ch/file/722398/1.1/gLite-3-UserGuide.html
12. Koblitz, B., Santos, N., Pose, V.: The amga metadata service. Journal of Grid Computing (2007) doi 10.1007/s10723-007-9084-6
13. GridWay 5.2 Documentation: User Guide:
    http://www.gridway.org/documentation/stable/userguide
14. Vázquez-Poletti, J.L., Huedo, E., Montero, R.S., Llorente, I.: A comparison between two grid scheduling philosophies: EGEE WMS and GridWay. Multiagent and Grid Systems 3(4), 429–439 (2007)
15. Neocleous, K., Dikaiakos, M.D., Fragopoulou, V., Markatos, E.: Failure management in grids: The case of the EGEE infrastructure. Technical Report TR-0055, CoreGRID (2006), http://www.coregrid.net/mambo/images/stories/TechnicalReports/tr-0055.pdf

A Framework for Interactive Parameter Sweep Applications

Adianto Wibisono[1], Zhiming Zhao[1], and Adam Belloum[-], and Marian Bubak[1,2]

[1] Informatics Institute, University of Amsterdam, Amsterdam, the Netherlands
{wibisono,zhiming,adam,bubak}@science.uva.nl
[2] Institute of Computer Science AGH, Krakow, Poland
bubak@agh.edu.pl

Abstract. This paper describes ongoing efforts on adding interactivity for performing parameter sweep experiments. The literature study and analysis of requirements gathered from use cases in various scientific domains indicate that interactivity is needed but not fully supported by most of existing frameworks designed to support parameter sweep applications. Based on this study we identify the requirements for interactivity during execution of parameter sweep experiments and the type of interactive actions needed to steer parameter sweep execution. Preliminary design of a framework that would support interactivity is presented and it will be analyzed further with Model Driven Architecture modelling approach and ORC to formally analyze grid service interaction used in this framework. The implementation of this framework will be based on existing components from WS-VLAM project.

Keywords: e-Science, parameter sweep, interactivity, workflows, virtual laboratory.

1 Introduction

Applications that can benefit from the computational power offered by Grid based systems are science and engineering simulations, many of which can be structured as loosely coupled *Parameter Sweep Applications* (PSA) [3]. A typical PSA is an application which has to be executed independently large number of times, to evaluate either a single or a multi dimensional parameter space of the experiments to locate a particular point in the parameter space that satisfies certain criteria. These applications are very often developed and run as workflows [7].

Albeit simple in terms of computational model, parameter sweep studies occurs frequently in scientific computation across a broad range of scientific disciplines. For example this application model is needed in bio-medical [10], bio informatics [8], bio-physics domain [11], data mining [13] and many other scientific domains. Recognizing that science has an exploratory and evolutionary nature, dynamic and interactive behaviour need to be supported [7]. However, existing frameworks for performing parameter sweep such as Nimrod-G [1] or

M. Bubak et al. (Eds.): ICCS 2008, Part III, LNCS 5103, pp. 481–490, 2008.

APST [2] focus mainly on the scheduling and the management of the computing resources required for the execution of PSAs. Performing parameter sweep experiments based on these frameworks typically involve multiple batch submissions of parameter ranges. Scientists often prefer to guide the parameter space exploration based on some heuristics, the current way of performing parameter sweep experiments does not allow such an action. In this paper we are aiming to provide interactivity features to support the exploratory scientific process in parameter sweep applications. This interactivity will allow more efficient ways of exploring parameter space, thus increasing the productivity of performing parameter sweep experiments.

The rest of this paper is organized in the following way: in Section 2 we discuss the basic processes involved in parameter sweep experiments derived from example use cases. A critical overview of currently available parameter sweep frameworks is presented in Section 3. Section 4 discussed goals and requirements to support interactive parameter sweep experiments. A design to satisfy these requirement is presented in Section 5 along with current prototype status. We conclude in Section 6 with summary and future research directions.

2   Basic Processes in Parameter Sweep Experiments

In this section we describe two use cases to analyze the basic lifecycle of parameter sweep experiments and then distinguish its basic processes. The use cases will also be used to illustrate issues that have to be addressed in developing framework to support interactive parameter sweep.

The first use case is the *ADDA Distributed Light Scattering*. Discrete Dipole Approximation (DDA) [11] is a general method to calculate scattering and absorption of electromagnetic waves by particles of arbitrary geometry and composition. Amsterdam DDA (ADDA) is used to study light scattering pattern in red blood cells (RBC). Mature red blod cells are modeled as biconcave disk in DDA simulations, and the results of this simulation are compared against experimental results based on scanning flow cytometer of actual RBC. There are 6 variables that defines the parameter space in ADDA experiments, which characterize the modelled RBC such as hemoglobin concentration, biconcave disk dimensions which includes height, width, diameters and angle of rotation. In this experiments a parameter sweep is conducted to find a set of parameter related to modelled RBC which will minimize the error when comparing simulation with experimental results. Currently, the approach of finding appropriate parameters are by performing 40000 light scattering patterns simulation for RBC randomly varying the 6 parameters. Result of this simulation is then used to solve inverse light scattering problems using χ^2 test based on direct comparison of experimental LSP and simulated ones. Each of these simulations take in average 10 to 15 minutes, and performing possible combinations of the 6 parameters requires a couple of processing months.

The second use case is about *Functional Magnetic Resonance Imaging (fMRI) Applications*. FMRI is a modality that enables the observation of brain activity during physical or cognitive simulation. Acquired MRI scans containing time series of 3D datasets during brain rest and stimulation phases need to be processed to generate brain activation maps. The analysis of fMRI data is performed with fMRIB software library and statistical parametric mapping tools [12]. Although these packages hide much of the image analysis complexity, the choice of parameters still plays an important role in fMRI analysis. Most researchers adopt standard parameter values for spatial smoothing, delay hemodynamic response function (HRF) and image registration. However these default values not always are the appropriate ones for a number of cases. The search for the optimal parameter requires investigating a large parameter space. Previous sweep experiments which varies the HRF delay parameters [10] using ranges from 4s up to 8s with steps 0.25 seconds applied to 22 different data sets generates 374 jobs. Repeating this experiments up to 6 times with job completion varies from 30 to 140 minutes takes 416.7 hours total experiment times. This search can be optimized if the experts can interact at run time with the system.

The experiment scenarios of the two use cases consist of many independent tasks, each of which corresponds to the evaluation of one point in multi-dimensional parameter space. Each of these tasks are the basic unit of application used in parameter sweep which usually consists of legacy application or existing library used such as the ADDA library, or FSL library for the fMRI experiments. The independent nature of the tasks composing the two applications makes them suitable for a parameter sweep approach and opens up possibility to grid enabled computing resources. Unlike the case for tightly coupled parallel applications which has communication overhead trade offs when using distributed computing resources, the search the optimal paremeters in the two use cases presented in this section does not involve any communication among the different runs.

In both scenarios, we can distinguish a number of basic steps. First, a preliminary run is done based on an initial guess of interesting parameter ranges. This preliminary run is used to determine the predicted behaviour of the modeled system (parameter space landscape) under limited input conditions. Based on analysis of this landscape, a process of gradual refinement is started [4]; in each subsequent runs, the parameter space is reduced by the user and/or through the parameter optimization method employed. Based on the analysis performed after each run, a different region of parameter space as well as a different optimization method can be chosen.

3   Overview of Parameter Sweep Frameworks

A support environment for parameter sweep experiments, namely parameter sweep framework, allows a scientist to efficiently perform and manage different steps of a parameter sweep experiment. A number of parameter sweep frameworks have been developed e.g Nimrod-G [1], AppLes Parameter Sweep Template (APST) [2], P-Grade [9], Virtual Instrument [4], Science Experimental Grid Laboratory (SEGL) [5] and SimX [16]. These systems aim to support the basic processes in parameter sweep experiments and provide user front end to describe, execute and monitor the experiments. Nimrod-G [1] provide a tool for parameterized simulation on distributed systems designed for large widely distributed computational systems. APST [2] focuses on adaptive scheduling of parameter sweep applications on the Grid. P-Grade [9] provide an integrated portal and workflow based approach for parameter sweep applications. Science Experimental Grid Laboratory (SEGL) [5] aims at management of complex and dynamic parameter studies in grid environments, by automating the creation of complex modules for the computation and providing dynamic control of the study, utilizing the use of results from previous computational stages. These existing frameworks focus more on scheduling and management of computing resources to satisfy computational demand of parameter sweep applications. The essential feature to monitor and to intervene with running parameter sweep simulation is something which is missing from these existing frameworks, which we will address in our research. A related research that have similar goal with our efforts is Virtual Instrument [4] which allows user directed parameter space search on distributed computing resources, specifically targetted for M-Cell application. Another related research that allow parameter space adjustment during experiment execution is SimX [16].

Table 1 summarizes the support from existing environment for performing parameter sweep. From this table we can see that most of existing systems does support basic job farming of tasks, provide basic handling of experimental results, but only few of them attempted to provide interactivity i.e Virtual Instrument [4] and SimX [16]. The former is a completed project which managed to add interactivity by assigning priority to interesting region of parameter space. The later is an ongoing effort which is aiming to add interactivity by

Table 1. Features supported by existing parameter sweep environment

	Experiment description	Monitoring	Scheduling Strategy	Result Management	Emphasizes
Nimrod-G	Parametric Modelling Language	MDS	Deadline and Budget Constraint	File Based	Computational Economy
APST	APST-XML	MDS, NWS, Ganglia	Adaptive, XSufferage	SRB, GridFTP	Scheduling Heuristics
P-Grade	PS-Workflow (DAG)	MDS, BDII	Condor Dagman based	-	Workflow of Parameter Sweep
Virtual Instrument	Model Description Language	APST Based	APST Based	VI Database	Interactive Computational Steering
Sim-X	Scirun dataflow	-	Greedy, Fairshare	Shared Object Layer (SISOL)	Engineering Optimization
SEGL	Task/ data flow	-	Unicore Based	OODB	Dynamic Parameter Studies

allowing user to change optimization strategy during experiments. Other type of interactivity is not yet studied and fully supported by available systems.

4 Requirements of Framework for Interactive Parameter Sweep

A support environment for interactive parameter sweep experiments, allows a scientist to efficiently perform and manage different steps of a parameter sweep experiment. The goal of performing parameter sweep experiments is to locate some particular point in the parameter space that satisfies some subjective criteria. Back to our examples, in the ADDA experiments the criteria is to minimize the error compared to the experimental light scattering pattern of RBC cells, and in the fMRI applications this criteria is to observe the effect of delay in hemodynamic response function parameters in the analysis results. User must be able to guide the search away from certain regions of parameter space to be explored once it is indicated that intermediate results for this region is not satisfiying. User can then concentrate on other regions based on intermediate application results. This requires that the framework allows the creation and cancellation of application tasks on the fly.

Framework for interactive parameter sweep (FRIPS) extends normal parameter sweep framework by adding user support for interactive visualization of computing results, parameter space choosing, decision making, and provenance of computation. FRIPS benefits the state of the art results of existing framework but have new requirements on system design. In this section, we discuss those requirements from user perspective i.e the scientist/expert who wish to exploit existing parameter sweep applications, application developers perspective i.e the developer of parameter sweep applications and from the problem solving environments perspective.

Domain scientists and application developers are basically two types of users of FRIPS. The domain scientists are experts of the application model and are interested in using to determine most interesting parameter regions for an application. From the point of view of domain scientists, the details and complexity of underlaying grid environment should be hidden so that they can focus on their main concern on performing their experiments. They also need to be able to monitor the job execution, and view intermediate results. For the interactivity they need to be able to change parameter space or set new policy for execution at runtime.

Application developers use this framework to develop experiment for specific applications. From the point of view of application developer, porting legacy application should be possible with minimal effort. The framework should be flexible enough to allow for different sampling policy or optimizer for experiments. From the perspective of a problem solving environment, FRIPS provide high level user seamless access to underlaying computing environments. In general the following characteristics are desired scalability, adaptability, platform independence and fault tolerance.

5   Design Considerations

In this section we describe the design of interactive parameter sweep framework which will addresses the requirements mentioned in the previous section. The main challenges in such framework lies on the dynamic nature of both the execution environment and the way scientist conduct such experiments. Therefore the framework need to be adaptive, both to the changes in the resources, and to parameter space steering initiated by user.

5.1   Functional Components

We distinguish a number of functional components of FRIPS, a front end, visualizer, a coordinator for parameter sweep experiment, an execution manager, worker and a result repository, as shown in Fig. 1.

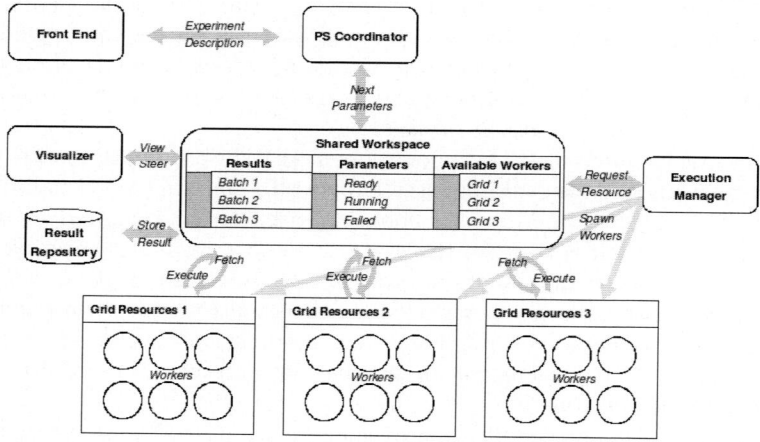

Fig. 1. Main modules of interactive parameter sweep framework

Using the *Front End*, a user can describe experiments, view computing results, and steer the execution at runtime. Description of the experiments includes parameter space, sampling strategy, and observation space description. User would be able to made such intervention based on feedback that he sees on the visualizer. He might make decisions such as modification of parameter space, abort the experiment if intermediate results looks not promising etc.

Visualizer is required to allow monitoring of the progress in an ongoing parameter sweep experiments. Depending on the user configuration at the beginning of the experiment, intermediate result visualizer might visualize individual experimental results or an aggregated result of previous set of parameter spawned by parameter sweep coordinator.

Parameter Sweep Coordinator (PS-Coordinator) is the main component which is responsible for managing parameter space region which is currently being explored during the experiments. This component determines the next set of interesting parameter to be explored, cancel pending parameter set which is considered no longer interesting by the user, and manage the results obtained from past execution of parameter spaces. The decision taken by parameter sweep coordinator can be based on interactive steering from user, or based on sampling strategy defined by user at the beginning of the study. Once the parameter sweep coordinator made decision about the next set of parameter to be explored, it will contact resource manager for mapping each of parameter to available resources.

Execution Manager is responsible for spawning workers responsible for executing each individual experiments to appropriate grid resources. We will possibly reuse existing resource manager from our previous work [15] since this component is concern only with appropriate resource allocation for workers.

Results Repository is needed to keep track of results from previous runs of experiments. This components will also be responsible for the provenance of the experiments to ensure that current experiment could be reproduced.

Shared Workspace is component responsible form maintaining a state of running parameter sweep experiments. It is a tuple space like workspace, which will be central communication points between components used in the framework.

Worker is the components that wraps real applications used to perform parameter sweep, to allow the legacy application to be interfaced with our framework. It is responsible for fetching parameters from shared workspace, executing the application and storing results back to the workspace.

5.2   Current Prototype

We have started to develop a prototype for the proposed framework. Currently we are focusing on the PS Coordinator, execution manager, and management of experiment status. We consider these three components as the main building blocks on top of which the interactivity features is built. One of the most important issue to develop these components is to provide a platform to manage the state of the parameters. All three components (PS Coordinator, execution manager, and management of experiment status) need to have access to the state of the parameter to perform their task. In the current prototype the state of a running parameter sweep experiment is maintained by a tuple space like workspace. Shared workspace allows concurrent access to experiment status such as active parameters, intermediate results and available workers.

The PS Coordinator, execution manager and later on the visualizer can concurrently access or modify the status of the experiment stored in the shared workspace. First PS Coordinator stores the next set of parameters to be executed into the workspace. This set is determined by initial sampling strategy, or on later stage based on feed back from user interaction. The visualizer retrieves the available intermediate results, and presents them to the end-user allowing him to follow the progress of the experiment and interact with the system to make the adequate changes. Finally, the execution manager accesses the experiments to provide information about available workers.

In the current prototype each worker is spawned by the execution manager, and provided with the following elements: the location of the workspace, and the experiments it is associated with. The worker is responsible for retrieving the parameters needed to execute the experiment, performs the actual execution of the experiment and stores back results in the shared workspace. If the size of results is very large only a reference to the actual location of the results is stored.

Since the worker is an active agent pulling parameters from the workspace, we do not have problem with restriction from common grid infrastructure where inbound connectivity to worker nodes is prohibited. Using this pull model the framework is more scalable, any number of workers can be added as soon as resources are available to accommodate the worker.

The current prototype supports limited interactivity features. At this stage the user can cancel at will unprocessed parameters and/or replacing with new set of parameters, stopping the experiment if the scientist consider no further exploration is necessary and also requesting execution manager to spawn more workers in resources if needed. For monitoring purposes, users can view active parameters that are being processed, available workers and intermediate results. Further interactivities will be supported at the next development stage of this framework.

5.3   Modelling of the System

We have presented our approach for designing and prototyping the interactive parameter sweep framework. We will perform further analysis based on Model Driven Architecture (MDA) [6], an architectural framework based on computational independent model, platform independent model and platform specific model that will enables rapid development of new system specification.

We will use available grid services as a building block for this framework. We consider that it is necessary to formally analyze the interaction among this basic grid services within our framework. For analysing the interaction among basic grid services used in the framework, we are considering the use of orchestration language ORC [14]. Grid service interactions including resource discovery, resource contract negotiation, job placement is essential in the development of this interactive framework. It has been shown in [14] that ORC can provide the means to describe these essential features of grid management.

6   Summary

In this paper we use two use cases to analyze basic scenarios in typical parameter sweep applications, and then a review on existing parameter sweep frameworks is given. From the research, we learned that including human knowledge in the runtime loop of parameter sweep application is essential to improve the efficiency of performing parameter sweep experiments. We argued that the batch based execution in existing frameworks hampers the inclusion of human expertise in runtime loop of parameter sweep, and we propose a new framework for interactive parameter sweep.

We identify the requirements for interactivity during execution of parameter sweep experiments. We present our initial design supporting these requirements and describe the modelling approach that we will use to further analyze this design. A prototype that supports basic interactivity, has been developed.

In this paper we have focused mainly on achieving interactivity on performing parameter sweep application. In future work, we plan to focus our research on the adaptive execution of parameter sweep experiments. The added interactivity poses additional challenge, since changes now come from both dynamic grid resource environment, and interactive behaviour of the user.

Acknowledgements. This work was carried out in the context of the Virtual Laboratory for e-Science project (www.vl-e.nl). Part of this project is supported by a BSIK grant from the Dutch Ministry of Education, Culture and Science (OC&W) and is part of the ICT innovation program of the Ministry of Economic Affairs (EZ). The authors of this paper would like to thank all the members in the VL-e SP2.5 especially to Vladimir Korkhov, Dmitry Vasyunin and Silvia Ollabariaga for fruitful discussion.

References

1. Buyya, R., Abramson, D., Giddy, J.: Nimrod/G: An Architecture for a Resource Management Scheduling System in a Global Computational Grid. In: 4th International Conference on High-Performance Computing in the Asia-Pacific Region, pp. 283–289. IEEE Computer Society, Los Alamitos (2000)
2. Casanova, H., Berman, F., Obertelli, G., Wolski, R.: The AppLeS Parameter Sweep Template: User-Level Middleware for the Grid. Sci. Program. 8, 111–126 (2000)
3. Casanova, H., Bartol, T., Berman, F., Stiles, J.: Distributing MCell Simulations on the Grid. Int. J. High Perform. Comput. Appl. 15, 243–257 (2001)
4. Casanova, H., Berman, F., Bartol, T., Gokcay, E., Sejnowski, T., Birnbaum, A., Dongarra, J., Miller, M., Ellisman, M., Faerman, M., Obertelli, G., Wolski, R., Pomerantz, S., Stiles, J.: The Virtual Instrument: Support for Grid-Enabled Mcell Simulations. Int. J. High Perform. Comput. Appl. 18, 3–17 (2004)
5. Currle-Linde, N., Kuester, U., Resch, M., Risio, B.: Science Experimental Grid Laboratory (SEGL) Dynamic Parameter Study in Distributed Systems. In: Joubert, G.R., Nagel, W.E., Peters, F.J., Plata, O.G., Tirado, P., Zapata, E.L. (eds.) PARCO 2005, vol. 33, pp. 49–56. Central Institute for Applied Mathematics, Jülich, Germany (2005)
6. Djuric, D., Devedzic, V.: Model Driven Architecture and Ontology Development. Springer, Heidelberg (2006)
7. Gil, Y., Deelman, E., Ellisman, M., Fahringer, T., Fox, G., Gannon, D., Goble, C., Livny, M., Moreau, L., Myers, J.: Examining the Challenges of Scientific Workflows. Computer 40, 24–32 (2007)
8. Inda, M., Belloum, A., Roos, M., Vasyunin, D., de Laat, C., Hertzberger, L.O., Breit, T.M.: Interactive Workflows in a Virtual Laboratory for e-Bioscience: the SigWin-Detector Tool for Gene Expression Analysis. In: 2nd IEEE International conference on e-Science Grid computing, pp. 19–19. IEEE Computer Society, Washington (2006)

9. Kacsuk, M.: Parallel program development execution in the grid. In: International Conference on Parallel Computing in Electrical Engineering, pp. 131–138. IEEE Computer Society, Washington (2002)

10. Olabarriaga, S., Nederveen, A., O'Nuallain, B.: Parameter Sweeps for Functional MRI Research in the Virtual Laboratory for e-Science Project. In: 7th IEEE International Symposium on Cluster Computing and the Grid, pp. 685–690. IEEE Computer Society, Washington (2007)

11. Penttila, A., Zubko, E., Lumme, K., Muinonen, K., Yurkin, M., Draine, B., Rahola, J., Hoekstra, A., Shkuratov, Y.: Comparison between discrete dipole implementations exact techniques. J. Quant. Spectros. Radiat. Transf. 106, 417–436 (2007)

12. Smith, S., Jenkinson, M., Woolrich, M., Beckmann, C., Behrens, T., Johansen-Berg, H., Bannister, P., De Luca, M., Drobnjak, I., Flitney, D., Niazy, R., Saunders, J., Vickers, J., Zhang, Y., De Stefano, N., Brady, J., Matthews, P.: Advances in Functional Structural MR Image Analysis and Implementation as FSL. NeuroImage 23, 208–219 (2004)

13. Stankovski, V., Dubitzky, W.: Special section: Data mining in grid computing environments. Future Gener. Comput. Syst. 23, 31–33 (2007)

14. Stewart, A., Gabarro, J., Clint, M., Harmer, T.J., Kilpatrick, P., Perrott, R.: Managing Grid Computations: An ORC-Based Approach. In: Guo, M., Yang, L.T., Di Martino, B., Zima, H.P., Dongarra, J., Tang, F. (eds.) ISPA 2006. LNCS, vol. 4330, pp. 278–291. Springer, Heidelberg (2006)

15. Wibisono, A., Vasyunin, D., Korkhov, V., Zhao, Z., Belloum, A., de Laat, C., Adriaans, P., Hertzberger, L.O.: WS-VLAM: A GT4 Based Workflow Management System. In: Shi, Y., van Albada, G.D., Dongarra, J., Sloot, P.M.A. (eds.) ICCS 2007. LNCS, vol. 4489, pp. 191–198. Springer, Heidelberg (2007)

16. Yau, S.M., Grinspun, E., Karamcheti, V., Zorin, D.: Sim-X: Parallel System Software for Interactive Multi-experiment Computational Studies. In: 20th International Parallel and Distributed Processing Symposium, pp. 10–10. IEEE Press, New York (2006)

Comparative Studies Simplified in GPFlow

Lawrence Buckingham, James M. Hogan, Paul Roe, Jiro Sumitomo,
and Michael Towsey

Queensland University of Technology
GPO Box 2434, Brisbane QLD 4001, Australia
{l.buckingham,j.hogan,p.roe,j2.sumitomo,m.towsey}@qut.edu.au

Abstract. We present a novel, web-accessible scientific workflow system
which makes large-scale comparative studies accessible without programming
or excessive configuration requirements. GPFlow allows a workflow defined on
single input values to be automatically lifted to operate over collections of input
values and supports the formation and processing of collections of values without the need for explicit iteration constructs. We introduce a new model for
collection processing based on key aggregation and slicing which guarantees
processing integrity and facilitates automatic association of inputs, allowing
scientific users to manage the combinatorial explosion of data values inherent in
large scale comparative studies. The approach is demonstrated using a core task
from comparative genomics, and builds upon our previous work in supporting
combined interactive and batch operation, through a lightweight web-based user
interface.

Keywords: Comparative Studies, Collection Processing, eScience.

1 Introduction

Collection of large quantities of data is perhaps the defining characteristic of
eScience, one especially noticeable in the earth and biological sciences through high
throughput remote sensing and sequencing machines. Collected data is typically
stored in a database and subsequently analyzed and visualized to address particular
scientific questions or discover new knowledge. In the latter case, data, as opposed to
theory or experiment, is starting to drive science.

The quantities of data now being collected preclude manual analysis, and a computational approach has been inherent from the start. In consequence, much of the 'low
hanging fruit' from these data sets is now routinely picked, and attention has turned to
more sophisticated questions and less obvious signals, often based on fine-grained
comparison across large sets of data records. Such comparative studies are essential to
the new science, and scientific workflow systems must evolve to support this new reality through: (i) integrated collection and input association processing; (ii) mechanisms for result traceback and ensuring processing integrity; and (iii) tools for result
filtering and aggregation. This paper is motivated by large-scale comparative genomics, but the work is applicable across many disciplines, with the management of input
collections and their combinatorial output values a common theme.

M. Bubak et al. (Eds.): ICCS 2008, Part III, LNCS 5103, pp. 491–500, 2008.

Bioinformatic sequence data is typically analyzed through a pipeline of different tools, perhaps to align sequences and search for motifs. Tool pipelines are either realized manually or through some kind of script or workflow system. The explosive increase in the number of genomes available has made single sequence analyses almost obsolete. Bioinformaticians now wish to compare and analyze multiple versions of similar sequences, and the greater statistical significance afforded by automated comparisons is vital to scientific investigation.

Unfortunately, existing automation tools make such studies difficult; typically they require some level of programming and provide limited support for experimentation. Moreover, responsibility for managing the combination and selection of constituent data streams remains largely with the user, as does the task of ensuring traceable associations between input data tuples and result values. The next generation of workflow systems should support large scale collection processing in a manner transparent to the scientific user. Such a user should be able to analyze a single data point, e.g. a sequence, to automatically lift the analysis to operate across a set of data points, and subsequently to apply synthesis operations to the resulting collection of values. Few available systems transparently support both these properties without requiring an element of programmer intervention. Interactive experimentation in comparative studies is poorly supported. An interactive map-reduce [1] workflow paradigm is required.

In this paper we present a novel workflow system for undertaking comparative studies, which supports interactive experimentation, automatic lifting to collection oriented computation, and automatic input association and synthesis of collections. This work extends our previously reported GPFlow system [2] – which leverages a commercial workflow system – by carefully applying the principles of structure data flow to produce a system focused on comparative studies.

The next section describes approaches taken to collection oriented computation within scientific workflow, and describes several other systems which support comparative studies through this approach. Sec. 3 presents our model for collection oriented workflow, with an illustration of the approach through the phylogenetic tree inference workflow following in Sec. 4. We conclude by examining the future directions for this work and its applications.

2 Background and Related Work

Numerous workflow systems have now been developed to support scientific research. Most have excellent support for integrating legacy tools, data and web services, but few cater for experimentation or provide an elegant model for comparative studies, if they directly support comparative studies at all. Fox and Gannon [3] observe that scientific workflow is strongly influenced by earlier work in the areas of data flow programming and distributed parallel programming. In this section we focus on collection processing in a data flow context. To that end, we first review models of data flow programming that relate to collection processing. We subsequently examine collection processing in two extant scientific workflow systems, namely Kepler [4], [5] and Taverna [6].

2.1 Data Flow Programming Models

In a data flow programming environment [7] a program is considered to be a directed graph in which nodes represent instructions while edges represent data transfer connections. Data arrives at a node on incoming edges (input ports) and leaves via outgoing edges (output ports). A node becomes eligible to fire when a value is available at each of its input ports. When a node fires, it removes values from the input ports, performs a computation, and places the resulting values in its output ports.

Most existing data flow programming environments are based on the token stream model. In this model, values are carried along the edges of the graph by tokens. Nodes are stream processors, and edges are realized by queues of tokens. When a node fires, it removes one token from each incoming queue and generates a new token for each output port. At each output port, which may be the origin of one or more edges, a distinct copy of the resulting token is appended to each queue connected to the port.

Since collection processing is fundamentally an exercise in the structuring of data, it is salutary to examine data structures in data flow environments. Davis and Keller [8] proposed two approaches that might be taken to implement data structures. The token-based tuple model introduces tuple-type tokens, which in turn contain other tokens, including nested tuple tokens where necessary. A tuple would correspond to an array in a procedural language, with elements accessible by index selection.

An alternative to the token stream model is also described in [8]. In the structure model, a data structure is incrementally constructed on each edge of the data flow graph. Several interpretations would be available for this structure, including that of a token stream, a tuple, a tree or a scalar value. During a computation, the entire history is retained in the form of the accumulated collection of data structures. In traditional data flow research this model has found little support, chiefly due to the potentially burdensome memory requirement. However, in scientific workflow often as not the history of a computation is as important as the final outcome. With the shift away from data flow at the microprogramming level, where the entire history needed to remain within physical memory, to coarse grain data flow with ready access to massive external storage, this is a less prohibitive consideration.

2.2 Collection Processing in Scientific Workflow Systems

In this section we describe collection processing in two widely used scientific workflow systems, Kepler [3], [5] and Taverna [6]. We summarize by positioning collection processing in GPFlow relative to these two systems and the overarching data flow theory.

Kepler is an extension of Ptolemy II [9], a mature and powerful visual data flow programming system in which nodes are called actors, edges are called channels, and actors communicate by sending messages along channels, following the token stream data flow model. Ptolemy II exhibits many features to be found in a general purpose programming language: conditional execution, iteration and procedural abstraction. Moreover, Ptolemy II provides a broad range of data flow execution modes.

Kepler implements collection processing by extending the component library rather than modifying the underlying language. A system that allows for hierarchical structures to be encoded and streamed through the actor pipeline is described in detail in

[3]. The presence of a collection is indicated by special delimiter tokens which mark the beginning and end of a collection. Collections may be nested arbitrarily using balanced begin-end tokens. Collection-aware actors are able to detect the delimiter tokens and act on them in a manner akin to SAX XML parsing. Collection-aware composite actors introduced in [5] allow conventional actors to be encapsulated and lifted to a "collection-aware" mode. The collection-aware composite actor uses scope expressions to characterize those parts of the token stream that are of interest and to specify how they are to be processed, including explicit support for iteration.

Taverna [6] is the scientific workflow management tool for the myGrid project [10]. Workflows are internally specified using the Simple conceptual unified flow language (Scufl), and enacted by the Execution Flow layer, which provides limited collection processing. Taverna provides a light-weight internal object model that permits the representation of lists and trees and the attachment of mime types to objects flowing through the system, exhibiting some properties of the structure model of data flow. Taverna insulates the user from explicit computational detail by providing implicit data-driven iteration over collections. A component, designed to read and write single values, is able to be lifted in a map-like way and applied to each element of a list, producing a list as the result. However, processing is less convenient if a component receives two or more lists in place of the singleton input values it was designed to accept. Here, user intervention is required to determine the association mode between the lists. Moreover, processing is constrained to operate over either the cross or dot product of the two lists.

Philosophically, GPFlow lies closer to Taverna, shielding users from computational aspects of workflow and freeing them to concentrate on their data and the scientific question at hand. However, in GPFlow collection processing is an integral facet of the execution regime, based on a variant of the structure data flow model. As in Taverna, a map-like operation lifts the workflow to operate over sets of input values, but GPFlow uses the data flow graph to determine automatically the association mode to be employed for converging set-valued input streams, and introduces a novel key aggregation and slicing mechanism to facilitate selection and ensure processing integrity. Collection processing is further enhanced by providing a simple operation that allows the user to gather disparate values to form an array for subsequent processing. Our model is described in the following section.

3   Collection Processing in GPFlow

GPFlow provides an interactive web-based workflow environment which allows the user to construct workflows from scientifically meaningful components without programming. The system implements a structured data flow model, in which a cumulative data structure is created over the lifetime of a computation. A GPFlow workflow presents as an acyclic data flow graph, yet provides powerful iteration and collection formation capabilities. The remainder of this section describes collection processing in GPFlow, in three stages: a conceptual data model; the iteration model supported by GPFlow; and finally the rules used to create collections from disparate data items.

3.1 Implementing the Data Model

A step in a scientific workflow is the execution of a computational tool. In GPFlow, these tools are wrapped and exposed to users as *Components* and a GPFlow workflow is the execution of a sequence of these components. This sequence is represented by organizing components as vertices in an acyclic data flow graph. The use of acyclic graphs enables the entire execution history of a workflow to be captured using only a simple data structure. In the graph, the termination of an edge at a particular node represent component inputs, and the origin of edges represent component outputs. We refer to the binding sites of edges as *channels*.

If a component takes input parameters, then the value of each parameter can be either be *bound* to the output of another component, shown as a directed edge between nodes, or be *unbound* and provided directly by the user through the user interface.

If the user specifies a collection through the UI, or binds a collection of values to a component's input channel, then the component will be executed for each element in the collection. Each execution is represented by a *Processor*, which encapsulates the result of the execution as well as the input value used to obtain the result. Collectively, the set of processors form the component's *result set*.

The ability to bind collections to channels means that each component output channel implicitly defines an output collection, which consists of the set of corresponding outputs from the Processors in the result set. When a component's input channel is bound to the output channel of another component, and that output channel produces a collection, then that component in turn will iterate over the collection to itself produce collection of outputs.

3.2 Data Driven Iteration

At the most basic level, a Component iterates over the Cartesian product of its input collections and queues one Processor for each combination. This works well if the data flow graph is a simple pipeline or strictly divergent tree structure, but if the data flow graph contains a fork-merge sub-graph, the simple iteration model breaks down by introducing spurious computations that could never have occurred if the user inputs were each entered in manually. Consider the workflow illustrated in Fig. 1.

Fig. 1 depicts a hypothetical workflow containing three components, each representing a step in a user's workflow. These are denoted by boxes labelled *C1...C3*. C1 is set to iterate over a collection of user supplied inputs, represented by the multi-document icon labelled *U*. Result sets generated by components in the workflow are represented by multi-document icons labelled *A, B, C*. Here, component 1 has no bound input channels. Component 2 has an input channel bound to the sole output channel of component 1. Component 3 has two input channels, bound to the outputs of components 1 and 2.

If the input collection U, contained only one element, $u_1 \mid u_1 \in U$ i.e. $U=\{u_1\}$ then the workflow would execute as follows:

1. Step 1 executes f_1, consuming the single supplied value u_1 and producing a single result $a_1 = f_1(u_1)$, where $a_1 \in A$, $A=\{a_1\}$
2. Step 2 executes f_2, consuming the single supplied output value o_2 and producing a single result $b_1 = f_2(a_1)$, where $b_1 \in B$, $B=\{b_1\}$
3. Finally, step 3 executes f_3, producing $c_1 = f_3(a_1, b_1)$, where $c_1 \in C$, $C=\{c_1\}$

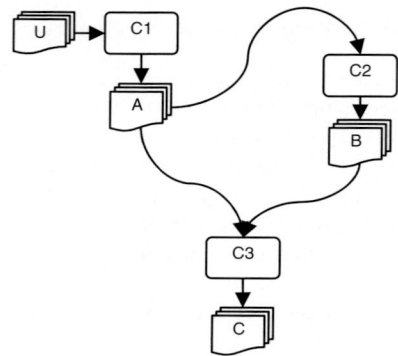

Fig. 1. Fork-merge sub-graph

Note that the cardinality of A, B and C is 1 when the cardinality of U is 1. On the other hand, if U contained multiple elements, data-driven iteration lifts the workflow to operate over the Cartesian product of the input collections. This gives rise to the following sets of output values:

- $A = \{f_1(u) \mid u \in U\}$
- $B = \{f_2(a) \mid a \in A\}$
- $C = \{f_3(a, b) \mid a \in A; b \in B\}$

If we examine the pairs of output values $\{(a_i, b_j) \mid a_i \in A; b_j \in B\}$ consumed as inputs by component 3, we see that although the component definition does not express an explicit association between its input channels, an implicit association is introduced by the workflow wiring structure. The only elements from A and B that may be paired are those derived from a common value of u. For example, given $U = \{u_1, u_2\}$:

- $A = \{a_1, a_2\}, a_1 = f_1(u_1), a_2 = f_1(u_2)$
- $B = \{b_1, b_2\}, b_1 = f_2(f_1(u_1)), b_2 = f_2(f_1(u_2))$

In other words, if a user were to manually input the two elements in U, running the workflow each time, (which is the kind of labour that we are trying to automate), Step 3 would execute twice as $f_3 (a_1, b_1)$ and $f_3 (a_2, b_2)$. It would not execute four times using the Cartesian product of all possible pairs of inputs from A and B. We prevent this spurious execution by introducing key-based association.

   Key-based association exploits the fact that every data value in a user input collection has a unique and well-defined address, from which the originating component, collection and position within that collection can be deduced. We use the user input addresses to form a key for each result in the system. The key of a user input value is its own address. Any value derived, directly or indirectly, from a user input value, contains the address of that value as part of its key. Thus a key is a list of user input addresses which encodes the provenance of each derived value.

   We consider two user input addresses to be comparable if and only if they originate in the same channel of the same component. Two keys k_1 and k_2 are said to be associated if and only if:

- No address a_1 in k_1 is comparable to an address a_2 in k_2, or
- For every address a_1 in k_1 that is comparable to an address a_2 in k_2, $a_1 = a_2$.

That is, two values are associated if and only if they derive from completely distinct lineages, or in the case that they are derived from overlapping sets of channels, they are derived from the same value in each of those channels.

By ensuring that a Processor is only queued for execution if its input values are mutually associated, we preserve the structural integrity of the workflow when it is iterated. We also remove the need for user intervention to specify the association mode.

3.3 Collection Formation: Aggregation and Key-Slicing

GPFlow provides two ways to form a collection: aggregation and key-slicing. An example of aggregation is where we wish to merge the elements of two parallel arrays to form a single array of 2-dimensional vectors. An example of key-slicing is where we wish to gather all values produced by a component to perform some synthesis or summarizing operation.

If an input field has type "Array of T" for some type T, it may be connected to one or more output channels, provided their types are "T" or "Array of T". When a value is assembled for such an input field, a single sequence is constructed. This sequence contains all constituent elements of the nominated antecedent output channels, subject to the key association criterion described above. This extension alone is sufficient to enable aggregation.

Key-slicing is based on the observation that the collected keys of the result set generated by a component form a discrete hypercube, with dimensionality defined by the set of keys that index the elements of the result set. An output value is associated with each point in this hypercube. If we remove a key field, we project onto a hypercube of lower dimension, each point of which indexes a collection of values, namely those distinguished by the value of the key removed. Intuitively, we take a slice through the result cube.

To implement this in GPFlow, we permit the user to nominate one or more unbound input variables to be removed from the key for a particular input channel. Sliced inputs may belong to the component that contains the output channel or to any of its antecedents. Any sliced input values are selectively ignored when the association test is applied during input value assembly. This provides the user with an intuitive way to specify collections.

4 Case Study: A Phylogenetic Tree Inference Workflow

This paper is focused on the integration of collection processing into scientific workflow systems to enable comparative parametric studies and data synthesis. A good example of the latter is provided by the workflow fragment described below, which mirrors the illustrative case study presented in [3] and provides a clear opportunity for comparison of collection processing as seen by users in GPFlow and Kepler.

The workflow contains the following steps:

1. Get Homologs - user supplies the locus tag of one or more reference genes, and a list of comparison genomes. BLAST [11] is used to identify homologs of the reference genes in the comparison genomes.
2. Get NCR - extracts a DNA sequence adjacent to each of the homologous genes.
3. Align - uses ClustalW [12] to align the DNA sequences produced by step 2, forming a multiple alignment.
4. Phylip Pars - takes the multiple alignment emitted by step 3 and executes the Pars program [13] once for each of a set of distinct seed values. The result set consists of a set of trees.
5. Consensus Tree - uses the key-slice operator to gather the trees inferred by Pars into a collection as required by the Phylip Consense [13] program, which produces a single consensus tree.

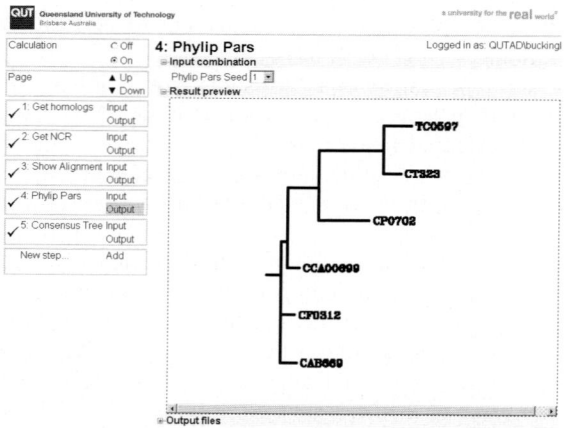

Fig. 2. Consensus tree workflow

Fig. 2 shows one of the trees produced by Phylip Pars resulting from a gene drawn from the Chlamydia trachomatis genome, indexed according to the seed value supplied to Pars. The user may readily view and compare alternate results by means of the drop down menu in the top section of the screen. The navigation component at left allows the user to move between components, adjusting parameters as necessary and triggering restarts of components whose input state has changed.

Although it is not immediately apparent in Fig. 2, this workflow contains an implicit loop in step 4 to generate a forest of distinct trees, which are gathered to form a collection in step 5 by clicking a single check box in the user interface. Use of data driven iteration and key-slicing allows us to hide detail from the scientist, who sees this as a simple linear pipeline of components, each of which has a readily identifiable role in the experiment. The workflow can be described in the language of the scientific domain without recourse to the language of computer science. This stands in contrast with the workflow described in detail in [3], in which the user encounters components such as "TextFileReader", "ExceptionCatcher", "NexusFile-Composer" and "TextFileWriter", and where explicit control flow primitives potentially introduce further confusion. The actions covered in steps 4 and 5 of the GPFlow workflow require 10 steps in Kepler. Furthermore, a special kind of actor had to be included in the Kepler solution to permit formation of a collection, where this situation is elegantly handled by the use of key-slice aggregation in GPFlow.

The next step for the user, having prototyped this workflow fragment, might be to execute the entire workflow over a range of input genes. This is accomplished by selecting two or more reference genes in the Get Homologs input panel. The workflow result set is then automatically recalculated, with the single consensus tree produced at the end of the run replaced by a corresponding collection of trees. This collection of trees would be available for individual analysis or further aggregated processing. Doing this manually would require the user to laboriously run through the entire workflow for each input gene. To achieve the same result in Kepler would require coding of another explicit loop.

5 Conclusions and Further Work

There is a widely acknowledged need for better eScience tools which enable scientists to concentrate on their research rather than the technology. We have presented a model and system enabling sophisticated comparative studies, vital in eScience, to be devised and undertaken without the need for programming or scripting, or for explicit management of the collection and data associations. A GPFlow workflow defined on a single value may be lifted to operate on a collection of values with no change required to the workflow. The model supports collective operations on aggregated data sets, through workflows encompassing a map reduce style of dataflow, and manages dependencies through a novel mechanism based on key aggregation and slicing. Thus, GPFlow differs from existing approaches in automatically determining the association mode for combinations of collections from the dataflow graph.

As in earlier versions of our system, the workflow is interactive; workflows can be interrupted and changed on the fly, thereby supporting experimentation. Workflows can be published and shared, thus enriching the research environment across a community.

Further work will address integration with Grid computing so that grid services may be easily invoked – currently the system can support arbitrary web services but has no specific facilities for grid computing. A longer term investigation is underway to raise the level of abstraction of the workflow further to enable high level questions and queries to be posed, ultimately including hypothesis generation. We envisage that this will form a layer on top of the current model. Finally we wish to apply the system to other fields, notably the analysis of environmental sensor network data.

The system will be made freely available under a BSD style license. A demonstration system can be accessed from the project home page (http://www.mquter.qut.edu.au).

Acknowledgements. We gratefully acknowledge the support of Microsoft Research, the Queensland Government and QUT.

References

1. Dean, J., Ghemawat, S.: MapReduce: Simplified Data Processing on Large Clusters. In: Sixth Symposium on Operating System Design and Implementation OSDI 2004, USENIX Association, San Francisco, CA (2004)
2. Rygg, A., Roe, P., Wong, O.: GPFlow: An Intuitive Environment for Web Based Scientific Workflow. In: Fifth International Conference on Grid and Cooperative Computing Workshops, pp. 204–211. IEEE Computer Society, Los Alamitos (2006)
3. Fox, G.C., Gannon, D.: Special Issue: workflow in Grid Systems (Editorial). Concurrency and Computation: Practice and Experience 18(10), 1009–1019 (2006)
4. McPhillips, T., Bowers, S.: An Approach for Pipelining Nested Collections in Scientific Workflows. ACM SIGMOD Record, ACM SIGMOD/PODS 34(3), 12–17 (2005)
5. McPhillips, T., Bowers, S., Ludäscher, B.: Collection-Oriented Scientific Workflows for Integrating and Analyzing Biological Data. In: Leser, U., Naumann, F., Eckman, B. (eds.) DILS 2006. LNCS (LNBI), vol. 4075, pp. 248–263. Springer, Heidelberg (2006)
6. Oinn, T., Greenwood, M., Addis, M., Alpdemir, M.N., Ferris, J., Glover, K., Goble, C., Goderis, A., Hull, D., Marvin, D., Li, P., Lord, P., Pocock, M., Senger, M., Stevens, R., Wipat, A., Wroe, C.: Taverna: lessons in creating a workflow environment for the life sciences. In: Concurrency and Computation: Practice and Experience, vol. 18, pp. 1067–1100. Wiley InterScience, Chichester (2006)
7. Johnston, W.M., Hanna, J.R.P., Millar, R.J.: Advances in dataflow programming languages. ACM Computing Surveys 36(1), 1–34 (2004)
8. Davis, A.L., Keller, R.M.: Data flow program graphs. IEEE Computer 15(2), 26–41 (1982)
9. Hylands, C., Lee, E., Liu, J., Liu, X., Neuendorffer, S., Xiong, Y., Zhao, Y., Zheng, H.: Overview of the Ptolemy Project. Technical Memorandum UCB/ERL M02/25, University of California, Berkeley (2003)
10. Stevens, R., Robinson, A., Goble, C.A.: myGrid: Personalized Bioinformatics on the Information Grid. Bioinformatics, Oxford Journals 19(suppl. 1), i302-i304 (2003)
11. Altschul, S.F., Gish, W., Miller, W., Myers, E.W., Lipman, D.J.: Basic local alignment search tool. J. Mol. Biol. 215(3), 403–410 (1990)
12. Thompson, J.D., Higgins, D.G., Gibson, T.J.: CLUSTAL W: improving the sensitivity of progressive multiple sequence alignment through sequence weighting, position-specific gap penalties and weight matrix choice. Nucleic Acids Res. 22, 4673–4680 (1994)
13. Felsenstein, J.: PHYLIP (Phylogeny Inference Package) version 3.6. Distributed by the author. Department of Genome Sciences, University of Washington, Seattle

Resource Discovery Based on VIRGO P2P Distributed DNS Framework

Lican Huang

Institute of Network & Distributed Computing,
Zhejiang Sci-Tech University, Hangzhou, P.R.China
huang_lican@yahoo.co.uk

Abstract. Resource discovery plays an important role on scientific workflows in large scalable network environment. This paper presents a resource discovery framework based on VIRGO P2P Distributed DNS. With the convention of resource name as the format– *functionscheme":" global-hier-part"/"local-name*, this framework supports flexible queries using partial keywords and wildcards, and range queries by a SQL-like query statement. The global-hier-part is managed by registers the same as DNS servers except the extension of RRs. The DNS servers construct n-tuple overlay virtual hierarchical overlay network of VIRGO. With cached addresses of DNS servers, the overload of traffic in tree structure can be avoided. The time complexity, space complexity and message-cost of lookup with this framework is $O(L)$, where L is the number of sub domains in Domain Name.

1 Introduction

Resource discovery plays an important role on scientific workflows in large scalable network environment. Before submitting a job to require a QoS performance, a suitable workflow engine host node with a set of attributes (such as CPU type, memory, operating system type) among lots of workflow engines is required to find for executing the job. Scientific workflow often requires dynamic selection of workflow routines, Web Services or workflow engines. In some case, we need to find a suitable Web Service from multiple copies of semantic Web Services to achieve better performance [1][2]. The most suitable workflow engines are required to be discovered before submitting sub workflows to these engines to execute in parallel [3]. In other cases, the Services with the same functions but with different algorithms (for example, data clustering services implemented by using the algorithms of Neural Network or SVM) may be only chosen at the just pre-execution point according to the intermediate results of workflow execution [4]. All the above cases illustrate the importance of resource discovery(the resources may be static and/or dynamic. The dynamic resources are meant both in dynamic change of attributes or dynamic exist in the network).

When there are more and more computers connected in Internet and more and more users use Internet, resources in Internet will be more and more. In this large,

M. Bubak et al. (Eds.): ICCS 2008, Part III, LNCS 5103, pp. 501–509, 2008.

decentralized, distributed resource sharing environments today or in the future, the efficient discovery of static and/or dynamic resources is a big challenge.

The resource discovery based on Client/Server is not suitable for huge amount of resources. There are many approaches for distributed resource discovery based on P2P technologies[5]. The un-structural P2P technology such as Freenet[6] using flooding way has shortage of heavy traffic and un-guaranteed search. The structural P2P technology using DHT such as Chord [7] loses semantic meanings of discovered objects. Resource discovery based on Virtual and dynamic hierarchical architecture(VDHA) [8] has shortage of high traffic load in root nodes and possible split of virtual trees. VIRGO[9][10] P2P network hybrids structural and unstructured P2P technologies by merging n-tuple replicated virtual tree structured route nodes and randomly cached un-structured route nodes(LRU and MinD) to solve the above problems.

The resources can be classified as hierarchical Domains. It is nature way to use DNS to discover resources. There are DNS-based resource discovery approached [11] by using the existing and mature DNS protocols[12]. The main advantage is the use of existing, widely accepted and consolidated DNS technologies and APIs to accomplish resource advertisement and dynamic discovery. Although DNS implementation today is scalable for translating Domain Names into IP addresses, it may encounter problem if using it to discover huge amount resources today and even more amount resources in the future. The framework proposed here based on VIRGO P2P technology is suitable for the huge amount of resources' discovery.

This paper presents a framework for distributed resource discovery based on VIRGO P2P technologies. The resources are classified as multi-layer hierarchical catalogue domains according to their semantic meaning. We then name resources as format– *functionscheme ":"global-hier-part"/"local-name* [13]. The global-hier-part part in this format is taken as Domain name. The authoritative DNS servers controlling these Domain names manage these resources. These DNS servers construct n-tuple overlay virtual hierarchical overlay network. With cached addresses of DNS servers, the overload of traffic in tree structure can be avoided. We here change some places of VIRGO to suit the distributed DNS implementation[14]. The lookup protocols of distributed DNS is similar as the protocols in VIRGO[9], which is illustrated in detail in the paper[10]. The resource discovery is effective and guaranteed. The time complexity, space complexity and message-cost of lookup protocol is $O(L)$, where L is the number of sub domains in Domain Name.

The structure of this paper is as follows: section 2 describes VIRGO network structure for distributed DNS; section 3 presents the Convention for Universal Resource Name; section 4 discusses Distributed Resource Discovery Framework; and finally we give out conclusions.

2   VIRGO Network Structure for Distributed DNS

VIRGO [9] is a domain-related hierarchical structure hybridizing un-structural P2P and structural P2P technology. VIRGO consists of prerequisite virtual group tree, and cached connections . Virtual group tree is similar to VDHA [8] , but with multiple gateway nodes in every group. Virtual group tree is virtually hierarchical, with one root-layer, several middle-layers, and many leaf virtual groups. Each group has N-tuple gateway nodes. In VIRGO network, random connections cached in a node's route table are maintained. These cached connections make VIRGO a distributed network, not just a virtual tree network like VDHA. With random cached connections, the net-like VIRGO avoids overload in root node in virtual tree topology, but keeps the advantage of effective message routing in tree-like network. As the change of contents in route table, VIRGO uses different lookup protocol and maintenance protocols from VDHA.

In VIRGO, user uses client to access VIRGO via access point node(called as entrance node). All users are managed by their owner nodes. Some nodes (called Gateway node)take route functions in several different layers of virtual groups. Gateway node is the node which is not only in low-layer group but also in up-layer group.

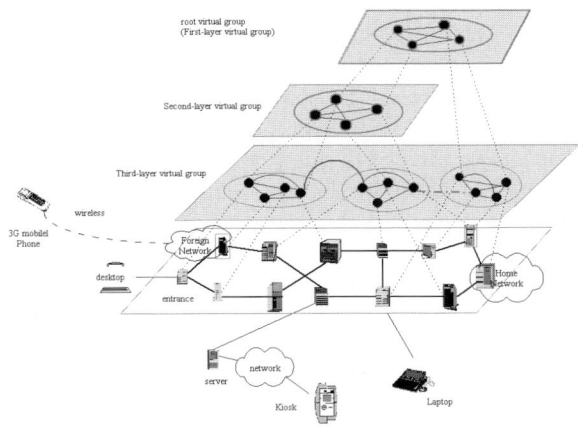

Fig. 1. Two-tuple Virtual Hierarchical Tree Topology

Fig. 1 shows two-tuple virtual hierarchical tree topology (the nodes in different layers connected with dash line are actually the same node). In Figure 1, from the real network, three virtual overlay groups are organized. From these virtual groups two nodes per virtual group are chosen to form the upper layer virtual group.

3   Convention for Universal Resource Name

In Internet, every resource needs an uniform name. There are RFCs URI [15] ,URL[16] and URN[17]. Assuming there are so huge number of resource names, how to name these resources as to easily resolve is a big issue. Here we argue that the resources can be classified into a hierarchical structure.

Therefore, the convention for resource name is formatted as global unique hierarchical domains like domain names in DNS, and a local name to specify the local attributes of resource name. The format of convention of resource names is as follows:

$$functionscheme" : "global - hier - part"/"local - name$$

Here,the functionscheme is type or function or schema for the resource; the global-hier-part is the global hierarchical name similar as domain name and the local-name is the local name for the resource.

For example,

$blog : Britney.popular.music/storyofBritney$ is a resource name for a blog article about the story of Britney Spears. Whereas, $song : Britney.popular.music/Lonely$ is a resource name for the song Lonely by Britney Spears. Here, Britney Spears is classified as Britney.popular.music.

The local-name part in the format can be used for the description of resource semantic meanings. It can also a URL for the other repository such as UDDI , Grimoires[18] which contains the resource metadata and/or other information.

The global-hier-part is formed as Domain Names. The resources are classified into different domains according to their semantic meaning. The top domains may be governed by International trustee such as The Internet Corporation for Assigned Names and Numbers (ICANN). Other level domains may be controlled by virtual organizations organized by the experts with specific knowledge of these domains. For example, in Domain Name www.Madonna.popular. music, music Domain is managed by music virtual organization, which reports to ICANN to approve, popular.music Domain is controlled by popular music virtual organization,which reports to music virtual organization to approve. The request of Domain Name www.Madonna.popular.music is sent to virtual organization of popular music to be approved. In Domain Name www.Beethoven.classic.music, classic.music Domain is controlled by classic music virtual organization, which reports to music virtual organization to approve. The request of www.Beethoven. classic.music is managed by virtual organization of classic music.

4   Distributed Resource Discovery Framework

We treat global-hier-part in the invention of resource name as domain name. We also use the similar concept of DNS zone for DNS server to manage the domain names. For example. if a DNS server's zone is Science.Biology.Bioinformatics, then this DNS server manages all resource information under the catalogue of Science.Biology.Bioinformatics. In other case, if a DNS server's zone is Science.Biology, then this DNS server manages all resource information under the catalogue of Science.Biology, which may includes sub domains of BioInformatics, botany, zoology,etc.

Resource names are published in their authoritative DNS server. The extension RRs in DNS server contains the parts of the functionscheme and local-part of conventions of resource name. The local-part of resource can be a URL for

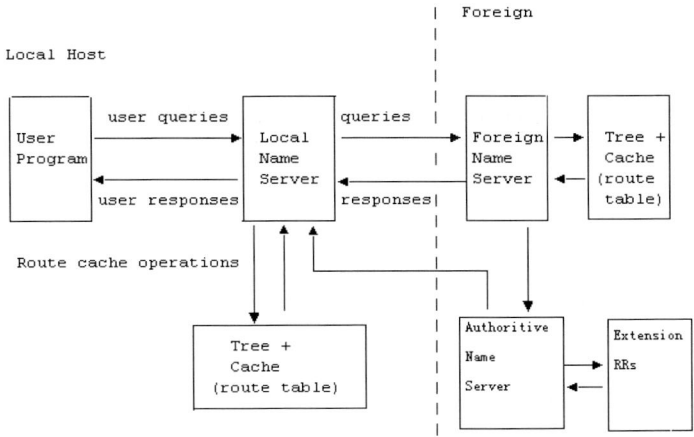

Fig. 2. Distributed Resource Discovery Framework

the other repository such as UDDI, Grimoires[18] which contains the resource metadata and/or other information.

Every DNS server is the same but some coexist in more than one layer. Every DNS Server maintains a route table and RRs related to its authoritative domain zone. Route table includes addresses of Foreign Name Servers which are prerequisite for Virtual Hierarchical Overlay Network and cached addresses of Foreign Name Servers which are refreshed by TTL rule, etc.

The DNS servers join VIRGO network by placing themselves in virtual groups according to their authoritative domain names(The only root domain- "root" is anonymous placed at the top point of all the domains). When a DNS server joins, its route table will add some other nodes' entities, and some other nodes will update their route table by adding the joining node's entity. The detail of the algorithm can be in the Internet draft[14].

The query process is as the following(see figure 2): User program sends QUERY MESSAGE, which contains SQl-like statement in the following subsection, to Local Name Server. If Local Name Server is the authoritative Domain Name Server, then the Local Name Server will check its RRs to resolve the request resource. Otherwise, The Local Name Server will route to the Foreign Name Server which is closer to the authoritative Domain Name Server by calculating theoretical hops. Then the Foreign Name Server routes to the even closer Foreign Domain Name Server. Repeat this process, until the authoritative Domain Name Server has been found. Finally, the authoritative Domain Name Server resolves request resource by check its RR record, and responses to the Local Name Server. The latter will forward the response to the User Program. The details of the algorithm can be found in [14].

For example, to discover resource chloroP, we first locate the DNS authoritative server for domain-Science.Biology.Bioinformatics, then in this server, we check its extension RRs for the resource, which gives URL of other repository.

Finally, we find resource information such as URL, metadata ,etc.in the item of localdirectore/chloroP/ in the repository.

4.1   SQL-Like Query Language and Maintenance Language

The local-part part in resource convention can be formed as entities with attributes in the domain defined by global-hier-part. We use SQL-like query language to query resources. The SQL-like query language is formed as the following:

$query ::= SELECT\{SCHEMA|SUBDOMAIN|ENTITY|ATTRIBUTE|*$
$|expr[[AS]c\_alias]\{, expr[[AS]c\_alias]...\}\}$
FROM $domainref[$ WHERE $search\_condition];$

Here, SCHEMA is for querying functionschema for a Domain; SUBDOMAIN is for querying subdomain for a given super Domain; ENTITY is for querying all entities for a given search condition ; ATTRIBUTE is for querying all attributes for a given search condition ; expr is for the attribute queried, which may be a set of attributes; [AS] c_alias is the alias name for expr; domainref is for Domain Name, in which we can use * to indicate all sub domains; search_condition is similar to SQL statement in Database.

In the following, there are several examples of query operations.

Operation 1

The form "?:global-hier-part" is to query all functionschema of the global-hier-part domain. For example, "?:Britney.popular.music" queries all functionschema of Britney.popular.music. We can use the following Statement:

$SELECT\ SCHEMA\ FROM\ Britney.popular.music;$

Operation 2

The form "functionschema:global-hier-part/?" is to query all resources with the functionschema in the global-hier-parts domain. For example, "song:Britney. popular.music/?" queries all songs by Britney Spears. We can use the following Statement:

$SELECT\ *\ FROM\ Britney.popular.music\ WHERE\ SCHEMA\ ='song';$

Operation 3

Suppose that global-hier-part can be expressed as leafDomain.superDomain. The form "functionschema:?.superDomain/" is to query all leafDomains. This is possible because of the hierarchical structure and the protocols. For example, "song:?.popular.music" queries all popular singers. We can use the following Statement:

$SELECT\ SUBDOMAIN\ FROM\ *.popular.music\ WHERE\ SCHEMA\ ='song';$

Operation 4 The form "functionscheme :global-hier-part/local-name" is to query specific resource. For example, "song:Britney.popular.music/Lonely" queries the information of the song-Lonely. We can use the following Statement:

$SELECT\ Lonely\ FROM\ Britney.popular.music\ WHERE\ SCHEMA\ ='song';$

Operation 5

The form "functionscheme :global-hier-part/expression" is to query the resources which satisfy the conditions of the expression indicated. Expression can be scope, maximum, minimum, where condition like SQL statements. For example, "song:Britney.popular.music/where year between 2006 and 2007" queries all songs by Britney Spears which are produced between 2006 and 2007. We can use the following Statement:

$SELECT$ * $FROM$ $Britney.popular.music$ $WHERE$ $SCHEMA$ $='song'AND$ $year \geq 2006$ AND $year \leq 2007;$

Another example is the case for selecting an optimal service. We first use Ganglia to get performance attributes of all machines in a given domain ,and select an optimal machine by calculating maximum value of the formula –"CPU speed*CPUnumber/CPUload"; we can use the following statement:

$SELECT$ $ServiceAddress$ $FROM$ $cluster.computer.IT$ $WHERE$ $SCHEMA$ $='hardware'$ AND $max(ganglia.CPUspeed*ganglia.CPUnumber/$
$ganglia.CPUload);$

The resources are maintained by the SQL-like maintenance language. Insert statement registers new resource into the domain. Delete statement drops the resource registration from the domain. Update statement changes the information registered in the domain. The formats of these statements are as the following:

$DELETEFROM$ domainref [WHERE $search\_condition$];

$INSERTINTO$ domainref[(entity.attribute){, $entity.attribute...$})]
$VALUES(expr|NULL\{, expr|NULL...\});$

$UPDATE$ domainref SETentity.attribute $= expr|NULL\{,$
$entity.attribute = expr|NULL...\};$

For example, the following statement deletes all songs by Britney Spears which are produced between 2006 and 2007.

$DELETE$ $FROM$ $Britney.popular.music$ $WHERE$ $SCHEMA$ $='song'AND$
$year \geq 2006$ AND $year \leq 2007;$

4.2   Complexity

Because the DNS server nodes are virtually organized as a tuple virtual tree, every DNS server has a route table which includes prerequisite DNS servers' IP addresses for Tree Paths (TREE portion) and cached DNS servers' IP addresses (CACHED portion).

Because the message is routed according to the minimum of theoretical distance from destination node , and the route table contains TREE portion, every hop reduces the distance from destination node by at least one hop, Therefore,

$hops(a, b) < length(a) + length(b) - 1$ (1)

Where, hops(a,b) is for the hops from node a to node b; length(a),length(b) are for node a domain name lengths and node b domain name lengths respectively. For example, the length of www.nic.fr is 3. So,

$timecomplexity = O(L)$ (2)

$message\_cost = O(L) (3)$,

where L is the length of domain name.

Because the route table of the virtual gateway nodes virtually existed from root layer to bottom layer groups has the maximum route items of nodes's information, we have:

$MaxItems = L * N\_tuple * nvg + Max\_Cached$ (4),

where L is the length of domain name., N_tuple is multiplicity of gateway nodes for virtual tree, nvg is number of virtual groups, Max_Cached is the maximum number of cached records in the route table

Therefore,

$SpaceComplexity = O(L)$ (5)

5   Conclusion

Resource discovery plays an important role on large scalable scientific workflows. We here presents a resource discovery framework based on a novel distributed DNS. The proposed distributed DNS framework is based on VIRGO P2P technologies. The resources are classified into hierarchical domains, which can be managed by registers the same as DNS servers except the extension of RRs. The DNS servers construct n-tuple overlay virtual hierarchical overlay network of VIRGO. With cached addresses of DNS servers, the overload of traffic in tree structure can be avoided. The time complexity, space complexity and message-cost of lookup with this framework is $O(L)$, where L is the length of domain name. With the convention of resource name as the format– *functionscheme ":"   global-hier-part"/"local-name*, this framework supports flexible queries using partial keywords and wildcards, and range queries. The framework presented here uses the similar message of DNS. Therefore, we can use existing, widely accepted and consolidated DNS technologies , APIs and open source codes to implement the framework. Meanwhile, the framework with VIRGO P2P technology is scalable for discovery of huge amount of resources today and even more amount resources in the future. We plan to implement this framework by merging the source codes of VIRGO project[19] and dnsjava[20].

Acknowledgements. This paper is funded by "Qianjiang Rencai" funds by Government of Zhejiang Province(No:2007R10013).

References

1. Huang, L., Akram, A., Allan, R.J., Walker, D.W., Rana, O.F., Huang, Y.: A Workflow Portal Supporting Multi-Language Interoperation and Optimisation. Journal Concurrency and Computation: Practice and Experience 19(12), 1583–1595 (2007)
2. Walker, D.W., Huang, L., Rana, O.F., Huang, Y.: Dynamic Service Selection in Workflows Using Performance Data. Journal of Scientific Programming 15(4), 235–247 (2007)

3. Huang, L.: Framework for Workflow Parallel Execution in Grid Environment. In: Shi, Y., van Albada, G.D., Dongarra, J., Sloot, P.M.A. (eds.) ICCS 2007. LNCS, vol. 4489, pp. 228–235. Springer, Heidelberg (2007)
4. Huang, L.: Production Rule Based Selection Decision for Dynamic Flexible Workflow. In: 3rd IEEE International Conference on e-Science and Grid Computing, pp. 627–632. IEEE Press, New York (2007)
5. Foster, I., Iamnitchi, A.: On Fully Decentralized Resource Discovery in Grid Environments. In: Lee, C.A. (ed.) GRID 2001. LNCS, vol. 2242, pp. 51–62. Springer, Heidelberg (2001)
6. Clarke, I., Sandberg, O., Wiley, B., Theodore, W., Hong, T.W.: Freenet: A distributed anonymous information storage and retrieval system. In: Federrath, H. (ed.) Designing Privacy Enhancing Technologies. LNCS, vol. 2009, pp. 46–66. Springer, Heidelberg (2001)
7. Stoica, I., Morris, R., Karger, D., Kaashoek, F.M., Balakrishnan, H.: Chord: a scalable peer-to-peer lookup service for internet applications. In: 2001 conference on Applications, technologies, architectures, and protocols for computer communications, pp. 149–160. ACM Press, New York (2001)
8. Huang, L., Wu, Z., Pan, Y.: Virtual and Dynamic Hierarchical Architecture for e-Science Grid. International Journal of High Performance Computing Applications 17(3), 329–347 (2003)
9. Huang, L.: VIRGO: Virtual Hierarchical Overlay Network for Scalable Grid Computing. In: Sloot, P.M.A., Hoekstra, A.G., Priol, T., Reinefeld, A., Bubak, M. (eds.) EGC 2005. LNCS, vol. 3470, pp. 911–921. Springer, Heidelberg (2005)
10. Huang, L.: A P2P service discovery strategy based on content catalogues. Data Science Journal 6, S492–S499 (2007)
11. Giordano, M.: DNS-Based Discovery System in Service Oriented Programming. In: Sloot, P.M.A., Hoekstra, A.G., Priol, T., Reinefeld, A., Bubak, M. (eds.) EGC 2005. LNCS, vol. 3470, pp. 840–850. Springer, Heidelberg (2005)
12. Mockapetris, P.: DOMAIN NAMES - IMPLEMENTATION AND SPECIFICATION. Specification, RFC1035 (1987), http://www.ietf.org/rfc/rfc1035.txt
13. Huang, L.:Distributed Universal Resource Name Resolution based on Distributed DNS(Internet Draft) (2008), http://www.tools.ietf.org/html/draft-licanhuang-dnsop-urnresolution-01
14. Huang, L.: Distributed DNS Implementation in IpV6 (Internet Draft) (2007), http://tools.ietf.org/id/draft-licanhuang-dnsop-distributeddns-02.txt
15. Berners-Lee, T. , Fielding, R., Irvine, U.C., Masinter, L.: Uniform Resource Identifiers (URI): Generic Syntax, RFC 2396 (1998), http://www.ietf.org/rfc/rfc2396.txt
16. Berners-Lee, T. , Masinter, L. , McCahill, M.: Uniform Resource Locators (URL), RFC 1738 (1994), http://www.faqs.org/rfcs/rfc1738
17. Daigle, L., Gulik, D. v., Iannella, R., Faltstrom, P.: URN Namespace Definition Mechanisms, RFC 2611 (1999), http://www.ietf.org/rfc/rfc2611.txt
18. Wong, S.C., Tan, V., Fang, W., Miles, S., Moreau, L.: Grimoires: Grid Registry with Metadata Oriented Interface: Robustness, Efficiency, Security. IEEE Distributed Systems Online 6(10) (2005)
19. VIRGO: HomePage of VIRGO, http://virgo.sourceforge.net/
20. DNSjava: HomePage of DNSjava, http://www.dnsjava.org/

Securing Grid Workflows with Trusted Computing

Po-Wah Yau, Allan Tomlinson, Shane Balfe, and Eimear Gallery

Information Security Group
Royal Holloway, University of London
Egham, Surrey TW20 0EX, UK
{p.yau,allan.tomlinson,s.balfe,e.m.gallery}@rhul.ac.uk

Abstract. We propose a novel scheme that uses Trusted Computing technology to secure Grid workflows. This scheme allows the selection of trustworthy resource providers based on their platform states. The integrity and confidentiality of workflow jobs are provided using cryptographic keys that can only be accessed when resource provider platforms are in trustworthy states. In addition, platform attestation is used to detect potential workflow execution problems, and the information collected can be used for process provenance.

1 Introduction

Grid computing [1] is a distributed computing paradigm which seeks to exploit the synergies of technology and social collaboration to solve data or computation-intensive problems. Solving such problems requires the management of multiple tasks, their relationships and execution to produce valid and reliable results — this is the focus of Grid workflow research [2]. Although the authorisation and authentication of Grid jobs has been extensively studied [3] [4], the management of Grid workflows introduces additional security issues. The risk to a user's data and results is dramatically increased when using workflows, because the entire dataset is exposed to the Grid.

The use of historical information has been proposed to schedule jobs so that 'untrusted' nodes can be avoided [5], but such information may be incorrect or open to manipulation. Moreover, this approach cannot prevent, detect, or react to single job manipulation, the effects of which would propagate throughout the associated workflow. The provision of a data provenance service, which records how data has been collected and processed [6] [7], helps to address this problem, but only forms part of a reactive solution to detect problems after running a workflow.

The application of Trusted Computing (TC) technology is emerging as a potential solution to a number of Grid security problems [8] [9], and this paper investigates how this technology may be applied to secure Grid workflows. Section 2 provides a brief summary of Grid workflows, and outlines a set of security requirements. Section 3 gives an overview of TC and its application to Grid security. Section 4 describes our proposal for securing Grid workflows, and Section 5 contains an analysis. Final remarks are given in Section 6.

M. Bubak et al. (Eds.): ICCS 2008, Part III, LNCS 5103, pp. 510–519, 2008.

2   Grid Workflows

A workflow defines a logical ordering of tasks to be completed, and can be represented as a directed acyclic graph or flowchart with parallel, sequential and choice branches and loops [10]. Each workflow task can operate on either a new set of data or the results from a parent task, i.e. intermediate data. Thus, data is input from storage resources that may be either internal or external to where the computation is taking place.

Typically, abstract workflow specifications are passed to a Workflow Resource Broker (WRB), i.e. a workflow execution engine/system, which maps workflow tasks onto physical jobs that will be submitted to a Grid. The creation of a physical workflow of Grid jobs requires the WRB to select Grid resource providers and schedule jobs to be submitted to them. The system will select resource providers that meet static and dynamic workflow requirements, for example the availability of software applications and libraries or, indeed, specific security policies.

During workflow execution, data can be moved using one of three approaches — centralised, mediated or peer-to-peer [10]. Centrally managed data movement is the easiest to implement, as all data is transferred via a central point. Mediated data movement involves a distributed management system with synchronised replication catalogue services. Finally, using a peer-to-peer method involves transferring data directly between resource providers.

2.1   Workflow Security

Trust in the WRB is critical, as it is relied upon by a user to ensure that their workflow will be executed as expected, and thus produce valid results. A user delegates control to a WRB to map workflow tasks to jobs, which must then be submitted to the appropriate resource providers. The WRB is also trusted not to divulge workflow information that would allow an attacker to coordinate attacks on the workflow. The compromise of a single job might not reveal any sensitive information, whereas an attack on several jobs might. Therefore, it is essential to maintain confidentiality of the locations to which workflow jobs are submitted.

Resource providers might be selected based on direct experience and/or other indirect metrics, such as reputation or trust measurements based on provenance services [5] [6]. However, there is a risk that this information is unreliable, incorrect or out-of-date. Thus, a WRB needs to be able to reliably determine if it can trust a resource provider to behave as expected before sending it a workflow job.

Many observers have commented on the vulnerabilities surrounding Grid middleware and the subsequent risk of job execution compromise [9] [11] . Therefore, it is also necessary for the assurances determined during the selection process to hold true until job execution has finished. This includes the protected transport of output data to the required destination, be it the WRB or another resource provider. If the integrity of the job execution platform is not maintained, then the WRB must be alerted to the potential compromise so that it can react accordingly. An undetected compromise could mean that resources are wasted on executing the rest of the workflow using incorrect data.

Finally, audit trail information must be reliably collected. As stated above, provenance information, or the procedure for collecting provenance information itself, could be flawed and a mechanism is required to detect when this is the case. Audit information will also assist the debugging of workflows, as confidence in the resource providers will help to eliminate a large potential source of errors.

3   Trusted Computing

A trusted platform is one that behaves in a particular manner for a specific purpose. Such a platform can be built following the Trusted Computing Group's[1] Trusted Platform Module (TPM) specifications [12] [13] [14]. These specifications describe a tamper-resistent device with cryptographic coprocessor capabilities. This device provides the host platform with a number of services including: special purpose registers for recording platform state; a means of reporting this state to remote entities; and asymmetric key generation, encryption and digital signature capabilities. TC also encompasses new processor designs [15] and OS support [16] which facilitate software isolation. These concepts are examined in more detail elsewhere — see, for example [17] [18]. For the purposes of this paper we examine four TC-related concepts: integrity measurement, TPM keys, sealing and platform attestation.

3.1   Integrity Measurement

An integrity measurement is the cryptographic hash of a platform component (i.e. a piece of software executing on the platform) [16]. For example, the integrity measurement of a program can be calculated by computing a cryptographic digest of a program's instruction sequence, its initial state and its input. Integrity measurements are stored in special purpose registers within the TPM called Platform Configuration Registers (PCRs).

3.2   TPM Keys

A TPM can generate an unlimited number of asymmetric key pairs. For each of these pairs, private key use and mobility can be constrained. Key use can be made contingent upon the presence of a predefined platform state (as reflected in the host platform's TPM PCRs). Additionally, a private key can be migratable, non-migratable or certifiable migratable.

A non-migratable key is inextricably bound to a single TPM instance, and is known only to the TPM that created it. A certificate for a non-migratable key and its security properties may be created by the TPM on which it was generated. A certifiable migratable key (CMK) can be migrated but also retains properties which can be certified by the TPM on which the CMK was generated. When a CMK is created, control of its migration is delegated to a migration (selection) authority. In this way, controlled migration of the key is made possible, so that an entity other than the TPM owner helps to decide where the CMK can be migrated. This ensures that the certified security properties of the key are maintained.

---

[1] https://www.trustedcomputinggroup.org

3.3   Sealing

This is the process by which data is encrypted and associated with a set of integrity measurements representing a particular platform configuration. The protected data can only be decrypted and released for use by a TPM when the current state of the platform matches the integrity measurements to which the data was sealed.

3.4   Platform Attestation

Platform attestation enables a TPM to reliably report information about the current state of the host platform. On request from a challenger, a TPM provides signed[2] integrity measurements reflecting (all or part of) the platform's software environment. The challenger can use this information to determine whether it is safe to trust the platform and its software environment. This involves validating the received integrity measurements against a set of values it believes to be trustworthy, possibly provided by a trusted third party such as a software vendor.

However, there are potential issues surrounding the binary representation of software components — such a representation is static and inflexible; program behaviour has to be inferred; upgrades and patches are difficult to deal with; and revocation is problematic [19]. In order to overcome these problems, the concept of property-based platform state representation has been proposed [19] [20], in which a platform's state is represented by a set of high-level security properties. Using such techniques, migratable and certifiable migratable keys can be generated such that private key use is bound to properties, data can be sealed to properties, and the TPM can attest to platform properties, rather than specific software integrity measurements.

3.5   Application of TC to Grid Security

A number of authors have considered how Trusted computing could be applied to Grid Computing [9] [11] [21] [22]; the main goal of much of this prior art is to prevent or detect resource provider misbehaviour. Mao et al. [23] propose Daonity, a system which establishes a relocatable key enabling controlled group sharing of encrypted content. Löhr et al. [24] propose a scheme in which resource providers publish *attestation tokens*, which contain public keys from non-migratable TPM key pairs and the platform states to which private key use is bound. Each token is signed by the TPM to prove that it was produced by an authentic TPM.

4   Securing Grid Workflows

We now describe how Trusted Computing may be used to provide the following security services to Grid workflows:

1. Trusted Resource Provider Selection;
2. Confidentiality of job information;

---

[2] Using a private attestation signing key.

3. Integrity of job information; and
4. Audit data for process provenance.

Job information can include a job script, any executables, and input and output data. TC can be used to provide strong assurances to the Grid user that a workflow has executed correctly, and that the data was protected from malicious entities.

4.1 Assumptions

In order to fully utilise TC in Grid computing, the supporting TC architecture must be integrated into Grid environments. The proposal in this paper operates on the following assumptions:

Trusted Computing prevalence: There exists a Workflow Resource Broker (WRB) that is equipped with a trusted platform, as described in Section 3. A subset of Grid resource providers will also have trusted platforms installed; the scheme only uses such providers to process workflow jobs.
Resource broker verification service: This is provided by a trusted third party, and will be used to determine whether or not a WRB is trustworthy. This is achieved by verifying the platform state attested to by a WRB against known trusted states.
Public keys: All entities involved will have a certified copy of the chosen WRB's public signature verification key. Conversely, the WRB will have the public signature verification keys of all entities.

The underlying assumption is that trusted platforms exist within a Grid network, supported by the TC infrastructure. With major backing from hardware and software vendors, TC is becoming more pervasive, which will lead to the greater availability of TC supporting entities such as migration authorities.

4.2 The Scheme

A user relies on a trusted resource broker verification service to determine the trustworthiness of a WRB, using platform attestation (see Section 3). A workflow specification tool is used to create an abstract workflow of Grid tasks that is passed to the WRB, together with an encompassing security policy. The WRB maps the workflow tasks to a set of jobs, that are scheduled for submission to selected resource providers meeting the user's security requirements. To achieve this, the WRB may have to translate high-level user requirements into low-level platform state requirements. Workflow execution is then protected using TC services, as we next describe.

Key Distribution. Consider a sequence of jobs $a_0, a_1, ..., a_n$ that make up a user's workflow. For each job a_i, the WRB matches the user's high-level security requirements to a private key SK_i, whose use is contingent on the selected resource provider's platform satisfying low-level state information α (see Section 3). A resource provider could use either of the following two methods to obtain a private key in our framework:

1. The private keys can be created *a priori* or dynamically by the WRB as certifiable migratable keys for each of the jobs in the user's workflow, with the WRB specifying itself as the migration (selection) authority. The WRB specifies the states to which the private keys are bound prior to their migration to the selected resource providers.

2. The resource providers themselves each create a non-migratable private key bound to a specific platform state; this state and the corresponding public key are advertised as part of an attestation token [25]. The WRB pulls the attestation tokens from a service register and uses them to select appropriate resource providers.

The result is that the WRB can seal data that a resource provider can only access when it is in a trusted state. This allows the workflow to be protected, as described below.

Protecting the Workflow. Once the private keys have been provisioned, the WRB creates a symmetric key K_i for each job a_i, and generates a set of information to send to each chosen resource provider RP_i:

$$\text{WRB} \rightarrow \text{RP}_i : ID_W || r_i || g_{K_i}(a_i || r_i) || e_{PK_i}(K_i) || IP_{i+1} || PK_{i+1} ||$$
$$ID_{RP_{i-1}} || VK_{RP_{i-1}} || \alpha_{i-1} || \sigma \tag{1}$$

where:

- ID_W contains the identifiers of the workflow and the WRB;
- r_i is a random nonce chosen by the WRB;
- g is the generation-encryption function of an agreed authenticated encryption scheme [26] [27] — $g_{k_i}(a_i || r_i)$ generates the ciphertext and message authentication code for the concatenation of the job and nonce;
- $e_{PK_i}(K_i)$ is the key K_i encrypted using RP_i's public key PK_i;
- IP_{i+1} is the address to which any job output should be sent — this could be either the WRB or the next resource provider in the workflow RP_{i+1}, either for storage or further processing;
- PK_{i+1} is the public key used to encrypt job output;
- $ID_{RP_{i-1}}$ is the identifier of the preceding resource broker;
- VK_{i-1} is the public verification key used to verify messages from RP_{i-1} (see Section 4.1);
- α_{i-1} is the platform state that RP_{i-1} had to be in in order to process a_{i-1};
- σ is the digital signature of the WRB on the entire message.

Note that in the case of RP_0, RP_{i-1} would be the WRB. Thus, the state α_{i-1} sent to RP_0 would be the platform state of the WRB; this information can be used for auditing purposes (see below).

Executing the workflow. The following is the process of workflow execution at an arbitrary RP_i, after receiving message 1 (see Section 4.2). We assume that each message also contains both the identifier of its originator and a digital signature.

$$\text{RP}_{i-1} \rightarrow \text{RP}_i : ID_W \| \text{ready} \qquad\qquad (2)$$

$$\text{RP}_i \rightarrow \text{RP}_{i-1} : ID_W \| C(r_{RP_i}) \qquad\qquad (3)$$

$$\text{RP}_{i-1} \rightarrow \text{RP}_i : ID_W \| \alpha_{i-1}(r_{RP_i}) \| g_{K_i'}(R(a_{i-1})) \| e_{PK_i}(K_i') \qquad\qquad (4)$$

$$\text{RP}_i \rightarrow \text{WRB} : ID_W \| r_{RP_i} \| \alpha_{i-1}(r_{RP_i}) \qquad\qquad (5)$$

For the above interaction, the following are the steps taken by RP_i to execute a_i upon receiving message 2:

1. Verify σ from message 1.
2. Use the private key SK_i to decrypt the symmetric key K_i.
3. K_i is passed to the appropriate Grid application, which decrypts a_i and verifies its data integrity.
4. Generate a random nonce r_{RP_i} and send an attestation challenge $C(r_{RP_i})$ to RP_{i-1} (message 3);
5. Compare the response $\alpha_{i-1}(r_{RP_i})$ from message 4 with α_{i-1} from message 1;
6. The results of the comparison are sent to the WRB for auditing (see message 5). If the check has failed, then RP_i waits for further instructions from WRB, which raises an exception.
7. Otherwise, the symmetric key K_i' from message 4 is decrypted and used to recover the results $R(a_{i-1})$ of the previous job. K_i' would have been generated by RP_{i-1} (see step 9).
8. Job a_i is processed using $R(a_{i-1})$.
9. Once a_i has completed, RP_i creates a fresh symmetric key K_{i+1}', generates $g_{K_{i+1}'}$ $(R(a_i))$ and encrypts the key $e_{PK_{i+1}}(K_{i+1}')$.

5   Security Analysis

Establishing trust in the WRB is a fundamental precursor to our scheme. It cannot be expected that a standard Grid user will be able to interpret attestation integrity measurements, hence we require a trusted third party to perform this task on behalf of the user. From this, we have a basis for determining if the results of a workflow can be secured and, indeed, trusted.

Part of this trust is formed from assurances that workflow jobs were executed correctly and not compromised in any way. This requires protection in two directions. In the forward direction, it is necessary to ensure that only trusted resource providers are selected to process workflow jobs. These jobs, together with input and output data, should have confidentiality and integrity protection so that only authorised resource providers can process them. In the reverse direction it is essential to determine whether or not the selected resource providers were compromised when processing their allocated jobs. The rest of this analysis focuses upon how well the proposed scheme provides these security services, with references back to the messages and steps described in Sections 4.2 and 4.2.

5.1 Trusted Resource Provider Selection

Job protection is achieved using private keys that have been sealed to particular platform states that match the user's security requirements. During job scheduling, the WRB only considers resource providers that can provide trusted platforms in the required states (message 1). This means that if a resource provider deviates from the predefined state, either by accident or due to malicious attack, then that resource provider will be unable to access the private key to decrypt job information (step 2), and intermediate (input) data encrypted by preceding resource providers (step 7).

5.2 Confidentiality and Integrity of Job Information

Job information and workflow results are protected using authenticated encryption which provides both confidentiality and integrity services (messages 1 and 4). This requires the use of symmetric keys, which are generated by the WRB and resource providers. In turn, the symmetric keys are encrypted with the private keys procured before workflow execution. This is a standard key management technique, utilising the speed of symmetric cryptography to protect the large quantities of data, and the key distribution advantages of public key cryptography to protect the less bandwidth-intensive symmetric keys.

5.3 Process Provenance

Platform attestation is used to reliably collate audit data for process provenance. Before accepting intermediate data from a resource provider, our scheme requires that the resource provider attests to its platform state post-job execution (steps 4–6). This provides three advantages. Firstly, any compromise can be detected immediately — if this occurs near the beginning of the workflow then considerable resources are saved from unnecessarily processing the rest of the workflow using incorrect data. Secondly, it allows the WRB to react by rescheduling the job to another resource provider. Thirdly, a record of attestation results are kept to provide a detailed audit trail.

Since the WRB requires resource providers on which workflows terminate to attest to their platform states, and input data for the workflow resides on trusted storage nodes, our proposal provides a complete audit trail to augment any additional provenance system being used. Thus, the proposed scheme enables the detection of any resource providers that may have compromised workflow results. An additional system will be required to manage the audit information collected, and this will be explored in future work.

6 Final Remarks

Grid workflows provide significant advantages when completing highly complex computations if strong assurances that participating entities will behave as expected can be provided. This requires both the judicious selection of trustworthy Grid resource providers, and a means to determine whether or not this trust still holds after job processing. This trust is built using Trusted Computing technology — there exists challenges in the implementation, as discussed in [28], and this will be the focus of future work. We have presented a novel scheme that enables trusted resource provider selection, protects the integrity and confidentiality of jobs within a workflow and provides data for process provenance. The provision of these security services enables Grid users to derive confidence in the execution of their workflows, and from this establish trust in workflow results.

Acknowledgements. The first and second authors are sponsored by the Engineering and Physical Sciences Research Council (EPSRC) UK e-Science programme of research (EP/D053269). The third author is sponsored by the U.S. Army Research Laboratory and the U.K. Ministry of Defence (Agreement Number W911NF-06-3-0001). The forth author is sponsored by the Open Trusted Computing project of the European Commission Framework 6 Programme. Many thanks to Professor Chris Mitchell for his comments.

References

1. Foster, I., Kesselman, C.: The Grid 2: Blueprint for a New Computing Infrastructure, 2nd edn. Morgan Kaufmann Publishers, San Francisco (2004)
2. Taylor, I.J., Deelman, E., Gannon, D.B., Shields, M. (eds.): Workflows for e-Science: Scientific Workflows for Grids. Springer, Heidelberg (2007)
3. Chadwick, D.: Authorisation in Grid Computing. Information Security Technical Report 10(1), 33–40 (2005)
4. Foster, I., Kesselman, C., Tsudik, G., Tuecke, S.: A security architecture for computational grids. In: Proceedings of the 5th ACM conference on Computer and Communications Security, San Francisco, California, United States, November 2–5, pp. 83–92. ACM Press, New York (1998)
5. Song, S., Hwang, K., Kwok, Y.-K.: Risk-resilient heuristics and genetic algorithms for security-assured grid job scheduling. IEEE Transactions on Computers 55(6), 703–719 (2006)
6. Rajbhandari, S., Wootten, I., Ali, A.S., Rana, O.F.: Evaluating provenance-based trust for scientific workflows. In: Proceedings of the Sixth IEEE International Symposium on Cluster Computing and the Grid, Singapore, May 2006, pp. 365–372. IEEE Press, Los Alamitos (2006)
7. Simmhan, Y.L., Plale, B., Gannon, D.: A survey of data provenance in e-Science. ACM SIGMOD Record 34(3), 31–36 (2005)
8. Mao, W., Martin, A., Jin, H., Zhang, H.: Innovations for grid security from trusted computing — protocol solutions to sharing of security resource. In: Proceedings of the 14th Int. Workshop on Security Protocols, Cambridge, UK, March 2006. LNCS, Springer, Heidelberg (to appear)
9. Martin, A., Yau, P.W.: Grid security: Next steps. Information Security Technical Report 12(3), 113–122 (2007)
10. Yu, J., Buyya, R.: A taxonomy of scientific workflow systems for grid computing. ACM SIGMOD Record 34(3), 44–49 (2005)
11. Cooper, A., Martin, A.: Towards a secure, tamper-proof grid platform. In: Proceedings of the 6th IEEE International Symposium on Cluster Computing and the Grid, May 2006, pp. 373–380. IEEE Press, Los Alamitos (2006)
12. Trusted Computing Group: TPM Main Part 1 Design Principles Specification Version 1.2 Revision 94 (2006)

13. Trusted Computing Group: TPM Main Part 2 TPM Data Structures Version 1.2 Revision 94 (2006)

14. Trusted Computing Group: TPM Main Part 3 Commands Specification Version 1.2 Revision 94 (2006)

15. Intel: LaGrande Technology Architectural Overview. Technical Report 252491-001, Intel Corporation (2003)

16. Peinado, M., England, P., Chen, Y.: An Overview of NGSCB. In: Mitchell, C.J. (ed.) Trusted Computing. IEE Professional Applications of Computing Series, vol. 6, pp. 115–141. The Institute of Electrical Engineers (IEE), London (2005)

17. Mitchell, C.J.: Trusted Computing. IEE Professional Applications of Computing, vol. 6. IEE Press, London (2005)

18. Pearson, S. (ed.): Trusted Computing Platforms: TCPA Technology in Context. Prentice Hall, Englewood Cliffs (2003)

19. Haldar, V., Chandra, D., Franz, M.: Semantic remote attestation - A virtual machine directed approach to Trusted Computing. In: Proceedings of the 3rd USENIX Virtual Machine Research & Technology Symposium (VM 2004), San Jose, CA, USA, May 6-7, 2004, pp. 29–41. USENIX (2004)

20. Sadeghi, A.R., Stüble, C.: Property-based attestation for computing platforms: Caring about properties, not mechanisms. In: Proceedings of the 2004 Workshop on New Security Paradigms (NSPW 2004), Nova Scotia, Canada, September 20-23, 2004, pp. 67–77. ACM Press, New York (2004)

21. Cooper, A., Martin, A.: Towards an open, trusted digital rights management platform. In: Proceedings of the ACM workshop on Digital rights management (DRM 2006), Alexandria, Virginia, USA, October 30, 2006, pp. 79–88. ACM Press, New York (2006)

22. Yau, P.W., Tomlinson, A.: Using trusted computing in commercial grids. In: Akhgar, B. (ed.) Proceedings of the 15th International Workshops on Conceptual Structures (ICCS 2007), Sheffield, UK, July 22-27, 2007, pp. 31–36. Springer, Heidelberg (2007)

23. Chen, H., Chen, J., Mao, W., Yan, F.: Daonity — Grid security from two levels of virtualisation. Information Security Technical Report 12(3), 123–138 (2007)

24. Löhr, H., Ramasamy, H.V., Sadeghi, A.R., Schulz, S., Stüble, C.: Enhancing Grid Security Using Trusted Virtualization. In: Xiao, B., Yang, L.T., Ma, J., Muller-Schloer, C., Hua, Y. (eds.) ATC 2007. LNCS, vol. 4610, pp. 372–384. Springer, Heidelberg (2007)

25. Löhr, H., Ramasamy, H.V., Sadeghi, A.R., Schulz, S., Schunter, M., Stüble, C.: Enhancing Grid security using trusted virtualization. In: Proceedings of the 1st Benelux Workshop on Information and System Security (WISSEC 2006), Antwerpen, Belgium, November 8-9, 2006, COmputer Security and Industrial Cryptography (COSIC), K.U. Leuven, ESAT/SCD (2006)

26. Dent, A.W., Mitchell, C.J.: User's guide to cryptography and standards. 1st edn. Artech House (2005)

27. International Organisation for Standardization: ISO/IEC 19772: Information technology – Security techniques – Authenticated encryption (2007)

28. Balfe, S., Gallery, E.: Mobile agents and the deus ex machina. In: Proceedings of the 21st International Conference on Advanced Information Networking and Applications (AINA 2007), Niagara Falls, Canada, May 21-23, 2007, pp. 486–492. IEEE Press, Los Alamitos (2007)

DaltOn: An Infrastructure for Scientific Data Management

Stefan Jablonski[1], Olivier Curé[2], M. Abdul Rehman[1], and Bernhard Volz[1]

[1] University of Bayreuth, Bayreuth, Germany
{Stefan.Jablonski,Abdul.Rehman,Bernhard.Volz}@uni-bayreuth.de
[2] Université Paris Est, S3IS/IGM France
ocure@univ-mlv.fr

Abstract. It is a common characteristic of scientific applications to require the integration of information coming from multiple sources. This aspect usually confronts end-users with data management issues which involve the transportation of data from one system to another as well as the syntactic and semantic integration of data, i.e. data come in different formats and have different meanings. In order to deal with these issues in a systematic and well structured way, we propose a sophisticated framework based on process modeling. In this paper, we present the three major conceptual architectural abstractions of the system and detail its execution.

1 Introduction

Information integration aims to enable the rapid development of new applications requiring information from multiple sources. This task is becoming a critical issue for both businesses and individuals. Its complexity is mainly due to the uprising of data volumes and the proliferation of sources, types of information. For instance, in other papers we report from a medical application [4] which enables patients to perform self-medication safely and efficiently. This web-based application supports patients with many services including a drug proposition service. The quality of this service partly depends on the ability to integrate coherently valuable information sources (e.g. databases) and thus to answer patient's inquiries in a consistent manner. This integration is not limited to be a syntactical issue (i.e. format alignments) but essentially a semantic issue, i.e. the information contained in the sources may not agree on a common semantics and hence produce inconsistent medical conclusions.

Furthermore data integration is not a "one shot approach" as it may be needed to integrate new sources and to take care of the schema evolution of already integrated sources. Thus we promote to incorporate data integration in the information system underpinning an application. We consider a process-based information system to be the ideal candidate for such incorporation. We call this approach Data Logistics (DaLo, [9]) and summarize it in Section 2. Section 3 presents the framework responsible for the data management issues within the DaLo, named DaltOn, and emphasizes all execution aspects on a concrete micrometeorology example. Section 4 relates our research to other approaches and in Section 5 we conclude this paper.

M. Bubak et al. (Eds.): ICCS 2008, Part III, LNCS 5103, pp. 520–529, 2008.

2 Workflow Management and Data Logistics

As described in several publications [5] [9] we pursue a special approach to workflow (process) management in medical and scientific applications which is called Data Logistics (DaLo). This approach supports a very flexible execution of workflows as compared to the traditional way of prescribed workflow execution in conventional workflow management systems [8]. DaLo-workflows propose many interesting features but for the purpose of this publication only the focus on modeling data specific issues is relevant. What does this mean? In a first phase of the workflow specification the principle structure of a workflow is defined. This is the usual way of specifying what steps have to be executed in what order, what people (or systems) are responsible to perform those steps and what applications have to be called when a workflow step has to be executed. Next, in a second phase of specification, data management issues are focused. This means that issues related to data management must be defined. We will explain this feature with an example in Section 3.

We call each issue that has to be dealt with when modeling and executing a workflow "perspective". Thus we differentiate at least the following perspectives: The functional perspective describes what step has to be executed; the behavioral perspective defines the execution order of work steps, the organizational perspective determines agents eligible and responsible to perform a work step and the operational perspective describes tools (applications) used when a certain work step is executed. All the data management related issues are dealt by the data perspective. The underlying meta model for this perspective oriented approach is called Perspective Oriented Process Management (POPM) which is presented in [10].

Fig. 1 shows a process model taken from a real world scenario of a scientific workflow from micrometeorology research. Before we go into more detail of the process and its execution, we want to describe the structure of the process according to POPM. Each step of a process depicted as a rectangle shows the functional perspective. The small rectangle underneath the right side of a process step denotes the operational perspective. A black arc and a small rectangle over the arc (showing data items) is considered as the data perspective which deals with both the description of input and output data of work steps and data flow between work steps as well. The gray arcs depict the behavioral perspective which realizes execution dependencies. It is out of the scope of this paper to discuss all the perspectives and their implementation, instead our focus is on the implementation of the data perspective. In the next sections we will show how our framework implements the data perspective and copes with data management issues.

As already mentioned, the workflow depicted in Fig. 1 is taken from micrometeorology research where weather data is acquired and then analyzed. In data acquisition phase, scientists collect datasets from diverse sensors at different locations, prepare them, dump the prepared datasets at an intermediate place - so called "FileServer", and finally move them from the FileServer to a central repository (EcoDatabase). In the first step "GetData", data are extracted from sensors and then moved to the next step "ValidateData" which uses the in-house built application "ValApp" for validating the extracted data. Then after validation the (extracted and validated) data are dumped into the FileServer at a different location. In order to move the data from the FileServer to the central repository, the "DataSelection" step extracts the data and moves it to the next step "StoreSensorData", in case interpolation is not required. If it is required then the step "Interpolation" will be executed; otherwise the data will be moved directly to "StoreSensorData". The step "StoreSensorData" is then responsible for storing the data into EcoDatabase using the "DBEco" application.

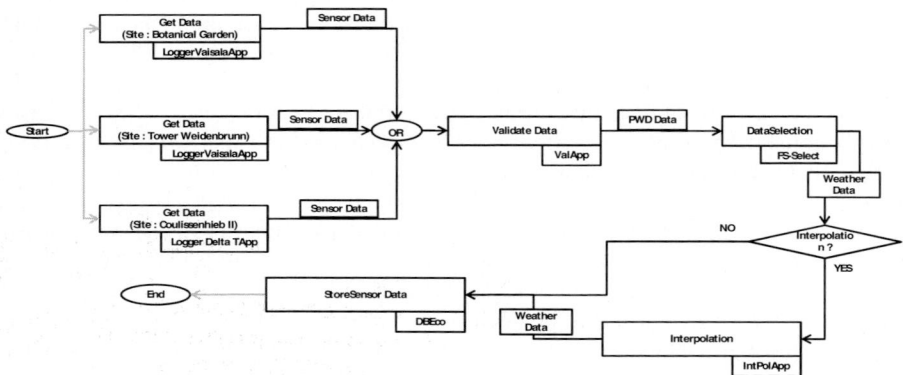

Fig. 1. An example workflow from the micrometeorology domain

It is worth to consider the overall architecture of our workflow management system before starting a detailed discussion of those parts that deal with data management issues. In principle, the architecture comprises one component for each perspective in use. In the context of this paper, the implementation of the data perspective is of most interest. Data perspective of the POPM approach is implemented by DaltOn (Data Logistics and Ontology based integration).

Whenever a data transmission between work steps of a workflow takes place, DaltOn is called. A data transmission is always taking place in between two work steps regardless whether data must be moved physically or not. Data are exchanged between data sources which are associated with regarding work steps. Due to different notions and namespaces of participating applications of a DaLo workflow, format, terminology and ontology transformations are needed. The constitutive idea of this data integration task has been published in [5]; in this paper we contribute the architecture of the DaltOn framework. Section 3 presents this architecture in more detail; especially its components are introduced and their orchestration is depicted.

3   Architecture of DaltOn

3.1   Introductory Example

DaltOn is a framework which deals with data management issues such as data exchange, semantic and syntactic data integration in well structured and transparent way. Thus it facilitates domain users (especially scientists in scientific workflows) in managing domain related processes by allowing DaltOn to care about data complexity. Fig. 2 depicts a part of the example workflow shown in Fig. 1. It highlights a scenario of the two work steps "DataSelection" and "StoreSensorData", especially it is focusing on the data transfer between these steps.

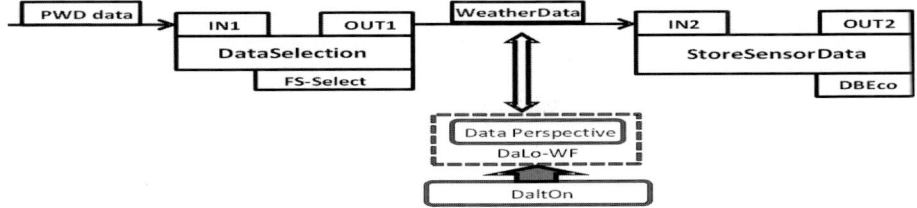

Fig. 2. Focusing data transfer between work steps

In Fig. 2 we want to zoom into the data perspective, i.e. into the overall transmission of data between work steps. Each work step consumes input data (INx) and produces output data (OUTx). Therefore the input and output of the work step "Data-Selection" in Fig. 2 are IN1 and OUT1 and those for the work step "StoreSensorData" are IN2 and OUT2 respectively. The output of the "DataSelection" step (OUT1) is a document containing weather data (single items are separated by spaces) taken from sensors in a proprietary format called "PWD". The schema is defined in terms of an ObjectID (OID) which identifies each data item uniquely, a TimeStamp (TS) that gives the time at which data has been recorded, a HardwareError code (HE) which shows the status of the (sensor) hardware, the VisibilityPerOneMinute value (VOM) which reflects the visibility, an InstantPresentWeather code (IPW) that provides NWS codes and WaterIntensity value (WI) which is showing the intensity of water at one minute average. The input of the work step "StoreSensorData" (IN2) is again a document that contains weather data but this time the format is not PWD but XML. Furthermore the weather data requires filtering since records with a VOM greater than 2000 are not integrated into the database and thus can be excluded.

Also a terminological transformation is required since codes in the OUT1 dataset will be integrated with values in the IN2 dataset. The schema of IN2 is defined as ObjectID (OID), TimeStamp (TS), Status, Visibility, InstantPresentWeather (IPW) and WaterIntensity (WI). Obviously, OUT1 must then somehow be related to IN2 since the output of "DataSelection" should become the input of "StoreSensorData". Here some major conceptual questions arise: Is OUT1 syntactically and semantically compatible with IN2? If some incompatibility shows up, how can data transformations are performed? And how are the data actually transported? We introduce "WeatherData" as a kind of common data structure compatible to OUT1 and IN2, respectively, and the data that have to be transported from "DataSelection" to "Store-SensorData". Then the tasks to be performed by DaltOn are first to convert OUT1 into "WeatherData", transport it and finally convert "WeatherData" into IN2.

3.2 Architectural Components

Fig. 3 depicts the architecture of DaltOn and its associated components. DaltOn has three major conceptual architectural abstractions, namely Data Provision, Data Integration and the internal Repository.

Data Provision aims at data exchange between data producing steps (sources) and data consuming steps (sinks). It consists of two components, namely Data Transportation (DT) and Data Selection/Filtering (DSF). DT handles the physical data transportation between sources and sinks by utilizing wrapper objects which encapsulate each source and sink. DSF is responsible for extracting the dataset based on end-users' selection and filtering criteria through configuration data at modeling time, hence only the data which are required by sink are extracted. Wrappers are supporting the communication with the sources; they provide a uniform interface for the access of each source/sink and extract or insert the data using the source's/sink's proprietary format and access method (e.g. SQL statements in case the source/sink is a relational database).

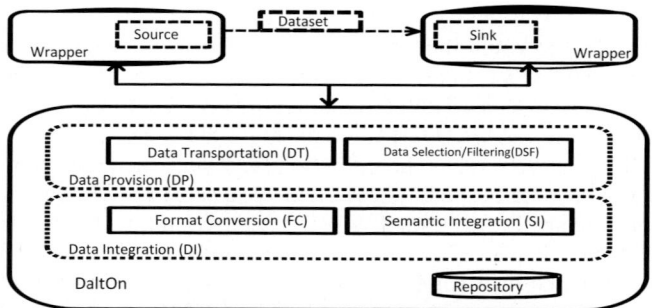

Fig. 3. DaltOn's architectural components

Here, it is worth to mention that DaltOn is not restricted to be used solely in the context of workflow management. Source and sink are two abstract components which represent a data producer and a data consumer, respectively. In our context, both data producers and data consumers are work steps; however, they can represent conventional applications in other contexts as well. Therefore data wrappers work as adapters to the data sources in order to provide a uniform access since each (type of) source usually has its own interface and is accessed differently.

Data Integration instead aims at syntactic and semantic transformations. Accordingly, this module encircles two components, namely Format Conversion (FC) and (Ontology-based) Semantic Integration (SI). FC is not only responsible for converting data formats between sources and sinks but can also be used by other components of the DaltOn framework. For instance, the SI component consumes and produces the data only in XML format. This representational format can totally be independent from the formats issued by sources and required by sinks. The FC component then takes care about these types of format conversions.

SI deals with data integration using a (semantic) mediation mechanism based on ontologies. In order to operate properly, it requires the following from a DaLo workflow: (i) a reference ontology, (ii) local ontologies associated to both applications and whose elements (i.e. concepts and properties) relate to elements of the reference ontology (iii) schema for the IN and OUT data and (iv) mappings between elements of the schemata and their respective local ontology. The semantic mediation is performed via matching concepts [15] from one local ontology to another. Since we discussed the theory and the algorithms of the SI component in detail in [5], interested readers should refer to this publication in order to get more details on this component.

Another conceptual abstraction in the DaltOn framework is the Repository. This encompasses information about all operations of DaltOn performed on each data item. It also contains information needed by every components of DaltOn, in particular SI with its reference and local ontologies, document schemata, mappings between schema and ontology elements, instance documents and ontology alignments.

3.3 Execution Semantics

DaltOn is invoked by the workflow management system (WfMS) each time when a work step is ready to be executed. Thus DaltOn prepares and moves the data necessary for performing that particular work step of a process.

We will now demonstrate how DaltOn components interact with each other and with WfMS using the example shown in Fig. 2. The logical sequence of messages occurring during this exemplary run of the DaltOn framework is depicted using sequence diagram in Fig. 4. Data Integration operations performed by DaltOn can be divided into three phases as show in the diagram. The first phase is responsible for extracting the data from the source (in our example the work step "DataSelection") and converting it into the format that can be acceptable by the SI component later on. The second phase performs semantic integration of the data taken from step "DataSelection" in order to make it compatible with step "StoreSensorData" (the description of the example showed that there is some semantic integration necessary). Finally the third phase is performed which converts the data into a format that is understood by the "StoreSensorData" and transports the converted data to the location where "StoreSensorData" expects it to be. When DaltOn is invoked by the WfMS it stores all the information necessary for the whole integration process into its local repository (we did not depict these intermediate storage tasks/messages as they would have overcrowded Fig. 4). This information comprises the local ontologies of source and sink ("DataSelection" and "StoreSensorData"), the reference ontology that connects local ontologies, information about formats of source and sink ("PWD" and "XML") and references where data can be found (data wrapper for "FS-Select"), where it is to be placed (data wrapper for "DBEco") and a selection criteria that determines which subset of data should be used ("VOM not greater than 2000").

In the first phase, the data extraction from "DataSelection" is started by DaltOn with the "PrepareSelectionStatement" call to the DSF. The DSF then reads the information stored in the repository and creates a statement that is valid for the data source ("DataSelection") and that reflects the selection criteria provided by the user during modeling time of the process. After the statement has been prepared, DaltOn asks the DT component to move the data from the source ("DataSelection") into the internal repository (message "MoveData") using the selection criteria generated in the previous call. As DT component cannot access the data directly, it uses a data wrapper for the source to extract the selected subset (message "GetData"). After the data has been extracted and moved into the repository, DaltOn triggers FC for format conversion since data extracted from "DataSelection" is in a proprietary format ("PWD") but the SI component expects it to be in XML. Thus the FC component converts the data from PWD to a XML representation in reaction to the "Convert" message received from DaltOn and also stores the resultant data into repository. The second phase, the semantic integration, can then be started. DaltOn thus calls the SI component (message "Integrate"). In order to perform the semantic integration, the SI component extracts the data, all the relevant ontologies (two local ontologies and the reference ontology) and some mapping information from the repository. After the integration has been finished, the (converted) data is put again into the repository. This also demarks the end of the second phase.

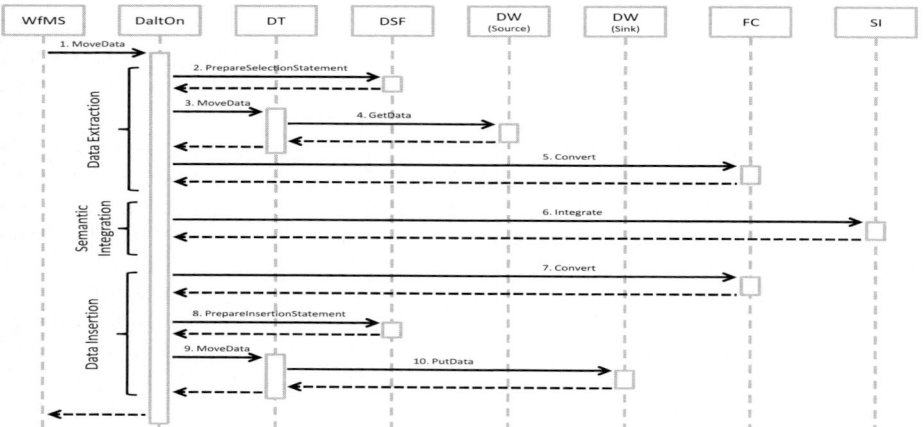

Fig. 4. Sequence diagram showing the interaction of DaltOn components

The third and last phase, the transportation of the data, is initiated by DaltOn with a call to the FC component (message "Convert") since the format of the output of the SI component may differ from that which is expected by the step "StoreSensorData". After the conversion DaltOn instructs the DSF component to prepare an insertion statement (message "PrepareInsertionStatement"). Because a work step may consume data combined from several sources, special statements are necessary to tell clearly which part of the combined data item has to be stored. After this insertion statement has been prepared, DaltOn calls the DT component for transporting the data to a place where the work step "StoreSensorData" expect it (message "MoveData"). Again DT uses a data wrapper – this time for the sink, i.e. "StoreSensorData".

It is noteworthy to mention that all the data exchange in between components is performed through the DaltOn Repository. This allows the Repository to track down all operations performed on the data and record them for later use (Data Provenance). DaltOn does not need to systematically carry out all integration steps; e.g. if both source and sink, use the same format and ontology, only data transportation is performed. Thus all the calls for the FC and SI components can be omitted.

The sequence diagram of Fig. 4 also shows the design rationale of DaltOn: every sub-component (e.g. SI, DSF, FC, DT) and the Repository are in principle independent of each other. This allows for exchanging and re-combining the sub-components easily in the future and reduces the resources necessary for maintaining the system. This design fosters the sustainability of the DaltOn system.

4 Related Work

We want to relate our approach to two domains, data integration solution provided by scientific workflows and data integration systems in scientific applications.

Kepler [11] is a scientific workflow system which has been developed for scientists like ecologist, geologist and biologists and it is built on Ptolemy II, a PSE from electrical engineering [14]. Kepler uses a semantic mediation based approach to integrate heterogeneous data [3]. Concerning the semantic data integration, the approaches of Kepler and DaltOn's SI are quite similar as they both aim to transform data semantically. This transformation uses ontologies and mappings, called registration mappings, to semi-automatically generate mappings between services with heterogeneous schemas, named structural types. Although the objectives of these systems are almost identical, the design of the solution is different as Kepler does not use the semantics of the ontologies to generate the mappings. In fact, end-users define registration mappings in the form of correspondences between queries on the service output/input, named ports, and contextual paths on the ontologies. The system then treats the paths to generate mappings between ports. We argue that SI is a real semantic approach as it exploits concept definitions to perform the alignment of ontologies. These ontologies are represented in Description Logics [1] and thus propose some associated reasoning procedures (concept subsumption, instance checking, knowledge base consistency) that are used during ontology matching [15].

Regarding syntactic data integration and data transportation, in Kepler the end-user has to introduce explicitly predefined and specialized actors for format conversion and for data transportation. Taverna [7] is a scientific workflow management system like Kepler; it is part of the myGrid project [12]. The Taverna workbench allows users to graphically build, edit, and browse workflows. It started life in bioinformatics applications and workflows are mostly used for the specification and execution of ad hoc in silico experiments using bioinformatics resources. These resources might include information repositories or computational analysis tools providing a Web Service based or custom interface. The workflows are enacted by the FreeFluo [6] engine and can be monitored within the Taverna workbench. In order to convert data formats, Taverna follows almost the same approach as Kepler by introducing so-called "Shims". Shims are little services (local workers, beanshell processors/code snippets, and in some cases, nested workflows) which transform one format to another. DaltOn approach differs from these systems in the way that it handles format conversion (syntactical conversion) implicitly by its so called FC component. Taverna doesn't support semantic data integration directly [13], instead the end-user typically needs some kind of specialized translation services which transform the schemas and perform the mappings as well. Thus for semantic data integration, the end-user needs to access a service that knows how to map schemas and ontologies. Whereas DaltOn approach provides a system which deals with this type of semantic integration issue in a systematic way so that end-user doesn't need to introduce such type of services.

BACIIS [2] (Biological and chemical information integration system) is an on-demand information integration system for life science web-databases. The architecture of BACIIS is based on the mediator-wrapper approach augmented with a knowledge base. The approach is similar to DaltOn approach as it follows a semantic mediation for data integration and extracts data via implementation of specific wrappers. It provides web based interface on which end-users define queries to diverse web databases. BACIIS aims at integrating life science data sources and produces results as an integrated view on a web-based interface, whereas DaltOn aims at data exchange between diverse heterogeneous data sources. In addition DaltOn provides a conceptual separation between data retrieval (data provision) and integration mediation. Also in BACIIS, the concepts underlying Data Logistics have not been considered. In BACIIS, mediator transform data from its source database format to the internal format used by the integration system, while in DaltOn, format conversion is dealt with in a systematic way.

5   Conclusion

We are stressing in this paper that data integration is a critical issue for information management and processing. We argue that a novel and valuable approach to ensure proper communication between systems and applications is to incorporate the data integration and exchange solution within a perspective oriented process modeling. DaltOn is such a framework and implements the data perspective of the POPM approach. The most important advantages to adopt such an approach are the following: The first advantage is an increased readability of the scientific workflows. In conventional approaches (cf. Section 4) data management steps (syntactic and semantic conversions, filtering etc.) are mixed up with "real" workflow steps which describe the scientific analysis to be performed. Through the integration of POPM with DaltOn these two issues can adequately be separated (cf. Section 2); thus work steps for scientific analysis are not disguised by work steps that deal with data management issues. Second there is a clear modularization inside the DaltOn framework. This software engineering principle nicely fosters the adaptation of DaltOn to varying application scenarios. For example, if a new format needs to be integrated, only the FC component of DaltOn needs to be adjusted. The same applies to the incorporation of new data sources (e.g. a new type of sensors); here only a new or adjusted wrapper component must be provided to DaltOn. Third, especially in the scientific domain, users are interested in information about what happened to their data during the execution of a workflow. This well recognized request is often referred to as data provenance. The DaltOn repository holds all information about tasks performed on the data inside a workflow and can provide it to the user– according to the demand of data provenance. Scientists are then able to retrace every single operation performed on data. Another major advantage is the possibility to integrate scientists in the workflow, i.e. the workflow management system can directly interact with users. This is possible because POPM fosters the inclusion of organization models into the organizational perspective. Thus scientists that are developing a process can clearly define which step(s) must be executed by a person rather than a computerized agent. This is helpful in scenarios where decisions are based on users' experiences which determine the continuation of a scientific analysis.

Last but not least re-use of existing concepts such as processes, data, selection criteria and applications is fostered. The modeling environment for POPM provides libraries for each type of concept that store the definitions of these and allow sharing information across many applications. DaltOn can be improved in many different ways. For instance, we aim to extend the relation between the internal Repository, the data integration and the data provenance issue.

References

1. Baader, F., et al.: The Description Logic Handbook: Theory, Implementation, and Applications. Cambridge University Press, New York (2003)
2. Ben-Miled, Z., Li, N., Baumgartner, M., Liu, Y.: A decentralized approach to the integration of life science web databases. Informatica (Slovenia) 27(1), 3–14 (2003)
3. Bowers, S., Ludäscher, B.: An Ontology-Driven Framework for Data Transformation in Scientific Workflows. In: Rahm, E. (ed.) DILS 2004. LNCS (LNBI), vol. 2994, pp. 1–16. Springer, Heidelberg (2004)
4. Curé, O.: Semi-automatic Data Migration in a Self-medication Knowledge-Based System. In: Althoff, K.-D., Dengel, A., Bergmann, R., Nick, M., Roth-Berghofer, T.R. (eds.) WM 2005. LNCS (LNAI), vol. 3782, pp. 373–383. Springer, Heidelberg (2005)
5. Curé, O., Jablonski, S.: Ontology-based Data Integration in Data Logistics Workflows. In: Parent, C., Schewe, K.-D., Storey, V.C., Thalheim, B. (eds.) ER 2007. LNCS, vol. 4801, Springer, Heidelberg (2007)
6. Freefluo Workflow Enactment Engine (visited: November 27, 2007), http://freefluo. sourceforge.net/
7. Hull, D., et al.: Taverna: a tool for building and running workflows of services, NAR (web service issue) (2006)
8. Jablonski, S., Bussler, C.: Workflow Management – Modeling Concepts, Architecture and Implementation. International Thomson Computer Press, London (1996)
9. Jablonski, S.: Process Based Data Logistics: Data Integration for Healthcare Applications. In: 1st European Conference on eHealth (ECEH 2006), Fribourg, Switzerland (October 2006)
10. Jablonski, S., Volz, B.: Database Based Implementation of Meta Modeling Concepts (in German). Datenbank Spektrum 24 (to appear 2008)
11. Ludäscher, B., et al.: Scientific Workflow Management and the Kepler System. Concurrency and Computation: Practice & Experience 18(10), 1039–1065 (2006)
12. myGrid Website (visited November 27, 2007), http://www.mygrid.org.uk
13. Oinn, T.: Taverna: Lessons in creating a workflow environment for the life sciences in Concurrency and Computation: Practice and Experience. Grid Workflow Special Issue 18(10), 1067–1100 (2005)
14. Potolemy II Website, http://ptolemy.eecs.berkeley.edu/ptolemyII
15. Shvaiko, P., Euzenat, J.: A Survey of Schema-Based Matching Approaches. In: Spaccapietra, S. (ed.) Journal on Data Semantics IV. LNCS, vol. 3730, pp. 146–171. Springer, Heidelberg (2005)

Workshop on Intelligent Agents and Evolvable Systems

Intelligent Agents and Evolvable Systems

Krzysztof Cetnarowicz[1], Robert Schaefer[1], Bojin Zheng[2], Maciej Paszyński[1],
and Bartlomiej Śnieżyński[1]

[1] Department of Computer Science
AGH University of Science and Technology, Cracow, Poland
[2] College of Computer Science,
South-Central University for Nationalities, Wuhan, China

The workshop *Intelligent Agents and Evolvable Systems* has originated as the fusion of two former workshops: *Intelligent Agents and Evolvable Systems* and *Algorithms and Evolvable Systems*. It has been organized since 2004 within the frame of the *International Conference on Computational Science*. Mathematical considerations, algorithmic aspects and designing computer applications supporting the development and management of systems for the large scale computation have been emphasized.

Two contributions are related to the formal description of multi agent systems. The first one delivered by Smołka, concerns agents that carry tasks of distributed parallel computation. The results ensure existence of time-optimal scheduling and locate the optimal scheduling rules among static ones. The second paper written by Capkovič, is devoted to the modeling inter-agents relations by Petri Nets. Main results have been obtained by using the discrete event systems control theory and illustrated on several examples.

The next three papers contain new agent-based concepts of intrusions detection (infections) in computer systems. Byrski has introduced the system which is appropriate for the specific communication in Mobile Ad-hoc Networks in his paper. Immune-inspired mechanisms allow detecting anomalies coming from communication threads. Prusiewicz has proposed in his paper another system detecting the security policy violations in computer networks. Cetnarowicz et.al. have proposed a social layer that may be used for evaluation of agents behavior. Agents learn patterns of behaviors, which may be used to detect intruders and eliminate them from the system.

The following group of papers have dealt with applications of multi-agents paradigm for solving various, complex engineering problems. Barbucha et al. have presented application of agents to creation of asynchronous team which is able to realize an optimization process and giving accessibility by WEB. Al-Kanhal et al. have proposed an application of multi-agent systems to the optimization of dynamic uncertain process occurring when optimal uncertain manufacturing process are created. Problems of uncertain, human-originated knowledge grounding, accomplished with the use of ontologies in particularly applying uncertainties and several ontologies have been discussed by Szymański et al. Turek has presented an agent-based architecture, which should provide a multi-agent system that enables management of a multi-robot team and its simple extensibility. An interesting application of multi-agent system has been

M. Bubak et al. (Eds.): ICCS 2008, Part III, LNCS 5103, pp. 533–534, 2008.

proposed by Srovnal et. al. It is an embedded system for hydrogen powered car control. Multi-agent architecture with dynamic mutual negotiation is applied. A new interesting research area is the application of graph grammars for modeling evolvable systems. The graph grammar can be effectively utilized for modeling self-adaptive PDE solvers, as it has been presented by Paszyńska et al. The graph representation of a problem can be stored in distributed manner and maintained by intelligent agents, as it has been proposed by Kotulski.

Some papers are focused on theoretical aspects of evolutionary computation and multi-agent systems and are related to real-world applications. Mirabedini et al. have proposed an ant-based routing algorithm, inspired by swarm intelligence and enhanced by fuzzy systems, to solve the routing problem in networks. In their system, multiple constrains can be considered in a simple and intuitive way. Qin et al. have constructed an adapted dynamic vehicle routing problem model based on multi-objective optimization. Then, they have proposed a hybrid multi-objective ant colony algorithm to solve this problem. The hybrid algorithm employs an EA to speed up the convergence of the algorithm. Khelil et al. have proposed the cancer diagnostic system to classify patients who may be affected by cancer. Fang et al. have analyzed the advantage of partnership selection based on grey relation theory and ant colony algorithm. The pre-election process can help to reduce problem search space and complexity. Ciepiela et al. have presented the hierarchical approach to multi-objective optimization problem based on Hierarchical Genetic Strategy (HGS). Barabasz et al. have presented hp-HGS, an enriched version of the HGS algorithm, to solve parametric inverse problems. Furthermore, they have analyzed the asymptotic behavior of their algorithm. Byrski at al. have combined evolutionary algorithms and multi-agent systems to form evolutionary multi-agent systems. The parameters of such systems can be tuned on-line. Application of agent-based co-evolutionary approach to the financial-decision support enabling financial investor planning process by generation of possible investments strategies have been discussed by Dreżewski et al. The contribution delivered by Fidanova discusses special kind of the evolvable systems – the ant colony and its application to finding quasi-optimal solution to Global Positioning System surveying problems.

Several papers have touched the problem of learning in multi-agent systems. Gehrke at al. have presented an application of rule induction algorithm for traffic prediction. Agents use knowledge generated from historical traffic data for route planning. The learning process in this work is performed outside of the multi-agent system by AQ21 program. The paper presented by Śnieżyński has proposed an architecture to design an agent, which has several learning modules for a number of aspects of its activity and every module can use different learning strategy. The architecture is tested on Fish-Banks game simulator. Other papers cover the development of self-learning agents systems. The application of agents with built-in machine learning routines designed to retrieve and analyze data coming from civil engineering data-bases has been presented by Kasperkiewicz et al. The paper by Jurek focuses on self-learning schemes in agents with application to the syntactic pattern recognition.

Task Hibernation in a Formal Model of Agent-Oriented Computing Systems

Maciej Smołka

Institute of Computer Science, Jagiellonian University, Kraków, Poland
smolka@ii.uj.edu.pl

Abstract. The paper contains recent enhancements of a formal model of agent-based computing systems. In such systems a computational task together with its data is enveloped in a shell to form a mobile agent. The shell carries the agent's logic, ie. abilities to make decisions about whether to migrate to a less loaded machine, split oneself or continue the task. The model describes an agent-based computational application as a controlled Markov chain. In this paper the operation of the agent hibernation, which is the last resort in the case of a server overload, is included in the model. This modification has an influence on the form of the state equations as well as the form of admissible control strategies.

1 Introduction

The multi-agent paradigm is already a classical design approach in a wide variety of domains (cf. [1], [2]), which can take advantage of the idea of mobile intelligent autonomous application unit. Computing systems are seldom considered as one of these domains. However, the concept of self-organising computational application composed of mobile tasks, which can move between interconnected computers according to a scheduling policy in order to find a better environment for executing themselves is well-known for several years. The article [3] describes such a system with a scheduling policy based on the heat conduction phenomenon.

A step forwards has been to put a task together with its data into an agent box, give the agent the ability to migrate, to communicate with other agents and to split itself (first of all its task) into two child agents (typically equal). In such a case a scheduling strategy may be incorporated in all agents' (distributed) intelligence. The strategy tells an agent in what order it should perform its activities (computations, migration, partitioning) to achieve its goals, which include typically the finishing of computations in the shortest possible time.

A multi-agent computational system of this type has been developed for the last several years (cf. [4]). It exploits a scheduling policy based on the phenomenon of the molecular diffusion in crystals (cf. [5], [6]). In this policy an agent makes decisions about actions to perform resting on its knowledge of the load of its computational node as well as the load of the node's direct neighbours. When an agent wants and is able to migrate, it chooses the least loaded neighbour as the target. The application of the multi-agent paradigm together with

M. Bubak et al. (Eds.): ICCS 2008, Part III, LNCS 5103, pp. 535–544, 2008.

local, diffusion-based agent scheduling strategies provide us with a relatively simple decentralised management of large-scale distributed computations.

Many theoretical questions has been raised during the development of the above mentioned multi-agent system (MAS). The fundamental one is whether the used heuristic scheduling strategy is in any sense optimal or quasi-optimal. This in turn requires a precise definition of the optimality of a strategy. If the answer to the first question is negative, another problem is whether there exists an optimal policy at all and, if so, what are its characteristics.

To address these (and other) questions a formal model of computing multi-agent systems has been proposed (cf. [6,7,8]). The model is based on the stochastic optimal control theory. The model is still under development and its last version is described in [9]. It already provides us with a precise definition of the optimal task scheduling in our context as well as results on the existence of optimal scheduling strategies and optimality conditions. The above mentioned MAS has served as an example for the development of the model, but the latter has proven more general (and complicated).

In this paper the model is extended by considering the operation of the agent hibernation. Agents are hibernated eg. in an early stage of migration (cf. [4]), but migrations has been already considered in our model (cf. [9]). The interesting case of the hibernation occurs when a computational node is overloaded and an agent cannot find any neighbouring node to emigrate. The local MAS server will then serialize the agent to the disk and deserialize it when the load is sufficiently low. We modify our state equations to allow such hibernations.

Finally we present results on the existence and the characterisation of optimal scheduling strategies, which are adapted to the modified model.

2   System Architecture

First let us introduce to the reader the architectural principles of our exemplary computing MAS. They constitute the foundations and the starting point for the development of the mathematical model. Here we recall only the features most important for the model, for a more complete description and some implementation details we refer the reader to [4] and [5].

The suggested architecture of the system is composed of *a computational environment* (MAS platform) and *a computing application* being a collection of mobile agents called *Smart Solid Agents* (SSA). The computational environment is the triple $(\mathbf{N}, B_H, perf)$, where:

$\mathbf{N} = \{P_1, \ldots, P_N\}$, where P_i is a MAS server called *a Virtual Computational Node* (VCN). Each VCN can maintain a number of agents.

B_H is the connection topology $B_H = \{\mathcal{N}_1, \ldots, \mathcal{N}_N\}, \mathcal{N}_i \subset \mathbf{N}$ is a direct neighbourhood of P_i (including P_i as well).

$perf = \{perf_1, \ldots, perf_N\}, perf_i : \mathbb{R}_+ \to \mathbb{R}_+$ is a family of functions where $perf_i$ describes the relative performance of VCN P_i with respect to the total memory request M_{total}^i of all agents allocated at the node.

The MAS platform is responsible for maintaining the basic functionalities of the computing agents. Namely it delivers the information about the local load concentration $\mathbf{L}_j$ and Q_j (see (2) and (3) below), it performs agent destruction, hibernation, partitioning and migration between neighbouring VCN's and finally it supports the transparent communication among agents.

We shall denote an SSA by A_i where index i stands for an unambiguous agent identifier (possibly a UUID). Each A_i contains its computational task and all data necessary for its computations. Every agent is also equipped with a shell which is responsible for the agent logic. At any time A_i is able to denominate the pair (E_i, M_i) where E_i is the estimated remaining computation time measured in units common for all agents of an application and M_i is the agent's RAM requirement in bytes. An agent may undertake autonomously one of the following actions:

- *continue* executing its internal task,
- *migrate* to a neighbouring VCN or
- decide to be *partitioned*, which results in creating two child agents $\{A_{i_j} = (T_{i_j}, S_{i_j})\}, j = 1, 2.$

We assume that in the case of the agent partitioning the following conditions hold:
$$E_i > E_{i_j}, \quad M_i > M_{i_j}, \quad j = 1, 2.$$

The parent SSA disappears after the partition.

A computing application may be characterised by the triple $(\mathbf{A}_t, G_t, Sch_t)$, $t \in [0, +\infty)$ where $\mathbf{A}_t$ is the set of application agents active at the time t, G_t is the tree representing the history of agents' partitioning until t. All agents active till t constitute the set of its nodes $\bigcup_{s=0}^{t} \mathbf{A}_t$, while the edges link parent agents to their children. All information on how to rebuild G_t is spread among all agents such that each of them knows only its neighbours in the tree. $\{Sch_t\}_{t \in [0, +\infty)}$ is the family of functions such that $Sch_t : \mathbf{A}_t \rightarrow \mathbf{N}$ is the current schedule of application agents among the MAS platform servers. The function is defined by the sets ω_j containing indices of agents allocated on each $P_j \in \mathbf{N}$. Every ω_j is locally stored and managed by P_j.

Each server $P_j \in \mathbf{N}$ asks periodically all local agents (allocated on P_j) for their requirements and computes the local load concentration

$$L_j = \frac{E_{total}^j}{perf_j(M_{total}^j)} \text{ where } E_{total}^j = \sum_{i \in \omega_j} E_i \text{ and } M_{total}^j = \sum_{i \in \omega_j} M_i. \tag{1}$$

Then P_j communicates with neighbouring servers and establishes

$$\mathbf{L}_j = \left\{ (L_k, E_{total}^k, M_{total}^k, perf_k) : P_k \in N_j \right\} \tag{2}$$

as well as the set of node indices

$$Q_j = \{ k \neq j : P_k \in N_j, L_j - L_k > 0 \}. \tag{3}$$

3   Global State of a Computing Application

In this section we introduce some key concepts appearing in our formal model of computing multi-agent systems. The detailed description of the model may be found in [9].

The key idea behind the model is to abandon the considering of single agents' behaviour in favour of observing a global quantity characterising the state of a computational application in an appropriate way. To this end we have introduced the notion of the *vector weight of an agent*, which is the mapping

$$w : \mathbb{N} \times \mathbf{A} \longrightarrow \mathbb{R}_+^M$$

with $M \geq 1$. The weight has at least one *positive* component as long as the agent is *active* (ie. its task is being executed) or *hibernated*. In other words the equality

$$w_t(A_i) = 0 \tag{4}$$

means that the agent A_i does not exist yet or already. Note that we observe the application state in discrete time moments, so the set of times is $\mathbb{N}$. We assume that the dependency of the total weight of child agents after partition upon their parent's weight before partition is well-known and linear, i.e. there is a matrix $\mathbf{P} \in \mathbb{R}_+^{M \times M}$ such that in the case of partition $A \rightarrow \{A_1, A_2\}$ we have

$$w_{t+1}(A_1) + w_{t+1}(A_2) = \mathbf{P} w_t(A). \tag{5}$$

The single agent weight is only an auxiliary notion needed to define what shall be one of the main observed global quantities, ie. the *total weight of all agents allocated on a virtual node P at any time t*, which is

$$W_t(P) = \sum_{Sch_t(A)=P} w_t(A).$$

In previous papers W_t was the system state, but as we want to differentiate between active and hibernated agents, we shall split the total weight into two terms

$$W_t(P) = W_t^a(P) + W_t^h(P),$$

where $W_t^a(P)$ is the total weight of *active* agents and $W_t^h(P)$ is the total weight of *hibernated* agents. We shall observe the evolution of both quantities separately.

If the components of w include E_i and M_i defined in Sec. 2 (as in [7], [8]), then:

- $M_i > 0$ for active agents and $M_i = 0$ for hibernated ones,
- E_i is positive till the agent destruction and does not change as long as the agent is hibernated.

In this case obviously E_{total}^i and M_{total}^i will be among the components of W. But in general it may be convenient to find other *state variables* (see [9] for considerations on that topic). In any case both E^i_{total} and M^i_{total} should remain *observables* of our system.

In the sequel we shall assume that the number of virtual nodes

$$\sharp \mathbf{N} = N$$

is *fixed*. We could also assume that it is *bounded*, but this is not a big generalisation. For the sake of conciseness we introduce the notation

$$W^{a,j}_t = W^a_t(P_j), \quad W^{h,j}_t = W^h_t(P_j)$$

for $j = 1, \ldots, N$. Then W^a_t and W^h_t may be interpreted as vectors from $\mathbb{R}^{MN}_+$ or, if it is more convenient, as nonnegative $M \times N$ matrices.

According to the interpretation of (4) the equality $W_t = 0$, which is equivalent to $W^a_t = W^h_t = 0$, means that at the time t there are neither active nor hibernated agents, ie. the computations are finished. In other words 0 is the final state of the application's evolution.

4   Global State Evolution

In this section we shall formulate the equations of the evolution of our state variables, ie. W^a_t and W^h_t. They are expected to be a generalisation of the state equations presented in [9], so they should reduce to those equations in the absence of hibernations. Consequently they shall be stochastic difference equations, which means that the pair (W^a_t, W^h_t) shall form a discrete stochastic process.

First of all let us recall what has been called the 'established' evolution equation. It has been the equation showing the evolution of an application in the absence of agent migrations and partitions and it has the following form.

$$W_{t+1} = F(W_t, \xi_t) \tag{6}$$

where F is a given mapping and $(\xi_t)_{t=0,1,\ldots}$ is a given sequence of random variables representing the background load influence. We assume that ξ_t are mutually independent, identically distributed and have a common finite set of values Ξ, which is justified in many natural situations (cf. [8], this paper also considers more general assumptions on the background load).

In our case, we extend the meaning of the 'established' evolution by excluding hibernations as well. This results in the following conditions

$$W^h_t = 0, \quad W^a_t = W_t.$$

Thus we can rewrite (6) to obtain

$$\begin{cases} W^a_{t+1} = F(W^a_t, \xi_t) \\ W^h_{t+1} = 0. \end{cases} \tag{7}$$

Since 0 has to be *an absorbing state* of $\overline{W}_t = (W^a_t, W^h_t)$, we need to assume that for every t

$$F(0, \xi_t) = 0 \tag{8}$$

with probability 1. To guarantee reaching the final state we need also another assumption stating that there exists $t > 0$ such that for every initial condition $\widehat{W}$ and W_t evolving according to (6) we have

$$\Pr\left(W_t = 0 \mid W_0 = \widehat{W}\right) > 0. \tag{9}$$

It is easy to see that a similar condition holds for $\overline{W}_t$ and the natural initial state $(\widehat{W}, 0)$. A desired consequence of (9) (cf. [10]) is that with any initial condition our application will eventually finish the computations, which makes the assumption quite useful.

The equations of migration and partition are almost the same as in [9], the only difference is that they describe the behaviour of W_t^a (W_t^h does not change during a migration or a partition).

In the case of sole hibernations and dehibernations at node P_j (and no migrations, partitions or 'established' evolution) the system shall behave according to the following equation.

$$\begin{cases} W_{t+1}^{a,j} = \left(I - \text{diag}(u_t^{h,j}(W_t^a))\right) W_t^{a,j} + \text{diag}(u_t^{a,j}(\overline{W}_t))W_t^{h,j} \\ W_{t+1}^{h,j} = \left(I - \text{diag}(u_t^{a,j}(\overline{W}_t))\right) W_t^{h,j} + \text{diag}(u_t^{h,j}(W_t^a))W_t^{a,j} \\ W_{t+1}^{a,i} = W_t^{a,i} \\ W_{t+1}^{h,i} = W_t^{h,i} \qquad \text{for } i \neq j. \end{cases} \tag{10}$$

$u_t^{h,j} : \mathbb{R}_+^{MN} \to [0,1]^M$ and $u_t^{a,j} : \mathbb{R}_+^{MN} \times \mathbb{R}_+^{MN} \to [0,1]^M$ are the proportions of components of the total weights of agents, respectively, hibernated and activated (dehibernated) to the corresponding proportions of the total weights of all agents allocated at node P_j at the moment t. We assume that the decision of hibernating some agents is the result of a resource shortage. As hibernated agents do not make use of any resources interesting from our point of view (we assume that server disks are large enough) this decision is based on the weight of active agents only. In other words u_t^h does not depend on W_t^h. In contrast the decision of reactivating some hibernated agents may depend on some of their features (eg. even if some RAM is free, every single hibernated agent may be too big to be activated), so u_t^a depends on both weight components.

Now we are in position to present the complete state equations. They are an extension of the state equations presented in [9] and they are constructed in a similar way, ie. as a combination of the above mentioned simplified equations reducing to these equations in their described particular context. We propose the following combination.

$$\begin{cases} W_{t+1}^{a,i} = g^i\left(F^i(\widetilde{W}_t^a, \xi_t), \ \mathbf{P} \, \text{diag}(u_t^{ii}(W_t^a)) \, W_t^{a,i}, \ \sum_{j \neq i} \text{diag}(u_t^{ji}(W_t^a)) \, W_t^{a,j}, \right. \\ \qquad \left. \text{diag}(u_t^{a,i}(\overline{W}_t)) \, W_t^{h,i}\right) \\ W_{t+1}^{h,i} = (I - \text{diag}(u_t^{a,i}(\overline{W}_t))) \, W_t^{h,i} + \text{diag}(u_t^{h,i}(W_t^a)) \, W_t^{a,i} \\ W_0^{a,i} = \widehat{W}^i \\ W_0^{h,i} = 0 \end{cases} \tag{11}$$

for $i = 1, \ldots, N$, where

$$\widetilde{W}_t^{a,i} = \left(I - \sum_{k=1}^N \mathrm{diag}(u_t^{ik}(W_t^a)) - \mathrm{diag}(u_t^{h,i}(W_t^a))\right) W_t^{a,i}$$

and $\widehat{W}$ is a given initial state. For (11) to reduce to simplified equations we need an assumption on g, which may be eg. $g(s,0,0,0) = s$, $g(0,p,0,0) = p$, $g(0,0,m,0) = m$, $g(0,0,0,h) = h$. Note that in [8] and earlier papers we used a stronger condition, ie. $g(s,p,m,h) = s + p + m + h$.

It follows that W_t is a *controlled stochastic process* with a *control strategy*

$$\pi = (\mathbf{u}_t)_{t \in \mathbb{N}}, \quad \mathbf{u}_t = (u_t, u_t^a, u_t^h) : \mathbb{R}_+^{MN} \longrightarrow U. \tag{12}$$

The control set U contains such elements $\mathbf{a} = (\alpha, \alpha^a, \alpha^h)$ from $[0,1]^{M(N \times N)} \times [0,1]^{MN} \times [0,1]^{MN}$ that satisfy at least the following conditions for $m = 1, \ldots, M$.

$$\alpha_m^{ij} \cdot \alpha_m^{ji} = 0 \text{ for } i \neq j, \quad \alpha_m^{i1} + \cdots + \alpha_m^{iN} + \alpha_m^{h,i} \leq 1 \text{ for } i = 1, \ldots, N. \tag{13}$$

The first equation in (13) can be interpreted in the following way: *at a given time migrations between two nodes may happen in only one direction*. The second equality means that *the number of agents leaving a node, partitioned or hibernated at the node must not exceed the number of agents active at the node*.

Remark 1. It is easy to see that the control set U defined by the conditions (13) is compact (and so are of course its closed subsets).

As in [9] we do not take the whole $\mathbb{R}_+^{MN} \times \mathbb{R}_+^{MN}$ as the state space. Instead we choose finite subsets S^a, S^h of $\mathbb{N}^{MN}$ both containing 0 and $S = S^a \times S^h = \{s_0 = (0,0), s_1, \ldots, s_K\}$ shall be the state space. Consequently we assume that F and g have values in S^a. Additionally, we need also to assume that for every $t \in \mathbb{N}$ and $\overline{W} = (W^a, W^h) \in S$

$$u_t(\overline{W}) \in U_{\overline{W}} = \{\mathbf{a} \in U : G^a(\overline{W}, \mathbf{a}, \xi) \in S^a, \ G^h(\overline{W}, \mathbf{a}) \in S^h \text{ for } \xi \in \Xi\}$$

where G^a and G^h denote the right hand sides of, respectively, the first and the second equation in (11). The above equality implies that $G^a(\overline{W}, 0, \xi) = F(W^a, \xi) \in S^a$ and $G^h(\overline{W}, 0) = W^h \in S^h$ for any $\overline{W}$ and ξ, which means that $(0,0) \in U_{\overline{W}}$, therefore $U_{\overline{W}}$ is nonempty for every $\overline{W} \in S$. On the other hand we have $G^a(0, \mathbf{a}, \xi) = F(0, \xi) = 0$ and $G^h(0, \mathbf{a}) = 0$, i.e. 0 is an absorbing state of $\overline{W}_t$ independently of a chosen control strategy.

Remark 2. Similarly to [9] it remains true that $\overline{W}_t$ is a *controlled Markov chain* with transition probabilities $p_{ij}(\mathbf{a}) = \mathrm{Pr}(G(s_i, \mathbf{a}, \xi_0) = s_j)$ for $i, j = 0, \ldots, K$, $\mathbf{a} \in U_{s_i}$, $G = (G^a, G^h)$. The transition matrix for the control $\mathbf{u}$ is $P(\mathbf{u}) = [p_{ij}(\mathbf{u}(s_i))]_{i,j=0,\ldots,K}$.

5   Optimal Scheduling Problem

Let us now recall (after [7]) the definition of the optimal scheduling for a computing MAS in terms of the stochastic optimal control theory. We have already the state equations (11), so we need also a cost functional and a set of admissible controls.

The general form of considered cost functionals is

$$V(\pi; s) = E[\sum_{t=0}^{\infty} k(\overline{W}_t, \mathbf{u}_t(\overline{W}_t))] \qquad (14)$$

where π is a control strategy (12) and $s = (s^a, 0)$ is the initial state of $\overline{W}_t$, i.e. $\overline{W}_0 = s$. Since 0 is an absorbing state we shall always assume that remaining at 0 has no cost, i.e. $k(0, \cdot) = 0$. This condition guarantees that the overall cost can be finite.

The form of the set of admissible strategies is a modification of the one used in [9], namely

$$\mathbf{U} = \left\{\pi : \ \mathbf{u}_t(\overline{W}) \in U_{\overline{W}}, \ t \in \mathbb{N}\right\}.$$

Now we can formulate the *optimal scheduling problem*. Namely given an initial configuration $(\widehat{W}, 0)$ we look for a control strategy $\pi^* \in \mathbf{U}$ such that

$$V(\pi^*; (\widehat{W}, 0)) = \min\{V(\pi; (\widehat{W}, 0)) : \ \pi \in \mathbf{U}, \ \overline{W}_t \text{ is a solution of (11)}\}. \quad (15)$$

In other words an *optimal scheduling* for W_t is a control strategy π^* realising the minimum in (15).

Our main general tool for proving the existence of optimal scheduling strategies is [9, Prop. 4]. Its key assumptions are (R1) and (R2). They are expressed in terms of special properties of the transition matrix $P(\mathbf{u})$, but they mean that K-step (for (R2) n-step for some n) probability of reaching 0 from *every* initial state is positive provided we use:

- for (R1) any stationary strategy $\pi = (\mathbf{u}, \mathbf{u}, \dots)$;
- for (R2) one particular stationary strategy.

In (R2) we have to impose a stronger assumption on the one-step cost k, nevertheless it is much easier to check than (R1). In our case thanks to (9) the assumption (R2) is satisfied eg. for the zero control strategy $\pi = (0, 0, \dots)$.

For this reason in the sequel we shall consider only costs which satisfy the second part of (R2), ie.

$$k(s, \mathbf{a}) \geq \varepsilon > 0, \quad \text{for } s \neq 0, \ \mathbf{a} \in U_s. \qquad (16)$$

Among cost functionals presented in [9] only the expected total time of computations V_T satisfies automatically (16). Let us recall that it has the following form.

$$V_T(\pi; s) = E\big[\inf\{t \geq 0 : W_t = 0\} - 1\big]. \qquad (17)$$

Remaining two functionals (V_L and V_M) do not satisfy the assumption (16),
but we can reduce the problem by adding a term cV_T with (maybe small) $c > 0$
to both of them.

First of the two (the one promoting good load balancing) after such a modification has the following form.

$$V_{LT}(\pi; s) = E\left[\sum_{t=0}^{\infty}\sum_{i=1}^{N}(L_t^i - \overline{L}_t)^2\right] + cV_T(\pi; s), \quad \overline{L}_t = \frac{1}{N}\sum_{i=1}^{N}L_t^i \quad (18)$$

where L_t^i is the load concentration (1) at P_i at the moment t. These quantities
are well defined because we have assumed that E_{total}^i and M_{total}^i are observables
of our system.

The last cost functional from [9] (penalising migrations) after the modification
has the following form.

$$V_{MT}(\pi; s) = E\left[\sum_{t=0}^{\infty}\sum_{m=1}^{M}\sum_{i\neq j}\mu_m^{ij}(u_{m,t}^{ij}(W_t^a))\right] + cV_T(\pi; s). \quad (19)$$

$\mu_m^{ij} : [0, 1] \rightarrow \mathbb{R}_+$ allows us to take into account the *distance* between P_i and P_j.

Considerations accompanying (15) along with [9, Prop. 4] result in the following corollary.

Corollary 3. *Problem* (15) *for* V_T, V_{LT} *or* V_{MT} *has the unique solution.*

Finally, let us recall the optimality conditions presented in [9, Prop. 6] and
rewrite it in our context in the following corollary.

Corollary 4. *Let* V *denote any of the functionals* V_T, V_{LT}, V_{MT}. *The optimal
solution of* (15) *is a stationary strategy* $\pi^* = \mathbf{u}^\infty = (\mathbf{u}, \mathbf{u}, \dots)$ *and it is the
unique solution of the equation*

$$V(\pi^*; s) = \min_{\mathbf{a}\in U_s}\left[\sum_{j=1}^{K}p_{ij}(\mathbf{a})V(\pi^*; s_j) + k(s, \mathbf{a})\right]. \quad (20)$$

The solution of (20) *exists and it is the optimal solution of* (15).

(20) can be solved by means of an iterative procedure such as Gauss-Seidel.
The complexity of the computational problem depends first of all on the size of
the state space S, which in general is expected to be quite big. To make things
better in a typical situation many states are inaccessible from one another so
the matrix p_{ij} is sparse. The complexity is increased on the other hand by the
minimisation and depends on the size (and the structure) of control sets U_s.

6   Conclusions

The presented MAS architecture along with diffusion-based agent scheduling
strategies form a relatively easy to manage and quite efficient framework for
large-scale distributed computations. The presented mathematical model provides us with a precise definition of optimal task scheduling in such an environment. It also gives us some useful results concerning the existence of optimal

scheduling strategies (Cor. 3) as well as the optimality conditions (Cor. 4). The latter show that the choice of scheduling strategies utilised during tests (cf. [5]) was proper, because it is a stationary strategy and according to Cor. 4 the optimal strategy belongs to that class. In this paper the formal model has been extended to consider agent hibernations neglected so far. Also the existence results and the optimality conditions has been adapted appropriately. It appears that it is an important extension and omitting hibernations in some cases might result in a wrong observation of the state of a computing application. Further plans concerning the model include detailed studies on the form of the optimal strategies and, on the other hand, finishing some experiments (and starting some new ones) expected to extend the model's empirical basis.

References

1. Bradshaw, J.M. (ed.): Software Agents. AAAI Press, Menlo Park (1997)
2. Wooldridge, M.: An Introduction to Multi-agent Systems. Wiley, Chichester (2002)
3. Luque, E., Ripoll, A., Cortés, A., Margalef, T.: A distributed diffusion method for dynamic load balancing on parallel computers. In: Proceedings of EUROMICRO Workshop on Parallel and Distributed Processing, San Remo, Italy, pp. 43–50. IEEE Computer Society Press, Los Alamitos (1995)
4. Uhruski, P., Grochowski, M., Schaefer, R.: Multi-agent computing system in a heterogeneous network. In: Proceedings of the International Conference on Parallel Computing in Electrical Engineering (PARELEC 2002), Warsaw, Poland, pp. 233–238. IEEE Computer Society Press, Los Alamitos (2002)
5. Grochowski, M., Schaefer, R., Uhruski, P.: Diffusion Based Scheduling in the Agent-Oriented Computing System. In: Wyrzykowski, R., Dongarra, J., Paprzycki, M., Waśniewski, J. (eds.) PPAM 2004. LNCS, vol. 3019, pp. 97–104. Springer, Heidelberg (2004)
6. Grochowski, M., Smołka, M., Schaefer, R.: Architectural principles and scheduling strategies for computing agent systems. Fundamenta Informaticae 71(1), 15–26 (2006)
7. Smołka, M., Grochowski, M., Uhruski, P., Schaefer, R.: The dynamics of computing agent systems. In: Sunderam, V.S., van Albada, G.D., Sloot, P.M.A., Dongarra, J. (eds.) ICCS 2005. LNCS, vol. 3516, pp. 727–734. Springer, Heidelberg (2005)
8. Smołka, M., Schaefer, R.: Computing MAS dynamics considering the background load. In: Alexandrov, V.N., van Albada, G.D., Sloot, P.M.A., Dongarra, J. (eds.) ICCS 2006. LNCS, vol. 3993, pp. 799–806. Springer, Heidelberg (2006)
9. Smołka, M.: A formal model of multi-agent computations. LNCS. Springer, Heidelberg (to appear, 2008)
10. Kushner, H.: Introduction to Stochastic Control. Holt, Rinehart and Winston (1971)

Synthesis of the Supervising Agent in MAS

František Čapkovič*

Institute of Informatics, Slovak Academy of Sciences
Dúbravská cesta 9, 845 07 Bratislava, Slovak Republic
Frantisek.Capkovic@savba.sk
http://www.ui.sav.sk/home/capkovic/capkhome.htm

Abstract. The systematic approach to the synthesis of the agent-supervisor supervising a group of agents is presented. It is based on results of the DES (discrete event systems) control theory. The agents as well as the agent-supervisor are understood to be DES and they are modelled by Petri nets (PN). The method based on PN place invariants is used in order to synthetise the supervisor. The illustrative example is introduced to explain the elementary steps of the proposed technique. A counter-example illustrating the limits of the approach is introduced too.

Keywords: Agent, invariants, MAS, modelling, Petri nets, supervisor.

1 Introduction

Agents are usually understood to be persistent (not only software but also material, personal, etc.) entities that can perceive, reason, and act in their environment, and communicate with other agents. Hence, multiagent systems (MAS) can be apprehended as a composition of collaborative agents working in shared environment. The agents together perform a more complex functionality. Communication enables the agents in MAS to exchange information. Thus, the agents can coordinate their actions and cooperate with each other. However, an important question arises here: What communication mechanisms enhance the cooperation between communicating agents?

In this paper the synthesis of the supervising agent is presented. It utilizes some results of the DES (discrete event systems) control theory, especially those concerning the theory of place invariants [8,6] and the supervisor [9] and/or controller synthesis based on this theory. Mathematical and/or graphical description of the agents as well as the global MAS will be delt on the background of place/transition Petri nets (P/T PN) described e.g. in [8].

Consider two simple agents A_1, A_2 representing, respectively, two processes P_1, P_2 given in Fig. 1a). While the resources used by the agents are sufficient, the agents can work autonomously. However, when the resources are limited a coordinated action is necessary. In such a case a supervisor or a controller is necessary. The procedure of the mutex (<u>mut</u>ual <u>ex</u>clusion) frequently occurs in

---

* Partially supported by the Slovak Grant Agency for Science (VEGA) under grant # 2/6102/26.

M. Bubak et al. (Eds.): ICCS 2008, Part III, LNCS 5103, pp. 545–554, 2008.

different systems when it is necessary to solve conflicts among parallelly operating subsystems. It is very frequent also in DES like flexible manufacturing systems (e.g. a robot serves two transport belts), communication systems (two users want to use the same channel), transport systems (two trains want to enter the same segment of the railway), etc. The PN model of the simplest form of the mutex - the mutex of two processes - is given in Fig. 1b). It is not very complicated to synthesize such a supervisor in this simple case. Contrariwise, in case of more agents and more complicated conditions for co-

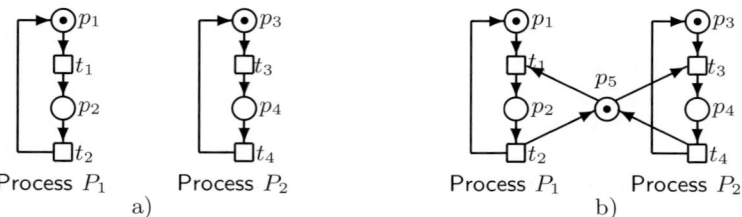

Fig. 1. The PN-based model of the a) two simple processes and b) their mutex

operation among them, founding the supervisor is by far more intricated. Consider the group of 5 autonomous agents $Gr_A = \{A_1, A_2, A_3, A_4, A_5\}$ given in Fig.2. Consider e.g the situation when it is necessary to ensure that only one agent from the subgroup $Sgr_1 = \{A_1, A_4, A_5\}$ and only one agent from the subgroup $Sgr_2 = \{A_2, A_4, A_5\}$ and only one agent from the subgroup $Sgr_3 = \{A_3, A_4, A_5\}$ can simultaneously cooperate with other agents from Gr_A. In other words, the agents inside the designated subgroups must not work simultaneously. Even, the agents A_4 and A_5 can work only individually (any cooperation with other agents is excluded). However, the agents A_1, A_2, A_3 can work simultaneously. In such a situation the synthesis of the supervisor realizing such a mutex (i.e. deciding which agents can simultaneously work and when) is not so simple. Consequently, a systematic (orderly, not stray) approach has to

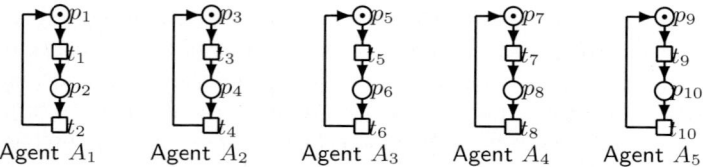

Fig. 2. The PN-based model of the group of 5 simple agents

be found. There exists the DES control synthesis method based on the PN place invariants (P-invariants) in DES control theory [11,7,2,4]. Its principle can be utilised for our purposes.

2   The Supervisor Synthesis

Before starting the supervisor synthesis it is necessary to introduce the mathematical model of P/T PN. Namely, the ability to express the states of agents, the ability to test properties by means of the PN invariants and the reachability graph (RG) as well as the ability to synthesize the supervisory controllers by means of the invariants and RG [3,5] utilizing linear algebra predestinate PN to be frequently used. The model has the form as follows

$$\mathbf{x}_{k+1} = \mathbf{x}_k + \mathbf{B}.\mathbf{u}_k \quad , \quad k = 0, N \tag{1}$$

$$\mathbf{B} = \mathbf{G}^T - \mathbf{F} \tag{2}$$

$$\mathbf{F}.\mathbf{u}_k \leq \mathbf{x}_k \tag{3}$$

where k is the discrete step of the dynamics development; $\mathbf{x}_k = (\sigma_{p_1}^k, ..., \sigma_{p_n}^k)^T$ is the n-dimensional state vector of DEDS in the step k; $\sigma_{p_i}^k \in \{0, c_{p_i}\}$, $i = 1, ..., n$ express the states of the DEDS elementary subprocesses or operations - 0 (passivity) or $0 < \sigma_{p_i} \leq c_{p_i}$ (activity); c_{p_i} is the capacity of the DEDS subprocess p_i as to its activities; $\mathbf{u}_k = (\gamma_{t_1}^k, ..., \gamma_{t_m}^k)^T$ is the m-dimensional control vector of the system in the step k; its components $\gamma_{t_j}^k \in \{0, 1\}$, $j = 1, ..., m$ represent occurring of the DEDS elementary discrete events (e.g. starting or ending the elementary subprocesses or their activities, failures, etc. - 1 (presence) or 0 (absence) of the corresponding discrete event; $\mathbf{B}$, $\mathbf{F}$, $\mathbf{G}$ are structural matrices of constant elements; $\mathbf{F} = \{f_{ij}\}_{n \times m}$, $f_{ij} \in \{0, M_{f_{ij}}\}$, $i = 1, ..., n$, $j = 1, ..., m$ expresses the causal relations among the states of the DEDS (in the role of causes) and the discrete events occuring during the DEDS operation (in the role of consequences) - 0 (nonexistence), $M_{f_{ij}} > 0$ (existence and multiplicity) of the corresponding causal relations; $\mathbf{G} = \{g_{ij}\}_{m \times n}$, $g_{ij} \in \{0, M_{g_{ij}}\}$, $i = 1, ..., m$, $j = 1, ..., n$ expresses very analogically the causal relations among the discrete events (causes) and the DEDS states (consequences); because $\mathbf{F}$ and $\mathbf{G}$ are the arcs incidence matrices the matrix $\mathbf{B}$ is given by means of them according to (2); $(.)^T$ symbolizes the matrix or vector transposition.

The main idea of the approach to the supervisor synthesis consists in imbedding of additional PN places (slacks) and finding the structural interconnections between them and the original PN places. Consequently, a new PN subnet representing the supervisor and its interface with the original system are found and added to the PN model. Thus, the desired behaviour of agents is forced. The definition of the P-invariant $\mathbf{v}$ of PN in general is the following

$$\mathbf{B}^T.\mathbf{v} = \emptyset \tag{4}$$

with $\mathbf{v}$ being n-dimensional vector and $\emptyset$ is m-dimensional zero vector. However, usually there are several invariants in PN models. Hence, the set of the P-invariants of PN is created by the columns of the $n \times n_x$-dimensional (n_x expresses the number of invariants) matrix $\mathbf{V}$ being the solution of the equation

$$\mathbf{V}^T.\mathbf{B} = \emptyset \tag{5}$$

In PN with the slacks mentioned above we have to use the following structure with augmented matrices. Consequently, the equation (5) has to be written in the form

$$[\mathbf{L}, \mathbf{I}_s].\begin{bmatrix} \mathbf{B} \\ \mathbf{B}_s \end{bmatrix} = \emptyset$$

where $\mathbf{I}_s$ is $(n_x \times n_x)$-dimensional identity matrix and $\mathbf{L}$ describes conditions (as to marking of some places in the PN models of agents) of the desired MAS behaviour. While $\mathbf{B}$ is the known structural matrix of the agents (without the supervisor), $\mathbf{B}_s$ is the supervisor structure which is searched by the synthesis process. Thus,

$$\mathbf{L}.\mathbf{B} + \mathbf{B}_s = \emptyset; \qquad \mathbf{B}_s = -\mathbf{L}.\mathbf{B}; \qquad \mathbf{B}_s = \mathbf{G}_s^T - \mathbf{F}_s \qquad (6)$$

where the actual structure of the matrix $\mathbf{L}$ has to be respected. The augmented state vector and the augmented structural matrices (of the original system together with the supervisor) are the following

$$\mathbf{x}_a = \begin{bmatrix} \mathbf{x} \\ \mathbf{x}_s \end{bmatrix}; \quad \mathbf{F}_a = \begin{pmatrix} \mathbf{F} \\ \mathbf{F}_s \end{pmatrix}; \quad \mathbf{G}_a^T = \begin{pmatrix} \mathbf{G}^T \\ \mathbf{G}_s^T \end{pmatrix}$$

where the submatrices $\mathbf{F}_s$ and $\mathbf{G}_s^T$ correspond to the interconnections of the incorporated slacks with the actual (original) PN structure.

2.1   The Illustrative Example

It can be seen that in the above mentioned case of the five agents the form of $\mathbf{L}$ follows from the imposed conditions prescribing the limited cooperation of agents expressed mathematically as

$$\sigma_{p2} + \sigma_{p8} + \sigma_{p10} \leq 1 \qquad (7)$$
$$\sigma_{p4} + \sigma_{p8} + \sigma_{p10} \leq 1 \qquad (8)$$
$$\sigma_{p6} + \sigma_{p8} + \sigma_{p10} \leq 1 \qquad (9)$$

The form of the augmented matrix $[\mathbf{L}, \mathbf{I}_s]$ follows from the modified conditions (introducing the slacks) which are given mathematical terms as

$$\sigma_{p2} + \sigma_{p8} + \sigma_{p10} + \sigma_{p11} \qquad\qquad = 1 \qquad (10)$$
$$\sigma_{p4} + \sigma_{p8} + \sigma_{p10} \qquad + \sigma_{p12} \qquad \doteq 1 \qquad (11)$$
$$\sigma_{p6} + \sigma_{p8} + \sigma_{p10} \qquad\qquad + \sigma_{p13} = 1 \qquad (12)$$

where p_{11}, p_{12}, p_{13} are the additional places - i.e. the slacks s_1, s_2, s_3 (in general, the number i of such slacks s_i can also be different from $i = 3$, of course) - that ensure the desired eliminations of the agents cooperation activities. The number of the slacks depends on the number of conditions and on their form. As it is clear from the equations (10)-(12), only one place from the set of four places

p_2, p_8, p_{10}, p_{11} can be active while the rest ones have to be passive. Analogically, the same is valid for the sets p_4, p_8, p_{10}, p_{12} and p_6, p_8, p_{10}, p_{13}.

$$
\mathbf{F} =
\begin{pmatrix}
\mathbf{F}_{A_1} & 0 & 0 & 0 & 0 \\
0 & \mathbf{F}_{A_2} & 0 & 0 & 0 \\
0 & 0 & \mathbf{F}_{A_3} & 0 & 0 \\
0 & 0 & 0 & \mathbf{F}_{A_4} & 0 \\
0 & 0 & 0 & 0 & \mathbf{F}_{A_5}
\end{pmatrix}
;\;
\mathbf{G} =
\begin{pmatrix}
\mathbf{G}_{A_1} & 0 & 0 & 0 & 0 \\
0 & \mathbf{G}_{A_2} & 0 & 0 & 0 \\
0 & 0 & \mathbf{G}_{A_3} & 0 & 0 \\
0 & 0 & 0 & \mathbf{G}_{A_4} & 0 \\
0 & 0 & 0 & 0 & \mathbf{G}_{A_5}
\end{pmatrix}
\tag{13}
$$

$$
\mathbf{F}_{A_i} =
\begin{pmatrix} 1 & 0 \\ 0 & 1 \end{pmatrix}
;\;
\mathbf{G}_{A_i} =
\begin{pmatrix} 0 & 1 \\ 1 & 0 \end{pmatrix}
;\;
\mathbf{B}_{A_i} =
\begin{pmatrix} -1 & 1 \\ 1 & -1 \end{pmatrix}
;\; i = 1, ..., 5
\tag{14}
$$

$$
\mathbf{L} =
\begin{pmatrix}
0 & 1 & 0 & 0 & 0 & 0 & 0 & 1 & 0 & 1 \\
0 & 0 & 0 & 1 & 0 & 0 & 0 & 1 & 0 & 1 \\
0 & 0 & 0 & 0 & 0 & 1 & 0 & 1 & 0 & 1
\end{pmatrix}
;\;
\mathbf{B} =
\begin{pmatrix}
-1 & 1 & 0 & 0 & 0 & 0 & 0 & 0 & 0 & 0 \\
1 & -1 & 0 & 0 & 0 & 0 & 0 & 0 & 0 & 0 \\
0 & 0 & -1 & 1 & 0 & 0 & 0 & 0 & 0 & 0 \\
0 & 0 & 1 & -1 & 0 & 0 & 0 & 0 & 0 & 0 \\
0 & 0 & 0 & 0 & -1 & 1 & 0 & 0 & 0 & 0 \\
0 & 0 & 0 & 0 & 1 & -1 & 0 & 0 & 0 & 0 \\
0 & 0 & 0 & 0 & 0 & 0 & -1 & 1 & 0 & 0 \\
0 & 0 & 0 & 0 & 0 & 0 & 1 & -1 & 0 & 0 \\
0 & 0 & 0 & 0 & 0 & 0 & 0 & 0 & -1 & 1 \\
0 & 0 & 0 & 0 & 0 & 0 & 0 & 0 & 1 & -1
\end{pmatrix}
\tag{15}
$$

$$
\mathbf{B}_s = -\mathbf{L}.\mathbf{B} =
\begin{pmatrix}
-1 & 1 & 0 & 0 & 0 & 0 & -1 & 1 & -1 & 1 \\
0 & 0 & -1 & 1 & 0 & 0 & -1 & 1 & -1 & 1 \\
0 & 0 & 0 & 0 & -1 & 1 & -1 & 1 & -1 & 1
\end{pmatrix}
\tag{16}
$$

Hence, by wirtue of (6), we have

$$
\mathbf{F}_s =
\begin{pmatrix}
1 & 0 & 0 & 0 & 0 & 0 & 1 & 0 & 1 & 0 \\
0 & 0 & 1 & 0 & 0 & 0 & 1 & 0 & 1 & 0 \\
0 & 0 & 0 & 0 & 1 & 0 & 1 & 0 & 1 & 0
\end{pmatrix}
;\;
\mathbf{G}_s^T =
\begin{pmatrix}
0 & 1 & 0 & 0 & 0 & 0 & 0 & 1 & 0 & 1 \\
0 & 0 & 0 & 1 & 0 & 0 & 0 & 1 & 0 & 1 \\
0 & 0 & 0 & 0 & 0 & 1 & 0 & 1 & 0 & 1
\end{pmatrix}
\tag{17}
$$

Thus, we obtain the structure of MAS given in Fig. 3 where the supervisor is represented by the structure consisting of the places p_{11}, p_{12}, p_{13} together with the interconnections joining them with the elementary agents. Such MAS exactly fulfils the prescribed conditions. The reachability graph (RG) [8] of the system is given in Fig. 4. Its nodes are the states reachable from the initial state $\mathbf{x}_0$ and its edges are weighted by corresponding PN transitions enabling the state-to-state transit. The control synthesis based on RG is described in the author's works [3,5]. Consequently, this problem is not discussed here. In the light of the above introduced facts we can illustrate also the systematic synthesis of the simplest mutex presented in Fig. 1. Here we have

$$
\sigma_{p_2} + \sigma_{p_4} \leq 1
\tag{18}
$$

$$
\sigma_{p_2} + \sigma_{p_4} + \sigma_{p_5} = 1
\tag{19}
$$

$$\mathbf{L} = (0, 1, 0, 1); \quad \mathbf{B} = \begin{pmatrix} -1 & 1 & 0 & 0 \\ 1 & -1 & 0 & 0 \\ 0 & 0 & -1 & 1 \\ 0 & 0 & 1 & -1 \end{pmatrix}; \quad \mathbf{L.B} = (1, -1, 1, -1) \quad (20)$$

$$\mathbf{B}_s = -\mathbf{L.B} = (-1, 1, -1, 1); \quad \mathbf{F}_s = (1, 0, 1, 0); \quad \mathbf{G}_s^T = (0, 1, 0, 1) \quad (21)$$

Of course, the result is the same like that given in the Fig. 1.b) - compare (21) with Fig. 1b).

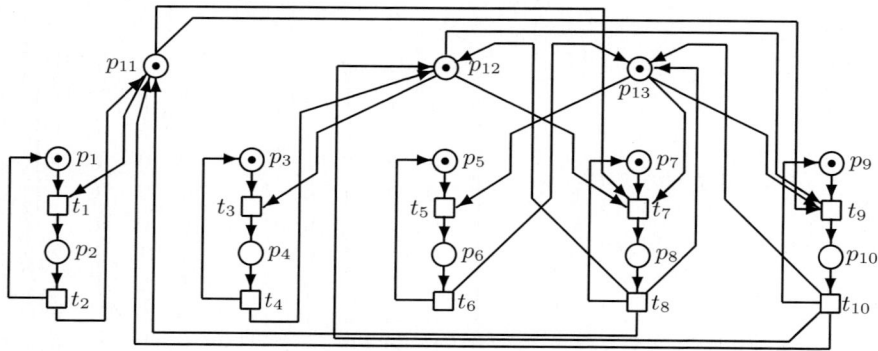

Fig. 3. The PN-based model of the controlled cooperation of the agents

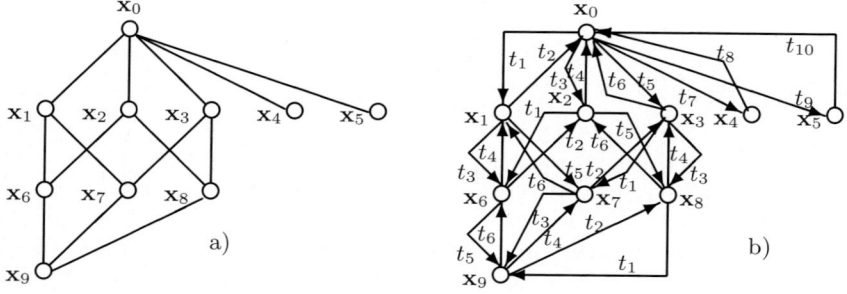

Fig. 4. The RG of the five cooperating agents a) in the form where the edges are oriented bidirecitonally, but weights of the elementary edges are not introduced because they are different depending on the orientation; b) in the form where both the orientation of edges and their weights are introduced implicitly

3   Applicability of the Approach

The method described above is suitable for the so called pure PN. P/T PN is pure when there is no 'loop' between a place and a transition - i.e. it is the net without the loops of the form given in Fig. 5a). Namely, in such a case the matrices $\mathbf{F}$, $\mathbf{G}^T$ have nonzero element equal to 1 in the same place - in general, when there exists the 'loop' between p_i and t_j the element $f(i.j)$ of the matrix $\mathbf{F}$ is equal to the element $gT(i,j)$ of the matrix $\mathbf{G}^T$. The effect is, that the difference $\mathbf{G}^T$ - $\mathbf{F} = \mathbf{B}$ gives us a deformed information about the system structure, because the elements $f(i,j)$, $gT(i,j)$ eliminate each other (namely, $gT(i,j)$ - $f(i,j) = 0$). Fortunately, in PN theory a simple solution of such a problem is possible. It is sufficient to replace the transition in question by the fictive structure introduced in Fig. 5b). The auxiliary place p' together with its input and output transitions t' and t'' replace the original transition t. It means that the impureness in the form of a 'loop' can be replaced (without changing any dynamic property of the net) very simply by the replacement of the transition in question by two fictive transitions and a fictive place set between them. Such a replacement is invariant. After removing the 'loop' problems the original method of the supervisor synthesis can be utilized also for the systems modelled by the 'impure' P/T PN.

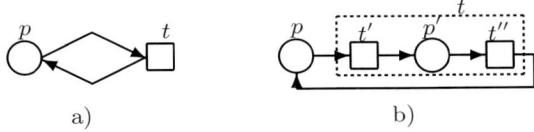

Fig. 5. The replacement of the loop. The original loop is on the left. The structure replacing the loop is given on the right.

3.1   Dealing with Problems of Mutual Interactions Among Agents

Many times, another form of demands on the cooperation among agents occurs. Especially, the situation - see e.g. [1] - when there exist mutual interactions among the agents to be supervised. In such a case a particular marking (i.e. the state of a concrete place) of one agent affects (e.g. stops) firing a transition in another agent. The situation is illustrated in Fig. 6a), where the transition ^{A_1}t of the agent A_1 is prevented from firing under marking the place ^{A_2}p of the agent A_2. As we can see in that picture, a figment that the place p' represents the supervising agent A_3 (or, better said, its part) can be accepted and consecutively, the above described approach to the supervisor synthesis can be utilized. Namely, the transition ^{A_1}t can be substituted by means of the auxilary fictive place p'' together with its fictive input and output transitions t', and t'' analogically to that displayed in Fig. 5b). Hence, the new situation - i.e. new structure - is displayed in Fig.6b). It can be seen that it is the classical mutex. Such a substitution can be utilized in synthesis of the supervisor in case of agents with more complicated structure. For example, the well known kind of the reader-writer cooperation named as the Owicki/Lamports mutex (defined e.g. in [10]) can be automatically synthetised using this substitution.

However, it is necessary to say, that there exist cases where the synthesis of the supervisor based on invariants is not so simple (i.e. fully automatized) or even, it is impossible. Here, the experience and invention of the creator of the

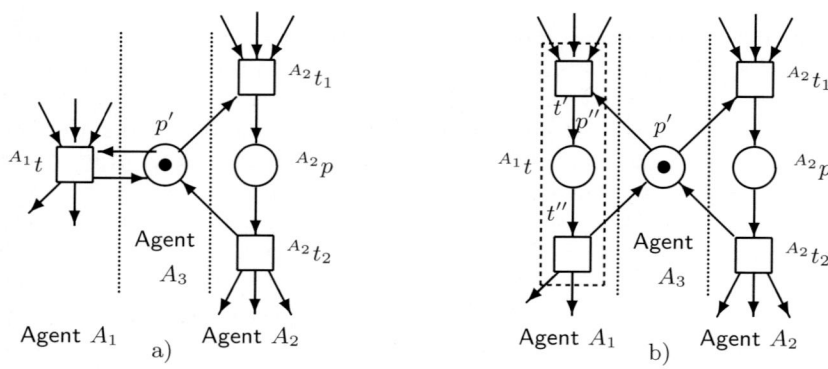

Fig. 6. The prevention of the transition ^{A_1}t (the part of the Agent A_1) from firing under the marking ^{A_2}p (the part of the Agent A_2) by means of p' - the part of the supervisor, i.e. the Agent A_3. a) The original schema; b) The substitutive schema.

agents supervisor has to be used. In the next illustrative example such a system is presented and described.

3.2   The Counter-Example

In order to illustrate such a counter-example let us consider the so called Peterson's mutex algorithm [10] given in Fig. 7. This system consists of two agents A_1 (writer) and A_2 (reader). The PN places describe the following activities: p_1 - pending 1 of A_1; p_2 - pending 2 of A_1; p_3 - critical A_1; p_4 - finished A_1; p_5 - pending 0 of A_1; p_6 - quiet A_1; p_7 - pending 1 of A_2; p_8 - pending 2 of A_2; p_9 - critical A_2; p_{10} - finished A_2; p_{11} - pending 0 of A_2; p_{12} - quiet A_2; p_{13} - at A_1 (left); p_{14} - at A_2 (right).

The state p_4 signals to A_2 that A_1 is presently not striving to become critical. This allows A_2 to "easily" access its critical region, by the action represented by the transition t_{12}. Likewise, the state p_{10} allows A_1 to access its critical state, by the action represented by the transition t_5. The shared token alternates between the states p_{13} and p_{14} The step from p_1 to p_2 results in the token in p_{13}: by action expressed by the transition t_2 in case A_1 obtains the token from p_{13}, or by action expressed by the transition t_3 in case at A_1 held the token anyway. The step from p_7 to p_8 likewise results in the token in p_{14}. Hence, the token is always at the site that executed the step from 'pending 1' to 'pending 2' most recently.

After leaving p_6 along the quiescent action represented by the transition t_7, the A_1 takes three steps to reach its critical state p_3. In the first step, the fair action represented by the transition t_6 brings A_1 from p_5 to p_1 and removes the state p_4. Fairness of the transition t_6 is local, because the transition t_6 is local to A_1, with p_4 the only forward branching the place in the transition t_6, which is connected to the A_2 by the loop (p_4, t_{12}). The second step, from p_1 to p_2, results in the shared token in the place p_{13}, as described above. The third step brings A_1 to p_3, with action expressed by the transition t_5 in case A_2 signals with p_{10},

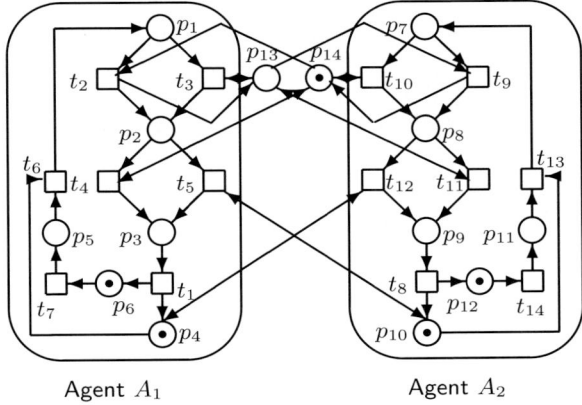

Agent A_1 Agent A_2

Fig. 7. The PN-based model of the two agents communication - the writer (agent A1) and the reader (agent A2) - in the form of the Peterson's mutex

that it is presently not interested in going critical, or with action represented by the transition t_4 in case A_2 more recently executed the step from p_7 to p_8.

The algorithm's overall structure guarantees that one of p_{10} or p_{14} will eventually carry a token that remains there until eventually either the event represented by t_5 or t_4 occurs. The two agents are structurally symmetrical, but the initial state favors A_2 (if the place p_{14} contains the token) or A_1 (if the place p_{13} contains the token).

In this example the invariants-based approach to the supervisor synthesis fails. Namely, any satisfying matrix **L** cannot be found. Searching for reasons it was detected that the problem consists in the specific structure of the P-invariants of the global system presented in Fig. 7. As we can see from the following matrix of the global system P-invariants the invariants of the agents A_1, A_2 are mutually disjoint

$$
\mathbf{V}^T = \begin{pmatrix}
0\,0\,0\,0\,0\,0\,1\,1\,1\,0\,1\,1\,0\,0 \\
0\,0\,0\,0\,0\,0\,1\,1\,1\,1\,0\,0\,0\,0 \\
0\,0\,0\,0\,0\,0\,0\,0\,0\,0\,0\,0\,1\,1 \\
1\,1\,1\,0\,1\,1\,0\,0\,0\,0\,0\,0\,0\,0 \\
1\,1\,1\,1\,0\,0\,0\,0\,0\,0\,0\,0\,0\,0
\end{pmatrix}
$$

4   Conclusion

The method of the DES control theory based on PN place invariants was utilized for the synthesis of the agent (with prescribed properties) supervising a group of agents in MAS. The illustrative example documenting the soundness of the method was introduced. It was shown that the method is simply applicable for autonomous agents modelled by pure PN. Additionally, it was described how the applicability can be extended for the agents modelled by impure PN. Namely,

554   F. Čapkovič

a simple substitution of the transition in the loops of the PN model is sufficient to remove the impureness. As a result of the research it can be said that the method is very useful for the synthesis of the agents supervising a group of automous agents. In case of a group of cooperating agents some problems can occur. Therefore, the counter-example was presented. It was shown that e.g. for the reader-writer in the form of the so called Peterson's mutex the supervising agent cannot be synthetised by the proposed method. Hence, there exists a challenge for the future investigation on this way, especially, to define general conditions for the validity of the approach.

References

1. Bordbar, B., Giacomini, L., Holding, D.J.: UML and Petri Nets for Design and Analysis of Distributed Systems. In: IEEE International Conference on Control Applications (CCA 2000) and IEEE International Symposium on Computer-Aided Control System Design (CACSD 2000), pp. 610–615. IEEE Press, Piscataway (2000)
2. Buy, U., Darabi, H.: Sidestepping Verification Complexity with Supervisory Control. In: Workshop on Software Engineering for Embedded Systems: From Requirements to Implementation, 8 pages (2003),
http://www.cs.uic.edu/~shatz/SEES/Schedule.htm
3. Čapkovič, F.: An Application of the DEDS Control Synthesis Method. Journal of Universal Computer Science 11, 303–326 (2005)
4. Čapkovič, F.: DES Modelling and Control vs. Problem Solving Methods. Int. J. Intelligent Information and Database Systems 1, 53–78 (2007)
5. Čapkovič, F.: Modelling, Analysing and Control of Interactions among Agents in MAS. Computing and Informatics 26, 507–541 (2007)
6. Martinez, J., Silva, M.: A simple and fast algorithm to obtain all invariants of generalized Petri net. In: Application and Theory of Petri Nets, vol. 52, pp. 301–311. Springer, New York (1982)
7. Moody, J.O., Antsaklis, P.J.: Supervisory Control of Discrete Event Systems Using Petri Nets. Kluwer Academic Publishers, Boston (1998)
8. Murata, T.: Petri Nets: Properties, Analysis and Applications. Proceedings IEEE 77, 541–580 (1989)
9. Ramadge, P.J., Wonham, W.M.: Supervisory Control of a Class of Discrete Event Processes. SIAM Journal on Control and Optimization 25, 206–230 (1987)
10. Reisig, W.: Elements of Distributed Algorithms. Modeling and Analysis with Petri Nets. Springer, Berlin (1998)
11. Yamalidou, K., Moody, J.O., Antsaklis, P.J., Lemmon, M.D.: Feedback Control of Petri Nets Based on Place Invariants. Automatica 32, 15–28 (1996)

Grounding of Human Observations as Uncertain Knowledge

Kamil Szymański and Grzegorz Dobrowolski

AGH University of Science and Technology, Cracow, Poland
camel_sz@go2.pl, grzela@agh.edu.pl

Abstract. The article presents some methods of uncertain, human-originated knowledge grounding, accomplished with the use of ontologies with uncertainities and several ontology ABox joining techniques. Several people with uncertain knowledge about some domain write down their domain knowledge in a form of simple text in english. The goal is to obtain uncertain ontology ABoxes, representing their knowledge, and to join them, resulting in total, grounded, uncertain domain knowledge. The joining process takes care of different facts description details, possible facts conflicts as well as varied facts certainity degrees and people credibility.

Keywords: uncertain knowledge, uncertain knowledge grounding, ontology ABoxes joining, human-originated knowledge.

1   Introduction

The paper is a proposition of solving the task of uncertain, human-originated knowledge grounding. Several people with uncertain knowledge about some domain write down uncertain domain facts in a form of simple text in english. The goal is to provide one, total uncertain domain knowledge, based on mentioned information pieces, that can vary in described facts, their certainities and description details. It is also necessary to take into consideration different people credibilities and possible facts conflicts. The solution presented in the paper is based on automated building of several domain ontologies [4] with uncertainities from the text (ABoxes only, with constant TBox supplied externally). Subsequently, ABoxes of the ontologies are joined [2,12] in a way that reflects total knowledge the most adequately. The resulting ontology contains evaluated certainity degrees of every sentence as well, depending on subjective certainity degrees of knowledge contributors and their credibilities.

2   Solution Concept

The whole process of knowledge grounding presented in this paper is shown in Fig. 1. The abstract TBox, called also the base TBox, is a kind of upper ontology [9] for all possible domain TBoxes. It contains several basic concepts:

M. Bubak et al. (Eds.): ICCS 2008, Part III, LNCS 5103, pp. 555–563, 2008.

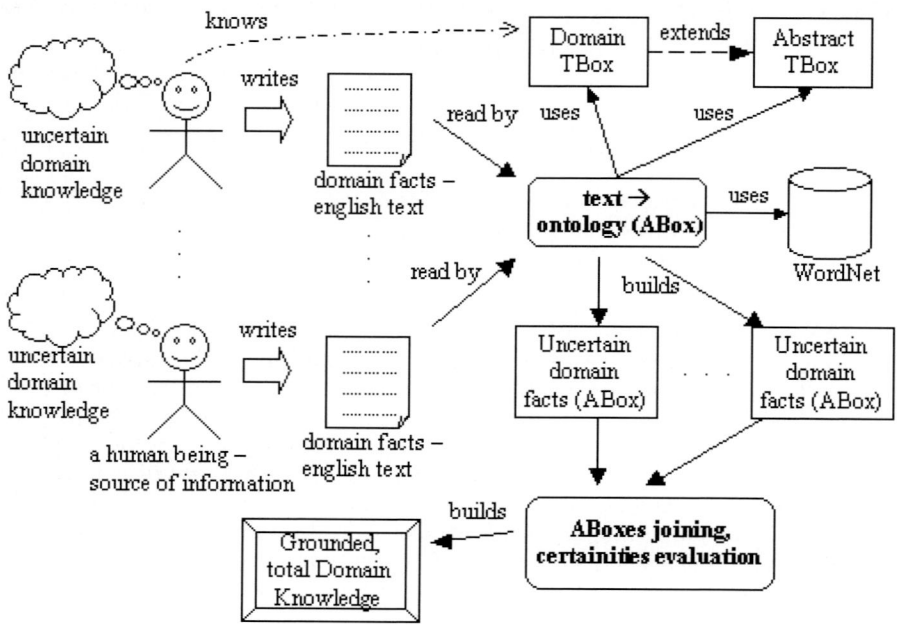

Fig. 1. Solution concept

Entity — *Existence*, for representing classes of objects, Relation — *Relation*, for representing relations between two entities, a subject and an object, and Property — *Property*, which represents entity attributes. Among relations, there are *actions* and *states*. In the abstract TBox it is also modelled that each action can have many following actions. It is important for events chronology. Among properties, there are numerical properties, *NumericalProperty*, and enumerative properties, *EnumProperty*, the latter having a finite set of values.

The domain TBox contains concrete, domain terminology. All its concepts should extend a proper base concept from the abstract TBox. Domain concepts, like *a car* or *a person* should be subclasses of *Existence*, while concrete action *hitting* and state *has color* should inherit abstract *action* and *state* relations respectively. *Speed*, *Height* or *Age* could serve as domain examples of *NumericalProperty*.

There are several human beings, sources of information, who have knowledge about some domain. The knowledge is uncertain as well as different people have varied credibilities. To model those uncertainities in ontologies [13,14], which were chosen for knowledge representation, several certainity factors are introduced, with values within $\langle -1..1 \rangle$, where 1 means total certainty and -1 — total contradiction. There are three kinds of relation certainities:

- *relationCertainity* — performing action certainty
- *relationSubjectCertainity* — the action subject certainty
- *relationObjectCertainity* — the action object certainty

A similar certainty degree for entity attributes can be defined — let it be called *propertyCertainity*. All degrees mentioned above are related to subjective opinions of a source of information and can be assigned to each sentence. Additionally each person has yet another certainty degree assigned that reflects his credibility. Each certainty value present in a sentence must be multiplied by the credibility factor of the information source.

It is assumed that people know domain terminology — the domain TBox. They write down their domain knowledge in a form of a piece of text in english. This is where knowledge processing, described in the article, begins. Firstly, all pieces of text are parsed in order to extract domain ABoxes — known domain facts.

Then, previously obtained ABoxes are joined using several techniques, like identifier-based joining, pattern-based joining, action sequences joining (used in domains describing events with many actions, where chronology is an important factor) or closure-based joining. For output ontology, although grounded, after using mentioned methods, is still uncertain, proper certainty evaluation formulas needed to be suggested as well. All phases of the process and used methods are more precisely described in the following sections.

The implemented solution with text parsing is temporary as it has a few disadvantages. Firstly, a person writing the text, must know what sentences are permissible, in what order they should be written for the resulting ontology to be correct and logical and what domain aspects are served. Those requirements are hard to fulfill in real conditions. Secondly, there is no way to assure that all needed information were written — the writer could still forget to mention some facts he knows about. Those flaws will greatly be limitted when a domain-aware agent is introduced. The agent will lead the dialog with a person, traversing the domain ontology and available terminology, thus restricting the role in the whole process of a human being to giving the answers to concrete, precise questions. This intelligent agent will have the ability of asking proper questions, because of its converstation history and context awareness. The abilities of the agent are thought to be similar to those described in [6].

3   Domain ABoxes Building

At first, the text is formatted and all unnecessary phrases and prepositions are deleted for every sentence in the text to be identified. Then, semantic analysis of each sentence takes place. It is done using WordNet project [1]. A list of the most probable parts of speach, like a noun or a verb, is created for every word in each sentence. The building module, basing on a list of available sentence schemes in the form of

$$\langle subject, relation, object, sentence\ certainity\ phrase\rangle,$$

builds the domain ontology connected with the concrete person, contributing to the knowledge. Several associate lists, which need to be provided externally, are used, such as

⟨certainity phrase, certainity value⟩, eg. ⟨'maybe', 0.4⟩, ⟨'certainly', 1.0⟩,

or base speach forms to a corresponding class in ontology mapping, like

⟨'getting closer', class:Approaching⟩,
⟨('Toyota','Avensis'), class:ToyotaAvensisCar⟩.

The 'it' preposition is also served. This way, the subject of the current sentence does not need to be repeated in the following one as long as it stays the same. Instances are identified and created in the knowledge base as well.

4   Domain ABoxes Joining

Having all domain ABoxes, forming knowledge from several human beings, the joining process [3,8] can be started to obtain aggregate, complete knowledge. Different credibilities of sources of information or points of view and possible missing facts must be taken into account while performing ontology ABoxes joining.

4.1   Identifier-Based Joining

The idea of performing joining based on identifiers has its origin in assumption that in real world there are certain entity kinds for which an identification system was created, like unique car licence plates. Two objects with the same identifiers are found to be synonims. The big advantage of the method is high matching accuracy, but from the other hand the amount of such entity types is rather small.

4.2   Pattern-Based Joining

An instance can be described by a set of numerical properties, like speed or height, and discrete properties with finite number of possible values, such as colours. If two individuals represent the same class and their corresponding properties' values are similar, there is a chance for these two instances to be synonims. A pattern is a set of elements of the following two types:

– A discrete role — its values must be equal in comaparing instances, like sex,
– A numerical role with fault margin — certain differences in values are permitted

All patern conditions have to be met to make two instances synonims. This method works the best if the domain is not too broad and the number of actors is not large.

4.3   Action Sequences Joining

A specification of events chronology in events-oriented domains is one of the most important things to be taken care of. Matching action types and succession is an important factor in synonims searching between corresponding actions and their actors. The analysis of possible successive actions for every action in domain ABoxes results in directed graph of action succession creation. Then, the action succession graph need to be linearized, using the breadth-first searching algorithm. For example, after linearization of the graph from the Fig. 2 (left side), the following action sequences will be created: (A, B, E, F), (A, C, E, F), (A, D, F). After all graphs linearization is complete, a comparison between each pair of

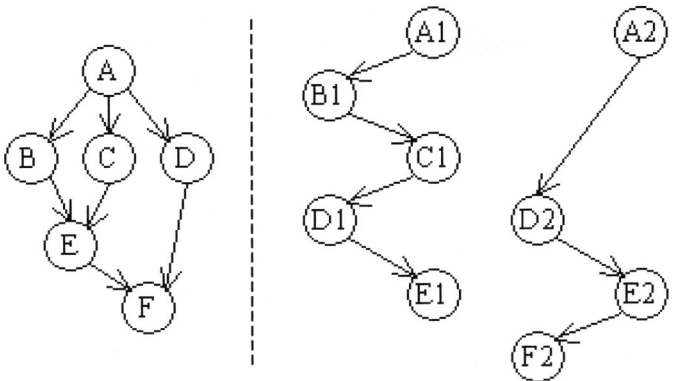

Fig. 2. An example action graph (left), Action sequences (right)

sequences from different ABoxes is performed. The comparison results in *correspondence sequence* detection, which consists of potentially synonimical actions. For an action pair to be accepted to the correspondence sequence, both actions have to be of the same kind. Additionally, the actions' subjects or objects have to be of the same class or classes near each other in class hierarchy (if the classes represent detailed concepts and only for certain, configurable cases).

Compared action sequences can, and probably will be only similar, not equal, because different descriptions of a complex event may vary in details. That is why not all actions from the sequence have to form a correspondence sequence — only matching ones. For example, in case of action sequences from Fig. 2 comparison, a correspondence sequence ((A1, A2), (D1, D2), (E1, E2)) will be created. From all correspondence sequences found for a certain pair of ontologies, the longest sequence is chosen. All its paired actions are considered synonims if the sequence length is more than three.

4.4   Closure-Based Joining

Ontologies can be treated like a connection graph with individuals or actions as vertices and *hasSubject* or *hasObject* relations as edges. The main idea behind closure-based joining is an assumption that if, for example, two individuals (verticies in the connection graph) are synonims, their adjoining vertices (for example two other individuals) could be synonims as well unless they are not similar or have different types.

Two kinds of closures are defined:

- action closure — if two actions have the same type and their subjects and objects are synonims, the actions are synonims as well,
- subject (object) closure (entity closure) — if two actions are synonims and their objects (subjects) are synonims, their subjects (objects) are synonims as well.

These closures are applied multiple times so as synonims discovery propagates through the connection graph.

4.5   Certainities Evaluation

If two facts do not collide with each other, there is not any problem with them — both are taken into the joined ABox as they were in domain ABoxes. Undoubtly there will be some colliding facts too, and not colliding facts, but with different certainity factors. In the first case, a fact with the highest certainity value must be taken, in the latter — although there is no argue concerning the fact itself, its final certainity degree needs to be calculated.

A fact with different certainity values. Let the fact in question has certainity values $p_1, p_2, \ldots, p_n$, coming from information sources $1..n$. These values already include credibilities $w_1..w_n$ of sources ($p_i = w_i * q_i$, where q_i is a subjective opinion of the ith source about the fact). The final certainity p is the mean of p_i values.

Colliding facts. In this case one of the facts need to be taken and its final certainity value must be evaluated. There are two possibilities, depending on property kind:

- discrete properties — a property value with the highest certainity degree need to be chosen. Let property values $a_i, (i = 1..n)$ have among n people certainities $p_{i,1}..p_{i,k(i)}, (1 \leq k(i) \leq n)$, and these certainities include sources credibilities (the same as previously). Property value a_x with the highest mean certainity value $p_x = \sum_{j=1..k(x)} p_{x,j}/k(x)$ is taken into the joined ABox. The value p_x is an output certainity value of the fact as well.
- numerical properties with fault margin — in this case two values are considered not equal if their difference is greater than the fault margin. Otherwise, the means of the values and their certainities (p_i) are calculated and treated like a common value. The rest of the procedure is analogous to the previous case, barring the difference that now the calculations can be performed many times for different mean combinations (eg. when a and b are within the fault margin, b and c, too, but a and c not any more, both the mean of a and b and the mean of b and c must be taken into account, in different iterations). When the means are evaluated, the number of elements forming the means have to be remembered, because those values must have proportionally higher wage in following evaluations.

5   Experiments

Road crashes ontology was designed for experiments. It includes, among other things, cars and road signs hierarchies — both concepts inherit *Existence* class from base TBox. Actions include cars approaching, hitting, turn signalling and road line crossing, to name a few. A car can have colour, speed and licence plates. Four crash witness reports were created and then joined.

Report 1 (witness credibility — 0.3):
I was driving a red Opel Vectra. Its license plates are ABA2010. I approached Toyota Avensis. Its license plates are ABA2557. I indicated left. I crossed the dashed line. I was driving 60 km/h. I hit Citroen C3. Its license plates are ABA3232. It was driving 100 km/h.

Report 2 (witness credibility — 0.6):
I was driving green Toyota Avensis. Opel Vectra approached me. It indicated left. Maybe it crossed the dashed line. It hit Citroen C3.

Report 3 (witness credibility — 0.9):
I was watching Opel Vectra with license plates ABA2010. No way it indicated left. It hit Citroen C3. Its license plates are ABA3232.

Report 4 (witness credibility — 0.3):
I was driving silver Citroen C3. Its license plates are ABA3232. I was driving 60 km/h. Red Opel Vectra crossed continuous line. Its license plates are ABA2010. I hit it.

The result of joining witnesses' reports is as following:
Red Opel Vectra with licence plates ABA2010 was riding 60km/h. Silver Citroen C3 with licence plates ABA3232 was riding 80km/h. The Opel Vectra maybe approached green Toyota Avensis with licence plates ABA2557. It is very doubtful that the Opel Vectra turned on its signalization. It maybe overran a dashed line. It probably hit the Citroen C3.

It can be spotted that information coming from a witness with high credibility (report 3) has clearly higher final certainty factor (minimal certainity of Opel Vectra driver's turn signaling, greater certainty of Opel Vectra and Citroen C3 crash). Synonims among car instances were detected because of non-conflicting car identifiers (licence plates). The joined knowledge is a kind of all reports sum — contains even those facts that were not mentioned by all information sources, like colours of cars. The speed of Citroen C3 (numerical property), 80 km/h, was calculated as the mean of 100 km/h and 60 km/h. It is worth noting that both values had the same certainity degrees in corresponding domain ABoxes, so the waged mean was not necessary here. Witness 2 credibility was high enough that even if he was not totally sure that Opel Vectra crossed the dashed line, it outclassed a certain sentence of low credible witness 4, who was saying the line was continous — there is a dashed line in the output ontology. Moreover, that version was confirmed by witness 1.

It is worth noting that testing presented process of uncertain knowledge joining (grounding) is a complex task which cannot be truely automated. A proper domain ontology must be manually created, basing on provided abstract TBox, and several additional mappings, dependant on used domain, need to be provided as well — they were mentioned in the text. Every ontology is different in its structure, too, and that greatly influences the outcome of the used algorithms. It was assumed that automatically obtained results would be checked by confronting them with anticipated ones. It is subjective to each person reading them wether they are satisfying or not.

6   Conclusions and Future Work

The article presented the methodology of human-origined, uncertain knowledge grounding with the use of several ABoxes with uncertainities joining methods. Knowledge gaining was based on domian ABoxes with uncertainity values building. Each ABox was built from the text which contained domain knowledge of some particular information source. Several joining methods, like identifier-based joining, pattern-based joining, closure-based joining and action sequences joining, together with sentence certainity values evaluation, resulted in domain knowledge joining in such a way that maximized knowledge credibility and gave broader view on problem domain. Presented methodology supports different fact certainity degrees as well as credibilities of information sources. Possible fact conflicts and differently detailed event descriptions are also taken care of. The experiments from the previous section were to show effectiveness of used methods.

Future work will focus on knowledge gaining method extension. Inclusion of domain-aware agent [6] that will lead the dialogue with people, asking them proper questions, related to dialog context and history, is planned. That approach will eliminate some current disadvantages, like demanding knowledge about domain and system features from people.

Acknowledgements. We would like to thank Grzegorz Twarduś and Michał Pelczar for their help in writing this paper and system implementation.

References

1. Cognitive Science Laboratory, Princeton University: WordNet - a lexical database for the English language, http://wordnet.princeton.edu/
2. Euzenat, J., Le Bach, T., Barrasa, J., Bouquet, P., De Bo, J., Dieng, R., Ehrig, M., Hauswirth, M., Jarrar, M., Lara, R., Maynard, D., Napoli, A., Stamou, G., Stuckenschmidt, H., Shvaiko, P., Tessaris, S., Van Acker, S., Zaihrayeu, I.: State of the art on ontology alignment. Knowledge Web Deliverable, Technical Report, INRIA (2004)

3. Fridman, N., Musen, M.: SMART: Automated Support for Ontology Merging and Alignment. In: Twelth Workshop on Knowledge Acquisition, Modeling and Management, Banff, Canada (1999)
4. Gruber, T.: What is an Ontology,
   http://www-ksl.stanford.edu/kst/what-is-an-ontology.html
5. Haase, P., Motik, B.: A Mapping System for the Integration of OWL-DL Ontologies. In: IHIS 2005, Bremen (November 2005)
6. Josyula, D.P., Fults, S., Anderson, M.L., Wilson, S.: Application of MCL in a dialog agent
7. Kalfoglou, Y., Schorlemmer, M.: Ontology Mapping: The State of the Art. The Knowledge Engineering Review 18(1), 1–31 (2003)
8. Luszpaj, A., Szymański, K., Zygmunt, A., Koźlak, J.: The Process of Integrating Ontologies for Knowledge Base Systems. In: 7th Software Engineering Conference, Cracow (2005)
9. Niles, I., Pease, A.: Towards a Standard Upper Ontology. In: Proceedings of the 2nd International Conference on Formal Ontology in Information Systems. FOIS-2001 (2001)
10. Pan, J.Z., Stamou, G., Tzouvaras, V., Horrocks, I.: f-SWRL: A Fuzzy Extension of SWRL
11. Pazienza, M.T., Stellato, A., et al.: Ontology Mapping to support ontclogy-based question answering. In: 4th International Semantic Web Conference (ISWC-2005), Galway, Ireland (November 2005)
12. Pinto, H.S., Martins, J. P.: Some Issues on Ontology Integration. Portugal (2001)
13. Stoilos, G., Stamou, G., Tzouvaras, V., Pan, J.Z., Horrocks, I.: Fuzzy OWL: Uncertainty and the Semantic Web
14. Straccia U.: Answering Vague Queries in Fuzzy DL-Lite

Application of Multi-agents in Control of Hydrogen Powered Car to Optimize Fuel Consumption

Bohumil Horak, Jiri Koziorek, and Vilem Srovnal

VSB - Technical University of Ostrava, FEECS, Department of Measurement and Control,
17. listopadu 15, 708 33 Ostrava-Poruba, Czech Republic
{bohumil.horak,jiri.koziorek,vilem.srovnal}@vsb.cz

Abstract. Mobile embedded systems belong among typical applications of the distributed systems control in real time. An example of a mobile control system is the hydrogen powered prototype car control system. The design and realization of such distributed control system represent demanding and complex task of real time control for minimization of race car fuel consumption. The design and realization of distributed control system, mention above, is prepared for testing as a complex laboratory task. The control system software uses multi-agent technology with dynamic mutual negotiation of mobile system parts. The real hardware and software model is also important motivation for extended study.

Keywords: Vehicle Control System, Distributed Control System, Multi-agents, Learning.

1 Introduction

A team of several specialists and students of Department of Measurement and Control, VSB-Technical University of Ostrava has designed and realized a prototype of hydrogen powered car based on fuel cell technology and electrical DC drive. The project is called HydrogenIX (Figure 1 shows a car) and the works and testing activities came through between October 2004 and today.

The motivations for the project are following:

- The development of mentioned race car is the first application of mobile system with the fuel cell in "Laboratory of Fuel Cells".
- Activate the interest of students, Ph.D. students, specialists and public in renewable and alternative energy sources.
- Enlarge cooperation between university and external subjects in the field of renewable and alternative energy sources and related technologies.
- Demonstrate results of the project in energy consumption economization at mobile vehicles.

The Shell Eco-Marathon competition is organized by Shell Company and takes place at the race circuit in Nogaro, France. Teams from all over the Europe compete to have lowest consumption of the fuel. Even if the majority of teams use petrol engines in their vehicles, there are also vehicles powered by diesel, LPG, CNG, hydrogen and

M. Bubak et al. (Eds.): ICCS 2008, Part III, LNCS 5103, pp. 564–573, 2008.

Fig. 1. The HydrogenIX car

other alternative energies. The results are obtained by the calorific value recalculating for each types of fuel. Therefore, it is possible to compare different types of fuel.

2   Control System

The vehicle powered by electricity generator with hydrogen fuel cell needs the electronic control system that provides control for all its parts. The complex control is necessary for basic vehicle operations and for many subsystems that have to be coordinated and controlled [6]. The control system realizes especially following tasks:

- Control of fuel cell operations – hydrogen input valve control, combustion products output valve control, fuel cell fan control, coupling of produced electrical energy to electric DC-drive system.
- DC drive control – motor current control, speed control.
- Safety and security of the car – safety of the fuel cell system and drive system, processing of hydrogen detector information, temperature measuring, etc.
- Driver control panel – complete interface to pilot that allows controlling the car – start/stop, speed set point, time measuring, emergency buttons and indicators.
- Data archives with saved process variables – saving important process data to archives for next export and analyze.
- Display actual data in car – display panel in the car is a "process" visualization system. All-important data are displayed online.
- Communication with PC monitoring station – the control system sends data and receives commands from PC monitoring station using wireless communication system.

The race car embedded control system uses the microcontrollers Freescale HCS12. The control system has distributed architecture that is divided into five main hardware parts:

- Race car pilot cockpit with pilot control panel - steering wheel, optimal track guidance control display and voice communication,
- Control block of the electricity generator with hydrogen fuel cell,
- Control block of electricity accumulator units with super capacitors,
- Control block of the power DC drive system,
- Interference block of physical environment conditions.

All parts of control system are connected via CAN communication network. The wireless communication between the car and PC monitoring station is realized by data link used mixed GPRS voice and data communication technologies.

The car control system as well as the PC monitoring station are equipped by GSM terminal and differential GPS station. The PC monitoring station provides a process visualization application that is realized by SCADA system Promotic. The process visualization displays all parameters measured during the car operation, all system states, alarms, make possible to display trends of required values, and log measured data in data archives.

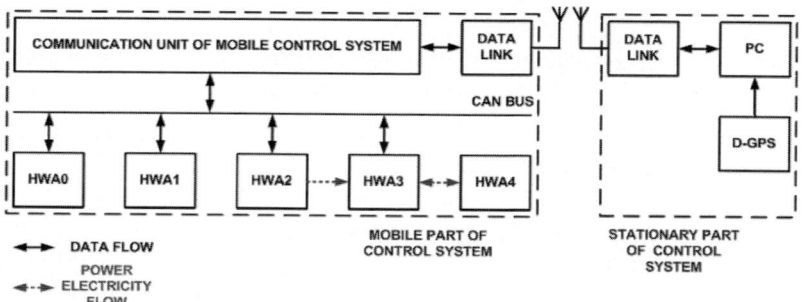

Fig. 2. The block scheme of the race car stationary and mobile part of control system, HWA0 - pilot, cockpit control unit, HWA1 - physical condition description unit, HWA2 - generator of electric power, HWA3 - energy accumulator unit, HWA4 - unit of power drive, recuperation

The complete block diagram of the car control system is demonstrated in figure 2 and a laboratory realization in figure 3.

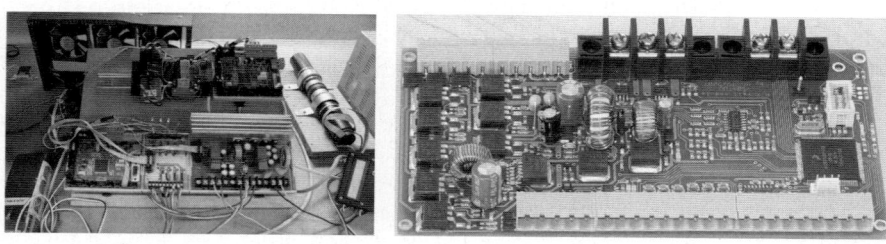

Fig. 3. The HydrogenIX car control electronic testing workplace (on the left) and final realization of control unit

2.1 Operating Values Monitoring

The car control system monitors many variables [3]. Some of these variables are used for basic control activities, the others are used for optimization of operation. The measured variables are following:

- Electrical variables – fuel cell voltage and current, super capacitors voltage and current, DC drive voltage and current and on-board battery voltage and current.

- Non-electrical variables – temperatures (exterior air, cockpit, cooling air, fuel, and fuel canister), pressures (air, fuel), car speed, wind speed and wind direction, air humidity, race track position.

In figure 5, are shown graphs of chosen quantities on time archived by control system during testing runs. Testing run 1 (Fig. 4a) shows the case of periodic changing of the velocity by driver. Testing run 2 (Fig. 4b) shows a start with maximum power and then velocity set to constant value.

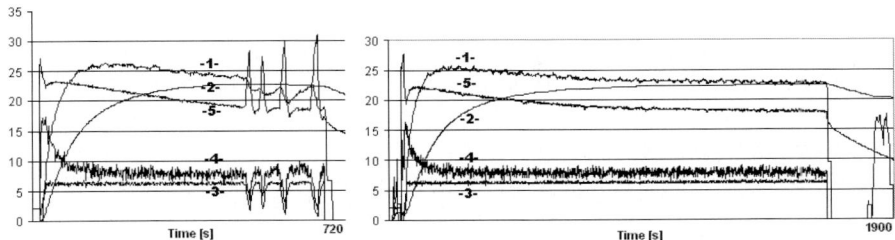

Fig. 4. a) Testing run 1. Instant velocity [km/h] (1), average velocity [km/h] (2), Fuel cell current [A] (3), Motor current [A] (4), Fuel cell voltage [V] (5)   b) Testing run 2. Instant velocity [km/h] (1), Average velocity [km/h] (2), Fuel cell current [A] (3), Motor current [A] (4), Fuel cell voltage [V] (5).

The vehicle is also equipped by bio-telemetry system that makes possible to monitor biological functions of the pilot. For biophysical monitoring was chosen pulse frequency, body temperature, respiration frequency, electrocardiography – ECG.

3   Concept of Multi-agent Control System

Optimization task of the racetrack passage is quite complex, the laboratory car has to consummate a minimal energy controlled in real time. Range of inputs and outputs of the control system, communication flows and safety of all operations require good level of adaptability with environment changes and event situations – base of the strategic control.

The strategic control includes number of components. The related subject should know what result is preferred. It should be able to react at intentions and actions of other participants of the situation. It should be capable of cooperation and compromise searching where it is rational [1] and [2].

Basic subject of strategic control is called an agent. The agent is software and/or hardware entity created with the aim of autonomous solving of tasks with respect at environment in which they occur. His functionality – internal action depends at communication with neighbourhoods (e.g. with sensors, actuators or other agents). Functionality is therefore given by its tasks, targets, behaviour and states. An intelligence of the agent represents often the capability of conformity, development of interactions with neighbourhoods, quick learning, data accumulation and self analysis. Multi-agent systems (MAS) increase flexibility, efficiency and provide learning capability of new reactions and behaviour. In comparison with classic technologies the learning is a „new feature" [4]. In learning process is possible distinguish methods on-line and off-line learning. During off-line learning process is possible to set-up of parameters and databases of system before ride. During on-line learning process learn the agents during ride. Agents can modify own behaviour by method test/mistake. Problem is in acquisition of right training set, that cover all possible situations.

Among expected basic properties of proposed MAS belong a strategic, targeted system behaviour, robustness and adaptability at environment changes. This can be provided by decentralization of control activities in the control system, by distribution of functions and by modularity based on fundamental elements – agents [1].

The agent system has a common target - safe realization of seven rounds through racing circuit in total time near of 50 minutes with minimal fuel consumption. For successful assertion of the race strategy the extraction and knowledge of changeable environment and learning capabilities are very important.

Planning of actions for accomplishment of common target is based on information about track - racing circuit and about internal state of individual agents. The influence at the fuel consumption have track geometry (length, inclination / declination, cross fall), weather (temperature and air-pressure, force and wind-direction, air-humidity, waterless / damp / soaked track), car (aerodynamic resistance, rolling and friction-resistance).

It is possible to characterize targets of individual hardware agents:

- HWA0 – pilot (maintaining of the drive direction, passive car brake, start/stop of the race car).
- HWA1 – physical condition description unit (measure the physical quantities of environment).
- HWA2 – generator of electric power (fuel-saving generation of electricity).
- HWA3 – energy accumulator (optimal select of energy source for consumption, optimal refill of sources with electric power).
- HWA4 – unit of power drive and energy recuperation (optimal control of electricity consumption).

4   MAS Structure Description

The MAS structure block scheme is shown at figure 5. The higher level of control system is represented by a personal computer. Differential GPS positioning system represents the relative coordinate system of environment – allow to precise of the position of the race car on the circuit. The GPRS communication modem is connected at the output – data link, which transmits commands for race car embedded mobile control system.

The separate task is the transformation which converts the global digital data (in-clination, declination, wind speed and wind direction) and digital data of the position into the object coordinates (car position on the circuit), which are saved in the data-base of the circuit. This database is common for all agents in the control system.

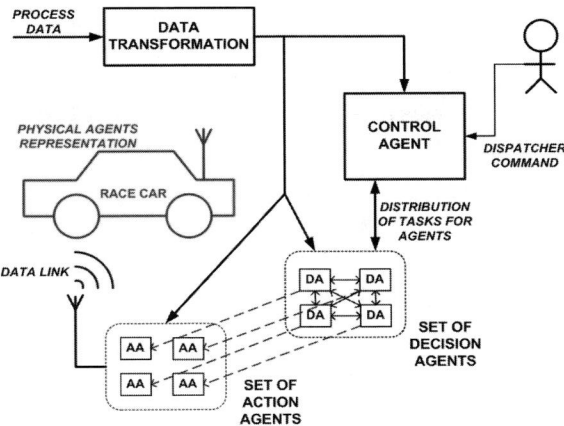

Fig. 5. Block scheme of Multi-Agent control system concept

Each agent access to the all actual data and is capable to control its behaviour in the qualified way. The basic characteristic of a control algorithm of a subordinate agent is independence on the number of decision making agents for car on the circuit.

Main architecture of such hybrid agent system is characterized by:

- Layered control. An agent is described by number layers of abstraction and complexity.
- Layered knowledge base.
- Bottom-up activating.
- Top-down execution.

The algorithm of agent's cooperation was proposed by the control agent on a higher level. The control agent (Fig. 6) goals (minimize fuel consumption) can be achieved by the correct selection of cooperating agents (division of tasks among agents). The decision making agents are able to select correct tasks for themselves and further to select executive agents.

The sensation module evaluates the actual state of the data. The data does not describe the position of car but gives the information about the capability of car to move with optimal trajectory and optimal velocity. In the case that the dynamic control and decision making system will not find the solution for a certain situation, then it will try to find the partial solution from several situations. The controller selects the relevant tasks for the agents for these situations. The controller saves or updates the information about the scene state and control in the relevant database which can be used for static control.

The main task of the decision agent (Fig. 7) is to schedule the relevant task. The agent will schedule the tasks in relation to the environment and the internal state of the relevant hardware unit. The module of perception provides the information about the environment.

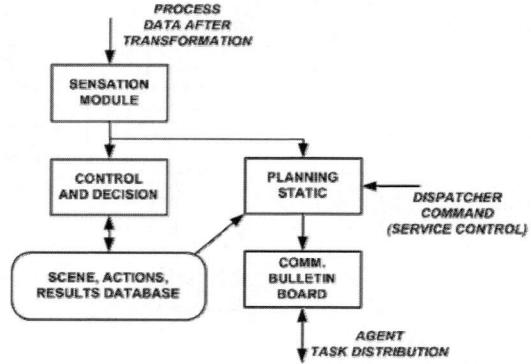

Fig. 6. Control agent scheme

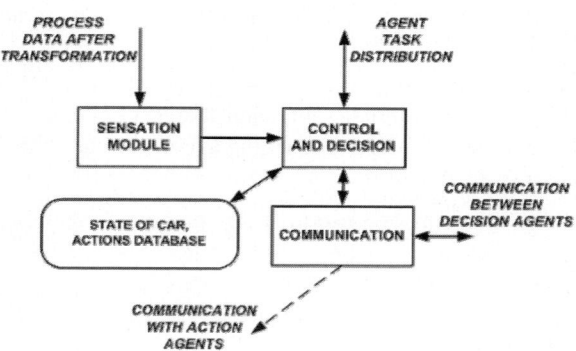

Fig. 7. Decision agent scheme

In the case that more than one agent will select the same task the decision agent must evaluate the probability of the action's success. The agent with the highest probability of success will schedule the actual task. The task selected is handed over to the executive agent who realizes it. The executive agent will receive also the information from cooperating agents for optimization of the car movement.

The activity of the action agent (Fig. 8) is simple. The agent moves with the race car from an actual position to a new position. The information concerning the actual position is obtained from the sensation module.

A new position for further moment (for instance another frame of the GPS data) is evaluated by the control module. The new position is calculated on the basis of a precalculated trajectory. After completion of the trajectory calculation the agent will determine the position for next frame (new position) and will transmit it to the action module.

Agents are connected with the environment through interface realized by sensors, actuators and communication module units (HWA0-4, D-GPS). Control is divided to the layers they use information from knowledge bases [2].

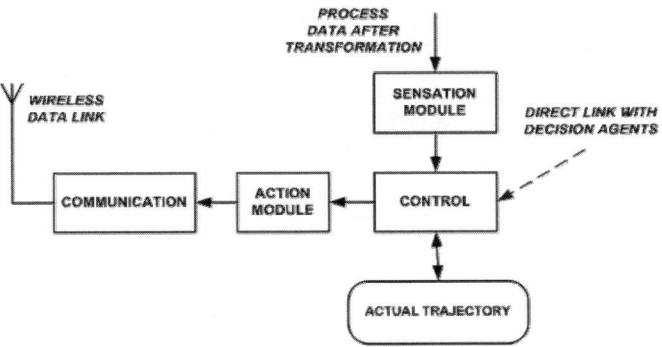

Fig. 8. Action agent scheme

Local planning layer. Some situations cannot be solved by execution of template action like an answer to stimulation from environment only, but they require certain deliberation level. A function of plans creation for solving of the targets performs the layer of local planning. Local planning layer has such fundamental data structures:

- Targets – are defined by state sets. The sets are characterized by attributes that are fulfilled at reaching targets.
- Planning – principles of planning. Sets of plans are predefined as data structures – plans library. Target sets are mapped into the plans library. For each target is possible to assign the plan for its reaching.
- Plans library – contains the plans for reaching of agent targets.
- Scheduling – secures time-limited plans stratification. Created plan schedules are executed like the step sequences.

Cooperative planning layer. A basic control cycle of cooperative planning layer is creation, interpretation, decision making and execution of local plans. In first phase the reports from nearby layers are processed. Reactive layer sends requests to solve new task or status of executed behaviour templates. Schedules of active plans are actualized. Subsequently the status of reactive layer executed procedures is checked.

In case of successful procedures finalization the plan is erased from accumulator. Reports from highest layer are related to creation or cancellation of commitment for the plan execution at local base or plan evaluation. In case of plan execution request or its cancellation the accumulator of active plans is actualized.

The plan availability is a result of difference between its relative value for the agent and its execution costs. The plan value is derived from target value that is possible to reach by plan. The plan costs are determined by function that assigns to every plan a real number calculated at basis of its fundamental action costs according to specific rules.

In gradual hierarchy of targets, it is not possible to realize the target without coordination with other agents. The purpose of multi-agent planning is in "unlocking" such target through building of common plan. Unlocking the target doesn't explicit it's reaching. It's sufficient if agent can continue after execution of common plan by tracking of own local plans. Common plans contain actions of all agents without synchronizations that are not directly executable. It is important to transform common plans into executable plans for individual agents.

Reactive layer. It is responsible for adequate reactions to the stimulations from environment that requires the immediate reaction and execution of called procedures from local planning layer. Fundamental characterization of such layer is:

- Use of effective algorithm of comparison with patterns of behaviour. Serve to pick-out of the actual situations.
- Situation description for timely actual reactions at received stimulus.
- Hard-wired links. Recognized situations are fix-connected with targets for reactive behaviour. Immediate execution of program actions.
- Solution request of situations not–corresponding with couple's situation-action is transmitted in local planning layer.
- Execution liability is coming from local planning layer to activate procedures of reactive layer patterns of behaviour.

4   Conclusions

The algorithm of the control system should be proposed in a way to ensure the requirements for the immediate response of control so that the system of race car would be controlled in real-time. That is why, it is very important that the algorithm for the critical highest speed, the critical average speed, the dead-line time and the fuel (stored energies) consumption would be optimized. The system response should be shorter than the sensitivity (position error) and/or time between two data frames from a GPS station. In the case that this limit is exceeded, the control quality may be decreased.

The main features of algorithm adjustment are following:

-Dynamic control in the control and decision module of a control agent.
-The control and decision modules and communication protocol of the decision agents.
-The strategy of planning in the control model of the action agent.
-Learning of a race strategy and using the extraction results for decision rules generation as a part of the rules decision database of a decision agent.

Acknowledgments. The Grant Agency of Czech Academy of Science supplied the results of the project No. p. 1ET101940418 with subvention.

References

1. Pavliska, A., Srovnal, V.: Robot Disassembly Process Using Multi-agent System. In: Dunin-Keplicz, B., Nawarecki, E. (eds.) CEEMAS 2001. LNCS (LNAI), vol. 2296, pp. 227–233. Springer, Heidelberg (2002)
2. Srovnal, V., Horak, B., Bernatik, R., Snasel, V.: Strategy Extraction for Mobile Embedded Control Systems Apply the Multi-agent Technology. In: Bubak, M., van Albada, G.D., Sloot, P.M.A., Dongarra, J. (eds.) ICCS 2004. LNCS, vol. 3038, pp. 631–637. Springer, Heidelberg (2004)

3. Garani, G., Adam, G.: Qualitative Modelling of Manufacturing Machinery. In: Proceedings of the 32nd IEEE Conference on Industrial Electronics, pp. 3591–3596. IEEE, Paris (2006)
4. Kaelbling, L.P.: Learning in Embedded Systems. MIT Press, Cambridge, Mass (1993)
5. Vilaplana, M.A., Mason, O., Leith, D.J., Leithead, W.E.: Control of Yaw Rate and Sideslip in 4-wheel Steering Cars with Actuator Constraints. In: Murray-Smith, R., Shorten, R. (eds.) Switching and Learning 2004. LNCS, vol. 3355, pp. 201–222. Springer, Heidelberg (2005)
6. Turek, W., Marcjan, R., Cetnarowicz, K.: Agent-Based Mobile Robots Navigation Framework. In: Alexandrov, V.N., van Albada, G.D., Sloot, P.M.A., Dongarra, J. (eds.) ICCS 2006. LNCS, vol. 3993, pp. 775–782. Springer, Heidelberg (2006)

Extensible Multi-Robot System*

Wojciech Turek

AGH University of Science and Technology, Krakow, Poland
wojciech.turek@agh.edu.pl

Abstract. In recent decades many important problems concerning mobile robots control algorithms have been solved. In order to integrate those solutions into complex, reliable, production-quality systems, an effort must be made to define abstract methodologies, architectures and frameworks, that would provide common functionalities and support features desirable in production systems: scalability, extensibility, flexibility and durability. This paper presents an agent-based architecture, which should provide extensibility of Multi-Robot Systems.

1 Solved Problems in Mobile Robotics

In recent years research into mobile robotics received significant attention. Constant development of hardware and communication technologies increases range of potential applications of systems using robots. Theoretically such systems should be able to successfully fulfil tasks like guarding, cleaning or all types of flexible transportation in warehouses, offices, etc. However, in case of mobile robots the distance between theoretical suitability and practical, reliable and worthwhile applications is very large, mainly due to enormous complexity of required software modules regarding many different domains.

This complexity caused separation of several well-defined problems concerning robotics, which became subjects of research. One of the most important of those was the localization and mapping problem, or SLAM [1], which aimed at the issue of map building and navigation of one or more robots in an unknown environment. The SLAM is now considered solved. Another popular problem regards algorithms of reactive navigation and collision avoidance, which must be applied in unknown environments, where global path planning cannot be applied. Several interesting solutions have been proposed [2]. Of course all of those have common limitations caused by limited knowledge.

When a map is available, the problem of path planning and motion coordination arises. In complex system with large number of robots operating on limited space a robust solution of this issue is crucial. The complexity of real-time coordination problem makes it impossible to calculate one correct solution, therefore motion coordination is often considered optimization problem [3].

Complex, separated operations performed by single or multiple robots are called tasks. If several tasks are to be performed simultaneously, a task scheduling

---

* This work has partly been supported by the grant MNiSzW Nr 3 T11C 038 29.

M. Bubak et al. (Eds.): ICCS 2008, Part III, LNCS 5103, pp. 574–583, 2008.

and assignment algorithm is required. Several approaches designed for different constraints can be found in literature [4].

Most of basic issues associated with mobile robots have already been addressed, and (better or worse) solutions are available. However, there are hardly any propositions concerning methods for integrating these solutions into robust systems, that could claim production quality. A detailed review of MRSs (Multi-Robot Systems) architectures was presented by Farinelli et al. in [5]. Most of those solutions were designed for solving particular types of problems, without paying any attention on features of the software, which are mandatory in production systems: **scalability, extensibility, flexibility and durability**. It seems justified to say, that a system, that cannot provide those features will never be introduced into large scale applications. Therefore an effort must be made to define abstract methodologies and frameworks, that would support the features and provide basic services required by most MRSs: universal navigation, flexible traffic control, sustainability and abstract task management and scheduling.

2   Extensibility of Systems

There are numerous works trying to define extensibility and identify methods of achieving it [6]. Extensibility of a system is typically defined as a presence of mechanisms dedicated for extending the system, and measured as effort required to implement extensions. The effort includes time needed for implementing new functions as well as the hardware that need to be added or modified. Other features taken into considerations can include the influence that new functionality has on former features of the system, or ability of introducing the changes without altering normal runtime. In the domain of software engineering, extensibility is one of the key aims. Lack of prepared mechanisms may result in huge looses for the institution which faces sudden need of extending functionality of its crucial software.

Supporting software extensibility includes creation of interfaces and hooks, which should be as flexible as possible, to allow new functionalities to access existing system resources without altering its internal structure. The interfaces should not assume which features will be added, but should rather consistently cover whole functionality which can be exposed. Ability of introducing new features without halting the system or causing risks of unexpected behaviours is also very desirable. It can be achieved by using distributed architectures, which separate runtime process of different components.

It seems that the multi-agent design and runtime paradigm is a good basis for building extensible software system. Features like parallel and asynchronous execution, inherent distribution, interaction based on messages or runtime code inclusion make it possible to implement safe functionality addition mechanisms. What needs to be done is implementation of flexible protocols used by crucial agents in the system, which would allow new agents to integrate. Another feature that might be useful is a replacement mechanism, which would support changing agent's code on runtime, without loosing agent's internal state.

3   Extensibility in Multi-Robot Systems

It is highly probable, that within few decades mobile robots will be commonly used as physical tools of many systems. Applications can include performing tasks like cleaning, guarding, all types of transportation, guiding and providing information, and many other unpredictable uses, which are the most important in context of this paper. It is obvious that complex systems, which use mobile robots, should also support functionality modification and extension abilities.

Identification of general methods of supporting extensibility in a Multi-Robot System or MRS requires defining abstract architecture of such a system. Without loosing generality it is safe to say, that most of the following elements will be present in a typical MRS:

- set of mobile robots and communication infrastructure,
- map and localization methods,
- path planning and motion coordination algorithms,
- task scheduling and execution algorithms,
- sustainability methods and infrastructure (robots charging, maintenance).

Several general directions of extension can also be identified. An extensible multi-robot system should support all of those with no need of altering normal functionality.

1. It should be possible to add a new type of task, which can be solved by robots already present in the system.
2. Addition of a new type of robot, which would be used for solving tasks already known in the system, should also be possible. The software responsible for solving the tasks should be able to make use of the new hardware without any modification.
3. Robot of a type which is new to the system may have different maintenance requirements – the system must allow addition of maintenance functionality.
4. New types of tasks and robots may require modifications in coordination algorithms, including task scheduling policies.
5. Expansion of working environment, which includes elements unknown to the system, should be supported. For example, if a second floor of a warehouse is to be included, robots must become able of using an elevator.
6. New types of tasks, robots and new management requirements may require modifications in motion coordination algorithms.

Before describing the system that will support the requirements, methods of making a MRS inextensible should be identified. It may seem surprising that one of the most popular methods of extensibility removal is related to application of software agents. The issue is caused by a very popular observation of similarities between features attributed to a software agent and an autonomous mobile robot (autonomy, mobility, communication and cooperation). The similarities encourage creation of an autonomous robot-agent, which is a software component embedded into a hardware robot. Typically the agent implements

algorithms for navigation and solving several individual tasks and a communication protocol, which is used for performing cooperative tasks. If these features
are compared against the extensibility scenarios listed above, it is obvious that
the approach is unacceptable. The same effect can be achieved by using task-
based coordination protocols, which are typically forced by too strong autonomy
of the robots.

Another method of reducing extensibility is application of fully autonomous
navigation without any remote management or coordination mechanisms. Fully
autonomous mobility may seem very advanced, but can cause a lot of trouble
when a new type of robot is introduced to the system, and even more when
working environment is to be extended.

It is impossible to foresee all possible applications of a complex robot during
system design process. However, the ability of performing calculations on-board
often leads to creation of high-level asynchronous interfaces, usually hiding basic
functions of a robot. As a result, the system owns a robot unable of performing
simple tasks, despite no hardware limitations present. The general conclusion
is: too much autonomy given to hardware robots is not the best idea in the
context of system extensibility. In the following section a MRS will be described,
which uses agent design and runtime paradigm and meets identified extensibility
requirements.

4   Agent-Based Extensible Multi-Robot System

Every complex system requires different types of modules with different amount
of autonomy or proactivity included. The architecture will use an agent-based
design paradigm, which usually involves a discussion about different definitions
of a software agent. Therefore at the very beginning of this section several assumptions concerning modules naming is needed:

- Master - a proactive module with a separated execution thread or threads,
  responsible for performing high-level tasks. Its activities will include decision making, creation and destruction of other elements, management and
  coordination.
- Worker - a reactive module with a separated execution thread. It will perform
  orders received by asynchronous messages, according to a deterministic and
  known algorithm. Its functionality may include immediate request-response
  operations as well as long-lasting tasks.
- Server - a reactive module with no separated execution thread. It will respond
  immediately to all requests received, using a deterministic algorithm.

From a design point of view probably only first or first two of those should be
called agents. But as the implementation will use an agent platform, all of those
will be agents on runtime, differing in amount of proactivity shown. Therefore the
names used for particular modules will end with "MasterAgent", "WorkerAgent"
or "ServerAgent" suffix.

The architecture presented here can be implemented on a group of complex robots performing all computations and communication using on-board hardware, as well as robots using external communication and computation infrastructure. In both cases it is assumed, that there is an agent framework available, which is logically separated from the robots. The framework is supposed to support basic messaging and management services, like for example those defined by FIPA [7] standards. Obviously there are going to be differences in implementation of both cases, but at the level of an abstract architecture all assumptions and conclusions are very similar. Modules of the architecture may be divided into four groups, responsible for robots, navigation sustainability and task execution.

4.1   Robot

The most basic module that need to be defined is a robot controller. Following the conclusion of the previous section and the definitions presented above, a robot controller should be defined as a RobotWorkerAgent or RWA – a reactive module, able of performing long-lasting operations and offering a deterministic, detailed interface. The interface is divided into two groups: common and specialized functions. The common functions, that must be implemented by all the robots in the system, contain getters for identification information, state (position, orientation) and information about specialized services.

The specialized RWA's interface should contain at least a complete wrapper to all low-level hardware functions. In addition it can contain any higher level methods, which implement most common algorithms. In particular, every mobile robot will include basic functions for setting velocities of its engines, and optionally several more complex functions concerning mobility, like for example reactive navigation methods. Methods in specialized interfaces are grouped into services, which are used for selecting proper robots for particular types of tasks.

If a robot is able of hosting a part of the agent platform, then the RWA is supposed to be located at the robot, where it can have the most direct access to the hardware. Otherwise the RWA is placed on a remote device (computer or other robot), which is equipped with communication hardware able to manage the robot. The first scenario guarantees better performance, but can be difficult to implement on small or simple robots. Other elements of the system must take into consideration, that the RWA is strongly connected with a hardware unit, and therefore can fail and disappear from the system without any warning or notification.

4.2   Navigation Subsystem

The navigation subsystem described in this section is an extension of the solution proposed in [8]. The basic idea is to divide the environment into two types of fragments: rooms, where autonomous, reactive navigation should work fine for all types of robots, and areas, where remote assistance and motion coordination is required. The areas may include narrow passages or junctions but also doorways or elevators. An example of such division is shown on the left side of Fig. 1.

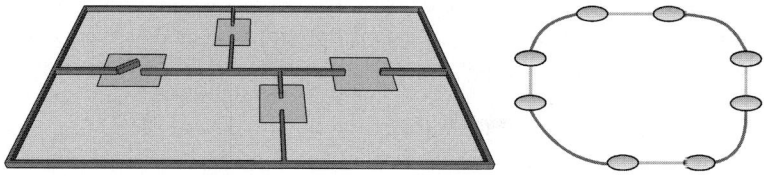

Fig. 1. Fragment of environment and a graph created for it

The graph on the right side is generated directly from the layout of rooms and areas in the environment. Edges of the graph represent rooms and areas, while nodes are intersections of those. The graph is maintained by a MapServerAgent or MSA, which supports path planning functionalities. Its interface must contain methods for modifying the structure of the environment and a method for finding the fastest path for a given robot between any two locations in the environment. Each edge associated with a room has a constant weight representing distance between connected areas. Weights of the edges associated with areas are calculated dynamically for a particular robot by an agent which manages the area, called AreaMasterAgent or AMA. Each AMA may have a different weight calculation strategy, which can be used for preventing high traffic. It can also return an infinite value, which forbids a particular robot crossing the area. Calculation of edge weight is the simplest of responsibilities incumbent on an AMA. Its basic task is to coordinate safe motion of robots within the area. The coordination algorithm used by AMAs will differ depending on type of the area and current traffic.

4.3   Sustainability Subsystem

Each type of robot may have different maintenance requirements. Typically these will include battery charging, in some cases more complicated operations might be performed automatically. Performing service operations requires specialized hardware devices installed in selected locations of the environment, called depots and managed by DepotMasterAgents or DMAs. When a new robot is introduced into the system, it is manually assigned to selected DMA, which becomes robot's basic Master and caregiver. It is supposed to maintain the robot working properly or report failures to a human operator. While registering, new robot sends available services descriptors, which contain service name and characterizing parameters. The name defines unambiguously associated parameters and a set of methods available in RWA's specialized interface. Although service name can and should be meaningful in a natural language, it is not intended to be semantically analysed by any element of the system.

Mobility services can be used as a good example. Three services could be created, named for example: ,,basic_mobility", ,,reactive_mobility" and ,,intelligent_mobility". Basic should contain on method: move(linear_velocity, angular_velocity); and several parameters describing robots features, like maximum velocities, accelerations, turning radius etc. Reactive mobility service should implement one method: go(x,y, velocity); and one parameter describing maximum velocity. Intelligent mobility service, which would use MapServerAgent, would implement a method go(x,y), which would return estimated time of arrival or information about destination inaccessibility.

Besides maintaining robots, DMA acts as a robot rental for modules responsible for task execution. It implements a ,,Query Service" method, which takes service name or robot type as a parameter and returns names, services and availability information of all robots supporting given condition. It also supports a ,,Request Robot" method, which allows the caller to master requested robot for specified period of time. To provide tasks management and priorities in the system, methods of DMAs must recognize priority of a caller - robots are shown as available only when no requests were made, or requests were made by callers with lower priority.

It should be pointed out, that a DMA can, but does not have to be associated with a specified location in the environment. It can but does not have to move all idle mobile robots to that location. For example there could be a DMA, which would be responsible for providing naming and rental services for several industrial manipulators mounted at fixed locations.

4.4   Task Execution Subsystem

Addition of a new type of task will probably be the most often extension in every Multi-Robot System, therefore lots of attention should be put to simplify this operation. Two types of agents must be implemented to add a new type of task:

- TaskExecutorMasterAgent (TExMA), which will use a group of robots located near the task to fulfil it,
- TaskExecutorFactoryServerAgent (TExFSA), which will support information about TExMA requirements (robots count, services and estimated execution time), and will be able to create an instance of TExMA.

The most proactive element of the system, is a TaskSchedulerMasterAgent or TSMA. It is fully independent on type of tasks performed, therefore it does not have to be modified when a new type of task is added. It can be created (by a user interface or by automated task detection mechanisms) for a group of tasks,

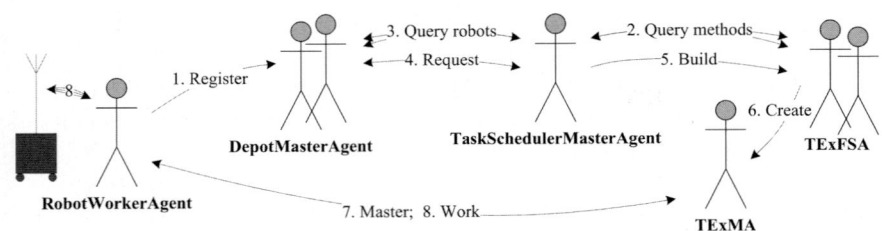

Fig. 2. The most important steps of task execution process

as well as for a single task. There can be different implementations present in the system to enable different scheduling strategies. Each TSMA has a priority value assigned, which it uses in communication with DMAs. A sequence of the most important steps in tasks execution process is shown in Fig. 2.

At the first glance this architecture might seem too complicated, but it can be justified by analysing required execution threads in considered process. If there are multiple tasks to be synchronized, there must be a supervising entity (TSMA) which will watch and manage progress. If a task involves several cooperating robots, there must be a supervising entity (TExMA) which will coordinate actions of particular hardware devices. As it was pointed out before, Server Agents (TExFSA) do not require separated execution threads – are agents only from implementation point of view.

4.5 Features of the Architecture

The approach presented here was designed to meet MRS extensibility requirements described earlier in this paper. The most important of those was support for addition of new types of tasks to the running system, which was described in previous section. Algorithms of task execution are separated in agents, therefore relatively small amount of code must be implemented to add new type of task. Many coexisting task scheduling agents are allowed, implementing different algorithms – implementations can be added or removed on runtime.

A new type of robot can also be easily added to the system without altering its functioning. If the robot does not have any special service requirements (different that provided by the system already), the only think that must be done is implementation of a new RobotWorkerAgent, which will be created for the robot and registered in selected existing DepotMasterAgent. Otherwise a new DepotMasterAgent must be implemented and deployed as well. All of these operations can be performed on runtime.

The navigation subsystem is built of separated agents responsible for managing traffic in fragments of the environment. Global map representation implemented by a MapServiceAgent is not dependent on type or implementation of AreaMasterAgents; the only think required is proper reaction for several types of messages – new types of environment fragments can be easily added.

The architecture has other desirable features, of which the most important is scalability – the only central elements of the architecture are the MapServiceAgent and internal services of agent platform used. Other components are created as needed and operate only in particular fragment of the system, using limited resources. Agent paradigm allows transparent distribution of the system among multiple computers, which makes the system easily scalable. It is worth pointing out, that the subsystems hardly depend on each other, making the architecture very flexible. It could even be used as a robot-soccer controller implementation, where particular plays are implemented by TExMAs, and a TSMA is responsible for detecting situations opportune for performing a play. Navigation subsystem is not used at all, while one DepotMasterAgent is used to move robots to idle positions in a pitch. Obviously there are better approaches to the particular problem; the point is that the architecture allows addition of significantly different functionalities to existing systems.

5 Implementation and Tests

All components of the architecture were successfully implemented and tested. The agent platform was JADE [9], one of the most popular FIPA-compliant platforms, which is very suitable for experiments concerning agents interoperability, provided that the scale of the system is limited. Unfortunately it is not feasible to get into implementation details due to limited length of this paper. Therefore only general functionality and conclusions of the example will be described here. To make the example simpler, the navigation subsystem was disabled. Its implementation and tests involving complex environments were described in [8].

First version of the system was supposed to support box moving functionality in a simple warehouse (figure 3). There were three identical robots used, each equipped with a two-degrees-of-freedom gripper. Each robot implemented basic and reactive mobility services and a 'gripper' service, which contained a method for setting gripper's position in both axis, and higher level 'grab' and 'drop' methods. There was one type of task defined, characterized by a source and destination locations of a box, and one TExMA using 'grab' and 'drop' methods.

The implementation was tested using the RoBOSS [10] simulation system. First new requirement added to the system was moving a box initially placed on

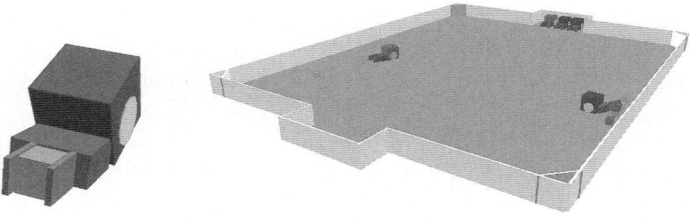

Fig. 3. Simulation model of the robots and the environment used in the example

another box. Former TExMA would have failed to fulfil this type of task, because 'grab' method always picked boxes from the ground – a new type of task and a new TExMA (using lower level gripper control) was created and introduced into the running system. Next innovation in the system were very heavy boxes, which could not be lifted by available robots. Solving the issue required adding new type of robots, which supported mobility and gripper service like the others, and a new 'stronger_gripper' service. New type of task, and a new TExFSA-TExMA pair were created and included. As expected, new robots were used for solving both new and old types of tasks – all possible services were supported.

It is easy to notice, that if robots services and the first task in the system were properly characterized (by 3D initial location and maximum weight of a box), new types of tasks and TExMAs would not have been necessary. The most interesting conclusion is, that new functionalities can be added successfully despite serious lacks and mistakes in existing implementation of a system.

6   Conclusions

It seems justified to say, that agent design and runtime paradigm is a good approach to creating extensible software systems, which use robots as hardware effectors. An approach proposed in this paper can be a good basis for building extensible multi-robot systems. Obviously not all applications of robots can and should be fitted into this architecture, however some ideas and solutions presented here may help increasing extensibility of complex multi-robot systems.

More reliable verification of the approach will require tests of more 'real-life', larger scale scenarios. Further investigation into different task scheduling strategies and failures handling methods must be carried out to provide reliability, robustness and flexible high level management methods. Tests on different types of hardware robots are also planned.

References

1. Smith, R., Self, M., Cheeseman, P.: Estimating Uncertain Spatial Relationships in Robotics. In: Autonomous robot vehicles, pp. 167–193. Springer, Heidelberg (1990)
2. Minguez, J., Montano, L., Santos-Victor, J.: Reactive navigation for non-holonomic robots using the ego kinematic space. In: Proc. if Int. Conf. on Robotics and Automation, Washington, USA, pp. 3074–3080 (2002)
3. Bennewitz, M., Burgard, W., Thrun, S.: Finding and Optimizing Solvable Priority Schemes for Decoupled Path Planning Techniques for Teams of Mobile Robots. Robotics and Autonomous Systems 41(2), 89–99 (2002)
4. Farinelli, A., et al.: Task assignment with dynamic perception and constrained tasks in a multi-robot system. In: Proc. of the IEEE Int. Conf. on Robotics and Automation (ICRA), pp. 1535–1540 (2005)
5. Farinelli, A., Iocchi, L., Nardi, D.: Multirobot systems: A classification focused on coordination. IEEE Transactions 34(5), 2015–2028 (2004)
6. Nurnberg, P.: Extensibility in component-based open hypermedia systems. Journal of Network and Computer Applications 24, 19–38 (2001)
7. O'Brien, P.D., Nicol, R.C.: FIPA – Towards a Standard for Software Agents. BT Technology Journal 16(3), 51–59 (1998)
8. Ambroszkiewicz, S., Cetnarowicz, K., Turek, W.: Multi-Robot Management Framework based on the Agent Dual-Space Control Paradigm. In: Proc. of the AAAI 2007 Fall Symposium, Arlington, Virginia, USA, pp. 32–37 (2007)
9. Bellifemine, F., Poggi, A. and Rimassa, G.: JADE – A FIPA-compliant agent framework. In: Proc. of the PAAM 1999, London, UK, pp. 97–108 (1999)
10. Turek, W., et al.: RoBOSS - an universal tool for robots modelling and simulation. In: Proc. of Computer Methods and Systems, Krakow, Poland, pp. 347–354 (2005)

Agent-Based Immunological Intrusion Detection System for Mobile Ad-Hoc Networks

Aleksander Byrski[1] and Marco Carvalho[2]

[1] AGH University of Science and Technology, Kraków, Poland
olekb@agh.edu.pl
[2] Institute for Human and Machine Cognition, Pensacola, U.S.A.
mcarvalho@ihmc.us

Abstract. Mobile Ad-hoc Networks are known to bring very special challenges to intrusion detection systems, mostly because of their dynamic nature and communication characteristics. In the last few years, several research efforts have proposed the use of immune-inspired systems for intrusion detection in MANETs. In most cases, however, only low-level pattern construction and matching have been considered, often customized to specific routing strategies or protocols. In this paper we present a more general, agent-based approach to the problem. Our approach proposes the use of artificial immune systems for anomaly detection in a way that is independent of specific routing protocols and services. After introducing the problem and the proposed system, we describe our proof-of-concept implementation and our preliminary experimental results over NS-2 simulations.

1 Introduction

Anomaly and Intrusion Detection Systems (IDS) have long been proposed in support of security strategies for computer networks. Most commonly applied in the context of enterprise networks, conventional IDS generally relies on a number of detection elements (sensors) and some (often centralized) components that correlate information among sensors to identify anomalies. Such components are responsible for learning how to identify and differentiate normal (self) patterns, from abnormal (non-self) traffic or system patterns.

Mobile Ad-hoc Networks (MANETs) are characterized by their lack of a fixed support infrastructure and their transient nature. Together, these characteristics lead to a very challenging environment for IDS implementation. Frequent changes in topology and communication patterns in MANETs require the use of specialized protocols and strategies for routing, transport and security. In particular, the use of autonomous agents performing the duties of a single security detector and being able to communicate with neighboring agents to share information and inferences is well suited for IDS implementation in MANETs.

Biologically-inspired approaches for anomaly detection systems have proven to be very interesting, often yielding very effective results [1] for some applications.

M. Bubak et al. (Eds.): ICCS 2008, Part III, LNCS 5103, pp. 584–593, 2008.

In particular, immune systems-based detection and defense mechanisms seem to provide a good analogy to the requirements and capabilities expected from an IDS for these kinds of environments. The approach proposed in [2], however, like most others, is defined for a specific routing protocol. In the paper we propose a more general approach to the problem. We first provide a brief introduction on the state-of-the-art in Intrusion Detection Systems for MANETs. After presenting the architecture and core components of the proposed systems, we introduce and discuss the modelling of behavior patterns. We conclude the work with by presenting and discussing our preliminary experimental results and our conclusions.

2   Intrusion Detection System for MANET

The primary goal of an Intrusion Detection Systems (IDS) is to detect the unauthorised use, misuse and abuse of computer systems and networks resources. The earlier research on IDS dates back from the 80's [3], when it basically aimed on providing auditing and surveillance capabilities to computer networks. Following that idea, the first generic intrusion detection model [4] was proposed in 1987.

The basic implementation of that model consisted in a real-time expert system whose knowledge was derived from statistical inference based on the audit trails of users or system resources. It stored characteristics describing the normal behavior of subjects with respect to objects and provided the signature of abnormal behaviors - a statistical metric and model were used to present profiles. As a subject, an individual system user, a group of system users or the system itself can be considered, while objects can be files, programs, messages, records, terminals etc. When a subject acts upon a specific object, it usually generates an event, which alters the statistical metric state of both subject and object. A knowledge base contains activity rules to be fired for updating profiles, detecting abnormal behaviour, and producing reports. An inference engine works by triggering rules matching profile characteristics. Since then, various IDS have been developed and a number of intrusion detection systems have directly employ this model e.g. [5] [6].

Mobile ad-hoc networks (MANET) are self organized networks without any predefined structure (other that the end users are equipped with radio-based networking interfaces). Communication beyond the transmission range is made possible by having all the nodes serve as routers. They should participate in common routing protocol (such as AODV). This makes these networks very difficult to perform monitoring, because of dynamical reorganization of the topology. In classical (non ad-hoc) networks possible reasons for node misbehavior may be caused by faulty software or hardware, sometimes caused by a human intruder. Other treats arise for ad-hoc networks, i.e. misuse of the routing protocols [7] [8].

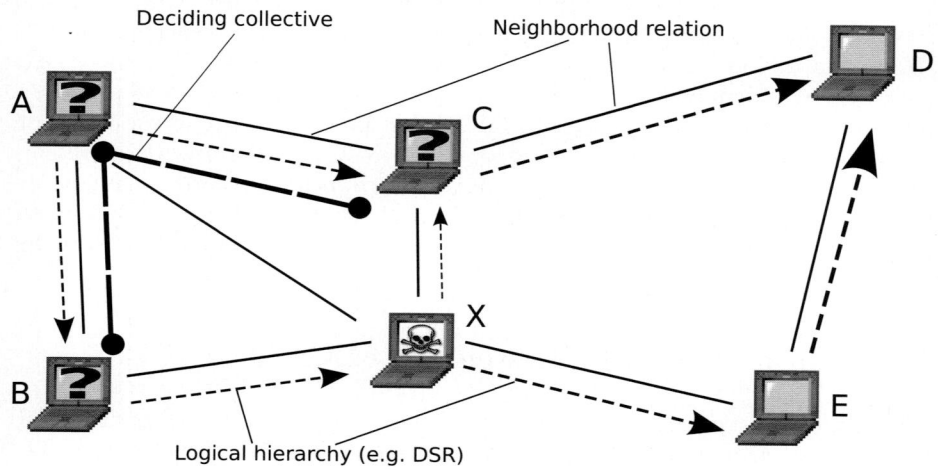

Fig. 1. Structure of distributed IDS

MANETs seem to pose special requirements for IDS, because of [7]:

- Mobility and dynamism – nodes in MANET are highly mobile and topology changes in sometimes unpredictable manner.
- Lack of fixed traffic points – there are no firewalls or routers as in classical computer networks, all nodes are used as routers.
- Limitations of host-resident network intrusion detection – detectors may also become the target of an attack per se, or by distracting of their communication protocol.
- Wireless communication – RF medium is susceptible to eavesdropping, jamming, interference and many other MAC threats what may effect in loss of packets and intermittent connectivity.
- Resource constraints – the resources vital co communicating in MANET environment are limited, e.g. energy (battery operated nodes), varying throughput because of dynamic topology configuration.

So the IDS for MANETs must be decentralized, with some level of data aggregation and information sharing – e.g. the detectors may consult themselves in order to evaluate the accuracy of detection and provide better responses.

Sterne e.a. [7] propose a reasonable solution to the problem based on the hierarchical organization of detectors. The dynamic nature of MANETs, however, tend to complicate the creation of (virtual) dynamic structure, often compromising this kinds of approaches. A far more simple and yet effective approach for detecting unfavourable behaviors might be considered by using non-hierarchical approaches similar to ethically-social mechanisms of decision undertaking, proposed in [9].

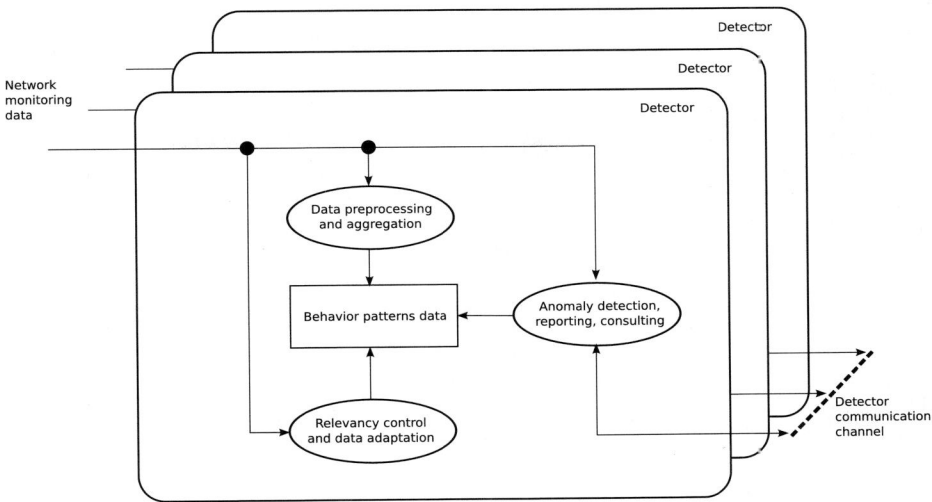

Fig. 2. Detection algorithm

3 Agent-Based IDS for MANET

3.1 System Structure Overview

Proposed IDS consists of a set of detectors (that may be perceived as intelligent agents, because of their autonomy [10]) introduced into the system (i.e. several nodes that take part in normal routing of the packets are considered as detectors). After sensing some kind of disturbance in the behavior of certain nodes, the detectors should try to reach neighboring detectors and communicate with them, in order to consult their observation. Then the decision of raising an alarm may be undertaken.

In this way constructed system allows to preserve no hierarchy and to undertake the decision based on asking several (possible) neighboring detectors for an opinion (see Fig. 1) what ensures reliability of the approach (even if the detectors cannot maintain contact among themselves, they still may react to the behavior they sense).

The behavior of the intruder node X is evaluated by its neighbors. Based on the overhearing of the X's transmission, the decision algorithms implemented inside nodes A, B and C, after consulting among them any of these nodes may report the invader to specific authorities (what will usually involve sending a message to the administrator).

3.2 Detection Algorithm Overview

The main task of the detectors is to perform monitoring of the routed and overheard packets (that are received by neighboring nodes) and build a certain model of normal (or abnormal, depending on the actual detection algorithm used) behavior of the system. Then, current behavior of certain nodes is evaluated based on the model. When a sudden change of certain node's behavior occurs, the alarm or need for consulting is raised.

Specific algorithm of adapting the collected normal behavior should be also considered. Generally the course of the algorithm should be optimized in order to sense fast changes in the behavior of neighboring nodes, and to adapt to the slow ones.

In Fig. 2 the structure of the detector is shown. General aspects of the algorithm must be supplemented with specific anomaly detection algorithm that would be able to construct the behavior model and to perform certain reasoning in order to classify unknown behavior of the neighbors.

4   Behavior Model and Anomaly Detection

4.1   Behavior Pattern Model

One of the most important thing in IDS is to propose specific behavior pattern creation what would let to evaluate neighbor behavior. Le Boudec and Sarafijanovic propose the approach based on classification of aggregated count of packets overheard during specific period of time [2].

The approach however considers only one routing protocol (AODV), what makes their approach improper for the other popular protocols (e.g. OLSR or ZRP). Le Boudec and Sarafijanovic use extensively biological inspiration, though the universal approach should be independent of the detection algorithm. Besides, following algorithm should allow to consider any routing protocol in order to create more adaptable and universal IDS. In this section, the algorithm for constructing behavior patterns for the nodes in MANET will be presented.

In order to capture the behavior in a certain period of time, first, specific packet signature is constructed. Packet signature is a way of describing certain number of similar packets overheard in the network. Packet signature may be described as a vector of values

$$PS = ATR^k \tag{1}$$

where k is length of the packet signature and ATR is one of the spaces described below (in fact the contents of this Cartesian product may be further adapted and extended according to the specific type of network):

- SRC, DST – source and destination identification, may be IP address, MAC address or other unique ID ($SRC, DST \subseteq \mathbb{N}$).
- DIM – distance mark describing how far (e.g. in hops) are interlocutors (when the protocol allows to get this information) ($DIM \subseteq \mathbb{N}$).
- PTF, PTT – port number from (to) describing the range of the ports that the packet is sent from ($PTF, PTT \subseteq \mathbb{N}$).
- PYS – payload size ($PYS \subseteq \mathbb{N}$)).

- PYT – payload type ($PYT \subseteq \mathbb{N}$)).

E.g. packet signature may look as follows:

$$PS_1 = (atr_1, atr_2, \ldots, atr_7) = (10, 12, 5, 1003, 1005, 128, 'CBR') \qquad (2)$$

being a vector described in the following packet signature space:

$$PS = SRC \times DST \times DIM \times PTF \times PTT \times PYS \times PYT \qquad (3)$$

In order to capture the behavior during specific time, packet signatures are aggregated based on the receiver's ID and presented in the following form:

$$B_1 = \{(PS_1, NO_1), (PS_2, NO_2), \ldots\} \qquad (4)$$

where B_i is behavior of the node i and NO_i is number of PS_i gathered in a specific period of time (it may be also frequency or value any other function dependent on the number of packet signatures). B_i is in fact a vector described in the following space:

$$B = APS^k = (PS \times \mathbb{R})^k \qquad (5)$$

where:

- k is maximal number of packet signatures aggregated in one behavior pattern.
- APS is aggregated packet signature (value describing number of packet signatures is added at the end of the vector).

Packets which are aggregated into a specific group being the part of the behavior based on certain similarity measure:

$$SIMAPS : APS^2 \to \mathbb{R} \qquad (6)$$

Range of this function may be constrained (e.g. to the interval $[0, 1]$) in order to clearly state the maximal, minimal and medium values of similarity. This similarity function depends on the following similarity measure used to discover whether two attributes are similar:

$$SIMATR : ATR^2 \to \mathbb{R} \qquad (7)$$

In order to evaluate the similarity of the behavior patterns (what is needed to implement several detector algorithms, e.g. immunological–based ones) similar function should be defined:

$$SIMB : B \times B \to \mathbb{R} \qquad (8)$$

Range of this function may also be constrained (e.g. to the interval $[0, 1]$) for the same reason as mentioned above.

4.2   Anomaly Detection Algorithm

Although any anomaly detection algorithm may be employed by detector, for the current prototype implementation and generation of experimental results, immune-based anomaly detection algorithm was used. Based on the several similar approaches presented i.a. by [1] negative selection algorithm was used.

Negative selection requires construction of self and non-self behavior patterns. Self patterns are constructed in a way described in 4.1. Every detector maintains a dataset with the collection of self patterns (normal behavior) collected during normal course of network operation, and non-self patterns (anomalous behavior) which are generated randomly with use of specific similarity measure. I.e. the non-self set of behavior patterns contains only these patterns that are not similar to any of self patterns (by the means of similarity function described by equation 8). One of possible implementation of this similarity function may look as follows:

$$SIMB(bp_1, bp_2) = \frac{\sum_{aps_1 \in bp_1, aps_2 \in bp_2} SIMPS(aps_1, aps_2)}{\#bp_1 \cdot \#bp_2} \qquad (9)$$

where:

- $bp_1, bp_2 \in B$
- $\#bp_1$ is count of elements in the set bp_1 (count of aggregated packet signatures).

Using this equation the similarity of the two behavior patterns may be determined. The denominator was introduced in order to scale the output to the interval $[0, 1]$, so, for the same patterns the function will return value 1. In order to complete the definition, $SIMPS$ function must be stated, e.g. as follows:

$$SIMAPS(aps_1, aps_2) = \frac{1}{k+1} \cdot \#S \qquad (10)$$

where:

- $S = \{(atr_{1i}, atr_{2i}) | SIMATR(atr_{1i}, atr_{2j}) > t\}$ – is a set of tuples containing corresponding attributes of aps_1 and aps_2 (the same value of index i),
- $t \in [0, 1]$ is a similarity threshold,
- $i \in \mathbb{N}$.

After collecting of the self behavior patterns the detector starts to monitor the communication of the neighboring nodes and report the anomalous behavior (behavior that is similar to one of its non-self pattern) to neighboring detectors (consulting) or to the end-user (alarm).

During the consulting some of the non-self patterns may be exchanged among the detectors, in order to spread the knowledge about behavior throughout the detectors set.

4.3   Collective Decision

After stating that the behavior of the observed neighbor is unfavourable, the procedure of collective decision is started, that consists in consulting of the neighboring detectors, when detector discovers that the observed behavior of a neighbor is anomalous. When the answer to the question is returned, the detector includes it into consideration. Then the possible action of the detector may be determined using different collective intelligence managing techniques (e.g. Winner Takes All, when the decision of the most reliable neighbor is the most important in consultation, or Winner Takes Most where the average decision of the neighbors is taken into consideration) inspired by [9].

The decision undertaken by a detector should however base not only on the collective voting techniques because in the environment where no-one has found any intruders, the collective will never decide to raise an alarm when one of the members will find an intruder. Another thing is the autonomy of the agents-detectors, which relieves them from relying completely on external information. Instead, the detector should maintain a database describing the behavior of its observed neighbors. It must be of course dynamically modified because of the changes of the network topology. The information contained there will be volatile. Anyway, after spotting several subsequent unfavorable activities of the neighbor, the detector should raise an alarm without consulting the collective.

Voting-based techniques require some sort of global control mechanism (that should assign the weights to the detectors), which is undesirable in this kind of distributed environment. The one rational possibility is to introduce second level of detectors, so called „super-detectors", that should maintain a database of their neighboring detectors and apply specific reliability weight that might be used in order to help the collective to undertake the decision. In this way hierarchical structure of the detectors will be introduced, however it should not be strict because of the dynamic nature of the network. Instead, the super-detector should be chosen collectively from the group of neighboring detectors and their function might not rest forever (they might be reduced in the future to the role of simple detector). The reliability weight of the certain detector might be changed only by the super-detector. These ideas are now considered as subjects of further research.

4.4   Adaptation and Pattern Exchanging

Detector maintains his own measurements of the collective agreement (e.g. in a very simple case it stores the information, how many times his own decision was similar to the decision of the collective). After observing the measure of the collective decision and changes in the input data (using specific self–pattern matching measures) it may try to send or acquire some of behavior patterns and broadcast an offer to perform such an exchange to neighboring detectors. The detector may also decide to drop some of its patterns and regenerate them from scratch in order to adapt to the changes present in the environment.

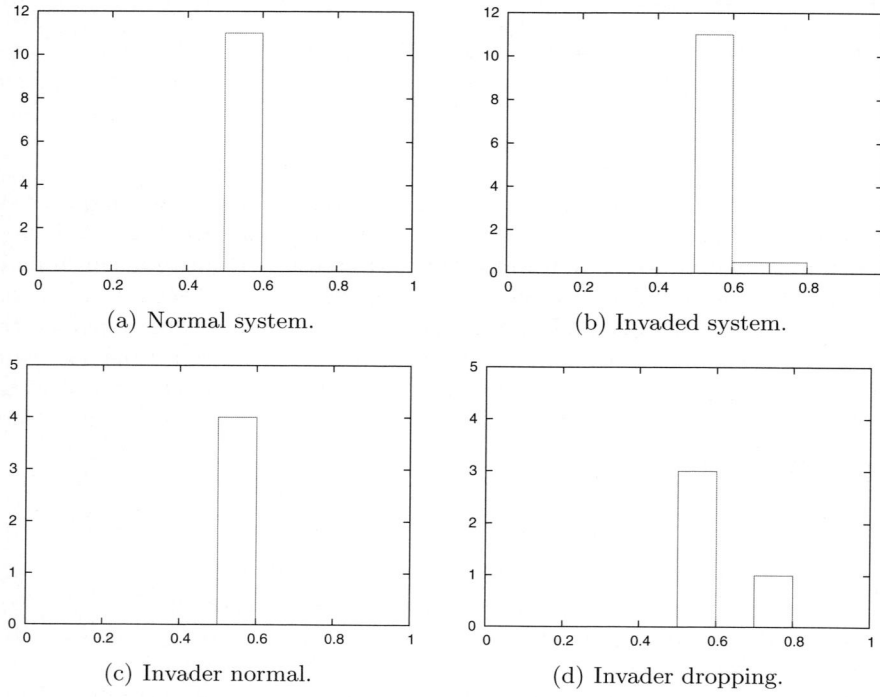

Fig. 3. Count of the test patterns highly similar to non-self patterns (Y-axis) in the similarity range (X-axis) for the whole system (a,b) and invader (c,d)

5   Preliminary Experimental Results

The simulation was performed with using NS-2 network simulator. MANET routing protocol AODV was used along with 802.11 wireless communication. Specific simulation environment consisted of 30 agents organized in three concentric circles, rotating in different directions. There was one node (detector) in the center that received the information sent from one node located outside the circles. Normal behavior of the environment consisted in observing the transmission by the detector, building a behavior model during 100 s of simulation. In order to simulate anomalous behavior, one node from the most central circle stopped forwarding (started dropping) the packets after 50 s of simulation. The behavior pattern results were collected and displayed in tables and histograms presented below. Histograms presented In Fig. 3 show the number of the non-self matching of the behavior patterns collected in the system with normal and anomalous behavior. The data was collected for two observed nodes. The examined test results were gathered for the whole system before (see Fig. 3(a)) and after intrusion (see Fig. 3(b)). The graph changes, there are more patterns similar to non-self patterns during the intrusion. Then the evaluation of single intruder node was performed. Comparing Fig. 3(c) and Fig. 3(d) yields, that higher matching among test patterns and non-self patterns occurs during the intrusion.

6   Conclusion

In this paper we have introduced and discussed an agent-based architecture of IDS for MANETs. In our approach, an intelligent agent-based system is augmented with an immune system-based anomaly detection algorithm. Our preliminary NS-2 based experimental results were encouraging, and seem to indicated that the proposed system can be effectively used to detect abnormal behavior in MANET environments. In continuation of this effort, we will expand our simulation analysis to include more complex network scenario and traffic patterns.

References

1. Dasgupta, D.: Artificial Immune Systems and Their Applications. Springer, New York (1998)
2. Boudec, J.Y.L., Sarafijanovic, S.: An artificial immune system approach to misbehavior detection in mobile ad hoc networks. In: Ijspeert, A.J., Murata, M., Wakamiya, N. (eds.) BioADIT 2004. LNCS, vol. 3141, pp. 396–411. Springer, Heidelberg (2004)
3. Anderson, J.: Computer security threat monitoring and surveillance. Technical report, James P. Anderson Co., Fort Washington, PA (1980)
4. Denning, D.E.: An intrusion-detection model. IEEE Trans. Softw. Eng. 13(2), 222–232 (1987)
5. Ilgun, K., Kemmerer, R., Porras, P.: State transition analysis: A rule-based intrusion detection approach. Software Engineering 21(3), 181–199 (1995)
6. Jackson, K., DuBois, D., Stallings, C.: An expert system application for detecting network intrusion detection. In: Procedings of the 14th National Computer Security Conference, pp. 215–225 (1991)
7. Sterne, D., et al.: A general cooperative intrusion detection architecture for manets. In: IWIA 2005: Proc. of the Third IEEE Int. Workshop on Information Assurance (IWIA 2005), pp. 57–70. IEEE Computer Society, Los Alamitos (2005)
8. Drozda, M., Szczerbicka, H.: Artificial immune systems: Survey and applications in ad hoc wireless networks. In: Proc. of the 2006 Int. Symposium on Performance Evaluation of Computer and Telecommunication Systems (SPECTS 2006), Calgary, Canada, pp. 485–492 (2006)
9. Rojek, G., Cieciwa, R., Cetnarowicz, K.: Algorithm of behavior evaluation in multi-agent system. In: Sunderam, V.S., van Albada, G.D., Sloot, P.M.A., Dongarra, J. (eds.) ICCS 2005. LNCS, vol. 3516, pp. 711–718. Springer, Heidelberg (2005)
10. Bradshaw, J.M. (ed.): Software Agents. AAAI Press/The MIT Press (1997)

Social Layers in Agents' Behavior Evaluation System

Krzysztof Cetnarowicz[1], Renata Cięciwa[2], and Gabriel Rojek[3]

[1] Institute of Computer Science
AGH University of Science and Technology
Al. Mickiewicza 30, 30-059 Kraków, Poland
cetnar@agh.edu.pl
[2] Department of Computer Networks
Nowy Sącz School of Business — National-Louis University
ul. Zielona 27, 33-300 Nowy Sącz, Poland
rcieciwa@wsb-nlu.edu.pl
[3] Laboratory of Computer Science
AGH University of Science and Technology
Al. Mickiewicza 30, 30-059 Kraków, Poland
rojek@agh.edu.pl

Abstract. Behavior evaluation is an approach to a security problem in a multi-agent system that reflects security mechanisms in a human society. The main idea of this approach is behavior evaluation of all agents existing in society that is done autonomously by every agent belonging to that society. All autonomous behavior evaluations have to be collected and processed in order to create a collective decision of a society of agents. This approach reflects security mechanisms existing in a small society in which every human being has enough possibilities to observe and evaluate all other members of the society. This results in large computational complexity. In this paper a modification to behavior evaluation is presented which involves two simple social layers. Social layers are characteristic for more complex and larger societies and could be a means of lower computational complexity.

1 Introduction

Ethically-social mechanisms play a key role in everyday life of every member in a society. These mechanisms enable to find dishonest and undesirable humans on the basis of continuous observation and evaluation of their behavior, or results of that behavior. In a small society every individual observes and evaluates the behavior of all other observable people. The results of autonomous behavior evaluations form one decision of the whole society e.g. decision to exclude somebody from a group. Security mechanisms in society have the decentralized and distributed character — all individuals make their own autonomous evaluations. The results of those evaluations form one decision of the entire society.

In order to design security mechanisms in a multi–agent system that are similar to those functioning in small human societies, two base problems have to be

M. Bubak et al. (Eds.): ICCS 2008, Part III, LNCS 5103, pp. 594–603, 2008.

solved. The first problem is the design of evaluation mechanisms with which every agent will be equipped. These mechanisms should enable an agent to evaluate the behavior of another agent functioning in society. The results of an agent's behavior are actions which are perceived in a computer system as objects. These objects registered in a certain period of time create a sequence of actions that could be processed in order to qualify whether it is a *good* or a *bad* acting agent in this particular system, in which evaluation takes place. Another problem is management, collecting and processing of results of autonomously made behavior evaluations in order to state if a particular agent, which is possibly variously evaluated by different agents, is generally *good* or *intruder* (also called a *bad* agent).

The solutions of the base problems mentioned in the above paragraph were presented in our earlier work (e.g. [5]). After implementing and testing characteristic of security mechanisms for small societies, 3 problems concerning ethically-social approach emerge:

- an agent with undesirable behavior could be unidentified by a society of evaluating agents as *bad* and in consequence this agent would not be excluded from that society,
- a *good* agent could be mistakenly treated as an *intruder*,
- computational complexity of security mechanisms is too high because of the base nature of ethically-social mechanisms.

However, we could limit the disadvantageous phenomena with the use of e.g. actions sampling (as presented in [1]), or earlier results collection (as presented in [2]), some additional mechanisms which are noticed in societies are still possible to use and implement in the ethically-social behavior evaluation. The idea of dividing all members of society into two (or even more) groups called social layers seems very useful. The individuals in one social layer could evaluate the behavior of all members of the society and decide which individual is an *intruder*. The individuals belonging to the other social layer could not have direct mechanisms to discriminate and remove *intruders*. Clarifying, the first presented layer will be called the remove layer (because of the possibility of direct removing some *bad* agents) and the second presented layer will be called the subordinate layer (because this layer does not have a direct impact on *intruders*' removing).

In the above paragraph only some main assumptions about the idea of social layer are presented. To implement this idea the main criterion for belonging to the remove layer has to be decided. Before presenting those details, the main mechanisms of ethically-social security solutions have to be presented: in Sect. 2 behavior evaluation mechanisms that are built into agents and in Sect. 3 mechanisms of management, collecting and processing of results of behavior evaluations. In Sect. 4 the details of the idea of social layers are presented, which is the main topic of this article. The presented theoretical assumptions are next tested and results of these tests are presented in Sect. 5. The main conclusions of this paper are stated in Sect. 6.

2   The Division Profile

All security mechanisms, which enable an agent to make behavior evaluations are named the *division profile*. Algorithms of the division profile are inspired by immunological mechanisms of T cells generation, which enable to detect some anomalies. In a case of behavior evaluation, immunological intruders detection mechanisms have to operate on observed actions made by evaluated agent. This approach is opposite to the one proposed in e.g. [3,4] in which immunological mechanisms operate on the structure of resources. Another difference between artificial immunology and ethically–social approach is the autonomy of a process (an agent) in a secured system — in artificial immunology approach one detection system is considered for a particular computer system (or sub-system). In ethically–social approach every agent autonomously uses his own instantion of detection mechanisms, what induces the necessity of application of some additional algorithms in order to agree collective decision of all agents.

According to immunological mechanisms of T cells generation the division profile has three stages of functioning: creation of collection of *good* (*self*) sequences of actions, generation of detectors and behavior evaluation stage. In further subsections some key aspects of three mentioned stages are presented. More precise description of the division profile functioning is presented in e.g. [5,6].

2.1   Collection of Good Sequences of Actions

The collection W of *good* sequences of actions of an agent consists of sequences of actions undertaken by this agent. The length of a sequence is fixed to l. Presuming there are stored h last actions undertaken by every agent, own collection W will contain $h - l + 1$ elements. An agent in order to generate the collection W should collect information representing actions undertaken by him in the past. But, on the other hand, an agent in order to evaluate behavior of an other agent has to collect information representing actions undertaken by the evaluated agent. So an agent should have information about all actions made in the system. This information is stored in the table of actions, in which every agent is equipped. In the table of actions there are stored last h actions of every visible agent.

2.2   Generation of Detectors

The generation of detectors of an agent happens when an agent first 'knows' his last h actions, so after this agent has undertaken h actions. The algorithm of detectors generation uses the negative selection — from set R_0 of generated sequences of length l those matching with any sequence from collection W are rejected. At the start of presented process set R_0 contains every possible sequence. Sequence matching means that elements of those sequences are the same. Sequences from set R_0 which will pass such a negative selection create a set of detectors R.

2.3 Behavior Evaluation Stage

Once detectors of an agent have been generated, this agent can evaluate behavior of an other agent. The result of behavior evaluation process of an evaluating agent a is a coefficient attributed to an evaluated agent k. This coefficient marked as m_a^k is a number of counted matches between:

 - detectors of the agent a which evaluates behavior,
 - sequences of actions undertaken by the agent k (this sequences of actions are taken from the table of actions of the agent a).

Marking the length of a detector as l and the number of stored actions as h, the coefficient m_a^k is a number from a range $\langle 0, h - l + 1 \rangle$. The maximum of counted matches is equal $h - l + 1$, because every fragment of sequence of actions, which has length equal to the length of a detector, can match only one detector.

3 Mechanisms of Distributed Evaluations Agreement

An algorithm of agents evaluations management, collection and processing is used to agree one common decision of all agents, which belong to the remove layer. The difficulty in this agreement is caused by the fact that an agent could be differently evaluated by various agents. The discussion of this problem is presented in [5], in this section are presented only key information essential to discuss the main topic of this article.

Each action undertaken by an agent may cause change of the results of behavior evaluations that are done by other agents in the system. This approach lets us formulate *the algorithm of evaluation management* as follows: If an agent k belonging to any social layer undertakes an action, a request of evaluation of the agent k is sent to all agents (except the agent k) in the remove layer, which have direct impact on agent removing.

An agent a in case of receiving a request of evaluation of an agent number k sends back only the coefficient o_a^k in the range $0 \le o_a^k \le 1$. The coefficient o_a^k is given by function:

$$o_a^k = \left(\frac{m_a^k}{h - l + 1} \right)^4 \tag{1}$$

where $h - l + 1$ is the maximum of counted matches of agent a. The power function of evaluation behavior increases a weight of high coefficient m_a^k (the exponent was set empirically).

In order to decide if the agent k is in general *good* or *bad* the environment uses *the algorithm of evaluation's collecting and processing*, which consists of following actions:

1. All results of behavior evaluations are stored (that results are sent by agents in response to the request of evaluation of the agent k).

2. Gained coefficients are summed and then this sum is divided by the number of agents which got the request of evaluation. If this obtained number is greater than $\frac{1}{2}$ agent k is eliminated.

4   Social Layers

The presented work focuses on the idea of two coexisting social layers: the remove layer and the subordinate layer. Agents belonging to the remove layer have direct impact on removing of all agents (of all social layers) existing in the environment. Agents belonging to the subordinate layer can only evaluate behavior, but do not have a possibility of presenting their results in order to remove the agents.

In order to make this idea implementable in our ethically-social security system, the criterion for agents to belong to the remove layer has to be stated. Hypothetically, the criterion could be 'experience' of an agent — agents which have been in the secured system long enough could have the right to evaluate and, on this basis, eliminate other agents from their environment. Another criterion is also possible — only those agents are chosen to the remove layer, which evaluations results mostly conform to the opinion of the whole society of agents. Checking this criterion and changing of agents between social layers can be done permanently after some constant time periods. The short discussion presented here does not include all variants of possible criteria. The research presented in this article focuses on the second mentioned criterion.

Analyzing presented ideas, *the algorithm of determination which agent will belong to the remove layer* could be presented as follows:

1. During the first $h + 1$ constant time periods Δt all agents belong to the remove layer.
2. Afterwards, during the next *checking_time* the agents observe each other's evaluation results. If an agent's opinion is the same as the opinion of the whole society it increases its social rank by 1 point.
3. In each $h + 1 + n * checking\_time$ (n=1,2,..) constant time period Δt 25 per cent of agents with the highest social ranks are chosen. Only these agents form the remove layer. The opinions of individuals belonging to this social layer are taken into consideration in the process of distinction between *good* entities and *intruders*.
4. The social ranks of all agents are reset. The society of agents acts in accordance with the algorithms of management, collecting and processing of results of behavior evaluations presented in Sect. 3. Nonetheless, in the randomly chosen time periods the whole society is requested to evaluate the behavior of an agent a with the purpose of establishing the social ranks of all agents existing in the society. If it happens that any agent belonging to the remove layer is deleted, an agent from the subordinate layer with the highest social rank is moved to the remove layer.
5. The steps number 3. and 4. are repeated.

In the research presented below tests are performed in which *checking_time* is equal to 10, 100 constant time periods Δt. Moreover, the subordinate layer is not relieved from the duty of constant behavior evaluation of all other entities in the society despite the fact that their opinions are not taken into consideration in the process of making decision of an agent removing. Consequently, such an approach to the algorithm presented above does not decrease the computational complexity of the security mechanisms, but lets us choose the agents to the remove layer with a high degree of precision.

5   Results of Experiments

A multi–agent system was implemented in order to test the security mechanisms existing in a society divided into social layers . The environment of designed system has two types of resources: type A and type B. Resources are used by agents independently, but refilling of all resources is only possible when every type of resources reaches the established low level. The researched system reflects operations in a computer system which should be executed in couples e.g. opening / closing a file. There are a lot of attack techniques that are limited to only one from a couples (or trios...) of obligatory operations (e.g. SYN flood attack [7]). The simulated system has three types of agents:

- *type 50/50 agents* – agents which take one unit of randomly selected (A–50%, B–50%) resource in every full life cycle; only this type of agents needs resources to refill its energy (if energy level of a 50/50 agent wears off, this agent will be eliminated)
- *type 80/20 agents* – agents which take one unit of randomly selected (A–80%, B–20%) resource in every full life cycle; type 80/20 agents should be treated as intruders because the increased probability of undertaking actions of one type can block the system;
- *type 100/0 agents* – agents which take one unit of A resource in every full life cycle; type 100/0 agents are also called intruders.

To some degree, the behavior of 80/20 agents is similar to the behavior of 50/50 agents but is undesirable in the secured system like intruders behavior. In all experiments presented here there are initially 80 agents of type 50/50, $10 - 80/20$ agents and $10 - 100/0$ agents. All agents in the system are equipped with the division profile mechanisms with parameters $h = 18$ and $l = 5$. The simulations are run to 1000 constant time periods Δt and 20 simulations were performed. Diagrams presented in the next paragraphs show the average taken from 20 simulations.

In the experiments presented below we compare the results obtained in simulations of a homogeneous society and societies divided into two social layers with different *checking_time* fixed at 10 constant time periods Δt in one case and 100 in the other one.

5.1   The Phenomenon of Self–destruction

In particular situations a *good* agent could be mistakenly treated as an *intruder*. Such a problem is a consequence of the random choice of undertaken actions. As a result, some sequences of actions of *good* agents can be similar to actions of *bad* agents. This phenomenon has been named *the phenomenon of self–destruction*.

Several tests were performed in order to check what is the level of mentioned phenomenon depending on the *checking_time* of remove layer of agent society. Afterwards, the results of these experiments were compared with the results obtained for a homogeneous society. The diagram in Fig. 1 shows the average number of agents type 50/50 in separate time periods.

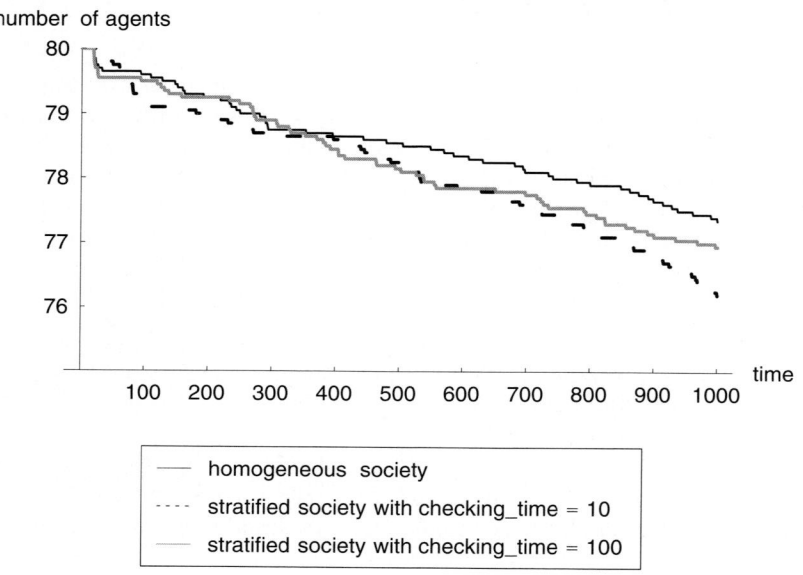

Fig. 1. Number of type 50/50 agents in separate time periods

In the homogeneous society the level of self–destruction was equal 3,31% which means that on average not more than 3 *good* agents were mistakenly treated as *intruders* and removed from the implemented system. The simulations of the society divided into two social layers showed that the level of the self–destruction phenomenon slightly increased – 4,75% in case of *checking_time* set to 10 and 3,81% for *checking_time* set to 100 constant time periods Δt. However, the rate of agents' removing tends to be higher during the early stage of the system with *checking_time* equals 10. Such a problem could stem from the fact that with so short *checking_time* it is very difficult to differentiate *good* agents from *type 80/20 agents*. Therefore, *intruders* can be chosen to the remove layer and, consequently, have the direct impact on removing of other agents.

The presented research indicate that the social layers have not significant effect on the phenomenon of self–destruction. Nevertheless, the *checking_time* of the society should be carefully chosen in order to recognize *intruders* more precisely.

5.2   The Rate of Intruders Detection

In every security system, it is crucial to recognize *bad* entities as soon as possible
and remove them from the environment. In some cases an agent with undesirable
behavior could be not identified by a society of evaluating agents as *bad* and, as
a result, this agent would not be excluded from that society.

In our simulations *type 100/0 agents* were detected during the first 28 constant
time periods Δt. Thus, when the system achieved the behavior evaluation stage
all *type 100/0 agents* were identified properly and eliminated from the system
when they tried to undertake actions.

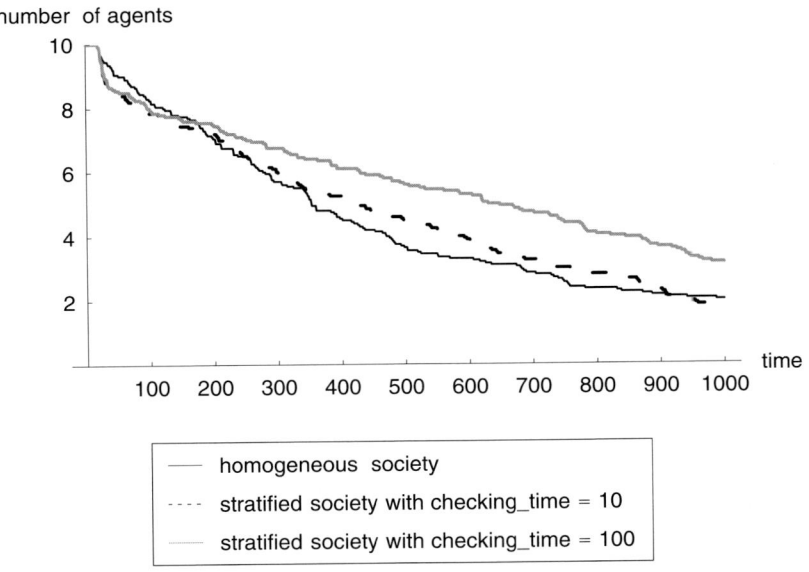

Fig. 2. Number of type 80/20 agents remained in the system in separate time periods

However, the precise recognition of *type 80/20 agents* is more difficult and
takes more time. The division between *50/50 agents* and *80/20 agents* is hin-
dered by random character of agents decision which resource to undertake (some
solutions of this problem were suggested in [2,6]). The diagram in Fig. 2 shows
the average number of agents type 80/20 remained in the system in separate
time periods.

In the homogeneous society the level of *intruders* elimination was equal 80%.
The simulations of the society divided into two social layers with *checking_time*
equals 10 showed that this level insignificantly increased to 82,5%. In the case
of the stratified society with *checking_time* equals 100 the level of *intruders*
detection decreased to 68,5%. During the initial stage of simulations the rate
of *bad* agents removing was similar to the results obtained in other mentioned
cases. However, this rate seems to be reduced at the moment of dividing the

society into two social layers. Such a problem could be caused by the fact that the *checking_time* is too long. Therefore, if the *type 80/20 agent* hadn't been recognized as an intruder before the division of the society, it could act almost not endangered during the next 100 constant time periods Δt due to the fact that the agents, which formed the remove layer probably do not possess detectors to identify its malicious behavior.

6   Conclusion

Some modifications in the ethically–social security approach were presented in this paper. The modifications are named social layers, because society of agents is divided into two coexisting groups: remove layer that consist of agents, which can make behavior evaluations and have direct impact on *intruder* removing and subordinate layer that consists of agents, which do not have direct impact on *intruder* removing (but can make behavior evaluations). The implementation of idea of social layers presented in the paper contains the criterion which agent should belong to the remove or subordinate layers. The researched criterion is connected with the opinion that only those agents should belong to the remove layer, which behavior evaluations are the closest to all evaluations undertaken in the secured society.

The main field of our interest was how the introducing of social layers would influence on the base problems of the ethically-social security mechanisms. The results of experiments were presented for three cases:

- all agents belong to the remove layer (called in this paper homogeneous society),
- only 25% of agents belong to the remove layer and the criterion of belonging to this layer is checking every 10 constant time periods Δt,
- only 25% of agents belong to the remove layer and the criterion of belonging to this layer is checking every 100 constant time periods Δt.

The obtained results indicate that the implementation of social layers in ethically–social security system does not have a significant effect on the self–destruction phenomenon and the rate of intruders detection. However, the time of checking which agent should belong to which layer should be carefully chosen in order to recognize *intruders* precisely and eliminate them from the system as soon as possible.

To conclude, the idea of social layers seems very interesting because it does not make security mechanisms worse and makes it possible to reduce the computational complexity. The computational complexity could be reduced due to the fact that the agents belonging to the subordinate layer do not have to make behavior evaluations every time, which could be the field of our future research.

Acknowledgments

This work is partially supported by the Ministry of Science and Higher Education of Poland, grant No. 3 T08B 042 29.

References

1. Cetnarowicz, K., Cięciwa, R., Rojek, G.: Behavior Evaluation with Actions' Sampling in Multi–agent System. In: Pěchouček, M., Petta, P., Varga, L.Z. (eds.) CEEMAS 2005. LNCS (LNAI), vol. 3690, pp. 490–499. Springer, Heidelberg (2005)
2. Cetnarowicz, K., Cięciwa, R., Rojek, G.: Behavior Evaluation with Earlier Results Collection in Multi–agent System. In: Proceedings of The Agent Days 2005, Malaga, July 7-8, 2005, pp. 77–84 (preprint, 2005)
3. Forrest, S., Perelson, A.S., Allen, L., Cherukuri, R.: Self-nonself Discrimination in a Computer. In: Proc. of the 1994 IEEE Symposium on Research in Security and Privacy, pp. 202–212. IEEE Computer Society Press, Los Alamitos (1994)
4. Hofmeyr, S.A., Forrest, S.: Architecture for an Artificial Immune System. Evolutionary Computation 7(1), 45–68 (2002)
5. Rojek, G., Cięciwa, R., Cetnarowicz, K.: Algorithm of Behavior Evaluation in Multi-agent System. In: Sunderam, V.S., van Albada, G.D., Sloot, P.M.A., Dongarra, J. (eds.) ICCS 2005. LNCS, vol. 3516, pp. 711–718. Springer, Heidelberg (2005)
6. Rojek, G., Cięciwa, R., Cetnarowicz, K.: Heterogeneous Behavior Evaluations in Ethically–Social Approach to Security in Multi-agent System. In: Alexandrov, V.N., van Albada, G.D., Sloot, P.M.A., Dongarra, J. (eds.) ICCS 2006. LNCS, vol. 3993, pp. 823–830. Springer, Heidelberg (2006)
7. Schetina, E., Green, K., Carlson, J.: Internet Site Security. Addison-Wesley Longman Publishing Co., Boston (2002)

Graph Transformations for Modeling *hp*-Adaptive Finite Element Method with Triangular Elements

Anna Paszyńska[1], Maciej Paszyński[2], and Ewa Grabska[1]

[1] Faculty of Physics, Astronomy and Applied Computer Science,
Jagiellonian University,
ul. Reymonta 4, 30-059 Kraków Poland
anna.paszynska@uj.edu.pl
[2] Department of Computer Science
AGH University of Science and Technology,
Al. Mickiewicza 30, 30-059 Kraków, Poland

Abstract. The paper presents composition graph (CP-graph) grammar, which consists of a set of CP-graph transformations, suitable for modeling triangular finite element mesh transformations utilized by the self-adaptive *hp* Finite Element Method (FEM). The *hp* adaptive FEM allows to utilize distributed computational meshes, with finite elements of various size (thus *h* stands for element diameter) and polynomial orders of approximation varying locally, on finite elements edges and interiors (thus *p* stands for polynomial order of approximation). The computational triangular mesh is represented by attributed CP-graph. The proposed graph transformations model the initial mesh generation, procedure of *h* refinement (breaking selected finite elements into son elements), and *p* refinement (adjusting polynomial orders of approximation on selected element edges and interiors). The graph grammar has been defined and verified by implemented graph grammar transformation software tool.

1 Introduction

The Composite Programable graph grammar (CP-graph grammar) was proposed in [1], [2], [3] as a tool for formal description of various design processes. The CP-graph grammar expresses a design process by graph transformations executed over the CP-graph representation of the designed object. In this paper, a new application of the CP-graph grammar is proposed, where the grammar is utilized to describe mesh transformations occuring during the self-adaptive *hp* Finite Element Method (FEM) calculations. The paper is an extension of the CP-graph grammar model introduced in [4], [5] for rectangular finite element meshes. The developed self-adaptive *hp* FEM [6], [7], [8], [9], [10] is an evolvable system that generates a sequence of optimal *hp* meshes, with finite elements of various size and polynomial orders of approximations varying locally on finite element edges, faces and interiors. The generated sequence of meshes delivers

M. Bubak et al. (Eds.): ICCS 2008, Part III, LNCS 5103, pp. 604–613, 2008.

exponential convergence of the numerical error with respect to the mesh size (number of degrees of freedom or CPU time). Each *hp* mesh is obtained by performing a sequence of *h* and *p* refinements on the predefined initial mesh. The *h* refinement consists in breaking selected triangular finite element into four son elements. The *p* refinement consists in increasing polynomial order of approximation on selected edges, faces and/or interiors. The process of generation of an

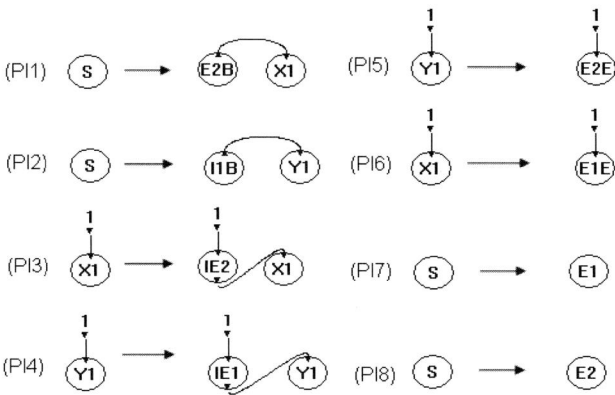

Fig. 1. Graph grammar productions responsible for an initial mesh generation

hp mesh is formalized by utilizing CP-graph grammar. This allows to express mesh transformation algorithms by means of graph transformations managed by control diagrams. This involves the generation of the initial mesh, as well as *h* and *p* refinements. The formalization of the process of mesh transformation allows to code the mesh regularity rules on the level of graph grammar syntax, which simplifies the algorithms and prevents computational mesh to be inconsistent.

The computational triangular mesh is represented by attributed CP-graph. The graph grammar consists of a set of graph transformations, called productions. Each production replaces a sub-graph defined on its left-hand-side into a new sub-graph defined on its right-hand-side. The left-hand-side and right-hand-side sub-graphs have the same number of free in/out bounds (bounds connected to external vertices). The corresponding free in/out bounds have the same number on both sides of a production. Thus, the embedding transformation is coded in the production by assuming the same number of free bounds on both sides of the production. The execution of graph transformations are controlled by control diagrams prescribing the required order of productions. The presented graph grammar has been defined and verified by the implemented software tool.

2   Graph Transformations for Modeling the Initial Mesh Generation

In this section, the subset of graph transformations, modeling generation of an arbitrary hp refined mesh, based on initial mesh with horizontal sequence of triangular finite elements is presented. The described graph transformations can be generalized into a case of arbitrary two dimensional initial mesh. The process of the initial mesh generation is expressed by the graph transformations presented in Fig. 1. There are two types of triangular finite elements, presented in Fig. 2. Each triangular finite element consists in three vertices, three edge nodes and one interior node.

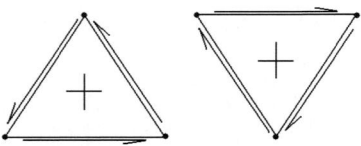

Fig. 2. Two types of triangular finite elements. Each element consists in 3 vertices, 3 edge nodes and one interior node.

Once the sequence of initial mesh elements is generated, the structure of each element is produced, as it is expressed by productions presented in Fig. 3. The V label stands for an element vertex, F stands for an element edge (face), $I1$ and $I2$ stand for interiors for two types of elements, respectively. If an element is adjacent to mesh boundary, then its free bounds are connected to B labeled graph vertex, denoting the boundary, as well as to FAL labeled vertex, denoting missed second father of the edge (edges located inside the domain have two father elements). There is the similar production for the second element type. There are also similar productions for elements $E1B$, $E2B$, $E1E$, $E2E$ located at the beginning and the end of the sequence.

We can identify common edges of adjacent finite elements. The production that is identifying two adjacent elements, and actually removing one duplicated edge, is presented in Fig. 4. The identification is possible, since we keep finite elements connectivity at the top level of the graph.

3   Graph Transformations for Modeling Mesh Refinements

Once the structure of triangular finite elements is generated, we can proceed with mesh refinements in the areas with strong singularities, where the numerical error is large. The decision about required refinements is made by knowledge driven artificial intelligence algorithm [6], [7], [8], [9], [10]. The selected finite elements can be either h, p or hp refined.

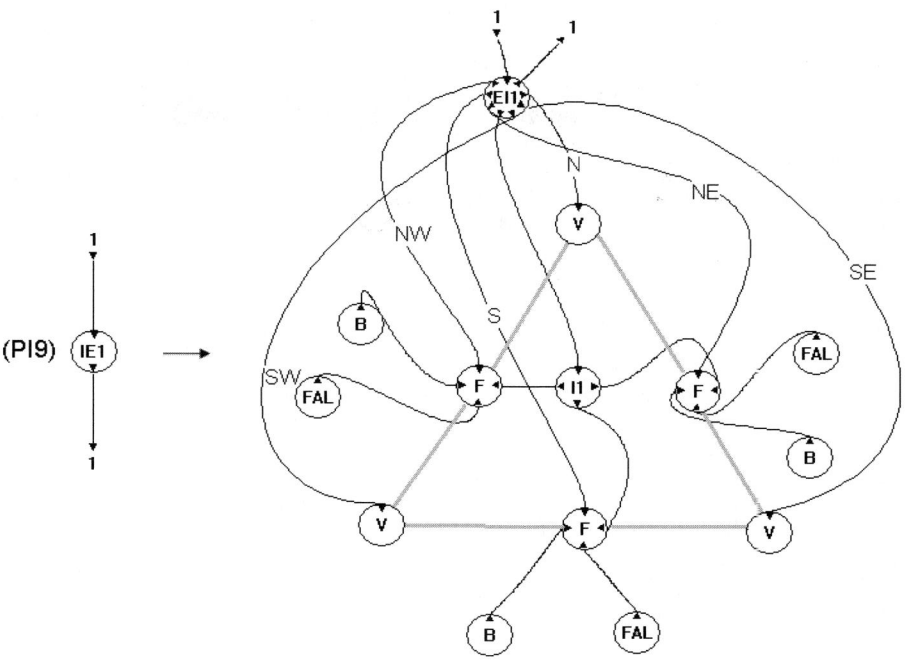

Fig. 3. Production for generating the structure of the first element type

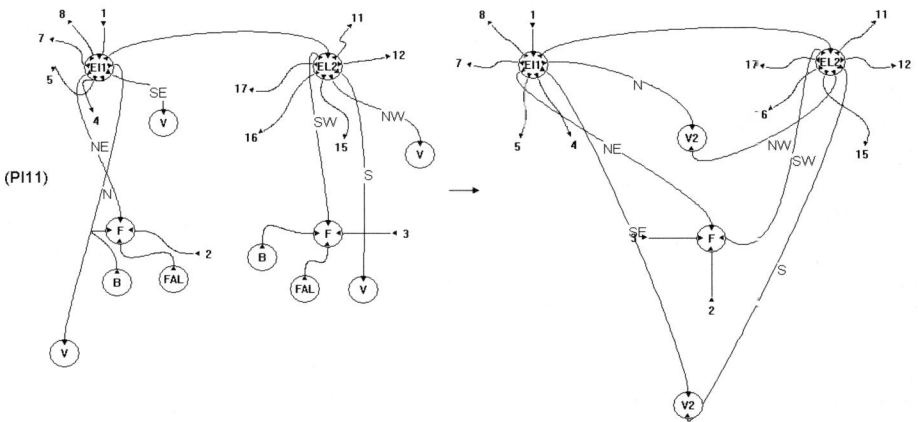

Fig. 4. Production for identifying the common edge of two adjacent elements

The *h* refinement is expressed by breaking element edges and interiors. To break an element interior means to generate four new element interiors, and three new edges, as it is illustrated on middle panel in Fig. 5. To break an element edge

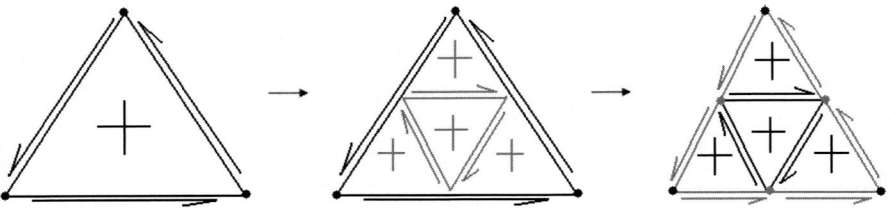

Fig. 5. The h refinement of a triangular element: breaking of the element interior followed by breaking of the element edges

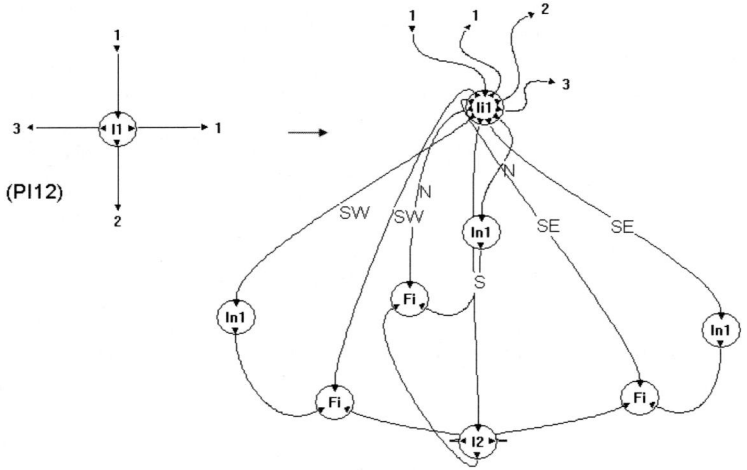

Fig. 6. Graph grammar production for breaking element interior, for the element of the first type

means to generate two new edge nodes and one new vertex, as it is illustrated on right panel in Fig. 6 for all three edges of the original triangular element. These procedures are expressed by (PI12-15) productions presented in Figs. 3-4. The newly created finite elements are never stored in the data structure. They are dynamically localized at the bottom level of generated refinement trees.

The following mesh regularity rules are enforced during the process of mesh transformation, see [6]. An element edge can be broken only if two adjacent interiors have been already broken, or the edge is adjacent to the boundary. This is expressed by (PI14) and (PI15) productions in Figs. 8 and 9. An element interior can be broken only if all adjacent edges are of the same size as the interior. This is expressed by (PI16) production in Fig. 10. The graph vertex JI4 is obtained after breaking all adjacent edges, and propagating adjacency information along the refinement tree, from the father node down to children nodes. The propagation of the adjacency data is obtained by executing six productions, related to three different edges of two types of elements. One of these productions is presented in Fig. 11. In other words, the history of adjacent edges refinements is coded within the label of graph vertex representing element interior. The interior can be broken only after breaking adjacent edges and propagating the adjacency information along the refinement trees. The goal of the mesh regularity rule is to

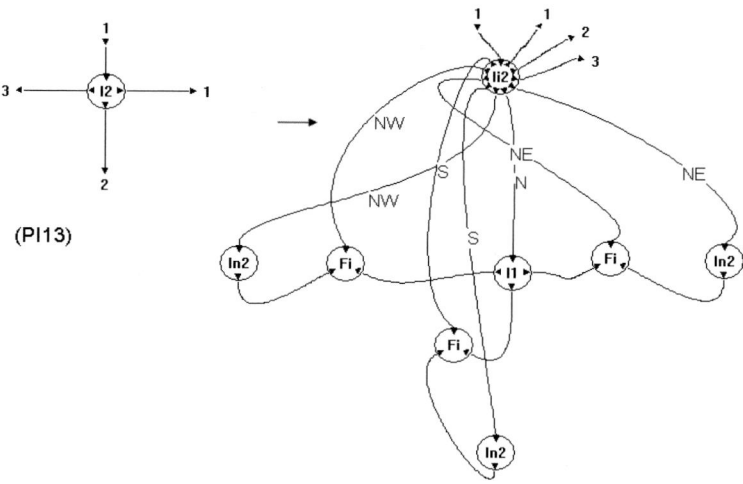

Fig. 7. Graph grammar production for breaking an element interior, for the element of the second type

avoid multiple constrained edges, which leads to problems with approximation over such an edge. The mesh regularity rule enforces breaking of large adjacent unbroken elements before breaking small element for the second time, which is illustrated in Fig. 12. The mesh regularity rules are enforced on the level of graph grammar syntax.

4 Numerical Example

We conclude the presentation with the sequence of triangular finite element meshes generated for the L-shape domain model problem [6]. The problem consists in solving the Laplace equation

$$\Delta u = 0 \text{ in } \Omega \tag{1}$$

over the L-shape domain Ω presented in Fig. 13. The zero Dirichlet boundary condition

$$u = 0 \text{ on } \Gamma_D \tag{2}$$

is assumed on the internal part of the boundary Γ_D. The Neumann boundary condition

$$\frac{\partial u}{\partial n} = g \text{ on } \Gamma_N \tag{3}$$

(Pl14)

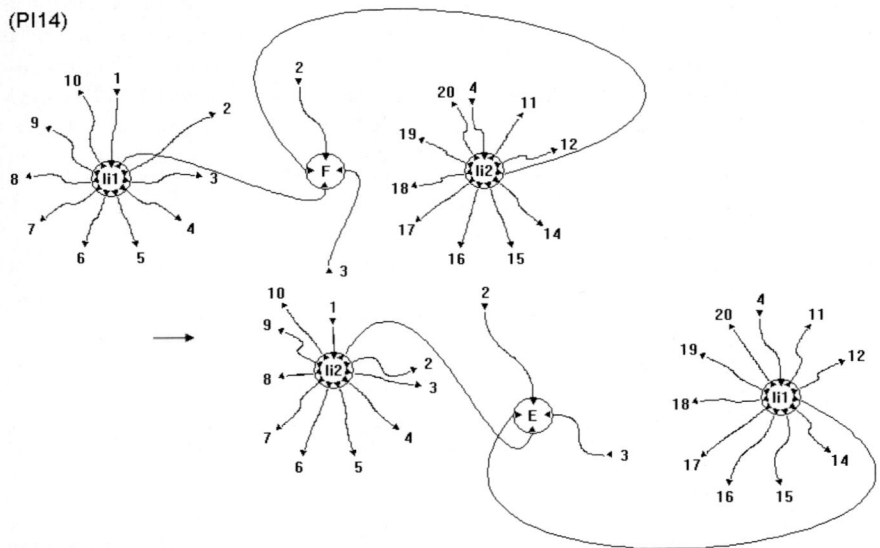

Fig. 8. Graph grammar production allowing for breaking an element edge, for the edge surrounded by two broken interiors

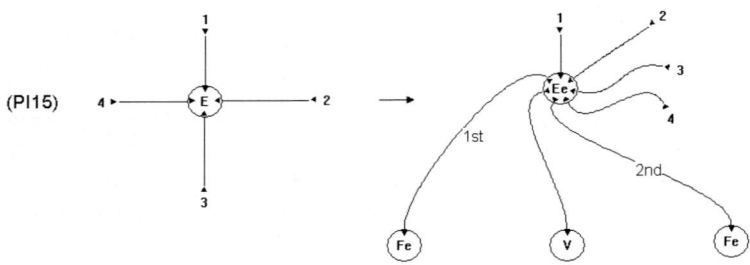

Fig. 9. Graph grammar production for actual breaking of an element edge

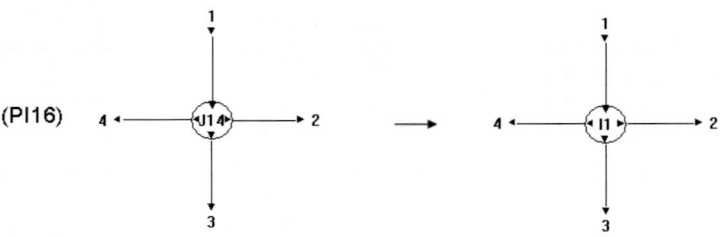

Fig. 10. Graph grammar production allowing for breaking an element interior

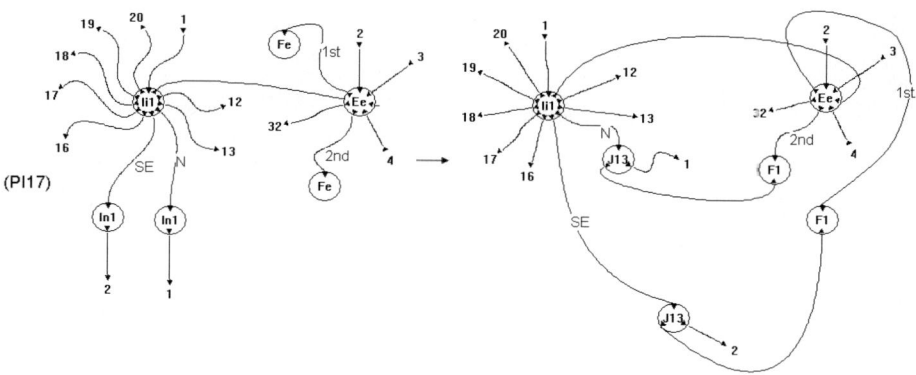

Fig. 11. One of productions executing propagation of the adjacency data along the refinement tree

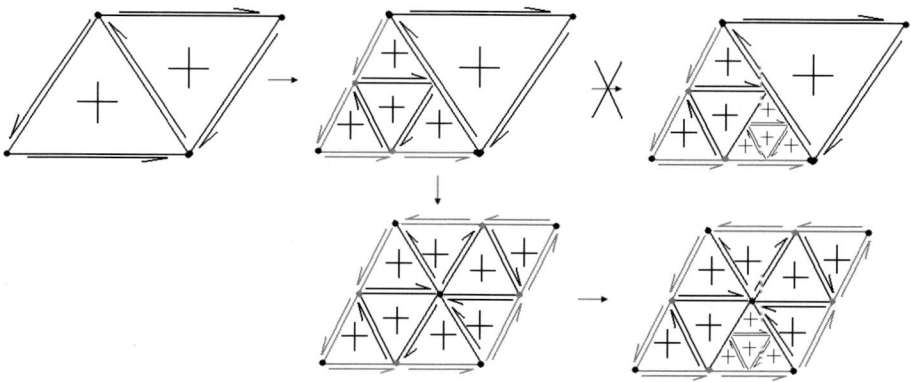

Fig. 12. Mesh regularity rule enforces breaking of large adjacent element before breaking of small element

is assumed on the external part of the boundary Γ_N. The temperature gradient in the direction normal to the boundary is defined in the radial system of coordinates with the origin located in the central point of the L-shape domain.

$$g\left(r, \theta\right) = r^{\frac{2}{3}} \sin \frac{2}{3} \left(\theta + \frac{\pi}{2}\right) . \tag{4}$$

The solution $u : R^2 \supset \Omega \ni u \mapsto R$ is a temperature distribution inside the L-shape domain.

The initial mesh consists in six triangular finite elements, presented on the first panel in Fig. 13. The self-adaptive *hp*-FEM code generates a sequence of meshes delivering exponential convergence of the numerical error with respect to

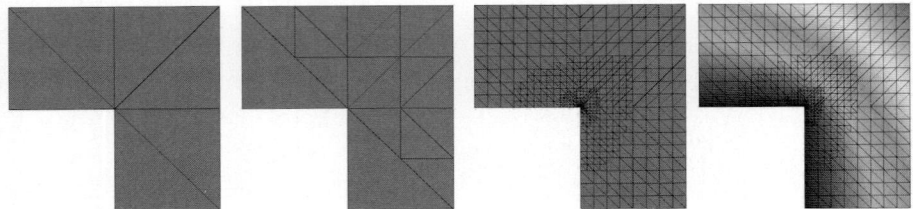

Fig. 13. The sequence of triangular finite element meshes for the L-shape domain problem

the problem size. The initial mesh, the second mesh, the optimal mesh delivering less then 5% relative error accuracy of the solution, and the solution over the optimal mesh are presented in Fig. 13.

5   Conclusions

The CP-graph grammar is the tool for a formal description of triangular mesh transformations utilized by adaptive FEM. It models all aspects of the adaptive computations, including mesh generation, h and p refinements, as well as mesh regularity rules, including elimination of multiple constrained nodes. The technical nightmare with implementing the mesh regularity rules has been overcome by including the mesh regularity rules within the graph grammar syntax. The graph grammar have been formally validated by utilizing graph grammar definition software [3]. The graph grammar can be easily extend to support anisotropic mesh refinements and three dimensional computations.

Acknowledgments. The work reported in this paper was supported by Polish MNiSW grant no. 3 TO 8B 055 29. The work of the first author has been also supported by the Foundation for Polish Science under Homming Programme.

References

1. Grabska, E.: Theoretical Concepts of Graphical Modeling. Part One: Realization of CP-Graphs. Machine Graphics and Vision 2(1), 3–38 (1993)
2. Grabska, E.: Theoretical Concepts of Graphical Modeling. Part Two: CP-Graph Grammars and Languages. Machine Graphics and Vision 2(2), 149–178 (1993)
3. Grabska, E., Hliniak, G.: Structural Aspects of CP-Graph Languages. Schedae Informaticae 5, 81–100 (1993)
4. Paszyński, M., Paszyńska, A.: Graph transformations for modeling parallel hp-adaptive Finite Element Method. In: Parallel Processing and Applied Mathematics, Gdańsk, Poland, September 2007. LNCS, vol. 4967, Springer, Heidelberg (in press, 2008)
5. Paszyński, M.: Parallelization Strategy for Self-Adaptive PDE Solvers. Fundamenta Informaticae (submitted, 2007)

6. Rachowicz, W., Pardo, D., Demkowicz, L.: Fully Automatic hp-Adaptivity in Three Dimensions. ICES Report 04-22, pp.1–52 (2004)
7. Demkowicz, L.: Computing with hp-Adaptive Finite Elements, vol. I. Chapman & Hall/Crc Applied Mathematics & Nonlinear Science, New York (2006)
8. Demkowicz, L., Pardo, D., Paszyński, M., Rachowicz, W., Zduneka, A.: Computing with hp-Adaptive Finite Elements, vol. II. Chapman & Hall/Crc Applied Mathematics & Nonlinear Science, New York (2007)
9. Paszyński, M., Kurtz, J., Demkowicz, L.: Parallel Fully Automatic hp-Adaptive 2D Finite Element Package. Computer Methods in Applied Mechanics and Engineering 195(7-8), 711–741 (2007)
10. Paszyński, M., Demkowicz, L.: Parallel Fully Automatic hp-Adaptive 3D Finite Element Package. Engineering with Computers 22(3-4), 255–276 (2006)

On Some Method for Intrusion Detection Used by the Multi-agent Monitoring System

Agnieszka Prusiewicz

Institute of Information Science & Engineering, Wrocław University of Technology, Poland
Wybrzeże Wyspiańskiego 27, 50-370 Wrocław, Poland
agnieszka.prusiewicz@pwr.wroc.pl

Abstract. In this paper an original method for intrusion detection used by the multi-agent monitoring system is proposed. Due to the complexity of the problem of the network security maintaining it is a good idea to apply the multi-agent approach. The multi-agent system is used in the tasks of computer network monitoring in order to detect the security policy violations. In this paper the algorithm for detecting some type of attack is proposed. This algorithm is based on the network traffic analysis.

1 Introduction

In this paper an original method for intrusion detection used by the multi-agent monitoring system is proposed. The problem of the network security is taken up since eighties [9] and is developed up today [5, 8, 20]. A comprehensive survey of anomaly detection systems is presented in [16].

A comparison of different approaches to intrusion detection systems is given in [4]. The general challenge of the current intrusion detection systems is to find the difference between the normal and abnormal user behaviour. Intrusion detection system (IDS) should not only recognise the previously knows patterns of attack, but also react in case of appearance of the new events that violate the network security policy. The first models of IDS systems were centralised namely data were collected and analyzed on a single machine. However, the distributed nature of the task of the network security monitoring requires applying of the distributed tools for network security maintaining. Many examples of the distributed IDS systems are given in [1].

The most important postulate addressed to the intrusion detection systems is that, such systems should automatically react in case of detecting the security policy breach to prevent the attack execution or to reduce the potential loss in the network systems. IDS systems should be equipped with the components responsible for the permanent observation of the states of monitored nodes and components that integrate the results of these observations and diagnose the security level of the system. It is reasonable to apply the agency for solving the tasks of the network security monitoring. A multi-agent approach in a network monitoring system was used in works [7, 3, 10, 19] where the general system architecture was proposed.

In our approach we also apply the agency to solve the tasks of network security monitoring. In work [11] the framework of an original proposal of the intrusion

M. Bubak et al. (Eds.): ICCS 2008, Part III, LNCS 5103, pp. 614–623, 2008.

detection system based on the multi-agent approach was presented. In particular, the architecture of such system and the task of agents were specified. Proposed ideas were further developed and in the work [14] the problem of anomalies detection on the basis of the nodes traffic analysis was discussed. Then in the work [17] the agents' knowledge organisation and the general idea of the algorithm for detecting the distributed denial of service (*DDoS*) attack was proposed. In this work we develop the algorithm for detecting the distributed denial of service attack.

2   The General Agents' Knowledge Structure

It is assumed that two types of agents are in the monitoring system: managing agents and monitoring agents. Each monitoring agent *ma* is responsible for one monitoring region *mr* consisted of the set of nodes *V*. Managing agents may modify the size of the monitoring regions by adding or removing nodes from the set *V* . It is assumed that the monitoring regions may overlap.

2.1   The Characteristic of the Monitoring Agents

The agent *ma* observes the states of the nodes from his monitoring region *mr* with a reference to the some properties from the set *P*. *ma* stores all the observations in his private database *DB*.

Definition 1. A single observation of agent *ma* is a tuple:

$$O(v,(p,x),t) \in DB \tag{1}$$

where $v \in mr$, $p \in P$, $t \in T$, T is the universe of the timestamps and *DB* denotes the database of the agent *ma*.

Single observation $O(v,(p,x),t)$ refers to the situation that at the timestamp t the agent *ma* has observed in the node v the value of the parameter p equals x [17].

DB consists of the *set of observations* (see definition 1) and *communication data*. *Communication data* consists of the communication matrix *CM* of the timestamps in which the communication between the nodes took place.

The second module embodied in the monitoring agent is *the anomaly detection module* that is supplied by the observations from *the set of observations*. This module is responsible for data analyzing to detect the security policy violation in the network system [17].

Definition 2. An *anomaly detection module* of agent *ma* is defined as a triple:

$$AD = \left\langle DB_{[t_{b'},t_{e'}]}, AT, AC \right\rangle \tag{2}$$

where:

– $DB[t_{b'},t_{e'}]$ is a subset of *DB* restricted to time interval $[t_{b'},t_{e'}]$ and is defined as:

$$DB[t_{b'},t_{e'}] = \{O(v,(p,x),t): p \in P, v \in mr; t \in [t_{b'},t_{e'}]\} \tag{3}$$

For a chosen time interval $[t_{b'}, t_{e'}]$ each tuple $O(v, (p, x), t) \in DB$ that fulfils the following condition: $t \in [t_{b'}, t_{e'}]$ is sent into *the anomaly detection module AD* .

–   *AT* is a table of discovered anomalies defined as:

$$AT = \{A(v, p, \alpha, [t_b, t_e]): p \in P, v \in mr; [t_b, t_e] \subseteq [t_{b'}, t_{e'}]\} \qquad (4)$$

A single anomaly $A(v, p, \alpha, [t_b, t_e]) \in AT$ is interpreted as the agent *ma* believes at the level α that in the time interval $[t_b, t_e]$ in the node v an anomaly connected with the property p occurred.

–   *AC* is an *anomaly archive* consisted of discovered anomalies that are sent from *AT* after attack detection. However only anomalies connected with discovered attack are sent to the archive *AC* . In this way the time interval of discovered anomalies is not a criterion of sending given tuples $A(v, p, \alpha, [t_b, t_e])$ into *anomaly archive*.

Anomaly table *AT* is created for each monitoring region independently. On agents' autonomy assumption it is quite possible that one agent may discover anomaly state of the node v with reference to some property's value, while other agent having the same node in its monitoring area will not observed any anomaly with reference to this property's value.

The third module closed in internal agents' knowledge structure is *ontology module* that consists of the attack patterns. Each monitoring agent subordinates only to one managing agent.

2.2   The Characteristic of the Managing Agents

Managing agents are responsible for coordination and management of the monitoring system. The internal structure of the managing agents is as follows. Each managing agent is equipped with: *data storage module, data analysis module, attack detection module, queue task module* and *communication module*. In *the data storage module* two types of data are stored: data received from the monitoring agents and knowledge about the pattern of attacks. Data sent by monitoring agents are derived from their *anomaly detection modules*. In *the data analysis module* conflicts of inconsistent knowledge incoming from distributed resources are solved applying consensus methods. Many a time managing agent after receiving the *warn* message from a one agent, asks other agents about their opinions. Also the new patterns of attacks are discovered in the data analysis module.

The attack detection module is responsible for the attack detections. In *the attack detection module* there are embodied two algorithms: the algorithm for determining of the sources attack and the algorithm for determining of the attack propagation scenarios. In this paper the attention to the first algorithm is paid, see Section 4 for details. *The Queue task module* consists of the tasks that must be executed or delegated by managing agent [17].

3   The Language of Communication

The agents communicate with each other to exchange knowledge about the states of monitored nodes or send the requests. They use a *communication language* compatible with standard ACL (Agent Communication Language) proposed by FIPA (Foundation of Intelligent Physical Agents) [15]. The ACL like KQML (Knowledge Query and Manipulation Language), developed by DARPA knowledge Sharing Effort, rely on speech act theory [2, 18]. These languages base on the performatives. The messages might be sent to one or more receivers. From the number of receivers point of view two types of performatives are considered: *peer-to-peer* and *multicast*. In the algorithm for determining the set of *Masters* following types of messages are used: *warn* messages, *multi_query_if* messages, *confirm* (*disconfirm*) messages and *multi_request* performatives with *inform* action execution. A monitoring agent *ma* sends the *warn* message to the managing agent as a warning of the anomaly detection in his monitoring region. There must exist a tuple $A(v, p, \alpha, [t_b, t_e])$ in the agent's *ma anomaly table*. The message with *query_if* performative is sent if the sender wants the receiver to ask about the anomaly appearance, i.e. the value of the parameter p in a given node v is higher than a given value x. *Multi_query_if* is used to send a query to a group of the agents. For example the managing agent may ask, if in the given time interval the value of the parameter p in a node v was abnormal i.e. higher than x: *multi_query_if*($A(v, (p > x), \alpha, [t_b, t_e])$).The *confirm* (*disconfirm*) message is send by the managing or monitoring agent if they want to confirm (disconfirm) the true of some proposition. If the managing agent sends *confirm* (*disconfirm*) message to more than only one agent then we have the *multi_confirm* (*multi_disconfirm*) message. The *multi_request* performative is sent by the managing agent to a group of monitoring agents. In this kind of message also the type of action must be specified. In the algorithm proposed in this paper an *inform* action is considered. In case of *multi_request* performatives with *inform* action execution the managing agent sends to the group of the monitoring agents a request of sending information about the discovered anomalies. If the managing agent requests of sending information about a state of a single node v with the reference to the one parameter p in the time interval $[t_b, t_e]$ then the message content is as follows: *send_information*($O(v, p, [t_b, t_e])$). The managing agent may also ask about the nodes that have been communicated wit a given node. In this case the *communication matrix* is analyzed.

4   The Algorithm for the DDoS Attack Detecting

The algorithm for the sources of *DDoS* attack detecting is embodied in the managing agent's *attack detection module*. It is assumed that network traffic should be analyzed in order to detect anomalies. The main purpose is to detect the attack before blocking the chosen goal. In order to detect the *DDoS* attacks the anomalies in traffic characteristics and anomalies in communication schemes must be observed. In the *DDoS* attack it is assumed that *Attacker* is the source of attack and initializes the overall attack in a computer network by putting a maliciously software in chosen computers (nodes). These nodes become *Masters*. In the next step each *Master* starts to infect other nodes that become *Demons*. Next *Demons* infect other nodes and those next ones etc. and in this way spreading stage is expanded. When sufficient number of nodes is infected or at the fixed time point final attack or on earlier chosen computer (attack goal) is carried out. Due to the numerous numbers of requests sent simultaneously from different *Demons* at the same time interval to the *attack goal* all his resources are allocated and blocked [13].

4.1   The Idea of the Algorithm

The goal of applying the algorithm for the sources attack detecting is to determine the node $v^k \in V$ that is an *Attacker*. This algorithm consists of two main steps. In the first step the set of *Masters* is detected and in the second one on the base of *Masters'* input traffic analysis the *Attacker* is trailed. In this work the first Step is discussed in detail. The idea of the algorithm is as follows [17]:

Step 1. Determine the set *H* of *Masters*.
Step 2. If the cardinality of the set *H* is higher than threshold value τ then go to Step
   3 else go to Step 1.
Step 3. Apply the *procedure of the Attacker detection*.

4.2   The Algorithm for Determining the Set of the Masters

Let us assume that the set of managing agents is one-element and we detect the anomaly with reference to one monitored property.

  The algorithm for determining the set of the *Masters* is divided into two parts: *Preliminary part* and *Principal part*.

Preliminary part
Step 1. The monitoring agent *ma* detects the anomaly in the node $v \in mr$ with the reference to the property $p \in P$. It means that there exists the tuple $(v, p, \alpha, [t_b, t_e])$
   in his anomaly table *AT*.
Step 2. The agent *ma* determines the node $v_k \in mr$, in which the anomaly of the value of the parameter *p* appeared at earliest.
Step 3. The monitoring agent *ma* affirms the *local anomaly* in his monitoring region, indicates the node v_k and sends a *warn message* to the managing agent.
Step 4. If the node v_k is monitored by the other monitoring agents, then Go to the Step 5, else Go to the Step 8.
Step 5. The managing agent asks other monitoring agents about their opinions about the state of the node v_k in a given time interval $[t_{b-\Delta}, t_e]$. The managing agent sends to them the message with *multi_request* performative with *inform action* execution.
Step 6. The managing agent collects the answers and obtains consistent opinion about the state of the node v_k. In this step, the procedure for verification of the node state is carried out, in which the consensus methods are applied. The aim of this procedure is to obtain the agreement of the node state.
Step 7. If anomaly exists then Go to Step 8 else send into the *queue task module* the task of the node state verification in some time.
Step 8. The managing agent initializes *Principal part* of the algorithm.

Principal part

Step 1. The managing agent asks monitoring agents which nodes communicated with the node v_k in the time interval $[t_{b-\Delta},t_b]$. The managing agent sends the message with *multi_request* performative with *inform action* execution.

Step 2. Monitoring agents on the basis of their communication matrixes determine the nodes that communicated with the node v_k in the time interval $[t_{b-\Delta},t_b]$ and send the answers.

Step 3. The managing agent obtains the set K of the nodes that communicated with the node v_k in the time interval $[t_{b-\Delta},t_b]$.

Step 4. The managing agent commissions the monitoring agents to verify if in the nodes from K the anomaly of the value of the parameter p in the time interval $[t_{b-\Delta},t_b]$ appeared. For each $v \in K$ the managing agent sends the message with *multi_guery_if* performative.

Step 5. Monitoring agents on the basis of their anomaly tables send the answers to the managing agent.

Step 6. The managing agent determines the set of infected nodes K'. If the nodes from K' are monitored by more than one monitoring agents, then the managing agent obtains the consensus opinion about the anomalies.

Step 7. If the set K' is not empty then go to Step 8 else go to Step 9.

Step 8. From the set K' managing agent selects the node v' that communicated with the node v_k at earliest and applies for the node v' the *Principal Part*. Go to the Step 1.

Step 9. The node v_k is a *Master*. Include v_k into the set H. Stop.

4.3   An Example

Fig. 1 shows the anomaly of the input and output traffic in the node $v_2 \in mr_3$. The node v_2 is monitored by the agent ma_3. This agent has observed the anomaly of the input traffic in the following time intervals: $[t_3,t_7]$, $[t_{11},t_{16}]$, $[t_{18},t_{20}]$ and $[t_{20},t_{25}]$ with the reference to the values of p_1 and p_2. In the time intervals: $[t_9,t_{14}]$ and $[t_{19},t_{23}]$ the anomaly of the output traffic has been observed. The nodes: v_3, v_{13}, v_{17}, v_{16}, v_{15} communicated with v_2 while v_2 communicated with v_8 and v_{19}.

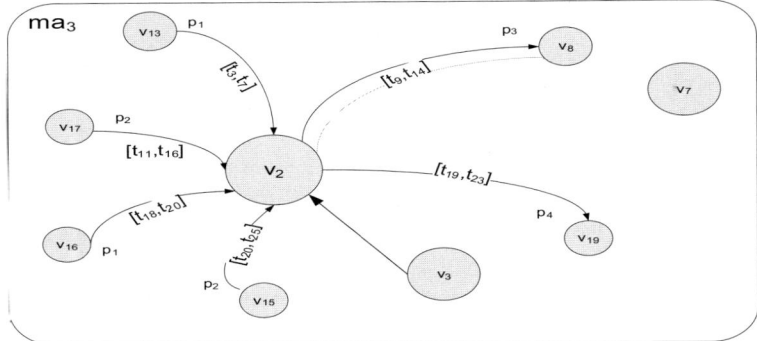

Fig. 1. Anomaly of the input and output traffic in the node v_2

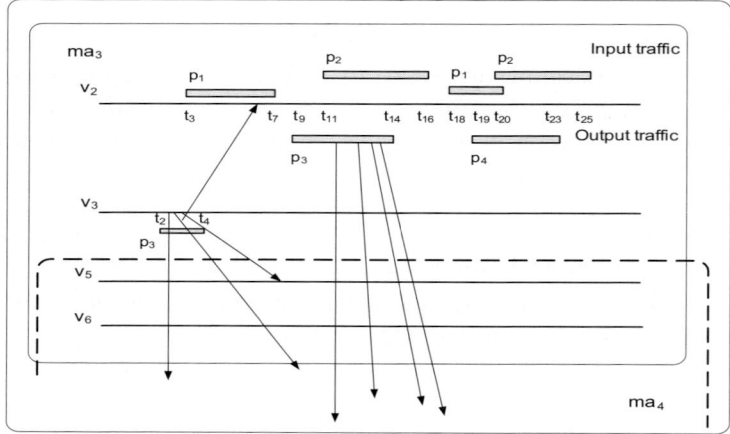

Fig. 2. Graphic representation of the piece of the anomaly table created by the ma_3.

- the anomaly of the parameter p_1 in the time intervals $[t_3,t_7]$ and $[t_{18},t_{20}]$
- the anomaly of the parameter p_2 in the time intervals $[t_{11},t_{16}]$ and $[t_{20},t_{25}]$
- the anomaly of the parameter p_3 in the time interval $[t_9,t_{14}]$
- the anomaly of the parameter p_4 in the time interval $[t_{19},t_{23}]$

Let us concentrate on the analyzing of the parameter p_3. On the basis of the time intervals in which this anomaly occurred we deduce that the node v_3 is the local source of the attack in the region mr_3. Let us assume the belief coefficient α equals 1 for all detected anomalies and the coefficient Δ equals 4.
The algorithm for determining the set of the *Masters* is as follows.

Preliminary part
Step 1. The monitoring agent ma_3 detects the anomaly in the node $v_2 \in mr_3$ with the reference to the property $p_3 \in P$. It means that there exists the tuple $(v_2, p_3, 1, [t_9, t_{14}])$ in his anomaly table AT.
Step 2. The agent ma_3 determines the node $v_3 \in mr_3$, in which the anomaly of p_3 appeared at earliest.
Step 3. The monitoring agent ma_3 sends w*arn message* to the managing agent.
Step 4. The node v_3 is monitored only by the one agent.
Step 8. The managing agent initializes *Principal part* of the algorithm.

Principal part
Step 1. The managing agent asks monitoring agents which nodes communicated with the node v_3 in the time interval $[t_0, t_4]$. The managing agent sends the message with *multi_request* performative with *inform action* execution.
Step 2. Monitoring agents on the basis of their communication matrixes send the answers.

Step 3. The managing agent receives the answers from two agents: ma_1 and ma_2 and determines the set $K = \{v_7, v_9, v_{12}\}$ (Fig 3).

Step 4. The managing agent commissions the monitoring agents to verify if in the nodes from K the anomaly of the value of the parameter p_3 in the time interval $[t_0, t_4]$ appeared. The managing agent sends the message with *multi_guery_if* performative.

Step 5. Monitoring agents on the basis of their anomaly tables send the answers to the managing agent.

Step 6. The managing agent determines the set K' of infected nodes: $K = \{v_7, v_{12}\}$

Step 8. From the set K' managing agent selects the node v_7 and applies for the node v_7 the *Principal Part*. Go to the Step 1.

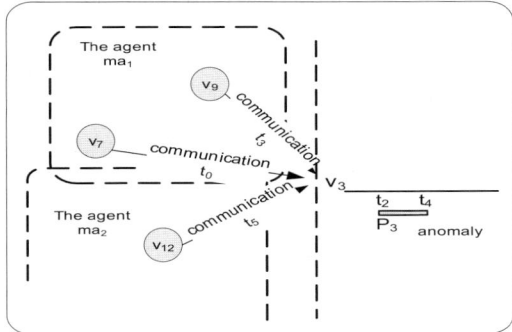

Fig. 3. The nodes v_7, v_9 and v_{12} have communicated with the node v_3 and one of them has infected the node v_3.

5   Conclusions

In this paper the problem of automatic intrusion detection was discussed. This paper is a continuation of the issues presented in previous works [11, 14]. In particular the ideas presented in [17] were here developed and the algorithm for automatic detecting of the *DDoS* attack was in detail described. This algorithm consists of two parts and the first part was here discussed. As a future work the second part of the algorithm must be developed and practical verification of presented results must be done.

Applying agency into the tasks of intrusions detection is the chance of building of the automatic intrusion detection systems that act in an incomplete, inconsistent and unpredictable environment. Agents as autonomous entities observe the nodes, take the decisions, communicate each other, integrate the results of their local activity and solve the global task of intrusion detection. In this paper we concentrate on the presentation of the algorithm for intrusion detection, and it was assumed that after discovering the anomaly of some parameter the agents apply the algorithm for the source of

DDOS attack detection. It must be pointed out that the anomalies discovering plays a crucial role in a whole system. This problem basis not only on measuring some parameters and comparing them with templates but also requires taking into account other individual nodes' characteristics. Undoubtedly the effectiveness of monitoring system depends not only on the applied intrusion detection algorithms but also on the input of these algorithms i.e. effectiveness of lower layers responsible for anomalies detecting.

References

1. Ajith, A., Thomas, J., Han, S.Y.: D-SCIDS: Distributed soft computing intrusion detection system. Journal of Network and Computer Applications 30(1), 81–98 (2001)
2. Austin, J.L.: How to Do Things with Words, Cambridge (Mass.) (1962), Harvard University Press, 2nd edn. (2005)
3. Balasubramaniyan, J.S., Garcia-Fernandez, J.O., Isacoff, D., Spafford, E., Zamboni, D.: An Architecture for Intrusion Detection Using Autonomous Agents. In: Proceedings of the 14th Annual Computer Security Applications Conference (1998)
4. Biermann, E., Cloete, E., Venter, L.M.: A comparison of Intrusion Detection systems. Computers and Security 20(8), 676–683 (2001)
5. Bejtlich, R.: Tao of Network Security Monitoring. In: The: Beyond Intrusion Detection, Addison-Wesley, Reading (2004)
6. Crosbie, A.M., Spafford, G.: Applying Genetic Programming to Intrusion Detection. In: Proceedings of the AAAI Fall Symposium on Genetic Programming, AAAI Press, Cambridge, Menlo Park (1995)
7. Crosbie, B.M., Spafford, G.: Defending a Computer System using Autonomous Agents. In: Proceedings of the 18th National Information Systems Security Conference (1995)
8. Dasgupta, D.: Immunity-Based Intrusion Detection System: A General Framework. In: Proceedings of the 22nd National Information Systems Security Conference, USA (1999)
9. Denning, D.E., Edwards, D.L., Jagannathan, R., Lunt, T.F., Neumann, P.G.: A prototype IDES: A real-time intrusiondetection expert system. Technical report, Computer Science Laboratory, SRI International, Menlo Park (1987)
10. Fenet, S., Hassas, S.: A distributed Intrusion Detection and Response System based on mobile autonomous agents using social insects communication paradigm, Electronic Notes in Theoretical Computer Science (2001)
11. Kołaczek, G., Pieczyńska, A., Juszczyszyn, K., Grzech, A., Katarzyniak, R., Nguyen, N.T.: A mobile agent approach to intrusion detection in network systems. In: Khosla, R., Howlett, R.J., Jain, L.C. (eds.) KES 2005. LNCS (LNAI), vol. 3682, pp. 514–519. Springer, Heidelberg (2005)
12. KQML Specification, http://www.cs.umbc.edu/kqml/
13. Mirkovic, J., Dietrich, S., Dittrich, D., Reiher, P.: Internet Denial of Service: Attack and Defense Mechanisms. Prentice Hall, Englewood Cliffs (2005)
14. Nguyen, N.T., Juszczyszyn, K., Kołaczek, G., Grzech, A., Pieczyńska, A., Katarzyniak, R.: Agent-based approach for distributed intrusion detection system design. In: Alexandrov, V.N., van Albada, G.D., Sloot, P.M.A., Dongarra, J. (eds.) ICCS 2006. LNCS, vol. 3993, pp. 224–231. Springer, Heidelberg (2006)
15. FIPA Specifications, http://www.fipa.org/

16. Patcha, A., Park, J.-M.: An overview of anomaly detection techniques: Existing solutions and latest technological trends. Computer Networks 51(12), 3448–3470 (2007)
17. Prusiewicz (Pieczyńska) A.: A Multi-agent System for Computer Network Security Monitoring. In: KES–AMSTA 2008, LNCS (LNAI), vol. 4953, pp. 842–849 Springer, Heidelberg (2008)
18. Searle, J.: Speech Acts. Cambridge University Press, Cambridge (1969)
19. Spafford, E., Zamboni, D.: Intrusion detection using autonomous agents. Computer Networks: The International Journal of Computer and Telecommunications Networking 34(4), 547–570 (2000)
20. Wilson, E.: Network Monitoring and Analysis: A Protocol Approach to Troubleshooting. Prentice Hall, Englewood Cliffs (1999)

Web Accessible A-Team Middleware

Dariusz Barbucha, Ireneusz Czarnowski, Piotr Jędrzejowicz,
Ewa Ratajczak-Ropel, and Izabela Wierzbowska

Department of Information Systems,
Gdynia Maritime University, Morska 83, 81-225 Gdynia, Poland
{barbucha,irek,pj,ewra,i.wierzbowska}@am.gdynia.pl

Abstract. The paper proposes a middleware called JABAT (Jade-based A-Team). JABAT allows to design and implement A-Team architectures for solving combinatorial optimization problem. JABAT is intended to become the first step towards next generation A-Teams which are fully Internet accessible, portable, scalable and in conformity with the FIPA standards.

Keywords: JABAT, A-Team, optimization, computionally hard problems, multi-agents systems.

1 Introduction

Last years a number of significant advances have been made in both the design and implementation of autonomous agents. A number of applications of agent technology is growing systematically. One of the successful approaches to agent-based optimization is the concept of an asynchronous team (A-Team), originally introduced by Talukdar [15].

A-Team is a multi agent architecture, which has been proposed in [15] and [16]. It has been shown that the A-Team framework enables users to easily combine disparate problem solving strategies, each in the form of an autonomous agent, and enables these agents to cooperate to evolve diverse and high quality solutions [14]. Acording to [15] an asynchronous team is a collection of software agents that solve a problem by dynamically evolving a population of solutions. Each agent works to create, modify or remove solutions from the population. The quality of the solutions gradually evolves over time as improved solutions are added and poor solutions are removed. Cooperation between agents emerges as one agent works on solutions produced by another. Each agent encapsulates a particular problem-solving method along with methods to decide when to work, what to work on and how often to work.

The reported implementations of the A-Team concept include two broad classes of systems: dedicated A-Teams and platforms, environments or shells used as tools for constructing specialized A-Team solutions. Dedicated (or specialized) A-Teams are usually not flexible and can be used for solving only particular types of problems. Among example A-Teams of such type one can mention the OPTIMA system for the general component insertion optimization problem

M. Bubak et al. (Eds.): ICCS 2008, Part III, LNCS 5103, pp. 624–633, 2008.

[13] or A-Team with a collaboration protocol based on a conditional measure of agent effectiveness designed for Flow optimization of railroad traffic [4].

Among platforms and environments used to implement A-Team concept some well known include IBM A-Team written in C++ with own configuration language [14] and Bang 3 - a platform for the development of Multi-Agent Systems (MAS) [12]. Some implementations of A-Team were based on universal tools like Matlab [16]. Some other were written using algorithmic languages like, for example the parallel A-Team of [5] written in C and run under PVM operating system.

The above discussed platforms and environments belong to the first generation of A-Team tools. They are either not portable or have limited portability, they also have none or limited scalability. Agents are not in conformity with the FIPA (The Foundation of Intelligent Psychical Agents) standards and there are no interoperability nor Internet accessibility. Migration of agents is either impossible or limited to a single software platform.

To overcome some of the above mentioned deficiencies a middleware called JABAT (Jade-based A-Team) was proposed in [11]. It was intended to become the first step towards next generation A-Teams which are portable, scalable and in conformity with the FIPA standards. JABAT allowes to design and implement an A-Team architecture for solving combinatorial optimization problems. In this paper we report on e-JABAT which is an extension of JABAT to become the fully Internet-accessible solution.

The paper is organized as follows: Section 2 gives a short overview of the JABAT features. Section 3 introduces the concept of e-JABAT and describes the required actions, which have to be carried out by the user wishing to use the Web-based JABAT interface to obtain a solution to the problem at hand. Section 4 offers more details on the e-JABAT architecture. Section 5 contains some comments on flexibility of e-JABAT. Finally, Section 6 contains conclusions and suggestions for future research.

2   Main Features of the JABAT Middleware

JABAT is a middleware supporting design and development of the population-based applications intended to solve difficult computational problems. The approach is based on the A-Team paradigm.

Main features of JABAT include:

- The system can solve instances of several different problems in parallel.
- The user, having a list of all algorithms implemented for the given problem may choose how many and which of them should be used.
- The optimization process can be carried out on many computers. The user can easily add or delete a computer from the system. In both cases JABAT will adapt to the changes, commanding the optimizing agents working within the system to migrate.

– The system is fed in the batch mode - consecutive problems may be stored and solved later, when the system assesses that there is enough resources to undertake new searches.

The use case diagram depicting the functionality of JABAT is shown in Fig. 1.

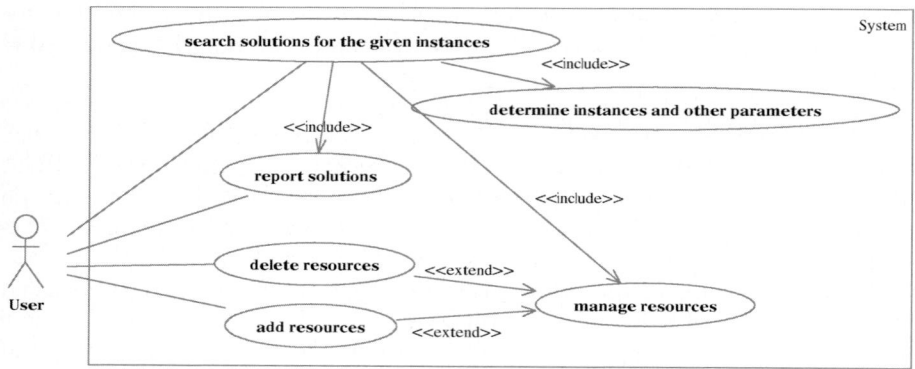

Fig. 1. Use case diagram of the functionality of JABAT

JABAT produces solutions to combinatorial optimization problems using a set of optimising agents, each representing an improvement algorithm. The process of solving of the single task (i.e. the problem instance) consists of several steps. At first the initial population of solutions is generated. Individuals forming the initial population are, at the following computation stages, improved by independently acting agents, thus increasing chances for reaching the global optimum. Finally, when the stopping criterion is met, the best solution in the population is taken as the result.

The way the above steps are carried out is defined by the "strategy". There may be different strategies defined in the system, each of them specyfying:

– how the initial population of solutions is created (in most cases the solutions are drawn at random),
– how to choose solutions which are forwarded to the optimizing agents for improvement,
– how to merge the improved solutions returned by the optimizing agents with the whole population (for example they may be added, or may replace random or worst solutions),
– when to stop searching for better solutions (for example after a given time, or after no better solution has been found within a given time).

To validate the system a number of experiments has been conducted. Experiments have involved a variety of combinatorial optimization problems. The results reported in [10], [2], [6], [7], [8] have proved ability and efectiveness of JABAT with regard to solving computationally hard problems.

3   e-JABAT

In this paper we propose an extension of JABAT making it accessible through a web interface, in which most of the original functionality is available for users from all over the world. The system with its interface is further on referred to as e-JABAT. A working copy of e-JABAT may be found at the address `http://jabat.wpit.am.gdynia.pl`.

3.1   Solving Tasks

To solve a task within the system, a user has to register at the e-JABAT website. Registered users obtain access to the part of the website in which tasks can be uploaded for solving. The uploaded tasks are sequenced and solved in the order of uploading. They may be solved in parallel (even the tasks of different problems). After a task has been solved the user can download the results saved in a text file in the user's space.

Thus, the user can:

- upload a file containing a task to be solved,
- observe the status of all tasks he has uploaded (waiting, being solved, solved),
- observe the logfiles in which some additional information or error messages may be found,
- download the files with solutions,
- delete his tasks.

The user willing to upload a task to the system must also:

- Choose from the available list the problem that the task belongs to. At present four different problems have been implemented in e-JABAT and are available for the users of the web interface. Theese problems are:
  - resource constrained project scheduling problem with single and multiple node (RCPSP, MRCPSP),
  - clustering problem (CP),
  - euclidean planar traveling salesman problem (TSP) and
  - vehicle routing problem (VRP).

  For each of these problems the format of the file containing the task to be solved is specified and published in the website.
- Choose which optimizing agents should be involved in the process of searching for the solution. Each of the optimizing agents within the system represents different optimization algorithm. For the problems implemented in the system there are available agents executing the following algorithms:
  - the local search algorithm, algorithm based on the simple evolutionary crossover operator and the tabu search algorithm for the RCPSP,
  - the Lin-Kerninghan algorithm and the evolutionary algorithm for the TSP,
  - the 2-optimum algorithm operating on a single route, the λ interchange local optimisation method, the evolutionary algorithm and the local search algorithms for the VRP,

- the random local search, the hill-climbing local search and the tabu search algorithms for the CP.
  For each of these algorithms a short description is available at the JABAT website.
– Choose a strategy from the list of available strategies and optionally define a set of options for this strategy, for example the size of the initial population or the length of time after which the search for better solutions stops.

For each optimizing agent the user may define the minimum and the maximum number of running copies of the agent. The system will initially use the minimal specified number and then the number will increase if there is enough computational resources available.

Fig. 2 shows the task upload screen, where the user's choices are shown.

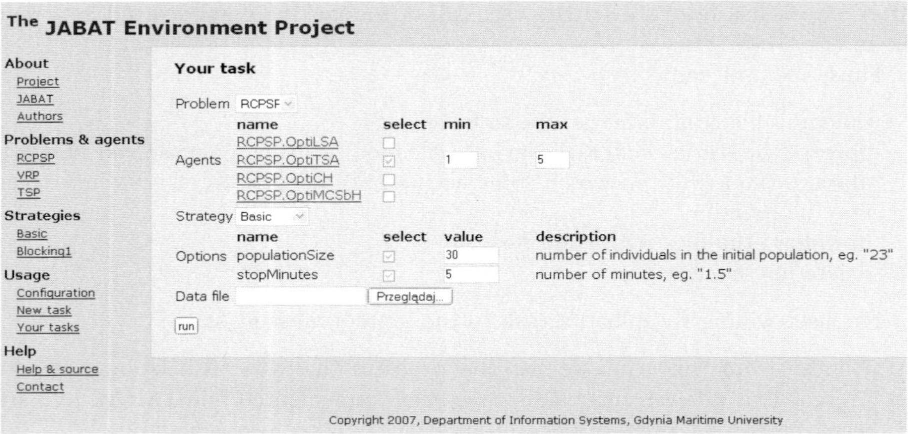

Fig. 2. Task upload screen

The report file created for each user's task includes the best solution obtained so far (that of the maximum or minimum value of fitness), average value of fitness among solutions from current population, the actual time of running and the time in which the best solution was last reached. The file is created after the initial population has been generated and then the next set of data is appended to the file every time the best solution in the population changes. The final results are added to the content of the file when the stopping criterion has been met. The report on the process of searching for the best solution may be later analysed by the user. It can be easily read into a spreadsheet and converted into a summary report with the use of the pivot table.

3.2 Adding/Deleting Resources

JABAT makes it possible for optimisation agents to migrate or clone to other computers. By the use of mobile agents the system offers decentralization of

computations resulting in a more effective use of available resources and reduction of the computation time. Each registered user may launch a JADE
container on his current host and attach it to the copy of JABAT running on
jabat.wpit.am.gdynia.pl.

4 e-JABAT Architecture

The system consists of two parts: JABAT engine, responsible for the actual
solving of computational task and web interface, in which a user can upload the
tasks and their parameters and download the results (Fig. 3).

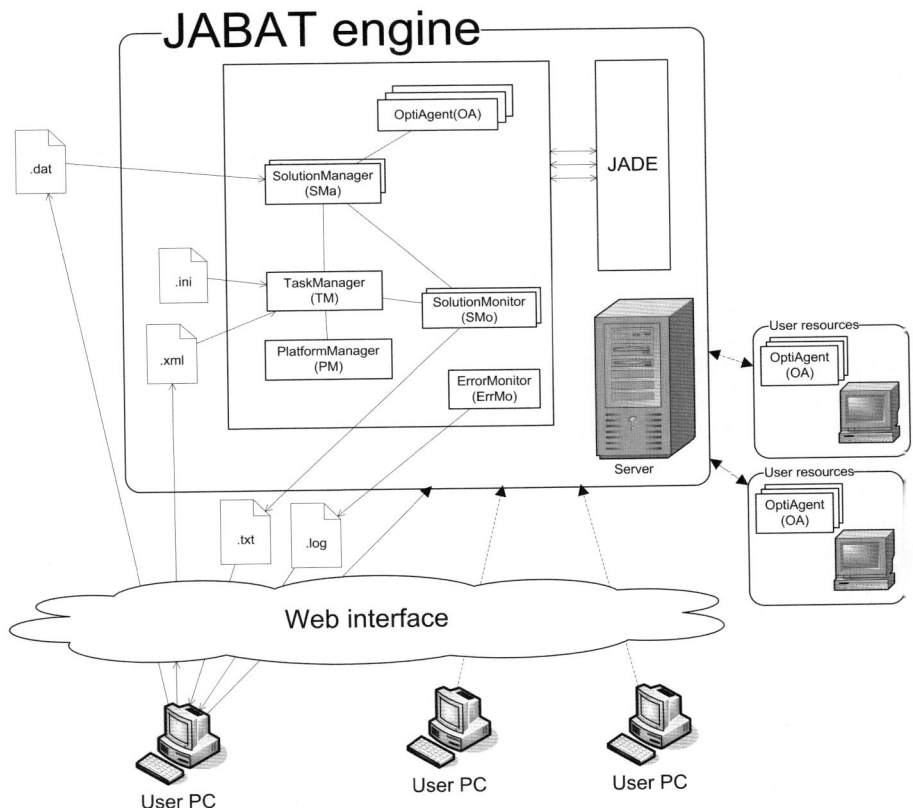

Fig. 3. e-JABAT architecture

4.1 Web Interface

Users obtain access to the JABAT engine through the web interface that has been
created with the use of Java Server Faces and Facelets technologies. The interface

allows the user to specify the task and to provide some additional information concerning details on how the task solving process should be carried. The task uploaded by the user is saved in the directory from which it can be later read by the JABAT engine. The information given by the user during the process of uploding the task are written in an XML file stored in the area called input directory from which JABAT can read it.

Each XML batch file stored in the input directory contains a single set of data provided by the user: the problem name, one or more instance data, list of optimising agents which should be run in order to solve the task, name of the selected strategy and additional options for this strategy. The XML input files stored in the initial directory are read by the JABAT engine and solved in the order of appearance. The results are stored in the output directory, from which they can be downloaded by the users to which they belong.

4.2   The JABAT Engine

The JABAT engine is built using JADE (Java Agent Development Framework), a software framework proposed by TILAB [17] for the development and run-time execution of peer-to-peer applications. JADE is based on the agents paradigm in compliance with the FIPA [9] specifications and provides a comprehensive set of system services and agents necessary to distributed peer-to peer applications in the fixed or mobile environment. It includes both the libraries required to develop application agents and the run-time environment that provides the basic services and must be running on the device before agents can be activated [3].

JADE platforms have containers to hold agents, not necessarily on the same computer. In JABAT containers placed in different platforms are used to run agents responsible for searching for optimal solutions using pre-defined solution improvement algorithms.

Within the JABAT engine the following types of agents are used:

- *OptiAgents* - representing the solution improvement algorithms,
- *SolutionManagers* - managing the populations of solutions,
- *TaskManagers* - responsible for initialising the process of solving an instance of a problem (it for example creates and deletes agents that are designated to the task)
- *SolutionMonitors* - recording the results,
- *PlatformManager* - organising the process of migration between different containers and
- *ErrorMonitor* - monitoring unexpected behavior of the system.

Agents Responsible for Solving a Task. The most important for the process of solving a task are *OptiAgents* and *SolutionManagers*. They work in parallel and communicate with each other exchanging solutions that are either to be improved when they are sent to *OptiAgents*, or stored back into the common memory when sent to *SolutionManager*.

Each *OptiAgent* is a single improvement algorithm. An *OptiAgent* can communicate with all *SolutionManagers* working with instances of the same problem.

An agent sends out the message about its readiness to work. Such message contains information about the number of solutions from the common memmory required to execute the improvement procedure. In response the *SolutionManager* sends the details of the task and appropriate number of solutions. The respective *OptiAgent* processes them and sends back the improved solution or solutions. The process iterates, until some stopping criterion is met.

Each *SolutionManager* is responsible for a single population of solutions created to solve a single task. Its actions include generation of the initial pool of solutions, sending solutions to the *OptiAgents*, merging improved solutions with the population storef in the common memory and deciding when the whole solution finding process should be stopped. All these activities are managed in accordance with the strategy that has been choosen for the particular task. This strategy is a part of the *SolutionManager* provided as one of the set of the agent parameters.

Apart from the above *SolutionManager* is also responsible for sending periodical reports on the state of computations to the *SolutionMonitor* monitoring the respective task. The *SolutionMonitor*, in turn, prepares and saves information on the results thus obtained in the report file available to the user.

Agents Responsible for Creating/Deleting/Relocating Other Agents.
There are two important agents categories managing the process of solving tasks: *TaskManagers* and *PlatformManager*. Their role is to create, delete, relocate or copy agents responsible for the actual problem-solving process.

There is only one *TaskManager* in the system, responsible for reading input data and creating or deleting all agents designated to the particular task. *TaskManager* may initialise the process of solving next task before the previous has stopped if there is any such task waiting and if the global system setting allows for that.

The *PlatformManager* manages optimization agents and system platforms. It can move optimization agents among containers and create (or delete) their copies to improve computations efficiency. The *PlatformManager* work is based on the following simple rules:

- the number of *OptiAgents* cannot exceed the maximum number and can not be smaller than the minimum number of *OptiAgents*, as specified by the user,
- if JABAT has been activated on a single platform (computer), then all OptiAgents would be also placed on this platform,
- if JABAT has been activated on multiple platforms, with main container placed on one computer and the remote joined containers placed on other computers, then OptiAgents are moved from the main container to outside containers to distribute the workload evenly.

5   Flexibility of e-JABAT

The JABAT engine has been designed in such a way, that it can be easily extended to solving new problems or solving them with new algorithms. The main idea is to reduce the amount of work of the programmer who wants to solve new problems or wishes to introduce new ways of representing tasks or solutions, new optimising algorithms or finally new replacement strategies. e-JABAT makes it possible to focus only on defining these new elements, while the processes of communication and population management procedures will still work. More detailed information about extending the functionality of JABAT can be found in [1].

6 Conclusions

The goal of the research presented in this paper was to propose a middleware environment allowing Internet accessibility and supporting development of A-Team systems. The solution - e-JABAT - has achieved this goal. Some of the advantages of e-JABAT have been inherited from JADE. The most important advantage seem to be e-JABAT ability to simplify the development of distributed A-Teams composed of autonomous entities that need to communicate and collaborate in order to achieve the working of the entire system. A software framework that hides all complexity of the distributed architecture plus a set of predefined objects are available to users, who can focus on the logic of the A-Team application and effectiveness of optimization algorithms rather than on middleware issues, such as discovering and contacting the entities of the system. It is believed that the proposed approach has resulted in achieving Internet accessible, scalable, flexible, efficient, robust, adaptive and stable A-Team architectures. Hence, e-JABAT can be considered as a step towards next generation A-Team solutions.

During the test and verification stages JADE-A-Team has been used to implement several A-Team architectures dealing with well known combinatorial optimization problems. Functionality, ease of use and scalability of the approach have been confirmed.

Further research will concentrate on extending scalability and efficiency features of e-JABAT. One of such features under current development is an extension of the system functionality to use the middleware through the Internet solely on a computer or computers directly controlled or owned by the user. Another development under way is the construction of the intelligent help which would provide guidance to users less advanced in using Java technologies.

References

1. Barbucha, D., Czarnowski, I., Jędrzejowicz, P., Ratajczak, E., Wierzbowska, I.: JADE-Based A-Team as a Tool for Implementing Population-Based Algorithms. In: Chen, Y., Abraham, A. (eds.) Intelligent Systems Design and Applications, IDSA, Jinan Shandong China, pp. 144–149. IEEE, Los Alamos (2006)
2. Barbucha, D., Jędrzejowicz, P.: An experimental investigation of the synergetic effect of multiple agents solving instances of the vehicle routing problem. In: Grzech, A. (ed.) Proceedings of the 16th International Conference on Systems Science, Wroclaw, vol. II, pp. 370–377 (2007)

3. Bellifemine, F., Caire, G., Poggi, A., Rimassa, G.: JADE. A White Paper, Exp 3(3), 6–20 (2003)
4. Blum, J., Eskandarian, A.: Enhancing intelligent agent collaboration for flow optimization of railroad traffic. Transportation Research Part A 36, 919–930 (2002)
5. Correa, R., Gomes, F.C., Oliveira, C., Pardalos, P.M.: A parallel implementation of an asynchronous team to the point-to-point connection problem. Parallel Computing 29, 447–466 (2003)
6. Czarnowski, I., Jędrzejowicz, P.: Implementation and Performance Evaluation of the Agent-Based Algorithm for ANN Training. In: Nguyen, N.T., Grzech, A., Howlett, R.J., Jain, L.C. (eds.) KES-AMSTA 2007. LNCS (LNAI), vol. 4496, pp. 131–140. Springer, Heidelberg (2007)
7. Czarnowski, I., Jędrzejowicz, P.: An agent-based approach to the multiple-objective selection of reference vectors. In: Perner, P. (ed.) MLDM 2007. LNCS (LNAI), vol. 4571, pp. 117–130. Springer, Heidelberg (2007)
8. Czarnowski, I., Jędrzejowicz, P.: An Agent-Based Algorithm for Data Reduction. In: Bramer, M., Coenen, F., Petridis, M. (eds.) Research and Development in Intelligent Systems XXIV and Applications and Innovations in Intelligent Systems XV, Proceedings of AI 2007, the Twenty-seventh SGAI International Conference on Artificial Intelligence, Springer, London (2007)
9. The Foundation for Intelligent Physical Agents, http://www.fipa.org/
10. Jędrzejowicz, P., Ratajczak-Ropel, E.: Agent-Based Approach to Solving the Resource Constrained Project Scheduling Problem. In: Beliczynski, B., Dzielinski, A., Iwanowski, M., Ribeiro, B. (eds.) ICANNGA 2007. LNCS, vol. 4431, pp. 480–487. Springer, Heidelberg (2007)
11. Jędrzejowicz, P., Wierzbowska, I.: JADE-Based A-Team Environment. In: Alexandrov, V.N., van Albada, G.D., Sloot, P.M.A., Dongarra, J. (eds.) ICCS 2006. LNCS vol. 3993, pp. 719–726. Springer, Heidelberg (2006)
12. Neruda, R., Krusina, P., Kudova, P., Rydvan, P., Beuster, G.: Bang 3: A Computational Multi-Agent System. In: Proceedings of the IEEE/WIC/ACM International Conference on Intelligent Agent Technology, IAT 2004 (2004)
13. Rabak, C.S., Sichman, J.S.: Using A-Teams to optimize automatic insertion of electronic components. Advanced Engineering Informatics 17, 95–106 (2003)
14. Rachlin, J., Goodwin, R., Murthy, S., Akkiraju, R., Wu, F., Kumaran, S., Das, R.: A-Teams: An Agent Architecture for Optimization and Decision-Support. In: Rao, A.S., Singh, M.P., Müller, J.P. (eds.) ATAL 1998. LNCS (LNAI), vol. 1555, pp. 261–276. Springer, Heidelberg (1999)
15. Talukdar, S.N., de Souza, P., Murthy, S.: Organizations for Computer-Based Agents. Engineering Intelligent Systems 1(2) (1993)
16. Talukdar, S., Baerentzen, L., Gove, A., de Souza, P.: Asynchronous teams: cooperation schemes for autonomous agents. Journal of Heuristics 4, 295–321 (1998)
17. Jade - Java Agent Development Framework, http://jade.tilab.com/

Multi-agent System for Dynamic Manufacturing System Optimization

Tawfeeq Al-Kanhal and Maysam Abbod

School of Engineering and Design,
Brunel University, West London, UK Uxbridge, UK. UB8 3PH
Tawfeeq.Al-kanhal@Brunel.ac.uk

Abstract. This paper deals with the application of multi-agent system concept for optimization of dynamic uncertain process. These problems are known to have a computationally demanding objective function, which could turn to be infeasible when large problems are considered. Therefore, fast approximations to the objective function are required. This paper employs bundle of intelligent systems algorithms tied together in a multi-agent system. In order to demonstrate the system, a metal reheat furnace scheduling problem is adopted for highly demanded optimization problem. The proposed multi-agent approach has been evaluated for different settings of the reheat furnace scheduling problem. Particle Swarm Optimization, Genetic Algorithm with different classic and advanced versions: GA with chromosome differentiation, Age GA, and Sexual GA, and finally a Mimetic GA, which is based on combining the GA as a global optimizer and the PSO as a local optimizer. Experimentation has been performed to validate the multi-agent system on the reheat furnace scheduling problem.

Keywords: GA, PSO, multi-agent system, reheat furnace, scheduling.

1 Introduction

Intelligent Manufacturing means the application of Artificial Intelligence (AI) and Knowledge-based technologies in general to manufacturing problems. This includes a large number of technologies such as machine learning, intelligent optimization algorithms, data mining, and intelligent systems modeling. Such technologies have so far proved to be more popular than AI Planning and Scheduling in such applications.

In this research, different types of intelligent optimization methodologies have been explored for the purpose of planning and scheduling with the emphasis on the application of the technology to reheat furnaces scheduling. An informal definition of the terms AI Planning and AI Scheduling, has to be defined as accepted in the manufacturing community which is as follows:

Planning: the automatic or semi-automatic construction of a sequence of actions such that executing the actions is intended to move the state of the real world from some initial state to a final state in which certain goals have been achieved.

M. Bubak et al. (Eds.): ICCS 2008, Part III, LNCS 5103, pp. 634–643, 2008.
© Springer-Verlag Berlin Heidelberg 2008

This sequence is typically produced in partial order, which is with only essential ordering relations between the actions, so that actions not so ordered appear in pseudo-parallel and can be executed in any order while still achieving the desired goals.

Scheduling: in the pure case, the organization of a known sequence of actions or set of sequences along a time-line such that execution is carried out efficiently or possibly optimally. By extension, the allocation of a set of resources to such sequences of actions so that a set of efficiency or optimality conditions are met.

Scheduling can therefore be seen as selecting among the various action sequences implicit in a partial-order plan in order to find the one that meets efficiency or optimality conditions and filling in all the re-sourcing detail to the point at which each action can be executed.

This paper addresses the issues involved in developing a suitable methodology for developing a generic intelligent scheduling system using a multi-agent architecture. The system includes a number of agents based on different intelligent techniques, such as Genetic Algorithms (GA) and its derivates, Particle Swarm Optimization (PSO), and hybridizations of the systems. Also, it must operate in an environment which requires the system to respond rapidly to complex, potentially real time response to a dynamic system. A metal reheating scheduling problem is chosen as the test bed.

2 Multi Agent System

Conceptually, multi-agent system architecture consists of a series of problem solving agents, and the control mechanisms. The agents are used co-operatively to solve a complex problem which can be solved by any of the agents individually. The subdivision of the system into agents increases the search space for a solution to the problem under investigation, which also facilitates the integration of other intelligent system components into the system structure. The agents are only allowed to communicate with each other via the system, a data structure which stores all the information which is either input or output from any of the agents. The purpose of the control mechanism is to decide at what time, and in which order, the agents are to be executed. At any one time, there may be many agents who are ready to execute, it being the role of the control mechanism to determine which of these agents will best meet the goals of the system and constrains set by the environment, such as fast or accurate solutions. Thus the system can be described as being examples of opportunistic reasoning systems [5]. In the following sections, the different agents used in the system are described.

2.1 Practical Swarm Optimization

Particle Swarm Optimization is a global minimization technique for dealing with problems in which a best solution can be represented as a point and a velocity. Each particle assigns a value to the position they have, based on certain metrics. They remember the best position they have seen, and communicate this position to the other members of the swarm. The particles will adjust their own positions and velocity based on this information. The communication can be common to the whole swarm, or be divided into local neighborhoods of particles [6].

2.2 Genetic Algorithms (GA)

GAs are exploratory search and optimization methods that were devised on the principles of natural evolution and population genetics [4]. Unlike other optimization techniques, a GA does not require mathematical descriptions of the optimization problem, but instead relies on a cost-function, in order to assess the fitness of a particular solution to the problem in question. Possible solution candidates are represented by a population of individuals (generation) and each individual is encoded as a binary string containing a well-defined number of chromosomes (1's and 0's). Initially, a population of individuals is generated and the fittest individuals are chosen by ranking them according to *a priori*-defined fitness-function, which is evaluated for each member of this population. In order to create another better population from the initial one, a mating process is carried out among the fittest individuals in the previous generation, since the relative fitness of each individual is used as a criterion for choice. Hence, the selected individuals are randomly combined in pairs to produce two off-springs by *crossing over* parts of their chromosomes at a randomly chosen position of the string. These new offspring represent a better solution to the problem. In order to provide extra excitation to the process of generation, randomly chosen bits in the strings are inverted (0's to 1's and 1's to 0's). This mechanism is known as *mutation* and helps to speed up convergence and prevents the population from being predominated by the same individuals. All in all, it ensures that the solution set is never naught. A compromise, however, should be reached between too much or too little excitation by choosing a small probability of mutation.

2.3 Age Genetic Algorithm (AGA)

The age GA emulates the natural genetic system more closely to the fact that the age of an individual affects its performance and hence it should be introduced in GAs. As soon as a new individual is generated in a population its age is assumed to be zero. Every iteration age of each individual is increased by one. As in natural genetic system, young and old individuals are assumed to be less fit compared to adult individuals [3]. The effective fitness of an individual at any iteration is measured by considering not only the objective function value, but also including the effect of its age. In GA once a particular individual becomes fit, it goes on getting chances to produce offspring until the end of the algorithm; if a proportional selection is used; thereby increasing the chance of generating similar type of offspring. More fit individuals do not normally die, and only the less fit ones die. Whereas in AGA, fitness of individuals with respect to age is assumed to increase gradually up to a pre-defined upper age limit (number of iterations), and then gradually decreases. This, more or less, ensures a natural death for each individual keeping its offspring only alive. Thus, in this case, a particular individual cannot dominate for a longer period of time. Rest of the process of evolution in AGA is same as that in GA.

2.4 Sexual Genetic Algorithm (SGA)

The selection of parent chromosomes for reproduction, in case of GA, is done using only one selection strategy. When considering the model of sexual selection in the area of population genetics it gets obvious that the process of choosing mating partners in natural populations is different for male and female individuals. Inspired by the idea of male vigor and female choice, Lis and Eiben [7] have proposed Sexual GA that utilizes two different selection strategies for the selection of two parents required for the crossover. The first type of selection scheme utilizes random selection and another selection strategy uses roulette wheel selection for the selection of two parents. Rest of the process is similar to that of GA.

2.5 Genetic Algorithm with Chromosome Differentiation (GACD)

In GACD [1], the population is divided into male and female population on the basis of sexual differentiation. In addition, these populations are made dissimilar artificially, and both the populations are generated in a way that maximizes the hamming distance between the two classes. The Crossover is only allowed between individuals belonging to two distinct populations, and thus introduces greater degree of diversity and simultaneously leads to greater exploration in the search space. Selection is applied over the entire population, which serves to exploit the information gained so far. Thus, GACD accomplishes greater equilibrium between exploration and exploitation, which is one of the main features for any adaptive system. The chromosomes in the case of GACD are different as it contains additional gene that helps in determining the sex of the individuals in the current population.

2.6 Mimetic Genetic Algorithms (MGA)

MGAs are inspired by the notions of a mime [2]. In MGA, the chromosomes are formed by the mimes not genes (as in conventional GA). The unique aspect of the MGA algorithm is that all chromosomes and offspring are allowed to gain some experience before being involved in the process of evolution. The experience of the chromosomes is simulated by incorporating local search operation. Merz and Freisleben [8] proposed a method to perform local search through pair wise interchange heuristic. The local neighborhood search is defined as a set of all solutions that can be reached from the current solution by swapping two elements in the chromosome.

In this research, the MGA local search engine is based on PSO. When the population is generated, it is passed to PSO for gaining some experience. The PSO will train the individuals to find local solutions to the problem within a constrained environment. Once the individuals are trained, they are passed back to the GA for performing the mating operations, and consequently finding solutions for the optimization problem.

3 Reheat Furnace Model

Metals reheating furnace scheduling is chosen as a test bed for the optimization algorithm. Fig. 1 shows a typical continuous annealing process known as the continuous annealing and processing line [10]. In this furnace, the material for annealing is a cold-rolled strip coil, which is placed on a pay-off reel on the entry side of the line. The head end of the coil is then pulled out and welded with the tail end of the preceding coil. Then the strip runs through the process with a certain line speed. On the delivery side, the strip is cut into a product length by a shear machine and coiled again by a tension reel. The heat pattern of the strip is determined according to the composition and the product grade of the strip. The actual strip temperature must be within the defined ranges from the heat pattern to prevent quality degradation. The value of the heat pattern at the outlet of the heating furnace is the reference temperature for the control. In most cases, the strip in the heating furnace is heated indirectly with gas-fired radiant tubes. The heating furnace is 400 to 500 m in strip length and is split into several zones. The furnace temperature and fuel flow rate are measured at each zone, while the strip temperature is measured only at the outlet of the furnace with a radiation pyrometer. It takes a few minutes for a point on the strip to go through the furnace.

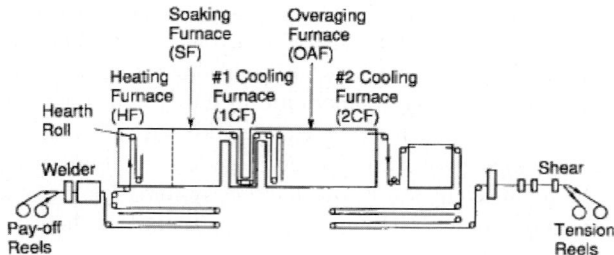

Fig. 1. Outline of a continuous annealing process [10]

For simplicity, a single heating furnace model is considered. The physical state of the steel piece annealing process is denoted by $z(t)$ and represents the temperature the metal as it evolves through the heating furnace. The metal piece temperature rise depends on its thickness, mass, and the furnace reference temperature F; which is pre-designed at a plant-wide planning level. The thermal process in the heating furnace can be represented by a nonlinear heat-transfer equation describing the dynamic response of each metal piece temperature so that the temporal change in heat energy at a particular location is equal to the transport heat energy plus the radiation heat energy as follows [9]:

$$\frac{dz(t)}{dt} = K_1 u + K_2 [F^4 - z(t)^4] \tag{1}$$

where $K_1 = \dfrac{f - z(t_0)}{L}$ and $K_2 = \dfrac{2\sigma_{sb}\phi_s}{60 d_s 10^{-3} \tau}$

and L is the furnace length (m); t_0 the heating start time; σ_{sb} is the Stefan–Boltzmann constant (= 4.88×10^{-8} kcal/m^2 h deg^4)); Φ_s is the coefficient of radiative heat absorption, $0 < \Phi_s < 1$ (assumed as 0.17); d_s is the strip specific heat (kcal/m^3 deg); τ is the metal thickness (mm). The heat energy equation is a nonlinear differential type and simulated in the following environment: L = 500 (m); d_s = 4.98×10^4 kcal/m^3 deg^4; Φ_s = 0:17; τ = 0:71 (mm), u = 100 (m/min) and $z(t0)$ = 30°C.

4   Optimization Results

4.1   Heating Schedules

Two types of scheduling problems were considered, the first consists of 5 jobs, while the second consists of 10 jobs. The scheduling problem is based on finding the best schedule to enter the metal pieces in sequence and to set the furnace temperature to the required setting for each piece. The objective function is to minimize the heating fuel consumption and the time to complete all the jobs. Table 1 shows the 5 (first 5 in the table) and 10 jobs heating temperature and time.

The 5 jobs problem has a search space of 5! = 120 solutions with a total time of 7400 sec. While the 10 jobs problem is more complicated and has a search space of 10! = 3,628,800 solutions with a total time of 16750 sec.

An unscheduled 10 jobs sequence simulation is shown in Fig. 2. Due to the large differences between the sequenced jobs temperature, the furnace temperature has to be raised and lowered to meet the required temperature for each piece. Since the furnace has to be heated and cooled to meet the required piece temperature, this will cause the process to take a long time and high energy consumption. The need for optimization the schedule for shortest time and lower energy consumption will be achieved through the multi-agent optimization system.

Table 1. Experimental jobs selections

Job no.	Temperature (°C)	Heating Time (sec)
1	800	1000
2	1200	2000
3	400	1500
4	600	1200
5	1000	1700
6	1400	1550
7	900	2200
8	700	800
9	1300	1900
10	400	3000

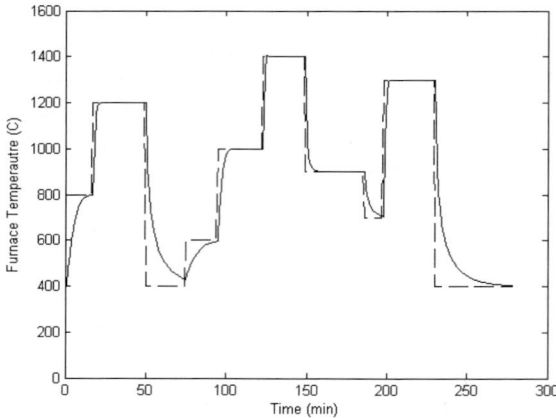

Fig. 2. Unscheduled 10 jobs heating sequence

The optimizers cost function is based on normalizing the fuel consumption and the time take for completing all the jobs in the sequence. Equal weighting has been given to both objectives (50% each). The final cost function is set by equation (2).

$$f = 0.5 \times (\text{norm. fuel}) + 0.5 \times (\text{norm. time}) \qquad (2)$$

4.2 PSO Schedule Optimization

The PSO algorithm was set to a population size of 100, while the inertial cognitive and social constants are as follows:

$W_{min} = 0.4$, $W_{max} = 0.9$, $c_1 = 1.4$, $c_2 = 1.4$, Velocity constraints $= \pm 1$,
No. of iterations = 200

Due to the fact that there are unfeasible schedule solutions that might be obtained by the PSO algorithm, a penalty was given to all unfeasible solutions. This step has been added to constrain the PSO in order not to search in the unfeasible solutions areas. The algorithm was run for 200 iterations on both schedules (5 and 10 jobs). The optimum solution is found after 15 iterations for the 5 and 10 jobs schedule. Fig. 3 shows the cost function minimization for both cases. The 5 jobs case solution was found after 15 iterations and it presents the optimum schedule. Similarly, the 10 jobs schedule, a minimum cost function was found after 15 iteration (f = 1583) which does not present the optimum cost function (f = 1210). Table 2 shows the best solutions found for both cases.

Table 2. PSO cost function minimization

Jobs type	Cost function	Iteration no.
5 jobs	621.36	15
10 jobs	1583.03	15

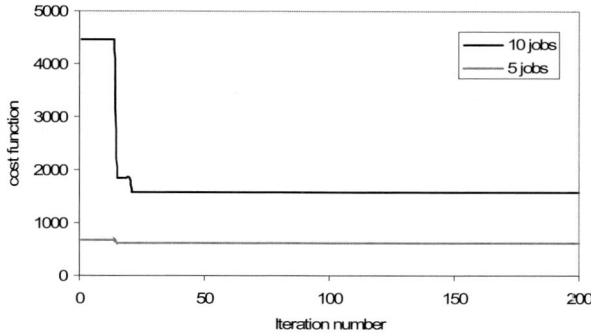

Fig. 3. PSO cost function minimization for 5 and 10 jobs schedule

4.3 GAs Schedule Optimization

The GA algorithm was set as a binary code of 4 bits for each of the numbers of the jobs in the schedule. The schedule of 10 jobs makes the chromosome $10 \times 4 = 40$ bits long. The 4 bit binary number maps to a search space 1 to 16 job selection. The GA was set with a mutation rate of 0.03 and a single point crossover at a rate of 0.9. However, the different derivations of the GAs will need different settings depending on the type of mating and selections procedures. Therefore it was necessary to experiment with all the algorithms separately to find the best setting for each type. Table 3 shows the best performance found by the GAs after many simulation runs.

Experimenting with the first type of scheduling (5 jobs) was simple as the number of solutions is limited (n! = 120 solutions) and the best solution can be found easily. The schedule optimization results are shown in Table 3 for the different GAs and Fig. 4 shows the cost function minimization. All the GAs types found the optimal solution (f = 586.2) which is the best solution. However, the MGA was the first to find the solution, in two iterations only. While SGA required 73 iterations for find the optimum solution. The 5 jobs optimum schedule obtained is [3 4 1 5 2].

Table 3. Parameters of best performing GAs

Algorithm	Cross over probability	Mutation probability	Cost function	Iteration no.
GA	0.90	0.03	586.2614	7
SGA	0.92	0.02	586.2614	73
GACD	0.95	0.03	586.2614	69
AGA	0.90	0.05	586.2614	46
MGA	0.99	0.01	586.2614	2

Experimenting with the second type of scheduling (10 jobs) was based on the same best GAs settings found during the 5 jobs experiments. The second type search space is very large (n! = 3,628,800 solutions). The schedule optimization results are shown in Table 4 for the different GAs and Fig. 5 shows the cost function minimization. The different GAs types found different optimal solution where the standard GA (f = 1209.8) is the best solution. However, the standard GA required 81 iterations to find the solution. Meanwhile MGA optimal cost function was not far from the best optimum GA, and it took 26 iterations. The 10 jobs optimum schedule obtained is [3 5 6 9 2 7 1 8 4 10].

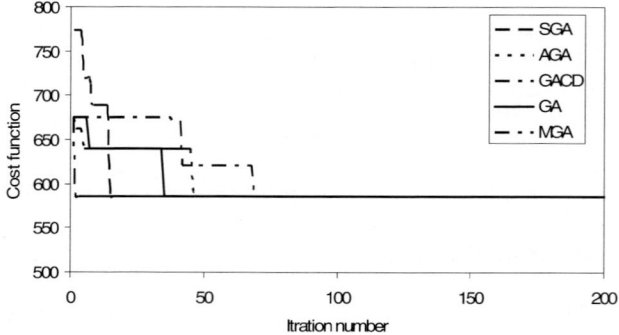

Fig. 4. GAs cost function minimization for 5 jobs schedule

Table 4. Cost function optimization algorithms for 10 jobs schedule

Algorithm	Cost function	Iteration no.
GA	1209.86	81
SGA	1210.20	73
GACD	1256.25	32
AGA	1261.43	235
MGA	1213.66	26

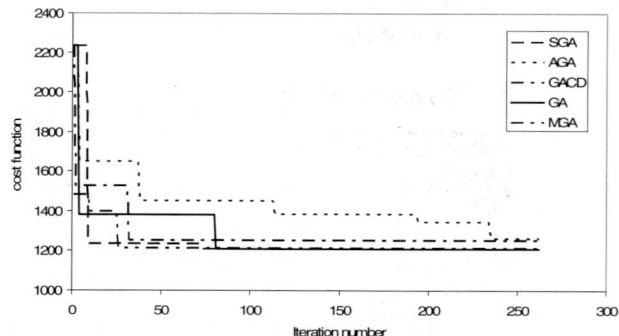

Fig. 6. GA cost function minimization for 10 job schedule

The search speed of the different GAs allow an interaction between the GAs generations. The fast divergence algorithms can provide good chromosomes to the more accurate slow algorithms via the multi agent system. This will be governed by the control system which should schedule the algorithms to run concurrently and at the same time communicate with each other.

5   Conclusions

In this paper a description of the multi-agent optimization algorithm has been given. Different intelligent optimization techniques have been utilized, such as GA and PSO. GAs are found to be a time consuming but robust optimization technique which can meet the requirements of manufacturing systems. GAs are capable to handle real world problems because the genetic representation of precedence relations among operations fits the needs of real world constraints in production scheduling. Moreover, GAs are applicable to a wide array of varying objectives and therefore they are open to many operational purposes.

The speed of GA can be improved by introducing fast algorithms, such as PSO, in order to find an initial population that advances the GA in finding the solutions in real time. Furthermore, using different types of GA can be beneficial in terms of finding an accurate solution; however, this has come to a price of being slow. Accurate GA takes longer time to converge, while less accurate GAs are much faster in converging. The multi-agent system architecture allows the communication between different agents, which in this case, at early stages, the fast and less accurate GA can pass its chromosomes to the slow and more accurate GA, which will benefit from the good chromosomes at an early stage.

References

1. Bandyopadhyay, S., Pal, S.K., Mulak, U.: Incorporating Chromosome Differentiation in Genetic. Information Science 104(8), 293–319 (1988)
2. Dawkins, R.: The Selfish Gene. Oxford University Press, Oxford (1976)
3. Ghosh, A., Tsutsui, S., Tanaka, H.: Individual Aging in Genetic Algorithms. In: Australian and New Zealand Conference on Intelligent Information Systems (1996)
4. Goldberg, D.E.: Genetic Algorithms in Search, Optimization, and Machine Learning. Addison-Wesley, Reading (1989)
5. Gonzalez, A.J., Dankel, D.D.: The Engineering of Knowledge-Based Systems - Theory and Practice. Prentice-Hall International, Englewood Cliffs (1993)
6. Kennedy, J., Eberhart, R.: Particle Swarm Optimization. In: Proc. IEEE Int'l. Conf. on Neural Networks, Perth, Australia, pp. 1942–1948 (November 1995)
7. Lis, J., Eiben, A.: A Multi-sexual Genetic Algorithm for Multi-objective Optimization. In: Fukuda, T., Furuhashi, T. (eds.) Proc. of 1996 International Conference on Evolutionary Computing, Nagoyo, Japan, pp. 59–64. IEEE, Los Alamitos (1996)
8. Merz, P., Freisleben, B.: A Genetic Local Search Approach to the Quadratic Assignment Problem. In: Back, C.T. (ed.) Proceedings of the 7th international conference on genetic algorithms, pp. 465–472. Morgan Kaufmann, San Diego (1997)
9. Watanapongse, C.D., Gaskey, K.M.: Application of Modern Control to a Continuous Anneal Line. IEEE Control System Magazine, 32–37 (1988)
10. Yoshitani, N., Hasegawa, A.: Model-Based Control of Strip Temperature for the Heating Furnace in Continuous Annealing. IEEE Transactions on Control Systems Technology 6(2), 146–156 (1998)

GRADIS – Multiagent Environment Supporting Distributed Graph Transformations

Leszek Kotulski

Department of Automatics, AGH University of Science and Technology
Al. Mickiewicza 30, 30 059 Krakow, Poland
kotuslki@agh.edu.pl

Abstract. Graph transformations are a powerful notation formally describing different aspects of modeled systems. Multiagent systems introduce distribution, parallelism and autonomous decision properties. In the paper a basic properties of the GRADIS agent's framework, joining of both approaches, are discussed. This framework supports splitting the graph, describing a problem, onto a few partial graphs, that can be maintained by different agents. Moreover, the multiagent's cooperation enables the application to the local graphs the graph transformation rules introduced for the centralizes graph; this permits us transfer all theoretical achievements of the centralized graph trans-formations to the distributed environment. The usefulness of the hierarchical graphs structure are and some examples of its usefulness are presented.

1 Introduction

Graphs are very useful formalism describing in a natural way a wide spectrum of problems. Unfortunately their use is limited with respect size and distribution. All the propositions mentioned in the three volume Handbook on Graph Grammars and Computing by Graph Transformations [1][2][3] remember the transformed graphs in one place. Moreover, the size of these graphs is limited due to a computational complexity of the parsers and the membership checkers. Even in the case of a few solutions, which offer the polynomial complexity of the solution of the mentioned problems (like $O(n^2)$ in [4]), this complexity permits us to think rather about hundreds or thousands nodes than about billions of ones. Let's note that billion of nodes is size of the small semantic web solutions.

The natural solution of the mentioned problem seems to be a graph distribution and a parallel transformation of these subgraphs. While the Multiagent Systems are characterized [5] by the assumption that:

- each agent has incomplete information or capabilities for solving the problem and, thus, a limited point of view,
- there is no global system control,
- data are decentralized,
- computations are asynchronous,

they seam to be the natural candidate for supporting distributed graph transformations.

M. Bubak et al. (Eds.): ICCS 2008, Part III, LNCS 5103, pp. 644–653, 2008.

The first approach, of the author, at solving the similar problem [6] also assumes creation a set of agents; each of them maintains a local graph, and transforms it (what sometimes needs an cooperation with another agent). The presented solution of the distributed control of the allocation state assumes that the rules of the agents cooperation are designed for the given set of graph grammar production (describing graphs transformations); any change of this grammar causes the redesigning of the agents cooperation rules.

The GRADIS agent's framework, presented in the paper, enables the cooperation of the agents (maintaining a local graph) in a way, that is independent from the graph grammar definition. What is more, the cooperation of the local graph transformations systems, that are founded upon different types of graph grammars, is also possible.

The scope of the paper is the following: in section 2 the basic model of the Multiagent GRADIS framework is introduced; in section 3 the theoretical foundation of the complementary graph construction (supporting the graph distribution) is presented; in section 4 the way of the cooperation of agents using different types of grammars is considered; finally, some concluding remarks are presented.

2   GRADIS Agent Model

The GRADIS framework (that is an acronym of GRAph DIStribution toolkit) makes possible the distribution of a centralized graph and the controlling its behavior with the help of concurrent processes. The proposed solution is based on Multiagent technology; an agent is responsible both for:

- a modification of the maintained (local) graph, in a way described by the graph transformation rules associated with it.
- a cooperation with other agents for the purpose of holding the cohesion of the whole graph system.

The GRADIS agent model assumes the existence of two types of agents, called: maintainers and workers.

The maintainer agent – maintains the local graph; the whole set of maintainers take care about the global system cohesion understood as a behavior equivalent to the graph transformations made over the centralized graph. Initially we assume, that at the beginning one maintainer controls the centralized graph, but it is able to split itself onto the set of maintainer agents controlling parts of the previous graph transformation. The cooperation of the maintainers is based on the exchanging information among the elements of the agent's local graph structure; the graph transformation rules are inherited from the centralized solution. The formal background of the maintainer's activities is presented in section 3.

The worker agents – are created: temporarily – for the purpose of realization of the given action (eg. for finding subpattern) or permanently - to achieve the more complex effect (eg. for detail designing of the element represented a the lower graph hierarchy). The worker graph structure is generated while its creation (by a maintainer or other worker agent) and is associated with some part of the parent's graph structure. However, this association in not the direct association among some nodes of maintained by these agents graph structure; we assume, that the parent worker association is made on the graph transformation level i.e. some worker's transformation enforces the realization of some graph transformation over the parent's graph structure. This problem will be in detail discussed in section 4.

The GRADIS agent's model supports the hierarchical model structure with the following limitation:

1. a maintainer agent can exist only at the first level of this structure,
2. a maintainer cannot split itself when it is the parent of some workers,
3. a worker cannot split itself,
4. workers cooperation is limited only to worker parent relations.

The 1, 3, 4 assumptions seam to be the fundamental one, and we have not the intention to modify them in the next time. We would like to cancel the second assumption, and the algorithm of the assurance correct cooperation of the worker with the maintainer agents, created by splitting worker's parent is in the final phase of the realization.

3   Complementary Graph as a Background for Maintainer Agent Work

The data structure, that is maintained and transformed by agents, has a form of labeled (attributed) graphs. Let Σ^v and Σ^e be a sets; the elements of Σ^v are used as node labels and the elements of Σ^e are used as edge labels. The graph structure are defined as follows:

Definition 3.1
A (Σ^v,Σ^e)-graph is a 3-tuple (V,D,v-lab) where V is nonempty set, D is a subset of $V\times\Sigma^e\times V$, and v-lab is a function from V into Σ^v.   ∎

For any (Σ^v,Σ^e)-graph G, V is set of nodes, D is set of edges and v-lab is a node labeling function. One can extend this graph definition eg. by introduction attributing functions both for nodes and edges, but these extensions will not influence on the rules of the centralized graph distribution and their transformation, because of that they will not consider here.

Our intention is splitting of the graph G onto a few parts and distribute them onto different locations. Transformation of each subgraph G_i will be controlled by some maintainer agent.

To maintain the compatibility of centralized graph with the set of split subgraphs some nodes (called border nodes) should be replicated and placed in the proper subgraph. Graphically, we will mark a border node by a double circle representation; we also introduce the set Border(G) to express that v is a border node in the graph G by a formulae $v\in$ Border(G). During the splitting of the graph we are interested in checking if the connection between two nodes crosses a border among the subgraphs; the function PathS(G, v,w) will return all sets of the nodes belonging to the edges creating a connection (without cycles) among v and w. For example for the graph G presented in figure 3.1 PathS(G,a,c)={{a,c},{a,b,c},{a,d,e,c},{a,d,f,g,e,c}}.

Definition 3.2

The set of graphs $G_i=(V_i, D_i, v\text{-lab}_i)$, for $i=1..k$, is a split of graph G to a complementary forms iff exist a set of injective homomorphisms s_i from G_i to G such as :

1) $\bigcup\limits_{i=1..k} s_i(G_i) = G$

2) $\forall i, j = 1..K \quad \left(s_i(V_i) \cap s_j(V_j)\right) = \left(s_i(Border(G_i)) \cap s_j(Border(G_j))\right)$

3) $\forall w \in V_i \quad \forall v \in V_j : \quad \exists p \in PathS(G, w, v) \implies \exists b \in Border(G_i) : s_i(b) \in p$

4) $\forall j = 1..k \quad v \in Border(G_j) \iff \left(\exists w \in G_j : w \leftrightarrow v\right)$ or $G_j = \{v\}$

where $\leftrightarrow$ means that the nodes are connected by an edge.   ∎

The introduced formal definition is difficult to use in practical construction of the complementary graphs, because of that we introduce an algorithm for splitting of the centralized graph.

Algorithm 3.1

Let H be a subgraph of G then two complementary graph H' and H'' are created in the following steeps:

1. initially H'=H and H''=G\H
2. for every $v \in H$ such that exist $w \in$ G\H if w is connected with v then v is replicated and:
   - v stays in H' but it is marked as a border node,
   - border node v', a copy of v, is attached to H'' with all edges connecting v with G\H,
   - proper references are added to v and v' (iff v has been a border node before this operation this references should also associate these node with another ones),
3. some reindexation of $V_{H'}$ and $V_{H''}$ should be made for optimizing local transformations.   ∎

The unique addressing of nodes in the glued graph (after 3-th steep of an algorithm or any sequences of local graph transformations) is guaranteed by the remembering their indices as a pair (local_graph_id, local_index). We also assume, that marking of the node as a border one is associated with designation for it of an unique index in the border nodes set (border_id is equal to 0).

Algorithm 3.2

The construction of G from the complementary graphs set $\{G_i\}$ is made in the following way:

- for the boarder nodes, one of the replicas, indexed as (0,glob_bord_node_index), is added to V,
- foe all normal (not border) nodes are added to V, with their local indexation,
- the edges in E are inherited from the local graphs (if one node of the edge is a border node in the final edge its global representative appears).
- The labeling function lab is the union of lab_i.   ∎

Splitting the graph G onto a few new complementary forms can be made by execution of the algorithm 3.1 on the already split graph H' or H" (and so on).

An example of G and set of two comple-mentary distributed graphs are presented appropriately on Fig. 3.1 and 3.2. The indexation inside G is consistent with one introduced in algorithm 3.2.

For any border node v in the graph G_i we can move boundary in such a way that, all nodes that are connected with v (inside another complementary graphs) are incorporated to G_i as a border nodes and the v node replicas are removed from another graphs (i.e. v stays a normal node). For graphs presented in Fig. 3.2 an incorporate((0,1),1) operation creates graphs presented in Fig. 3.3.

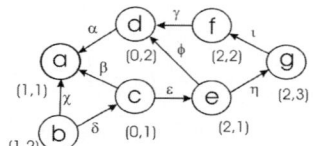

Fig. 3.1 Graph G

Let $\mathbb{G}$ be a class of graphs, then analogically as in [7], we define $\mathbb{R}$ as a class of rules and $\Rightarrow_r$ as a rule application operator, that for $r \in \mathbb{G} \times \mathbb{G}$ yielding a binary operation over the graphs. A pair (G,G'), usually written as G$\Rightarrow_r$G', establish a direct derivation from G to G' through r. The GRADIS framework associates with each distributed complementary graph the maintainer agent, that not only makes the local derivations possible, but also assures the coordination and synchronization of parallel derivations on the different

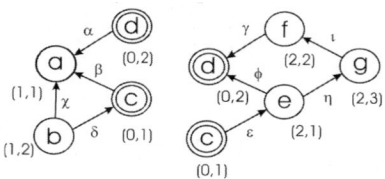

Fig 3.2 Complementary graphs

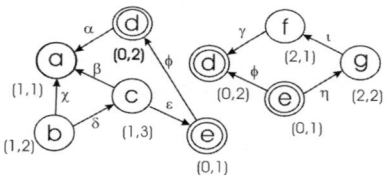

Fig. 3.3. Complementary graphs- 2

complementary graphs. Each maintainer agent is also able to gather (after some cooperation with other agents) a temporary information about the neighborhood of the replicas of the pointed boundary node; for any natural number k and any border node $v \in$ Border(G_i) i-th maintainer agent is able to return a graph B=k-distance_ neighborhood(v) that is a subgraph of the global graph G limited to the nodes distant from v no more then k.

Graph grammars provide a mechanism in which a local transformations on a graph can be modeled in a mathematically precise way. The main component of a graph grammars is a finite set of productions; a production is in general, a triple (L,R,E) where L and R (left- and right-hand graph of production, respectively) are a graph and E is some embedding mechanism. Such a production can be applied to graph G whenever there is an m occurrence of L in G. It is applied by removing m(L) from G, replacing it by (an isomorphic copy) of R, and finally using embedding mechanism E to attach R to the remainder G-m(L).

The general idea of the proposed application of the production L→(R, E) over a distributed graph is the following:

1. each of i-th maintainer agent autonomously decide to apply this production when one can find the m occurrence of L inside G_i , none of border nodes are removed and all removed edges belongs to G_i edges.
2. otherwise, i-th maintained agent needs the cooperation with the rest of the system, that will be made in three steep:
    2.1. the gathering an information: for all nodes v such that
        $v \in$ Border$(G_i) \cap$ Nodes(m(L)) we are looking such k that the graph B designnated as k-distance_ neighborhood(v) covers L and all removed nodes and edges.
    2.2. the preparing an environment: all nodes belongs to B and does not belongs to G_i are incorporated to the graph G_i by execution of sequence incorporate(…) operations made in the transactional mode.
    2.3. the applying of a production: a production $L \rightarrow (R,E)$ can be applied in a new created graph G_i' according to 1-th rule (local derivation).   ∎

The presented algorithm does not depend on the specific properties of the graph transformation mechanism like NLC embedding transformation (in case of the algorithmic approach) [8] or single- and double-pushout (in the case of the algebraic approach) [9]. In [10] the cooperation among the agents in the case of these types graph transformations is considered and a detail algorithms basing on these transformation properties are presented.

4   Cohesion Graph Grammar as a Background for Workers Agents Cooperation

The maintainer cooperation, based on complementary graphs, make possible to distribute some centralized solution onto the distributed agent's environment; each of agents maintains the "reasonable" part of the centralized graph, what makes the parallel computation effective. As it was mentioned in section 2, GRADIS framework offers more one type of agents – the worker agents. The worker agent is created by the maintainer agent temporarily or permanently to support the maintainer creativity. For a better understanding of the problem we consider some examples.

The first example is a part of the problem of finding subpatterns, presented in [11]. Let, set of maintainer agents remember a set of complementary graphs $\{G_i\}$ describing large pattern, and let the graph grammar SUB defines a set subpatterns; we would like to check iff a given node v of G_i graph is a starting point of any subpattern (defined by SUB) contained by H=$\bigcup G_i$.

For the simplicity we assume that in all subpatterns minimal distance between two node is less or equal to k. Usually, graph B=k-distance_ neighborhood(v) in H does not contain itself in the graph G_i. The sequence of incorporate operations can enlarge graph G_i to G_i' such that B$\subseteq G_i$', but this solution seems unlucky from two reasons:

- it causes growing the i-th maintainer structure and finally it leads to the centralized solution,

- the i-th maintainer agent should in parallel way service two types graph grammars transformations (inherited from centralized solution and defined by SUB graph grammar) what would enforce the need of these activities synchronization. This synchronization could not be independent from the definition of the mentioned grammars.
- It seams to be better replicate the structure of graph B, and create a worker agent that solves the membership problem parameterized by graph B, node v, and graph grammar SUB[1]. Let's consider maintainer-worker relationship:
- the worker is created only for realization of the particular task,
- the maintainer and worker structures are not synchronized,
- the worker after the task realization informs the maintainer about its work effects.

In GRADIS we assume that the communication among these agents is made at the graph transformation level, i.e. the worker recommends execution by the maintainer of some transformations depending on the final effect of its work.

The second example is inspired by a distributed adaptive design process [12]. This process composes from a few phases: firstly we create basic plans (eg. of a building design), next we can plan rooms arrangement and finally details of the furniture. Graph transformations seams to be very promising for this application [13].

We assume that, the building plain (in a graph form) is maintained for the set of maintainer agents, and each of maintainer agents takes care about the subgraph responsible for parameterization (allocation of walls, doors, windows, etc.) of one or more rooms. With each of the rooms it is associated a permanent worker agent that is responsible for this room arrangement. Let's note that, in this case, the maintainer-worker relationship is more complex then in the previous example. The moving walls or doors has a very strong influence on the room arrangement process and from the other hand putting of the wardrobe on the wall with door is possible only when these door will become transferred into other place. Thus the cooperation between graph transformation systems supported by maintainer and worker agents should be very close.

The formal background of the for the support above problem bases on a conjugated graph grammars theory [14][15]. In the conjugated graph a new type of nodes appears – remote nodes. Remote nodes represent the nodes appearing in other graph structures. In the conjugated graphs grammars we assume that, the P graph transformations (on the first agent graph structure) in which exist a remote node w is associated with the Q graph transformation (on the second agent graph structure), such that it modifies the neighborhood of the node represented by w. The pair P and Q are called conjugated transformations in context of remote node w. In order to synchronize the set of conjugated graph transformations we assume that, GRADIS assures that <u>both P and Q graph transformations will be successfully performed</u>.

To guarantee this we consider three types of the conjugated graph grammars:

- the strictly conjugated graph grammars – when the created conjugated graph structure guarantee, that application of P graph transformation forces the possibility of application Q graph transformation. In [15] there is proved that the fulfillment eight condition guarantee strictly conjugated model for double-pushout graph grammars.

---

[1] It is assumed that SUB graph grammar is one of that are able to solve the membership problem in a polynomial time complexity (like for example ETPL(k) graph grammar [16]).

- the conditional conjugated graph grammar – when both the transformations are performed, when some (earlier defined) conditions are fulfilled by the both graph structures.
- the transactional conjugated graph grammars – when both the production are performed in the transactional mode i.e. either all of these graph transformation can be performed without violating the conjugated graph cohesion or none of them will be applied.

Let's notice that in conditional and transactional models, there was not any assumption on the type graph transformation performed by agents. There is no objection to construct a system in which parent and son agents uses different graph transformation systems.

The level of a furniture design (supported by the worker created for the support the room arrangement) points out yet another problem. Assume that, we would like to furnish the house in the style Louise XVI and all chairs should be identical. The first part of the problem can be solved by preparing of the proper graph grammar. The second at the first sight demands the non-hierarchical worker's cooperation. Fortunately, it can be solved by introduction worker's equivalence, i.e. while a worker creation its parent can decide whether create a new agents instance or to use already created agent. The common agent react on demand any of its parents, but its reaction is broadcasted to all of them. This solves the problem of some subsystems unification; in our example when the one room arranger suggest made chairs more wide, then this modification (if accepted by the other arrangers) will be made for all chairs.

5   Conclusions

There was a few fundamental assumptions of the agents GRADIS framework:

- the introduction of the possibility graph transformation over the set of distributed graphs,
- the making use of the all earlier theoretical achievement of the centralized graph transformations in the multiagent's environment.
- the support the agents cooperation over the hierarchical graph structures.

The concept of the distribution (to specify a concurrent and distributed system) was initially considered in the algebraic approach. The distributed graph transformation [2] was developed with the aim to naturally express computation in system made of interacting parts. In general, it is assumed an existence of some global state, repressentted as a graph, that is used to synchronize execution of the amalgamated productions or synchronized productions. In [9], a distributed attributed graph (Distr(AGr) has two levels: a network and a local level. Network nodes are assigned to the local graphs, which represent their state. This graphs are transformed via distributed rules. They consist from a network rule and a set of local rules, one for each node of network rule. In addition to the usual, in the double push-out approach, *dangling* and *identification* conditions, two additional *locality* conditions should be verified [17][18]. The satisfying of the *locality* conditions by a local graph morphism assures that the push-out can be constructed component wise. There are a few examples of usefulness of

Distr(AGr) in the visual design of distributed system [17], the specifying integrated refactoring [19], the modeling a analysis of a distributed simulation protocols [20]. The problem of evaluation of the distributed rules and graphs in a real distributed environment is not up till now considered (the last implementation [20] is made in a centralized environment with help of AToM tool [21]).

The complementary graphs model, introduced for the worker agents in the GRADIS framework, assure that for each of the local graphs any graphs transformation are introduced locally using only these graph; what means that GRADIS does not influence of the properties of the graph transformation (considered in context of global graph generation) such as: confluention, Local Church-Roster problem; the explicit parallelism problem is also solved on condition, that concurrency instead real parallelism is possible, because we assume that, the evaluation of $m(L_i)$ is made in a critical section with respect of the nodes belonging to $m(L_i)$.

The conjugate graph grammars system, supporting the hierarchical graph transformation systems, assumes that each agent can be supported by another type of graph transformation. As a consequence, for each of the subproblem we can use the graph transformation system that describes it in the best way.

The different graph transformation systems, coexisting in a hierarchical graph structure, seams to be promising solution with respect the final computational effectiveness evaluation; it is obvious that the graph structure supported by the maintainer agents, with respect its size, should be managed by graph transformation systems with the polynomial time complexity. Unfortunately, such systems have too weak descriptive power to describe most of the considered problems. On the other hand, the maintainer usually keep an information about the structural properties of the modeled system; for such type of information exist at least one class of graph transformation systems (ETPL(k) graph grammars [16]) with enough descriptive power [22] and the polynomial time complexity of the parsing and membership checking. From the worker agents we need not pay such attention on the graph transformation complexity, when the size of supported graphs is limited to tens. Finally, the computational complexity of such a hierarchical system, constructed in such a way, can be acceptable.

One of the most important limitation of the current GRADIS framework is lack possibility of the maintainer's graph split, when some worker has been created by them (then more that some of this worker could be created as a permanent one). This problem is our preferential work, and an algorithm of the assurance correct cooperation of the worker with maintainer agents, created by splitting the worker's parent seams to in the final phase of the realization.

Reference

1. Rozenberg, G.: Handbook of Graph Grammars and Computing By Graph Trans-formation, Foundations edn., vol. I. World Scientific Publishing Co., Singapore (1997)
2. Ehrig, H., Engels, G., Kreowski, H.-J., Rozenberg, G.: Handbook of Graph Grammars and Computing By Graph Transformation, Application, Languages and Tools edn., vol. II. World Scientific Publishing Co, Singapore (1999)
3. Ehrig, H., Kreowski, H.-J., Montanari, U., Rozenberg, G.: Handbook of Graph Grammars and Computing By Graph Transformation, Concurrency, Parallelism, and Distribution edn., vol. III. World Scientific Publishing Co, Singapore (1999)

4. Flasiński, M.: Distorted Pattern Analysis with the Help of Node Label Controlled Graph Languages. Pattern Recognition 23(7), 765–774 (1990)
5. Sycara, K.P.: Multiagent Systems. AI Magazine, 79–92 (1998)
6. Kotulski, L.: Supporting Software Agents by the Graph Transformation Systems. In: Alexandrov, V.N., van Albada, G.D., Sloot, P.M.A., Dongarra, J. (eds.) ICCS 2006. LNCS, vol. 3993, pp. 887–890. Springer, Heidelberg (2006)
7. Kreowski, H.-J., Kuske, S.: Graph Transformation Units and Module. In: [2], pp. 607–640
8. Engelfriet, J., Rozenberg, G.: Node Replacement Graph Grammars. In: [1], 3–94
9. Ehrig, H., Heckel, R. Löwe, M., Ribeiro, L., Wagner, A.: Algebraic Approaches to Graph Transformation – Part II: Single Pushout and Comparison with Double Pushout Approach In: [1], pp. 247–312
10. Kotulski, L.: On the Distribution Graph Transformations, Preprint of Automatics Chair. AGH University of Science and Technology (January 2007); (submitted to the Theoretical Computer Science)
11. Kotulski, L.: Distributed Graphs Transformed by Multiagent System In: International Conference on Artificial Intelligence and Soft Computing ICAISC, Zakopane. LNCS(LNAI), vol. 5097. Springer, Heidelberg (2008) (accepted) (to be published)
12. Strug, B., Kotulski, L.: Distributed Adaptive Design with Hierarchical Autonomous Graph Transformation Systems. In: Shi, Y., van Albada, G.D., Dongarra, J., Sloot, P.M.A. (eds.) ICCS 2007. LNCS, vol. 4488, pp. 880–887. Springer, Heidelberg (2007)
13. Grabska, E., Strug, B.: Applying Cooperating Distributed Graph Grammars in Computer Aided Design. In: Wyrzykowski, R., Dongarra, J., Meyer, N., Waśniewski, J. (eds.) PPAM 2005. LNCS, vol. 3911, pp. 567–574. Springer, Heidelberg (2006)
14. Kotulski, L., Fryz, Ł.: Assurance of system cohesion during independent creation of UML Diagrams. In: Proceedings at the Second International Conference on Dependability of Computer Systems DepCoS - RELCOMEX 2007, Poland, June 14-16, 2007, pp. 51–58. IEEE Computer Society, Los Alamitos (2007)
15. Kotulski, L., Fryz Ł.: Conjugated Graph Grammars as a Mean to Assure Consistency of the System of Conjugated Graphs. In: Third International Conference on Dependability of Computer Systems DepCoS - RELCOMEX 2008 (accepted, 2008)
16. Flasinski, M.: Power Properties of NLC Graph Grammars with a Polynomial Mem-bership Problem. Theoretical Comp. Sci. 201(1-2), 189–231 (1998)
17. Fisher, I., Koch, M., Taentzer, G., Vohle, V.: Distributed Graph Transformation with Application to Visual Design of Distributed Systems. In: [3], pp. 269–337
18. Taentzer, G.: Distributed Graphs and Graph Transformation, Applied Categorical Structures. Special Issue on Graph Transformation 7(4) (December 1999)
19. Bottoni, P., Parisi Presicee, F., Taentzer, G.: Specifying Integrated Refactoring with Distributed Graph Transformations. In: Pfaltz, J.L., Nagl, M., Böhlen, B. (eds.) AGTIVE 2003. LNCS, vol. 3062, pp. 220–235. Springer, Heidelberg (2004)
20. de Lara, J., Taentzer, G.: Modelling and Analysis of Distributed Simulation Protocols with Distributed Graph Transformation. In: Proceedings of the Fifth international Conference on Application of Concurrency To System Design. ACSD, pp. 144–153. IEEE Computer Society, Washington (2005)
21. de Lara, J., Vangheluve, H.: AToM3: A Tool for Multi-Formalism Modeling and Meta-Modelling. In: Kutsche, R.-D., Weber, H. (eds.) FASE 2002. LNCS, vol. 2306, pp. 174–188. Springer, Heidelberg (2002)
22. Kotulski, L.: Graph representation of the nested software structure. In: Sunderam, V.S., van Albada, G.D., Sloot, P.M.A., Dongarra, J. (eds.) ICCS 2005. LNCS, vol. 3516, pp. 1008–1011. Springer, Heidelberg (2005)

User-Assisted Management of Agent-Based Evolutionary Computation

Aleksander Byrski and Marek Kisiel-Dorohinicki

AGH University of Science and Technology, Kraków, Poland
{olekb,doroh}@agh.edu.pl

Abstract. In the paper the need and general idea of user-assisted management of computational intelligence systems is discussed. First some methodological issues are presented with particular attention paid to agent-based approaches. These general considerations are supported by the case of evolutionary multi-agent system (EMAS) with immunological selection, applied to multi-modal function optimization subjected to a user-driven tuning procedure. Finally, preliminary experimental results considering the influence of selected parameters on the performance of the system are shown.

1 Introduction

Computational intelligence relies mostly on heuristics based on evolutionary, connectionist, or fuzzy systems paradigms, but also on other techniques, which used by themselves or in combination exhibit behavior that, in some sense, may be considered intelligent. One of the main advantages of such approaches is their ability to produce sub-optimal solutions even in cases, when traditional methods fail. One of the main drawbacks of these approaches is that they need to be adapted to each problem to be solved. Their configuration (a number of parameters in the simplest case) determines their behavior and thus often significantly influences their solving capabilities.

In practice the configuration of a particular technique appropriate for a given problem is usually obtained after a number of experiments performed by a user, which is often an unreliable and time-consuming task. Meanwhile the choice of specific mechanisms and their parameterization is crucial for the accuracy and efficiency of the examined technique. That is why various approaches were proposed to deal with the problem more reliably and possibly without involving a user directly. Yet automatic discovery of an acceptable system configuration by means of some optimization technique requires resolving some methodological issues, e.g. needs the formulation of the search domain and criterion, which may be difficult in many cases.

In the paper the problem of tuning of agent-based computational systems is considered. The abstraction of an agent provides a natural perspective for designing decentralized systems and may facilitate the construction of hybrid soft computing systems [1]. Immunological mechanisms introduced into agent-based evolutionary optimization may serve as an example of such approach [2]. At the same time, these immunological mechanisms may be perceived as a means for automatic tuning of evolutionary processes in the population of agents.

M. Bubak et al. (Eds.): ICCS 2008, Part III, LNCS 5103, pp. 654–663, 2008.

This paper starts with the discussion of the structure of management of computational system, aiming at discovering its acceptable configuration, with a particular emphasis on agent-based models. Next, as an example of such a system, the idea of *immunological evolutionary multi-agent system* is presented. The results of the user-assisted tuning of the system and their discussion conclude the paper.

2 User-Assisted Configuration of a Computational System

Most computational intelligence systems need a vast amount of work before they are able to provide acceptable solutions for a given problem. Hybrid systems, which put together elements of different techniques, by the effect of synergy often provide better solving capabilities and more flexibility [3], but they are obviously more difficult to configure. This is mainly because such systems are usually very hard to model and analyze. In many cases it may be even admitted that the principles of their work are not known, just like for many evolutionary algorithms: there were some approaches to model specific algorithms, yet they were successful only in very simple cases, like for a simple genetic algorithm [4].

When theoretical analysis of a computational model fails, its configuration needs to be decided experimentally. A result depends on the experience and patience of an expert, who is to perform many experiments to check accuracy and efficiency of the particular technique with different settings. Instead of an expert some automatic procedure may be applied to search for an optimal configuration of the computational system. In this case some criterion is indispensable to drive the search, which formulation may be really difficult in many cases. It must take into consideration ambiguity of the system evaluation based on its current state, or even based on gathered data from a single run, because of stochastic character of many soft computing techniques.

The task of automatic optimization of a complex computational system presents at least two more levels of problems for its designer. First, a particular class of optimization algorithms must be chosen that is suitable for that particular application. It is worth to remember that most optimization techniques are dedicated for parametric optimization, so a computational system must be described in terms of a fixed set of decision variables of acceptable types. Also the evaluation of each examined configuration may be a really time consuming job because it may require multiple system runs. Second, the configuration of the optimization algorithm needs to be tuned, so in fact the problem is repeated at the higher level of abstraction. Such approach in the field of optimization techniques is often called a *meta*-algorithm (compare the idea of meta-evolutionary approach [5]).

Indeed, it seems that automatic discovery of a system configuration for computational intelligence techniques is often an important but only auxiliary tool. The role of a human expert cannot be overestimated, because of a variety of available options and characteristics that describe the system behavior. Yet, it would be helpful, if an expert were equipped with an appropriately defined procedure and automatic system for testing, defining parameters, visualizing the results, lastly automatically searching for optimal parameters of the system.

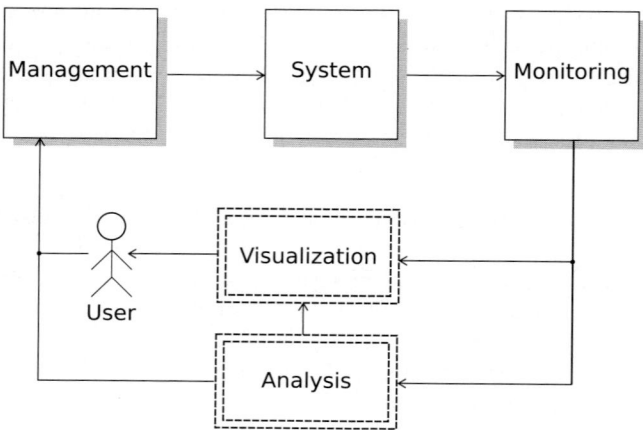

Fig. 1. A general scheme of user-assisted management of a computational system

In Fig. 1 a general structure of management of a computational system is presented. In the simplest case it may be realized as a set of scripts run by some engine, based on the user recommendations. Several characteristics of the system may be monitored and possibly stored into a database. Then gathered data may be visualized and analyzed by the user, or processed by some optimization algorithm to obtain some estimation of the system quality. Both the user and analyzer module amy use similar goal function in order to look for the best configuration of the system.

3   Management of an Agent-Based Computational System

Agent-based computational systems open possibilities of introducing on-line tuning strategies. These may be realized as specialized agents, which may perform analysis of the system behavior while it is still running. Of course because of assumed agents' autonomy they cannot directly control computing agents, but rather apply some indirect strategy based e.g. on resource management, which in turn may influence the behavior of computing agents.

For multi-agent systems building effective and efficient monitoring mechanisms is not an easy task. This is mainly because of the assumed autonomy of agents, but for computational systems also because of the number and variety of agents that produce huge amount of data, which quickly become out-of-date. Also problems of distribution and heterogeneity of agents are of vast importance. The proposed solution assumes local on-line processing of only required (subscribed) information via *monitoring services*, available both to the agents and external software tools via dedicated interaction protocols [6].

In a computational MAS acquisition of required information may be realized mostly by the core infrastructure, since it "knows" a lot about the agents' states. Thus in this case a monitoring subsystem should be tightly integrated with the agent platform. Fig. 2 shows a typical structure of a computational MAS together with a monitoring services

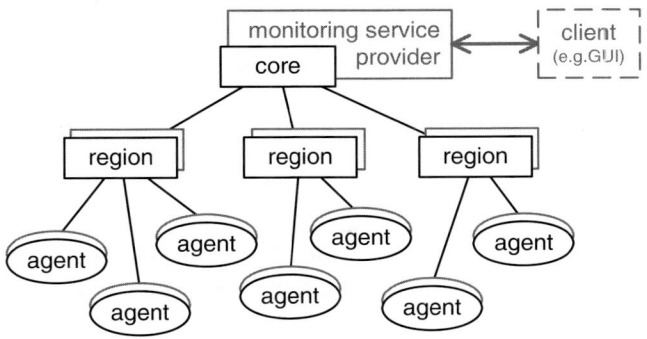

Fig. 2. Monitoring infrastructure for computational MAS

provider, which is a local authority responsible for management of all monitoring resources in a particular location (computing node). Since some directory of monitoring resources is indispensable to support processing and delegation of monitoring services, the monitoring services provider also delivers appropriate interfaces for agents of the system and external clients to facilitate identification of agents, their properties, and actual state.

The prototype implementation of the monitoring subsystem for a computational MAS was realized for *AgE* platform[1] — a software framework facilitating agent-based implementations of distributed (mostly evolutionary) computation systems. Monitoring services rely on the instrumentation of base classes equipped with the mechanism of properties and notifications, which may be used by monitoring services provider, according to *Observer* design pattern [7].

4   Evolutionary and Immunological Multi-Agent Systems

The idea of agent-based evolutionary optimization most generally consists in the incorporation of evolutionary processes into a multi-agent system at a population level. In its fine-grained model (EMAS – *evolutionary multi-agent systems*) it means that besides interaction mechanisms typical for agent-based systems (such as communication) agents are able to *reproduce* (generate new agents) and may *die* (be eliminated from the system). Inheritance is accomplished by an appropriate definition of reproduction (with mutation and recombination), which is similar to classical evolutionary algorithms. Selection mechanisms correspond to their natural prototype and are based on the existence of non-renewable resource called *life energy*, which is gained and lost when agents perform actions [8].

This shortly described approach proved working in a number of applications, yet it still reveals new features, particularly when supported by specific mechanisms, like immunological approach proposed as a more effective alternative to the classical energetic selection (iEMAS – *immunological* EMAS). In order to speed up the process of

---

[1] http://age.iisg.agh.edu.pl

selection, based on the assumption that "bad" phenotypes come from the "bad" genotypes, a new group of agents (acting as lymphocytes) may be introduced [2]. They are responsible for recognizing (assuming some predefined affinity function) and penalizing (by decreasing agent's energy or even removing an agent) agents with genotypes similar to the pattern possessed.

More thorough description of these ideas may be found e.g. in the referenced papers [8] [2]. Important here is that lymphocytes may be considered as autonomous agents performing on-line analysis of the system behavior (by searching for "bad" agents) and influencing computational agents (by penalizing them) in order to speed up the process of selection (see 3). Thus original concept of EMAS is not modified, instead newly introduced agents may be perceived as a means for automatic adaptation of computation.

It should be also stressed that the variety of parameters describe the system behavior in both EMAS and iEMAS – among the most important one may distinguish:

- number of demes (islands),
- initial number of agents on a deme,
- initial agents' energy,
- death energy level,
- duration of negative selection,
- agent rendezvous period,
- agent–agent evaluation rate,
- lymphocyte–agent similarity function coefficient,
- lymphocyte–agent penalization rate,
- lymphocyte prize for "bad" agent discovery.

In fact over 30 parameters need to be established for each particular run of the system, and many of them may prove important for the quality of the solutions obtained.

5 Experimental Results

The most important characteristics of the system under consideration is as follows:

- There are three fully connected demes, in every deme there are 20 individuals in the initial population.
- Agents contain real-valued representation of ten dimensional search space.
- Two variation operators are used: discrete crossover and normal mutation with small probability of macro-mutation.
- Lymphocyte contains a mutated pattern of the late agent and use the similarity function based on computing differences for corresponding genes to discover "bad" agents.
- Lymphocytes are rewarded for finding "bad" agents, and removed after a longer time of inactivity.

Below, the results of both introducing immunological mechanisms into EMAS and a user-driven tuning of EMAS and iEMAS parameters are presented. The tuning was

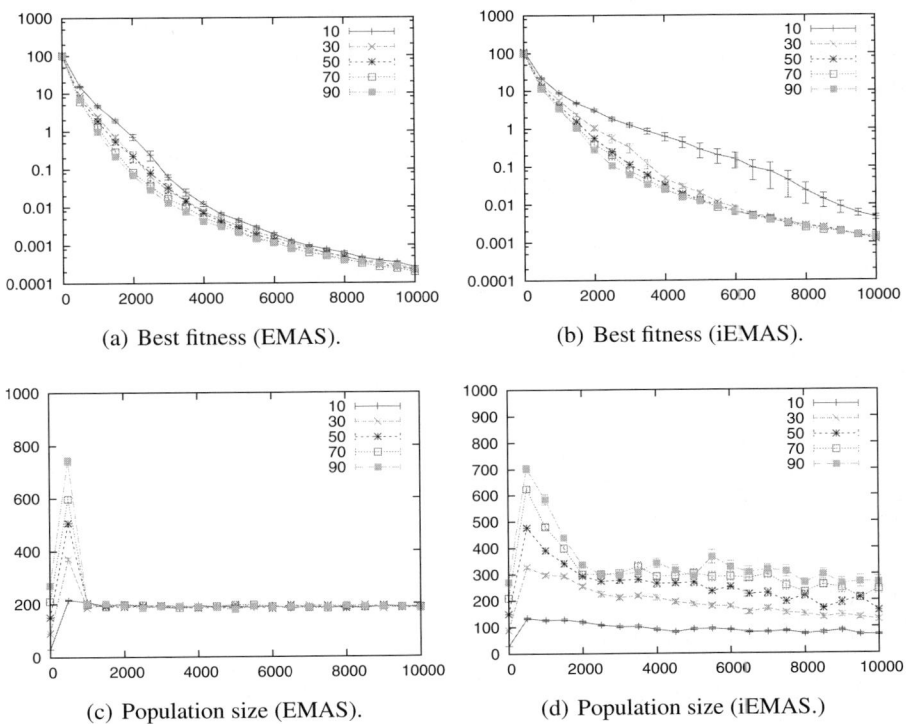

(a) Best fitness (EMAS).

(b) Best fitness (iEMAS).

(c) Population size (EMAS).

(d) Population size (iEMAS.)

Fig. 3. Best fitness and population size (depending on step of systems' work) in EMAS and iEMAS for different initial number of agents (10–90)

realised using a set of scripts parameterized by the user. Every experiment was repeated 10 times. As the test optimization problem the Rastrigin function in 10 dimensions was chosen.

One of the main goals of the research was to check, how does the computation efficiency and accuracy in EMAS differ from iEMAS depending on different parameters. The efficiency was described by several important characteristics such as population size and the number of fitness function calls, and the accuracy was expressed in terms of the best fitness value. The tuning was based on arbitrarily chosen parameters: initial population size, rendezvous rate (EMAS and iEMAS), and negative selection period (only iEMAS).

First of all, parameters common to EMAS and iEMAS were explored. The results gathered for different values of initial population size are shown in Fig. 3 It is easy to see, that this parameter does not affect in a significant way the work of EMAS – the value of both best fitness and population size are very similar for all the values of the tested parameter. This is very important for stating, that there is no need to introduce hundreds of agents into a subpopulation in order to get reliable results.

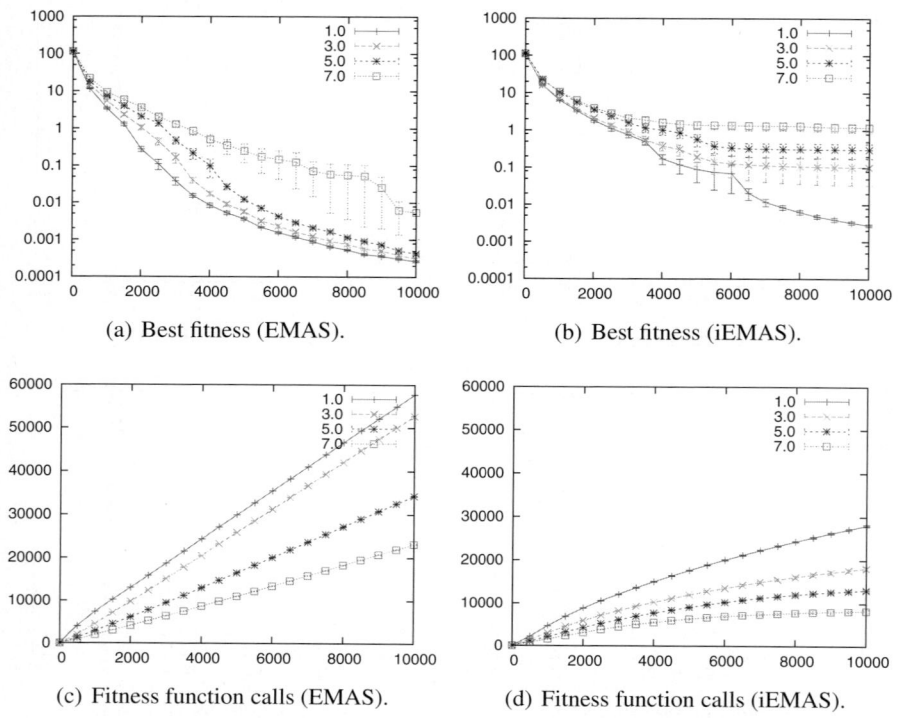

Fig. 4. Best fitness and number of fitness function calls (depending on step of systems' work) in EMAS and iEMAS for different rendezvous period (1-7)

However, for iEMAS the situation is quite different. It may be clearly observed, that for high values of the parameter, iEMAS reaches better suboptimal solution of the problem. Characteristics of population size for EMAS and iEMAS differ, because the population in iEMAS is affected by the actions of lymphocytes, so it is important to optimize their parameters, otherwise they may remove promising solutions from the population.

The second parameter that was chosen for presentation is the length of rendezvous period (agents meet and evaluate themselves in intervals of such length). Best fitness and the number of fitness function calls for EMAS and iEMAS were examined (see Fig. 5). Comparing the graphs shown in Fig. 4(a) and Fig. 4(b) one can see, that introduction of lymphocytes leads to decreasing the efficiency of finding suboptimal solutions of the problem. Yet, the graphs shown in Fig. 4(c) and Fig. 4(d) prove, that though decreasing the accuracy, efficiency increases: there are fewer agents in iEMAS than in EMAS, therefore there are also fewer fitness function calls, the system seems to be more suitable for solving problems with complex fitness functions.

Frequent evaluations lead to increasing efficiency (better suboptimal solutions were found) in both systems (see Fig. 4(a), Fig. 4(b)), but the number of fitness function

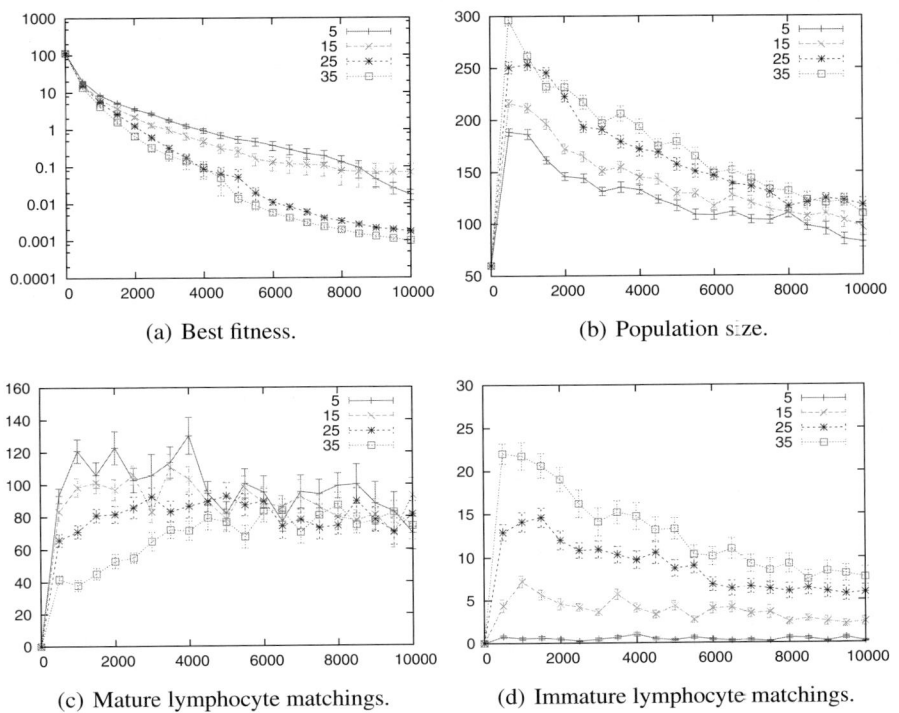

(a) Best fitness.

(b) Population size.

(c) Mature lymphocyte matchings.

(d) Immature lymphocyte matchings.

Fig. 5. Best fitness, population size and lymphocyte matchings (depending on step of system's work) for different lengths of negative selection (5-35)

calls in this case was high (see Fig. 4(c), Fig. 4(d)), so there is a clear tradeoff between efficiency and accuracy. Usually such deliberation lead to assign medium values to the coefficients affecting these characteristics, yet a better answer could be given based e.g. on multi-criteria tuning strategy [9].

After stating that iEMAS may help in reaching better efficiency, and their parameters are crucial, search for optimal parameters of iEMAS was undertaken. One of the vital parameters of iEMAS is the length of negative selection. The results were shown in Fig. 5. Best fitness characteristic is significantly affected by changing the value of the observed parameter (see Fig. 5(a)) obtaining better values for longer periods. It is, because lymphocytes that are longer verified, do not remove good agents from the population. It may be verified by looking on Fig. 5(c) and Fig. 5(d) where count of mature and immature lymphocytes matchings was shown. Longer trained lymphocytes are often removed after meeting high energy agent, so the count of immature matchings is high. However mature matchings are almost independent on the length of negative selection, so the optimal value of this parameter should be medium, because lymphocytes should affect the population of agents, but they should not cause extinction.

6   Conclusion

In the paper problems of configuring computational intelligence systems were discussed, with particular attention paid to agent-based systems. Agent paradigm may facilitate the construction of hybrid systems, which was illustrated by the ideas of evolutionary multi-agent systems with immunological selection applied. These general considerations were supported by an experimental study on tuning of selected parameters of EMAS and iEMAS.

The process of system parameters tuning in the presented agent-based systems is performed on two levels:

- Inner, where certain subset of system's parameters is adjusted in the runtime. In iEMAS, specialized group of agents (lymphocytes) is introduced into the original population of agents. They influence the computation, affecting the structure of the population.
- Outer, where complete set of system's parameters is adjusted, according to the user's preferences and passed to the system before running. It may be done manually by the user, or automatically by some *meta*–approach, such as evolutionary algorithm.

In the course of the user-driven tuning of EMAS and iEMAS, several conclusions were drawn, which may help in future development of such systems:

- Change of starting population size did not affect the accuracy of EMAS, but it was important for iEMAS, which modifies the structure of the population. Better results were found in iEMAS for larger populations, but this parameter must be adjusted carefully, because too many agents in population will lead to high computation cost. There is apparently a tradeoff between accuracy and efficiency in iEMAS, and the lymphocytes clearly affect these aspects.
- In iEMAS because of introduction of lymphocytes, there was much fewer evaluations of fitness function, than in EMAS, so this approach seems good for the optimization problems with costly fitness evaluation. Observed parameters (rendezvous period) is closely bound to the cost of the computation in these systems.
- Negative selection length affected greatly the probability of mature or immature matchings among lymphocytes and agents. This parameter is crucial for immunological selection, and assigning improper value may lead to degeneration of the computation (too high value of this parameter de facto transforms iEMAS into EMAS, because lymphocytes do not have the possibility of modifying the agent population).

These conclusions were followed by more observations and experiments that could not be included in the paper, because of editorial limit. However they will constitute a starting point for further research, which should lead to proposing new strategies for on-line tuning of computational ingelligence systems.

References

1. Kisiel-Dorohinicki, M., Dobrowolski, G., Nawarecki, E.: Agent populations as computational intelligence. In: Rutkowski, L., Kacprzyk, J. (eds.) Neural Networks and Soft Computing. Advances in Soft Computing, pp. 608–613. Physica-Verlag (2003)

2. Byrski, A., Kisiel-Dorohinicki, M.: Agent-based evolutionary and immunological optimization. In: Shi, Y., van Albada, G.D., Dongarra, J., Sloot, P.M.A. (eds.) ICCS 2007. LNCS, vol. 4488, pp. 928–935. Springer, Heidelberg (2007)
3. Bonissone, P.: Soft computing: the convergence of emerging reasoning technologies. Soft Computing 1(1), 6–18 (1997)
4. Vose, M.D.: The Simple Genetic Algorithm: Foundations and Theory. MIT Press, Cambridge (1998)
5. Freisleben, B.: Metaevolutionary approaches. In: B"ack, T., Fogel, D.B., Michalewicz, Z. (eds.) Handbook of Evolutionary Computation, IOP Publishing and Oxford University Press (1997)
6. Kisiel-Dorohinicki, M.: Monitoring in multi-agent systems: Two perspectives. In: Dunin-Keplicz, B., Jankowski, A., Skowron, A., Szczuka, M. (eds.) Monitoring, Security, and Rescue Techniques in Multi-Agent Systems, pp. 563–570. Springer (2005)
7. Gamma, E., Helm, R., Johnson, R., Vlissides, J.: Design patterns: elements of reusable object-oriented software. Addison-Wesley Professional (1995)
8. Kisiel-Dorohinicki, M.: Agent-oriented model of simulated evolution. In: Grosky, W.I., Plášil, F. (eds.) SOFSEM 2002. LNCS, vol. 2540, Springer, Heidelberg (2002)
9. Deb, K.: Multi-Objective Optimization using Evolutionary Algorithms. John Wiley & Sons (2001)

Generating Robust Investment Strategies with Agent-Based Co-evolutionary System

Rafał Dreżewski, Jan Sepielak, and Leszek Siwik

Department of Computer Science, AGH University of Science and Technology, Kraków, Poland
{drezew,siwik}@agh.edu.pl, sapielak@gmail.com

Abstract. Agent-based co-evolutionary systems seem to be convenient for modeling market-related behaviors, since they allow for defining competing and/or co-operating agents which can interact and communicate with each other and influence the environment and other agents. In the course of this paper the idea of utilizing of agent-based co-evolutionary approach for supporting decisions of financial investor through generating possible investment strategies is presented and experimentally verified.

1 Introduction

Almost simultaneously with the growing maturity of financial markets, international stocks, etc., researchers paid their attention to mathematical models, formulas and theorems supporting investment decisions. It is enough to mention in this place such models supporting building effective portfolio as Nobel prize winner Harry Markowitz' Modern Portfolio Theory (MPT) proposed in 1952 [8] [9], or its extension proposed in 1958 by James Tobin [15]. Although, mentioned theory lays the foundations of modern capital investments practically it is nowadays rather only historically-important method of assets pricing. Next, Capital Asset Pricing Model (CAPM) was proposed by J. Traynor [16], J. Lintner [7], J. Mossin and formalized by W. Sharpe [14]—and it was based of course on previous work of Markowitz and his MPT theory. On the basis of the critique of CAPM (e.g. so called Roll's Critique)—Arbitrage Pricing Theory (APT) was proposed by Stephen A. Ross in mid-1970s [13]. In 1990s so-called Post Modern Portfolio Theory (PMPT) was proposed. The notion of PMPT was used for the first time probably by B. M. Rom and K. W. Ferguson in 1993 [12].

The real and wide support of the investments-related decision making was possible along with the use and applying computational units and dedicated computational methods and algorithms. Because of the complexity and difficulty of market-related problems attempts of applying computational intelligence heuristics rather than accurate numerical methods was natural and justified. However, observing market situations and behaviors—entrepreneurs, small and medium enterprises (SMEs), and corporations—all of them all the time have to be more innovative, cheaper, more effective etc. than the others in order to maximize their profits. That is why, all the time some enterprises introduce some organizational, financial or technological innovations and the rest of the market-game participants have to respond to such changes. All the time we are eye

M. Bubak et al. (Eds.): ICCS 2008, Part III, LNCS 5103, pp. 664–673, 2008.

witnesses of a peculiar "arms race", which can be compared to "arms races" caused by co-evolutionary interactions. The range of dependencies that can be seen on the market can be pretty wide—from co-operation, through competition until antagonism. It is obvious however, that all activities of each participant of the market game are conformed to one overriding goal—to survive and to maximize profits (analogies to Darwin's "survival of the fittest"). From the interactions with another enterprises point of view it can be realized by: eliminating from the market as many (weak) rivals as possible and taking over their customers, products, delivery channels etc. (analogy to "predator-prey" interactions), by sucking out of another ("stronger") enterprises' customers, technologies, products etc. (analogy to "host-parasite" interactions), by supplementing partners' portfolio with additional products, technologies, customers etc., and by co-operating with other actors of the market (analogies to living in "symbiosis"). It is seen clearly that one of the most important activity of all market-game participants is co-existence with co-development—and from the computational intelligence point of view we would say—co-evolution (this subject is wider covered in [4]).

Because (generally speaking of course and under additional conditions) participants of the market game are autonomous entities (from the computational intelligence point of view we would say—"agents"), they are distributed, they act asynchronously, and they interact with another entities and with the environment to achieve their individual goal—maximizing profit—in the natural way, applying co-evolutionary multi agent systems seems to be the perfect approach for modeling such phenomenons and environments. In this paper we will present exemplary agent-based co-evolutionary approach for supporting investment decisions.

In the literature there can be found some attempts for applying evolutionary-based approaches for supporting investing-related decisions. S. K. Kassicieh, T. L. Paez and G. Vora applied the genetic algorithm to making investment decisions problem [6]. The tasks of the algorithm included selecting the company to invest in. Their algorithm operated on historical stock data. O. V. Pictet, M. M. Dacorogna, R. D. Dave, B. Chopard, R. Schirru and M. Tomassini ([10]) presented the genetic algorithm for automatic generation of trade models represented by financial indicators. Three algorithms were implemented: genetic algorithm, genetic algorithm with fitness sharing technique developed by X. Yin and N. Germay ([17]), and genetic algorithm with fitness sharing technique developed by the authors themselves in order to prevent the concentration of individuals around "peaks" of fitness function. Proposed algorithms selected parameters for indicators and combined them to create new, more complex ones. F. Allen and R. Karjalainen ([1]) used genetic algorithm for finding trading rules for S&P 500 index. The algorithm could select the structures and parameters for rules.

Creating a system which uses co-evolutionary and agent-based approach to investment strategy generation is interesting because such systems probably could better explore solution space and slow down a convergence process. However, to our best knowledge there have been no attempts of implementation of such systems so far. In the next sections the realization of agent-based co-evolutionary system for generating investment strategies is presented and its properties are experimentally verified.

The paper is organized as follows: in the Section 2 the general idea and architecture of evolutionary system for generating investment strategies is presented. The architecture of the system has been already described in detail in [3], so in this paper we will only focus on the algorithms used in the system. In the Section 3 some selected results obtained with the use of the proposed approach are presented. As compared to the results presented in [3], now we deeper analyze the proposed agent-based co-evolutionary algorithm for generating investment strategies. Finally this work is briefly summarized and concluded in the Section 4.

2   Agent-Based Co-Evolutionary System for Generating Investment Strategies

As it was said, the architecture of the evolutionary system for generating investment strategies has been described in [3], so in this section we will mainly focus on the algorithms used as the computational components. The algorithms used in the system include: evolutionary algorithm (*EA*), co-evolutionary algorithm (*CCEA*), and agent-based co-evolutionary algorithm (*CoEMAS*).

The system has component architecture ([3]) what implies that we can easily replace the given element of the system with another one. The most important parts of the system are computational component, historical data component, visualization component, and strategies evaluation component.

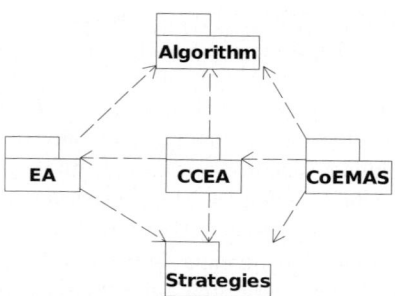

Fig. 1. Dependencies between the evolutionary algorithms packages

The classes implementing the algorithms used as computational components was organized into packages shown in the Fig. 1. Among the most important packages the following should be mentioned:

1. *Algorithm*—it is the most general package containing classes, which are the basis for other classes of strategies generation algorithms;
2. *Strategies*—all classes connected with the representation of strategies are placed within this package;
3. *EA*—it contains the implementation of the evolutionary algorithm;
4. *CCEA*—it contains the implementation of the co-evolutionary algorithm;
5. *CoEMAS*—which contains the implementation of the agent-based co-evolutionary algorithm.

In all three algorithms the strategy is a pair of formulas. First formula indicates when one should enter the market and the second indicates when one should exit the market. Each formula, which is a part of a strategy, can be represented as a tree, which leafs and nodes are functions performing some operations. Such tree has a root and any quantity of child nodes. The function placed in the root always returns logical value. The Fig. 2 shows the tree of exemplary formula, which can be symbolically written in the following way: $MACD > Mov(MACD, 6, EXPONENTIAL)$.

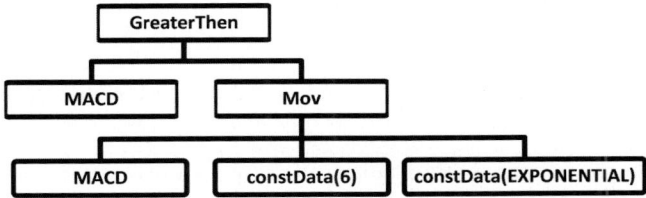

Fig. 2. The tree of exemplary formula

Functions which formulas are built of (there are altogether 93 such functions in the system) are divided into four categories: functions returning stock data (e.g. Close (returns close prices)), mathematical functions (e.g. Sin (sine function)), tool functions (e.g. Cross (finds a cross point of two functions)), and indicator functions (e.g. AD, which calculates the Accumulation/Distribution indicator).

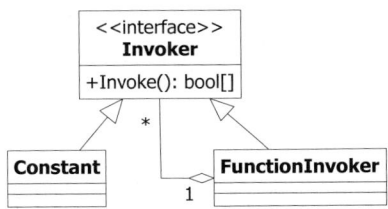

Fig. 3. Class diagram for formula representation

The Fig. 3 shows how formula tree is represented. There are three main classifiers. The interface *Invoker* allows to treat all tree nodes and leaves in the same manner. Class *FunctionInvoker* represents a node of the tree if it wraps function with arguments or represents leaf if it wraps function without arguments. Class *Constant* wraps constant value and it can be also treated as a function which returns always the same value.

2.1   Implemented Algorithms

In the *evolutionary algorithm* the genotype of individual contains two formulas. One is treated as exiting the market formula and the second as entering the market formula.

Fitness estimation is performed with the use of historical stock session data. As a result of formula execution two arrays with logical values are created. One of them concerns entering the market actions and the second one exiting the market actions. The algorithm which computes the profit processes these arrays and determines when the *buy* and when the *sell* actions occur. If the system is outside of the market and in the *buy* array is the *true* value then entering the market occurs. If the system is inside the market and in the *sell* array is the *true* value then exiting the market occurs. In the other cases no operation is performed.

Equation (1) shows the manner of the profit computation:

$$PR = \sum_{i=1}^{t} P_{en} - C_{en} + P_{ex} - C_{ex} \tag{1}$$

where PR is the profit earned as a result of the strategy use, t is the number of transactions, P_{en} and P_{ex} are respectively the prices of entering and exiting the market, C_{en} and C_{ex} are respectively commissions paid at entering and exiting the market.

During the computing of the fitness, besides the profit, also another objectives are taken into consideration: formula complexity—to complicated formula could cause computations slow down, and transaction length which depends on user's preferences. For the sake of simplicity weighted multi-objective optimization was used.

Equation (2) shows how average transaction length is computed:

$$TL = \frac{\sum_{i=1}^{t} BN_i}{t} \tag{2}$$

where TL is the average length of the transaction, and BN_i is the number of the bars (daily quotations) the transaction spans.

Equation (3) is used to calculate the fitness:

$$FV = W_{pr} \cdot PR - W_{fc} \cdot FC_b - W_{fc} \cdot FC_s + W_{tl} \cdot TL \tag{3}$$

where FV is the fitness value, W_{pr} is the weight of the profit, W_{fc} is the weight of the formula complexity, FC_b is the complexity of the buy formula, FC_s is the complexity of the sell formula, W_{tl} is the weight of the transaction length.

In the evolutionary algorithm the tournament selection mechanism [2] was used. Three kinds of recombination were used in the evolutionary algorithm:

- *return value* recombination—it requires two functions in the formulas with the same return type. The functions with their arguments are swapped between individuals.
- *argument* recombination—it requires two functions in the formulas with the same arguments. The arguments of the functions are swapped between individuals.
- *function* recombination—it requires two functions in the formulas with the same arguments and with the same return type. The functions are swapped between individuals and the arguments stay at its original places.

Two kinds of mutations were used:

- *function arguments* mutation—it requires constant value argument. Such constant is changed to another from the allowed range.
- *function* mutation—it replaces the function with another one. There are three variants of this mutation: the function is replaced by another one with the same parameter and the same return value, the function is replaced by another one with the same return type, but parameters are created randomly, the function is replaced by another one with the same return type, but parameters are preserved if it is possible—the parameters which cannot be taken from the function being replaced are created randomly.

In the *co-evolutionary algorithm* the co-operative approach was used ([11]). In this model of co-operative co-evolution the problem is divided into sub-problems which are solved by sub-populations (species). Each sub-population is processed by the evolutionary algorithm. Individuals from the different populations interact only during the fitness estimation stage. There are two kinds of species in the CCEA: entering the market and exiting the market. Each individual has one formula encoded in its genotype. This formula is treated as formula of entering the market or as formula of exiting the market—depending on to which sub-population (species) the given individual belongs. Individual of the given species was evaluated in the group of individuals. Such group consists of evaluated individual and the chosen representative individual coming from the other sub-population. At the initialization stage of the CCEA algorithm the representative individual is chosen randomly. After the initialization stage the representative individual is the best individual from the last generation. Genetic operators and selection mechanism used in the CCEA are the same as in case of the evolutionary algorithm.

   Co-evolutionary multi-agent algorithm algorithm is the agent-based version of the *CCEA* algorithm. Now, each co-evolutionary algorithm is an agent which independently performs computations. The populations of each co-evolutionary algorithm also consist of agents-individuals. The populations exist within the environment which has the graph-like structure with computational nodes ("islands") and paths between them. In each node of the environment there exist resources. Such resources circulate between the agents and the environment. During the evolution process agents give back certain amount of the resources to the environment. Resources are also given to the agents by the environment in such a way that "better" (according to the fitness function) agents are given more resources. Agents with the resource amount less than the value of the parameter specified by the user die. Such mechanism plays the role of selection—thanks to this the agents neither live too long and keep the resources unnecessarily nor they cause the stagnation of the evolutionary processes.

   The fitness estimation and the genetic operators are generally the same as in the case of previously described algorithms. Quite different is the reproduction mechanism, which is now based on the resources. During the reproduction process the parents are chosen on the basis of the amount of resources they possess. After the recombination the parents give the children certain amount of their resources. During the mutation phase the amounts of the resources possessed by the individuals do not change. The agents-individuals can migrate between the nodes of the environment. During the migration of the individual its resource amount is decreased by the certain constant—the resources are given back to the environment.

3   Experimental Results

In [3] the implemented algorithms were analyzed taking into consideration their efficiency, convergence properties and generalization properties. In this paper we deeper analyze properties of the agent-based co-evolutionary algorithm, for example the influence of some parameters' values on the best individual's fitness, and the ability of rules generalization. We also additionally verify the properties mentioned above.

During the experiments all the algorithms worked through 500 generations. They generated strategies for 10 stocks and historical stocks data coming from the period of 5 years were taken into consideration. The session stock data came from the WIG index ([5]) and the period from 2001-09-29 to 2006-09-29.

In the case of analyzing additional aspects of CoEMAS, the system worked through 300 generations, and generated strategies for 3 stocks. Data came from the period of 5 years, whereas while analyzing the ability of the generalization the algorithm worked through 500 generations and the data came from a randomly selected period of at least 5 years. Strategies were generated for 4, 7, 10, 13, and 16 securities chosen randomly.

All the experiments were carried out with the population size equal to 40. In the CoEMAS that population size was achieved using environment with the resource level equal to 980 (CoEMAS is the algorithm with variable population size, which depends on the total amount of the resources within the system). Presented experiments were carried out on the PC with one AMD Sempron 2600+ processor.

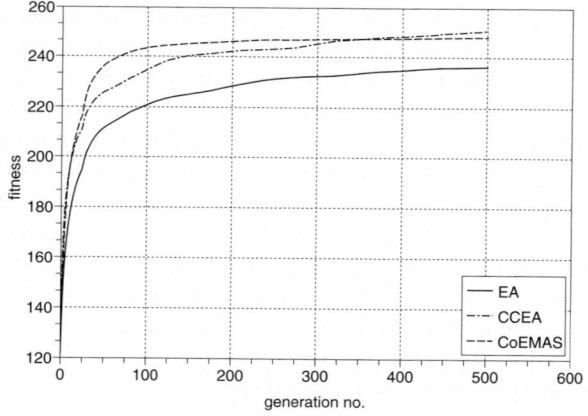

Fig. 4. Fitness of the best individuals in the population (average values from 20 experiments)

Both, Table 1 and Fig. 4 show that CCEA has the best efficiency—the fitness value is the highest in this case. In the case of CCEA high convergence does not cause the fitness worsening. The EA algorithm reveals the worst efficiency and also the highest convergence. The CoEMAS algorithm obtained slightly worse fitness then CCEA, but has

Table 1. The comparison of efficiency, work time and convergence of the algorithms

Algorithm	The best fitness in last generation	Convergence in last generation	Algorithm work time (m:s)
EA	236.171	59.481	1:57.27
CCEA	251.052	58.792	12:9.341
CoEMAS	248.222	44.151	14:49.048

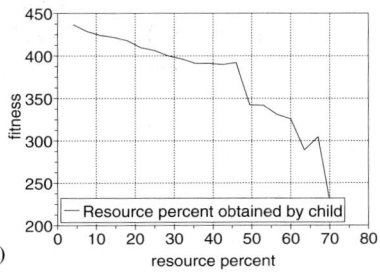

a)
b)

Fig. 5. The minimal amount of resource needed for individuals to survive in CoEMAS (each point is the average of 10 experiments) (a), and the percentage of resources transferred from parental agents to their children during recombination (b)

the smallest convergence. The Fig. 4 shows that the best fitness of the EA during the entire evolution process is much worse than the best fitness of the remaining two algorithms. The best fitness of the CCEA until 350 generation is worse than the best fitness of the CoEMAS, but later (after 350 generation) CCEA obtained a little better results.

Providing the high quality results is costly. The cost is manifested by the computational time. The work time of both CCEA and CoEMAS is considerably longer (six and seven times respectively) than this of EA.

Search for optimal values of the parameters of all algorithms was performed by using back to back experiments. For a given value of the parameter each algorithm was launched a few times and then the average results was computed. Then the value of the particular parameter was modified and efficiency was estimated again. The conclusions coming from such experiments are as follows: for most of the parameters no regularity was found. But there are some exceptions. The most interesting plots/diagrams are presented in the Fig. 5.

Table 2. Results of examining the CoEMAS capability of generalization

Number of stocks in group (n)	Profit (%) per year	Profit (%) per year for random stocks
4	83.71	5.62
7	132.71	23.64
10	48.79	27.11
13	35.33	26.20
16	100.83	25.93

The Fig. 5a shows the influence of the minimal amount of resource needed for individuals to survive on CoEMAS efficiency. At the beginning, with the growth of the value of this parameter the efficiency of the CoEMAS slightly increases. But when the resource amount exceeds 10 units the same efficiency systematically decreases.

The Fig. 5b presents the percentage of the resources transferred from parental agents to their children during the recombination in CoEMAS. It indicates that transferring a large amount of the resource causes a decrease in the efficiency of the algorithm. Whereas transferring a small amount of the resource causes increase in the algorithm efficiency.

As it was mentioned, CoEMAS was also analyzed from the ability for generalization point of view. In the Table 2 the results of experiments concerning generalization are presented. The results show that the number of stocks in a group does not influence the results of strategy generation for the given stocks. Moreover, small amount of stocks in a group (less than 7) causes obtaining poor results for random stocks along with the small capability for generalization. When the number of stocks taken for strategy generating is greater or equal to 7, then the algorithm gives good results in spite of the random securities.

4   Summary and Conclusions

In the course of this paper three evolutionary based approaches for supporting investors with their decisions related to entering (purchasing stocks) and exiting (selling stocks) the market were presented and experimentally verified.

As it was mentioned in the Section 1, using co-evolutionary based approaches seems to be a natural way of modeling market-related interactions, relations and behaviors. In this paper such an approach was used for generating robust investment strategies.

As it can be observed in the Section 3 preliminary experiments and obtained results are very promising since the co-evolutionary algorithms (both "classical" and agent-based) allow for obtaining much more valuable results than compared evolutionary algorithm. During the experiments it was possible to distinguish such parameters of the proposed CoEMAS approach that influence the system and obtained results most significantly and in most regular way.

Authors are obviously still researching and experimenting with the proposed algorithms and the future works will include wide range of additional experiments (also with data coming not only from Warsaw Stock Exchange). On the basis of obtained results further modification and tuning up of proposed approaches will be performed. Also comparisons of obtained results with strategies generated not only on the basis of another (evolutionary based) heuristics but also with the use of classical models, systems and algorithms which are being used by investors for supporting their financial decisions are included in the future research plans. Additionally, as it was mentioned in the Section 1, since the goal of this paper was to experimentally verify the proposed agent-based co-evolutionary approach—it is natural that more interesting were comparisons of efficiency, and the influence of some parameters' values on the obtained results than analyzing particular strategies obtained during experiments. It explains the way of presenting obtained results in the Section 3 but in the future work not only quantitative but also qualitative comparisons of generated strategies will be performed.

References

1. Allen, F., Karjalainen, R.: Using genetic algorithms to find technical trading rules. Journal of Financial Economics 51(2), 245–271 (1999)
2. Bäck, T., Fogel, D., Michalewicz, Z.: Handbook of Evolutionary Computation. IOP Publishing and Oxford University Press (1997)
3. Dreżewski, R., Sepielak, J.: Evolutionary system for generating investment strategies. In: Giacobini, M. (ed.) EvoWorkshops 2008. LNCS, vol. 4974, pp. 83–92. Springer, Heidelberg (2008)
4. Dreżewski, R., Siwik, L.: Co-evolutionary multi-agent system for portfolio optimization. In: Brabazon, A., O'Neill, M. (eds.) Natural Computation in Computational Finance, pp. 273–303. Springer, Heidelberg (in print, 2008)
5. Historical stock data, http://www.parkiet.com/dane/dane_atxt.jsp
6. Kassicieh, S.K., Paez, T.L., Vora, G.: Investment decisions using genetic algorithms. In: Proceedings of the 30th Hawaii International Conference on System Sciences, vol. 5. IEEE Computer Society, Los Alamitos (1997)
7. Lintner, J.: The valuation of risk assets and the selection of risky investments in stock portfolios and capital budgets. Review of Economics and Statistics 47, 13–37 (1965)
8. Markowitz, H.: Portfolio selection. Journal of Finance 7(1), 77–91 (1952)
9. Markowitz, H.: The early history of portfolio theory: 1600-1960. Financial Analysts Journal 55(4), 5–16 (1999)
10. Pictet, O.V., Dacorogna, M.M., Dave, R.D., Chopard, B., Schirru, R., Tomassini, M.: Genetic algorithms with collective sharing for robust optimization in financial applications. Technical Report OVP.1995-02-06, Olsen & Associates Research Institure for Applied Economics (1995)
11. Potter, M.A., De Jong, K.A.: Cooperative coevolution: An architecture for evolving coadapted subcomponents. Evolutionary Computation 8(1), 1–29 (2000)
12. Rom, B., Ferguson, K.: Post-modern portfolio theory comes of age. The Journal of Investing (Winter, 1993)
13. Ross, S.: The arbitrage theory of capital asset pricing. Journal of Economic Theory 13(3) (1976)
14. Sharpe, W.F.: Capital asset prices: A theory of market equilibrium under conditions of risk. Journal of Finance 19(3), 425–442 (1964)
15. Tobin, J.: Liquidity preference as behavior towards risk. The Review of Economic Studies 25, 65–86 (1958)
16. Treynor, J.: Towards a theory of market value of risky assets (unpublished manuscript, 1961)
17. Yin, X.: A fast genetic algorithm with sharing scheme using cluster analysis methods in multimodal function optimization. In: Forrest, S. (ed.) Proceedings of the Fifth International Conference on Genetic Algorithms. Morgan Kaufman, San Francisco (1993)

A Hybrid Multi-objective Algorithm for Dynamic Vehicle Routing Problems

Qin Jun, Jiangqing Wang, and Bo-jin Zheng

College of Computer Science, South-Central University for Nationalities, Wuhan 430074, Hubei, China
wrj_qj@hotmail.com, wjqing2000@yahoo.com.cn, wjqing2000@yahoo.com.cn

Abstract. This paper analyzes firstly the limitation of traditional methods when used to solve Dynamic Vehicle Routing Problem (DVRP), and then constructs an adapted DVRP model named DVRPTW based on Multi-objective Optimization. In this model, we consider two sub-objectives such as vehicle number and time cost as an independent objective respectively and simultaneously to coordinate the inherent conflicts between them. Also, a hybrid Multi-objective ant colony algorithm named MOEvo-Ant is proposed and some crucial techniques used by MOEvo-Ant algorithm are discussed too. In our ant colony algorithm, an EA is introduced into our ant colony algorithm to increase pheromone update. The main reason of the introduction is that we try to take advantage of the outstanding global searching capability of EA to speed up the convergence of our algorithm. Simulating experiments demonstrate that no matter when compared with the known best solutions developed by previous papers or when use it to solve dynamic vehicle routing problems generated randomly, our algorithm illustrates pretty good performance.

1 Introduction

Because of combining theoretical research and practical application characteristic together, Dynamic Vehicle Routing Problem (DVRP) demonstrates to be an active issue. The researchers also have achieved huge progress [1-4]. The objective of DVRP is how to find out a perfect route for loaded vehicles when customers' requirements or traffic information keep changing, which means to minimize the total cost of all routes with minimum number of vehicles without violating any constraints. Traditional algorithms produce a single solution by combining all of the objectives together in the way of traditional weight sum technique[56]. With the increasing development of logistics management field, the disadvantages of those traditional algorithms illustrate obviously as below:

- Only unique solution provided, but in some cases, more than one solution or a solution set is preferred. Thus, a decision maker can choose the best one to satisfy his/her own requests from the solution set.
- Various sub-objectives such as vehicle number, total distance, customers' waiting time, etc. are so different in meanings or order of value that it's not

M. Bubak et al. (Eds.): ICCS 2008, Part III, LNCS 5103, pp. 674–681, 2008.
© Springer-Verlag Berlin Heidelberg 2008

suitable to combine them into single objective with any weight sum techniques. Furthermore, Each sub-objective usually depends on the others. One optimized sub-objective is gained often at the cost of another sub-objective.

This paper takes advantages of both global searching ability of evolutionary algorithms and local searching capability of Ant Colony algorithm. We propose a new algorithm—MOEvo-Ant algorithm to solve DVRP. In order to satisfy personal requirements of users and coordinate conflicts between each sub-objective, our algorithm treats each sub-objective as an independent optimal objective and optimizes them simultaneously. As a result, we treat the traditional single objective optimization DVRP as a multi-objective optimization problem in this paper. This paper is organized as follow: In Section 1, the multi-objective optimization model of DVRP with time windows (DVRPTW) is given. Then MOEvo-Ant algorithm is given in Section 2. In Section 3, the algorithm's crucial techniques are discussed. In Section 4, some initial simulations are given. An analysis of the algorithm's performance is illustrated in Section 5. In the last section, some reviews are summarized.

2   Multi-objective Optimization Model of DVRPTW

In this paper, we consider the DVRPTW problem with dynamic requirements under dynamic network environment. The characteristics of the problem can be described in terms of the depot, the sort of the requirement (delivery or pick-up), the vehicle capacity, and the time-dependent route between requirement nodes. All vehicles used for service have to return to the depot before the end of the day. Every requirement node has its own time window. We consider only soft time window constraints in this paper. A vehicle is allowed to arrive at a requirement node (a delivery node or a pick-up node) outside of the time interval defined for service. However, there would be a penalty when the arriving time of a vehicle violates the time window.

At the beginning of a workday, we define the initial schedule of vehicle route, the task of it contains the requirements which were not completed the day before the workday, and today's customers' requirements. The real-time customers' requirement will be allowed given at any time. For those customers who have pick-up requirement, the destinations are the depot.

Our model considers two optimal objectives: minimize vehicle numbers and minimize the weighted sum of vehicle's traveling time, customers' waiting time and vehicles' waiting time . The objective functions are as follows.

$$\min(f_1, f_2)$$
$$f_1 = K$$
$$f_2 = \sum_{k \in K} (\alpha_2 \sum_{p=1}^{m_t} T_{i_{p-1}^t, i_p^t}^k + \alpha_3 \sum_{p=0}^{m_t} (T_{start_{i_p^t}} - T_{i_p^t}^k)^+ + \alpha_4 \sum_{p=0}^{m_t} (T_{i_p^t}^k - T_{end_{i_p^t}})^+)$$

$$(x - y)^+ = \max\{0, x - y\}$$

where:

K : the number of vehicles used in the system.

$\alpha_2, \alpha_3, \alpha_4$: the weights used for the traveling times, customers' waiting times and vehicles' waiting times at the moment of T.

$\{i_0^k, i_1^k, ..., i_{m_t}^k\}$: the sequence of requirement nodes of vehicle k planning to visit at the moment of T_t

$[T_{start_i}, T_{end_i}]$: the time window of requirement node i

T_i^k : the planned moment of planned route designed at the moment of T when vehicle k visits requirement node i_j.

Apparently, there are some conflicts between the two optimal objectives, therefore, there is not a single solution satisfying most every objective, but a group of solutions (called Pareto Optimization Set). Tan and Chew [7] analyzed the relationship between VRPTW problems' objectives using the testing data of Solomon, and indicated that the relationship between them can be variable. Sometimes these objectives are to some cases, consistent each other, but sometimes they are totally in conflict with each other. To solve the latter cases, it is obvious that the conception of Pareto Set should be used.

3   Algorithm Design

Currently, there are a large number of multi-objective optimal evolutionary algorithms (EA) based on Pareto conception. Those EA-based techniques have attracted lots of interest of academic world and more and more good performance algorithms [89] have been designed. Those algorithms can acquire not bad results in numerical value experiments, but still, they all share a common deficiency: poor performance in local searching. To solve the problem, we propose a new hybrid algorithm named MOEvo-Ant algorithm. The algorithm combines Ant Colony algorithm (ACO) and evolutionary algorithm. The main reason why we combine those two kinds of algorithms lies in that the ACOs have the characteristic of good local searching capability while the EAs have fairly good global searching performance. Rank strategy is adopted when evaluating an individual: each individual in the population has a rank value, representing the probability of the individual belongs to the Pareto set. The algorithm is described as follow:

```
void procedure MOEvo-Ant()
{
input:AImax; // the maximum iteration number
input: AI; // the iteration number of ACO
input:EI; // the iteration number of EA
input:AM; // the population size of ACO
input:EM; // the population size of EA
ParetoSet ⟵ NULL;
Pheromatrix ⟵NULL
EAo(Pheromatrix); // initialize the pheromone matrix by EA
Update(ParetoSet); //update Pareto candidate solution set
repeat
{
```

AntColony(Pheromatrix);// update the pheromone matrix by ACO
Update(ParetoSet);
EAo(Pheromatrix); // optimize pheromone matrix by EA
Update(ParetoSet);
}
until satisfying the stopping criterion
}

4 Crucial Techniques

4.1 Pheromone Updating

There are two ways used by MOEvo-Ant algorithm to update pheromone. The first way is to optimize the pheromone matrix using EA and record current best solution to construct status variable $X_k = (\vec{\tau}(k) R(K)), k = 0, 1, \ldots$.

The second way occurs during the iteration process of Ant Colony algorithm. In this paper we mainly consider the second case. Update of pheromone will be conducted by a process of global update given as follow:

$$
\tau_{ij}(t+1) =
\begin{cases}
\rho \cdot \tau_{ij}(t) + \sum_{k=1}^{m} \Delta\tau_{ij}^{k}, & if \; \rho \cdot \tau_{ij}^{k} + \sum_{k=1}^{m} \Delta\tau_{ij}^{k} \geq \tau_{\min} \\
\tau_{\min} & otherwise
\end{cases}
$$

where $0 \leq \rho \leq 1$ is the trail persistence$m$the number of ants$\Delta\tau_{ij}^{k} = Q/L_k$ the amount of pheromone laid by the k-th ant on edge (ij). $Q$is a constant$L_k$is the objective value of the k-th ant.

4.2 Evaluation Strategy of EA

Coding of evolutionary algorithm is based on pheromone matrix. When evaluating individual Ind_iin the algorithm, we generate AM ants, and calculate the pareto-dominate relationship between ants and the set of Pareto candidate solutionΩ. The process is given below:

1)Rank=0; int count[2]; count[1]=count[2]=0;
  2)for(i=1; i<= $|\Omega|$; i++)
if $\Omega_i \prec ant$ then count[1]++;
else if $\Omega_i \succ ant$ then count[2]++;
3)if count[1] ==0 then rank=1
else rank=max(2count[1]-count[2]).
The fitness calculation of Ind_i is defined as

$$
fitness(Ind_i) = \frac{1}{\sum\limits_{j=1}^{AM} rank_{ant_j} - AM + 1}
$$

4.3   Update Strategy of Pareto Candidate Solution Set

When an ant is generated, no matter generated by evolutionary algorithm or generated during the iterating process of ant colony algorithms, the updating strategy of Pareto candidate solution set remains the same. That is, if this ant is not dominated by any individual in the set, and the Pareto candidate solution set is not full, add it in to the set; otherwise, if this ant is not dominated but the set is full, it will be replaced with the closest candidate solution from this ant by Hamming distance. Pareto candidate solution set update process is described as follow:

1)Int count=0;
2)for(i=1;i<= $|\Omega|$;i++)
If $\Omega_i \prec ant$ then count++;
else if $\Omega_i \succ ant$ then delete Ω_i;
3)if count ==0 then
if $|\Omega| < Size$ then $\Omega \leftarrow \Omega \cup ant$;
else {Int dis=the hamming distance between Ω_1 and ant;
Int min=1;
for(i=2;i<= $|\Omega|$;i++)
{tmp= the hamming distance between $<= |\Omega|$ and ant;
if tmp<dis then min=i;
}
$\Omega \leftarrow \Omega \cup ant - \Omega_{min}$; }

5   Emulating Experiment

Currently, there is not a general benchmark used for DVRPTW. So we use the Dynamic Vehicle Routing Problem simulator (DVRPSIM) [10] designed by us to test the capability of the algorithm. Fig. 1 is the system status at scheduling timeT_i, 14~30 are requirement nodes produced dynamically. In this system, we suppose that two vehicles are assigned and planed routes of each vehicle are shown in Fig.1. Pick-up nodes are colored by gray, and delivery nodes colored by white. Parameters of algorithm are as follow:

Fig.2 is a Pareto optimality solution chosen randomly from the Pareto optimality set acquired by one running. The total cost is 9982. It demonstrates the availability and efficiency of the algorithm when used to solve the multi-objective problems.

6   Performance Analysis

In order to test the performance of our algorithm, twelve data sets generated by Solomon [11] are used, namely C1-01C1-09C2-04C2-08R1-05R1-07R2-03R2-10RC1-06RC1-08RC2-02RC2-06. Because MOEvo-Ant algorithm is a multi-objective optimal algorithm, in order to test the performance of algorithm

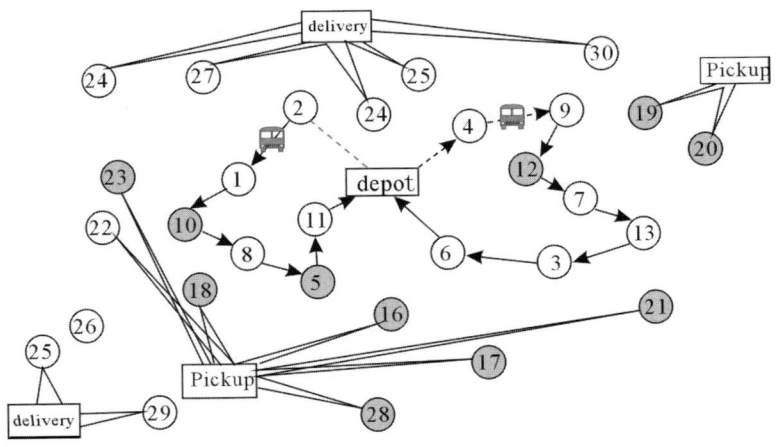

Fig. 1. System status at scheduling time T_i

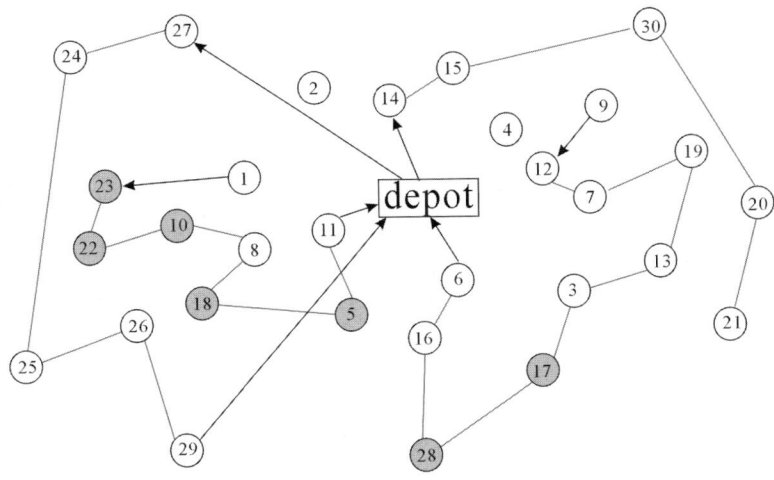

Fig. 2. Planned route using 4 vehicles chosen randomly from the Pareto optimality set

conveniently, we only consider whether the Pareto best solution set generated by our algorithm could cover the current known best solution. Table 1 gives out a comparison of computational results of our algorithm and known best solutions. Table 2 is the analysis of our algorithm.

Table 1. Comparison of MOEvo-Ant algorithm

Solomon Problem Category	Known best solutions	MOEvo-Ant
C1-01	10/828.94	10/828.94/→
C1-09	10/828.94	10/828.94/→
C2-04	3/590.60	3/590.60/→
C2-08	3/588.32	3/588.32/→
R1-05	14/1377.11	14/1377.11/→
R1-07	10/1104.66	11/1101.59/↑
R2-03	3/942.64	4/921.75/↑
R2-10	3/939.37	3/988.52/↓
RC1-06	11/1427.13	12/1416.28/↑
RC1-08	10/1142.66	10/1142.66/→
RC2-02	3/1377.089	4/1265.39/↑
RC2-06	3/1153.93	3/1367.91/↓

Table 2. Analysis of the MOEvo-Ant algorithm

Percent of cover	Percent of superior	Percent of failure
83%	33%	17%

In this table:

$$\text{Percentage of cover} = \frac{\text{number of the known-best solutions covered by the algorithm}}{\text{number of the solutions}}$$

$$\text{Percentage of superior} = \frac{\text{number of the solutions better than the known-best solutions}}{\text{number of the solutions}}$$

$$\text{Percentage of fail} = \frac{\text{number of the solutions worse than the known-best solutions}}{\text{number of the solutions}}$$

From the Table. 2, we can find out that, our algorithm is fairly competitive when used to solve DVRPTW. However, the results, when the sets R2-10RC2-06 are used, are not so good. The main reason, according to the related work of Tan and Lee[7], could be due to the specified feature existed in the testing data, Tan and Lee analyzed the data and find out that, there exists both harmony and conflict between objective data of vehicle number and that of total costs. Thus, when we use multi-objective optimal algorithm to conduct on these two data sets, the computational difficulty of our algorithm will increase dramatically. Another reason would be the stopping criterion, the algorithm will stop after 200 generations, the potential capability of our algorithm has not been illustrated thoroughly yet.

7   Conclusion

In this paper, we utilize the theory of multi-objective optimization to research the multi-objective optimization algorithm to deal with dynamic vehicle routing problems. The achievements of this paper are: 1) we adapt the traditional single optimization objective into two objectives – minimizing the number of vehicle and minimizing the total consumed time. 2) we introduce the concept of 'Pareto candidate solution', aiming to the defaults of traditional multi-objective optimization algorithms. 3) we design a new multi-objective optimization algorithm MOEvo-Ant combined with EA and ACO. This algorithm simultaneously evolves the number of vehicle and consumed time separately to harmonize the conflicts between those parameters. Further research plans contain: 1) the analysis of sensitive of algorithm parameters; 2) the influence of relationship between each objective to Pareto solution set.

Acknowledgement

This work is partially supported by the National Grand Fundamental Research 973 Program of China under Grant No. 2004CB719401 and the National Natural Science Foundation of China under Grant No. 60603008.

References

1. Fan, J.H., Wang, X.F., Ning, H.Y.: A multiple vehicles routing problem algorithm with stochastic demand [C]. In: Proceedings of the 6th World Congress on Intelligent Control and Automation, pp. 1688–1692 (2006)
2. Alvarenga, G.B., Silva, R.M.A., Mateus, G.R.: A hybrid approach for the dynamic vehicle routing problem with time windows [C]. In: Proc. 5th International Conference on Hybrid Intelligent Systems (2005)
3. Swihart, M.R., Papastavrou, J.D.: A stochastic and dynamic model for the single-vehicle pick-up and delivery problem [J]. European Journal of Operations Research 114(3), 447–464 (1999)
4. Secomandi, N.: Comparing neuro-dynamic programming algorithms for the vehicle routing problem with stochastic demands [J]. Computers and Operations Research 27(11), 1201–1225 (2000)
5. Wijeratne, A.B., Turnquist, M.A., Mirchandani, P.B.: Multiobjective routing of hazardous materials in stochastic networks [J]. European Journal of Operational Research 65, 33–42 (1993)
6. Wang, H., Chen, Y.: Multi-objective vehicle routing with time-window using genetic algorithms [J]. Computer application 24(9), 144–146 (2004)
7. Tan, K.C., Lee, T.H., Chew, Y.H., Lee, L.H.: A multiobjective evolutionary algorithm for solving vehicle routing problem with time windows [C]. In: IEEE International Conference on Systems, Man and Cybernetics, vol. 1, pp. 361–366 (2003)
8. Knowles, J.D., Corne, D.W.: Approximating the nondominated front using Pareto archived evolutionary strategy [J]. Evolutionary Computation 8(2), 149–172 (2000)
9. Deb, K.: Multi-Objective Optimization Using Evolutionary Algorithms, pp. 34–76. John Wiley & Sons ltd, Chichester (2001)
10. Wang, J.Q., Kang, L.S.: Design and implementation of dynamic vehicle routing problem simulator [J] 9 (2007)
11. Solomon, M.M.: Algorithms for vehicle routing and scheduling problems with time window constraints [J]. Operations research 35(2) (1987)

Asymptotic Behavior of hp–HGS (hp–Adaptive Finite Element Method Coupled with the Hierarchic Genetic Strategy) by Solving Inverse Problems

Robert Schaefer[1] and Barbara Barabasz[2]

[1] Department of Computer Science
[2] Department of Modeling and Information Technology,
AGH University of Science and Technology,
Al. Mickiewicza 30, 30-059 Cracow, Poland,
schaefer@agh.edu.pl
barabasz@metal.agh.edu.pl

Abstract. The new hp–HGS multi-deme, genetic strategy for economic solving of the parametric inverse problems is introduced. The inverse problems under consideration are formulated as the global optimization ones, where the objective express the discrepancy between the computed and measured energy. The efficiency of the proposed strategy results from the coupled adaptation of the accuracy of solving optimization problem and the accuracy of hp–FEM direct problem solver. The asymptotic analysis allows to estimate the expected computational cost of hp–HGS and to show its advantage over the single population SGA algorithm as well as over the HGS strategy without the FEM error scaling.

1 Introduction

The class of direct problems under considerations is defined by the abstract variational equation describing the sample physical phenomena (e.g. the variational equations of the linear elasticity described in [3]):

$$\begin{cases} u \in u_0 + V \\ b\,(d; u, v) = l\,(v) \;\; \forall v \in V \end{cases} \tag{1}$$

where u_0 is the shift of the Dirichlet boundary conditions and V is the proper Sobolev space. The form of functionals b, l depend of the physical phenomena and of its parameter $d \in \mathcal{D}$, where $\mathcal{D}$ is the regular compact in $\mathbb{R}^N$, $N < +\infty$. For symmetric and positively defined b, the variational problem (1) can be formulated as the minimization one

$$\begin{cases} u \in u_0 + V \\ \min\{E\,(d; u)\}, \end{cases} \tag{2}$$

where the functional $E(d; u) = \frac{1}{2} b\,(d; u, u) - l\,(u)$ stands for the total energy of the modeled system.

M. Bubak et al. (Eds.): ICCS 2008, Part III, LNCS 5103, pp. 682–691, 2008.

One of the most effective numerical method of solving the above problem is the hp–adaptive Finite Element Method (hp–FEM) which approximates its solution by the sequence of finite dimensional problems (see Demkowicz [4]). The coarse mesh solution $u_{h,p} \in V_{h,p}$ that satisfies

$$\begin{cases} u_{h,p} \in u_0 + V_{h,p} \\ b\left(d; u_{h,p}, v_{h,p}\right) = l\left(v_{h,p}\right) \ \forall v_{h,p} \in V_{h,p} \end{cases} \qquad (3)$$

is computed in each step of this strategy. Moreover the fine mesh solution $u_{\frac{h}{2},p+1} \in V_{\frac{h}{2},p+1}$ that satisfies the equation similar to (3) in the space $V_{\frac{h}{2},p+1}$ is computed. Notice that both approximate solution spaces satisfy $V_{h,p} \subset V_{\frac{h}{2},p+1} \subset V$. Formally, the solutions to (1), (2), (3) depend also on the parameter $d \in \mathcal{D}$.

Based on the relative hp–FEM error analysis (see Demkowicz [4])

$$err_{FEM}(d) = \left\| u_{h,p}(d) - u_{\frac{h}{2},p+1}(d) \right\|_E \qquad (4)$$

the final solution of the hp–FEM step is established. It persists the modification introduced by $u_{\frac{h}{2},p+1}$ only in elements, where this error is large. The norm $\|\cdot\|_E$ defined on the space V expresses the energy of its argument (see e.g. Ciarlet [3]).

We assume that all above problems (1), (2), (3) are well posed. i.e. they posses the unique solution for all possible values of the admissible parameter $d \in \mathcal{D}$.

The inverse problem under consideration leads to encountering the unknown parameter $\hat{d} \in \mathcal{D}$ while the energy $J(\hat{d}) = E(\hat{d}; u)$ of the exact solution $u \in V$ to (1) is known (e.g. it is measured during the laboratory test). It can be formulated as follows:

Find $\hat{g} \in \mathcal{D}$ such that :

$$\lim_{h \to 0, \, p \to +\infty} \left| J_{h,p}\left(\hat{g}\right) - J(\hat{d}) \right| \leq \lim_{h \to 0, \, p \to +\infty} \left| J_{h,p}\left(g\right) - J(\hat{d}) \right| \qquad (5)$$

The above problem is the global optimization one with the admissible set of solutions $\mathcal{D}$. The quantities $\hat{g}, g$ represent approximated parameters and $\hat{d}$ the exact parameter of the inverse problem which we are looking for. Moreover $J_{h,p}(g) = E(g; u_{h,p}(g))$ is the energy of solution $u_{h,p}(g)$ obtained by hp–FEM. We assume moreover, that the sufficient conditions are satisfied (see Demkowicz [4]) that the expression $\left| J_{h,p}(g) - J(\hat{d}) \right|$ has a finite limit for all $g, \hat{d} \in \mathcal{D}$, while $h \to 0$, $p \to +\infty$.

One of the main difficulties of the above optimization problem is caused by presence of more than one global extreme or local extremes with the objective very close to the minimal one. Moreover, the cost of the objective evaluation $\left| J_{h,p}(g) - J(\hat{d}) \right|$ is large, because of the large cost of iterative solving of direct problem. Please, notice that this cost strongly depends on the required accuracy of the direct problem solving. The global search with the exceptionally low computational cost (counting in the number of objective evaluations) is then desirable for solving the inverse problem (5).

We propose the new strategy hp–HGS which is based on the economic Hierarchic Genetic Strategy (HGS) (see Schaefer, Kołodziej [10]) with the adaptive solution accuracy. It offers relatively small number of fitness function calls and ability of global search of solution, especially in case of many local extremes. The proposed strategy additionally decreases the total computational cost by scaling the accuracy of the direct problem solving with respect to varying accuracy of the inverse problem solution.

The HGS proceeds tree-structured, dynamically changing set of dependent demes. The depth of HGS tree is limited by $m < +\infty$. All demes work asynchronously and are synchronized by the message-passing mechanism if necessary. The evolution of each deme is governed by the separate instance of the Simple Genetic Algorithm (SGA) (see Vose [15]).

The low-order demes (closer to the root) perform more chaotic search with the lower accuracy, while the demes of higher order perform the more accurate, local search. The various search accuracy is obtained by the various encoding precision and by the different length binary strings as the genotypes in demes of different order. The unique deme of the first order (root) utilizes the shortest genotypes, while the leafs utilizes the longest ones. To obtain the search coherency for demes of different order the special kind of hierarchical, nested encoding is used. First the densest mesh of phenotypes in $\mathcal{D}$ for the demes of m-th order is defined. Next the meshes for the lower order demes are recursively defined by selecting some nodes from the previous ones. The maximum diameter of the mesh δ_j associated with the demes of the order j determines the search accuracy at this level of the HGS tree. Of course $\delta_1 >, \ldots, > \delta_m$.

Each deme expecting leaf-demes sprouts the new child-deme after the constant number of genetic epochs K called the *metaepoch*. The child-deme is activated in the promising region of the evolutionary landscape surrounding the best fitted individual distinguished from the parental deme at the end of the metaepoch.

HGS implements also two mechanisms that allow to reduce the search redundancy. The first one called *conditional sprouting* disable to sprout new deme in the region already occupied or explored by the brother-deme (another child-deme of the same order sprouted by the same parent). The second mechanism called *branch reduction* reduces the branches of the same order that perform the search in the common landscape region or in the region already explored.

The HGS details and various implementations of this strategy as well as its high efficiency by solving global optimization problems are presented in [6], [11], [8], [10], [16], [7].

2   Relation Between the hp–FEM Error and HGS Objective Function Error

Let us apply HGS for solving the inverse problem (5). The fitness function for the particular deme should be based on the energy error

$$e_{h,p}(g) = \left| J_{h,p}(g) - J(\hat{d}) \right| \tag{6}$$

computed by using hp–FEM which approximate the objective function of the global optimization problem (5) for the particular values of h and p. As previously, $\hat{d} \in \mathcal{D}$ denotes the exact parameter value and $J(\hat{d})$ the known, exact energy of the exact solution while $J_{h,p}(g)$ the approximated value of energy computed by hp–FEM with respect to the parameter value $g \in \mathcal{D}$ obtained from the HGS individuals' genotype.

Let us assume for a while that g is constant and represents the parameter value decoded from the genotype that appears in the HGS deme of the j-th order, $j \in \{1, \ldots, m\}$. The regression of the error (6) while improving the FEM approximation may be evaluated as follows (see Lemma 2 in [12]):

$$e_{\frac{h}{2},p+1}(g) \leq \frac{1}{2}\|u_{\frac{h}{2},p+1}(g) - u_{h,p}(g)\|_E^2 + |J_{h,p}(g) - J(\hat{d})| \qquad (7)$$

where $e_{\frac{h}{2},p+1}(g) = \left|J_{\frac{h}{2},p+1}(g) - J(\hat{d})\right|$. Further using the formula (8) in [12] we may obtain

$$e_{\frac{h}{2},p+1}(g) \leq \|u_{\frac{h}{2},p+1}(g) - u_{h,p}(g)\|_E^2 + \|u(g) - u_{h,p}(g)\|_E^2 + L|g - \hat{d}| \qquad (8)$$

where L stands for the Lipshitz constants of the functional J and $|g - \hat{d}|$ is the error of the inverse problem solution that characterizes the individuals belonging to the HGS demes of the j-th order. It is easy to observe, that $|g - \hat{d}|$ corresponds to δ_j. The above formula shows, that the error of the energy evaluation over the fine FEM mesh is restricted by the relative FEM error on the coarse FEM mesh solution with respect to the fine mesh solution plus the absolute FEM error over the coarse FEM mesh plus the accuracy of the proper HGS branch.

3 hp–HGS Definition

The main idea of hp–HGS is to adjust dynamically the accuracy of the objective computation to the particular value of the parameter g encoded in the individuals' genotype as well as for the inverse problem error that characterizes the current HGS branch. It may be obtained by balancing the components of the FEM error given by the right hand side of the formula (8), assuming δ_j as the accuracy of inverse problem solving by the branch of j-th order. We will perform then the hp-adaptation of the FEM solution of the direct problem while the quantity $\frac{err_{FEM}}{\delta_j}$ is greater then the assumed *Ratio*, which stands for the parameter of this strategy.

Notice, that no matter how the fitness of the individual i is computed by iterative process of the hp–FEM adaptation, the fitness function f_j is well defined for all individuals in branches of j-th order (it is not a random variable).

The draft of the single hp–HGS deme activity is stressed in the pseudo-code Algorithm 1. The function *branch_stop_condition*(P) returns *true* if it detects the lack of evolution progress of the current deme P. The separate module continuously checks whether the satisfactory solution was found or hp–HGS could not find more local extremes. If yes, the *global_stop_condition* signal is send to all computing demes. The *conditional sprouting* mechanism is implemented as follows. Each branch excepting root computes the average of its phenotypes and send it to its parental deme. These values are analyzed by the *children_comparison(x)* procedure and compared with the phenotype of the best fitted individual x distinguished from the parental deme. This procedure returns *true* if x is sufficiently close to the existing child-demes. The *branch reduction* mechanism is omitted in Algorithm 1 for the sake of simplicity. All inter-deme messages are send and received asynchronously using buffers.

1: **if** $(j = 1)$ **then**
2:      initialize the root deme;
3: **end if**
4: $t \leftarrow 0$;
5: **repeat**
6:      **if** (*global_stop_condition* received) **then**
7:          STOP;
8:      **end if**
9:      **for** $(i \in P^t)$ **do**
10:          solve the direct problem for $g = code(i)$ on the coarse and fine FEM meshes;
11:          compute $err_{FEM}(g)$ according to the formula (4);
12:          **while** $(err_{FEM}(g) > Ratio * \delta_j)$ **do**
13:              execute one step of hp adaptivity;
14:              solve the problem on the new coarse and fine FEM meshes;
15:              compute $err_{FEM}(g)$ according to the formula (4);
16:          **end while**;
17:          compute fitness $f_j(i)$ using the FEM mesh finally established;
18:      **end for**
19:      **if** $(j > 1)$ **then**
20:          compute the phenotypes' average and send it to the parental deme;
21:          **if** $(branch\_stop\_condition(P^t))$ **then**
22:              STOP;
23:          **end if**
24:      **end if**
25:      **if** $(((t \bmod K) = 0) \wedge (j < m))$ **then**
26:          distinguish the best fitted individual x from deme P^t;
27:          **if** $(\neg children\_comparison(x))$ **then**
28:              sprout;
29:          **end if**
30:      **end if**
31:      perform proportional selection, obtaining multiset of parents;
32:      perform SGA genetic operations on the multiset of parents;
33:      $t \leftarrow t + 1$;
34: **until** $(false)$

Algorithm 1: Pseudo-code of the j-th order deme P in the hp–HGS tree

4 hp–HGS Asymptotic

The main goal of the asymptotic analysis presented below is to evaluate the efficiency of hp–HGS and compare its computational cost with the cost of two other strategies of solving inverse problem (5). The first strategy is the coupling of HGS with the same SGA engines in each branch as in hp–HGS, but with the fitness function f_m computed as in hp–HGS leafs (e.g. by solving the direct problem with the maximum accuracy) and then induced to all branches of lower order. Notice, that such induction is well defined because of the nested HGS encoding (all phenotypes in branches of the order j are also phenotypes in branches of the $j + 1$ order). The second strategy is the single population SGA with the same fitness f_m as previously. The size of the SGA population ensures the same initial local coverage of the admissible domain $\mathcal{D}$ by the SGA individuals as by individuals of each hp–HGS leaf.

We will intensively use the theory of the SGA heuristic (genetic operator) and its fixed points developed by Vose [15] as well as the convergence results of SGA sampling measures (see Schaefer [11], Chapter 4). Let $G_j : \Lambda^{r_j-1} \to \Lambda^{r_j-1}$ be the genetic operator (heuristic) of all branches (SGA demes) of the order j. It depends only on the number of genotypes r_j, fitness function f_j and the genetic operations applied in branches of the order j. Moreover the unit simplex $\Lambda^{r_j-1} \subset \mathbb{R}^{r_j}$ stands for the set of frequency vectors of all possible demes of the order j. We assume that each genetic operator G_j has the unique fixed point z_j in Λ^{r_j-1} that represent the limit population (i.e. the infinite cardinality population after the infinite number of genetic epochs). Moreover z_j stands for the global attractor of G_j on Λ^{r_j-1} (i.e. $\forall x \in \Lambda^{r_j-1} \lim_{t \to +\infty} (G_j)^t (x) = z_j$).

Each deme $x \in \Lambda^{r_j-1}$ of the order j may induce the probabilistic measure on $\mathcal{D}$ given by the density $\rho_x \in L^p(\mathcal{D}), p \geq 1$, so we may establish the mapping $\Psi_j : \Lambda^{r_j-1} \to \mathcal{M}(\mathcal{D})$ such that $\Psi_j(x)(A) = \int_A \rho_x(\xi)d\xi$ for each measurable set $A \subset \mathcal{D}$. The space $\mathcal{M}(\mathcal{D})$ collects all probabilistic measures over the admissible set $\mathcal{D}$. The densities $\rho_x, x \in \Lambda^{r_j-1}$ are piecewise constant on some subsets surrounding the phenotypes induced by the encoding of j-th order (e.g. on the Voronoi neighborhoods of phenotypes). Details of this construction may be found in Schaefer [11]. Let us denote $\rho_j = \rho_{z_j}$ and $\psi_j = \Psi_j(z_j)$ for the sake of simplicity.

We assume that SGA governing the evolution of the hp–HGS branches of j-th order $j \in \{1, \ldots, m\}$ are *well tunned* (see Schaefer [11], Definition 4.63) i.e. the densities ρ_j dominates on some closed sets $C^j \subset \mathcal{D}$ with the strictly positive Lesbegue measure (not necessary connected). Each set C^j surrounds the local extremes to the objective of the inverse problem (5) and is contained in the basins of attraction of these extremes. Moreover we assume that $C^1 \supset C^2 \supset ,\ldots, \supset C^m$.

The analogous assumptions are made for the strategy in which the fitness f_m is implemented in all HGS branches. The resulting quantities will be denoted as $\tilde{\rho}_j, \tilde{\psi}_j, \tilde{C}^j, j = 1, \ldots, m$ in this case.

Theorem 1. *Given the above assumptions, the hp–HGS deme of the order* $j_0 = 2, \ldots, m$ *survives with the probability not greater than* κ^{j_0} *given by the formula:*

$$\kappa^{j_0} = \psi_1(C^1) \prod_{j=2}^{j_0} \frac{\psi_j(C^j)}{\psi_j(C^{j-1})}.$$

□

Sketch of the proof: Similarly to the proof of Hypothesis 1 in [10], it may be easily checked, that if the hp–HGS branch is sprouted from C^{j-1} to C^j, then the probability of its surviving may be approximated by

$$\psi_1(C^1)(1 - \eta_1) \prod_{j=2}^{j_0} \frac{\psi_j(C^j)}{\psi_j(C^{j-1})}(1 - \eta_j)$$

where $1 - \eta_j$ is the probability that the branch of the order j survives if it is sprouted inside C^j. If the deme of the order j is sprouted outside C^j, then the probability of its surviving is η_j. This value becomes arbitrary small after the sufficient number of genetic epochs which is the issue of well tuning of this branch, and the inclusion $C^j \subset C^{j-1}$. □

Lemma 1. *Under the conditions similar to those assumed in the Theorem 1 the probability of surviving the deme of the order* $j_0 = 2, \ldots, m$ *in the HGS tree is not greater then*

$$\tilde{\kappa}^{j_0} = \tilde{\psi}_1(\tilde{C}^1) \prod_{j=2}^{j_0} \frac{\tilde{\psi}_j(\tilde{C}^j)}{\tilde{\psi}_j(\tilde{C}^{j-1})}$$

for the strategy in which the fitness f_m *is implemented in all HGS branches.* □

Lemma 1 generalizes the Hypothesis 1 in [10] to the case of HGS branches of the arbitrary order $j_0 = 2, \ldots, m$. Its proof is analogous to the proof of the theorem 1 above.

Corollary 1. *Let* $a_1 < a_2 <, \ldots, < a_m$ *are the averaged costs of solving direct problems for the individuals in hp–HGS branches of the orders* $1, \ldots, m$. *The computational cost of the single genetic epoch in all hp–HGS branches may be approximated by*

$$\mu_1 a_1 + \sum_{j=2}^{m} \sharp \Omega_{s_{j-1}} \kappa^j \mu_j a_j.$$

□

Proof: Using the theorem 1 we can evaluate the expected computational cost of the single genetic epoch of the hp–HGS branch of degree $j > 1$ which equals: $\sharp \Omega_{s_{j-1}} \kappa^j$ – the expected number of demes of the order j times the constant cardinality of each deme μ_j times the averaged cost a_j. The thesis is obtained by summing over all degrees of the hp–HGS tree. □

The above results allow to formulate three practical corollaries which try to evaluate the efficiency of the new introduced strategy.

Corollary 2. *Let hp–HGS satisfies all assumptions of the theorem 1 and SGA is applied to solving the same inverse problem with the fitness f_m. The expected number of individuals in hp–HGS leafs is smaller than the expected number of individuals in SGA, so the computational cost of the single genetic epoch in the SGA search is greater then the analogous cost of processing hp–HGS leafs.* □

Proof: The expected number of individuals in the hp–HGS leafs is $\sharp\Omega_{s_{m-1}}\,\kappa^m\,\mu_m$. The local initial coverage of $\mathcal{D}$ by the SGA individuals will be the same as the initial coverage by the hp–HGS leaf individuals if the size of the SGA population equals $\sharp\Omega_{s_{m-1}}\,\mu_m$. Because all hp–HGS branches are well tunned then $\kappa^m < 1$.

□

Corollary 3. *If the hp–HGS and HGS branches are similarly well tuned i.e. $\tilde{\psi}_j(\tilde{C}^j) \cong \psi_j(C^j)$, $j = 1,\ldots,m$, then the cost of the single genetic epochs in hp-HGS branches is smaller then the analogous cost in HGS branches.* □

Proof: We can easily observed, that the assumption of the corollary implies $\tilde{\kappa}^j \cong \kappa^j$. The cost of the single genetic epoch of the HGS root is $\mu_1\,a_m$ while in the hp–HGS root equals $\mu_1\,a_1$. The cost of the single genetic epoch of the HGS branch of the order $j > 1$ equals $\sharp\Omega_{s_{j-1}}\,\mu_j\,\tilde{\kappa}^j\,a_m$ while in the hp–HGS branch is $\sharp\Omega_{s_{j-1}}\,\mu_j\,\kappa^j\,a_j$. The thesis holds because $a_j < a_m$ for $j = 1,\ldots,m-1$. □

For sample, more detail comparison of HGS and hp–HGS computational costs the progression of the averaged costs of solving direct problems a_j by the growing branch order j will be evaluated. Here we assume the particular regression of the inverse problem error $\delta_j = \alpha^{j-1}\delta$, where δ and $\alpha < 1$ are some positive parameters. Moreover we assume that the hp–HGS error by solving direct problem decreases exponentially with respect to n_j – the number of degrees of freedom that defines the final approximate space $V_{h,p}$. This assumption is motivated by theoretical considerations (see Babuška [1], [2]) and many test computations (see e.g. [4], [5], [14], [13]).

Corollary 4. *Given the above assumptions the average cost a_j of solving direct problem for individuals of the hp–HGS deme of j-th order is*

$$\mathcal{O}\left((\theta\,(j-1) + \beta)^{3\gamma}\right)$$

where the constants $\theta > 0$, $\beta \geq 0$ and $\gamma > 1$ depend on the inverse problem under consideration. □

Proof: >From the definition of hp–HGS algorithm (see Algorithm 1) $err_{FEM} = \delta_j\,Ratio = \alpha^{j-1}\,\delta\,Ratio$ if we solve the direct problem for the individual in the branch of the j-th order, $j = 1,\ldots,m$. Assuming the exponential decrement of the hp–FEM error, err_{FEM} may be also expressed as $err_0 \exp\left(-C\,(n_j)^{\frac{1}{\gamma}}\right)$ where $err_0 > 1$ is the maximum error larger or equal then one established for the fitness computation in the root deme and $C > 0$, $\gamma > 1$ are some constants. Comparing both expressions we have $(\theta\,(j-1) + \beta)^{\gamma} = n_j$ where $\theta = -\frac{\ln\alpha}{C}$

and $\beta = \frac{\ln err_0 - \ln(\delta\, Ratio)}{C}$. The first constat θ is stritcly positive, because $C > 0$ and $\alpha < 1$ while $\beta \geq 0$ follows from the condition of the hp–HGS strategy (see Algorithm 1, line 12). Finally, the well known dependency $a_j = \mathcal{O}(n_j^3)$ (see e.g. Demkowicz [4]) motivates the thesis.   □

5   Conclusions

– Solving the inverse parametric problems is usually a difficult and time consuming numerical task. The sophisticated global optimization strategies have to be applied in order to find multiple solutions, keeping the memory and computational costs at the acceptable level.

– The hp–HGS strategy presented in this paper offers two ways to decrease the computational and memory costs by solving inverse parameter problems. Firstly it is obtained by decreasing of the number of the objective calls by using the adaptation of the inverse problem accuracy (HGS strategy). Secondly, the cost of the direct problem solution which is necessary for computing the particular value of the objective function is decreased by the proper scaling of the FEM error using the hp adaptation technique.

– The theoretical results of the hp–HGS analysis allow to evaluate the expected computational cost of this strategy (Corollary 1). Moreover it was shown that its computational cost is less then the cost of the single population SGA algorithm (Corollary 2) as well as the cost of the HGS strategy without the FEM error scaling (Corollary 3). The sample formula that allow the more detailed cost comparison was drown for the second case (Corollary 4).

– No matter how SGA with the binary encoding is sometimes criticized as a tool for solving optimization problems in the continuous domains, it is used in each branch of the first version of hp–HGS mainly because the theoretical results that characterize its asymptotic behavior are available. We plan to design the next version of hp–HGS which will be based on the hierarchic genetic strategy with the real number encoding (see [11], [16]).

Acknowledgements. The part of the work was performed within COST action P19 and supported by the MNiSzW, project nr. COST/203/2006.

References

1. Babuška I., Guo, B.: The hp-version of the finite element method, Part I: The basic approximation results. Comput. Mech. 1, 21–41 (1986)
2. Babuška, I., Guo, B.: The hp-version of the finite element method, Part II: General results and applications. Comput. Mech. 1, 203–220 (1986)
3. Ciarlet, P.: The Finite Element Method for Elliptic Problems. Society for Industrial & Applied Mathematics, Philadelphia (2002)

4. Demkowicz, L.: Computing with hp-Adaptive Finite Elements. One- and Two-Dimensional Elliptic and Maxwell Problems, vol. I. Chapmann & Hall / CRC Applied Mathematics and Nonlinear Science (2006)

5. Demkowicz, L., Kurtz, J., Pardo, P., Paszyński, M., Rachowicz, W., Zdunek, A.: Computing with hp-Adaptive Finite Elements. Frontiers: Three-Dimensional Elliptic and Maxwell Problems with Applications, vol. II. Chapmann & Hall / CRC Applied Mathematics and Nonlinear Science (2007)

6. Kołodziej, J.: Hierarchical Strategies of the Genetic Global Optimization. PhD Thesis, Jagiellonian University, Faculty of Mathematics and Informatics, Kraków, Poland (2003)

7. Kołodziej, J., Jakubiec, W., Starczak, M., Schaefer, R.: Identification of the CMM Parametric Errors by Hierarchical Genetic Strategy Applied. In: Burczyński, T., Osyczka, A. (eds.) Solid Mechanics and its Applications, vol. 117, pp. 187–196. Kluwer, Dordrecht (2004)

8. Momot, J., Kosacki, K., Grochowski, M., Uhruski, P., Schaefer, R.: Multi-Agent System for Irregular Parallel Genetic Computations. In: Bubak, M., van Albada, G.D., Sloot, P.M.A., Dongarra, J. (eds.) ICCS 2004. LNCS, vol. 3038, pp. 623–630. Springer, Heidelberg (2004)

9. Schaefer, R., Barabasz, B., Paszyński, M.: Twin adaptive scheme for solving inverse problems. In: X Conference on Evolutionary Algorithms and Global Optimization, pp. 241–249. Warsaw Technical University Press (2007)

10. Schaefer, R., Kołodziej, J.: Genetic search reinforced by the population hierarchy. In: De Jong, K.A., Poli, R., Rowe, J.E. (eds.) Foundations of Genetic Algorithms 7, pp. 383–399. Morgan Kaufman Publisher, San Francisco (2003)

11. Schaefer, R. (with the chapter 6 written by Telega H.): Foundation of Genetic Global Optimization. Springer, Heidelberg (2007)

12. Paszyński, M., Barabasz, B., Schaefer, R.: Efficient adaptive strategy for solving inverse problems. In: Shi, Y., van Albada, G.D., Dongarra, J., Sloot, P.M.A. (eds.) ICCS 2007. LNCS, vol. 4487, pp. 342–349. Springer, Heidelberg (2007)

13. Paszyński, M., Demkowicz, L.: Parallel Fully Automatic hp-Adaptive 3D Finite Element Package. Engineering with Computers 22(3-4), 255–276 (2006)

14. Paszyński, M., Kurtz, J., Demkowicz, L.: Parallel Fully Automatic hp-Adaptive 2D Finite Element Package. Computer Methods in Applied Mechanics and Engineering 195, 711–741 (2006)

15. Vose, M.D.: The Simple Genetic Algorithm. MIT Press, Cambridge (1999)

16. Wierzba, B., Semczuk, A., Kołodziej, J., Schaefer, R.: Genetic Strategy with real number encoding. In: VI Conference on Evolutionary Algorithms and Global Optimization, pp. 231–237. Warsaw Technical University Press (2003)

Traffic Prediction for Agent Route Planning

Jan D. Gehrke[1] and Janusz Wojtusiak[2]

[1] Center for Computing Technologies (TZI)
University of Bremen, 28359 Bremen, Germany
jgehrke@tzi.de
[2] Machine Learning and Inference Laboratory
George Mason University, Fairfax, VA 22030, USA
jwojt@mli.gmu.edu

Abstract. This paper describes a methodology and initial results of predicting traffic by autonomous agents within a vehicle route planning system. The traffic predictions are made using AQ21, a natural induction system that learns and applies attributional rules. The presented methodology is implemented and experimentally evaluated within a multiagent-based simulation system. Initial results obtained by simulation indicate advantage of agents using AQ21 predictions when compared to naïve agents that make no predictions and agents that use only weather-related information.

Keywords: Traffic Prediction, Intelligent Agents, Natural Induction.

1 Introduction

The importance of information technology (IT) in logistics has increased remarkably. IT systems support or take responsibility for logistics planning of a forwarder's whole motor pool or single vehicles. The just-in-time paradigm demands for high robustness to situation changes and real-time coordination abilities of all participants in the logistics network. These requirements are facilitated by pervasive mobile communication networks and devices as well as intelligent systems processing incoming information [1]. Due to the fierce competition in the logistics service market, companies are searching for new technologies that advance quality of service (e.g., promptitude and robustness to disturbances) and cost of carriage. Thus, even seemingly small differences may bring about a considerable competitive advantage and high economic impact.

The agent-based approach to intelligent logistics systems delegates the planning and decision making from central planning systems to single logistic entities, such as trucks and containers, that decide autonomously and locally. With reduced complexity for (re-)planning, this decentralized approach aims at increased robustness to changes such as new transport orders or vehicle breakdown. Approaches for agent-based transport planning and scheduling have been proposed by, e.g., Dorer and Calisti [2] and Bürckert et al. [3].

The ability of autonomous systems to react to situation changes calls for situation awareness provided by sensors, other agents, or external sources such as

M. Bubak et al. (Eds.): ICCS 2008, Part III, LNCS 5103, pp. 692–701, 2008.

databases. To make use of this information the agent needs to know its relevance, i.e., how information will influence its cost or utility function during planning of actions. Furthermore, the agent has to consider the spatio-temporal scope where some information is valuable. Because the scope of agent planning is the near future, agents also need prediction abilities.

In this paper we examine the influence of environmental knowledge in vehicle route planning in simulation experiments. In particular, we examine how road traffic predictions used in route planning affect vehicles' performance measured as the time needed to reach their destination. In order to do this, we compare performance of ignorant agents, with agents that are provided with weather information and agents that use rules induced from historical traffic data for predictions. The rules used in the latter approach are learned using the *natural induction* system AQ21.

2 Situation-Aware Vehicle Route Planning

The performance of an intelligent planning systems depends on a proper analysis of the current situation and prediction of the future. Depending on the domain, the planning scope may cover a few hours for regional transports or some days for international or overseas routes. Thus, an important requirement for intelligent planning systems is situation awareness [4] as well as situation prediction.

In order to investigate the impact of such abilities in logistics and to understand the spatial and temporal constraints we reduce the problem to a simplified single vehicle transportation scenario. An agent represents a truck that aims at finding the fastest route in a graph-based highway grid. This becomes a complex problem because there are highly dynamic environmental conditions that may enforce or reduce the speed at which the vehicle may travel. The examined conditions include traffic density and weather that vary in space and time.

In this study, the vehicle route planning applies an A* search algorithm with cost function (1) for reaching destination d when using (partial) route r at departure time t_{dep}:

$$f(r, d, t_{\mathrm{dep}}) = g(r, t_{\mathrm{dep}}) + h(end_r, d) \tag{1}$$

with g as the *estimated* driving time for r and h as the estimated driving time from r's endpoint end_r to d. Heuristics h is calculated as driving time at straight line distance from end_r to d at maximum vehicle speed. The route r consists of n consecutive edges (i.e., roads) $e_i \in r$ with $0 \leq i < n$. The route segment of first k edges is denoted by r_{k-1}. The driving time g on route r is calculated by:

$$g(r, t_{\mathrm{dep}}) = \sum_{i=0}^{n-1} \frac{length(e_i)}{v_{\mathrm{est}}(e_i, t_{\mathrm{dep},e_i})} \tag{2}$$

with v_{est} as the estimated vehicle speed on an edge which depends on the vehicle agent implementation (Sect. 3). Edge departure time for $i > 0$ is defined by

$$t_{\mathrm{dep},e_i} = g(r_{i-1}, t_{\mathrm{dep}}) \tag{3}$$

Because this setting ensures the criteria for the A* algorithm (non-negative costs and optimistic heuristics) it guarantees the optimal solution. However, the found route is optimal only provided that knowledge about the environment used in the cost function is complete and correct. But assumptions on *future* road conditions are possibly wrong because the environment continuously changes in a way that cannot be precisely predicted.

3   Simulation Model

In order to evaluate the impact of environmental knowledge and driving time prediction abilities we set up experiments with the multiagent-based simulation system *PlaSMA*[1]. The system applies discrete event, distributed simulation using software agents as logical simulation processes. Thus, PlaSMA provides a natural way to evaluate software agents behavior and interactions.

For the purpose of vehicle route planning, the simulation model includes two *world agents* and multiple *vehicle agents*. The world agents simulate weather and traffic within the simulation environment. The vehicle agents can be categorized in three classes: the *ignorant* agent, the *weather-aware* agent, and the *predictive* agent. While all vehicle agents use A* search with time cost function for routing, they use different knowledge in planning and thereby provide a comparative measure to evaluate the impact of their knowledge.

3.1   Weather and Traffic

For the logistics scenario we set up a road network as a 6×6 grid graph with a set $\mathcal{E}$ of directed edges (i.e., unidirectional roads) of 100 km length each. This idealized grid structure ensures that results are not biased by a special network structure that may have no implications for the general problem. Nevertheless, experiments can be conducted with other network structures as well. The two world agents generate simulation events that affect this graph with respect to traffic density $dens(e, t)$ and maximum safe speed $v_{\text{safe}}(e, t)$ for each edge $e \in \mathcal{E}$ at simulation time t. $dens(e, t)$ is a linear traffic *quality* value that is normalized to 1 and indicates the ability of a vehicle to drive at a reference speed v_{ref}, i. e., maximum speed is

$$v_{\max}(e, t) = v_{\text{ref}} \cdot dens(e, t) \tag{4}$$

For our experiments we use $v_{\text{ref}} = 130$ km/h for all roads. This value corresponds to the recommended speed on German Autobahn roads for passenger cars.

Speed v_{safe} depends on the current weather on each road. Different roads may have different weather but the weather within the area of each road is homogeneous. The simulation model is designed for a set of qualitative weather types $\mathcal{W} = \{VeryBad, Bad, Moderate, Good\}$. $weather(e, t)$ describes the weather at an edge e at time t. Each weather type $w \in \mathcal{W}$ corresponds to a truck-oriented maximum safe speed $v_{\text{safe}}(w)$ when facing that weather:

$$v_{\text{safe}}(VeryBad) = 35 \text{ km/h}, \ v_{\text{safe}}(Bad) = 65 \text{ km/h}$$
$$v_{\text{safe}}(Moderate) = 80 \text{ km/h}, \ v_{\text{safe}}(Good) = 100 \text{ km/h}$$

---

[1] Available from http://plasma.informatik.uni-bremen.de/

Then, the maximum safe speed for each edge $e \in \mathcal{E}$ at time t is defined by

$$v_{\text{safe}}(e, t) = v_{\text{safe}}(weather(e, t)) \tag{5}$$

The minimum of $v_{\max}(e, t)$ and $v_{\text{safe}}(e,t)$ determines the maximum possible average speed $v_{\text{avg}}(e, t)$ of a vehicle on edge e.

Weather Generation. The weather agent updates weather in the interval $\Delta t_{\mathcal{W}}$ separately for each edge. The new weather type $w \in \mathcal{W}$ on each edge e depends on the previous weather at e and the basic probability distribution $\mathbf{P}(\mathcal{W})$ for weather which is assumed to be location-independent.

The next weather is determined by the weather w randomly drawn from $\mathcal{W}$ according to $\mathbf{P}(\mathcal{W})$ but changes may be constrained depending on current weather to avoid sudden changes as determined by transition probability model $\mathbf{P}(\mathcal{W}_{t+\Delta t_{\mathcal{W}}} | \mathcal{W}_t)$. The actual distributions for weather and length of $\Delta t_{\mathcal{W}}$ are subject of the experimental setup (Sect. 6).

Traffic Generation. Similarly to the weather agent, the traffic simulation agent updates traffic density on each edge in interval $\Delta t_{\mathcal{T}}$. Though the generated traffic density is a real number, the traffic generator is based on a qualitative traffic model with a set of traffic classes $\mathcal{T}$ covering disjoint intervals of traffic density. Similarly to *levels of service* A to F in US Highway Capacity Manual [5], the model distinguishes six traffic classes $\mathcal{T} = \{VeryLow, \ Low, \ Medium, \ High, \ VeryHigh, \ Jam\}$ and the following traffic density intervals:

$$VeryLow = [0.0, 0.1), \ \ Low = [0.1, 0.25), \ \ Medium = [0.25, 0.4)$$
$$High = [0.4, 0.6), \ \ VeryHigh = [0.6, 0.85), \text{ and } Jam = [0.85, 1.0]$$

$\mathcal{T}_{\mu}$ denotes the mean value of a traffic class. The traffic density on an edge is determined by an edge-specific density matrix for time of day and day of week. For this purpose, we analyzed traffic volume data from German (BASt) and Austrian (ASFINAG) agencies that count traffic on national highways. According to the combined days and hours in aggregated agency models we set up the basic model matrix depicted in Table 1 that determines the basic mean traffic density value $\mathcal{T}_{\mu}(t)$. The basic model is used for all edges but each edge e has an additional and time-independent traffic bias function $\Delta dens(e)$ that shifts the basic model to higher or lower traffic densities.

The actual values for the density bias function $\Delta dens$ depend on the experimental setup (see Sect. 6). The traffic generation world agent calculates the density $dens(e, t)$ with Gaussian distribution $N(\mu, \sigma)$:

$$dens(e, t) \sim N\left(\mathcal{T}_{\mu}(t) + \Delta dens(e), \sigma\right) \tag{6}$$

3.2   Vehicle Agents

There are three kinds of vehicle agents. The *ignorant agent* is not aware of any environmental information and has no predictive abilities. Like all other vehicle

Table 1. Basic traffic density matrix

	Mo	Tu-Th	Fr	Sa	Su
Morning, 6am-10am	High	Medium	Medium	VeryLow	VeryLow
Noon, 10am-3pm	Low	Low	High	Low	Low
Afternoon, 3pm-7pm	Medium	Medium	High	Medium	Medium
Evening, 7pm-10pm	Low	Low	Medium	Low	Low
Night, 10pm-6am	VeryLow	VeryLow	Low	VeryLow	Low

agents it uses the algorithm described in Sect. 2. Because this agent assumes that all roads allow equal speed its dominant planning criterion is distance. Within the road grid the ignorant agent chooses a route that most closely matches the straight line, disregarding possible bad weather or traffic conditions.

The *weather-aware agent* is a vehicle agent that acquires status information on weather for relevant locations. The agent uses this information to determine speed $v_{\mathrm{est}}(e, t_{\mathrm{dep},e})$ for route planning (Sect. 2). It does not attempt to predict any traffic or changes of weather but naïvely assumes that there is no (relevant) traffic or change of weather w.r.t current time t_{cur}, i. e.

$$v_{\mathrm{est}}(e, t_{\mathrm{dep},e}) = v_{\mathrm{safe}}(e, t_{\mathrm{cur}}) = v_{\mathrm{safe}}(weather(e, t_{\mathrm{cur}})) \tag{7}$$

The status information is provided by the weather agent. Which locations are considered interesting depends on the current vehicle location, its destination, and a lookahead parameter λ. λ specifies the spatial range and the area for which the agent acquires environmental status information. The actual lookahead distance used in this study is $\lambda \cdot 100$ km. Because the agent assumes the weather to be static when planning the fastest route, it may turn out to be wrong when reaching edges that are far ahead. However, vehicle agents may reconsider their route at each junction, i.e., wrong decisions can be corrected if there are new, better alternatives.

The *predictive agent* is an extension of the weather-aware agent. It also assumes that weather at edge e will not change until arrival at e. But it also predicts traffic at e for $t_{\mathrm{dep},e}$ to determine $v_{\mathrm{est}}(e, t_{\mathrm{dep},e})$ (Sect. 2). The predictions rely on previous experiences that are used as data to learn situation- and edge-specific rules for expected speed (Sect. 5). Besides time of day and day of week these predictions may depend on weather and thus on λ, too. Edges that are not within λ distance are assumed to have the most likely weather, e.g. *Moderate*.

4   Natural Induction

In contrast to most methods known in the literature, including different forms of statistical learning, the approach used in this study puts an equal importance to accuracy and interpretability of learned models. While the importance of the former does not require any justification, the latter may not be clear, especially in the area in which learned knowledge is used by autonomous agents. Models learned by many methods can be regarded as a "black box" which may give very good predictions, but it is hard to understand and often impossible to validate or refine by human experts. Models described in a language that is easy to understand, for example, natural language or easy to interpret rules, can be modified by experts to reflect their background knowledge and improve predictions. In this study we use *natural induction* [6] that is an approach to inductive learning whose goal is to achieve high understandability of derived knowledge. It uses a highly expressive language *attributional calculus* (AC) that combines predicate, propositional and multi-valued logics. Because all forms of knowledge used in AC corresponds to different constructs in natural language, its expressions are easy to understand.

4.1 Knowledge Representation

The main form of knowledge in AC is an attributional rule. Attributional rules, which follow the general *if ... then ...* schema, are more general than those learned by most learning programs. This is because attributional rules use more expressive language which allows creating simpler descriptions than normal rules. A basic form of attributional rules is (8).

$$\text{CONSEQUENT} \Leftarrow \text{PREMISE} \tag{8}$$

Here, CONSEQUENT and PREMISE are conjunctions of attributional conditions. In this study we used attributional conditions in the form $[L \text{ rel } R]$ where L is an attribute; R is a value, a disjunction of values, or a conjunction of values if L is a compound attribute; and rel is a relation that applies to L and R. Other forms of attributional conditions may involve count attributes, simple arithmetical expressions, conjunctions and disjunctions of attributes, comparison or attributes, etc. [7].

4.2 The AQ21 Machine Learning System

A simple form of natural induction is implemented in the currently developed AQ21 system [7]. Given input data, problem definition, and optional background knowledge, AQ21 induces rules in the form (8), or in more advanced forms describing one or more class in the data. A set of rules constituting a description of a given class is called a *ruleset*. By repeating learning for all classes defined by values of an output attribute, AQ21 generates a *classifier*.

In order to learn rules for a given class AQ21 starts with one example, called a *seed*, belonging to the class. It generates a star, which is a set of maximally general rules that cover the seed and do not cover any examples from other classes. This is done by repeating an *extension-against* operation that generalizes the seed against examples not belonging to the concept being learned. Results of applying the extension-against are intersected and the best rules are selected according to user-defined criteria. If selected rules do not cover all examples belonging to the class, another seed is selected (from the not covered examples) and additional rules are learned. The process is repeated until all examples of the class are covered by the learned rules. AQ21 implements several modifications to the above basic algorithm as described, for example, in [7].

5   Learning and Applying Traffic Models

In the presented study traffic models were learned using simulated data collected over 15 years of simulation time. The training data consisted of 131,488 examples for each type of edge. To learn traffic models we used the AQ21 system (Sect. 4.2). It was executed with different settings of parameters, from which we selected the best according to predictive accuracy on testing data and simulation results. We applied the program in two modes, *theory formation* (TF) and *approximate theory formation* (ATF). The TF mode learns complete and consistent rulesets w. r. t. the training data, while the ATF mode allows partial inconsistence and incompleteness (e.g. in the presence of noise) by optimizing the $Q(w)$ quality measure [8]. In this study the weight, w, of completeness against consistency gain was 0.1.

Many training examples are ambiguous, meaning that for identical values of Day, Time, and Weather, different values of Speed are assigned. Among several methods for solving ambiguity available in AQ21 we investigated two. The first method, here called *majority*, assigned the most frequent class to all ambiguous examples, and the second method, here called *pos*, always treats ambiguous examples as positive examples. Details of the methods are in [9].

Application of learned models to classify new examples is done by executing AQ21's testing module. The program is provided with input files consisting of a testing problem description, learned models and one or more testing examples. In the presented experiments we tested two methods of evaluating rules, namely *strict* in which an example either matches a rule or not, and *flexible* in which the program calculates the degree to which examples match rules. In the latter case the degree of match equals the number of matching conditions in the rule to the total number of conditions in the rule. Detailed description of the parameters and other rule matching schemas are in [6,9]. Two methods of resolving imprecise classifications are tested here. The first method is pessimistic, namely it assumes the worst of the given answers (the lowest predicted speed). The second method is based on frequency of classes. For instance, if an example matches two classes "slow" and "very slow," and the former is more frequent in the training data, it is reasonable to assume that the prediction should be "slow."

Average predictive accuracies and precisions [9] of models learned and applied with different AQ21 parameters on testing data with variances 0.001 and 0.01 for $dens(e,t)$ ranged from 91% to 100% and 71% to 100%, respectively. The best results were obtained by AQ21 in the theory formation mode, with ambiguities treated as majority and flexible rule interpretation. There were 95% predictive accuracy, and 100% precision. All presented values are averaged for all types of roads in the simulation model. As an example, the following rule has been learned using the TF mode. It predicts that expected maximum speed for a given road is 60 kmph on Monday mornings provided moderate to good weather conditions.

```
[Speed=SPEED_60]
   <= [Day=Mo] & [Time=morning] & [Weather=moderate..good]
```

6   Experimental Evaluation

For evaluation we conducted experiments using the simulation system PlaSMA. Each experiment is specified by a simulation model parameter setting and 9 participating vehicle agents. There is one ignorant (named "IA"), one weather-aware ("WA"), and seven traffic-predicting agents ("AQ"). The latter are differing in the applied prediction rulesets for each edge (Sect. 5) as indicated by their index.

The agent named AQ_M is a traffic-predicting agent whose rules have been created manually *knowing* the actual traffic simulation model. Thus, this agent should provide results close to the achievable optimum for the applied algorithm. All agents try to optimize the driving time of a 1000 km trip in the road network. Simulation results are provided as an average driving time and its standard deviation for each vehicle agent and parameter setting. For statistical significance experiments were repeated between 4200 and 4800 times with each run corresponding to one trip. With significance level $\alpha = 0.05$ stated average values do not differ more than 0.02 hours from the actual value.

Most of the model parameters have been examined in prior studies [10]. For all experiments presented in this paper we set parameters to $\Delta t_T = 1$h, $\Delta t_W = 3$h, and $\lambda = 2$. Weather probabilities are set to

$$\mathbf{P}(\mathcal{W}) = \{P(\mathcal{W} = VeryBad) = 0.05, P(\mathcal{W} = Bad) = 0.2,$$
$$P(\mathcal{W} = Moderate) = 0.55, P(\mathcal{W} = Good) = 0.2\}$$

The edge-specific density bias $\Delta dens(e)$ was examined in a setting that is characterized by a fairly well-shuffled distribution of slower and faster edges (see Fig. 1). As an exception, the grid also includes three edges (dashed line) that form a fast partial route. These edges are not part of routes that most closely match the theoretical straight line route. Hence, ignorant vehicles that are dominated by the straight line heuristics and always choose such centered routes will not be affected. Weather-aware agents should not benefit as much as traffic predicting agents because only the latter will actually realize the traffic properties and consider them in planning.

Table 2 shows the simulation results for traffic density setting depicted in Fig. 1. With variance $\sigma_T^2 = 0.001$, the ignorant agent needs 13.31 ± 0.87 hours driving time on average. The weather-aware agent needs 12.61 ± 0.78 h, hence, it is 0.7 h or 5.3% faster than the ignorant agent. The prediction ruleset AQ_3 (AQ21 mode: TF, majority, strict) performs best with 12.47 ± 0.60 h (0.84 h, 6.3% faster). The standard deviation is significantly lower than that of IA and WA, too. AQ_3 and AQ_4 even come close to the reference ruleset AQ_M. With

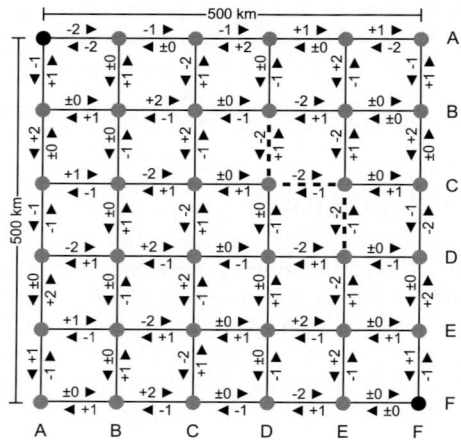

Fig. 1. Road grid annotated with $\Delta dens(e) \cdot 10$

$\sigma_T^2 = 0.01$, AQ_3 needs 0.16 hours longer due to more wrong predictions and all agents using TF mode perform equally good. The changes are rather small and all situation-aware agents still clearly outperform the ignorant agent. Thus, these agents can be considered sufficiently robust to less predictable environments.

Table 2. Average vehicle driving time in hours

σ_T^2	IA	WA	AQ_M	AQ_1 TF pos,str	AQ_2 TF pos,flx	AQ_3 TF maj,str	AQ_4 TF maj,flx	AQ_5 ATF pos,str	AQ_6 ATF pos,flx	AQ_7 ATF maj,str	AQ_8 ATF maj,flx
0.001	13.31 ±0.87	12.61 ±0.78	12.49 ±0.60	12.51 ±0.64	12.56 ±0.63	12.47 ±0.60	12.50 ±0.61	12.62 ±0.60	12.63 ±0.64	12.54 ±0.64	12.57 ±0.65
0.01	13.44 ±0.92	12.74 ±0.86	12.64 ±0.68	12.63 ±0.68	12.63 ±0.68	12.63 ±0.66	12.63 ±0.66	12.75 ±0.67	12.74 ±0.64	12.64 ±0.66	12.69 ±0.71

7   Conclusions

This paper presented an approach to the problem of predicting traffic for vehicle routing. Autonomous agents make predictions based on rules induced by the AQ21 system from historic data. By inducing attributional rules AQ21 realizes natural induction, that is an inductive learning process whose results are both accurate and easy to understand by people. Experiments performed within the PlaSMA multiagent-based simulation system indicated advantage of agents that use AQ21 predictions over naïve agents that consider only distance to the destination and agents that use only weather-related information.

Future research includes the comparison with different learning methods, investigation of the effect of the amount of historic data on the learning results and predictions, and investigation of the effects of changes of the environment on the predictions. Other important research directions include learning individualized models for each agent, based on the agent's experience and preferences, and application of the system in a real, not simulated, environment.

Acknowledgments. The authors thank Abdur Chowdhury for his contributions to experiment design and Jarek Pietrzykowski for his comments that helped to improve this paper. This research was partially funded by the German Research Foundation (DFG) within Collaborative Research Centre 637 (SFB 637). Development of the AQ21 system was partially funded by the National Science Foundation Grants IIS 9906858 and IIS 0097476.

References

1. Scholz-Reiter, B., Windt, K., Freitag, M.: Autonomous logistic processes: New demands and first approaches. In: Monostri, L. (ed.) Proceedings of the 37th CIRP International Seminar on Manufacturing Systems, pp. 357–362 (2004)
2. Dorer, K., Calisti, M.: An adaptive solution to dynamic transport optimization. In: Proceedings of AAMAS 2005, pp. 45–51 (2005)
3. Bürckert, H., Fischer, K., Vierke, G.: Holonic transport scheduling with teletruck. Journal of Applied Artificial Intelligence 14, 697–725 (2000)
4. Endsley, M.R.: Theoretical Underpinnings of Situation Awareness: A Critical Review. In: Endsley, M.R., Garland, D.J. (eds.) Situation Awareness, Analysis and Measurement, Lawrence Erlbaum Assoc., Mahwah (2000)
5. Transportation Research Board: Highway Capacity Manual (HCM) (2000)
6. Michalski, R.S.: Attributional Calculus: A Logic and Representation Language for Natural Induction. Tech. Report MLI 04-2, MLI, George Mason University (2004)
7. Wojtusiak, J., Michalski, R.S., Kaufman, K., Pietrzykowski, J.: The AQ21 Natural Induction program for pattern discovery: Initial version and its novel features. In: Proceedings of the 18th IEEE International Conference on Tools with Artificial Intelligence, Washington, DC (2006)
8. Kaufman, K., Michalski, R.S.: An Adjustable Rule Learner for Pattern Discovery Using the AQ Methodology. Journal of Intelligent Information Systems 14, 199–216 (2000)
9. Wojtusiak, J.: AQ21 User's Guide. Tech. Report MLI 04-3, MLI, George Mason University (2004) (updated, September 2005)
10. Gehrke, J.D., Wojtusiak, J.: A natural induction approach to traffic prediction for autonomous agent-based vehicle route planning. Tech. Report MLI 08-1, MLI, George Mason University (2008)

Agents for Searching Rules in Civil Engineering Data Mining

Janusz Kasperkiewicz and Maria Marks

Institute of Fundamental Technological Research,
Polish Academy of Sciences,
21 Świętokrzyska Str., 00-049 Warsaw, Poland
{J.Kasperkiewicz,M.Marks}@ippt.gov.pl

Abstract. The software agents are applied for a remote search of information. It seems natural that to analyse such information machine learning routines should be built-in into an agent system. After finding and processing the data the generated rules will be evaluated by means of so called *interestingness measures*, and only the best rules should be returned to the user.

The paper presents situation in civil engineering data processing, as a suggestion for designers of intelligent software tools, to work out difficult but much needed procedures that should be implemented into autonomous agent system, intended for retrieving special kind of information searched for example by materials technologists.

A simple architecture for an agent system is suggested without, however, getting into any technical details on how the elements of such system should be constructed.

Keywords: machine learning, data mining, software agents, engineering databases.

1 Introduction

Generally speaking the agents are to reduce the workload of their users, but there is a diversity of kinds of information interesting for different users. Many users are satisfied only with a task of simply finding appropriate documents, like text files, Internet sites, data tables, images, audio or video files, which are dedicated to specific topics. Introducing agents able not only to find but also to process the data might immensely increase the possibilities of remote knowledge search.

An agent is intelligent, autonomous code that can be send out on a mission, [1]. It can work in a static way, being located at the local system, or it may be mobile, working after being embedded in a remote machine. Agents may be static or mobile, acting in a collaborative way or separately, but their crisp definition is difficult and a classic paper of 1996, [2], was pointing out a general lack of precision in software agents taxonomy. It seems that the situation in this respect is not much different today. In the description which follows the case of static agent is assumed.

M. Bubak et al. (Eds.): ICCS 2008, Part III, LNCS 5103, pp. 702–711, 2008.

Machine learning (ML) programs are not agents, but it seems natural that they might and should be built in into an agent system. ML programs usually are constructed for purposes more sophisticated than simple finding of information sources. In our investigations we are using them mainly for searching rules representing knowledge about technological processes under inspection. An example here can be a process of selecting proper mix of components in a composite material.

A concept of an agent for engineering data search will be explained on an example. An agent could be a personification of an Internet user, who from the computer terminal is searching, e.g. around WWW, or some protected network, or the whole Internet, or only the single computer of the user, for a particular kind of information. After the information source is found, (e.g. a report with experimental results), it is usually downloaded by the user from the remote system, before it can be analysed to extract the knowledge in form of statements, (hypotheses, rules, generalisations). Generation of rules is performed by the ML programs. A rule of practical importance for a civil engineer can be for example an information, in form of a relation, what amounts of certain components of a concrete mix correspond to a desired quality of the resulting hardened concrete material.

In our group at IPPT[1] we are using various kinds of AI tools, such as ANNs (Artificial Neural Networks), but especially ML methods for prediction of rules. Important in our investigations were the methods developed by Ryszard Michalski and his team at George Mason University; programs: AQ15, AQ19, AQ21. We were also comparing them with various other ML algorithms.

The present paper is to indicate certain important characteristics of agent type programs that might bring most needed results. It is assumed that the agent will generate the rules concerning certain process; the details of creation of association rules, of the decision trees, etc., are not discussed in this text. Also technical questions such as the protection of the data or the robustness of the agent, etc., are not discussed here.

2   Machine Learning Tools

Learning systems have the ability of setting values of their certain internal parameters, of some preset, abstract algorithm, in a way that will minimize difference between the internal results and the external control values. The values can be scalars, vectors or categories. If during the training the configuration of the internal parameters tends to a certain steady state, such state represents the knowledge of the system about the process under consideration.

A generalization of the learned knowledge can be a black-box type tool, like in case of ANNs, (Artificial Neural Networks), or a set of various concepts, (rules, hypotheses), expressed in form of logical expressions, such as a proposition: $A \rightarrow B$.

---

[1] IPPT - Institute of Fundamental Technological Research, Polish Academy of Sciences.

This expression represents an inference containing an *antecedent* (A) and the *consequent* (B). The antecedent A may represent a larger set of simple conditions on the components of the input and B represents output conditions, typically it is a designation of the class, of a certain category.

The symbol A represents some concept, which is a conjunction, (or a *complex*), of simple, or atomic conditions, called *selectors*, described by numerical or nominal attributes of the dataset. The selectors are marked by square brackets, $([\dots])$. A few examples of selectors are: $[x \leq 12]$, $[y > 2.37]$, $[z \in (1.30, 1.77\rangle]$, in case of only numerical variables, and $[v \in 'red', 'blue', 'yellow']$, $[w \notin 'A', 'D', 'K', 'L', 'R']$ – where only nominal variables appear.

For any record in the training or testing datasets the hypothesis in form of the proposition $A \rightarrow B$ can be either true or false. In rare cases it can also be undefined.

A task for ML programs is to propose to the user rules of the type described above, selected from the collection of the trainee set of training hypotheses, [3]. The selection should correspond to a possibly low error rate of the power of a concept in predicting its consequences, characterizing the dataset under consideration. A finally selected rule might be for example:

if $[y > 2.37]$ **and** $[[v =' red']$ **or** $[v =' yellow']]$ **then** $[class = ClassIII]$

The requirements concerning the formatting of the database are slightly different for different ML programs, but it is relatively easy to translate an input script created for one system to another one.

There is a number of different Machine Learning programs available. Many of these are more or less for free, like AQ15, AQ19, AQ21, C4.5, WinMine, Gradestat, WEKA, Rosetta, (a rough sets toolbox), [4, 5, 6, 7].

There are many commercial data mining programs, as was described e.g. in [8, 9]. Examples are: DataCruncher, IBM Intelligent Miner, MineSet, CART, See5, WizRule, Magnum Opus, etc. Various elements of rules searching from database examples are also available now in large commercial packages dedicated to statistics and data mining, like Oracle, Statistica, SPSS or SAS.

In civil engineering there is a hypothetical possibility of combining various soft computing tools into a one system, [10, 11, 12], but so far it is still only the concept that needs much further programming work.

3   Civil Engineering Data Bases

Civil engineering experimental data are scattered around various sources, such as technical papers, reports, books, standards and instructions, also special databases, which are properties of laboratories or production managers. Many of such sources are available only in a paper form, so to transfer them into electronic form considerable additional work would be needed. It is assumed in this paper that there is a certain digital data environment accessible, in which the agent will function. This may be simply a selected directory in the computer of the user, the contents of a network of connected computers, or any library that may be opened in the Internet.

There are different types of engineering data. These are descriptive databases concerning composition of concrete and similar composite materials, data concerning various diagnostic problems, data enabling only the quantitative predictions, data allowing only for classification, etc. For example the properties of concrete depend on composition of the original concrete mix, on certain elements of its mixing technology, on its curing and protection, its age, sometimes on its whole history. The properties of concrete can be characterized by attributes in form of numbers, (like: strength, density, amounts of cement, water or additives), and in form of categories, (like: 'basalt', 'limestone', 'PFA1', 'PFA2', etc.)

The fields of investigation may be properties of materials, diagnostics of the quality of materials, diagnostics of whole engineering structures. In any case for processing by ML algorithms, (*Machine Learning*), the data must be properly formatted, so that the program would not encounter any unannounced values.

A very simple example of a structure of a database prepared for generation of rules for civil engineering purposes is shown in Table 1. The attributes of the records can, in the simplest case, be of two different types: *numerical* or *nominal*. There is number of different other attribute types, such as *date* or *string*, which are not discussed here. Any attribute in a database record may also have an *unknown value*, (a symbol '?' is usually applied).

Table 1. An example of a simple structure of a formatted database

name	type	action	min	max
			or - list of legal values	
No	numeric	ignore	1	455
nrpomiaru	numeric	ignore	1	502
lzdH	numeric	active	394	660
lzdM	numeric	active	461	833
lzdL	numeric	active	886	1027
senH	numeric	active	35	71
senM	numeric	active	62	286
senL	numeric	active	351	606
sazH	numeric	active	14	33
sazM	numeric	active	23	178
sazL	numeric	active	103	248
phase	nominal	ignore	cp, a, v	
HV	numeric	ignore	56	1298
class	nominal	active	A0, A15K, A15T, A30K, A30T	

A formatted database can be conceived as a matrix in which the rows represent records and the columns represent attributes.

All the information on the structure of a database like in this example are needed to organise properly the ML processing of the dataset. Not all the attributes must be taken into account during the final rule generation. Working on a number of attributes bigger than the necessary minimum might result in an unnecessary noisiness of the data. The designation 'ignore' in the example above, ('ignore', as opposite to 'active'), concerns the attributes that were excluded from the calculations in the example in Chapter 5.

There are various actions that an agent should do. A difficult and important task to be fulfilled by an agent is after identification of the source, (e.g. finding in the Internet a paper dedicated to some type of concrete), to evaluate this source from the point of view of "useful data", (i.e. whether it contains the kind of data that this particular agent will be able to recognise and format), to decide how the data will be procured and transformed, and - if needed - also imported to the agent's home site.

Not discussed here are quite obvious but more particular tasks for agent system, for example optical character recognition in case of the PDF documents available in form of images, recognition of different decimal systems, unification of descriptions, identification, proper understanding and translation in case of foreign languages, etc.

4   The Concept of Interestingness

The ML programs often produce for a given dataset many, even hundreds of particular rules. For a user who is looking for rules as simple as typical empirical formulae proposed over the ages by the human experts, the rules produced typically by a ML program are of unequal value, many practically useless. Worth further attention may be only some of them – those "most interesting" ones.

The concept of *interestingness* has appeared in data mining literature mainly in the last decade, (perhaps one of the first uses of the term was in 1995, by Silberschatz and and Tuzhilin, as cited in [13]). It concerns the relative importance of any rules conceived by people, but mainly of the rules generated by machine learning algorithms. The meaning of the term *interestingness* is slightly imprecise, because the value of a rule depends naturally on the point of view of the user.

There are more than 30 different interestingness measures discussed in the literature; cf. for example [13, 14, 15, 16]. They are mostly constructed by algebraic operations on a set of primary measures of amounts of records in a database under consideration, (in certain texts, depending on the field of interests of the author, instead of the term *records* used is the word *transactions*; e.g. [17]). And some investigators apply their own measures concerning the quality of the rules, without referring at all to the notion of *interestingness*, (e.g. [5, 7, 18]).

With association rules in form of $A \rightarrow B$, (the antecedents in A and consequent or target class B, being conjunctions of simple rules or selectors of attributes), and n being a total number of records in a database, the primary measures are: n_a, n_b, n_{ab} and $n_{a\neg b}$, meaning numbers of records matching the conditions, respectively, of A, B, $A \cap B$, $A \cap \neg B$, (here $\neg B$ means: NOT B).

Examples of four simplest and most typical interestingness measures are: *Support, Confidence, Conviction* and *Lift*[2]. They are, respectively, defined by formulae:

---

[2] the same name *Lift* corresponds to a different operator in *See5*, although there is close linear correlation between the results of the both formulae.

$$support = (n_a - n_{a \neg b})/n, \quad confidence = 1 - n_{a \neg b}/n_a,$$
$$conviction = n_a n_{\neg b}/(n n_{a \neg b}), \quad lift = n(n_a - n_{a \neg b})/(n_a n_b).$$

Among the interestingness measures discussed in the literature rarely mentioned is a measure characterizing the simplicity of a rule. For example very important may be the difference whether the rule is composed of only 1, 2 or 3 selectors, and not – e.g. – 20 selectors or more. The issue is treated in AQ21, where a special function is introduced that can be used to minimize the complexity of learned rules.

In the present paper it is proposed to introduce an additional interestingness measure called *Simplicity*, calculated as an inverse of the number of selectors on the antecedents' side of the rule in question.

In this way a vector of 5 interestingness measures designated respectively as: *sup.*, *conf.*, *conv.*, *lift* and *simpl.*, presents a compact set of parameters, enabling the user a quick evaluation and comparison of the applied ML procedures.

5 Experiments

The particular experiments with ML procedures discussed in what follows were dedicated to a case of data collected during the microindentation tests on hardened concrete, using a Vickers indenter, and recording acoustic emission signals, which were subsequently processed by wavelet transformation, [19]. The database of about 300 records contained, among others, 9 columns of numbers characterising the acoustic emission signal, (AE), on selected frequency and magnitude levels. The structure of the database was presented in the Table 1, above.

The task of ML programs were to discover from the AE data the rules allowing recognizing which records correspond to concrete containing certain additives, (like fly ash or PFA), and which ones are without those additives. The issue may be of importance in case of forensic analysis problems in construction of questionable quality. The presence of additives in the experiment was identified by a class code, (there were three levels of the additive content: 0, 15 and 30 percent, and two additive sources: K and T; cf. the bottom raw of the Table 1).

In search of the rules applied were mainly four ML tools: See5, AQ19, WEKA and AQ21, [5, 6, 7, 18]. A number of different rules were obtained from different methods. In many cases the primary measures concerning numbers of records supporting rules generated by the system could be evaluated directly from the accuracy statistics indices built-in into programs. In other cases they were calculated manually using MS Excel.

Selected results in form of rules, the primary measures and the corresponding interestingness measures are presented in Table 2.

The database from which the numbers were taken in the Table 1 was of 239 records, $(n = 239)$, with the numbers, (n_b), of records in five different classes A0, A15K, A15T, A30K and A30T being, respectively, 80, 60, 62, 63 and 64. The numbers n_a, n_{ab} and $n_{a \neg b}$ were either taken from the ML programs or were counted in Excel.

Table 2. Examples of interestingness measures obtained using four ML programs (the cases of *conv=max* correspond to rules in which there were no erroneous predictions)

	n_a	n_{ab}	$n_{a\neg b}$	sup	conf	conv	lift	simpl
Rules according to See5								
[senH <= 47] → Class = A0	229	78	151	**0.24**	0.34	1.15	1.06	1.00
[sazH > 18][sazL <= 167] → Class = A30T	108	41	67	0.12	0.38	1.30	1.57	0.50
[senL > 400][sazH <= 17] → Class = A15T	16	14	2	0.04	**0.88**	**6.49**	**3.77**	0.50
[senH > 47] → Class = A15K	100	39	61	0.12	0.39	1.34	1.75	**1.00**
[lzdH > 486][senH > 54] → Class = A15K	35	16	19	0.05	0.46	1.51	2.05	0.50
[senH > 47][senH <= 54] → Class = A30K	62	21	41	0.06	0.34	1.22	1.43	0.50
Rules from AQ19								
[lzdH = 510..578][lzdM = 697..800] [lzdL = 976..1007, 1027][senH = 39..42] [senL = 351..393] → Class = A0	51	35	16	0.11	0.69	2.41	2.14	0.20
[lzdH = 486..555, 571][lzdM = 655..748, 807] [lzdL = 950..1005][senH = 39..41] [senL = 379..507] → Class = A15T	28	21	7	0.06	**0.75**	3.25	3.23	0.20
[lzdH = 394..531, 633][lzdL = 950..1006] [senH = 42..57][senM = 99..286][sazM = 47..148] [sazL = 113..145, 177..195] → Class = A30T	19	16	3	0.05	0.84	5.10	**3.49**	0.17
Rules from WEKA, (classifier PART)								
[senH > 47][lzdL <= 942] [lzdH > 465] → Class = A15K	9	8	1	0.02	0.89	**7.36**	3.99	0.33
[senH > 47][lzdL > 974][sazH > 21] → Class = A30T	10	10	0	0.03	**1.00**	**max**	**4.14**	0.33
Rules from AQ21								
[lzdH = 522..568][lzdM = 680..744] [lzdL = 976..1000][senH <= 45] [senM = 73..103][sazH <= 18][sazM >= 28] [sazL = 106..138] → Class = A0	16	16	0	0.05	**1.00**	**max**	3.11	0.13
[senH = 45..66] → Class = A15K	135	48	87	0.15	0.36	1.27	1.59	**1.00**
[sazH <= 17] → Class = A15T	132	44	88	0.13	0.33	1.22	1.44	1.00

The numbers in bold print in the last columns of the Table 2 correspond to best results from the point of view of a given interestingness measures. The limit values are 1 in case of *Support, Confidence* and *Simplicity* measures. There are no practical limits in cases of *Conviction* and *Lift*, or, to be more precise the maximum number that can appear there depends on the size of the database under investigation.

As can be seen the rules obtained by the programs have limited effectiveness. The rules are either of low support even being quite reliable, or are of low accuracy, (many false predictions), or are very complicated, (e.g. 8 selectors in a rule).

It should be added, that the recognition of the data by ML programs was generally rather good. It is obvious that effectiveness of the whole set of rules generated by any ML program used together as a whole collection of formulae may be very effective. For example by applying a whole set of 60 rules obtained by one of the programs, (it was *See5*), quite a satisfying confusion matrix was obtained, as shown in Fig. 1-a. Confusion matrix displays how the ML system assigns the records to their respective classes. When in the next run selected was the *Boost-10 trials* option the results were almost ideal – Fig. 1-b.

Similar results were obtained also using the other ML programs. The good recognition, however, is an effect of applying a kind of voting procedures, and

```
     (a)    (b)    (c)    (d)    (e)    <-classified as
     ----   ----   ----   ----   ----
      74            4             2     (a): class A0
       5     49     2      2      2     (b): class A15K
       7      1    53             1     (c): class A15T
       1      5     4     52      1     (d): class A30K
       8      1            2     53     (e): class A30T
```
(a)

```
     (a)    (b)    (c)    (d)    (e)    <-classified as
     ----   ----   ----   ----   ----
      80                                (a): class A0
             60                         (b): class A15K
                    62                  (c): class A15T
              1            62           (d): class A30K
                                 64     (e): class A30T
```
(b)

Fig. 1. Confusion matrices obtained on the same dataset with program See5:   (a) – default settings, (b) – after selecting the option: *Boost (10 trials)*

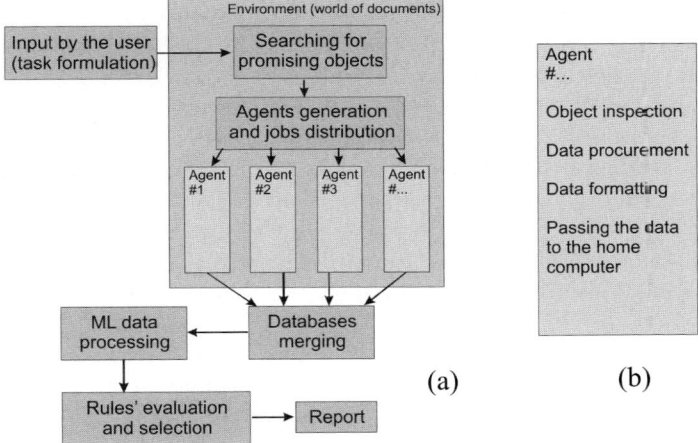

(a)   (b)

Fig. 2. Proposed architecture – (a) of the agent system, (b) of the sub-agent

there are no simpler formula resulting from such calculation. And these possibly simple formulae are what is really needed by engineers.

The selected rules presented in the first column of Table 2 were obtained at a default settings in cases of See5 and AQ19, applying the classifiers PART in case of WEKA, and applying the PD mode, (*Pattern Discovery*), in case of AQ21.

6 Conclusions

It seems that the ambitions presented in this paper seem to be rather far-reaching. It would be however really advantageous to strive towards creating an agent system able to realize proceedings described above. Or at least some elements of it, as these also might be helpful to a human researcher.

After being activated the agent, (agents), should start looking for appropriate data sources, to process what is found, and return finally a number of rules, each accompanied with an interestingness measures vector, as defined in the previous chapter. For the beginning, however, it would be important to have at least one agent that could identify which papers, from a list of hundreds of titles, contain the tables of interesting experimental results.

In Fig. 2 shown is what seems to be a possible architecture for such an agent system.

The input by the user would be some characterization of the field of search, for example by typing-in a number of keywords, (e.g.: "concrete, additives, admixtures, silica, PFA"; as previously this is an example from the field of concrete-like composite materials), and a proposal of a list of expected headings of the columns in the tables presenting the features aimed at during the search. Such list would have to contain as many alternatives as the user can think of. For example: "w, water, c, cement, silica, silica fume, sup, superplasticiser, aggregate, CA, FA, air entrainment, strength, compression strength, density, fc28", (as can be seen the same attributes may have different symbols in different data sources).

After receiving the results from such system the user is thereupon to decide which results are better from others. The rules obtained by different procedures are mutually supplementing, to combine the results, however, a human action would be needed, so the question of how to combine the rules is not discussed here further.

There is a general observation from the experiments performed on actual experimental data using different ML tools that very often the resulting set of rules was either too large, (e.g. 100 rules), or the rules were too complex, (more than $3 \div 4$ selectors), or they had too low the support or too low the accuracy. The results like those in Table 2 will allow the user to concentrate on only the most important results of the search.

The problem of finding optimal rule is a really multi-criteria task. The user in the presented example could support his or her estimate of the resulting rulsets by the five components of the interestingness vector. Later on the user will be able to work out the position to recognise the value of the search by the first look into the results.

Acknowledgments. This work was supported by Projects No. R04 013 01 and No. 4 T07E 036 30, sponsored by The Ministry of Science and Higher Education, Warsaw, Poland, (MNiSW), to which the authors are grateful.

References

[1] Ouderkirk, J.: Technical services task assignment: from macros to collection management intelligent agents. The Journal of Academic Librarianship 25, 397–401 (1999)
[2] Nwana, H.S.: Software agents: An overview. Knowledge Engineering Review 11(2), 205–244 (1995)
[3] Cichosz, P.: Learning systems. WNT, Warszawa (in Polish) (2000)

[4] Gradestat - a statistical program for exploratory data analysis, http://gradestat.ipipan.waw.pl/

[5] Michalski, R.S., Kaufman, K.A.: The AQ19 system for machine learning and pattern discovery: A general description and user's guide (2001)

[6] Witten, I.H., Frank, E.: Data Mining: Practical Machine Learning Tools and Techniques, 2nd edn. Morgan Kaufmann Series in Data Management Sys. Morgan Kaufmann, San Francisco (2005)

[7] Wojtusiak, J.: AQ21 user's guide. Technical Report MLI 04-3, P 04-5 (2004)

[8] Elder, J.I., Abbott, D.: A comparison of leading data mining tools. In: Fourth International Conference on Knowledge Discovery & Data Mining (1998)

[9] King, M., Elder, I.J., Gomolka, B., Schmidt, E., Summers, M., Toop, K.: Evaluation of fourteen desktop data mining tools. In: IEEE International Conference on Systems, Man, and Cybernetics, vol. 3, pp. 2927–2932 (1998)

[10] Alterman, D., Kasperkiewicz, J.: Evaluating concrete materials by application of automatic reasoning. Bulletin of the Polish Academy of Sciences. Technical Sciences 54(4), 352–362 (2006)

[11] Alterman, D.: Evaluation of concrete materials by automatic reasoning. IPPT PAN, Polish Academy of Sciences (in Polish) (manuscript, 2005)

[12] Kasperkiewicz, J., Alterman, D.: Holistic approach to diagnostics of engineering materials. Computer Assisted Mechanics and Engeneering Sciences 14, 197–207 (2007)

[13] Vaillant, B.: Mesurer la qualité des regles d'association. Etudes formelles et expérimentales. l'École Nationale Supérieure des Télécommunications de Bretagne en habilitation conjointe avec l'Université de Bretagne Sud (2006)

[14] Huynh, X., Guillet, F., Briand, H.: ARQAT: an exploratory analysis tool for interestingness measures. In: Janssen, J. (ed.) 11th international symposium on Applied Stochastic Models and Data Analysis (ASMDA 2005), Brest, France, May 17-20, 2005, pp. 334–344 (2005)

[15] Lallich, S., Vaillant, B., Lenca, P.: A probabilistic framework towards the parameterization of association rule interestingness measures. Methodology and Computing in Applied Probability 9(3), 447–463 (2007)

[16] Lenca, P., Meyer, P., Vaillant, B., Lallich, S.: On selecting interestingness measures for association rules: User oriented description and multiple criteria decision aid. European Journal of Operational Research 127(2), 610–626 (2008)

[17] McGarry, K.: A survey of interestingness measures for knowledge discovery. Knowl. Eng. Rev. 20(1), 39–61 (2005)

[18] Rulequest research 2007, See5: An informal tutorial http://www.rulequest.com/see5-win.html

[19] Kasperkiewicz, J.: On a possibility of structure identification by microindentation and acoustic emission. In: de Miguel, Y., Porro, A., Bartos, P. (eds.) 2nd International Symposium on Nanotechnology in Construction, NICOM 2, Bilbao, Spain, pp. 151–159. RILEM Publications S.A.R.L (2006)

Grammatical Inference as a Tool for Constructing Self-learning Syntactic Pattern Recognition-Based Agents

Janusz Jurek

IT Systems Department, Jagiellonian University
Straszewskiego 27, 31-110 Cracow, Poland
jjurek@wzks.uj.edu.pl
http://www.wzks.uj.edu.pl/ksi/jwj

Abstract. Syntactic pattern recognition-based agents have been proven to be a useful tool for constructing real-time process control intelligent systems. In the paper the problem of self-learning schemes in the agents is discussed. Learning capabilities are very important if practical applications of the agents are considered, since the agents should be able to accumulate knowledge about the environment and flexible react to the changes in the environment. As it is shown in the paper, the learning scheme in the agents can be based on a suitable grammatical inference algorithm.

1 Introduction

Syntactic pattern recognition-based agents have been proven to be a useful tool for constructing real-time process control intelligent multi-agent systems [6]. In [2,3] we describe an example of such system: a multi-agent system implemented for the purpose of the on-line monitoring, diagnosing, and controlling of a very complex industrial-like equipment (a high energy physics detector). The system had to perform four different tasks:

- monitoring and identifying the equipment behaviour,
- analysing and diagnosing the equipment behaviour,
- predicting consequences of the equipment behaviour,
- controlling, i.e. taking proper actions (eg. setting operating modes for equipment components).

The decision about the multi-agent architecture of the system has been caused by the fact that only fully-decentralised, distributed intelligent system consisting of autonomous components acting in a parallel way can meet the hard real-time constraints: the necessity of performing four mentioned above tasks simultaneously and on-line with respect to hundreds of elementary components of the equipment. The system has been built of different types of agents correspondingly to the layers of analysis (i.e. the whole equipment layer, modules (subdetectors) layers, and components layers). We have chosen the syntactic pattern recognition

M. Bubak et al. (Eds.): ICCS 2008, Part III, LNCS 5103, pp. 712–721, 2008.

approach as a base for constructing lowest layer agents of reactive type. They are responsible for monitoring and recognition of the behaviour of particular components.

The main part of syntactic pattern recognition agents is a parser. The parser performs syntax analysis of the symbolic description of a component behaviour. The knowledge about the possible behaviour of a component is remembered in the agent in the form of a formal string grammar. The grammar is used to construct a control table for the parser. The advantages of using the syntactic pattern recognition approach consist in very good computational properties (it is possible to implement an advanced parser of the linear, $O(n)$, computational complexity) and the capability to analyse and recognise even exceedingly complex patterns representing a process in time series. These two advantages make the approach better in case of constructing real-time process control intelligent systems than many other computationally inefficient classical artificial intelligence methods [10].

However, after several years of practical experiments with syntactic pattern recognition-based agents we have identified one important problem of the implementation of such agents: the difficulty in providing a self-learning feature. Learning capabilities are very important if we consider practical applications of the agents, since the agents should be able to accumulate knowledge about the environment (i.e. the possible behaviour of a component which is being monitored) and flexible and autonomously react to the changes in the environment. The learning scheme in the agents can be based on a suitable grammatical inference algorithm (i.e. the algorithm of automatical definition of a grammar from the sample of a pattern language). Unfortunately grammatical inference is still one of the main open issues in the syntactic pattern recognition area.

The research into grammatical inference of string grammars started almost forty years ago. The first, basic results were published in the seventies. During next years the methods of grammatical inference were improved in the aspects of the "quality" of generated grammar, and the computational efficiency. Although many different results have been already achieved (especially in case of *regular* grammars) [4,5], there is still lack of models of inferencing *context-sensitive* grammars, that is grammars of big generative/discriminative power. Most methods already developed are based on the identification of structures in words in a sample and the definition of the sequence of the structures. The dependencies between *numbers of symbols* in words are not examined. This results from the fact that the discriminative power of grammars we are able to inferred is too weak to reflect such dependencies. The only one practical method that allows to consider such quantitative information (in the context of syntactic pattern recognition) has been developed by Alquézar and Sanfeliu [1]. They have proposed a new type of grammars (ARE grammars, characterised by a big discriminative power) and a proper grammatical inference algorithm. Although the method is very innovative, it has some disadvantages. In particular, the time complexity of the grammatical inference is exponential, which makes the model unsuitable for the application in a real-time environment.

In the paper we present the recent results of the research into constructing a self-learning algorithm for syntactic pattern recognition-based agents. In section 2 we describe the general scheme of the agents. Section 3 contains the definitions corresponding to the string grammars which we use as a knowledge base in the agents. The description of the model of the self-learning of the agents (via grammatical inference) is included in section 4. Section 5 contains presentation of some experimental results. Conclusions are included in the final section.

2   Syntactic Pattern Recognition-Based Agents

There are three main elements of syntactic pattern recognition-based agents: a knowledge base (in the form of a formal grammar), a parser (a control table for the parser is constructed on the base of a grammar being the knowledge base), and a self-learning module (based on a grammatical inference algorithm used to update the grammar). The general scheme of a syntactic pattern recognition-based agent is shown in Fig. 1.

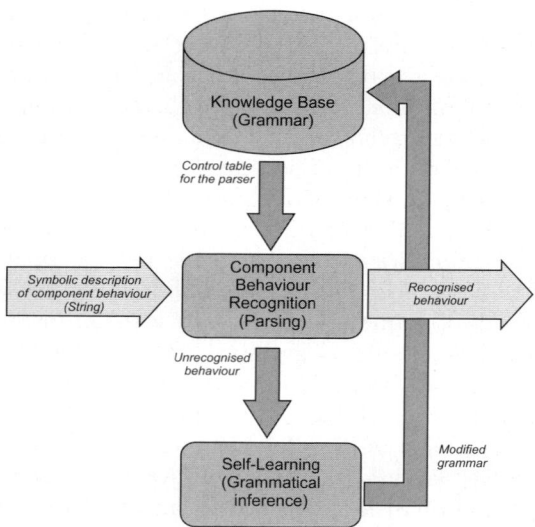

Fig. 1. A general scheme of a syntactic pattern recognition-based agent

A syntactic pattern recognition-based agent is of reactive type: it should be able to respond timely to changes in its environment (changes in the behaviour of a component which is being monitored) and to acquire knowledge about the environment. The agent receives symbolic data (a string of symbols) representing characteristics of a component behaviour. The analysis of the behaviour is made by the parsing of a received string. The results are outputted to the higher layer agents with different architecture which are beyond the scope of the paper [2].

When a received string cannot be recognised by the parser, then the self-learning activity in the agent should take place. The inability to recognise the string pattern means that the current grammar should be replaced by a broaden one which generates the missing pattern. Self-learning can be done via grammatical inference. The result is an updated grammar which becomes a new knowledge base.

The most important decision concerning the implementation of the agent has been the choice of GDPLL(k) grammars [8] as a formal string grammar which is used to store knowledge about a component behaviour and to construct a control table for the parser. The definition of GDPLL(k) grammar and its brief characteristics are included in next section.

3   Quasi Context Sensitive GDPLL(k) Grammars

Let us introduce two basic definitions [3,8].

Definition 1.   A *generalised dynamically programmed context-free grammar* is a six-tuple: $G = (V, \Sigma, O, P, S, M)$, where V is a finite, nonempty alphabet; $\Sigma \subset V$ is a finite, nonempty set of terminal symbols (let $N = V \setminus \Sigma$); O is a set of basic operations on the values stored in the memory; $S \in N$ is the starting symbol; M is the memory; P is a finite set of productions of the form: $p_i = (u_i, L_i, R_i, A_i)$ in which $\mu_i : M \longrightarrow \{TRUE, FALSE\}$ is the predicate of applicability of the production p_i defined with the use of operations ($\in O$) performed over M; $L_i \in N$ and $R_i \in V^*$ are left- and right-hand sides of p_i respectively; A_i is the sequence of operations ($\in O$) over M, which should be performed if the production is to be applied.   □

Definition 2. Let $G = (V, \Sigma, O, P, S, M)$ be a generalised dynamically programmed context-free grammar. The grammar G is called a *Generalised Dynamically Programmed LL(k) grammar*, GDPLL(k) grammar, if: 1) the LL(k) condition of deterministic derivation is fulfilled, and: 2) the number of steps during derivation of any terminal symbol is limited by a constant.   □

As we mentioned in Section 2, GDPLL(k) grammars have been chosen to store knowledge about components behaviour in the agents. The choice has resulted from two main features of GDPLL(k) grammars:

1. GDPLL(k) grammars are characterised by very good discriminative properties since their generative power is stronger than context-free grammars and "almost" as big as context-sensitive grammars [8]. It means that the grammars can store knowledge even about highly complicated patterns representing a component behaviour.
2. It is possible to implement a parser for the languages generated by GDPLL(k) grammars which performs syntax analysis in the linear time. This feature is crucial if we consider embedding of the agents in real-time applications.

In case of GDPLL(k) grammars these two contradictory requirements are balanced. However, as we noticed in Section 2, there is one more requirement concerning a class of grammars to be used as a knowledge base in the agents: the ability to construct a grammatical inference algorithm (preferably of polynomial time complexity) needed to provide self-learning feature in the agents. Grammatical inference is a well-known open problem in the syntactic pattern recognition area. It is especially difficult in case of grammars stronger than than context-free ones. We have done the research into the problem of the grammatical inference of GDPLL(k) grammars for a few years [7,9]. The algorithms developed during last years have been mostly dedicated to particular applications. Recently, we have achieved results that allow us to implement a self-learning system which can be used more generally: in many possible practical applications.

4   Grammatical Inference

Firstly, let us present main definitions needed to discuss the grammatical inference algorithm [9].

Definition 3.   Let A is a set of all (terminal) symbols which appear in the sample, Z set of integer variables. *Polynomial specification of a language* is of the form: $L_p(A, Z) = S_i^{p_j(n_k)}$ where: p_j is a polynomial of a variable $n_k \in Z$; variable n_k can be assigned only values greater or equal 1; S_i, called *polynomial structure*, is defined in a recursive way:

a) $S_i = (a_{i_1}...a_{i_r})$, where $a_{i_j} \in A$. (S_i is a *basic* polynomial structure), or:
b) $S_i = (S_{i_1}^{p_{i_1}(n_{i_1})}...S_{i_r}^{p_{i_r}(n_{i_r})})$, where S_{i_k} is defined as in a) or b). (S_i is a *complex* polynomial structure).

Extended polynomial specification of a language $L_{ep}(A, Z)$ is defined in the following way:

a) $L_{ep}(A, Z) = L_p(A, Z)$ or:
b) $L_{ep}(A, Z) = L_{ep}^1(A_1, Z_1) L_{ep}^2(A_2, Z_2) ... L_{ep}^z(A_z, Z_z)$, or:
c) $L_{ep}(A, Z) = L_{ep}^1(A_1, Z_1) + L_{ep}^2(A_2, Z_2) + ... + L_{ep}^z(A_z, Z_z)$,

if proper conditions of the unambiguity of the specification are fulfilled in case of all Z_i and A_i (see: [9]).   □

Example 1. Let $A = \{a, b, c\}$ be a set of terminal symbols, $Z = \{n, m\}$ be a set of integer variables. Then: $L_p(A, Z) = ((ab)^{2n+1}c^{n^2}(ba)^{2m})^{n+2}(ab)^{m^3}$ is an example of polynomial specification of a language. The polynomial structures in the specification are the following:

$S_1 = ((ab)^{2n+1}c^{n^2}(ba)^{2m})^{n+2}(ab)^{m^3}$	$p_1 \equiv 1$
$S_{1_1} = (ab)^{2n+1}c^{n^2}(ba)^{2m}$	$p_{1_1}(n) = n + 2$
$S_{1_2} = ab$	$p_{1_2}(m) = m^3$
$S_{1_{1_1}} = ab$	$p_{1_{1_1}}(n) = 2n + 1$
$S_{1_{1_2}} = c$	$p_{1_{1_2}}(n) = n^2$
$S_{1_{1_3}} = ba$	$p_{1_{1_3}}(m) = 2m$

Extended polynomial specification of a language is built as a catenation (case "b" in Definition 3) or alternatives (case "c" in Definition 4) of polynomial specifications. Then: $L_{ep}(A, Z) = ((ab)^{2n+1} + cb^{2m})ab^{n^2}$ can be an example of extended polynomial specification of a language. □

Our approach to the inferencing GDPLL(k) grammars is based on the following method. The input is the sample of a language (containing a new string pattern representing an unrecognised behaviour). We divide the inference process into two phases. The first one is responsible for extraction of the features of the sample and generalisation of the sample. The result is the extended polynomial specification of a language. In the second phase, a GDPLL(k) grammar is generated on the basis of the extended polynomial specification of the language.

Now, let us present the first phase: the method of creating the extended polynomial specification of a language from the sample. Let us notice that our method is canonical. It can be used for the detection of basic features only, and for relatively simple generalisation of the sample. The canonical character of the method may broaden the area of its possible applications.

Let the sample of a language be a set of m words: $Sample = \{w_1, ..., w_m\}$. Let Σ be a common alphabet for all words in the sample. Words belonging to the sample will be written in the form: $w = a_1^{n_1} a_2^{n_2} ... a_k^{n_k}$.

Definition 4. Let us consider two words: $w_1 = a_1^{p_1} a_2^{p_2} ... a_k^{p_k}$ and $w_2 = b_1^{q_1} b_2^{q_2} ... b_l^{q_l}$. We say that words w_1 i w_2 are *sequentially equivalent*, if the following conditions are fulfilled: $k = l$ and $a_i = b_i$ for $i = 1, ..., k$. *The template of an extended polynomial specification* is an expression constructed accordingly to the definition of extended polynomial specification L_{ep}, which is built with sets of sequentially equivalent words instead of polynomial specifications L_p. □

Example 2. Let $A = \{a, b, c\}$ be a set of terminal symbols. Two words: $a^2b^4a^2c^7$ and $a^6b^2a^3c^5$ are sequentially equivalent, since both words are built with the same sequence of different symbols: a, b, a, c. □

The algorithm of the construction of the extended polynomial specification of a language can be divided into three steps. In the first step the template of the extended polynomial specification is built. In the second step each set of sequentially equivalent words (in the template) is generalised in the form of a polynomial specification. In the last step the extended polynomial specification is generated (by applying the results of step 2 to the template).

Algorithm 1. The algorithm of the construction of the extended polynomial specification of the language from $Sample = \{w_1, ..., w_m\}$ consists of three steps:

Step 1. We construct the template of the extended polynomial specification. Let $V\{v_1, ..., v_n\}$ be a set which is a variable in the template of an extended polynomial specification which is being constructed, where $v_1, ..., v_n$ are words or fragments of words belonging to *Sample*. Initially, let the template of an extended polynomial specification be the variable: $V\{w_1, ..., w_m\}$. While there is such a set V in the template, which is not the set of sequentially equivalent words, we transform the template according to two rules. The first one is: $V\{w_{s_1}a_1w_{r_1}, ..., w_{s_n}a_nw_{r_n}\} := V\{w_{s_1}, ..., w_{s_n}\}\, V\{a_1w_{r_1}, ..., a_nw_{r_n}\}$, if w_{s_i} are sequentially equivalent (for $i = 1, ..., n$) and there are such symbols a_i and a_j different than the last symbol of w_{s_1} that $a_i \neq a_j$. The second one is: $V\{a_1w_1, ..., a_nw_n\} := V\{a_{1_1}w_{1_1}, ..., a_{1_k}w_{1_k}\} + ... + V\{a_{s_1}w_{s_1}, ..., a_{s_k}w_{s_k}\}$, if for any two symbols $a_{x_i}, a_{x_j} : a_{x_i} = a_{x_j}$, and for any two symbols a_{x_i}, a_{y_j}, where $x \neq y : a_i \neq a_j$. The result is the template in which all V are sets of sequentially equivalent words.

Step 2. We generalise each set of sequentially equivalent words (in the template) in the form of a polynomial specification. Let $P = \{w_1, ..., w_q\}$ be a set of sequentially equivalent words. Let the word $w_i \in P$ be of the form: $w_i = a_1^{n_{i_1}} a_2^{n_{i_2}} ... a_k^{n_{i_k}}$. The canonical algorithm of the construction of polynomial specification $L_p(A, Z)$ for P is the following. For $L_p(A, Z)$ we define: $A = \{a_1, ..., a_k\}$, and $Z = \{x_1, ..., x_k\}$. We assume that $L_p(A, Z) = a_1^{p_1(x_1)} ... a_k^{p_k(x_k)}$. For each $j = 1, ..., k$ we search for the linear function of variable x_j, which describes dependencies between numbers n_{i_j} (where: $i = 1, ..., q$). Let the function be of the form: $p_j(x_j) = d_jx_j + r_j$. Then we look for linear dependencies between $x_j = x_1, ..., x_k$ allowing to reduce Z set. For each pair x_g and x_h (where $g = 1, ..., k$, $h = 1, ..., k$) we check whether $x_h = x_g + s$ for each words in P and a constant value s. If the result of the test is positive, we store a new function for the index h, and we delete variable x_h from Z set. As the outcome of this step we get polynomial specification $L_p(A, Z)$ such that all words from P belong to $L_p(A, Z)$, as well as all other words which are considered to be "similar".

Step 3. We define the extended polynomial specification by inserting polynomial specifications (generated in step 2) to the template of the extended polynomial specification (generated in step 1).   □

The algorithm above is of polynomial complexity. The extended polynomial specification being the result of the algorithm is the input for the second phase.

Now, let us present the algorithm of automatic generation of a GDPLL(k) grammar from polynomial specification of a language (as the main algorithm of the second phase). The algorithm can be divided into two steps. In the first step we define a separate grammar for each polynomial specification which is included in the *extended* polynomial specification. In the second step we construct a target GDPLL(k) grammar for extended polynomial specification on the basis of the grammars constructed in the first step.

Algorithm 2.   The algorithm of automatic generation of a GDPLL(k) grammar from polynomial specification of a language consists of two steps:

Step 1. We define a separate grammar for each polynomial specification which is included in the extended polynomial specification in the following way. Let $L_p(A, Z)$ be a polynomial specification of a language. We will construct a GDPLL(k) grammar $G = (V, \Sigma, O, P, S, M)$ generating the language. Let $\Sigma := A$. For each variable $n \in Z$ in $L_p(A, Z)$ we define two variables: v_n and d_n in the grammar memory M. For each polynomial structure S in $L_p(A, Z)$ we define: a nonterminal symbol $X_S \in V$, and two memory variables: c_S and e_S in M. For each $S^{p(n)}$ structure in $L_p(A, Z)$ we define a set of productions in P in the following way. We use memory variables to implement "loops" during derivation. Current value of n (in exponent expression) is stored in v_n. Variable c_S is a counter of repetitions of S structure. Variable e_S contains the current evaluation of exponent expression for S structure. Boolean variable d_n stores information whether n is fixed or not (i.e. if the value of n has been determined before). Operations on the memory defined for each production are responsible for "programming" proper number of repetitions during derivation. (Formal definition of the set of production is described in [7]).

Step 2. We construct a target GDPLL(k) grammar for extended polynomial specification in the following way. Let $L_{ep}(A, Z)$ be an extended polynomial specification of language L. Let us assume that $L_{ep}(A, Z)$ is built by m polynomial specifications $L_p^i(A^i, Z^i)$ (we will use L_p^i to denote $L_p^i(A^i, Z^i)$). Let us assume that for each L_p^i a grammar $G^i = (V^i, \Sigma^i, O, P^i, S^i, M^i)$ has been constructed (in step 1) in such a way that: $N^i \cap N^j = \emptyset$ and $M^i \cap M^j = \emptyset$ for $i \neq j$. Now we define: $V := \bigcup V^i$, $\Sigma := \bigcup \Sigma^i$, $P := \bigcup P^i$ and $M := \bigcup M^i$, for $i := 1, ..., m$. Then for each extended polynomial specification L_{ep}^j (where $j \geq m$) included in the construction of the specification $L_{ep}(A, Z)$ we define: if $L_{ep}^j \neq L_p^j$, then a new symbol E^j is added to N; if $L_{ep}^j = L_{ep}^{j_1} L_{ep}^{j_2} ... L_{ep}^{j_z}$, then a new production $p: E^j \longrightarrow E^{j_1} ... E^{j_z}$ is added to P, where symbols E^j represent adequate extended polynomial specifications of the language; if $L_{ep} = L_{ep}^1 + L_{ep}^2 + ... + L_{ep}^z$, then z new productions: $E^j \longrightarrow E^{j_1}$, ..., $E^j \longrightarrow E^{j_z}$ are added to P.   □

The result of Algorithm 2 is a new (updated) definition of a GDPLL(k) grammar being a new knowledge base for the agent. Both algorithms presented above are of polynomial computational complexity, so the whole grammatical inference process is also computationally efficient (the time complexity of the whole grammatical inference is $O(m^3 * n^3)$, where m is the number of words in a sample, n is the maximum length of a word in a sample).

5   Implementation and Experimental Results

All grammatical inference algorithms presented in previous section (being the realization of the learning self-learning scheme in the syntactic pattern recognition-based agents) have been implemented and tested. The experimental implementation has been prepared in C++ language with the use of the object-oriented approach.

Two aspects of the grammatical inference module functioning have been verified: the correctness of the results of the inference, and the time efficiency.

Table 1. Inference of the grammar generating language: $\{a^n b^n a^n : n = 1, 2, \ldots\}$

Sample	Number of words	Maximum word length	Time (s)
1	3	9	27×10^{-5}
2	7	30	28×10^{-5}
3	10	90	29×10^{-5}
4	50	900	33×10^{-5}
5	100	900	39×10^{-5}
6	1000	900	246×10^{-5}

In order to test the module we have used mainly such samples that describe context-sensitive languages (intuitively: samples in which context dependencies concerning the number of given symbols in a word can be observed).

A typical example of a strong context-sensitive language (a language which cannot be generated by a context-free grammar) is the language: $\{a^n b^n a^n : n = 1, 2, \ldots\}$. The context property of the language can be easily proven on the basis of the well-known pumping lemma. We have prepared a rich set of test data for this language. The set has included samples having from 3 to 1000 words. In all cases grammatical inference module has recognised proper dependencies inside words (the same number of a, b, and c symbols in a word) and dependencies between words (all words are of the same pattern $a^n b^n a^n$). Then a proper GDPLL(k) grammar has been automatically generated.

The results of time tests of the grammatical inference module are shown in Table 1. The tests have been performed on relatively old PC (processor: AMD Athlon XP 2600+ 2.1 GHz, Microsoft Windows XP Professional system).

All the experiments and tests confirm that the algorithms of grammatical inference are of good (polynomial) computational complexity. Usually, execution time is much better than predicted on the base of (pessimistic) estimation $O(m^3 * n^3)$, where m is the number of words in a sample, n is the maximum length of a word in a sample.

6   Concluding Remarks

In the paper we have presented the recent results of the research into construction of syntactic pattern recognition-based agents. Such agents have been designed as a tool for the implementation of real-time process control intelligent multi-agent systems. The process diagnostic and control is a natural application for agents, since process controllers are themselves autonomous reactive systems. On the other hand, the real-time requirements make the construction of the agents particularly difficult [11,12,13].

Although we have proven practical usefulness of the syntactic pattern recognition-based agents by the application in a real-time process control system [2,6], we have observed that there is a significant problem concerning the definition of the universal self-learning method in the agents. Let us notice, that learning capabilities are very important if we consider practical applications of the agents, since they should be able to gain knowledge about the changes in the environment. In the paper we have shown that the learning scheme in the agents can be based on a grammatical inference algorithm. If a new (unrecognised) pattern of the behaviour in the monitored process appears, an updated grammar (being the knowledge base in the agents) will be generated. We have developed the algorithm that can automatically produce the definition of a very strong (in the sense of discriminative power) quasi-context sensitive grammar. The algorithm is of polynomial computational complexity.

The method presented in the paper still needs a thorough practical evaluation. We are going to test it in the environment of several different applications where the need of a self-learning feature of the agents is particularly noticeable. The discussion of the results will be a subject of further publications.

References

1. Alquézar, R., Sanfeliu, A.: Recognition and learning of a class of context-sensitive languages described by augmented regular expressions. Pattern Recognition 30, 163–182 (1997)
2. Flasiński, M.: Automata-Based Multi-agent Model as a Tool for Constructing Real-Time Intelligent Control Systems. In: Dunin-Keplicz, B., Nawarecki, E. (eds.) CEEMAS 2001. LNCS (LNAI), vol. 2296, pp. 103–110. Springer, Heidelberg (2002)
3. Flasiński, M., Jurek, J.: Dynamically Programmed Automata for Quasi Context Sensitive Languages as a Tool for Inference Support in Pattern Recognition-Based Real-Time Control Expert Systems. Pattern Recognition 32, 671–690 (1999)
4. De La Higuera, C.: Current Trends in Grammatical Inference. In: Amin, A., Pudil, P., Ferri, F.J., Iñesta, J.M. (eds.) SPR 2000 and SSPR 2000. LNCS, vol. 1876, pp. 28–31. Springer, Heidelberg (2000)
5. De La Higuera, C.: A bibliographical study of grammatical inference. Pattern Recognition 38, 1332–1348 (2005)
6. Jurek, J.: Syntactic Pattern Recognition-Based Agents for Real-Time Expert Systems. In: Dunin-Keplicz, B., Nawarecki, E. (eds.) CEEMAS 2001. LNCS (LNAI), vol. 2296, pp. 161–168. Springer, Heidelberg (2002)
7. Jurek, J.: Towards grammatical inferencing of GDPLL(k) grammars for applications in syntactic pattern recognition-based expert systems. In: Rutkowski, L., Siekmann, J.H., Tadeusiewicz, R., Zadeh, L.A. (eds.) ICAISC 2004. LNCS (LNAI), vol. 3070, pp. 604–609. Springer, Heidelberg (2004)
8. Jurek, J.: Recent developments of the syntactic pattern recognition model based on quasi-context sensitive languages. Pattern Recognition Letters 26, 1011–1018 (2005)
9. Jurek, J.: Generalisation of a Language Sample for Grammatical Inference of GDPLL(k) Grammars. In: Computer Recognition Systems 2. Advances in Soft Computing series, pp. 282–288. Springer, Heidelberg (2007)
10. Negnevitsky, M.: Artificial Intelligence. A Guide to Intelligent Systems. Addison-Wesley, Reading (2002)
11. Niederberger, C., Gross, M.: Hierarchical and Heterogenous Reactive Agents for Real-Time Applications. Computer Graphics Forum 22, 323–331 (2003)
12. Russell, S.J., Norvig, P.: Artificial Intelligence: A Modern Approach, 2nd edn. Prentice-Hall, Englewood Cliffs (2002)
13. Soto, I., Garijo, M., Iglesias, C.A., Ramos, M.: An agent architecture to fulfill real-time requirements. In: Proceedings of the Fourth International Conference on Autonomous Agents, Barcelona, Spain, June 03–07, 2000, pp. 475–482 (2000)

An Architecture for Learning Agents

Bartłomiej Śnieżyński

AGH University of Science and Technology, Institute of Computer Science
Kraków, Poland
Bartlomiej.Sniezynski@agh.edu.pl

Abstract. This paper contains a proposal of an architecture for learning agents. The architecture supports centralized learning. Learning may be performed by several agents in the system, but it should be independent (without communication or cooperation connected with the learning process). An agent may have several learning modules for different aspects of its activity. Each module can use different learning strategy. Application of the architecture is studied on example of Fish-Banks game simulator.

Keywords: multi-agent systems, machine learning, agent architecture.

1 Introduction

Multi-agent systems often work in complex environments. Therefore it is very difficult (or sometimes impossible) to specify and implement all system details a priori.

Applying learning algorithms allows to overcome such problems. One can implement an agent that is not perfect, but improves its performance. This is why machine learning term appears in a context of agent systems for several years.

A lot of multi-agent systems, which are able to learn, have been built so far. But in these works authors use their own architectures for learning agents, specialized for the considered application domains. The universal model of the learning agent was missing. It should be general enough to use in every domain, cover as many learning methods as possible, but also it should be specific enough to help to develop learning multi-agent systems.

This paper contains a proposal of the architecture for learning agents. The proposed architecture supports centralized learning only. It means that all the learning process is performed by an agent itself. Learning may be performed by several agents in the system, but it should be independent (without communication or cooperation with other agents regarding the learning process).

In the following sections learning in multi agent systems is briefly discussed, the architecture of the learning agent is described, and its use in developed system is presented.

2 Learning in Multi-agent Systems

Machine learning focuses mostly on research on an isolated process performed by only one module in the whole system. The multi-agent approach concerns the

M. Bubak et al. (Eds.): ICCS 2008, Part III, LNCS 5103, pp. 722–730, 2008.

systems composed of autonomous elements, called agents, whose actions lead to the realization of given goals. In this context, learning is based on the observation of the influences of activities, performed to achieve the goal by an agent itself or by other agents. Learning may proceed in a traditional – centralized (one learning agent) or decentralized manner. In the second case more than one agent is engaged in one learning process.

In multi-agent systems the most common technique is reinforcement learning [1]. It allows to generate a strategy for an agent in a situation, when the environment provides some feedback after the agent has acted. Feedback takes the form of a real number representing reward, which depends on the quality of the action executed by the agent in a given situation. The goal of the learning is to maximize estimated reward.

Supervised learning is not so widely used in multi-agent systems. However there are some works using such strategies (e.g. [2,3]). Supervised learning allows to generate knowledge from examples. Using this method instead of reinforcement learning has several advantages, see [4].

Architecture for learning agent can be found in [5]. Unfortunately it fits mainly reinforcement learning.

3   The Learning Agent Architecture

In this paper we propose a learning agent architecture for centralized learning which allows to use several learning modules in an agent. The architecture is presented in Fig. 1.

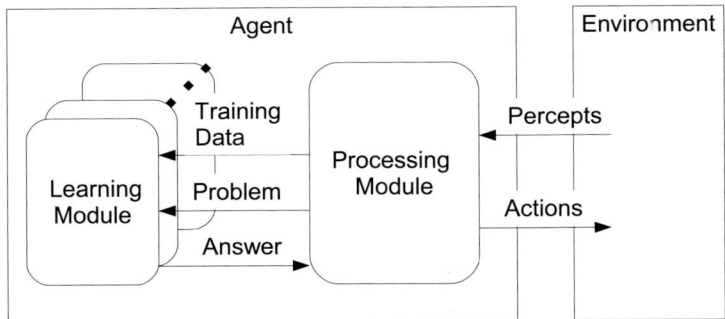

Fig. 1. Learning agent architecture

An agent gets percepts from an environment, and executes actions to interact with the environment. The main unit in an agent is a processing module, which is responsible for analyzing sensor input, and choosing appropriate action. It can be realized in a simple way, e.g. using reactive agent architecture, or in a complex way, e.g. using layered or BDI architectures. To improve the performance, agent can use learning modules. To learn, agent should provide a training data. After learning, the module can be used to get an answer for a provided problem. Therefore, the training data and the problem are inputs for the module, and the answer is an output. Several learning modules, using various learning strategies, can be used by one agent for different aspects of its activity.

Characteristics of the training data, the problem and the answer depend on the learning strategy used in the module. Therefore we can define a learning module as a four-tuple: (*Learning strategy, Training data, Problem, Answer*).

Details of the learning modules are domain-specific. *Learning strategy*, (knowledge representation, learning algorithm, and conditions for which learning is executed), structure and source of *Training data*, *Problem*, and *Answer* should be carefully designed by the system architect. Additional research is necessary to provide guidance in this aspect.

Two types of learning modules were developed and tested so far: reinforcement learning module and inductive rule learning (see section 4). Although, other learning methods can be also used. Below modules for three types of popular learning strategies are characterized.

3.1   Reinforcement Learning

As it was mentioned earlier, the most popular learning method in multi-agent systems is reinforcement learning. In this method, an agent gets description of the current state and using its current strategy chooses an appropriate action from a defined set. Next, using reward from the environment and next state description it updates its strategy. Several methods of choosing the action and updating the strategy have been developed so far. E.g. in Q-learning developed by Chris Watkins [6] action with the highest predicted value (Q) is chosen. Q is a function that estimates value of the action in a given state:

$$Q : A \times X \to \Re, \tag{1}$$

where A is a set of actions, and X is a set of possible states. Q function is updated after action execution:

$$Q(a, x) := Q(a, x) + \beta \Delta \tag{2}$$
$$\Delta = \gamma Q_{max} + r - Q(a, x) \tag{3}$$
$$Q_{max} = \max_a Q(a, x') \tag{4}$$

where $x, x' \in X$ are subsequent states, $a \in A$ is an action chosen, r is a reward obtained from the environment, $\gamma \in [0, 1]$ is a discount rate (importance of the future rewards), and $\beta \in (0, 1)$ is a learning rate. Various techniques are used to prevent from getting into a local optimum. The idea is to explore the solution space better by choosing not optimal actions (e.g. random or not performed in a given state yet) from time to time.

Reinforcement learning module can be responsible for managing all the agent activities or only a part of it (it can be activated in some type of states or can be responsible for selected actions only). The *Problem* definition that is provided consists of the description of the current state. The *Answer* is an action chosen using the current strategy (current Q function). *Training data* consists of the next state description (after executing action returned by the module), and a reward. The reward may be observed by the agent or may be calculated by the processing module using some performance measures.

3.2   Supervised Learning

Supervised learning allows to generate an approximation of a function $f : X \to C$ from labeled examples, which consist of pairs of arguments and function values. This approximation is called a hypothesis h. Elements of X are described by set of attributes $A = (a_1, a_2, \ldots, a_n)$, where $a_i : X \to D_i$. Therefore $x^A = (a_1(x), a_2(x), \ldots, a_n(x))$ is used instead of x.

Supervised learning module gets a *Training data*, which is a set $\{(x^A, f(x))\}$ and generates hypothesis h. *Problem* is a x^A, and the *Answer* is $h(x^A)$.

There are lots of supervised learning methods. They use various hypothesis representation, and various methods of hypothesis construction. One of the most popular algorithms is C4.5, inductive decision tree learning algorithm developed by Ross Quinlan [7]. It can be used if the size of the set C is small. In such a case we call C a set of classes, and hypothesis is called a classifier. C4.5 uses decision trees to represent h. The basic idea of learning is as follows. The tree is learned from examples recursively. If (almost) all examples in the training data belong to one class, the tree consisting of the leaf labeled by this class is returned. In the other case, the best attribute for the test in the root is chosen (using entropy measure), training examples are divided according to the selected attribute values, and the procedure is called recursively for every attribute test result with the rest of attributes and appropriate examples as parameters.

Another learning algorithm with broad range of abilities, which was used in the implemented system (see section 4) is AQ. It was developed by Ryszard Michalski [8]. Its subsequent versions are still developed. This algorithm also generates classifier from the training data, but h is represented by a set of rules, which have tests on attribute values in the premise part, and a class in a conclusion. Rules are generated using sequential covering: the best rule (e.g. giving a good answer for the most examples) is constructed by a beam search, examples covered by this rule are eliminated from a training set, and the procedure repeats.

Other methods, using different knowledge representation, such as support vector machines, Bayesian or instance-based models also fit the above specification. Similarly, learning module using artificial neural networks for classification or function approximation have the same input and output.

What is important, in the case of supervised learning, the processing module should provide in the training data a proper function value $f(x)$ for examples. If we are not able to provide this, inductive learning can not be used. However, if we have at least some qualitative information about $f(x)$ for given x^A, as we suggested in [9] we can build a classifier. Details of this work-around can be found in section 4.2.

3.3   Unsupervised Learning

In unsupervised learning the task of the learning module is to organize examples into groups called clusters, whose members are similar in a some way. Examples of this strategy are Kohonen neural networks and clustering. *Training data* have a form of example descriptions: $\{x^A\}$ (without any label). The *Problem* is an example description x^A, and the *Answer* is the example's cluster identifier.

This type of module was not tested yet. It is presented here to show that the framework proposed is general enough to cover this type of learning.

4   Application of the Architecture

In this section we present application of the proposed learning agent architecture. An multi agent system was built to simulate the Fish Banks game [10]. The game is a dynamic environment providing resources, action execution procedures, and time flow represented by game rounds. Each round consists of the following steps: ships and money update, ship auctions, trading session, ship orders, ship allocation, fishing, fish number update.

Agents represent players that manage fishing companies. Each company aims at collecting maximum assets expressed by the amount of money deposited at a bank account and the number of ships. The company earns money by fishing at fish banks. The environment provides two fishing areas: coastal and deep-sea. Agents can also keep their ships at the port. The cost of deep-sea fishing is the highest. The cost of staying at the port is the lowest but such ship does not catch fish. Initially, it is assumed that the number of fish in both banks is close to the bank's maximal capacity. Therefore, at the beginning of game deep-sea fishing is more profitable.

Usually, exploration of the banks by fishing is too high and after several rounds the number of fish decreases to zero. It is a standard case of "the tragedy of commons" [11]. It is more reasonable to keep ships at the harbor then, therefore companies should change theirs strategies.

Agents may observe the following aspects of the environment: arriving of new ships bought from a shipyard, money earned in the last round, ships allocations of all agents, and fishing results for deep sea and inshore area. All types of agents can execute the following two types of actions: order ships, allocate ships.

Three types of agents can play the game in the system: two types of learning agents using reinforcement learning and rule inductive learning, and a random agent.

Order ships action is currently very simple. It is implemented in all types of agents in the same way. At the beginning of the game every agent has 10 ships. Every round, if it has less than 15 ships, there is 50% chance that it orders two new ships.

Ships allocation action is controlled by a learning module or is done randomly. It is based on the method used in [12]. The allocation action is represented by a triple (h, d, c), where h is the number of ships left in a harbor, d and c are numbers of ships sent to a deep sea, and a coastal area respectively. Agents generate a list of allocation strategies for $h = 0\%, 25\%, 50\%, 75\%$, and 100% of ships that belong to the agent. The rest of ships (r) is partitioned; for every h the following candidates are generated:

1. All: $(h, 0, r), (h, r, 0)$ – send all remaining ships to a deep sea or coastal area,
2. Check: $(h, 1, r - 1), (h, r - 1, 1)$ – send one ship to a deep sea or coastal area and the rest to the other,
3. Three random actions: $(h, x, r - x)$, where $1 \leq x < r$ is a random number – allocate remaining ships in a random way,
4. Equal: $(h, r/2, r/2)$ – send equal number of ships to both areas.

The random agent allocates ships using one of the candidates chosen by random. Methods used by learning agents and their learning modules are described below.

4.1   Reinforcement Learning Agent

Reinforcement learning agent chooses action by random in the first round. In the following rounds, reinforcement learning module is used. In this module *Problem* is a pair (dc, cc), where $dc \in \{1, 2, \ldots 25\}$ represent catch in a deep-sea area, and $cc \in \{1, 2, \ldots, 15\}\}$ represents catch in a coastal area in the previous round. *Answer* is a triple representing ship allocation action (h, d, c), such that $h, d, c \in \{0\%, 25\%, 50\%, 75\%100\%\}, d + c = 1$. The *Training data* consists of a pair (dc', cc'), which is a catch in the current round, and a reward that is equal to the income (money earned by fishing decreased by ship maintenance costs). *Learning strategy* applied is the Q-Learning algorithm.

At the beginning Q is initialized as a constant function 0. To provide sufficient exploration, in a game number g a random action is chosen with probability $1/g$ (all actions have the same probability then). Therefore random or the best action (according to Q function) is chosen and executed.

4.2   Rule learning agent

Because agent has no information what action is the best in the given situation, it is not able to prepare a training data in the form of (*state*, f(*state*)). To overcome this problem the following work-around is used. Thank to comparison of income of all agents after action execution, the learning agent has information about quality of actions executed in the current situation. Leaning module is used to classify action in the given situation as good or bad. When it is learned, it may be used to give ranks to action candidates.

The *Problem* is defined as a five-tuple (dc, cc, h, d, c), it consists of catch in the both areas during the previous round and a ship allocation action parameters. The *Answer* is an integer, which represents the given allocation action rating. The agent collects ratings for all generated allocation action candidates and chooses the action with the highest rating.

Training examples are generated from agent observations. Every round the learning agent stores ship allocations of all agents, and the fish catch in the previous round. The action of an agent with the highest income is classified as

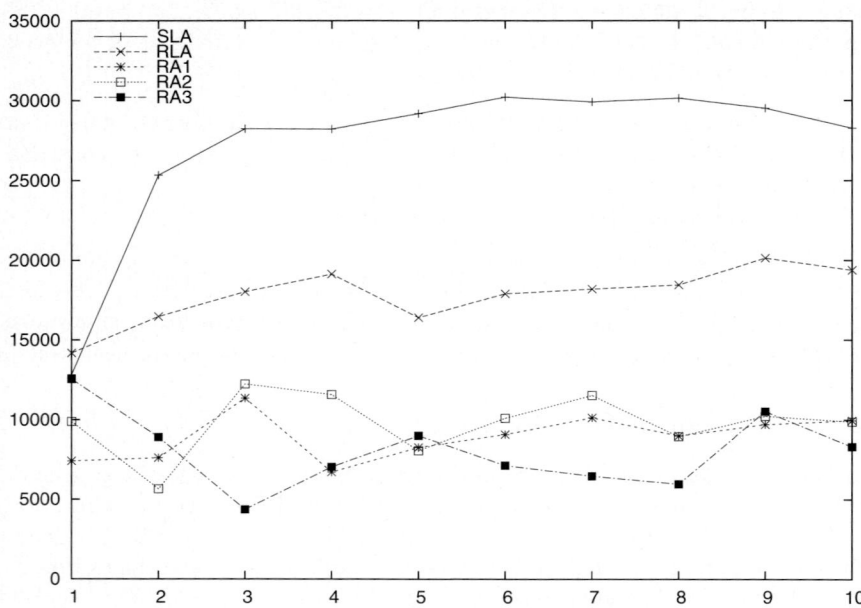

Fig. 2. The average performance of rule learning agent (SLA) reinforcement learning agent (RLA), and agents using random strategy (RA1, RA2, RA3)

good, and the action of an agent with the lowest income is classified as bad. If in some round all agents get the same income, none action is classified, and as a consequence, none of them is used in learning. *Training data* consists of the following pairs: $((dc, cc, h, d, c), q)$, where q is equal to *good* or *bad*. At the end of each game the agent uses training examples, which were generated during all games played so far, to learn a new classifier, which is used in the next game. *Learning strategy* used is AQ21 program, which is an implementation of the AQ algorithm [13].

Rating r of the action a is calculated according to the formula:

$$v(a) = \alpha \, good(a) - bad(a), \tag{5}$$

where $good(a)$ and $bad(a)$ are numbers of rules, which match the action and current environment parameters, with consequence *good* and *bad*, respectively, and α is a weight representing the importance of rules with consequence *good*.

4.3   Performance of the Agents

The average performance of agents presented above in 10 subsequent games is presented in Fig. 2. In these experiments there were three random agents and one reinforcement learning agent or one rule learning agent (α was equal to one). Performance was measured as a balance at the end of every game. In the figure the average performance from ten repetitions of the simulation is presented.

In the experiments the average balance of both types of learning agents increases with time. Reinforcement learning agent was worse than a rule learning agent, but tuning of its parameters and taking into account during learning actions of other agents should increase its performance. On the other hand, reinforcement learning works well even if the reward is delayed. More about comparison of these two learning strategies can be found in [4].

4.4   Two Learning Modules in One Agent

Currently, we are working on the version of the system, in which agent will be learning in two aspects: ship allocation and setting a catch limit. The former aspect will be the same as described above. The latter will be used to develop a strategy of limiting fishing in the areas with small number of fish. *Learning strategy* that is assumed is Q-Learning algorithm. Currently, *Problem* is defined as a fish catch in the previous round, *Answer* contain information if the limit proposal (which is constant) should be accepted or not. *Training data* consists of the fish catch information and a reward. The reward is equal to 0 in all rounds except the last one, when it is a balance of the agent.

5   Conclusion and Further Research

In the paper an architecture of learning agent is proposed. It was used in the description of learning agents in a Fish-Banks simulation system. Agents are learning ship-allocation strategy using reinforcement learning and rule induction.

The architecture is general enough to represent different approaches to learning. Applying the proposed model in a description of the system makes the description clearer. It also helps to develop learning agents and to add new learning modules to existing agents. It may be considered as a tool for learning agents design.

Currently the architecture supports centralized learning only. In the future it should be extended to cover distributed learning (cooperation and communication during learning). Also agents with more then one learning module should be studied and the possibility of interaction between modules in the same agent should be examined.

Acknowledgments. The author is grateful to Arun Majumdar, Vivomind Intelligence Inc. for providing Prologix system (used for implementation), and for help with using it, Janusz Wojtusiak, MLI Laboratory for AQ21 software and assistance, and last but not least Jaroslaw Kozlak, AGH University of Science and Technology for help with the Fish Bank Game.

References

1. Sen, S., Weiss, G.: Learning in multiagent systems. In: Weiss, G. (ed.) A Modern Approach to Distributed Artificial Intelligence. MIT Press, Cambridge (1999)
2. Sugawara, T., Lesser, V.: On-line learning of coordination plans. In: Proceedings of the 12th International Workshop on Distributed Artificial Intelligence, pp. 335–345, 371–377 (1993)

3. Śnieżyński, K.J.: Learning in a multi-agent approach to a fish bank game. In: Pěchouček, M., Petta, P., Varga, L.Z. (eds.) CEEMAS 2005. LNCS (LNAI), vol. 3690, pp. 568–571. Springer, Heidelberg (2005)

4. Śnieżyński, B.: Resource Management in a Multi-agent System by Means of Reinforcement Learning and Supervised Rule Learning. In: Shi, Y., van Albada, G.D., Dongarra, J., Sloot, P.M.A. (eds.) ICCS 2007. LNCS, vol. 4488, pp. 864–871. Springer, Heidelberg (2007)

5. Russell, S., Norvig, P.: Artificial Intelligence – A Modern Approach. Prentice-Hall, Englewood Cliffs (1995)

6. Watkins, C.J.C.H.: Learning from Delayed Rewards. PhD thesis, King's College, Cambridge (1989)

7. Quinlan, J.: C4.5: Programs for Machine Learning. Morgan Kaufmann, San Francisco (1993)

8. Michalski, R.S., Larson, J.: Aqval/1 (aq7) user's guide and program description. Technical Report 731, Department of Computer Science, University of Illinois, Urbana (June 1975)

9. Sniezynski, B.: Rule induction in a fish bank multiagent system. Technical Report 1, AGH University of Science and Technology, Institute of Computer Science (2005)

10. Meadows, D., Iddman, T., Shannon, D.: Fish Banks, LTD: Game Administrator's Manual. Laboratory of Interactive Learning, University of New Hampshire, Durham, USA (1993)

11. Hardin, G.: The tragedy of commons. Science 162, 1243–1248 (1968)

12. Kozlak, J., Demazeau, Y., Bousquet, F.: Multi-agent system to model the fishbanks game process. In: The First International Workshop of Central and Eastern Europe on Multi-agent Systems (CEEMAS 1999), St. Petersburg (1999)

13. Wojtusiak, J.: AQ21 User's Guide. Reports of the Machine Learning and Inference Laboratory, MLI 04-3. George Mason University, Fairfax, VA (2004)

Partnership Selection of Agile Virtual Enterprise Based on Grey Ant Colony Algorithm

Y.D. Fang[1], L.H. Du[2], H. Chen[1], B. Sun[1], and Y.L. He[3]

[1] Jin Hua North Road 4#, The Institute of Mechanical and Electrical Engineer, Xi'an Technological University, Xi'an, Shaan Xi, P.R.C
tomfangok@126.com, Chenhua126@163.com, sunbomm@163.com
[2] Cui Hua South Road 44#, School of Information,Xi'an University of Finance and Economics, Xi'an, Shaan Xi, P.R.C
dulh06@126.com
[3] You Yi western Road 127 #, The Key Laboratory of Contemporary Design and Integrated Manufacturing Technology, Northwestern Polytechnic University, Xi'an, Shaan Xi, P.R.C
hey1@nwpu.edu.cn

Abstract. The paper analyzes the art of partner selection, and enumerates the advantage of partnership selection based on grey relation theory and ant colony algorithm. Furthermore, evaluation framework of agile virtual enterprise (AVE) partner pre-election is analyzed and designed based on the characteristics of AVE. According to grey relation theory, considerable candidate enterprises are selected to reduce the size of problem. Lastly, cooperative enterprise selecting path is decided by making use of ant colony algorithm in terms of transportation cost.

Keywords: partner selection, grey relation theory, ant colony algorithm, agile virtual enterprise.

1 Introduction

Agile Virtual Enterprise (AVE) is a dynamic organization on the web environment, which is composed of a number of independent enterprises connected by information technology based on various hierarchy resources integration and sharing [1]. It's important to get the optimum cooperative enterprise scheme by analyzing, arranging and evaluating character information of candidate enterprise for AVE. At present, partner selection of the AVE always utilizes integer programming, fuzzy evaluation, multiple objectives programming, grey relation theory and so on. For example, P.Gutpa [2] introduces partner selection method in the distributed manufacturing environment according to evaluation of product manufacturing capability; Kasilingam [3] discusses partner selection problem in terms of cost by mixed integer programming method; Hinkle [4] selects partner according to cluster analysis; Siying [5] discusses partner selection problem by neural networks method; N.Q. Wu [6] puts forward partner selection algorithm based on graph method; P.J.Ma [7] introduces partner selection by fuzzy analysis hierarchy process. These methods don't consider scheduling relationship among manufacturing tasks, and it's difficult to evaluate various indices synthetically by the methods.

M. Bubak et al. (Eds.): ICCS 2008, Part III, LNCS 5103, pp. 731–739, 2008.

Grey relation analysis belongs to grey system theory put forward by Professor Deng Ju-long [8] in 1982, and it mainly researches quantification analysis problem of system state development. In grey relation theory, the geometry curve constructed by several stat. data is more similar, the relation degree is bigger. The relation sequence reflects approximate sequence of each project to objective project, and the project of maximal grey relation degree is best one. Ant colony algorithm [9] has character of positive feedback, distributing compute and heuristic search, and it is successfully applied in NP-hard problem (such as, traveling salesman problem (TSP), scheduling problem and job-shop problem). To typical TSP problem about Oliver30、Eil50 and Ei175, ant colony algorithm, genetic algorithm and simulated annealing algorithm are compared in optimization quality and convergence speed by simulation analysis in reference [10], and the experiment result shows that ant colony algorithm is most satisfied. Partner selection of AVE is directed graph in the nature, which is similar to TSP. The paper hence adopts nature heuristic approach, which is ant colony algorithm, to resolve partner selection of AVE driven by working procedure. What's more, grey relation theory is applied into the problem at each task vertex to reduce problem resolved space.

2   Grey Relation Approaches to AVE Partner Pre-election

2.1   Evaluation Framework of AVE Partner Pre-election

AVE partner is mainly pre-elected from time, cost, quality and service. The pre-selection evaluating index of AVE partner is made up of cost evaluating index, history evaluating index, time evaluating index and general evaluating index [11], and the evaluation system is shown as Fig.1

- Cost evaluating index (L_1)

The cost index information includes biding price (L_{11}) and transporting fee (L_{12}). L_{11} indicates that candidate enterprise put forward processing fee to finish assigned manufacturing task; L_{12} is transporting expenses from candidate enterprise to core enterprise.

- History evaluating index (L_2)

History evaluating index is made up of credit for collaboration (L_{21}), capability of disposing exception (L_{22}), quality of after service (L_{23}) and production quality grade (L_{24}). The digital rule of above index is: AAAAA(5)、AAAA(4)、AAA(3)、AA(2)、A(1).

- Time evaluating index (L_3)

Time evaluating index is made up of production completion time (L_{31}) and project postponing dateline (L_{32}).

- General evaluating index (L_4)

General evaluating index estimates candidate enterprises from its scope (L_{41}), enterprise important degree (L_{42}) and equipment capability (L_{43}). Moreover, the numeralization rule of general evaluating index includes: enterprise important level is divided into monopolization status (5), domination status (4), leading status (3), participation status (2), and obedience status (1); Equipment capability is made up of international leading level (5), international general level (4), internal leading level (3), internal general level (2), and behindhand status (1).

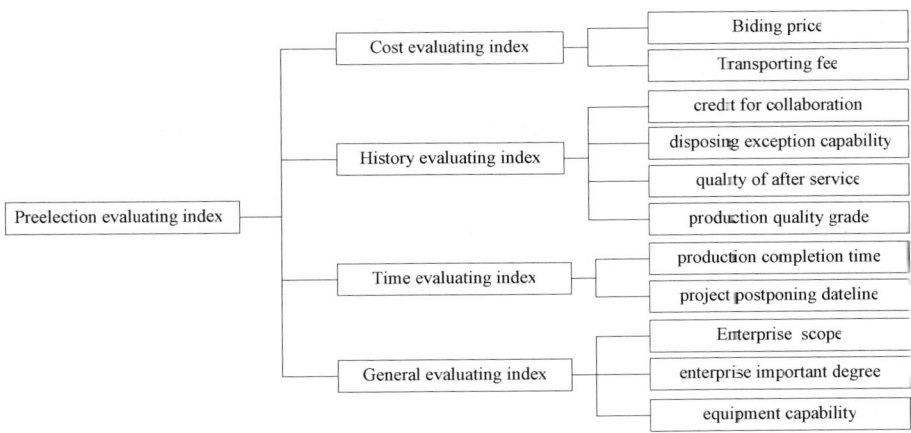

Fig. 1. Evaluation index system of AVE partner pre-election

2.2 Instance Analysis of Partner Selection Based on Grey Relation

Optimum set (reference sequence) is firstly determined in partner grey selection [12]. Cooperative manufacturing partner is filtered by acquiring grey relation coefficient of three hierarchies according to pre-election evaluating index system discussed in Section 2.1. The paper discusses AVE partner pre-election problem by giving an example of working procedure about semi-finishing turning flank groove for one aeroplane engineer front casket. In terms of final grey relation coefficient, five enterprises to be satisfied are chosen from ten candidate enterprises, and the detail information of each candidate enterprises is shown as Table 1.

According to the principle of the lowest cost, shortest manufacturing times, production excellent quality and most quick response speed, the optimum set is: $T^*= \{T^{*1}, T^{*2}, T^{*3}, T^{*4}\} = \{\{1.8,0.23\},\{5,5,5,5\},\{13,2\},\{4,5\}\}$, and the precedence arrangement sequence of each evaluating index is shown as: $L_{11}>L_{12}$; $L_{21}>L_{24}>L_{22}>L_{23}$; $L_{31}>L_{32}$; $L_{43}>L_{41}>L_{42}$; $L_1>L_3>L_2>L_4$. The first and second hierarchy evaluating index judgment matrixes are:

$$E_1^2 = \begin{bmatrix} 1 & 4 \\ 1/4 & 1 \end{bmatrix}, \quad E_2^2 = \begin{bmatrix} 1 & 5 & 7 & 3 \\ 1/5 & 1 & 3 & 1/3 \\ 1/7 & 1/3 & 1 & 1/5 \\ 1/3 & 3 & 5 & 1 \end{bmatrix} \quad E_3^2 = \begin{bmatrix} 1 & 4 \\ 1/4 & 1 \end{bmatrix} \quad E_4^2 = \begin{bmatrix} 1 & 5 & 1/5 \\ 1/5 & 1 & 1/9 \\ 5 & 9 & 1 \end{bmatrix}$$

$$E^1 = \begin{bmatrix} 1 & 5 & 3 & 7 \\ 1/5 & 1 & 1/3 & 3 \\ 1/3 & 3 & 1 & 5 \\ 1/7 & 1/3 & 1/5 & 1 \end{bmatrix}.$$

We can get weight vector of multi-hierarchy judgment matrixes by data pretreatment and plus multiply disposal, vector: $W_1^2=\{0.75,0.25\}$; $W_2^2=\{0.74, 0.21,$

Table 1. Detail information table of candidate enterprise

Candidate Set	L_{11}	L_{12}	L_{21}	L_{22}	L_{23}	L_{24}	L_{31}	L_{32}	L_{41}	L_{42}	L_{43}
Enterprise I	2.2	0.52	3	4	5	2	16	3	0.02	4	4
Enterprise II	2.6	0.34	5	5	2	3	13	5	0.8	3	2
Enterprise III	3.1	0.23	2	3	4	2	22	4	0.12	2	3
Enterprise IV	2.1	0.45	1	1	3	5	24	6	1.58	2	1
Enterprise V	1.8	0.49	4	4	5	4	18	3	0.78	4	5
Enterprise VI	2.8	0.78	5	5	1	3	17	7	2.12	1	2
Enterprise VII	2.9	0.66	2	3	3	2	21	3	0.09	3	3
Enterprise VIII	1.9	0.32	4	2	4	5	26	5	0.56	4	4
Enterprise IX	2.7	0.41	5	4	2	4	20	3	1.92	2	5
Enterprise X	2.5	0.72	1	1	3	3	19	2	0.99	4	1

0.08, 0.43}; $W_3^2 = \{0.75, 0.25\}$; $W_4^2 = \{0.28, 0.06, 0.67\}$; $W^1 = \{0.51, 0.14, 0.30, 0.05\}$.
There need to pretreat the data of assessment index, before grey relation coefficient is determined. What's more, the rule of treatment for cost type index is T*(i)/T (i), and the first grade difference matrix is:
P= {P_1, P2, P3, P_4} =

$$\begin{bmatrix}
0.08 & 0.14 & 0.14 & 0.08 & 0.00 & 0.20 & 0.07 & 0.08 & 0.02 & 0.00 & 0.06 \\
0.14 & 0.08 & 0.00 & 0.00 & 0.19 & 0.13 & 0.00 & 0.14 & 0.14 & 0.09 & 0.19 \\
0.19 & 0.00 & 0.22 & 0.14 & 0.07 & 0.20 & 0.14 & 0.12 & 0.22 & 0.18 & 0.12 \\
0.06 & 0.12 & 0.28 & 0.28 & 0.12 & 0.00 & 0.16 & 0.15 & 0.06 & 0.18 & 0.25 \\
0.00 & 0.13 & 0.08 & 0.08 & 0.00 & 0.07 & 0.10 & 0.08 & 0.15 & 0.00 & 0.00 \\
0.17 & 0.18 & 0.00 & 0.00 & 0.24 & 0.13 & 0.09 & 0.16 & 0.00 & 0.27 & 0.19 \\
0.18 & 0.16 & 0.22 & 0.14 & 0.12 & 0.20 & 0.14 & 0.07 & 0.22 & 0.09 & 0.12 \\
0.02 & 0.07 & 0.08 & 0.22 & 0.07 & 0.00 & 0.18 & 0.14 & 0.17 & 0.00 & 0.06 \\
0.15 & 0.11 & 0.00 & 0.08 & 0.19 & 0.07 & 0.12 & 0.07 & 0.02 & 0.18 & 0.00 \\
0.13 & 0.17 & 0.28 & 0.28 & 0.12 & 0.13 & 0.11 & 0.00 & 0.12 & 0.00 & 0.25
\end{bmatrix}$$

In terms of reference [13], each grade grey relation space comparing average values are defined as:

$$\overline{\Delta}_1 = \sum_{i=1}^{10} \sum_{j=1}^{2} |P_{ij}| \Big/ (10 \times 2) = 0.1 \qquad ; \qquad \overline{\Delta}_2 = \sum_{i=1}^{10} \sum_{j=3}^{6} |P_{ij}| \Big/ (10 \times 4) = 0.1 \qquad ;$$

$$\overline{\Delta}_3 = \sum_{i=1}^{10} \sum_{j=7}^{8} |P_{ij}| \Big/ (10 \times 2) = 0.1 \; ; \; \overline{\Delta}_4 = \sum_{i=1}^{10} \sum_{j=9}^{11} |P_{ij}| \Big/ (10 \times 3) = 0.1 \cdot$$ According to formula of

$\Delta \min = \min_i \min_j |P_{ij}|$, $\Delta \max = \max_i \max_j |P_{ij}|$, the maximal value and minimal value of first grade evaluating element are: $\Delta \min_1 = \Delta \min_2 = \Delta \min_3 = \Delta \min_4 = 0$, $\Delta \max_1 = 0.19$, $\Delta \max_2 = 0.28$, $\Delta \max_3 = 0.18$, $\Delta \max_4 = 0.25$. The average value proportion coefficient is defined as: $\gamma = \overline{\Delta} / \Delta \max$, thus, $\gamma_1 = 0.53$, $\gamma_2 = 0.36$, $\gamma_3 = 0.56$, $\gamma_4 = 0.4$. According to definition III of reference [10], the distinguishing coefficients are respectively: $\sigma_1 =0.78$, $\sigma_2 =0.61$, $\sigma_3 =0.81$, $\sigma_2 =0.65$. We can get first grade relation matrix by Eq. 1.

$$\xi_i^{(j)}=\frac{\Delta\min+\sigma\Delta\max}{P_{ij}+\sigma\Delta\max}\quad . \tag{1}$$

$$\Omega=\begin{bmatrix}
0.65 & 0.51 & 0.55 & 0.68 & 1.00 & 0.46 & 0.68 & 0.65 & 0.89 & 1.00 & 0.73 \\
0.51 & 0.65 & 1.00 & 1.00 & 0.47 & 0.57 & 1.00 & 0.51 & 0.54 & 0.64 & 0.46 \\
0.44 & 1.00 & 0.44 & 0.55 & 0.71 & 0.46 & 0.51 & 0.55 & 0.42 & 0.47 & 0.58 \\
0.71 & 0.55 & 0.38 & 0.38 & 0.59 & 1.00 & 0.48 & 0.49 & 0.73 & 0.47 & 0.39 \\
1.00 & 0.53 & 0.68 & 0.68 & 1.00 & 0.71 & 0.59 & 0.65 & 0.52 & 1.00 & 1.00 \\
0.47 & 0.45 & 1.00 & 1.00 & 0.42 & 0.57 & 0.62 & 0.48 & 1.00 & 0.38 & 0.46 \\
0.45 & 0.48 & 0.44 & 0.55 & 0.59 & 0.46 & 0.51 & 0.68 & 0.42 & 0.64 & 0.58 \\
0.88 & 0.68 & 0.68 & 0.44 & 0.71 & 1.00 & 0.45 & 0.51 & 0.49 & 1.00 & 0.73 \\
0.50 & 0.57 & 1.00 & 0.68 & 0.47 & 0.71 & 0.55 & 0.68 & 0.89 & 0.47 & 1.00 \\
0.53 & 0.47 & 0.38 & 0.38 & 0.59 & 0.57 & 0.57 & 1.00 & 0.58 & 1.00 & 0.39
\end{bmatrix}$$

On the basis of multi-hierarchy grey relation selection model [15], the multi-hierarchy grey relation coefficient for AVE partner selection is:

$$C=W_i\times\Omega_i=\sum_{i=1}^{4}W^1(i)\cdot W_2^i\cdot\Omega_i=\{0.67,0.74,0.57,0.63,0.82,0.61,0.52,0.75,0.66,0.58\}.$$

For the manufacturing task about turning flank groove of one aeroplane engineer front casket, the candidate enterprises pre-elected include: enterprise V, enterprise VIII, enterprise II, enterprise I and enterprise IX.

3   Application of Ant Colony Algorithm in Selecting AVE Partner

Considering part manufacturing character, working procedures are arranged for manufacturing chain from the view of the aspect of manufacturing technics, and manufacturing enterprise need to be selected from node of manufacturing chain. Thus, the problem of AVE partner selection is considered as directed graph. The metal cutting working procedure (MCWP) in one aeroplane engineer front casket AVE includes: drilling and boring the hole of plane to be combined (MCWP1), rough milling exterior plane (MCWP2), finish milling the plane to be combined (MCWP3), finish grinding the plane to be combined (MCWP4), boring the plane to be combined (MCWP5), finish milling exterior plane (MCWP6), finish boring radial hole (MCWP7), milling groove of the plane to be combined (MCWP8) and finish turning fore-and-aft groove (MCWP9), and the directed graph of AVE partner selection is shown as Fig.2.

With the increment of working procedure node, the space of problem about partner selection expands rapidly, and it belongs to typical NP-Hard compounding optimization problem. On the basis of AVE partner pre-election in terms of grey relation theory, the paper makes use of ant colony algorithm to search manufacturing path of least cost consumed.

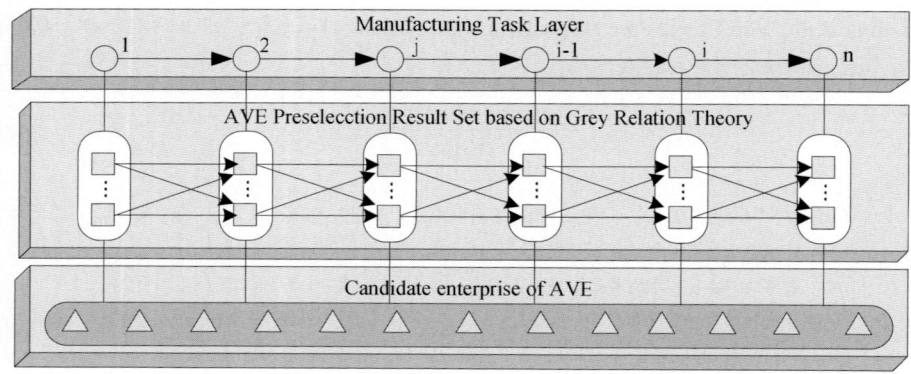

Fig. 2. Directed graph of AVE partner selection

Ant departs from vertex of in-degree zero, and its amount is the same as working procedure. It chooses the next vertex according to information consistency, and the travel is finished at the vertex of out-degree zero. The task of aeroplane engineer front casket includes nine working procedure, and the element of candidate enterprise set Ai ($i=1,2...m$) from each working procedure should be selected. What's more, the ant chooses the next intention enterprise at the t time, and arrives the destination at the t+1 time. All the ants travel $|Ai|$ steps at the interval of t and t+1 time, and it is thought that the algorithm finish one times iterative travel. If the algorithm realizes n times iterative travel, the ant colony algorithm is defined one times cycle, and update formula of information consistence is defined:

$$\tau_{ij}(t+n) = \rho \times \tau_{ij}(t) + \Delta\tau_{ij} .$$ (2)

Variable $\tau_{ij}(t)$ is denoted information consistency from enterprise i to enterprise j, and Variable ρ is indicated residual information consistency. When t equals to zero, the original information consistency $\tau_{ij}(0) = 0.6$; $\Delta\tau_{ij}$ is increment of information consistency, after the algorithm finishes one travel cycle.

$$\Delta\tau_{ij} = \sum_{k=1}^{m} \Delta\tau^k_{ij} .$$ (3)

$\Delta^k\tau_{ij}$ is indicated k ant leaves information element value from i enterprise to j enterprise at the internal of t and t+n, and it is calculated:

$$\Delta^k\tau_{ij} = \begin{cases} Q/L_k & ,\textit{if } k \textit{ ant travels from enterprise } i \textit{ to enterprise } j \\ 0 & , \textit{otherwise} \end{cases} .$$ (4)

Q is constant about information consistence that ant leaves, and its value impacts on the convergence speed of ant colony algorithm. The paper discusses how to select appropriate collaborating enterprise for each working procedure by utilizing ant colony algorithm. $P^k_{ij}(t)$ is indicated the probability of k ant traveling from enterprise i to enterprise j:

$$P_{ij}^k(t) = \begin{cases} \dfrac{[\tau_{ij}(t)]^\alpha \cdot [\eta_{ij}(t)]^\beta}{\sum\limits_{A_j}[\tau_{ij}(t)]^\alpha \cdot [\eta_{ij}(t)]^\beta} & if \quad j \in A_j \\ 0 & otherwise \end{cases}.$$ (5)

According to definition of arc visibility η_{ij} in typical TSP problem for ant colony algorithm, suppose $\eta_{ij}(t) = 1/c_{ij}(t)$. And $c_{ij}(t)$ is denoted the distance between enterprise i and j. Considering ant colony traveling rule in the process of AVE selection for manufacturing task shown as Fig.2, the transporting cost between vertex of directed graph can be expressed as: $Trans = \bigcup\limits_{m=1}^{10} T_m$,and $T_1 = Transport(E_{Sponsor}, E_j^{(1)})$, $T_i = Transport(E_j^{(i)}, E_k^{(i+1)})$, $T_{10} = Transport(E_j^{(10)}, E_{Sponsor})$, $i \in \{1,2,...,9\}$, $j \in \{1,2,...,5\}$, $k \in \{1,2,...,5\}$. The transporting cost matrix is acquired by calculating distance among various cities according to china railway passenger transport odograph:

$$T_2 = \begin{bmatrix} 1489 & 689 & 577 & 511 & 998 \\ 1807 & 1160 & 1195 & 1206 & 303 \\ 2000 & 1159 & 651 & 0 & 1509 \\ 547 & 1288 & 1802 & 2453 & 2577 \\ 2882 & 2042 & 1493 & 842 & 2351 \end{bmatrix} \quad T_3 = \begin{bmatrix} 1489 & 1807 & 547 & 2882 & 2033 \\ 689 & 1160 & 1288 & 2042 & 1463 \\ 577 & 1195 & 1802 & 1493 & 1498 \\ 511 & 1206 & 2453 & 842 & 1509 \\ 998 & 303 & 2577 & 2351 & 0 \end{bmatrix}$$

$$T_4 = \begin{bmatrix} 1187 & 577 & 1489 & 511 & 511 \\ 1182 & 1195 & 1807 & 1206 & 1206 \\ 3099 & 1802 & 547 & 2453 & 2453 \\ 1172 & 1493 & 2882 & 842 & 842 \\ 2185 & 1498 & 2033 & 1509 & 1509 \end{bmatrix} \quad T_5 = \begin{bmatrix} 3099 & 1172 & 1811 & 1182 & 1187 \\ 1424 & 1493 & 3336 & 842 & 2042 \\ 0 & 2882 & 741 & 1807 & 1489 \\ 2000 & 842 & 1159 & 1206 & 511 \\ 2000 & 842 & 1159 & 1206 & 511 \end{bmatrix}$$

$$T_6 = \begin{bmatrix} 2000 & 2033 & 547 & 2000 & 741 \\ 842 & 2351 & 3336 & 842 & 2042 \\ 1159 & 1463 & 1288 & 1159 & 0 \\ 1206 & 303 & 2277 & 1206 & 1160 \\ 511 & 998 & 2085 & 511 & 689 \end{bmatrix} \quad T_7 = \begin{bmatrix} 676 & 511 & 1206 & 2000 & 842 \\ 2185 & 998 & 303 & 2033 & 2351 \\ 3099 & 2085 & 2277 & 547 & 3336 \\ 676 & 511 & 1206 & 2000 & 842 \\ 1811 & 689 & 1160 & 741 & 2042 \end{bmatrix}$$

$$T_8 = \begin{bmatrix} 676 & 3099 & 676 & 1327 & 3099 \\ 511 & 1489 & 511 & 577 & 1489 \\ 1206 & 2085 & 1206 & 1807 & 2085 \\ 2000 & 0 & 2000 & 1424 & 0 \\ 842 & 2882 & 842 & 1493 & 2882 \end{bmatrix} \quad T_9 = \begin{bmatrix} 842 & 1159 & 2453 & 1509 & 0 \\ 2882 & 741 & 547 & 2033 & 2000 \\ 842 & 1159 & 2453 & 1509 & 0 \\ 1493 & 514 & 1802 & 1498 & 651 \\ 2882 & 741 & 547 & 2033 & 2000 \end{bmatrix}$$

$$T_1 = \begin{bmatrix} 1489 & 1807 & 2000 & 547 & 288 \end{bmatrix} \quad T_{10} = \begin{bmatrix} 2882 & 741 & 547 & 2033 & 2000 \end{bmatrix}^T$$

When one parameter is test, the other parameters are chosen default value to observe influence of the parameter to algorithm, and default value of respective parameter is: $\alpha = 1, \beta = 2, \rho = 0.9, Q = 100$. AVE partner selection is calculated 600

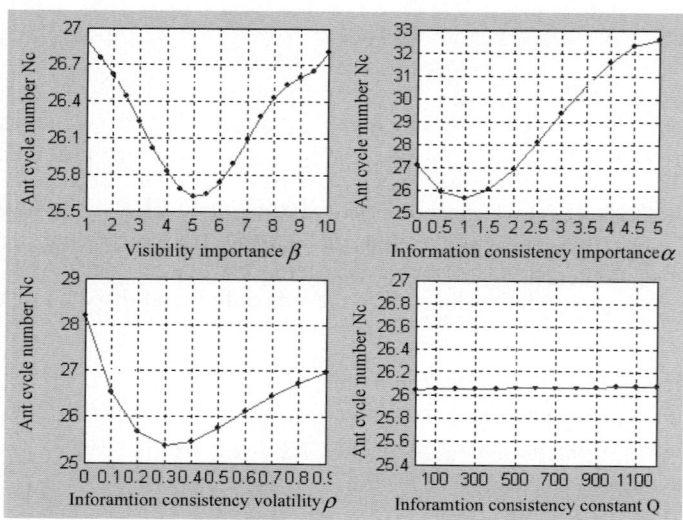

Fig. 3. Ant colony algorithm parameter optimum result analysis based on stat

times for each parameter, and average of experimentation result is computed. The value range of each parameter is: $\alpha \in [0,5]$, $\beta \in [1,10]$, $\rho \in [0,0.9]$, $Q \in [1,1100]$, and stat. analysis of experimentation is shown as Fig.3.

According to simulation result above, the optimization value range of each parameter is: $\beta \in (5-\varepsilon, 5+\varepsilon)$, $\alpha \in (1-\varepsilon, 1+\varepsilon)$, $\rho \in (0.3-\varepsilon, 0.3+\varepsilon)$, $Q \in [100, 1100]$. If β is too high, ant colony will select candidate enterprise of low transporting cost without considering information consistency. Otherwise, selection probability is influenced by information consistence too much, and it leads to algorithm converge too early. If α is too high, information consistence decides ant colony selection probability, and ant selects the path which the other ant have traveled. Otherwise, ant colony algorithm changes into classic greedy random algorithm. Volatility coefficient ρ reflects memory degree of ant colony for experience. If its value is too low, ant colony algorithm changes into greedy heuristic search, and algorithm cycle times is more. Otherwise, information consistence volatilizes too quickly, and transporting cost among enterprises in large determines selection probability. Information consistence const. Q influences ant colony algorithm search efficiency a little. On the basis of above analysis, the paper gives one group optimization parameter: $\alpha = 1, \beta = 5, \rho = 0.3, Q = 800$, and AVE partner selection result is shown as Table 2.

Table 2. AVE partner selection result based on ant colony algorithm

Calculation Num	1	2	3	4	5	6	7	8	9	10	Ave
Ant cycle Num	7	11	20	32	57	5	23	24	11	18	20.8
Running time (ms)	125	78	63	78	109	78	78	78	62	94	84.3

4 Conclusions

The bottleneck of AVE is how to manufacture high quality products at the right time and place by right manufacturing resource. The paper divided partner enterprise of AVE into two stages: In first stage, the partner pre-election of AVE problem is resolved by grey relation theory, ant it reduces problem resolved space and complexity. According to transport cost among candidate enterprises, partner selection is realized by ant colony algorithm on the basis of partner candidate set. Future research is to optimize resource allocation for plant of each AVE partner by grey relation and grey relation theory for the scheduling and dispatch of cooperative manufacturing.

Acknowledgments. The paper is supported by Research Fund of Xi'an Technological University (XAGDXJJ200706), Shaanxi Province Programs for Science and Technology Development (2007K05-11) and National Natural Science Foundation of China (50605051).

References

1. Yang, S.L., Li, T.F.: Agility evaluation of mass customization production manufacturing. Journal of Material Processing Technology 129, 640–644 (2002)
2. Gupta, P., Rakesh, N.: Flexible Optimization Framework for Partner Selection in Agile Manufacturing. In: Proceedings of the 4th Industrial Engineering Research Conference, pp. 691–700. Nashville, Tennesse (1995)
3. Raja, G.L.: Cheep: Selection of vendors- A Mixed-integer Programming Approach. Computers & Industrial Engineering 31, 347–351 (1996)
4. Hinkle, C.L., Robinson, P.J., Green, P.E.: Vendor Evaluation Using Cluster Analysis. Journal of Purchasing 29, 49–58 (1996)
5. W, S.Y.: A Supplier-Selecting System Using a Neural Network. In: IEEE International Conference on the Intelligent Processing Systems, vol. 40, pp. 468–471. IEEE Press, New York (1997)
6. Wu, N.Q.: An Approach to Partner Selection in Agile Manufacturing. Journal of Intelligent Manufacturing 10, 519–529 (1999)
7. Ma, P.J.: Study on Partner-Selecting Strategy in Agile Manufacturing. China Mechnical Engineer 10, 1176–1179 (1999)
8. Fang, Y.D.: Research of Manufacturing Resources Optimum Allocation Technology Supporting Rapid Extended Manufacturing. Northwestern Polytechnical University Philosophy Doctor Dissertation (2005)
9. Wang, S.R.: Research of Ant Colony Algorithm Optimization Theory Model and Job-shop Application. ZheJiang University Press, ZheJiang (2003)
10. Li, S.Y.: Ant Colony Algorithm and Application. Haerbing Technology Press, Haerbing (2004)
11. Fang, Y.D.: Research and Realization of Partner Selection System in Cooperative Manufacturing Based on Web. System engineer 23, 118–123 (2005)
12. Wen, Z.: The discussion about problems for grey relation theory. Data Stat. and management 18, 25–29 (1999)
13. Yang, D.: Research of Ant Colony Algorithm Applied in the Combination Optimization. Tianjing University Press, Tianjing (2004)

Hierarchical Approach to Evolutionary Multi-Objective Optimization

Eryk Ciepiela, Joanna Kocot, Leszek Siwik, and Rafał Dreżewski

Department of Computer Science, AGH University of Science and Technology,
Kraków, Poland
{siwik,drezew}@agh.edu.pl

Abstract. In this paper a new "hierarchical" evolutionary approach to solving multi-objective optimization problems is introduced. The results of experiments with standard multi-objective test problems, which were aimed at comparing "hierarchical" and "classical" versions of multi-objective evolutionary algorithms, show that the proposed approach is a very promising technique.

1   Introduction

The most natural process of decision making for human being consists in analyzing many—often contradictory—factors and searching for peculiar compromise among them. Such decisive process is known as a *multi-criteria decision making (MCDM)*. The most frequently, *MCDM* process is based on appropriately defined *multi-objective optimization problem (MOOP)*. Following [2]—*multi-objective optimization problem* in its general form is defined as minimizing/maximizing the set of objectives $f_m(\bar{x})$, *where* $m = 1, 2 \ldots, M$, taking into account the set of constraints, which define all possible (feasible) decision alternatives $(\mathcal{D})$.

Because there are many criteria, to indicate which solution is better than the other, so called dominance relation is used [2]. A solution of the multi-objective optimization problem in the Pareto sense means determination of all non-dominated alternatives from the set $\mathcal{D}$.

During over 20 years of research on evolutionary multi-objective algorithms (EMOAs) quite many techniques have been proposed. Generally all of these techniques and algorithms can be classified as elitist (which give the best individuals the opportunity to be directly carried over to the next generation) or non-elitist ones [2].

The Hierarchical Genetic Strategy (HGS) was introduced by Kołodziej and Schaefer [4] as one of the multi-deme, parallel genetic algorithms models. The main idea of HGS is running a set of dependent evolutionary processes in parallel. Its dependency relation has a tree structure with fixed depth. The tree nodes which are closer to the root perform chaotic search with low accuracy—they detect promising regions of the optimization landscape—while more accurate searching is done in further successor nodes.

M. Bubak et al. (Eds.): ICCS 2008, Part III, LNCS 5103, pp. 740–749, 2008.

In the course of this paper the new "hierarchical" approach to multi-objective optimization based on HGS model is presented and compared experimentally with "classical" EMOAs. "Hierarchical" in this context means that presented algorithm is trying to identify more and more precisely (in a hierarchical way) the more and more accurate approximation of non-dominated points.

The paper is organized as follows. First the HGS model is described with more details. Next the "hierarchical" approach to multi-objective optimization (MOHGS) is presented. The preliminary experimental results comparing "hierarchical" and "classical" versions of multi-objective evolutionary algorithms conclude the paper.

2   Hierarchical Genetic Strategy

As it was said already, the main idea of HGS is running in parallel a set of dependent evolutionary processes organized as a tree with more and more accurate searching done in the nodes located far and far away from the root of the tree [4]. The HGS node's individuals represent the solutions (phenotypes) with precision that increases with the node's level.

After a *metaepoch* which lasts a predefined number of iterations of evolutionary algorithm, each HGS node chooses the best individuals. Each of them constitutes a new "child" population. This procedure is called *sprouting operation* and is performed conditionally, according to the outcome of the *branch comparison operation*. It is reasonable as long as such a comparison prevents the same or similar individual from sprouting identical or similar populations in the child HGS nodes.

Avoiding exploration of the same regions of the optimization landscape is also supported by further operation called *reduction*. However, unlike *branch comparison operation*, *reduction* is performed after branches have been sprouted. *Reductions* can be performed both within the scope of sibling HGS nodes—then it is said to be local, and globally when sibling scope is exceeded. Details on how the above mentioned operations are carried out depend strongly on the implementation of HGS.

The individual being sprouted, has to be prepared to find a new HGS node of increased precision, what means that the individual (phenotype) has to be specified more precisely. This is accomplished e.g. by appending randomly least significant bits to the floating point number's binary representation.

3   Multi-Objective Hierarchical Genetic Strategy

The first implementations of HGS ([6]) used simple genetic algorithm (SGA) as the main optimization algorithm, therefore being limited to optimizing single-objective problems. Our goal in this paper is to introduce multi-objective optimization algorithms to HGS—thus developing multi-objective HGS (MOHGS)—and provide it with new features to take advantage of these algorithms' properties.

In our approach SGA algorithm was replaced by vector evaluated genetic algorithm (VEGA) [5], multiple objective genetic algorithm (MOGA) [3], strength pareto evolutionary algorithm (SPEA) [9] and non-dominated sorting genetic algorithm II (NSGA-II) [2]. However, any other multi-objective evolutionary algorithm could have been used as well.

3.1   Sprouting and Reduction Operators in MOHGS

Having more than one objectives to operate on, means that in the process of sprouting more than one criterion can be considered. Furthermore, these criteria do not have to be limited only to objectives, but other properties of the investigated population (generation) can be used—e.g. the domination level (see [2]). It can be also advisable to promote individuals which seem to differ from others in either the problem domain space or the objective space. It is also likely to take into consideration the criteria based on the ideas known from the multi-objective evolutionary algorithms, like niching or elitism (see [2]).

New HGS node sprouted with respect to one criterion creates a new population (generation), which neither can be compared with nor reduced to populations resulting from sprouting with regard to other criteria. In this way, HGS tree can be considered as a colored tree, where the node's color denotes the node sprouting criteria. Certainly, such an approach may cause existence of similar populations in HGS tree. However, it was found desirable as long as it ensures heavier computation of those populations that were promoted with respect to more than one criterion.

4   Selected Aspects of the System Realization

In the following section the key aspects of MOHGS implementation on the EA-AE platform ([1]) are described.

4.1   The EA-AE Platform

The EA-AE system is a runtime environment that provides the convenient approach to development, "composition", and running of different types of evolutionary algorithms. Its approach is based on the idea of multiple "processors" which, arranged in a sequence, create a single iteration of an algorithm. Each of the processors transforms the given generation of a population, in a way defined by the developer. A single processor can be for example responsible for performing mutation (or recombination) operation on the current generation of individuals.

Despite of being independent, the processors have to be able to exchange some information. As an example, the recombination based on the previously selected pool of individuals can be considered. If so, the information about pool has to be passed forward from pool selector processor. To enable such communication, the *individual properties* concept is introduced. Processors are able to annotate each individual of the generation they are transforming with "key-value" pairs. In this way, the information is exchanged in "per individual" context and is actually attached to the individual. These "key-value" pairs are called properties and are carried throughout the "evolution step" (see further in this section). The information the processor needs is expressed by property-existence requirements. Hence, the processors are characterized by properties they use and by those they provide.

Once the generation passes the chain of processors custom rankers rate (in the way defined by the developer) the generation's individuals. Each ranker keeps a ranking of the best individuals, to which the rated ones are compared. Depending on the comparison, they are inserted (or not) into the ranker's list.

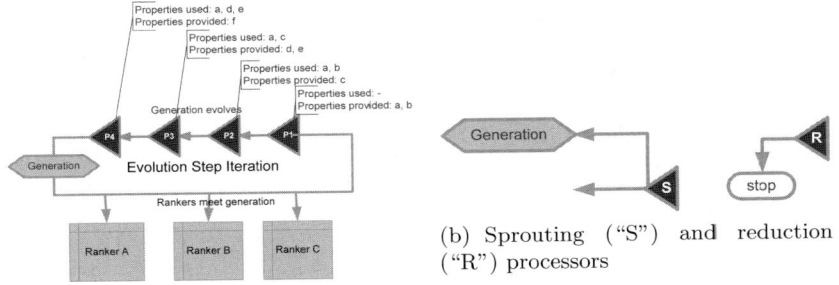

(b) Sprouting ("S") and reduction ("R") processors

Fig. 1. Principles of EA-AE platform functioning

The sequence of processors along with rankers constitute the *evolution step* (see Fig. 1a). Evolution step is performed in the context of the problem's objective function and the search region constraints defining a phenotype space.

The phenotype space is responsible for bidirectional mapping between individual's genotype and phenotype. Since within EA-AE phenotypes and genotypes are represented in a generic, normalized, and uniform way, generation processors remain universal and independent of phenotype space. The generic binary genotype is provided as well as floating point phenotype feature (decision variable), etc. What is more, since the objective function is defined using generic phenotypes, it is also separated from the phenotype space. Therefore, the orthogonal concerns such as: search region constraints along with genetic representation, individuals rating, and applied evolutionary algorithm are independent.

The evolution step constitutes the iteration of the evolutionary algorithm. Since the evolution step can be composed of processors and rankers chosen by developer, ideas from different algorithms may be combined in order to find the best heuristic for solving various optimization problems.

4.2 Multi-Objective HGS Meets EA-AE

The EA-AE platform was used for the implementation of the MOHGS algorithms. It was possible because of the EA-AE platform mechanisms and features, which include: the generation processors ability to spawn computation (which enables the HGS branch sprouting), and computation controller, which maintains the structure of the spawned computations and therefore enables HGS tree structure and HGS branch reduction operation. The implementation of MOHGS was reduced to providing HGS-specific generation processors.

As it was previously said, every generation from the HGS node is rated with regard to several criteria. For each of those, a certain number of best individuals is marked and can be used for sprouting new HGS nodes—each initially containing a population of one individual, copied from the initial population. A HGS node containing the derived population evolves in the same way as its parent node does, still, representing the individuals with higher precision. Simultaneously the initial population's node continues the evolutionary algorithm's transformations.

The sprouting process is performed by generic sprouting processor (see Fig. 1b) which designates individuals annotated by property with certain name to be sprouted.

The reduction operator is used when populations of two different HGS nodes are operating on a very similar area of decision variables space. The similarity of the HGS nodes defined as the similarity of the areas of decision variables space they operate on, was estimated by the Equation (1).

$$sim(P_i, P_j) = \frac{1}{|P_i| \cdot |P_j|} \sum_{ind^i \in P_i} \sum_{ind^j \in P_j} dist(ind^i, ind^j) \tag{1}$$

where: $sim(P_i, P_j)$ is the similarity measure of the two populations P_i and P_j (located in the i-th and j-th HGS nodes), $|P_i|$ is the number of individuals in the population P_i, ind^i is the individual from population P_i, $dist(ind^i, ind^j)$ is the distance between ind^i and ind^j individuals in the Euclidean metric ($ind \in \mathbb{R}^n$).

The nodes reduction operator is introduced as a custom generation processor (see Fig. 1b), which calculates the similarity at the same level of the HGS tree—not only locally between sibling nodes, but also globally. Since the measure (1) is symmetrical, in order to avoid computation redundancy as well as simultaneous reduction of two close populations, each node compares itself only to its ancestors. The node with a similarity measure value less than similarity threshold value does not continue computations. However, despite stopping the parent node, its children remains active.

To avoid the effectiveness losses, the reduction processor was placed before the processor responsible for sprouting in the HGS processor sequence. Otherwise, the amount of similar non-sibling child nodes would have been sprouted just before their parent nodes similarity evaluation.

Each of the HGS nodes is allowed to perform only a defined number of computing iterations. After that, like in the case of reduction, it is stopped. The whole strategy is completed when **all** the HGS nodes are stopped.

5 Experimental Results

The experiments were intended to compare "hierarchical" realizations of various multi-objective evolutionary algorithms with their "classical" versions. To achieve this the Zitzler-Deb-Thiele ZDT1, ZDT3, and ZDT6 problems ([8]) were used.

5.1 The Methodology of the Experiments

In order to ensure comparable results, both classical and HGS-based algorithms were allowed to evolve for the same overall number of iterations. The results' quality was compared using various metrics. The measured time of execution also gave an outlook on the overhead introduced by the HGS strategy.

For the experiments the following multi-objective evolutionary algorithms were chosen: vector evaluated genetic algorithm (VEGA), multiple objective genetic algorithm (MOGA), strength pareto evolutionary algorithm (SPEA) and non-dominated sorting genetic algorithm II (NSGA-II) [2].

The metrics used for measuring the quality of the obtained results are described in [7]. The first trivial metric that was taken into consideration was the *number of non-dominated individuals (solutions)*, which form the approximation of the Pareto frontier. Generally, the greater the value of this metric is, the better is the quality of the found solution.

The *Inferior Region (IR)* metric calculates the size of the area dominated by obtained non-dominated individuals. The greater is the Inferior Region Metric' value, the better is the approximation of the ideal Pareto frontier.

The *Dominant Region Metric (DR)* measures the size of the area that dominates the obtained non-dominated individuals. Generally the smaller is the Dominant Region Metric' value, the better is the approximation of the ideal Pareto frontier. However, its value decreases also in the case of increasing concentration of the Pareto frontier without moving towards the "good point", thus local variations of the metric' value are possible.

The *Accuracy Frontier Metric (AF)* calculates the size of the area that neither dominates the obtained individuals, nor is dominated by them. It can be observed that the value of this metric can be expressed by the formula $1 - DR - IR$, where DR and IR are values of the DR and the IR metrics, respectively. The smaller is the AF metric value, the more complete and concentrated the Pareto frontier approximation is. However, the Accuracy Frontier Metric actually does not measure the quality of the ideal Pareto frontier approximation.

The *Pareto Spread Metric* compares the range of the obtained Pareto frontier and the ideal Pareto frontier (approximated by "good point" and "bad point").

The tests for all implemented evolutionary algorithms were carried out on a "flat" HGS structure (without sprouting and reduction—see Section 3.1) as well as on a regular HGS tree. This way the equal conditions were assured for both "classical" evolutionary algorithm (the "flat" structure) and HGS using this algorithm.

To ensure that the number of iterations will be the same in both cases, after HGS execution, iterations on all its nodes were counted and the compared algorithm was iterated exactly the same number of times. The number of individuals per population was also exactly the same.

5.2   Discussion of the Results

The Figures 2, 3, and 4 present plots of the obtained solutions for ZDT-1 with VEGA, ZDT-3 with MOGA, and ZDT-6 with SPEA, respectively. These three were chosen from 20 experiment runs, each per every combination of the problem and the algorithm. Each of the presented experiments completed in 450 iterations.

Detailed comparison of the quality metrics results is shown in the Tables 1–5. The time of computation in each mentioned case was gathered in the Table 6.

On the basis of the experiment's results the HGS-based algorithms can be considered regarding to the following criteria:

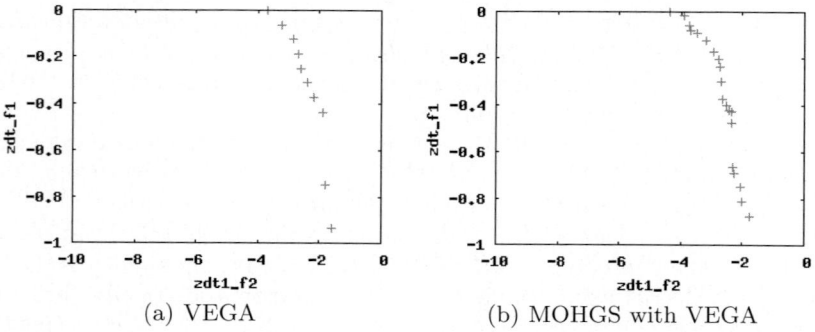

(a) VEGA   (b) MOHGS with VEGA

Fig. 2. The plot of obtained Pareto frontier for the ZDT-1 test problem

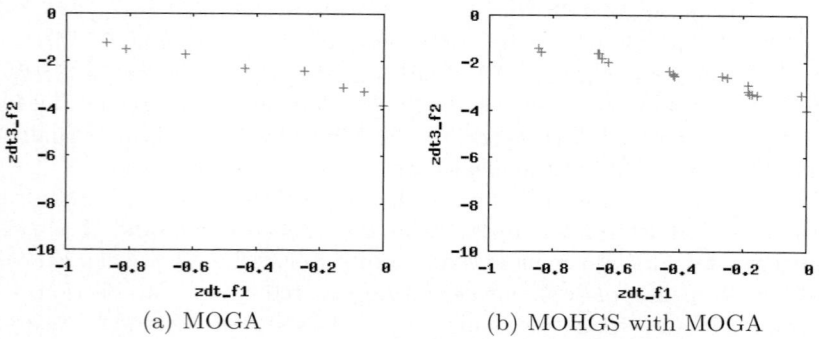

(a) MOGA   (b) MOHGS with MOGA

Fig. 3. The plot of obtained Pareto frontier for the ZDT-3 test problem

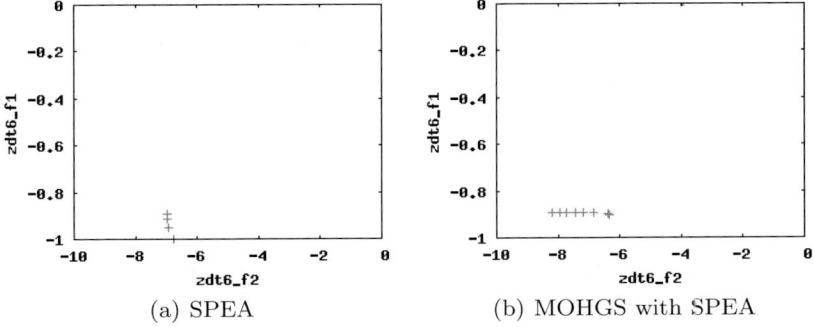

(a) SPEA   (b) MOHGS with SPEA

Fig. 4. The plot of obtained Pareto frontier for the ZDT-6 test problem

Table 1. Number of non-dominated individuals metric values comparison

test problem / algorithm	classic strategy	HGS
ZDT-1 / VEGA	10	21
ZDT-3 / MOGA	8	21
ZDT-6 / SPEA	7	8

1. Number of non-dominated individuals. In the case of HGS the number of non-dominated individuals is usually significantly greater than in the case of "classic" algorithm (see Table 1). It means that HGS offers a wider and more varied range of acceptable solutions.
2. Other Pareto frontier's quality metrics values. The "classic" strategies' Inferior Region metric values (Table 2) are in the most cases slightly better (the values are 1-5% higher). Similarly, the Dominant Region (Table 3) remains a little better (below 10%) in the case of "classic" algorithm. On the other hand, HGS obtains better Accuracy Frontier metric values (Table 4), which is implied by densely filled Pareto frontier. Moreover, the Overall Pareto Spread is more satisfactory in HGS (see Table 5). The differences in Accuracy Frontier and Overall Pareto Spread metrics decrease with the test problem complexity.
3. Efficiency. The overhead introduced by the MOHGS usually varies between 10-30% (see Table 6).

Table 2. Inferior Region metric values comparison

test problem / algorithm	classic strategy	HGS
ZDT-1 / VEGA	0.7741	0.7284
ZDT-3 / MOGA	0.7715	0.7673
ZDT-6 / SPEA	0.0323	0.0393

Table 3. Dominant Region metric values comparison

test problem / algorithm	classic strategy	HGS
ZDT-1 / VEGA	0.1981	0.2290
ZDT-3 / MOGA	0.1813	0.1981
ZDT-6 / SPEA	0.6958	0.7382

Table 4. Accuracy Frontier metric values comparison

test problem / algorithm	classic strategy	HGS
ZDT-1 / VEGA	0.0279	0.0426
ZDT-3 / MOGA	0.0472	0.0345
ZDT-6 / SPEA	0.2719	0.2225

Table 5. Overall Pareto Spread metric values comparison

test problem / algorithm	classic strategy	HGS
ZDT-1 / VEGA	0.1993	0.2283
ZDT-3 / MOGA	0.2276	0.2231
ZDT-6 / SPEA	0.0020	0.0019

Table 6. Time elapsed for computing [ms]

test problem / algorithm	iterations	classic strategy	HGS
ZDT-1 / VEGA	450	17906	23110
ZDT-3 / MOGA	450	27875	34188
ZDT-6 / SPEA	450	21766	19766

4. Even Pareto Spread. However not measured by any metric formula, worth taking into consideration is the observation that HGS much more evenly spreads the Pareto frontiers' individuals (compare Figures 2, 3, and 4).

While performing the experiments a strong and non-linear impact of HGS parameters such as node similarity threshold value, population size or metaepoch length on the strategy course was observed.

It is also worth mentioning that HGS enables distribution and parallelization of computations and does not cause high communication overhead, which involves only sending sprouted individual. However, the reduction operation in distributed environment remains a non-trivial and extremely important issue.

6   Concluding Remarks

In this paper the "hierarchical" approach to evolutionary multi-objective optimization was presented. The results of preliminary experiments show that still more research is needed on the proposed technique—however it seems to be a very promising approach especially in the case of difficult multi-objective problems, and the goal of this paper was to present the idea of MOHGS approach from the general point of view.

The future research could concentrate on additional verification of the proposed approach especially with the use of hard multi-objective problems and on the introduction of additional mechanisms (like niching and elitism) improving the obtained results. The agent-based realization of MOHGS seems to be the specially interesting direction of research. Such system will allow for introducing additional mechanisms—for example mechanisms of maintaining population diversity based on co-evolutionary interactions between evolving agents.

References

1. Ciepiela, E., Kocot, J., Siwik, L.: Composable runtime environment for building evolutionary algorithms. Technical report, Department of Computer Science, AGH University of Science and Technology (2006)
2. Deb, K.: Multi-Objective Optimization using Evolutionary Algorithms. John Wiley & Sons, Chichester (2001)
3. Fonseca, C., Fleming, P.: Genetic algorithms for multiobjective optimization: Formulation, discussion and generalization. In: Genetic Algorithms: Proceedings of the Fifth International Conference, pp. 416–423. Morgan Kaufmann, San Francisco (1993)
4. Schaefer, R., Kołodziej, J.: Genetic search reinforced by the population hierarchy. In: Foundations of Genetic Algorithms 7, pp. 383–399. Morgan Kaufman Publisher, San Francisco (2003)
5. Schaffer, J.D.: Some experiments in machine learning using vector evaluated genetic algorithms. PhD thesis, Vanderbilt University (1984)
6. Wierzba, B., Semczuk, A., Kołodziej, J., Schaefer, R.: Hierarchical genetic strategy with real number encoding. Technical report, Institute of Computer Science, Jagiellonian University (2003)
7. Wu, J., Azarm, S.: Metrics for quality assessment of a multiobjective design optimization solution set. Transactions of the ASME, Journal of Mechanical Design 123, 18–25 (2001)
8. Zitzler, E.: Evolutionary Algorithms for Multiobjective Optimization: Methods and Applications. PhD thesis, Swiss Federal Institute of Technology (2001)
9. Zitzler, E., Thiele, L.: An evolutionary algorithm for multiobjective optimization: The strength pareto approach. Technical Report 43, Swiss Federal Institute of Technology, Zurich, Gloriastrasse 35, CH-8092 Zurich, Switzerland (1998)

Author Index

Printing: Mercedes-Druck, Berlin
Binding: Stein+Lehmann, Berlin

Lecture Notes in Computer Science

Sublibrary 1: Theoretical Computer Science and General Issues

For information about Vols. 1– 4736
please contact your bookseller or Springer